HANDBOOK OF PATTERN RECOGNITION AND COMPUTER VISION

4th Edition

HANDBOOK OF PATTERN RECOGNITION AND COMPUTER VISION

4th Edition

editor

C H Chen

University of Massachusetts Dartmouth, USA

World Scientific

NEW JERSEY • LONDON • SINGAPORE • BEIJING • SHANGHAI • HONG KONG • TAIPEI • CHENNAI

Published by

World Scientific Publishing Co. Pte. Ltd.
5 Toh Tuck Link, Singapore 596224
USA office: 27 Warren Street, Suite 401-402, Hackensack, NJ 07601
UK office: 57 Shelton Street, Covent Garden, London WC2H 9HE

British Library Cataloguing-in-Publication Data
A catalogue record for this book is available from the British Library.

HANDBOOK OF PATTERN RECOGNITION AND COMPUTER VISION (4th Edition)

ISBN-13 978-981-4273-38-1
ISBN-10 981-4273-38-4

A Brief Introduction to the Handbook Series, by C.H. Chen

The Handbook Series consisting of four sister volumes in four editions published in 1993, 1999, 2005 and 2010 respectively is dedicated to the late Professor King-Sun Fu (1930-1985). The fourth volume is published to celebrate his 80th birthday in 2010.

I recall when I entered the Ph.D. program at Purdue University in September 1962, I was the first graduate student to work under Prof. King-Sun Fu in the pattern recognition area. He just came back from a sabbatical at MIT at that time and gradually shifted his research from learning control to pattern recognition. Two influential papers we read at that time were Marvin Minsky's Steps toward artificial intelligence, 1959 and C.K. Chow's paper on An optimum character recognition system using decision functions, 1957. Another influential paper at that time was the one by T. Marill and D.M. Green, on the effectiveness of receptors in recognition systems, 1963, that generated a lot of interest on feature selection using divergence measures. Subsequently there were influential papers on nearest neighbor decision rule and Parzen density estimates. The 1960s and early 70s were the time of great progress in statistical pattern recognition in theory and applications. I remember Prof. Fu was very active in pattern recognition applications ranging from fingerprint recognition, handprinted character recognition, remote sensing, seismic signal classification, speech recognition, etc. In 1968 he published the book, Sequential Methods in Pattern Recognition I did not realize until later that he was also working on syntactic pattern recognition during 1970s This is an area much dear to his research interest that he had laid down a nearly complete foundation of syntactic pattern analysis with every possible application to pattern recognition and scene analysis that can be made. His effort led to the book on Syntactic Methods in Pattern Recognition 1982. Among many other research efforts by him are the contextual pattern recognition, combined syntactic and statistical approach to pattern recognition, etc. Prof. Fu, in his twenty-five years of distinguished professional career, indeed has contributed and influenced the enormous progress in pattern recognition and computer vision in the past half century. A well written memorial to the late Prof. K.S. Fu was provided by Prof. C.C. Li in the second edition of the Handbook. I admire him not only on the academic achievements, but also on the inspirations he has provided to his Ph.D. students. He was always informal to his students who can see him whenever he was in office without prior appointments. He was very active and energetic in professional and lecture activities that required him to make almost one domestic trip each week and one overseas trip each month. Among his leadership activities is the establishment of IEEE Trans. on Pattern Analysis and Machine Intelligence, which has grown by page count from 1979 volume 1 of 417 pages to 2008 volume 30 of 2255 pages. Publication growth is only one measure of progress. The number and extent of pattern recognition and computer vision applications have grown far more than ten folds in the last 35 years and the impact of such applications to our daily lives is even more significant. It is indeed fortunate to see the continued and rapid progress in theory, methodologies, and applications of pattern recognition and computer vision.

The progress in pattern recognition since 1985 is particularly noted by the activities in combined and multiple classifiers, neural networks and support vector machines for pattern classifications, industrial applications of machine vision, and applications in biometrics, remote sensing, character recognition, etc. The dramatic and continued improvement in sensor and computation technologies plays a major role in the progress.

In the summer of 1990 I had a meeting with Dr. K.K. Phua, chairman of World Scientific Publishing in Singapore. He made an excellent suggestion for me to prepare a Handbook of Pattern Recognition and Computer Vision, which led to the publication of the first edition in 1993. The intent of the Handbook is not to serve as an encyclopedia or as a series of “Advances in PRCV” books. Unlike many engineering and software and hardware books, chapters in the handbook series have much longer lasting values. The basic problems and approaches to feature extraction, decision making, texture analysis, and clustering presented in the first edition remain to be the same. To a great extent each edition reflects some new emphasis in theory and applications. For examples: syntactic recognition was nicely presented in the first and second editions, human identification was well covered in the third edition, while life science areas and 3-D problems are more emphasized in the 4th edition. With enormous progress made even within a few years’ time, just by looking at the large number of papers in ICPR08, a handbook like this can provide only a very small representative articles of the areas considered. It is hoped that such small but careful selection can offer readers many promising directions for further pursuit. As Prof. Anil K. Jain nicely put it in his lecture at ICPR08, there are thousands of approaches but no silver bullet. Indeed in pattern recognition area alone, we are still looking for better features and feature space. There is endless progress in sight in pattern recognition until perhaps when machines can be designed to automatically look for the best features and feature space. In related areas, it is noted that signal processing can define new feature space and provide mathematical features and artificial intelligence can exploit higher level knowledge and they both will have continued impact on pattern recognition. On the application front, both patter recognition and computer vision should look beyond automation with accuracy and speed and seek for more useful solutions to environment, life sciences, energy and related problems.

In going over the Handbook series, I like to point out that Prof. Ching Y. Suen has provided the most consistent as well as important contributions to all four volumes of the handbook series. His long and pioneering research and education in pattern recognition and document understanding parallels to those of Prof. K.S. Fu. In addition to the special recognition of Prof. Suen's contribution, I like to take this opportunity to thank all contributors of the handbook series, and the past co-editors Profs. L.F. Pau and P.S.P. Wang, as well as the World Scientific Publishing, without whose effort and encouragement, this series would not become a reality. Finally I like to thank my wife, Wanda, for her continued support and understanding in over four decades of my publication and research activities.

The following is a list of table of contents for the third, second and first editions of the handbook series.

Handbook of Pattern Recognition and Computer Vision, vol. 3, World Scientific Publishing 2005

Contents — Page Numbers

Preface to the Third Edition

Part 1 Basic Methods in Pattern Recognition **1**

1.1 Statistical Pattern Recognition, by R.P.W. Duin and D.M.J. Tax 3

1.2 Hidden Markov Models for Spatial Temporal Pattern Recognition, by Brian C. Lovell and Terry Caelli 25

1.3 A New Kernel-based Formalization of Minimum Error Pattern Recognition, by Erik McDermont and Shigeru Katagiri 41

1.4 Parallel Contextual Array Grammars with Trajectories, by P. Helen Chandra, P. Helen Chandra, C. Martin-Vide, K.G. Subramanian, D. L. Van and P.S.P. Wang 55

1.5 Pattern Recognition with Local Invariant Features, by C. Schmid, G. Dorko, S. Lazebnik, K. Mikolajczyk and J. Ponce 71

Part 2 Basic Methods in Computer Vision **99**
2.1 Case-based Reasoning for Image Analysis and Interpretation, by Petra Perner 101
2.2 Projective Geometry and Computer Vision, by Quang–Tuan Luong 115
2.3 Skeletonization in 3D Discrete Binary Images, by Gabriella Sanniti di Baja and Ingela Nystrom 137
2.4 Digital Distance Transforms in 2D, 3D, and 4D, by Gunilla Borgefers 157
2.5 Computing Global Shape Measures, by Paul Rosin 177
2.6 Texture Analysis with Local Binary Patterns, by Topi Maenpaa and Matti Pietikainen 197
Part 3 Recognition Applications **217**
3.1 Document Analysis and Understanding, by Yuan Yan Tang 219
3.2 Chinese Character Recognition, by Xiaoqing Ding 241
3.3 Extraction of Words from Handwritten Sentences, by Incheol Kim and Ching Y. Suen 259
3.4 OCR Assessment of Print-fonts for Enhancing Human Vision, by Ching Y. Suen, Qizhi Xu and Cedric Devoghelaere 273
3.5 Clustering and Classification of Web Documents Using Graph Models, by Adam Shenker, Horst Bunke, Mark Last and Abraham Kandel 287
3.6 Automated Detection of Masses in Mammograms, by H.D. Cheng, X.J. Shi, R. Min, X.P. Cai and H.N. Du 303
3.7 Wavelet-based Kalman Filtering in Scale Space for Image Fusion, by His-Chin Hsin and Ching-Chung Li 325
3.8 Multisensor Fusion with Hyperspectral Imaging Data: Detection and Classification, by Su May Hsu and Hsiao-hua Burke 347
3.9 Independent Component Analysis of Functional Magnetic Resonance Imaging Data, by Vince D. Calhoun and Baoming Hong 365
Part 4 Human Identification **385**
4.1 Multimodal Emotion Recognition, by Nicu Sebe, Ira Cohen and Thomas S. Huang 387
4.2 Gait-based Human Identification from a Monocular Video Sequence, by Amit Kale, Aravind Sundaresan, Amit K. RoyChowdhury, Rama Chellappa 411
4.3 Palmprint Authentication System, by David Zhang 431
4.4 Reconstruction of High-Resolution Facial Images for Visual Surveillance, by Jeong-Seon Park and Seong-Whan Lee 445
4.5 Object Recognition with Deformable Feature Graphs: Faces, Hands, and Cluttered Scenes, by Jochen Triesch and Christian Eckes 461
4.6 Hierarchical Classification and Feature Reduction for Fast Face Detection, by Bernd Heisele, Thomas Serre, Sam Prentice, and Tomaso Poggio 481

Part 5 System and Technology **497**
5.1 Tracking and Classifying Moving Objects Using Single or Multiple Cameras, by Quming Zhou and J.K. Aggarwal 499
5.2 Performance Evaluation of Image Segmentation Algorithms, by Ziaoyi Jiang 525
5.3 Content-based Video Analysis for Knowledge Discovery, by Chia Hung Yeh, Shih-Hung Lee and C.-C. Jay Kuo 543
5.4 Object-Process Methodology and its Applications to Image Processing and Pattern Recognition, by Dov Dori 559
5.5 Musical Style Recognition – a Quantitative Approach, by Peter van Kranenburg and Eric Backer 583
5.6 Auto-detector: Mobile Automatic Number Plate Recognition, by Giovanni B. Garibotto 601
5.7 Omnidirectional Vision, by Hiroshi Ishiguro 619
Index 629

Handbook of Pattern Recognition and Computer Vision, vol. 2, World Scientific Publishing 1999

Contents (*indicates chapter reprinted or modified from vol. 1) — Page Numbers

Preface to the Second Edition
Preface
Foreword
A Memorial to the Late Professor King-Sun Fu
Part 1 Basic Methods in Pattern Recognition **1**
1.1 Cluster Analysis and Related Issues*, by Richard C. Dubes 3
1.2 Statistical Pattern Recognition*, by Keinosuke Fukunaga 33
1.3 Syntactic Pattern Recognition, by Kou-Yuan Huang 61
1.4 Neural Net Computing for Pattern Recognition*, by Yoh-Han Pao 105
1.5 On Multiresolution Wavelet Analysis using Gaussian Markov Random Field Models, by C.H. Chen and G.G. Lee 143
1.6 A Formal Parallel Model for Three-Dimensional Object Pattern Representation*, by P.S.P. Wang 183
Part 2 Basic Methods in Computer Vision **205**
2.1 Texture Analysis*, by Mihran Taceryan and Anil K. Jain 207
2.2 Model-Based Texture Segmentation and Classification*, by R. Challappa, R.L. Kashyap and B.S. Manjunath 249
2.3 Color in Computer Vision: Recent Progress, by Glenn Healey and Quang-Tuan Luong 283
2.4 Projective Geometry and Computer Vision, by Roger Mohr 313
2.5 3-D Motion Analysis from Image Sequences using Point Correspondences, by John J. Weng and Thomas S. Huang 339

2.6	Signal-to-Symbol Mapping for Laser Rangfinders, by Kenong Wu and Martin D. Levine	387
2.7	3-D Vision of Dynamic Objects, by S.Y. Lu and Chandra Kambhametu	425
Part 3	**Recognition Applications**	**453**
3.1	Pattern Recognition in Nondestructive Evaluation of Materials*, by C.H. Chen	455
3.2	Discriminative Training—Recent Prgress in Speech Recognition, by Shigeru Katagri and Erik Mcdermott	473
3.3	Statistical and Neural Network Pattern Recognition Methods for Remote Sensing Aplications, by Jon Atli Benediktsson	507
3.4	Multi-SEnsory Opto-Electronic Featiure Extraction Neural Associative Retriever, by H.K. Liu, Y.H. Jin, Neville I. Marzell and Shaomin Zhou	535
3.5	Classification of Human Chromosomes—A Study of Correlated Behavior in Majority Vote, by Louisa Lam and Ching Y. Suen	567
3.6	Document Analysis and Recognition by Computers, by Yuan Y. Tang, M. Cheriet, Jiming Liu, J.N. Said and Ching Y. Suen	579
3.7	Pattern Recognition and Visualization of Sparsely Sampled Biomedical Signals*, by Ching-Chang Li, T.P. Wang and A.H. Vagnucci	613
3.8	Pattern Recognition and Computer Vision for Geographic Data Analysis, by F. Cavayas and Y. Baudouin	625
3.9	Face Recognition Technology, by Martin Lades	667
Part 4	**Inspection and Robotic Applications**	**685**
4.1	Computer Vision in Food Handling and Sorting*, by Hordur Anarson and Magnus Asmundsson	687
4.2	Approaches to texture-Based Classification, Segmentation and Surface Inspection, by Matti Pietikainen, Timo Ojala and Olli Silven	711
4.3	Context Related Issues in Image Understanding*, by L.F. Pau	737
4.4	Position Estimation Techniques for An Autonomous Mobile Robot—A Review*, by Raj Talluri and J.K. Aggarwal	765
4.5	Computer Vision in Postal Automation, by G. Garibotto and C. Scagliola	797
4.6	Vision-Based Automatic Road Vehicle Guidance, by Dieter Koller, Quang-Tuan Luong, Joseph Weber and Jitendra Malik	817
Part 5	**Architecture and Technology**	**855**
5.1	Vision Engineering: Designing Computer Vision Systems*, by Rama Chellappa and Azriel Rosenfeld	857
5.2	Optical Pattern Recognition for Computer Vision*, by David Casasent	869
5.3	Infra-Red Thermography: Techniques and Applications, by M.J. Varga and P.G. Ducksbury	891
5.4	Viewer-Centered Representations in Object Recognition: A Computational Approach*, by Ronen Basri	925
5.5	Video Content Analysis and Retrieval, by Hongjiang Zhang	945

5.6 VLSI Architectures for Moments and Their Applications to Pattern Recognition, by Heng-da Cheng, Chen-Yuan Wu and Jiguang Li 979
Index 1003

Handbook of Pattern Recognition and Computer Vision, vol. 1, World Scientific Publishing 1993

Contents — Page Numbers

Preface
Part 1 Basic Methods in Pattern Recognition **1**
1.1 Cluster Analysis and Related Issues, by Richard C. Dubes 3
1.2 Statistical Pattern Recognition, by Keinosuke Fukunaga 33
1.3 Large-Scale Feature Selection, by Jack Skalansky et al. 61
1.4 Neural Net Computing for Pattern Recognition, by Yoh-Han Pao 125
1.5 Structural and Syntactic Pattern Recogniton, by Horst Bunke 163
1.6 A Formal Parallel Model for Three-Dimensional Object Pattern Representation, by P.S.P. Wang 211
Part 2 Basic Methods in Computer Vision **233**
2.1 Texture Analysis, by Mihran Tuceryan and Anil K. Jain 235
2.2 Model-Based Texture Segmentation and Classification, by R. Chellappa, R.L. Kashyap and B.S. Munjunath 277
2.3 Color in Computer Vision, by Quang-Tuan Luong 311
2.4 Projective Geometry and Computer Vision, by Roger Mohr 369
2.5 3-D Motion Analysis from Image Sequences Using Point Correspondences, by John J. Weng and Thomas S. Huang 395
2.6 Segmentation Tools in Mathematical Morphology, by Serge Beucher 443
2.7 Parallel Thinning Algorithm for Binary Digital Patterns, by Yung-Sheng Chen and Wen-Hsing Hsu 457
Part 3 Recognition Applications **491**
3.1 Pattern Recognition in Nondestructive Evaluation of Materials, by C.H. Chen 493
3.2 Pattern Recognition in Geophysical Signal Processing and Interpretation, by Yanda Li, Zhaoqi Bian, Pinfan Yan and Tong Chang 511
3.3 Signal Transient Analysis and Classification Techniques, by R. Craig Olson 541
3.4 Analysis of Economic and Business Information, by Claudio Gianotti 569
3.5 Optical Handwritten Chinese Character Recognition, by Jun S. Huang 595
3.6 Automatic Analysis and Understanding of Documents, by Yuan Y. Tang, Chang D. Yan, M. Cheriet and Ching Y. Suen 625
3.7 Pattern Recognition and Visualization of Sparsely Sampled Biomedical Signals, by Ching-Chung Li, T.P. Wang and A.H. Vagnucci 655

3.8	Understanding Microvessels in Two and Three Dimensions, by Carl E. Wick, Murray H. Loew and Joseph Kurantsin-Mills	667
Part 4	**Inspection and Robotics Applications**	**695**
4.1	Computer Vision in Food Handling and Sorting, by Hordur Arnarson and Magnus Asmundsson	697
4.2	Image Information Retrieval Systems, by Ang Yew-Hock, Arcot Desai Narasimhalu, and Saliman Al-Hawanamdeh	719
4.3	Context Related Sigues in Image Understanding, by L.F. Pau	741
4.4	Positional Estimation Techniques for An Autonomous Mobile Robot—A Review, by Raj Talluri and J.K. Aggrawal	769
Part 5	**Architectures and Technology**	**803**
5.1	Vision Engineering: Designing Computer Vision Systems, by Rama Chellappa and Azriel Rosenfeld	805
5.2	Optical Pattern Recognition for Computer Vision, by David Casasent	817
5.3	Spatial Knowledge Representation for Iconic Image Database, by Suh-Yin Lee and Fang-Jung Hsu	839
5.4	Viewer-Centered Representation for Object Recognition: A Computational Approach, by Ronen Basri	863
5.5	Connectionist Architectures in Low Level Image Segmentation, by W. Ekkehard Blanz, Charles E. Cox and Sheri L. Gish	883
5.6	Architectures for Image Processing and Computer Vision, by Giovanni Garibotto	921
5.7	Image Information Systems: Where Do We Go from Here?, by Shi-Kuo Chang and Arding Hsu	941
Subject Index		**967**

July 2009 C.H. Chen

Preface to the 4th Edition, by C.H. Chen

Largely motivated by the rapid progress in pattern recognition and computer vision, as shown by a large number of journal and conference publications and the program of the latest ICPR (International Conference on Pattern Recognition), the 4th edition of the Handbook is published one year ahead of six year interval from the 3rd edition. The year 2010 also marks the 80th birthday of the late Prof. King-Sun Fu, for which this Handbook series is dedicated to.

As in the previous editions, the book consists of five parts, with all thirty-three chapters being new. Part 1 Basic Methods in Pattern Recognition Part 2 Basic Methods in Computer Vision and Image Processing, Part 3 Recognition Applications Part 4 Computer Vision and Pattern Recognition in Life Sciences and Human ID Part 5. System and Technology.

Chapter 1.1 of **Part 1** by F. D. Torre presents a unified approach to component analysis for pattern recognition, making use of least-squares weighted kernel reduced rank regression (LS-WKRRR). Chapter 1.2 by V. Gunes, et al. introduces and explains several essential concepts and components for building multiple classifier systems. Chapter 1.3 by H. Bunke and K. Riesen deals with graph based object representation and introduces a novel approach for graph embedding in vector space. Chapter 1.4 by E. Granger, et al. shows that match tracking parameter of fuzzy ARTMAP neural networks with particle swarm optimization can produce a significantly lower generalization error than with other match tracking strategies. Chapter 1.5 by T.A. Duong, et al. presents a bio-inspired approach to object recognition by taking a global view of an object that considers the shape feature as the logical building block and the color feature, if available, as additional information. Though Part 1 covers disjoint methods in pattern recognition, it reflects the divergent approaches to pattern recognition as the present time.

Chapter 2.1 of **Part 2** by L. Zhang, et al. introduces the concepts and basic types of probabilistic graphical methods, which are shown to perform important computer vision tasks such as facial expression recognition and image segmentation. Chapter 2.2 by C. Lei and Y. H. Yang presents a region-tree based framework for general image discrete labeling problems with application to stereo matching and optical estimation.

Chapter 2.3 by T. Bouwmans, et al. has a comprehensive survey of the statistical background modeling in the context of moving object (foreground) detection. Chapter 2.4 by K. Kanatani, et al. presents a rigorous mathematical treatment of algorithms for 3-D reconstruction from two views. Chapter 2.5 by E. Rahtu and J. Heikkila considers affine invariance in feature extraction using normalization and invariant approaches with particular reference to multiscale framework that offers systematic tool to alleviate some of the basic problems encountered with traditional invariant techniques. Chapter 2.6 by N. Ohmishi, et al. presents an algorithm for detecting the dominant plane by using Independent Component Analysis for robot navigation. Chapter 2.7 by S. Tari considers creating a field within the shape domain with emergent structures capturing the parts automatically and the field is computed by minimizing an energy which captures both local and global as well as both region and boundary based interactions among shape points.

Thus Part 2 has covered an expanded set of topics in computer vision.

Chapter 3.1 of **Part 3** by G. Moser, et al. presents for the combined optical and synthetic aperture radar remote sensing image data a novel region-based semiparametric classification technique. Chapter 3.2 by L. Bruzzone and C. Persello has a comprehensive presentation of support vector machine for remote sensing classification. Chapter 3.3 by M. I. Shah and C.Y. Suen provides a comprehensive survey of word searching techniques to detect and locate instances of the given template/query images in the document image databases. Chapter 3.4 by C. L. He, et al. presents a novel rejection criterion that employs linear discriminant analysis to optimize the criterion for rejection in handwriting recognition. Chapter 3.5 by M. Liwicki and H. Bunke evaluates recognition techniques for whiteboard notes written in Roman scripts including the hidden Markov model (HMM)-based recognizer and the bidirectional long short-term memory networks. Chapter 3.6 by J. Kim and E. Andre investigates the automatic emotion recognition using physiological signals with physiological features extracted from various analysis domains for classification of four musical emotions using an extended linear discriminant analysis. Chapter 3.7 by F. Shih considers automatic solar flare detection with solar images observed by a telescope in the Big Bear Solar Observatory. Though Part 3 covers only a limited number of application areas it certainly reflects the emergence of many potential pattern recognition applications.

Chapter 4.1 of **Part 4** by Y. Feng and P.C. Yuen discusses the threats to biometric authentication systems and security enhancing schemes for biometric recognition systems. Chapter 4.2 by S. Sarker, et al. presents a framework of the problem of recognition of continuous sign language, i.e. signs in sentences and not isolated signs or finger-spelled signs, by using combined manual and non-manual information. Chapter 4.3 by P. Manandhar, et al. considers the automated intra-vascular ultrasound image segmentation (IVUS) for diagnosis of coronary heart disease by using active contour model algorithm with special reference to tracking the change in guidewire position such that its effect on IVUS can be minimized. Chapter 4.4 by V. Gervasio and J.A. Jorge employs the interactive technique to improve the capability of active contour model in the segmentation of medical images. Chapter 4.5 by V. Meas-Yedid, et al. addresses the issues of both cell segmentation and spot detection for the images of biological objects. Chapter 4.6 by E. Zaharia and D. Maroulis presents the evolutionary genetic algorithms to automate the procedures of gridding microarray images and segmenting microarray spots, which are both challenging problems normally requiring human intervention. Chapter 4.7 deals with the multi-class protein folds recognition by using error-correcting output coding and ensemble classifier (support vector machine or flexible neural nets). Part 4 thus presents largely chapters that employ pattern recognition and computer vision techniques in life science problems.

Chapter 5.1 of **Part 5** by M.D. Levine and Y. Yu describes a 3D facial reconstruction system with the ability to recover a 3D human face from a single 2D frontal natural or standard image by using separately computed stages of shape reconstruction and texture recovery. Chapter 5.2 by J. Zhou and H. Peng formulates the automatic annotation as a pattern recognition problem with special focus on annotating fruitfly gene expression pattern images during embryogenesis and considering applications for multi-objective vs. mutual-exclusive annotation, and ROI vs. entire image annotation. Chapter 5.3 by Z. Ying and R. Naidu presents the dual energy computed tomography technology and associated Automated Threat Detection techniques in the context of checked baggage screening. Chapter 5.4 by M. Sun, et al. deals with assessment of food intake and physical activity by addressing the issue in food dimension measurement using spectral sensors and image processing models and algorithms, and presenting an effective computational method to characterize walking and jogging, Chapter 5.5 by E. Vallasques, et al. has a very detailed survey of watermarking systems employing evolutionary computation techniques that automatically set user-defined parameters of watermarking tasks and a case study is also presented. Chapter 5.6 by B. Lu and X. Wang presents a framework of large scale classification problems, such as a large scale Japanese patent classification problem, that has three independent components: decomposing training data sets, training component or modular classifiers in parallel and combining trained component classifiers. Chapter 5.7 by G.B. Garibotto presents the main algorithms for human tracking around the 3D model based framework in an effort to implement an intelligent visual surveillance system. The topics covered in Part 5 again reflect the diversified activities in systems and technology.

The one sentence description of each chapter clearly is highly inadequate to present the excellent contributions by all authors We should let each chapter to speak for itself and thus the readers are strongly encouraged to go over individual chapters in detail. The handbook is indeed very rich in information about the vibrant activities in theory, applications and technology in pattern recognition and computer vision.

July 2009 C.H. Chen

Contents

A Brief Introduction to the Handbook Series (by C.H. Chen) v

Preface to the 4th Edition (by C.H. Chen) xiii

Contents xvii

Part 1. Basic Methods in Pattern Recognition **1**

Chapter 1.1 A Unification of Component Analysis Methods 3
F. De la Torre

Chapter 1.2 Multiple Classifier Systems: Tools and Methods 23
Veyis Gunes, Michael Ménard and Simon Petitrenaud

Chapter 1.3 On Dissimilarity Embedding of Graphs in Vector Spaces 47
Horst Bunke and Kaspar Riesen

Chapter 1.4 Match Tracking Strategies for Fuzzy ARTMAP Neural Networks 73
Eric Granger, Philippe Henniges, Robert Sabourin and Luiz S. Oliveira

Chapter 1.5 Shape and Color Features for Object Recognition Search 101
Tuan A. Duong, Vu A. Duong and Allen R. Stubberud

Part 2. Basic Methods in Computer Vision and Image Processing **127**

Chapter 2.1 Probabilistic Graphical Models and Their Applications in Computer Vision 129
Lei Zhang, Yan Tong and Qiang Ji

Chapter 2.2 A Region Tree Based Image Discrete Labeling Framework 157
Cheng Lei and Yee-Hong Yang

Chapter 2.3 Statistical Background Modeling for Foreground Detection: A Survey 181
Thierry Bouwmans, Fida El Baf and Bertrand Vachon

Chapter 2.4 Latest Algorithms for 3-D Reconstruction from Two Views 201
Kenichi Kanatani, Yasuyuki Sugaya and Yasushi Kanazawa

Chapter 2.5 Affine Invariant Features in Pattern Recognition 235
Esa Rahtu and Janne Heikkilä

Chapter 2.6 Statistical Aspects of Dominant Plane for Robot Navigation 257
Naoya Ohnishi and Atsushi Imiya

Chapter 2.7 Extracting Parts of 2D Shapes Using Local and Global Interactions Simultaneously 283
Sibel Tari

Part 3. Recognition Applications 305

Chapter 3.1 Contextual Classification of Multisensor Optical-SAR Remote Sensing Images 307
Gabriele Moser, Raffaele Gaetano, Giami Poggi, Giuseppe Scarpa and Sebastiano B. Serpico

Chapter 3.2 Approaches Based on Support Vector Machine to Classification of Remote Sensing Data 329
Lorenzo Bruzzone and Claudio Persello

Chapter 3.3 Word Spotting Techniques in Document Analysis and Retrieval — A Comprehensive Survey 353
Muhammad Ismail Shah and Ching Y. Suen

Chapter 3.4 Optimization of Rejection Parameters to Enhance Reliability in Handwriting Recognition 377
Chun Lei He, Louisa Lam, and Ching Y. Suen

Chapter 3.5 Recognition Techniques for Whiteboard Notes Written in Roman Script 397
Marcus Liwicki and Horst Bunke

Chapter 3.6 Detection of Affective Patterns in Physiological Signals Towards Improving Automatic Emotion Recognition 415
Jonghwa Kim and Elisabeth André

Chapter 3.7 Automated Detection and Classification of Solar Image Features 435
Frank Y. Shih

Part 4. Computer Vision and Pattern Recognition in Life Sciences and Human ID 453

Chapter 4.1 Biometric Template Protection: Towards a Secure Biometric System 455
Yicheng Feng and Pong C. Yuen

Chapter 4.2 Fusion of Manual and Non-Manual Information in American Sign Language Recognition 477
Sudeep Sarkar, Barbara Loeding and Ayush S. Parashar

Chapter 4.3 A Deformable Template Model with Feature Tracking for Automated IVUS Segmentation 497
Prakash Manandhar, Chi Hau Chen and Ahmet Umit Coskun

Chapter 4.4 User-Assisted Segmentation for Medical Images 517
Vasco Gervásio and Joaquim A. Jorge

Chapter 4.5 Automated Quantification of Cellular Processes with Applications to High-Content Screening 537
Vannary Meas-Yedid, Alexandre Dufour, Bo Zhang, Alexandre Grassart, Nathalie Sauvonnet and Jean-Christophe Olivo-Marin

Chapter 4.6 Advanced Evolutionary Algorithms for Intelligent Microarray Image Analysis 557
Eleni Zacharia and Dimitris Maroulis

Chapter 4.7 Computational Intelligence Approaches for Protein Fold Recognition 579
Yuehui Chen, Yaou Zhao and Jack Y. Yang

Part 5. System and Technology 593

Chapter 5.1 A Method for the Synthesis of a 3D Face Using a Single 2D Image 595
Martin D. Levine and Yingfeng Yu

Chapter 5.2 Automatic Annotation of Biological Image 625
Jie Zhou and Hanchuan Peng

Chapter 5.3 Dual Energy Computed Tomography for Checked Baggage Screening 645
Zhengrong Ying and Ram Naidu

Chapter 5.4 Assessment of Food Intake and Physical Activity: A Computational Approach 667
Mingui Sun, John D. Fernstrom, Wenyan Jia, Ning Yao, Steven A. Hackworth, Xiaoyu Liu, Chengliu, Qiang Liu, Yuecheng Li, Madelyn H. Fernstrom, Robert J. Sclabassi

Chapter 5.5 Intelligent Watermarking Systems: A Survey 687
Eduardo Vellasques, Eric Granger and Robert Sabourin

Chapter 5.6 A Parallel and Modular Pattern Classification Framework for Large-Scale Problems 725
Bao-liang Lu and Xiao-lin Wang

Chapter 5.7 3-D Computer Vision Modeling in Video Surveillance Applications 747
Giovanni B. Garibotto

Index 767

Part 1

Basic Methods in Pattern Recognition

CHAPTER 1.1

A UNIFICATION OF COMPONENT ANALYSIS METHODS

F. De la Torre

Robotics Institute. Carnegie Mellon University.
5000 Forbes av. 211 Smith Hall. Pittsburgh, PA 15213.
E-mail: ftorre@cs.cmu.edu

Over the last century Component Analysis (CA) methods such as Principal Component Analysis (PCA), Linear Discriminant Analysis (LDA), Canonical Correlation Analysis (CCA), k-means and Spectral Clustering (SC) have been extensively used as a feature extraction step for modeling, classification, visualization and clustering. CA techniques are appealing because many can be formulated as eigen-problems, offering great potential for learning linear and non-linear representations of data without local minima. However, the eigen-formulation often conceals important analytic and computational drawbacks of CA techniques, such as solving generalized eigen-problems with rank deficient matrices, lacking intuitive interpretation of normalization factors, and understanding relationships between CA methods.

This chapter proposes a unified framework to formulate many CA methods as a least-squares estimation problem. We show how PCA, LDA, CCA, k-means, spectral graph methods and kernel extensions correspond to a particular instance of a least-squares weighted kernel reduced rank regression (LS-WKRRR). The least-squares formulation allows better understanding of normalization factors, provides a clean framework to understand the communalities and differences between many CA methods, yields efficient optimization algorithms for many CA algorithms, suggest easy derivation for on-line learning methods, and provides an easier generalization of CA techniques. In addition, we derive weighted generalizations of PCA, LDA, SC and CCA (including kernel extensions).

1. Introduction

Over the last century, Component Analysis (CA) methods[1] such as Kernel Principal Component Analysis (KPCA),[2] Linear Discriminant Analysis (LDA),[3] Canonical Correlation Analysis,[4] and Spectral Clustering (SC)[5] have been extensively used as a feature extraction step in modeling, clustering, classification and visualization problems. The aim of CA techniques is to decompose a signal into *relevant* components that are optimal for a given task (e.g. classification, visualization). These components, explicitly or implicitly (e.g. kernel methods), define the representation of the signal. CA techniques are appealing for two main reasons. Firstly, CA models typically have a small number of parameters, and therefore can be estimated using relatively few samples. CA techniques are especially useful to handle high-dimensional data due to the *curse-of-dimensionality*, which usually requires a large number of samples to build accurate models. Secondly, many CA techniques can be formulated as eigen-problems, offering great potential for efficient learning of linear and non-linear models without local minima. The use of eigen-solvers to address statistical problems dates back to the 1930s, and since then, many numerically stable and efficient packages have been developed to solve eigen-problems. For these reasons, during the last century many computer vision, computer graphics, signal processing, and statistical problems were posed as problems of learning a low dimensional CA model.

Although CA methods have been widely used in many scientific disciplines, there is still a need for a better mathematical framework than the eigen-formulation to analyze and extend CA techniques. The least-squares unified framework proposed in this chapter provides a tool for analyzing, generalizing, and developing efficient algorithms to solve many CA methods. This chapter shows how Kernel PCA, Kernel LDA, Kernel CCA, k-means, and Normalized Cuts correspond to a particular instance of a least-squares weighted kernel reduced rank regression (LS-WKRRR) problem. This framework should provide researchers with a thorough understanding of a large number of existing CA techniques, and it may serve as a tool for dealing with novel least-squares (LS) problems as they arise. Preliminary versions of this work were published in a technical report.[6]

This paper recovers the spirit of three previous published papers seeking unified frameworks. Borga[7] showed how PCA, Partial Least Squares (PLS), Canonical Correlation Analysis (CCA) and Multiple Linear Regression (MLR) can be formulated as a generalized eigen-value problems (GEPs). He proposed a gradient-descent algorithm on the Rayleigh quotient to efficiently solve the GEP. Roweis and Ghahramani[8] showed how a Linear Dynamical System (LDS) is the generative model for Hidden Markov Models, Kalman Filter, vector quantization, Factor Analysis, and mixture of Gaussians. By introducing non-linearities in LDS, the authors[8] showed how Independent Component Analysis (ICA) can also be cast as an extension of a LDS. Yan *et al.*[9] have recently proposed a unifying view of PCA, LDA, LPP, Isomap, and LDA using a graph theoretical formulation. Additionally, the authors propose Marginal Fisher Analysis, a variant of non-parametric LDA.[10] This work differs from previous works in that we unified PCA, CCA, LDA, SC, and kernel generalizations with a LS-WKRRR. In addition, we propose new weighted extensions for PCA, CCA, LDA, and SC.

The rest of the chapter is organized as follows: Section 2 introduces the notation and formulates the typical covariance matrices of CA methods using a compact matrix formulation. Section 3 introduces the LS-WKRRR problem and derives the coupled generalized eigenvalue system resulting from solving it. Section 4 relates PCA, KPCA and weighted extensions to the LS-WKRRR. Section 5 shows how LDA, KLDA, CCA, KCCA and weighted extensions are a particular instance of LS-WKRRR. Section 6 shows the relationship between LS-WKRRR, k-means and spectral clustering. Section 7 finalizes the chapter presenting the conclusions.

2. Covariance Matrices in Component Analysis

Many CA methods can be formulated as generalized eigenvalue problems (GEPs). This section derives a compact matrix expression for most common covariance matrices used to solve CA methods through GEPs.

Let $\mathbf{D} \in \Re^{d \times n}$ (see notation [a]) be a matrix, where each column is a vectorized data sample from one of c classes. d denotes the number of features and n number of samples. Some of the most common CA covariance matrices can be conveniently expressed in matrix form as:[11]

$$\mathbf{S}_t = \frac{1}{n-1}\sum_{j=1}^{n}(\mathbf{d}_j - \mathbf{m})(\mathbf{d}_j - \mathbf{m})^T = \frac{1}{n-1}\mathbf{D}\mathbf{P}_t\mathbf{D}^T$$

$$\mathbf{S}_w = \frac{1}{n-1}\sum_{i=1}^{c}\sum_{\mathbf{d}_j \in C_i}(\mathbf{d}_j - \mathbf{m}_i)(\mathbf{d}_j - \mathbf{m}_i)^T = \frac{1}{n-1}\mathbf{D}\mathbf{P}_w\mathbf{D}^T$$

$$\mathbf{S}_b = \frac{1}{n-1}\sum_{i=1}^{c} n_i(\mathbf{m}_i - \mathbf{m})(\mathbf{m}_i - \mathbf{m})^T = \frac{1}{n-1}\mathbf{D}\mathbf{P}_b\mathbf{D}^T$$

where $\mathbf{m} = \frac{1}{n}\mathbf{D}\mathbf{1}_n$ is the mean vector, $\mathbf{m}_i$ is the mean vector for class i, n_i denotes the number of samples for class i, and $\mathbf{P}_i$ are projection matrices (i.e. $\mathbf{P}_i^T = \mathbf{P}_i$ and $\mathbf{P}_i^2 = \mathbf{P}_i$) with the following expressions:

$$\mathbf{P}_t = \mathbf{I}_n - \frac{1}{n}\mathbf{1}_n\mathbf{1}_n^T \qquad \mathbf{P}_w = \mathbf{I}_n - \mathbf{G}(\mathbf{G}^T\mathbf{G})^{-1}\mathbf{G}^T \qquad \mathbf{P}_b = \mathbf{G}(\mathbf{G}^T\mathbf{G})^{-1}\mathbf{G}^T - \frac{1}{n}\mathbf{1}_n\mathbf{1}_n^T$$

$\mathbf{G} \in \Re^{n \times c}$ is an indicator matrix such that $\sum_j g_{ij} = 1$, $g_{ij} \in \{0, 1\}$, and g_{ij} is 1 if $\mathbf{d}_i$ belongs to class j, and 0 otherwise. $\mathbf{S}_b$ is the between-class covariance matrix and represents the average distance between the means of the classes. $\mathbf{S}_w$ is the within-class covariance matrix that contains information about the average compactness of each class. $\mathbf{S}_t$ is the total covariance matrix. Using the previous matrix expressions, it is straightforward to show that $\mathbf{S}_t = \mathbf{S}_w + \mathbf{S}_b$. The upper bounds on the ranks of the matrices are $\min(c-1, d), \min(n-c, d), \min(n-1, d)$ for $\mathbf{S}_b, \mathbf{S}_w$, and $\mathbf{S}_t$ respectively.

3. A Generative Model for Component Analysis

This section introduces the formulation for the Least-Squares weighted kernel reduced rank regression (LS-WKRRR) problem. In the following sections, we will show how the LS-WKRRR is the generative model for many CA methods, including kernel PCA, kernel LDA, kernel CCA, k-means, and normalized cuts.

[a] Bold capital letters denote a matrix $\mathbf{D}$, bold lower-case letters a column vector $\mathbf{d}$. $\mathbf{d}_j$ represents the jth column of the matrix $\mathbf{D}$. All non-bold letters denote scalar variables. $\mathbf{d}^j$ is a column vector that represents the j-th row of the matrix $\mathbf{D}$. d_{ij} denotes the scalar in the row i and column j of the matrix $\mathbf{D}$. $\mathbf{1}_k \in \Re^{k \times 1}$ is a vector of ones. $\mathbf{I}_k \in \Re^{k \times k}$ denotes the identity matrix. $||\mathbf{d}||_2^2$ denotes the norm of the vector $\mathbf{d}$. $tr(\mathbf{A}) = \sum_i a_{ii}$ is the trace of the matrix $\mathbf{A}$ and $|\mathbf{A}|$ denotes the determinant. $vec(\mathbf{A})$ is a linear operator which converts a matrix $\mathbf{A} \in \Re^{m \times n}$ into a column vector $\mathbf{a} \in \Re^{mn \times 1}$. $||\mathbf{A}||_F^2 = tr(\mathbf{A}^T\mathbf{A}) = tr(\mathbf{A}\mathbf{A}^T)$ designates the Frobenius norm of a matrix. $\circ$ denotes the Hadamard or point-wise product, and $\otimes$ the Kronecker product.

3.1. *Least-Squares Weighted Kernel Reduced Rank Regression (LS-WKRRR)*

Since its introduction in the early 1950s by Anderson,[12,13] the reduced-rank regression (RRR) model has inspired a wealth of diverse applications in several fields such as signal processing[14,15] (also known as reduced-rank Wiener filtering), neural networks[16] (also known as asymmetric PCA), time series analysis,[12] and computer vision.[17] This section, extends previous work on RRR by introducing kernels and weights within a least-squares formulation, and it derives the system of GEPs resulting from solving it.

Given two data sets $\mathbf{X} \in \Re^{x \times n}$ and $\mathbf{D} \in \Re^{d \times n}$, the RRR model[12,15,16] finds a linear mapping, $\mathbf{T} \in \Re^{x \times d}$, that minimizes the LS error subject to rank constraints on $\mathbf{T}$. The RRR model minimizes $||\mathbf{D} - \mathbf{T}\mathbf{X}||_F^2$ subject to $rank(\mathbf{T}) = k$. The rank constraint is typically needed when $\mathbf{X}$ is high dimensional and the dimension of the features is larger than the samples ($x > n$).

The LS-WKRRR extends previous work on RRR on three aspects: (1) it explicitly parameterizes $\mathbf{T}$ as the outer product of two matrices of rank k, that is $\mathbf{T} = \mathbf{B}\mathbf{A}^T$, where $\mathbf{A} \in \Re^{x \times k}$ and $\mathbf{B} \in \Re^{d \times k}$, as has previously proposed;[15–17] (2) allows for non-linear regression. LS-WKRRR maps the input space of $\mathbf{D}$ and $\mathbf{X}$ to a feature space using kernel methods. That is, $\mathbf{\Gamma} = \phi(\mathbf{D}) = [\ \phi(\mathbf{d}_1)\ \phi(\mathbf{d}_2)\ \cdots\ \phi(\mathbf{d}_n)\ ] \in \Re^{d_d \times n}$ represents a mapping of $\mathbf{D}$. ϕ denotes a mapping from the d dimensional space to the feature space (d_d dimension). Similarly, $\mathbf{\Upsilon} = \varphi(\mathbf{X}) = [\ \varphi(\mathbf{x}_1)\ \varphi(\mathbf{x}_2)\ \cdots\ \varphi(\mathbf{x}_n)\ ] \in \Re^{d_x \times n}$ denotes the mapping for $\mathbf{X}$. ϕ, φ map the data to a (usually) higher dimensional space, where the data is more likely to behave linearly (if the right mapping is found). (3) The LS-WKRRR introduces weights for the features $\mathbf{W}_r \in \Re^{d_d \times d_d}$ and samples $\mathbf{W}_c \in \Re^{n \times n}$.

The LS-WKRRR problem minimizes the following expression:

$$E_0(\mathbf{A}, \mathbf{B}) = ||\mathbf{W}_r(\mathbf{\Gamma} - \mathbf{B}\mathbf{A}^T\mathbf{\Upsilon})\mathbf{W}_c||_F^2 \tag{1}$$

with respect to the regression matrices, $\mathbf{A} \in \Re^{d_x \times k}$ and $\mathbf{B} \in \Re^{d_d \times k}$. Typically $\mathbf{A}$ corresponds to the projection matrix (e.g. LDA), and $\mathbf{B}$ is a generative matrix for the column space of $\mathbf{\Gamma}$. $\mathbf{W}_r \in \Re^{d_d \times d_d}$ is a matrix that weights the contributions of features (e.g. PCA) or classes (e.g. LDA). Similarly, $\mathbf{W}_c \in \Re^{n \times n}$ weights the importance of each sample. In the following, we will assume that the weighting matrices are symmetric. The mappings ϕ and φ do not need to be explicitly computed and only the kernel between two samples needs to be defined. Kernel methods[18,19] make use of the *kernel trick* to implicitly define the mapping by means of a kernel function. A given function is a kernel if, and only if, the value it produces for two vectors corresponds to a dot product in some Hilbert feature space. This is the well-known Representer Theorem: *"Every positive definite, symmetric function is a kernel. For every kernel k, there is a function $\phi(\mathbf{x}) : k(\mathbf{d}_1, \mathbf{d}_2) = \langle\phi(\mathbf{d}_1), \phi(\mathbf{d}_2)\rangle$."*, where $\langle\rangle$ denotes dot product.

The necessary conditions on $\mathbf{A}$ and $\mathbf{B}$ for the critical points of Eq. (1) are:

$$\frac{\partial E_0}{\partial \mathbf{B}} = \mathbf{W}_r^2\mathbf{B}\mathbf{A}^T\mathbf{\Upsilon}\mathbf{W}_c^2\mathbf{\Upsilon}^T\mathbf{A} - \mathbf{W}_r^2\mathbf{\Gamma}\mathbf{W}_c^2\mathbf{\Upsilon}^T\mathbf{A} = \mathbf{0} \tag{2}$$

$$\frac{\partial E_0}{\partial \mathbf{A}} = \mathbf{\Upsilon}\mathbf{W}_c^2\mathbf{\Gamma}^T\mathbf{W}_r^2\mathbf{B} - \mathbf{\Upsilon}\mathbf{W}_c^2\mathbf{\Upsilon}^T\mathbf{A}\mathbf{B}^T\mathbf{W}_r^2\mathbf{B} = \mathbf{0} \tag{3}$$

Eq. (2) and Eq. (3) form a set of coupled equations that have solutions in terms of a GEP in either $\mathbf{A}$ or $\mathbf{B}$. Substituting the optimal $\mathbf{B} = \mathbf{\Gamma W}_c^2 \mathbf{\Upsilon}^T \mathbf{A}(\mathbf{A}^T \mathbf{\Upsilon W}_c^2 \mathbf{\Upsilon}^T \mathbf{A})^{-1}$ derived from Eq. (2) into Eq. (1) leads to the following expression:

$$E_0(\mathbf{A}) \propto tr\big((\mathbf{A}^T \mathbf{\Upsilon W}_c^2 \mathbf{\Upsilon}^T \mathbf{A})^{-1}(\mathbf{A}^T \mathbf{\Upsilon W}_c^2 \mathbf{\Gamma}^T \mathbf{W}_r^2 \mathbf{\Gamma W}_c^2 \mathbf{\Upsilon}^T \mathbf{A})\big) \quad (4)$$

Similarly, substituting the optimal value of $\mathbf{A} = (\mathbf{\Upsilon W}_c^2 \mathbf{\Upsilon}^T)^{-1} \mathbf{\Upsilon W}_c^2 \mathbf{\Gamma}^T \mathbf{W}_r^2 \mathbf{B}(\mathbf{B}^T \mathbf{W}_r^2 \mathbf{B})^{-1}$ from Eq. (3) into Eq. (1) leads to the following expression:

$$E_0(\mathbf{B}) \propto tr\big((\mathbf{B}^T \mathbf{W}_r^2 \mathbf{B})^{-1}(\mathbf{B}^T \mathbf{W}_r^2 \mathbf{\Gamma W}_c^2 \mathbf{\Upsilon}^T (\mathbf{\Upsilon W}_c^2 \mathbf{\Upsilon}^T)^{-1} \mathbf{\Upsilon W}_c^2 \mathbf{\Gamma}^T \mathbf{W}_r^2 \mathbf{B})\big) \quad (5)$$

Eq. (4) and Eq. (5) are the **fundamental equations** of CA methods. In the rest of the manuscript, we will show how to relate many CA methods to these two equations. Eq. (4) and Eq. (5) are standard Rayleigh quotients (i.e. $J(\mathbf{B}) = tr\big((\mathbf{B}^T \mathbf{S}_1 \mathbf{B})^{-1} \mathbf{B}^T \mathbf{S}_2 \mathbf{B})\big)$, and the solution is given by the following GEP, $\mathbf{S}_2 \mathbf{B} = \mathbf{S}_1 \mathbf{B} \Lambda$.[10] The solution of Eq. (4) is unique up to an invertible transformation $\mathbf{R}$, that is, $E_o(\mathbf{AR}) = E_o(\mathbf{A})$. Similarly for Eq. (5).

Recasting CA eigenvalue problems as a LS-WKRRR problem (E_0), eq (1), has a number of desirable benefits that will be illustrated throughout the chapter:

(1) E_0 provides a unifying expression for many CA methods. The commonalities and differences between the methods, as well as the intrinsic relationship, can be easily understood from E_0.
(2) The Least-Squares (LS) formulation provides an alternative and simple framework to understand normalization factors in CA methods. For instance, normalization terms in spectral graph clustering, or weighting factors in PCA/LDA.
(3) The surface of E_0, Eq. (1), has a unique local minimum,[20] so most optimization algorithms will exhibit almost global convergence properties. In general, optimization theory provides a solid framework for convergence analysis, and many optimization methods are suitable (section 3.2).
(4) Many numerical optimization methods (e.g. gradient descent, alternated least-squares) can be applied to solve E_0. Directly optimizing E_0 solves the small sample size (SSS) problem of standard GEPs when dealing with high dimensional data. Moreover, optimization-based algorithms are more efficient for large-scale problems, especially if matrices are sparse. In addition, using the least-squares formulation one can easily derive extensions for online component analysis methods.
(5) The least-squares formulation allows many straight-forward extensions of CA methods. It is unclear how to formulate these new extensions using an eigenvalue framework.

3.2. *Computational Aspects of LS-WKRRR*

This section proposes three optimization strategies to optimize the LS-WKRRR problem E_0, Eq. (1).

3.2.1. *Subspace Iteration*

Standard numerical packages to solve GEPs (i.e. $\mathbf{S}_1\mathbf{B} = \mathbf{S}_2\mathbf{B}\Lambda$) are not well suited to solve Eq. (4) or Eq. (5) for high-dimensional data, especially when the number of samples is less than the number of features (small size sample problem). In this case, methods that use iterative schemes for minimizing the Rayleigh quotient $\frac{\mathbf{x}^T\mathbf{S}_1\mathbf{x}}{\mathbf{x}^T\mathbf{S}_2\mathbf{x}}$[7,21] to obtain the largest/smallest eigenvalue, rely on deflation procedures in order to obtain several eigenvectors. Such a deflation process often breaks down numerically (especially when increasing the number of eigenvectors).[22] This section proposes a stable subspace iteration.

Given two covariance matrices, $\mathbf{S}_1 \in \Re^{d\times d}$ and $\mathbf{S}_2 \in \Re^{d\times d}$, and an initial random matrix $\mathbf{V}_0 \in \Re^{d\times q}$, the subspace iteration method[22] alternates the following steps:

$$\mathbf{S}_1\hat{\mathbf{V}}_{t+1} = \mathbf{S}_2\mathbf{V}_t \tag{6}$$

$$\mathbf{S} = \hat{\mathbf{V}}_{t+1}^T\mathbf{S}_1\hat{\mathbf{V}}_{t+1} \qquad \mathbf{T} = \hat{\mathbf{V}}_{t+1}^T\mathbf{S}_2\hat{\mathbf{V}}_{t+1} \tag{7}$$

$$\mathbf{SW} = \mathbf{TW\Delta} \tag{8}$$

$$\mathbf{V}_{t+1} = \hat{\mathbf{V}}_{t+1}\mathbf{W} \qquad \hat{\mathbf{V}}_{t+1} = \hat{\mathbf{V}}_{t+1}/||\hat{\mathbf{V}}_{t+1}||_F^2$$

The first step, Eq. (6), of the subspace iteration algorithm solves a linear system of equations to find $\hat{\mathbf{V}}_{t+1}$. In the second step, the data is projected onto the estimated subspace, Eq. (7). In order to impose the constraints that $\mathbf{V}_{t+1}^T\mathbf{S}_1\mathbf{V}_{t+1} = \mathbf{\Lambda}$ and $\mathbf{V}_{t+1}^T\mathbf{S}_2\mathbf{V}_{t+1} = \mathbf{I}_q$, a normalization is done by solving the following $q \times q$ generalized eigenvalue problem, $\mathbf{SW} = \mathbf{TW\Delta}$, Eq. (8). It can be shown[22] that as t increases, $\mathbf{V}_{t+1}$ will converge to the eigenvectors of $\mathbf{S}_1\mathbf{B} = \mathbf{S}_2\mathbf{B\Lambda}$ and $\mathbf{\Delta}$ to the eigenvalues $\mathbf{\Lambda}$. The convergence is achieved when $\frac{|\delta_i^{k+1}-\delta_i^k|}{\delta_i^{k+1}} < \epsilon\ \forall i$, where δ_i^k denotes the k-largest generalized eigenvalue. The subspace iteration algorithm converges linearly and the convergence rate is proportional to $\frac{|\delta_q|}{|\delta_{q+1}|}$. It is not critical that $\mathbf{V}_0$ does not have a projection onto the first q generalized eigenvectors, because numerical errors will provide such a projection.

The computationally intensive part of the subspace iteration algorithm is to solve the linear system of equations in Eq. (6) (especially for high dimensional data). To regularize the solution and improve efficiency, we approximate the covariance as $\mathbf{S}_1 \approx \mathbf{U\Lambda U}^T + \sigma^2\mathbf{I}_d$, where $\mathbf{U} \in \Re^{d\times k}$. The parameters σ^2, $\mathbf{U}$ and $\mathbf{\Lambda}$ can be estimated by minimizing:

$$E(\mathbf{U}, \mathbf{\Lambda}, \sigma^2) = ||\mathbf{S}_1 - \mathbf{U\Lambda U}^T - \sigma^2\mathbf{I}_d||_F^2. \tag{9}$$

The optimal parameters correspond to: $\sigma^2 = tr(\mathbf{S}_1 - \mathbf{U}\hat{\mathbf{\Lambda}}\mathbf{U}^T)/d - k$, $\mathbf{\Lambda} = \hat{\mathbf{\Lambda}} - \sigma^2\mathbf{I}_d$, where $\hat{\mathbf{\Lambda}}$ are the eigenvalues of the $\mathbf{S}_1$ and $\mathbf{U} \in \Re^{d\times k}$ the first k eigenvectors.

Once the factorization is done, inverting $\mathbf{S}_1$ can be done efficiently using the matrix inversion lemma[23] $(\mathbf{U\Lambda U}^T + \sigma^2\mathbf{I}_d)^{-1} = \frac{1}{\sigma^2}(\mathbf{I}_d - \frac{1}{\sigma^2}\mathbf{U}(\mathbf{\Lambda}^{-1} + \frac{\mathbf{I}_k}{\sigma^2})^{-1}\mathbf{U}^T)$.

3.2.2. *Alternated Least Squares (ALS)*

Solving the GEP resulting from the LS-WKRRR with a subspace iteration method or standard eigen-packages might be computationally intensive if the weighted covariance matrices do not have any special structure (e.g. sparse). For large amounts of high dimensional data previous approaches might not be efficient neither in space nor time, and Alternated Least-Squares (ALS) approaches might be more convenient.

ALS approaches alternate between solving for $\mathbf{A}$ with $\mathbf{B}$ fixed, and solving for $\mathbf{B}$ with $\mathbf{A}$ fixed. Each step can be computed in closed form as:

$$\mathbf{A} = (\mathbf{\Upsilon W}_c^2\mathbf{\Upsilon}^T)^{-1}\mathbf{\Upsilon W}_c^2\mathbf{\Gamma}^T\mathbf{W}_r^2\mathbf{B}(\mathbf{B}^T\mathbf{W}_r^2\mathbf{B})^{-1} \tag{10}$$

$$\mathbf{B} = \mathbf{\Gamma W}_c^2\mathbf{\Upsilon}^T\mathbf{A}(\mathbf{A}^T\mathbf{\Upsilon W}_c^2\mathbf{\Upsilon}^T\mathbf{A})^{-1} \tag{11}$$

After a few iterations, ALS strategies have shown slower convergence than gradient descent strategies.[24,25] In the case of kernel methods, the ALS procedure needs to re-parameterize $\mathbf{B}$, see section 4.2 for more details.

3.2.3. *Gradient descent*

For large amounts of high dimensional data, gradient descent approaches can provide a less computationally demanding alternative.[24,25] Eq. (2) and Eq. (3) suggests a simple gradient descent scheme:

$$\mathbf{A}^{t+1} = \mathbf{A}^t - \eta_a\frac{\partial E_0(\mathbf{A}^t)}{\partial \mathbf{A}} \qquad \mathbf{B}^{t+1} = \mathbf{B}^t - \eta_b\frac{\partial E_0(\mathbf{B}^t)}{\partial \mathbf{B}} \tag{12}$$

A major problem with the update of Eq. (12) is determining the optimal η. η can be found with a line search strategy,[24,26] or as an estimate on upper bound on the diagonal of the Hessian matrix.[25,27] Recently, Buchanan and Fitzgibbon[24] showed how a damped Newton algorithm on the joint pair $\mathbf{A}, \mathbf{B}$ (i.e. vec([$\mathbf{A}$; $\mathbf{B}$])) is more efficient than alternated least-squares algorithms to solve for $\mathbf{A}, \mathbf{B}$. Moreover, in the case of having missing data, the joint damped Newton algorithm is able to avoid local minima more often.

Finally, it is important to notice that both the ALS algorithm and the gradient descent algorithm will effectively solve the SSS problem of many common CA methods. This is another advantage of using optimization techniques on a LS formulation rather than solving the resulting eigen-problem. Moreover, the computational cost of the algorithms will be less than standard eigen-decompositions, $O(n^3)$ or $O(d^3)$ for $\mathbf{A}$ and $\mathbf{B}$, respectively.

4. PCA, KPCA, and Weighted Extensions

This section derives PCA, KPCA and weighted extensions as a particular case of the fundamental equation of CA methods (E_0), Eq. (1).

4.1. *Principal Component Analysis (PCA)*

PCA is one of the most popular dimensionality reduction techniques.[1,16,28,29] The basic ideas behind PCA date back to Pearson in 1901,[28] and a more general procedure was described by Hotelling[29] in 1933. PCA finds an orthogonal subspace ($\mathbf{B}$) that best preserves the covariance ($\mathbf{S}_t$) of the data $\mathbf{D}$. PCA maximizes:

$$J_1(\mathbf{B}) = tr(\mathbf{B}^T\mathbf{S}_t\mathbf{B}) \quad s.t. \quad \mathbf{B}^T\mathbf{B} = \mathbf{I}_k \tag{13}$$

where $\mathbf{B} \in \Re^{d \times k}$, being d the number of features, n number of samples, and k the dimension of the subspace. Typically $k \leqslant min(n, d)$. The columns of $\mathbf{B}$ form an orthonormal basis that spans the principal subspace of $\mathbf{D}$. PCA can be computed in closed-form by calculating the leading eigenvectors of the covariance matrix $\mathbf{S}_t$.[1,16] The PCA projections, $\mathbf{C} = \mathbf{B}^T\mathbf{D}(\mathbf{I}_n - \frac{1}{n}\mathbf{1}_n\mathbf{1}_n^T) \in \mathbf{R}^{k \times n}$, are decorrelated, that is $\mathbf{C}\mathbf{C}^T = \mathbf{\Lambda}$, where $\mathbf{\Lambda} \in \Re^{k \times k}$ is a diagonal matrix containing the eigenvalues of $\mathbf{S}_t$.

For large data sets of high dimensional data (d and n are large), minimizing a least-squares error function[30,31] is an efficient procedure (in both space and time) to compute the principal subspace of $\mathbf{D}$. There exist several least-squares error functions such that the stationary points are solutions of PCA. Consider the fundamental equation of CA, Eq. (1), where $\mathbf{\Upsilon} = \mathbf{I}_d, \mathbf{W}_r = \mathbf{I}_d, \mathbf{W}_c = \mathbf{I}_n, \mathbf{\Gamma} = \mathbf{D}$, and $\mathbf{D1} = \mathbf{0}$ (zero mean data):

$$E_1(\mathbf{B}, \mathbf{A}) = ||\mathbf{D} - \mathbf{B}\mathbf{A}^T||_F^2 \tag{14}$$

In this case, Eq. (4) and Eq. (5) transform to:

$$E_1(\mathbf{A}) \propto tr\big((\mathbf{A}^T\mathbf{A})^{-1}(\mathbf{A}^T\mathbf{D}^T\mathbf{D}\mathbf{A})\big) \tag{15}$$
$$E_1(\mathbf{B}) \propto tr\big((\mathbf{B}^T\mathbf{B})^{-1}(\mathbf{B}^T\mathbf{D}\mathbf{D}^T\mathbf{B})\big) \tag{16}$$

The $\mathbf{B}$ that maximizes Eq. (16) is given by the leading eigenvectors of covariance matrix ($\mathbf{D}\mathbf{D}^T$). Similarly, the optimal $\mathbf{A}$ corresponds to the eigenvectors of the Gram matrix ($\mathbf{D}^T\mathbf{D}$). Observe that the primal and dual formulation of PCA lead to a clean and direct connection with the estimates of the regression matrices $\mathbf{A}$ and $\mathbf{B}$.

Alternated least-squares (ALS) approaches to solve Eq. (14), alternate between solving for $\mathbf{A}$ while $\mathbf{B}$ is fixed and vice versa.[20,25,32,33] In the case of PCA, the ALS equation (Eq. (10) and Eq. (11)) reduce to solve the following systems of linear equations:

$$\mathbf{D}^T\mathbf{B} = \mathbf{A}\mathbf{B}^T\mathbf{B} \tag{17}$$
$$\mathbf{D}\mathbf{A} = \mathbf{B}\mathbf{A}^T\mathbf{A} \tag{18}$$

This optimization is equivalent to the Expectation Maximization (EM) algorithm in probabilistic PCA (PPCA)[31,34] when the noise becomes infinitesimal and equal in all directions. Once $\mathbf{A}$ and $\mathbf{B}$ are found, the unique PCA solution ($\hat{\mathbf{B}}$) can be obtained by finding an invertible transformation $\mathbf{R} \in \Re^{k \times k}$ that jointly diagonalizes $\hat{\mathbf{B}}^T\hat{\mathbf{B}}$ and $\hat{\mathbf{A}}^T\hat{\mathbf{A}}$, where $\hat{\mathbf{B}} = \mathbf{B}\mathbf{R}$ and $\hat{\mathbf{A}} = \mathbf{A}(\mathbf{R}^{-1})^T$. $\mathbf{R}$ has to satify the following simultaneous diagonalization:

$$\mathbf{R}^T\mathbf{B}^T\mathbf{B}\mathbf{R} = \mathbf{I} \qquad \mathbf{R}^T(\mathbf{A}^T\mathbf{A})^{-1}\mathbf{R} = \mathbf{\Lambda}^{-1}$$

where $\mathbf{\Lambda} \in \Re^{k \times k}$ is a diagonal matrix containing the eigenvalues of $\mathbf{S}_t$. $\mathbf{R}$ can be computed by solving the following $k \times k$ GEP $(\mathbf{A}^T\mathbf{A})^{-1}\mathbf{R} = \mathbf{B}^T\mathbf{B}\mathbf{R}\mathbf{\Lambda}^{-1}$.

Alternatively, PCA can also be derived from a least-squares optimization problem by considering E_0, Eq. (1), with the following values[30] : $\mathbf{\Gamma} = \mathbf{D}, \mathbf{W}_r = \mathbf{I}_d, \mathbf{W}_c = \mathbf{I}_n, \mathbf{A} = \mathbf{B}$:

$$E_2(\mathbf{B}) = ||\mathbf{D} - \mathbf{B}(\mathbf{B}^T\mathbf{D})||_F^2 \;\; s.t. \;\; \mathbf{B}^T\mathbf{B} = \mathbf{I}_k \tag{19}$$

However, Eq. (19) is more challenging to optimize because it is quartic in $\mathbf{B}$. Moreover, this formulation of PCA does not allow to incorporate robustness to intra-sample outliers[25].

4.2. *Kernel Principal Component Analysis (KPCA)*

Similar to PCA, KPCA can be derived from E_0, Eq. (1), by lifting the original data samples, $\mathbf{D}$, to a feature space, that is, $\mathbf{\Gamma} = \phi(\mathbf{D})$. The kernelized version of Eq. (14) can be written as:

$$E_3(\mathbf{B}, \mathbf{A}) = ||\mathbf{\Gamma} - \mathbf{B}\mathbf{A}^T||_F^2 \tag{20}$$

The optimal $\mathbf{A}$ can be obtained from one of the fundamental equations of CA, Eq. (4):

$$E_3(\mathbf{A}) \propto tr((\mathbf{A}^T\mathbf{A})^{-1}\mathbf{A}^T\mathbf{K}\mathbf{A}) \tag{21}$$

where $\mathbf{K} = \mathbf{\Gamma}^T\mathbf{\Gamma} \in \Re^{n\times n}$ is the standard kernel matrix. Each element $k_{ij} = k(\mathbf{d}_i, \mathbf{d}_j)$ of $\mathbf{K}$ represents the similarity between two samples by means of a kernel function. Observe that in the case of kernel methods, it is (in general) not possible to directly solve Eq. (5), because the covariance in the input space, $\mathbf{\Gamma}\mathbf{\Gamma}^T$, can be infinite dimensional.

The computational cost of the eigen-decomposition of $\mathbf{K}$ is $O(n^3)$ (no sparsity is assumed), where n is the number of samples. For large amounts of data (large n) an ALS or gradient-descent approach to computing KPCA is computationally more convenient (see Section 3.2.2). To apply the ALS method in the case of KPCA, a re-parameterization of $\mathbf{B}$ is needed. Recall that for KPCA, $\mathbf{B}$ can be expressed as a linear combination of the data in feature space $\mathbf{\Gamma}$;[18] that is, $\mathbf{B} = \mathbf{\Gamma}\boldsymbol{\alpha}$, where $\boldsymbol{\alpha} \in \Re^{n\times k}$. Substituting this expression into Eq. (20) results in:

$$E_3(\boldsymbol{\alpha}, \mathbf{A}) = ||\mathbf{\Gamma}(\mathbf{I}_n - \boldsymbol{\alpha}\mathbf{A}^T)||_F^2 \tag{22}$$

Assuming that $\mathbf{K}$ is invertible, similarly to the ALS-PCA, we can alternate between computing $\boldsymbol{\alpha}$ and $\mathbf{A} \in \Re^{k\times n}$ as:

$$\boldsymbol{\alpha} = \mathbf{A}^T(\mathbf{A}\mathbf{A}^T)^{-1} \quad \mathbf{A} = (\boldsymbol{\alpha}^T\mathbf{K}\boldsymbol{\alpha})^{-1}\boldsymbol{\alpha}^T\mathbf{K} \tag{23}$$

The computational cost of each iteration is $O(n^2k)$.

4.3. *Weighted Extensions*

In many situations, it is convenient to weight the features and/or samples. For instance, when modeling faces from images, it is likely that some pixels have more variance than others (e.g. pixels in the eye regions have more variance than pixels in the cheeks) and they could be weighted less in the model. It could also be the case that some face images are outliers, and we would like to reduce their influence in the subspace.

Eq. (4) and Eq. (5) provide a partial solution to the weighting problem. For instance, consider the weighted PCA case, with a matrix that weights rows ($\mathbf{W}_r$) and a matrix that weights columns ($\mathbf{W}_c$) in E_0, Eq. (1). The closed-form solutions for the weighted PCA is given by the fundamental equations of CA, Eq. (4) and Eq. (5):

$$E_0(\mathbf{A}) \propto tr\big((\mathbf{A}^T\mathbf{W}_c^2\mathbf{A})^{-1}(\mathbf{A}^T\mathbf{W}_c^2\mathbf{D}^T\mathbf{W}_r^2\mathbf{D}\mathbf{W}_c^2\mathbf{A})\big) \quad (24)$$

$$E_0(\mathbf{B}) \propto tr\big((\mathbf{B}^T\mathbf{W}_r^2\mathbf{B})^{-1}(\mathbf{B}^T\mathbf{W}_r^2\mathbf{D}\mathbf{W}_c^2\mathbf{D}^T\mathbf{W}_r^2\mathbf{B})\big) \quad (25)$$

Previous equations have a closed-form solution as a GEP. For $\mathbf{A}$, the GEP is $\mathbf{D}^T\mathbf{W}_r^2\mathbf{D}\mathbf{W}_c^2\mathbf{A} = \mathbf{A}\mathbf{\Lambda}_a$ and for the dual problem in $\mathbf{B}$, the GEP is $\mathbf{W}_r^2\mathbf{D}\mathbf{W}_c^2\mathbf{D}^T\mathbf{W}_r^2\mathbf{B} = \mathbf{W}_r^2\mathbf{B}\mathbf{\Lambda}_b$. The Generalized Singular Valued Decomposition (GSVD) provides[35,36] an alternative approach to solve the weighted PCA problem.

It is also possible to find a weighted KPCA solution for features and samples. Weighting the samples (i.e. $\mathbf{W}_c$) directly translates to weighting the kernel matrix $\mathbf{K}\mathbf{W}_c^2\mathbf{A} = \mathbf{A}\mathbf{\Lambda}_a$. If the weighting is in the feature space (e.g. Mahalanobis in feature space), the weighting can still be taken into account using the kernel trick.[37]

In general, for an arbitrary set of weights, the weighted PCA minimizes:

$$E_4(\mathbf{A}, \mathbf{B}) = ||\mathbf{W} \circ (\mathbf{D} - \mathbf{B}\mathbf{A}^T)||_F^2 \quad (26)$$

$\circ$ denotes the Hadamard or pointwise product. Observe that there is no closed-form solution in terms of GEP for the solution of Eq. (26).[32,36] Moreover, the problem of data factorization with arbitrary weights has several local minima depending on the structure of the weights.[24,38] Minimization of E_4, Eq. (26), has been typically used to solve PCA with missing data[24,32,38] or outliers in PCA[25,39] or LDA.[40] Recently, Aguiar et al.[41] have proposed a closed-form solution to the data factorization problem, when the missing data has a special structure.

5. LDA, KLDA, CCA, KCCA and Weighted Extensions

This section relates LDA, KLDA, CCA and KCCA to the LS-WKRRR problem of E_0, Eq. (1), and derives weighted generalizations.

5.1. *Linear Discriminant Analysis (LDA)*

Let $\mathbf{D} \in \Re^{d\times n}$ be a matrix, where each column is a vectorized data sample from one of c classes. d denotes the number of features and n number of samples. $\mathbf{G} \in \Re^{n\times c}$ is an indicator matrix such that $\sum_j g_{ij} = 1$, $g_{ij} \in \{0,1\}$, and g_{ij} is 1 if $\mathbf{d}_i$ belongs to class j, and 0 otherwise. LDA, originally proposed by Fisher[3] for the two-class case and later extended to the multi-class case,[10] computes a linear transformation ($\mathbf{A} \in \Re^{d\times k}$) of $\mathbf{D}$ that maximizes the Euclidian distance between the means of the classes ($\mathbf{S}_b$) while minimizing the within-class variance ($\mathbf{S}_w$). Rayleigh-like quotients are among the most popular LDA optimization criteria.[10] For instance, LDA can be obtained by maximizing:

$$J_2(\mathbf{A}) = tr((\mathbf{A}^T\mathbf{S}_1\mathbf{A})^{-1}\mathbf{A}^T\mathbf{S}_2\mathbf{A}) \quad (27)$$

where several combinations of $\mathbf{S}_1$ and $\mathbf{S}_2$ matrices lead to the same LDA solution (e.g. $\mathbf{S}_1 \in \{\mathbf{S}_w, \mathbf{S}_t, \mathbf{S}_w\}$ and $\mathbf{S}_2 \in \{\mathbf{S}_b, \mathbf{S}_b, \mathbf{S}_t\}$). The Rayleigh quotient of Eq. (27) has a closed-form solution in terms of a GEP, $\mathbf{S}_2\mathbf{A} = \mathbf{S}_1\mathbf{A}\Lambda_a$,[10] where Λ_a are the eigenvalues. In the case of high-dimensional data (e.g. images), the covariance matrices are likely to be rank-defficient due to lack of training samples, and standard eigen-solutions for LDA can be ill-conditioned. This is the well-known small sample size (SSS) problem. In recent years, many algorithms have been proposed to deal with the SSS problem, including PCA+LDA,[42,43] regularized LDA,[44–46] and many other methods that explore several combinations of the Null and Range spaces of the matrices $\mathbf{S}_1$ and $\mathbf{S}_2$.[47] See[48] for a unifying review of the optimal solution of Eq. (27) based on the analysis of the four fundamental spaces of $\mathbf{S}_1$ and $\mathbf{S}_2$.

LDA has been previously formulated as a regression problem for the two-class case (Fisher Discriminant),[49] and extended to the multi-class case using several approximations of the covariances.[44,50] Recently, Ye[51] extended the work of Hastie et al.[44] by finding the optimal indicator matrix $\mathbf{G}$ that corresponds to LDA. This section provides a simpler proof of the relation between regression and LDA using a convenient matrix formulation.[52] In the following, we will assume zero mean data ($\mathbf{D1} = \mathbf{0}$).

Consider E_0, Eq. (1), where $\boldsymbol{\Gamma} = \mathbf{G}^T$, $\boldsymbol{\Upsilon} = \mathbf{D}$, $\mathbf{W}_r = (\mathbf{G}^T\mathbf{G})^{-\frac{1}{2}}$, $\mathbf{W}_c = \mathbf{I}_n$, and $\mathbf{D1} = \mathbf{0}$, E_0 transforms to:

$$E_5(\mathbf{A}, \mathbf{B}) = ||(\mathbf{G}^T\mathbf{G})^{-\frac{1}{2}}(\mathbf{G}^T - \mathbf{B}\mathbf{A}^T\mathbf{D})||_F^2 \tag{28}$$

Considering $\mathbf{C} = \mathbf{A}^T\mathbf{D}$ and after eliminating $\mathbf{B}$, Eq. (28) can be re-written as:

$$E_5(\mathbf{A}) = ||(\mathbf{G}^T\mathbf{G})^{-\frac{1}{2}}\mathbf{G}^T(\mathbf{I}_n - \mathbf{C}^T(\mathbf{C}\mathbf{C}^T)^{-1}\mathbf{C})||_F^2 \propto \tag{29}$$

$$tr((\mathbf{A}^T \underbrace{\mathbf{D}\mathbf{D}^T}_{\mathbf{S}_t} \mathbf{A})^{-1}\mathbf{A}^T \underbrace{\mathbf{D}\mathbf{G}(\mathbf{G}^T\mathbf{G})^{-1}\mathbf{G}^T\mathbf{D}^T}_{\mathbf{S}_b} \mathbf{A}) \tag{30}$$

Eq. (30) is one of the standard Rayleigh quotients for LDA. Eq. (30) can also be derived from one of the fundamental equations of CA, Eq. (4). Recall that LDA is a supervised learning problem and the binary indicator matrix $\mathbf{G} \in \Re^{c\times n}$ is given. LDA can be understood as using RRR from the data samples ($\mathbf{D}$) to the labels ($\mathbf{G}$), weighted by $\mathbf{G}^T\mathbf{G}$ to compensate for unequal number of samples in each class. Observe, that directly optimizing Eq. (30) (e.g. gradient descent) with respect to $\mathbf{A}$ and $\mathbf{B}$ in Eq. 28 avoids the small sample size (SSS) problem and can be a numerically convenient algorithm for large amounts of high dimensional data.

5.2. *Kernel Linear Discriminant Analysis (KLDA)*

Kernel Linear Discriminant Analysis (KLDA)[53] can also be derived from E_0, Eq. (1). Consider E_0, Eq. (1), where $\boldsymbol{\Gamma} = \mathbf{G}^T$, $\boldsymbol{\Upsilon} = \varphi(\mathbf{D})$, $\mathbf{W}_r = (\mathbf{G}^T\mathbf{G})^{-\frac{1}{2}}$, $\mathbf{W}_c = \mathbf{I}_n$:

$$E_6(\mathbf{A}, \mathbf{B}) = ||(\mathbf{G}^T\mathbf{G})^{-\frac{1}{2}}(\mathbf{G}^T - \mathbf{B}\mathbf{A}^T\varphi(\mathbf{D}))||_F^2 \tag{31}$$

In this case, Eq. (4) translates to the following expression:

$$E_6(\mathbf{A}) \propto tr((\mathbf{A}^T\boldsymbol{\Upsilon}\boldsymbol{\Upsilon}^T\mathbf{A})^{-1}\mathbf{A}^T\boldsymbol{\Upsilon}\mathbf{G}(\mathbf{G}^T\mathbf{G})^{-1}\mathbf{G}^T\boldsymbol{\Upsilon}^T\mathbf{A}) \tag{32}$$

Using the Mercer theorem,[18,19] and assuming that the dimension of the feature space is larger than the number of samples, it can be shown that the solution to the KLDA problem can be expressed as $\mathbf{A} = \boldsymbol{\Upsilon}\boldsymbol{\alpha}$.[53] Using this fact, the KLDA can be found as the solution of the following GEP, $\mathbf{KG}(\mathbf{G}^T\mathbf{G})^{-1}\mathbf{G}^T\mathbf{K}^T\boldsymbol{\alpha} = \mathbf{K}^2\boldsymbol{\alpha}\boldsymbol{\Lambda}_{\boldsymbol{\alpha}}$, where $\mathbf{K} = \boldsymbol{\Upsilon}^T\boldsymbol{\Upsilon}$ is the kernel matrix and $\boldsymbol{\alpha}$ the eigenvectors of the GEP.

5.3. *Canonical Correlation Analysis (CCA) and Kernel CCA*

Canonical correlation analysis (CCA) is a technique to extract common features from a pair of multivariate data. CCA, first proposed by Hotelling in 1936,[4] identifies relationships between two sets of variables by finding the linear combinations of the variables in the first set ($\mathbf{D} \in \Re^{d_d \times n}$) that are most highly correlated with the linear combinations of the variables in the second set ($\mathbf{X} \in \Re^{d_x \times n}$). CCA has been used for matching sets of images in problems such as activity recognition from video,[54] robot navigation,[55] and pose estimation.[56]

Assuming zero mean data (i.e. $\mathbf{D1}_n = \mathbf{0}$, $\mathbf{X1}_n = \mathbf{0}$), CCA finds a combination of the original variables (i.e. $\hat{\mathbf{B}}^T\mathbf{D}$ and $\hat{\mathbf{A}}^T\mathbf{X}$) that maximize:[4]

$$J_3(\hat{\mathbf{A}}, \hat{\mathbf{B}}) = tr(\hat{\mathbf{B}}^T\mathbf{S}^{\mathbf{DX}}\hat{\mathbf{A}}) \quad s.t. \quad \hat{\mathbf{B}}^T\mathbf{S}_t^{\mathbf{D}}\hat{\mathbf{B}} = \hat{\mathbf{A}}^T\mathbf{S}_t^{\mathbf{X}}\hat{\mathbf{A}} = \mathbf{I} \tag{33}$$

where $\mathbf{S}_t^{\mathbf{X}} = \frac{1}{n-1}\mathbf{XX}^T, \mathbf{S}_t^{\mathbf{D}} = \frac{1}{n-1}\mathbf{DD}^T$, and $\mathbf{S}^{\mathbf{DX}} = \frac{1}{n-1}\mathbf{DX}^T$. The pair of canonical variates ($\hat{\mathbf{b}}_i^T\mathbf{D}$, $\hat{\mathbf{a}}_i^T\mathbf{X}$) is uncorrelated with other canonical variates of lower order. Each successive canonical variate pair achieves the maximum relationship orthogonal to the preceding pair. Observe that canonical correlations are invariant with respect to a full-rank affine transformation of $\mathbf{X}$ and $\mathbf{D}$. Eq. (33) has a closed-form solution as two symmetric GEPs:[4,57]

$$(\mathbf{S}_t^{\mathbf{X}})^{-1}\mathbf{S}^{\mathbf{XD}}(\mathbf{S}_t^{\mathbf{D}})^{-1}\mathbf{S}^{\mathbf{DX}}\hat{\mathbf{A}} = \hat{\mathbf{A}}\boldsymbol{\Lambda}_{\hat{a}} \tag{34}$$
$$(\mathbf{S}_t^{\mathbf{D}})^{-1}\mathbf{S}^{\mathbf{DX}}(\mathbf{S}_t^{\mathbf{X}})^{-1}\mathbf{S}^{\mathbf{XD}}\hat{\mathbf{B}} = \hat{\mathbf{B}}\boldsymbol{\Lambda}_{\hat{b}} \tag{35}$$

The number of solutions (canonical variates) is given by $\min{(d_x, d_d)}$.

Borga[7] proposed a unified eigen-framework for PCA, CCA, and Partial Least-Squares (PLS).[7] showed that the canonical factors can be obtained as the critical points of the following Rayleigh quotient:[7]

$$J_4(\mathbf{U}) = tr((\mathbf{U}^T \begin{pmatrix} \mathbf{0} & \mathbf{S}^{\mathbf{XD}} \\ \mathbf{S}^{\mathbf{DX}} & \mathbf{0} \end{pmatrix} \mathbf{U})^{-1}(\mathbf{U}^T \begin{pmatrix} \mathbf{S}_t^{\mathbf{X}} & \mathbf{0} \\ \mathbf{0} & \mathbf{S}_t^{\mathbf{D}} \end{pmatrix} \mathbf{U})) \quad s.t. \quad \mathbf{U} = \begin{bmatrix} \mathbf{A} \\ \mathbf{B} \end{bmatrix} \tag{36}$$

Previous eigen-formulation of CCA, Eq. (34), Eq. (35) and Eq. (36), is appealing from an analytical viewpoint; however, solving the eigen-system is not a numerically convenient method for large amount of high-dimensional data.[7] Alternatively, $\hat{\mathbf{A}}$ and $\hat{\mathbf{B}}$ can be obtained by performing gradient descent on Eq. (36),[7] or performing SVD on $(\mathbf{S}_t^{\mathbf{D}})^{-\frac{1}{2}}\mathbf{S}^{\mathbf{DX}}(\mathbf{S}_t^{\mathbf{X}})^{-\frac{1}{2}}$.[57,58]

In general, it is not clear how E_0, Eq. (1), can recover the canonical variates, because CCA treats both data sets $\mathbf{D}$ and $\mathbf{X}$ symmetrically, whereas LS-WKRRR only normalizes for $\mathbf{X}$. At this point, it is worth observing that if $\mathbf{X} = \mathbf{G}$ (the indicator matrix), the CCA solution of Eq. (35) is equivalent to the LDA solution, Eq. (27). Using our matrix notation, it is straightforward to show that, in this case, Eq. (35) in CCA reduces to $\mathbf{S}_w^{\mathbf{D}}\mathbf{B} = \mathbf{S}_t^{\mathbf{D}}\mathbf{B}\mathbf{\Lambda}$ (assuming zero mean data). Using this fact, we can interpret LDA as CCA. LDA finds the optimal linear subspace that makes $\mathbf{D}$ best correlated with the label matrix $\mathbf{G}$. Similar reasoning can be done for the case of Kernel CCA. Using this observation, it is simple to relate CCA to the fundamental equation of CA, Eq. (1). In order to treat all the variables symmetrically, we introduce weights in the predicted variable ($\mathbf{D}$) (as LDA), and show that the CCA solution can be recovered using the fundamental equation of CA, E_0. Consider E_0, Eq. (1), where $\mathbf{\Gamma} = \mathbf{D}$, $\mathbf{\Upsilon} = \mathbf{X}$, $\mathbf{W}_r = (\mathbf{D}\mathbf{D}^T)^{-\frac{1}{2}}$, and $\mathbf{W}_c = \mathbf{I}_n$.

$$E_7(\mathbf{A}, \mathbf{B}) = ||(\mathbf{D}^T\mathbf{D})^{-\frac{1}{2}}(\mathbf{D} - \mathbf{B}\mathbf{A}^T\mathbf{X})||_F^2 \quad (37)$$

After substituting these values into one of the fundamental equations of CA, Eq. (4), results in:

$$E_7(\mathbf{A}) \propto tr((\mathbf{A}^T\mathbf{S}_t^{\mathbf{X}}\mathbf{A})^{-1}\mathbf{A}^T\mathbf{S}^{\mathbf{XD}}(\mathbf{S}_t^{\mathbf{D}})^{-1}\mathbf{S}^{\mathbf{DX}}\mathbf{A}) \quad (38)$$

which corresponds to the GEP for CCA, Eq. (34). Similarly, Eq. (5) derives in:

$$E_7 \propto tr((\mathbf{B}^T(\mathbf{D}\mathbf{D}^T)^{-1}\mathbf{B})^{-1}\mathbf{B}^T(\mathbf{D}\mathbf{D}^T)^{-1}\mathbf{X}\mathbf{D}^T(\mathbf{D}\mathbf{D}^T)^{-1}\mathbf{D}\mathbf{X}^T(\mathbf{D}\mathbf{D}^T)^{-1}\mathbf{B}) \quad (39)$$

After a change of variable, $\mathbf{U} = \mathbf{B}(\mathbf{D}\mathbf{D}^T)^{-1}$, Eq. (39) can be re-written as:

$$E_7(\mathbf{U}) \propto tr((\mathbf{U}^T(\mathbf{S}_t^{\mathbf{D}})\mathbf{U})^{-1}\mathbf{U}^T\mathbf{S}^{\mathbf{DX}}(\mathbf{S}_t^{\mathbf{X}})^{-1}\mathbf{S}^{\mathbf{XD}}\mathbf{U}) \quad (40)$$

which is the same solution provided by CCA, Eq. (35).

There exist other least-squares formulations of CCA that are worth mentioning. To treat all the variables symmetrically, a LS function can be obtained by minimizing:

$$E_8(\mathbf{B}, \mathbf{A}) = ||\mathbf{B}^T\mathbf{D} - \mathbf{A}^T\mathbf{X}||_F^2 \quad s.t. \quad \mathbf{B}^T\mathbf{S}_t^{\mathbf{D}}\mathbf{B} = \mathbf{I}_d \quad \mathbf{A}^T\mathbf{S}_t^{\mathbf{X}}\mathbf{A} = \mathbf{I}_d \quad (41)$$

Assuming $\mathbf{S}_t^{\mathbf{D}} = \mathbf{D}\mathbf{D}^T$ is invertible, after optimizing w.r.t. the optimal $\mathbf{A} = \mathbf{S}_t^{\mathbf{D}^{-1}}\mathbf{D}\mathbf{X}^T\mathbf{B}$, Eq. (41) transforms to:

$$E_8(\mathbf{B}) = tr(\mathbf{B}\mathbf{S}_t^{\mathbf{D}}\mathbf{B}) - tr(\mathbf{B}^T\mathbf{S}^{\mathbf{DX}}\mathbf{S}_t^{\mathbf{X}^{-1}}\mathbf{S}^{\mathbf{XD}}\mathbf{B}) \quad s.t. \quad \mathbf{B}^T\mathbf{D}\mathbf{D}^T\mathbf{B} = \mathbf{I}_d \quad (42)$$

Changing $\mathbf{U} = \mathbf{S}^{\mathbf{D}^{\frac{1}{2}}}\mathbf{B}$, and using the cyclic permutation property of traces, it can be shown that the eigen-problem corresponds with the CCA solution, Eq. (35), that is:

$$E_8(\mathbf{U}) = -tr(\mathbf{U}^T\mathbf{S}_t^{\mathbf{D}^{-\frac{1}{2}}}\mathbf{S}^{\mathbf{XD}}\mathbf{S}_t^{\mathbf{X}^{-1}}\mathbf{S}^{\mathbf{DX}}\mathbf{S}_t^{\mathbf{D}^{-1^{-\frac{1}{2}}}}\mathbf{U}) \quad s.t. \quad \mathbf{U}^T\mathbf{U} = \mathbf{I}_d \quad (43)$$

It is interesting to point out that CCA can also be recovered using an unweighted regression. Yohai and Garcia[59] and Tso[60] have shown that the canonical variates minimize:

$$E_9(\mathbf{B}, \mathbf{A}) = |\mathbf{D} - \mathbf{B}\mathbf{A}^T\mathbf{X}| \quad s.t. \quad \mathbf{A}^T\mathbf{X}\mathbf{X}^T\mathbf{A} = \mathbf{I}_d$$

where $|.|$ denotes determinant. This is equivalent to minimizing Eq. (1) if $\mathbf{\Gamma} = \mathbf{D}$, $\mathbf{\Upsilon} = \mathbf{X}$, $\mathbf{W}_r = \mathbf{I}$, $\mathbf{W}_c = \mathbf{I}$ using the determinant instead of the trace as the loss function.

5.4. *Weighted Extensions*

Similarly to PCA and KPCA, for LDA and KLDA there are possible weighted extensions, consider Eq. (4) and Eq. (5) when $\mathbf{\Gamma} = \mathbf{G}^T$ and $\mathbf{W}_r = (\mathbf{G}^T\mathbf{G})^{-\frac{1}{2}}$:

$$E_0(\mathbf{A}) \propto tr\big((\mathbf{A}^T\mathbf{\Upsilon}\mathbf{W}_c^2\mathbf{\Upsilon}^T\mathbf{A})^{-1}(\mathbf{A}^T\mathbf{\Upsilon}\mathbf{W}_c^2\mathbf{G}(\mathbf{G}^T\mathbf{G})^{-1}\mathbf{G}^T\mathbf{W}_c^2\mathbf{\Upsilon}^T\mathbf{A})\big)$$
$$E_0(\mathbf{B}) \propto tr\big((\mathbf{B}^T(\mathbf{G}^T\mathbf{G})^{-1}\mathbf{B})^{-1}(\mathbf{B}^T(\mathbf{G}^T\mathbf{G})^{-1}\mathbf{G}^T\mathbf{W}_c^2\mathbf{\Upsilon}^T(\mathbf{\Upsilon}\mathbf{W}_c^2\mathbf{\Upsilon}^T)^{-1}\mathbf{\Upsilon}\mathbf{W}_c^2\mathbf{G}(\mathbf{G}^T\mathbf{G})^{-1}\mathbf{B}\big)$$

Previous equations extend work on weighted LDA approaches by allowing to weight the samples rather than the classes.[61] Similar expressions can be derived for weighted CCA and KCCA, exchanging $\mathbf{G}^T$ for $\mathbf{X}$, and we omit them in the interest of space.

6. K-means and Spectral Clustering

This section relates the LS-WKRRR to k-means, spectral clustering and proposes a new clustering method, Discriminative Cluster Analysis (DCA).

6.1. *k-means*

k-means[62,63] is one of the most popular unsupervised learning algorithms to solve the clustering problem. k-means clustering splits a set of n objects into c groups by minimizing the within-cluster variation. That is, k-means clustering finds the partition of the data that is a local optimum of the following energy function:[52,64–66]

$$J_5(\mathbf{b}_1, ..., \mathbf{b}_c) = \sum_{i=1}^{c}\sum_{j\in C_i} ||\mathbf{d}_j - \mathbf{b}_i||_2^2 \tag{44}$$

where $\mathbf{d}_j$ is a vector representing the jth data point, and $\mathbf{b}_i$ is the geometric centroid of the data points for class i. Eq. (44) can be rewritten in matrix form[52] as:

$$E_{10}(\mathbf{B}, \mathbf{A}) = ||\mathbf{D} - \mathbf{B}\mathbf{A}^T||_F^2 = tr(\mathbf{S}_w) \quad s.t. \quad \mathbf{A}\mathbf{1}_c = \mathbf{1}_n \quad and \quad a_{ij} \in \{0, 1\} \tag{45}$$

where $\mathbf{A} \in \Re^{n\times c}$ is the indicator matrix and $\mathbf{B} \in \Re^{d\times c}$ is the matrix of centroids. Recall that the equivalence between the k-means error function Eq. (44) and Eq. (45) is only valid if $\mathbf{A}$ strictly satisfies the constraints. Observe that Eq. (45) can be derived from the fundamental equation of CA, E_0, Eq. (1), where $\mathbf{\Upsilon} = \mathbf{I}_d, \mathbf{W}_r = \mathbf{I}_d, \mathbf{W}_c = \mathbf{I}_n, \mathbf{\Gamma} = \mathbf{D}$.

The k-means algorithm performs coordinate descent in $E_{10}(\mathbf{B}, \mathbf{A})$. Given the actual value of the centroids, $\mathbf{B}$, the first step finds for each data point $\mathbf{d}_j$, the $\mathbf{a}^j$ such that one of the columns is one and the rest 0, while minimizing Eq. (45). Recall that $\mathbf{a}^j$ refers to a column vector with the jth row of $\mathbf{A}$. The second step optimizes over $\mathbf{B} = \mathbf{D}\mathbf{A}(\mathbf{A}^T\mathbf{A})^{-1}$, which is equivalent to computing the mean of each cluster.

Eliminating $\mathbf{B}$, Eq. (45) can be rewritten as:

$$E_{10}(\mathbf{A}) = ||\mathbf{D} - \mathbf{D}\mathbf{A}(\mathbf{A}^T\mathbf{A})^{-1}\mathbf{A}^T||_F^2 =$$
$$tr(\mathbf{D}^T\mathbf{D}) - tr((\mathbf{A}^T\mathbf{A})^{-1}\mathbf{A}^T\mathbf{D}^T\mathbf{D}\mathbf{A}) \geq \textstyle\sum_{i=c+1}^{min(d,n)} \lambda_i \tag{46}$$

where λ_i are the eigenvalues of $\mathbf{D}^T\mathbf{D}$. The continuous solution of $\mathbf{A}$ lies in the $c-1$ subspace spanned by the first $c-1$ eigenvectors with largest eigenvalues of $\mathbf{D}^T\mathbf{D}$.[64,65] In this case, the error E_{10} is equal to the sum of the residual eigenvalues, i.e. $E_{10} = \sum_{i=c+1}^{min(d,n)} \lambda_i$. Observe that $tr((\mathbf{A}^T\mathbf{A})^{-1}\mathbf{A}^T\mathbf{D}^T\mathbf{D}\mathbf{A})$ is a Rayleigh quotient optimization problem with a closed-form eigenvalue solution. This is the spectral relaxation of the k-means algorithm.

6.2. *Normalized Cuts*

Recently, spectral graph methods for clustering have arisen as a solid approach to data clustering, and have grown in popularity.[65–69] Spectral clustering arises from concepts in spectral graph theory, where the connection between graphs and matrices provides powerful tools to tackle graph theoretical and linear algebra problems.

Spectral clustering, similarly to Laplacian Eigenmaps, constructs a weighted graph, $M(\mathbf{W}, Q)$, with n nodes $Q = [q_1, ..., q_n]$, where node i represents a sample $\mathbf{d}_i$, and each weighted edge, w_{ij}, measures the similarity between two samples, $\mathbf{d}_i$ and $\mathbf{d}_j$. Once the adjacency or affinity matrix (i.e. $\mathbf{W} \in \Re^{n\times n}$) is computed, the clustering problem can be seen as a graph cut problem,[70] where the goal is to find a partition of the graph that minimizes a particular cost function. A popular cost function is:

$$cut(M) = \sum_{q_i \in R, q_j \in Q-R} w_{ij} \tag{47}$$

where q_i denotes the i node of the Graph M, Q represents all the nodes and R is a subset of the nodes. Finding the optimal cut is an NP complete problem, and spectral graph methods use relaxations to find an approximate solution. However, minimization of this objective function, Eq. (47), favors partitions containing isolated nodes, and better measures such as normalized cuts[67] or ratio-cuts[71] have been proposed. Normalized cuts[67] finds a low dimensional embedding better suited for clustering by computing the eigenvector with the second smallest eigenvalue of the normalized Laplacian, $\mathbf{S}^{-\frac{1}{2}}\mathbf{L}\mathbf{S}^{-\frac{1}{2}}$, where $\mathbf{L} = \mathbf{S} - \mathbf{W} \in \Re^{n\times n}$, and $\mathbf{S}$ is a diagonal matrix whose elements are the sum of the rows of $\mathbf{W}$, that is, $s_{ii} = \sum_j w_{ij}$. Ratio-cuts[71] computes the second eigenvector of $\mathbf{L}$. See[69,72–74] for a comparison of different spectral clustering algorithms.

Recently,[66,75] established the connection between kernel k-means and normalized cuts, by means of kernel methods. In this section, we follow a simpler derivation of the same idea with our compact matrix notation, and linked to kernel PCA.[52] Consider E_0, Eq. (1), where $\mathbf{\Gamma} = \phi(\mathbf{D})$, $\mathbf{\Upsilon} = \mathbf{I}_n$, $\mathbf{W}_r = \mathbf{I}_d$, $\mathbf{W}_c = diag(\mathbf{\Gamma}^T\mathbf{\Gamma}\mathbf{1}_n)^{-\frac{1}{2}}$, the weighted kernelized version of k-means, Eq. (45), is:

$$E_{11}(\mathbf{B}, \mathbf{A}) = ||(\mathbf{\Gamma} - \mathbf{B}\mathbf{A}^T)\mathbf{W}_c||_F^2 \tag{48}$$

Recall that the weight matrix $\mathbf{W}_c$ weights each sample (columns of $\mathbf{\Gamma}$) differently. In this case, minimizing Eq. (48) is equivalent to maximizing Eq. (4), that is:

$$E_{11}(\mathbf{A}) \propto tr((\mathbf{A}^T\mathbf{W}_c^2\mathbf{A})^{-1}\mathbf{A}^T\mathbf{W}_c^2\mathbf{\Gamma}^T\mathbf{\Gamma}\mathbf{W}_c^2\mathbf{A}) \tag{49}$$

where $\mathbf{W} = \mathbf{\Gamma}^T\mathbf{\Gamma}$ is the standard affinity matrix in spectral graph methods. After a change of variable $\mathbf{Z} = \mathbf{A}^T\mathbf{W}_c$, Eq. (49) can be expressed as:

$$E_{11}(\mathbf{Z}) \propto tr((\mathbf{Z}\mathbf{Z}^T)^{-1}\mathbf{Z}\mathbf{W}_c\mathbf{\Gamma}^T\mathbf{\Gamma}\mathbf{W}_c\mathbf{Z}^T) \quad (50)$$

Eq. (50) is the same expression used in Normalized Cuts (Ncuts),[67,69] considering $\mathbf{W}_c = \mathbf{S}^{-\frac{1}{2}}$ and $\mathbf{W} = \mathbf{\Gamma}^T\mathbf{\Gamma} \in \Re^{n\times n}$, where $\mathbf{W}$ is the affinity matrix in Ncuts. Once again, with a LS view of Ncuts, the connection with kernel-PCA becomes evident, and normalization factors are easier to interpret. Moreover, the LS formulation is more general since it allows for different kernels and weights. For instance, the weight matrix could be used to reject the influence of a pair of data points with unknown similarity (i.e. missing data).

Typically, after the embedding is found, there are several multiway cut algorithms (directly splitting the samples into c clusters) to cluster in the embedded space.[72,76] See[77] for a review of rounding methods and more advanced rounding strategies. In related work, Rahimi and Recht[78] showed how Normalized Cuts,[67] originally presented as a graph-theoretic algorithm, can be framed as a regression problem, and also point out the problems of sensitivity to outliers. Zass and Shashua[73] showed the importance of normalization of the affinity matrix in spectral clustering. Important connections have also been made between clustering and manifold learning. Recently,[79] showed the connection between the continuous formulation of spectral embedding and Kernel PCA through learning eigenfunctions. Finally, similar relations could be derived for other spectral graph methods such as Ratio-cuts[71] or MinMaxCut.[80]

7. Conclusions

In this chapter, we have shown that the LS-WKRRR is a generative model for several CA methods. In particular, we have shown how the fundamental equation of CA E_0, Eq. (1), relates to PCA, LDA, CCA, k-means, spectral methods, and kernel extensions. We have derived the coupled symmetric system of eigen-equations to solve E_0, and showed several alternatives to solve the resulting GEP. The LS formulation of CA has several advantages: (1) provides a clean connection between many CA techniques. It allows understanding the communalities and differences between several CA methods, as well as the intrinsic relationships, (2) helps to understand *normalization* factors in CA methods, (3) suggests new optimization strategies, (4) yields efficient optimization algorithms to solve CA techniques that avoid typical problems when the covariance matrices are rank deficient (e.g. small size sample problem); (5) allows many straight-forward extensions of CA methods (e.g online learning versions). We have derived weighted extensions for PCA, LDA, CCA, and kernel extensions. Further work must be done to address the equivalence between methods when covariance matrices are rank-deficient and not invertible, and to relate other extensions of CA (e.g. Partial Least-Squares, maximum variance unfoldering) to this framework.

Acknowledgment

This material is based upon work supported by the U.S. Naval Research Laboratory under Contract No. N00173-07-C-2040. Any opinions, findings and conclusions or recommendations expressed in this material are those of the authors and do not necessarily reflect the views of the U.S. Naval Research Laboratory. Thanks to Louis-Phillippe Morency, Feng Zhou, Tomas Simon, Minyoung Kim, Karim Abou-Moustafa, Zaid Harchaoui, Chris Ding and Jordi Soler for helpful comments.

References

1. I. T. Jolliffe, *Principal Component Analysis*. (New York: Springer-Verlag, 1986).
2. B. Schölkopf, A. Smola, and K. Muller, Nonlinear component analysis as a kernel eigenvalue problem., *Neural Computation*. **10**, 1299–1319, (1998).
3. R. A. Fisher, The statistical utilization of multiple measurements, *Annals of Eugenics*. **8**, 376–386, (1938).
4. H. Hotelling, Relations between two sets of variates, *Biometrika*. **28**, 321–377, (1936).
5. B. Mohar, *Some applications of Laplace eigenvalues of graphs. In G. Hahn and G. Sabidussi (Eds.), Graph Symmetry: Algebraic Methods and Applications (Vol. NATO ASI Ser. C 497, p. 225-275)*. (Kluwer, 1997).
6. F. de la Torre. A least-squares unified view of pca, lda, cca and spectral graph methods. In *tech. report CMU-RI-TR-08-29, Robotics Institute, Carnegie Mellon University, May*, (2008).
7. M. Borga. Learning multidimensional signal processing. In *PhD Dissertation. Linkping University, Sweden*, (1998).
8. S. Roweis and Z. Ghahramani, A unifying review of linear gaussian models., *Neural Computation*. **11**(2), 305–345, (1999).
9. S. Yan, D. Xu, B. Zhang, and H. Zhang, Graph embedding: A general framework for dimensionality reduction, *PAMI*. **29**(1), 40–51, (2007).
10. K. Fukunaga, *Introduction to Statistical Pattern Recognition, Second Edition*. (Academic Press.Boston, MA, 1990).
11. F. de la Torre and T. Kanade. Multimodal oriented discriminant analysis. In *International Conference on Machine Learning*, pp. 177–184, (2005).
12. T. W. Anderson, Estimating linear restrictions on regression coefficients for multivariate normal distributions, *Ann. Math. Statist.* **12**, 327–351, (1951).
13. T. W. Anderson, *An Introduction to Multivariate Statistical Analysis*. (2nd ed. Wiley, New York, 1984).
14. S. S. Haykin, *Adaptive filter theory*. (Prentice-Hall, 1996).
15. L. Scharf, *The SVD and reduced rank signal processing. SVD and Signal Processing, II*. (Elsevier, 1991).
16. K. I. Diamantaras, *Principal Component Neural Networks (Therory and Applications)*. (John Wiley & Sons, 1996).
17. F. de la Torre and M. J. Black. Dynamic coupled component analysis. In *Computer Vision and Pattern Recognition*, pp. 643–650, (2001).
18. B. Scholkopf and A. Smola, *Learning with Kernels: Support Vector Machines, Regularization, Optimization, and Beyond*. (MIT Press, 2002).
19. J. Shawe-Taylor and N. Cristianini, *Kernel Methods for Pattern Analysis*. (Cambridge University Press, 2004).

20. P. Baldi and K. Hornik, Neural networks and principal component analysis: Learning from examples without local minima, *Neural Networks*. **2**, 53–58, (1989).
21. H. Murase and S. K. Nayar, Visual learning and recognition of 3D objects from appearance, *International Journal of Computer vision*. **1**(14), 5–24, (1995).
22. K. J. Bathe and E. Wilson, *Numerical Methods in Finite Element*. (Prentice-Hall. Englewood Cliffs, NJ., 1976).
23. G. Golub and C. F. V. Loan, *Matrix Computations*. (2nd ed. The Johns Hopkins University Press, 1989).
24. A. Buchanan and A. Fitzgibbon. Damped newton algorithms for matrix factorization with missing data. In *IEEE Conference on Computer Vision and Pattern Recognition*, (2005).
25. F. de la Torre and M. J. Black, A framework for robust subspace learning, *International Journal of Computer Vision*. **54**, 117–142, (2003).
26. R. Fletcher, *Practical methods of optimization*. (John Wiley and Sons., 1987).
27. A. Blake and A. Zisserman, *Visual Reconstruction*. (MIT Press series, Massachusetts, 1987).
28. K. Pearson, On lines and planes of closest fit to systems of points in space, *The London, Edinburgh and Dublin Philosophical Magazine and Journal*. **6**, 559–572, (1901).
29. H. Hotelling, Analysis of a complex of statistical variables into principal components, *Journal of Educational Psychology*. **24**, (1933).
30. E. Oja, A simplified neuron model as principal component analyzer, *Journal of Mathematical Biology*. **15**, 267–273, (1982).
31. S. Roweis. EM algorithms for PCA and SPCA. In *NIPS*, pp. 626–632, (1997).
32. K. R. Gabriel and S. Zamir, Lower rank approximation of matrices by least squares with any choice of weights, *Technometrics, Vol. 21, pp.* **21**, 489–498, (1979).
33. H. Shum, K. Ikeuchi, and R. Reddy, Principal component analysis with missing data and its application to polyhedral object modeling, *Pattern Analysis and Machine Intelligence*. **17**(9), 855–867, (1995).
34. M. Tipping and C. M. Bishop, Probabilistic principal component analysis, *Journal of the Royal Statistical Society B*. **61**, 611–622, (1999).
35. M. Irani and P. Anandan. Factorization with uncertainty. In *European Conference on Computer Vision*, pp. 539–553, (2000).
36. M. J. Greenacre, *Theory and Applications of Correspondence Analysis*. (Academic Press, London, 1984).
37. I. Tsang and J. Kwok. Distance metric learning with kernels. In *Proceedings of the International Conference on Artificial Neural Networks*, pp. 126–129 (June, 2003).
38. R. Hartley and F. Schaffalitzky. Powerfactorization: an approach to affine reconstruction with missing and uncertain data. In *Australia-Japan Advance Workshop on Computer Vision*, (2003).
39. D. Skocaj and A. Leonardis. Weighted and robust incremental method for subspace learning. In *International Conference on Computer Vision ICCV*, pp. 1494–1501, (2003).
40. S. Fidler, D. Skocaj, and A. Leonardis, Combining reconstructive and discriminative subspace methods for robust classification and regression by subsampling, *IEEE Transactions on Pattern Analysis and Machine Intelligence*. **28**(3), 337–350, (2006).
41. P. Aguiar, M. Stosic, and J. Xavier. Spectrally optimal factorization of incomplete matrices. In *IEEE Computer Vision and Pattern Recognition*, (2008).
42. A. Martinez and A. Kak, Pca versus lda, *IEEE Transactions on Pattern Analysis and Machine Intelligence*. **23**(2), 228–233, (2003).
43. P. Belhumeur, J. Hespanha, and D. Kriegman, Eigenfaces vs. fisherfaces: Recognition using class specific linear projection, *IEEE Transactions on Pattern Analysis and Machine Intelligence*. (19), 711–720, (1997).
44. T. Hastie, R. Tibshirani, and J. Friedman, *The Elements of Statistical Learning: Data Mining, Inference and Prediction*. (Springer, 2001).

45. W. Zhao. Discriminant component analysis for face recognition. In *ICPR*, pp. 818–821, (2000).
46. J. Hoffbeck and D. Landgrebe, Covariance matrix estimation and classification with limited training data, *IEEE Transactions on Pattern Analysis and Machine Intelligence.* **18**(7), 763–767, (1996).
47. J. Ye, Characterization of a family of algorithms for generalized discriminant analysis on undersampled problems, *The Journal of Machine Learning Research.* **6**(1), 483 – 502 (September, 2005).
48. S. Zhang and T. Sim, Discriminant subspace analysis: A fukunaga-koontz approach., *PAMI.* **29**, 1732–1745, (2007).
49. R. O. Duda, P. E. Hart, and D. G. Stork, *Pattern Classification.* (John Wiley and Sons Inc., 2001).
50. P. Gallinari, S. Thiria, F. Badran, and F. Fogelman-Soulie, On the relations between discriminant analysis and multilayer perceptrons, *Neural Networks.* **4**, 349–360, (1991).
51. J. Ye. Least squares linear discriminant analysis. In *ICML*, pp. 1087–1093, (2007).
52. F. de la Torre and T. Kanade. Discriminative cluster analysis. In *International Conference on Machine Learning*, vol. 148, pp. 241 – 248, New York, NY, USA (June, 2006). ACM Press.
53. S. Mika. Kernel fisher discriminants. In *PhD thesis, University of Technology, Berlin*, (2002).
54. T.-K. Kim, J. Kittler, and R. Cipolla., Discriminative learning and recognition of image set classes using canonical correlations, *IEEE Trans. on PAMI.* **29**(6), 1005–1018, (2007).
55. D. Skocaj and A. Leonardis. Appearance-based localization using cca. In *Computer Vision Winter Workshop*, (2000).
56. T. Melzer, M. Reiter, and H. Bischof, Appearance models based on kernel canonical correlation analysis, *Pattern Recognition.* **36**(9), 1961–1971, (2003).
57. K. Mardia, J. Kent, and J. Bibby, *Multivariate Analysis.* (Academic Press, London, 1979).
58. S. Cherry, Singular value decomposition analysis and canonical correlation analysis., *J. Climate.* (9), 2003–2009, (1997).
59. V. J. Yohai and M. S. Garcia, Canonical variables as optimal predictors., *The Annals of Statistics.* **8**(4), 865–869, (1980).
60. M. Tso, Reduced-rank regression and canonical analyis., *Journal of the Royal Statistical Society. Series B.* **43**(2), 183–189, (1981).
61. M. Loog, R. Duin, and R. Hacb-Umbach, Multiclass linear dimension reduction by weighted pairwise fisher criteria, *IEEE Transactions on Pattern Analysis and Machine Intelligence.* **7**(23), 762766, (2001).
62. J. B. MacQueen. Some methods for classification and analysis of multivariate observations. In *5-th Berkeley Symposium on Mathematical Statistics and Probability. Berkeley, University of California Press.*, pp. 1:281–297, (1967).
63. A. K. Jain, *Algorithms For Clustering Data.* (Prentice Hall, 1988).
64. H. Zha, C. Ding, M. Gu, X. He, and H. Simon. Spectral relaxation for k-means clustering. In *Neural Information Processing Systems*, pp. 1057–1064, (2001).
65. C. Ding and X. He. K-means clustering via principal component analysis. In *International Conference on Machine Learning*, vol. 1, pp. 225–232, (2004).
66. R. Zass and A. Shashua. A unifying approach to hard and probabilistic clustering. In *International Conference on Computer Vision. Beijing*, (2005).
67. J. Shi and J. Malik, Normalized cuts and image segmentation, *IEEE Transactions on Pattern Analysis and Machine Intelligence.* **22**(8), 888–905 (Aug., 2000).
68. A. Y. Ng, M. Jordan, and Y. Weiss. On spectral clustering: Analysis and an algorithm. In *Neural Information Processing Systems*, number 14, (2002).
69. S. Yu and J. Shi. Multiclass spectral clustering. In *Proceedings of International Conference on Computer Vision*, pp. 335–336 (Nice, France, October, 2003).
70. F. K. Chung, *Spectral Graph Theory.* (CBMS Regional Conference Series in Mathemtics, vol 92, American Mathematical Society, Providence, 1997).

71. L. Hagen and A. Kahng, New spectral methods for ratio cut partitioning and clustering., *IEEE. Trans. on Computed Aided Desgin.* (11), 1074–1085, (1992).
72. D. Verma and M. Meila. Comparison of spectral clustering methods. In *NIPS*, (2003).
73. R. Zass and A. Shashua. Doubly stochastic normalization for spectral clustering. In *Neural Information Processing Systems*, (2006).
74. M. Filippone, F. Camastra, F. Masulli, and S. Rovetta, A survey of kernel and spectral methods for clustering, *Pattern Recognition.* **41**(1), 176–190, (2008).
75. I. S. Dhillon, Y. Guan, and B. Kulis. A unified view of kernel k-means, spectral clustering and graph partitioning. In *UTCS Tech. Report TR-04-25*, (2004).
76. Y. Weiss. Segmentation using eigenvectors: a unifying view. In *ICCV*, (1999).
77. D. Tolliver. Spectral rounding and image segmentation. Technical report, doctoral dissertation, tech. report CMU-RI-TR-06-44, Robotics Institute, Carnegie Mellon University (August, 2006).
78. A. Rahimi and B. Recht. Clustering with normalized cuts is clustering with a hyperplane. In *Statistical Learning in Computer Vision*, (2004).
79. Y. Bengio, P. Vincent, J. Paiement, P. Vincent, and M. Ouimet, Learning eigenfunctions links spectral embedding and kernel pca, *Neural Computation.* (16), 2197–2219, (2004).
80. C. Ding, X. He, H. Zha, M. Gu, and H. Simon. A min-max cut algorithm for graph partitioning and data clustering. In *Proc. IEEE International Conference on Data Mining*, (2001).

CHAPTER 1.2

MULTIPLE CLASSIFIER SYSTEMS: TOOLS AND METHODS

Veyis Gunes

Institut FEMTO-ST / LMARC, Ecole Nationale Supérieure de Mécanique et des Microtechniques, 26 chemin de l'Epitaphe, F-25 000 Besançon, France
E-mail: v.gunes@laposte.net

Michel Ménard

Laboratoire d'Informatique, d'Image et d'Interaction, Université de La Rochelle, Avenue Michel Crépeau, F-17 042 La Rochelle Cedex 1, France

Simon Petitrenaud

Laboratoire d'Informatique de l'Université du Maine, Université du Maine, Avenue René Laënnec, F-72 085 Le Mans Cedex 9, France

As multiple experts can confront and exchange their ideas in order to improve the decision-making process, a pattern recognition system can use several classifiers in order to improve its recognition rate. Moreover, various decisions strategies, implying these classifiers in different ways, can contribute to a same recognition task. A first strategy consists in deciding using different opinions: this is the combination of classifiers. A second strategy consists in using one or more opinions for better guiding other classifiers in their training stages, and/or to improve the decision-making of other classifiers in the classification stage: this is the cooperation of classifiers. The third and last strategy consists in giving more importance to one or more classifiers according to various criteria or situations: this implies the selection of classifiers. Since time could be a selection criterion, the temporal aspect of pattern recognition, i.e. the possible evolution of the classes to be recognized, can be treated by the selection strategy.

1. Introduction

This chapter introduces and explains several essential concepts and components for building multiple classifier systems. Before explaining our motivations, it is necessary to place this work in the field of data fusion which is the fundamental framework to build and design such systems.

Klein[1] specifies a definition of data fusion, inspired by the Data Fusion Subpanel of the US Department of Defense: "data fusion is a multilevel, multifaceted process, dealing with the association, correlation, estimation, and combination of data and information from multiple sources …". Ng[2] divides data fusion into two stages: low-level and high-level[a] fusion processes. In Pattern Recognition, or PR, this chapter concerns mainly the first stage. This level of fusion uses data obtained previously by individual classifiers. Many authors have introduced different approaches for data fusion. The contexts are: tracking, data association, attribute data fusion, situation assessment, sensor management and control (Refs. 3, 4, 5 and 6). They explain system implementations with the Bayesian approach, the Dempster-Shafer Evidential reasoning, the Graph-Based Data Association, the Biologically Inspired Algorithms (neural networks, genetic algorithms, fuzzy logic, artificial immunology, and expert systems). This chapter explains fundamental architectures (or strategies) adapted to decision systems using multiple classifiers.

A *multiple classifier system* (MCS) is obtained by an association of decisions from several classifiers using mathematical techniques in order to create a single decision that will be more reliable, and thus, more useful for a human operator or computer tasks. Decision data which have been obtained using different classifier modalities exhibit diverse characteristics, dues to the classifier mathematical properties. The automated procedure of conveying all the meaningful decisions from the individual classifiers to a final decision is the objective of a MCS.

More precisely, in PR context, the aim of a MCS[b] is to obtain a highly powerful (in the sense of a high recognition rate) association of classifiers, based on the decisions of several classifiers (see Refs. 73 and 74). Another motivation for this type of system lies in the complexity of the forms of the classes involved; when the classes are multi-mode or when they overlap, it becomes difficult to model them, because their representation (for example, by centers, probability distributions or fuzzy sets) becomes complicated. Methods in which mixture models are used (in particular, see the work of McLachlan and Basford[44]) are suited to such classes, but whenever the size of training set is limited, this

[a] High-level fusion normally begins at the point where the objects are known.
[b] Other used terms are: multi-experts combination, committee of classifiers and fusion of classifiers.

modeling leads to an over-fitting with the available data, which do not necessarily represent the real classes accurately. By associating various classifiers (different algorithms and/or different constructions), one hopes to obtain better performances. Lastly, within the framework of the recognition of evolutionary objects, it is useful to determine the strategies of associations which are best adapted to the temporal aspect of dynamic classes.

The first allusions to *multiple classifier systems* or *systems of classifiers* (also called *ensemble* of classifiers), quoted by Ho,[30] go back to Nilsson[46] and Haralick[29] for the error analysis relating to a set of Bayesian classifiers. However, the MCS's were put into practice concretely only from the early 1990s, in particular within the framework of the recognition of words[30] and handwritten words.[59,60,73] Nowadays, they are also widely used in remote sensing (Ref. 72).

Currently, the most usual operation, which associates the various classifiers of a system, is the "combination" of outputs of these classifiers. Some researchers studied another kind of operation, called "adaptive selection of classifiers"[19] or "dynamic selection of classifiers".[31,65] Other authors used the outputs of individual classifiers as features for the input of another classifier (of "decision"), carrying out the training on the decisions of these classifiers.[66,62] This chapter deals with combination and these (others) classifier associations strategies. Thus, we introduce problems and solutions related to MCS design, fusion methodology and a generic architecture.

A method of automatic design of a MCS has been also proposed.[22] The proposal is to choose a set of classifiers of different methods and/or parameters among a greater set of classifiers and to select (with a validation set, different from the test set) the subset of the classifiers whose performances are high. The outputs of the selected classifiers are combined with the majority vote rule.

This chapter begins with some explanations about the advantages and the taxonomy of multiple classifier systems (Secs. 2 and 3 respectively). Then, the operations such as combination, cooperation and selection of classifiers are respectively detailed in Secs. 4–6. This is completed with a section about hybrid systems. Finally, Sec. 8 includes a conclusion and some prospects.

2. Advantages of Multiple Classifier Systems

Let $F=\{f_1, f_2, \ldots, f_j, \ldots, f_m\}$ be the feature space in which the f_i are different features and m is the number of available features. And let $\Omega=\{C_1, C_2, \ldots, C_i, \ldots, C_l\}$ be the decision space (also called frame or space of discernment), in which the C_i are the different classes or assumptions and l is the number of classes or assumptions. According to the situation, the MCS can have the following advantages:

- In the same way as an additional feature f_i makes it possible to a classifier to improve its decision (provided that this feature is complementary to the other features), an additional opinion (defined on the space of decision Ω) coming from another classifier makes it possible for a MCS to decide better (provided that this opinion is complementary to the other opinions).
- The association of classifiers allows a modular and dispatched/shared approach of PR.
- When, for a given PR problem, some features of F are real and the others are discrete or symbolic, it is often unavoidable to use two different kinds of classifiers. Thanks to a common space of discernment Ω, the opinions of the two classifiers can easily be combined.
- In certain cases, algorithmic complexity (in training and/or classification stage) can be reduced if the data processing is distributed on several levels.
- Lastly, for the evolutionary classes, it can be necessary to specialize several classifiers on different temporal intervals, and to use these classifiers according to time and/or according to the pattern to be classified.

3. Taxonomy of Multiple Classifier Systems

In addition to the *combination* of the outputs of classifiers, researchers introduced, quite recently, other types of operations. Gosselin[25] used the terms of *cooperation* and *combination* in the same way. Franke and Mandler[18] also used the terms in the same way, while proposing two different types of constructions; "combined outputs of cooperating classifiers being used as features for a classifier of higher level" or "combination of the outputs of classifiers being interpreted like votes of automatic specialists with each one having a certain point of view on the input pattern". We call these two types of associations respectively *cooperation* of classifiers and *combination* or *fusion* of classifiers. We propose to distinguish the systems of classifiers in three categories, according to the type of operation between the classifiers:

- combination of classifiers,
- cooperation of classifiers,
- selection of classifiers.

If a system operates with several types of associations, the system is known as hybrid or mixed. The first two types of associations also seem to be major concerns in the neural networks community. According to Sharkey *et al.*,[54] this community uses mainly two terms: *combination of an ensemble* (of classifiers) and *combination of modules* (of classification). Thus, the term *combination* is used here in the way of an association. According to these authors, "*in an* ensemble, *component nets are redundant in that they each provide a solution to the same task ... By contrast, under a modular approach, a task is decomposed into a number of subtasks, and the complete task solution requires the contribution of all the several modules ...*". It is clear that the ensemble approach is an expression of the combination of classifiers while the modular approach is an expression of the cooperation of classifiers. The authors also specify that the concept of selection of an *ensemble* members was approached in a certain number of publications, and propose to make use of it explicitly. The proposal is to determine the performances of different *sets* of classifiers (among a fixed higher number of classifiers), by testing them (method "test and select") on a validation set, different from the training set and the test set. But the disadvantages of this approach are that, on the one hand, the step of validation/constitution of an ensemble requires a high computing time and on the other hand, the obtained *ensemble* is fixed and cannot change according to a pattern to be classified.

Some authors, also interested in the complexity of the algorithms, proposed to carry out conditional combinations. Thus, Gosselin[25] proposed to classify the classifiers according to their performances and to treat an unknown pattern by a first classifier. Then, the proposal is to accept its decision if the pattern is not rejected. In the opposite case, the decision will be made using the combination of the outputs of the first and the second classifiers. The same reasoning can apply, until the pattern is classified or the outputs of all the classifiers are combined. This conditional combination makes it possible to reduce the execution times effectively. The disadvantage is the need for fixing multiple rejection thresholds associated with the various decisions.

Table 1 presents the related work in a synthetic way. This work is presented according to the above taxonomy as well as according to the types of specializations of the classifiers. The most common related works deal with the combination of classifiers using different features (in the first column, first line), whereas the most recent related researches involve rather the adaptive selection.

Table 1. Examples of Multiple Classifier Systems. To our knowledge, some solutions are not treated yet (empty cases)

Specialization of Classifiers by: Spaces ↓	Specialization of Classifiers by: Methods ↓	Combination of Classifiers	Co-Operation of Classifiers	Selection: Static	Selection: Adaptive
Features	*Heuristics and various*	Van Breukelen *et al.*[62] Chen *et al.*[80] Xu *et al.*[73], Ho[32]	Jaimes and Chang[37]	Sharkey *et al.*[54] "Test and Select"	Gosselin[25]
Decision	*Different principals*	Giacinto *et al.*[22] Ho *et al.*[30]	Giacinto and Roli[20]	Giacinto *et al.*[22]	Woods *et al.*[65]
Decision	*Different parameters* or *Different initializations*	Ng and Singh[45] Rogova[48] Cho and Kim[9]	Giacinto and Roli[20] Cao *et al.*[79]	Giacinto *et al.*[22] Sharkey *et al.*[54]	Cao *et al.*[79] "Multiple features and Multistage classifiers"
Representation	*Random selection* or *Local accuracy*	Wolpert[64] "Stacked generalization" Ho[33] "Random subspace"	Parikh *et al.*[42] Didaci and Giacinto[43]		Parikh *et al.*[42] Giacinto *et al.*[19]
Representation	*Clustering*	Gunes *et al.*[27] "3C"	Gunes *et al.*[27] "3C"		Gunes *et al.*[27] "3C"
Representation	*Temporal selection*	Gunes *et al.*[26]			

4. Combination of Classifiers

The advantages of combining classifiers instead of single classifier schemes are twofold: it helps reducing the computational requirements by using simple models and it can improve the classification skills. The encountered problems are those which can also be found in the following fields:

- combination of sensors,

- combination of expert opinions, and
- combination of data bases.

Data fusion techniques are employed in many applications: image processing (Refs. 77, 72, and many others), handwritten digit and word recognition,[34,59] face Recognition (detection),[66,83] speech recognition[68] and others. The information provided can come from various nature and type of sources (sensors, human experts, data bases, classifiers). Indeed, they can constitute a signal, a sound or an image, and be of numerical and/or symbolic type. Unfortunately, the sources are often sullied with inaccuracy and/or uncertainty and also with incompleteness and/or ambiguity. Figure 1 illustrates the possible interactions between the various modules of a MCS putting combination into practice. The connections in full line represent the features (with bars) or the decisions (with crosses). In this configuration, the chosen feature vectors x_1, x_2 and x_3 can be different, but they should concern the same pattern to be recognized. The figure shows a problem with l classes and d available features (thus, we have $d_1 \leq d$, $d_2 \leq d$ and $d_3 \leq d$). Generally, the classifiers e_1, e_2 and e_3 have a common decision space, i.e. the l classes of the various classifiers have the same significances.

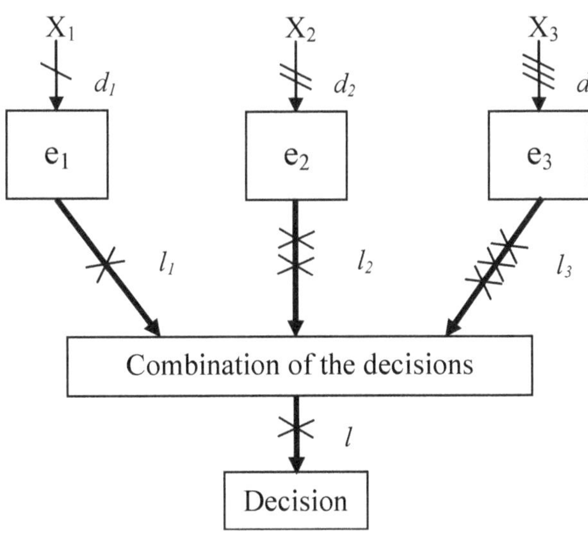

Fig. 1. A multiple classifier system carrying out the combination of classifiers. In most works, $l_1=l_2=l_3=\{C_1, C_2, \ldots, C_l\}$. The available features are $F=\{\{x_1\}, \{x_2\}, \{x_3\}\}$ and d_1, d_2 and d_3 are the dimensions of x_1, x_2 and x_3, respectively. The features of a classifier e_i can be different from the features of a classifier e_k but the tasks of the two classifiers can still be the same (e.g., these classifiers can use different features to recognize the same faces).

The theoretical and experimental results which are seen in the literature show clearly that the combination of classifiers can be effective only if the individual classifiers are "accurate" and "diverse",[53,61] i.e. if they have low error rates and if they induce different errors (this is also called complementarity). Two classifiers induce different errors when the patterns badly classified by the first are different from the patterns badly classified by the second. This latter condition means that the decisions or the outputs of the classifiers are decorrelated (each classifier specializes on a certain number of classes). It can be satisfied by using different feature spaces, different training sets or different (intrinsically) classifiers (various parameter settings or different types of classifiers). Some "diversity" measures are explained by Ranawana and Palade.[10]

Three types of combinations are possible. They come from the specialization of the classifiers according to:

- different features (by specialization on the feature space),
- different methods (by specialization on the decision space), or
- different learning sets (by specialization on the representation space).

By using various features, one specializes the classifiers in the feature space F and by using various methods (and/or parameter settings), one specializes the classifiers in the decision space. Indeed, to find the advantage of a classification method over another or a parameter setting against another, means to consider a decision against another or certain decisions against others. Lastly, by using various training sets, one specializes the classifiers in the space of representation. The majority of work already completed relates to specializations of the first type. However, the two other types should be considered as complementary.

The techniques of *Bagging*[78,47] and *Boosting*[52] implicitly specialize the various produced classifiers in the space of representation. The technique of *Bagging* re-defines the training set for each classifier, by independent and random selection (uniform probability distribution) of the patterns of an available training set, whereas that of *Boosting*, assigns to the patterns of the training set various weights for the various classifiers. In the latter technique, weights are also assigned to the classifiers, at the stage of combination of the decisions.

In this type of MCS, each classifier is applied in parallel and their outputs are combined to reach a consensus; these approaches use mainly the framework of the theory of vote,[41,73] the unanimous consensus,[31,73] methods of vote using heuristic rules of decision,[38] the Bayesian theory,[73] the Belief theory,[51,73] the Possibility theory[17] and the fuzzy sets theory.[69] The choice of the combination type, in particular, depends on the classifiers outputs type.

According to their type of outputs, the classifiers can be divided into three sets. Type 1 is defined by crisp decisions, type 2 is defined by ranked (and crisp) decisions and type 3 is defined by a similarity measure to each class (not crisp). Owing to the fact that the outputs of the type 3 are richer in information, our interests lie more in fusion for this type of outputs.

4.1. *Combinations inspired by the voting methods (output types: 1,2,3)*

Let us suppose that each classifier e_i carries out a "crisp" classification, assigning each vector of features (or pattern) to only one of the classes. A simple method of combination of the outputs of these classifiers is to interpret each output as "a vote" for one of the classes. The class which obtains a number of votes higher than a preset threshold is retained as a final decision. If this threshold is selected as being half of the number of voting classifiers, the rule is called combination with majority vote. Otherwise, the combination is done with the plurality method which can give, as results, one or more classes.

The advantage of this type of combination is that it can be used for any type of classifier, whatever is the type of outputs of these classifiers (i.e. "crisp" classification or not). If the outputs are not "crisp", a rank based method, the Borda[71] count can be applied. This method is a powerful tool that takes into account not only the first preferences of the classifiers but also all their other preferences. For the decision step, one can see the next section.

4.2. *Some usual and useful combinations (output type: 3)*

To generalize, let us define $v_{is}(x_k)$ as a similarity measure of the pattern x_k to the class C_i, calculated by the classifier e_s during the classification of the pattern and w_i as a similarity measure to the class C_i, calculated by the MCS. These measures can be, for example, probabilities or degrees of memberships. Let l be the number of possible assumptions (classes) and S the number of classifiers whose outputs enter into combination. Let l+1 be the class of rejection and T (with $0<T\leq 1$) be a chosen reject threshold. Usually, the most employed rules are:

The Maximum rule:

$$\forall i \in [1,l], \quad w_i(x_k) = \max_{s=1}^{S} \upsilon_{is}(x_k). \tag{1}$$

The Minimum rule:

$$\forall i \in [1,l], \quad w_i(x_k) = \min_{s=1}^{S} \upsilon_{is}(x_k). \tag{2}$$

The Mean rule:

$$\forall i \in [1,l], \qquad w_i(x_k) = \frac{1}{S}\sum_{s=1}^{S} \upsilon_{is}(x_k). \tag{3}$$

The Product rule:

$$\forall i \in [1,l], \qquad w_i(x_k) = \prod_{s=1}^{S} \upsilon_{is}(x_k). \tag{4}$$

Usually, the decision rule is defined by a function $SC(x_k)$ such as:

$$SC(x_k) = \begin{cases} C_j, & \text{if } w_j(x_k) = \max_{i=1}^{l} w_i(x_k) \quad \text{and} \quad w_j(x_k) \geq T \\ C_{l+1}, & \text{else.} \end{cases} \tag{5}$$

4.3. *Combination by the probability theory (output type: 3)*

Let $\{y_1, y_2, \ldots, y_s, \ldots, y_S\}$ be the decision vectors provided by the classifiers e_1, e_2, ..., e_s, ..., e_S respectively for the pattern x_k. In the probability model, the measure v_{is} calculated by the classifier e_s is represented by a probability, denoted as conditional probability according to class C_i:

$$\upsilon_{is}(x_k) = P(x_k/C_i), \qquad i = 1,\ldots,l. \tag{6}$$

The information combination concerning the class of x_k is represented by an *a posteriori* probability obtained with the rule of Bayes:

$$P(C_i/y_1, y_2, \ \ldots \ , y_S) = \frac{P(y_1, y_2, \ \ldots \ , y_S/C_i)P(C_i)}{P(y_1, y_2, \ \ldots \ , y_S)}, \tag{7}$$

in which $P(C_i)$ is the *a priori* probability of the class C_i. Under independence assumptions, i.e. independence of classifiers e_1, e_2, ..., e_S and conditional independence of e_1, e_2, ..., e_S to C_i, equation (8) can be simplified as follows:

$$P(C_i/y_1, y_2, \ \ldots \ , y_S) = \frac{P(C_i)\prod_{s=1}^{S} p(y_s/C_i)}{\sum_{n=1}^{l} P(C_n)\prod_{s=1}^{S} p(y_s/C_n)}. \tag{8}$$

Generally, the conditional probabilities are parameterized by some family distributions and all the parameters have to be estimated in a learning stage using methods such as the "maximum likelihood" and the so-called "expectation maximization" algorithms.[12] The determination of the *a priori* probabilities is not easy. In some related work, they are considered equiprobable and in others, they are estimated by the frequency of the patterns belonging to the class in the learning set (i.e. ratio of the number of patterns of a class over the total number of patterns for all the classes).

Several decision methods are available. They are all based on the *a posteriori* probability $P\ (Ci/y_1\ ,\ y_2\ ,\ \dots\ ,\ y_S)$. The most popular one is the MAP (maximum *a posteriori* probability):

$$SC(x_k) = \begin{cases} C_j, & \text{if } P(C_j / y_1, y_2, \ \dots\ , y_S) = \max_{i=1}^{l} P(C_i / y_1, y_2, \ \dots\ , y_S) \\ & \text{and } P(C_j / y_1, y_2, \ \dots\ , y_S) \geq T \\ C_{l+1}, & \text{else,} \end{cases} \tag{9}$$

in which $T \in\ ]0, 1]$ is a chosen reject threshold. Among other methods, one can state the "minimum expected risk", which generalizes the MAP algorithm, or the "maximum entropy". An established method for an evaluation of a combined classifier is measuring some aspect of its Receiver Operating Characteristic (ROC) curve, which graphs the trade-off between the conditional probabilities of detection and false alarm (Ref. 81).

Liggins *et al.*[5] introduce concepts of decision fusion, named distributed detection fusion, founded on the classical Bayesian likelihoods. They use, mainly, parallel fusion architecture. They design the system decisions fusion rules jointly to combine the local detector, describing an analysis of some of the common fusion rules, and how they are defined. Some other score level fusion techniques have been proposed in the literature. Ross *et al.*[8] propose three main categories: density-based, transformation-based and classifier-based schemes. The efficiency of the two first categories depends on the amount and quality of the available training data. As shown in sections 5, 6 and 7, other distributed design strategies can be implemented: cooperation, selection of classifiers, or hybrid methods. These can be a viable alternative when the number of training samples is limited. Schemes of classifier-based fusion consolidate the outputs of different matchers into a single vector of scores. The interest of the strategies is to focus on dependence relations between the classifiers and to allow a modular and distributed approach of decision.

4.4. *Combination by the Belief theory (output type: 3)*

In some cases, the use of this rule of Bayes can be unsuited to the combination of decisions of several classifiers. Indeed, the rule of Bayes requires that the measurements behave like probabilities. In the case of the decisions of experts, this condition is often impossible to respect,[21] because an expert can decide to give a certain credit to two classes (mixed assumption), which cannot be expressed simply with the probability theory. The Belief theory[51] or the

Transferable belief model[58] are one of the most known tools for the representation of uncertainty and inaccuracy in knowledge based systems. Compared to the statistical and fuzzy approaches, the Belief theory has the advantage of including a modeling of uncertainty and inaccuracy during the combination of classifiers and during the assignment of a pattern to a class. Masses are assigned to all the elements of the power set 2^{Ω}, rather than to the elements of the decision space Ω only. Bayesian fusion is considered as a particular case of the fusion in the framework of the Belief theory. The examples of cases where the Belief theory can be made profitable are:

- when the decision of a classifier is ambiguous, the theory takes this into account by giving masses to the union of classes,
- when the classifiers have different reliabilities; it is possible to take them into account by reducing or reinforcing their masses with a weighting,
- when the knowledge on the reliabilities of the classifiers depends on classes, this can be taken into account by modifying the masses according to this knowledge.

Let Ω be a limited set of l mutually exclusive assumptions ω_i, $i \in \{1, \ldots, l\}$ and 2^{Ω}, the power set of Ω. Let A be a subset of Ω and thus element of 2^{Ω} ($A \subseteq \Omega$, $A \in 2^{\Omega}$). The fundamental concept representing uncertainty is that of a distribution of belief by a function called function of mass allocation, defined as an application m, of the power set 2^{Ω} in the interval [0, 1], checking:

$$m(A) > 0 \ \ if \ \ and \ \ only \ \ if \ \ A \in 2^{\Omega}$$
$$and \ \ \sum_{A \subseteq \Omega} m(A) = 1. \tag{10}$$

The $m(A)$ quantity can be interpreted as a measurement of the belief assigned exactly to A and none of its subsets, in the absence of other information. In this theory uncertainty related to an event $A \in 2^{\Omega}$ is measured using the Belief (Bel) and Plausibility (Pl) functions.

The quantity Bel(A) is understood as a total belief assigned to A:

$$\begin{aligned} &\mathrm{Bel} : 2^{\Omega} \to [0,1] \\ &A \mapsto \mathrm{Bel}(A) = \sum_{B \subseteq A} m(B). \end{aligned} \tag{11}$$

The quantity Pl(A) is understood as a belief assigned to A. Indeed, 1 − Pl(A) is a measure of doubt in A. It is defined as follows:

$$\begin{aligned} &\mathrm{Pl}: 2^{\Omega} \to [0,1] \\ &A \mapsto \mathrm{Pl}(A) = \sum_{B \cap A = \varnothing} m(B) = 1 - \mathrm{Bel}(\overline{A}). \end{aligned} \tag{12}$$

The most significant operation concerning the distributions of belief is the combination (or rule) of Dempster.[11] Two distributions of Belief m_1 and m_2 are known as combinable if there are at least two nondisjoined subsets B and C such as $m_1(B)>0$ and $m_2(C)>0$. The orthogonal sum of m_1 and m_2, noted $m_{1,2}=m_1 \oplus m_2$ is then defined by:

$$m_{1,2}(A) = \frac{\sum_{B \cap C = A} m_1(B) m_2(C)}{1 - \sum_{B \cap C = \varnothing} m_1(B) m_2(C)}. \tag{13}$$

For S sources, owing to the fact that the rule of Dempster is associative and commutative, the combination is done by successive combinations (in an unspecified order) and makes it possible to obtain the final distribution of belief m which is noted:

$$m = \bigoplus_{i=1}^{S} m_i. \tag{14}$$

Several other combination rules have been proposed in the literature, particularly through a disjunctive combination.[16,56] Recently, new disjunctive and conjunctive rules suitable to combine belief functions induced by possibly overlapping bodies of evidence, and more or less reliable, have been proposed (see Ref. 84).

Finally, a rule of decision must be chosen and applied, usually in order to retain the most appropriate class or assumption C_i: a first method is to transform the mass in a convenient probability distribution and apply the MAP algorithm to this probability. In the Transferable Belief Model,[55,58] the authors suggest the use of what they call the *pignistic*[c] probability, denoted as P_{bet} and defined as:

$$\forall C_i \in \Omega,\ P_{\text{bet}}(C_i) = \sum_{A \ni C_i} \frac{m(A)}{|A|} \tag{15}$$

in which | A | is the cardinality of A. Then, the decision rule is defined as:

$$SC(x_k) = \begin{cases} C_j, & if \quad P_{\text{bet}}(C_j) = \max_{i=1}^{l} P_{\text{bet}}(C_i) \ \ and \ \ P_{\text{bet}}(C_j) \geq T \\ C_{l+1}, & else \end{cases} \tag{16}$$

in which $T \in]0, 1]$ is a chosen reject threshold.

---

[c] This term comes from "pignus": bet in Latin.

Other decision rules have been proposed, generally over simple assumptions[13,14]: for example, the maximum of plausibility, the maximum of credibility (Belief) and the rules using expected utility (see also Ref. 57).

4.5. *Other theoretical frameworks for the combination*

Other theoretical frameworks also make it possible to carry out combinations, often with equivalent results. As it is difficult to prove that such a classifier is better than a similar one for any PR problem, it is not easy to prove that such theoretical framework is more appropriate than for any PR problem. In particular, the Fuzzy Sets theory and the Possibility theory[17,70] constitute powerful theoretical frameworks. Verikas *et al.*,[63] among others, carried out combinations using the fuzzy integrals of Sugeno and Choquet. In these frameworks $v_{is}(x_k)$ represents the membership degree of a fuzzy set or a possibility distribution.

A global evaluation of the pattern x_k membership of the class C_i is then provided by the combination of $v_{is}(x_k)$, with $s \in [1,S]$. The advantage of these frameworks rely on the variety of combination operators[76,15]: some of them are disjunctive (sum, maximum, and more generally triangular co-norm operators). Others are conjunctive (product, minimum and more generally triangular norm operators) and some others are neutral, for example, the ordered weighted averaging operators.[67] After the combination step, as in the common cases, decision is usually done with the maximum of membership values.

More recently, some authors (Ref. 82) proposed a maximum entropy (maxent) approach to the fusion of classifier outputs. The maxent approach is quite versatile and allows expressing the reliability of the classifiers. The correlations between these classifiers can be easily integrated. Maximum entropy approach is also used by Fujino *et al.*[7] to combine trained models (naïve Bayes models) in order to improve classification performance. Maximum entropy does not introduce any additional ad-hoc information. With the available individual classifier decisions, the final decision is considered as the most honest characterization of the knowledge.

Problems involving more than two classes (polychotomous classification problem) can be solved directly using multiclass classifiers, or by combining several binary classifiers.[60] In the latter approach, complexity is reduced by decomposing a difficult task into simpler subtasks. Furthermore, some popular classifiers such as logistic regression,[36] support vector machines,[85] or binary classification trees,[75] are more adapted to two-class problems. When reducing a polychotomous classification problem into two-class problems, dichotomies can be obtained in various ways. Each class can be opposed to all others (one-against-all decomposition): l binary classifiers are trained, each one using all the training patterns.[24] Alternatively, each class can be opposed to each other (one-against-one or pairwise decomposition): in this case, $l(l-1)/2$ pairwise classifiers are trained using patterns from each pair (C_i, C_j) of classes.[39,60]

5. Cooperation of Classifiers

These methods use the decisions and/or the results of a classifier for better guiding one or more other classifiers in their decision-making. Generally, it is the decision vector which is transmitted between classifiers (in particular for Refs. 28, 34 and 35), because this type of information has less differences (in its form) from a type of classifier to another one (this leads to a more general method). This advantage leads to easier cooperation of various types of classifiers. However, other types of information can also be transmitted or received (e.g. a classifier could communicate the rejected classes).

The cooperation operation can induce either a reduction of the decision space,[79] or a reduction of the representation space, consequently, of the learning set,[27] or a reduction of the features space.[37] Thus, the classifiers entering into cooperation can be specialized in three different ways:

- in the feature space,
- in the decision space, and/or
- in the (features) representation space.

Figure 2 illustrates the possible interactions between the various classifiers of a MCS carrying out the cooperation operation. In this configuration, the vectors of features x_1 , x_2 and x_3 concern the same pattern to be classified, but they can be different from/to each other. The classifiers e_1 , e_2 and e_3 do not necessarily have a common decision space (as the case of hierarchical classifiers).

A particular case of a MCS carrying out cooperation operation is that of the multilevel (or hierarchical) classifiers put into practice primarily by the neural networks community. This type of system is a particular case of cooperation, since from Fig. 2, when one eliminates the connection between e_2 and e_3, and changes into one-way the connections between e_1 and e_2 on the one hand and between e_1 and e_3 on the other hand (directed towards the right-hand side), a system of hierarchical classifiers is obtained, in which e_1 is in the higher level and e_2 and e_3 are in the lower level. Generally, they make it possible to reduce the decision space, a classifier of the higher level supplying its outputs to a classifier of the lower level which is able to classify the pattern introduced in the system's input. This can also be seen as a serial architecture (as opposed to the combination or parallel architecture). The system suggested by Cao *et al.*[79] is of this type. It also makes it possible to carry out an adaptive selection (see this term in the following section). The reader can find in Table 1 the references of some work according to the operation type and the specialization type of the classifiers.

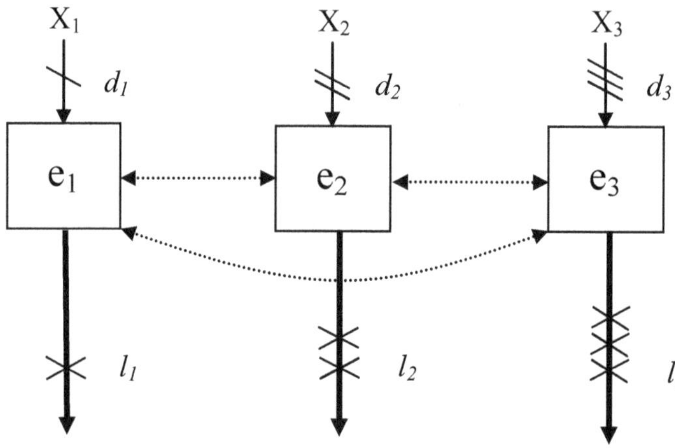

Fig. 2. System of classifiers carrying out the cooperation of classifiers. The vectors of features x_1, x_2 and x_3 concern the same pattern to be classified (but not necessarily, as opposed to the combination, can we have $l_1 \cap l_2 \cap l_3 = \varnothing$) and they can be different from one classifier to another. Furthermore, the results of a classifier can be used in another classifier (in the learning and/or classification stage) in different ways. For example, e_1 can supply the learning set of e_3.

6. Selection of Classifiers

In the *Machine Learning* community, most of the current studies relates to the comparison of various algorithms. The qualification criterion generally adopted is that of the performance of these algorithms on standard sets of data. As specified by Salzberg,[50] since the performances of the classifiers strongly depend on the specific field of application, such comparisons can lead to errors. This is why this community recommends the empirical evaluation of the algorithms in their application field. This led to the static selection of classifiers, quite naturally. However, within the framework of multiple classifier systems, the best classifier (or subset of classifiers) can be different from one pattern (to be classified) to another. This problem can be treated with adaptive selection of classifiers.

According to Giacinto *et al.*,[23] the approaches used by the neural networks community can be classified into two design strategies: the "direct" strategy and, referring to classifiers, the "over-produce and choose" strategy (see also Ref. 49). The first strategy consists in generating decorrelated neural networks directly (whose outputs can then be combined), while the second consists in creating a consequent initial set of networks and choosing the subsets of the decorrelated networks. The two strategies are, in fact, *static* selections of classifiers.

Within the framework of adaptive selection, several articles present methods of selection of a classifier using an estimate of local performances in small areas of the representation space, surrounding an unknown test sample.[65] However, only the output of the best classifier is used for the final decision-making.

Some authors consider that adaptive selection and dynamic selection have the same meaning[19,31] and we agree with this idea. In the case of the adaptive selection, various classifiers or subsets of classifiers are usually selected according to the pattern introduced in input of the system and several patterns, identical but resulting from different moments, lead to the selection of the same classifier(s). As in a dynamic classifier selection, adaptive classifier selection can also treat time explicitly and lead to different choices according to the pattern but also to the moment considered. This can be useful when the classes are evolutionary (e.g. when the class centers move as a course of time).

In a dynamic classifier selection (DCS), for a given pattern x, the PR system retains the outputs of one or more classifier(s) adapted to the pattern (i.e. with the numerical values of the features of the pattern treated), at its evolution moment. The behavior of the system can be periodical: for two patterns x_{t1} and x_{t2} respectively taken at the moments t_1 and $t_2=T+t_1$, in which T is the periodicity of the behavior of the process, the system must select, in both cases, the same classifier or subset of classifiers. In such cases, if the learning sets are constituted as a sum of all patterns, over a period of time, the static or the adaptive approaches can still be used. Thus, the dynamic classifier selection can also be seen as a special case of the adaptive one (see for example Ref. 43). Rightfully, Ranawana and Palade[10] call the two kind of selections *data dependent classifier selection*. In these conditions, we propose the following taxonomy:

Static classifier selection (SCS): For any pattern x_k, at any moment t, the classifier or the subset of classifiers retained by the MCS is that defined in the learning stage. The strategy "over-produce and choose" uses this type of selection. Giacinto *et al.*[22] and Sharkey *et al.*[54] also proposed this type of systems.

Adaptive classifier selection (ACS): For any pattern x, at any moment t, the system retains the outputs of the classifier(s) adapted to the values of the features of the pattern to be classified. The 3C algorithm which we proposed[27] is an example of this type of selection. Woods *et al.*[65] and Cao *et al.*[79] also proposed this type of systems.

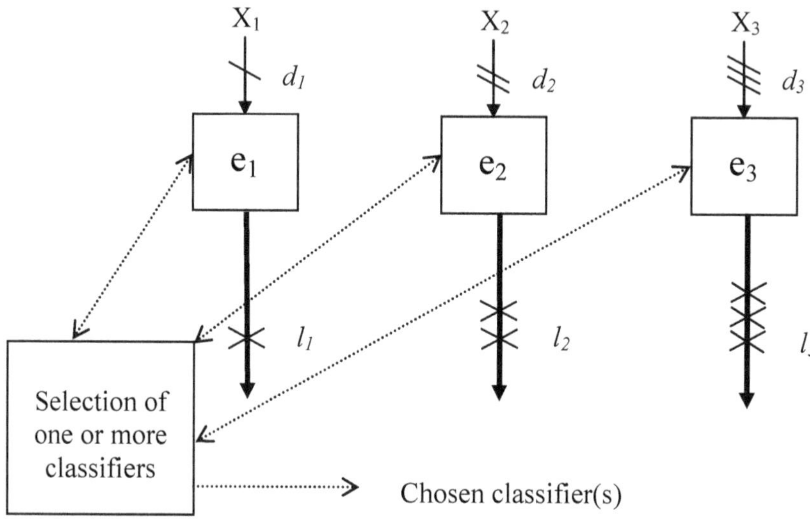

Fig. 3. System of classifiers carrying out the selection of classifiers. The selection module activates one or more classifiers according to the input pattern and/or according to time. The classifiers e_1, e_2 and e_3 do not necessarily have a common decision space. Thus, if only one classifier is selected at any time, we can have $l_1 \cap l_2 \cap l_3 = \varnothing$.

Figure 3 illustrates the possible interactions between the various classifiers of a system carrying out the selection operation. In this operation, the vectors of features x_1, x_2 and x_3 concern the same pattern to be classified, but they can be different from/to each other. The classifiers e_1, e_2 and e_3 do not necessarily have a common decision space. On the other hand, if the results of several classifiers are retained, it is necessary to respect the constraints of combination. Thus, one can associate with this operation, a combination and/or a cooperation of classifiers. There is then a *mixed* or *hybrid approach* of multiple classifier systems.

7. Hybrid Methods in Multiple Classifier Systems

Among the existing systems of classifiers, some of them carry out the operations of combination, cooperation and/or (static or adaptive) selection, at the same time, like the system suggested by Jaimes and Chang.[37] This system allows the combination and the cooperation of classifiers. The cooperation is done between hierarchical classifiers, i.e. the decisions of the classifiers of low levels are used as inputs for classifiers for higher levels who make decisions of higher levels.

These PR systems are called hybrid systems of classifiers and Fig. 1, 2 and 3, together, illustrate the possible interactions between the different classifiers of such a system (the outputs of the chosen classifiers undergo combination and decision). The connections in dotted lines, between the various "modules" represent various information. This information can be communicated during the training and the classification steps. In this way, the module of selection is an intelligent module which retains some information resulting in the training step and uses it in the classification step. A classifier can transmit its decisions to one or more other classifiers. Here are some examples of information exchange:

- The classifier e_1 selects the training set for classifier e_3 (in the training step, see Figure 2).
- The classifier e_1 informs the module of selection that it is necessary or not to take into consideration the outputs of the classifier e_3 (in the classification step, see Figure 3).
- The module of selection "requests an answer" from classifiers e_1 and e_2, concerning an unknown pattern available in input (i.e. selects the decision vectors of these classifiers for the combination step).

The dimension of the decision space Ω being l (combination output, i.e. the final decision vector), we have $\dim(l_1)\leq l$, $\dim(l_2)\leq l$ and $\dim(l_3)\leq l$. Thus, the classifiers e_1, e_2 and e_3 do not necessarily have a common decision space. When only one classifier is selected, there is no operation of combination. On the other hand, if we want to keep the results of several classifiers in order to combine them, it will then be preferable that the selected classifiers have a common decision space. If this latter constraint is not respected, for example, when a classifier decision is a compound class (two or more classes among the decision space), the combination is generally treated through the Belief theory.[51] This theory is well suited when the outputs of the classifiers are the richest in terms of information (type 3 outputs: they are not crisp). When the outputs are crisp, other methods, based on an ordering of all the classes are also available.

Example: If the selection recommends the choice of the results of the classifiers e_1 and e_2, it is preferable that the l classes or assumptions should be the same ones from one classifier to another, such as:

$$\begin{cases} \dim(l_1) = \dim(l_2) = l \\ \text{and } \ \Omega_1 = \Omega_2 = \Omega = \{C_1, C_2, \ \dots\ , \ C_l\} \end{cases} \tag{17}$$

The module of selection can be based on an algorithm of clustering/classification (it is the case for the 3C system, suggested in Ref. 27), while a classifier e_i can itself consist of several classifiers (i.e. of a system of classifiers). In the *conditional topology* approach,[40] the selection is based only on the reject option, i.e. a selection is carried out only if the first classifier(s) reject(s) the input pattern. However, according to the results of a first classifier (selection classifier), many choices are possible.

Some authors[27] have proposed a system which carries out the three types of operations (combination, cooperation and adaptive selection). Table 1 summarizes existing work and all the possibilities offered within the framework of the systems of classifiers. The hybrid systems are easily identifiable by the fact that one finds them or not in several columns (in this sense, the two columns of *selection* are considered as one column).

Designing high performance classifiers for structured data, consisting of multiple components, is an important and challenging research issue in the field of multimedia retrieval. Hybrid methods can be used in the framework of indexing and interactive retrieval. Indeed, the selection strategy of one or more classifiers can structure the representation space by using clustering algorithms. Clustering driven by queries about objects of interest can modify the feature space in a suited way and adapt the answer of classifiers to the query. The clustering algorithms are known to facilitate interactivity with, for example, semantic events as well as non-experts queries. In this way, selection modules allow users to formulate easily their queries.

8. Conclusion

In this article, we have given an overall view on *Multiple Classifier Systems* with different points of view over these systems and identified the different available topologies of such systems. We hope that this presentation can help readers for the design of their own topology of classifiers, according to their specific PR problem. In the literature, it can be seen that the operations of combination, cooperation and selection (static or adaptive) of classifiers are widely used in the *Systems of Classifiers* (i.e. in *Multiple Classifier Systems*) to recognize static classes. When the classes have complex forms, these operations generally make it possible to obtain better recognition rates, compared to one classifier only, especially if the diversity of classifiers is respected. If the temporal aspect of a PR problem isn't neglected, instead, taken into account explicitly, a dynamic selection of classifiers can be applied. This latter is, in fact, an adaptive or data dependent selection of classifiers.

It appears clearly that a formalization of the systems of classifiers is needed. Indeed, a wide variety of systems are proposed with various architectures, various terminologies and various validations. Therefore, the advantages and the drawbacks of the proposed approaches cannot be proved easily. Such a formalization, defined as a set of classifiers and a set of relations between these classifiers, would permit comparative studies.

References

1. L.A. Klein, *Sensor and Data Fusion Concepts and Applications*. (SPIE Optical Engineering Press, 1999).
2. G.W. Ng, *Intelligent Systems. Fusion, Tracking and Control*. CSI Series. (Research Studies Press LTD, England, 2003).
3. S. Blackman and R. Popoli, *Design and Analysis of Modern Tracking Systems*, (Artech House, Boston, London, 1999).
4. C.J. Lakhmi, N.M. Martin, Fusion of Neural Networks, Fuzzy Systems and Genetic Algorithms: Industrial Applications (CRC Press, 1998).
5. M.E. Liggins, D. Hall and J. Llinas, *Handbook of Multisensory Data Fusion. Theory and Practice*, (CRC Press, 2008).
6. E. Bossé, J. Roy and S. Wark, *Concepts, Models, and Tools for Information Fusion*, (Artech House, 2007).
7. A. Fujino, N. Ueda and K. Saito, "A classifier design based on combining multiple components by maximum entropy principle", *L. Notes in Computer Science* (Springer, Berlin/Heidelberg, 2005).
8. A.A. Ross, K. Nandakumar, A. K. Jain, *Handbook of multibiometrics*, (Springer, 2006).
9. S. B. Cho and J. H. Kim, "Combining multiple neural networks by fuzzy integral for robust classification", *IEEE Trans. SMC 25*, 2 (1995), p. 380.
10. R. Ranawana and V. Palade, "Multi-Classifier Systems: Review and a roadmap for developers", Int. J. Hybrid Intell. Syst. 3, 1 (2006), p. 35.
11. A. P. Dempster, "Upper and lower probabilities induced by a multivalued mapping", *Ann. Math. Statist.* 38 (1967), p. 325.
12. A. P. Dempster, N. M. Laird and D. B. Rubin, "Maximum likelihood estimation from incomplete data via the EM algorithm", *J. Roy. Statist. Soc., Series B 39* (1977), p. 1.
13. T. Denœux, "A k-nearest neighbor classification rule based on Dempster-Shafer theory", *IEEE Trans. SMC* 25, 5 (1995), p. 804.
14. T. Denœux, "Analysis of evidence-theoretic decision rules for pattern classification", *Patt. Recogn.* 30, 7 (1997), p. 1095.
15. D. Dubois and H. Prade, "A review of fuzzy sets aggregation connectives", *Inform. Sci.* 36 (1985), p. 85.
16. D. Dubois and H. Prade, "On the unicity of dempster rule of combination", *Int. J. Intell. Syst.* 1 (1986), p. 133.
17. D. Dubois and H. Prade, Possibility Theory: An Approach to Computerized Processing of Uncertainty (Plenum, New York, 1988).

18. J. Franke and E. Mandler, "A comparison of two approaches for combining the votes of cooperating classifiers", Proc. *11th Int. Conf. Pattern Rec.*, Vol 2, Netherlands (1992), p. 611.
19. G. Giacinto and F. Roli, "Adaptive selection of image classifiers", *L. Notes in Computer Science* 1310 (Springer, Berlin, 1997), p. 38.
20. G. Giacinto and F. Roli, "Ensembles of neural networks for soft classification of remote-sensing images", *Proc. Eur. Symp. Intelligent Techniques* (1997), p. 166.
21. D. Bahler and L. Navarro, "Methods for combining heterogeneous sets of classifiers", *17th Natl. Conf. Artificial Intelligence (AAAI 2000)*, 2000.
22. G. Giacinto and F. Roli, "An approach to the automatic design of multiple classifier systems", *Patt. Recogn. Lett.* 22 (2001), p. 25.
23. G. Giacinto, F. Roli and G. Fumera, "Design of effective multiple classifier systems by clustering of classifiers", *Proc. ICPR2000, 15th Int. Conf Pattern Recognition* (2000), p. 160.
24. R. Anand, K.G. Mehrotra, C.K. Mohan and S. Ranka. "Efficient classification for multiclass problems using modular neural networks". *IEEE Trans. on Neur. Netw.* 6(1), 1995, p. 117.
25. B. Gosselin, "Cooperation of multilayer perceptron classifiers", *Proc. 8th Works. Circuits Syst. Signal Proc.*, Netherlands (1997), p. 187.
26. V. Gunes, P. Loonis and M. Ménard, "A fuzzy petri net for pattern recognition: application to dynamic classes", *Knowledge And Information Systems* 4, 1 (2002), p. 112.
27. V. Gunes, M. Ménard and P. Loonis, "A multiple classifier system using ambiguity rejection for clustering-classification cooperation", *Int. J. Uncert. Fuzz. Knowledge-Based Syst.* 8, 6 (2000), p. 747.
28. D. Happel and P. Bock, "Overriding the experts: a stacking method for combining marginal classifiers", *Proc. 13th Int. FLAIRS Conf.*, (2000).
29. R. M. Haralick, "The table look-up rule", *Commun. Statistics — Theory and Methods* A5, 12 (1976), p. 1163.
30. T. K. Ho, "A theory of multiple classifier systems and its application to visual word recognition", Ph.D. Thesis, State University of New York at Buffalo, (1992).
31. T. K. Ho, J. J. Hull and S. N. Srihari, "Decision combination in multiple classifier systems", *IEEE Trans. Patt. Anal. Mach. Intell.* 16 (1994), p. 66.
32. T. K. Ho, "Random decision forests", Proc. 3rd Int. Conf. Document Analysis and Recognition (1995), p. 278.
33. T. K. Ho, "The random subspace method for constructing decision forests", *IEEE Trans. Patt. Anal. Mach. Intell.* 20, 8 (1998), p. 832.
34. T. S. Huang and C. Y. Suen, "Combination of multiple experts for the recognition of unconstrained handwritten numerals", *IEEE Trans. PAMI,* 17, 1 (1995), p. 90.
35. Y. S. Huang and C. Y. Suen, "A method of combining multiple classifiers – a neural network approach", *Proc. 12th Int. Conf. Pattern Recognition and Computer Vision* (1994), p. 473.
36. T. Hastie, R. Tibshirani and J. Friedman, The Elements of Statistical Learning: Data mining, Inference and Prediction (Springer, NewYork, 2001).
37. A. Jaimes and S. F. Chang, "Integrating multiple classifiers in visual objects detectors learned from user input", *4th Asian Conf. Computer Vision* (2000).
38. F. Kimura and M. Shridar, "Handwritten numerical recognition based on multiple algorithms", *Patt. Recogn.* 24, 10 (1991), p. 969.
39. S. Abe and T. Inoue, "Fuzzy support vector machines for multiclass problems", *Proceedings of ESANN*, 2002, p. 113.

40. L. Lam, "Classifier combinations: implementations and theoretical issues", *L. Notes in Computer Science* (Springer, Berlin, 2000), p. 77.
41. L. Lam and C. Y. Suen, "A theoritical analysis of the application of majority voting to pattern recognition", *Int. Conf. Pattern Recognition* (1994), p. 418.
42. D. Parikh, M. T. Kim, J. Oagaro, S. Mandayam, and R. Polikar, "Combining classifiers for multisensory data fusion", *Proc. of the IEEE Int. Conf. on Syst., Man and Cybern.*, Netherlands (2004), p. 1232.
43. L. Didaci and G. Giacinto, "Dynamic classifier selection by adaptive k-nearest-neighbourhood rule", *L. Notes in Computer Science* 3077 (Springer, Berlin, 2004), p. 174.
44. G. J. McLachlan and K. E. Basford, *Mixture Models: Inference and Applications to Clustering*, Marcel Dekker, NY, USA (1988).
45. G. S. Ng and H. Singh, "Data equalisation with evidence combination for pattern recognition", *Patt. Recogn. Lett.* 19 (1998), p. 227.
46. N. J. Nilsson, *Learning Machines*, McGraw-Hill, 1965.
47. J. R. Quinlan, "Bagging, boosting, and c4.5", *Proc. 13th Natl. Conf. Artif. Int.* (1996), p. 725.
48. G. Rogova, "Combining the results of several neural network classifiers", *Neural Networks* 7, 5 (1994), p. 777.
49. F. Roli and G. Giacinto, "Design of multiple classifier systems", *Hybrid Methods in Pattern Recognition*, Eds. H. Bunke and A. Kandel, (World Scientific, Singapore, 2002).
50. S. L. Salzberg, "On comparing classifiers: a critique of current research and methods", *Data Mining and Knowledge Discovery* 1 (1999), p. 1.
51. G. Shafer, *A Mathematical Theory of Evidence*, Princeton University Press, Princeton (1976).
52. R. E. Shapire, Y. Freund, P. Bartlett and W. S. Lee, "Boosting the margin: a new explanation for the effectiveness of voting methods", *Proc. 14th Int. Conf. Mach. Learn.* (1997).
53. A. J. C. Sharkey, "Multi-net systems", Combining Artificial Neural Nets, Ensemble and Modular Multi-Net Systems (1999), p. 1.
54. A. J. C. Sharkey and N. E. Sharkey, "The 'test and select' approach to ensemble combination", *L. Notes in Computer Science* 1857 (Springer, Berlin, 2000), p. 30.
55. P. Smets, "The combination of evidence in the transferable belief model", *IEEE Trans. Patt. Anal. Mach. Intell.* 12, 5 (1980), p. 447.
56. P. Smets, "Belief functions: the disjuntive rule of combination and the generalized Bayesian theorem", *Int. J. Approx. Reason.* 9 (1993), p. 1.
57. P. Smets, "Decision analysis using belief functions", *Int. J. Approx. Reason.* 4 (1994), p. 391.
58. P. Smets and R. Kennes, "The transferable belief model", *Artif. Intell.* 66 (1994), p. 191.
59. C. Y. Suen, C. Nadal, R. Legault, T. A. Mai and L. Lam, "Computer recognition of unconstrained handwritten numerals", *Proc. IEEE* 80 (1992), p. 1162.
60. F. Cutzu, "Polychotomous classification with pairwise classifiers: a new voting principle", *L. Notes in Computer Science* 2709 (Springer, Berlin, 2003), p. 159.
61. K. Tumer and J. Ghosh, "Error correlation and error reduction in ensemble classifiers", *Connect. Sci.* 8 (1996), p. 385.
62. M. Van Breukelen, R. P. W. Duin and D. M. J. Tax, "Combining classifiers for the recognition of handwritten digits", *1st IAPR TC1 Stat. Techn. in Pattern Rec.* (1997), p. 13.
63. A. Verikas, A. Lipnickas, K. Malmqvist, M. Bacauskiene and A. Gelzinis, "Soft combination of neural classifiers: a comparative study", *Patt. Recogn. Lett.* 20 (1999), p. 429.
64. D. H. Wolpert, "Stacked generalization", *Neural Networks* 5, 2 (1992), p. 241.

65. K. Woods, W. P. Kegelmeyer and J. R. K. Bowyer, "Combination of multiple classifiers using local accuracy estimates", *IEEE Trans. Patt. Anal. Mach. Intell.* 19, 4 (1997).
66. B. Achermann and H. Bunke, "Combination of classifiers on the decision level for face recognition", *Report IAM-96-002, Institut für Informatik*, (Univ. of Bern, Switzerland, 1996).
67. R. R. Yager, "On ordered weighted averaging aggregation operators in multi-criteria decision making", *IEEE Trans. SMC,* 18 (1988), p. 183.
68. K. Yu, X. Jiang and H. Bunke, "Combining acoustic and visual classifiers for the recognition of spoken sentences", *Proc. ICPR2000, 15th Int. Conf. Pattern Recognition* (2000), p. 491.
69. L. A. Zadeh, "Fuzzy sets", *Information and Control* 8 (1965), p. 338.
70. L. A. Zadeh, "Outline of a new approach to the analysis of complex systems and decision processes", *IEEE Trans. Syst. Man Cybern.* 3 (1973), p. 28.
71. J.C. Borda, "Mémoire sur les élections au scrutin", *Histoire de l'Académie royale des Sciences. Année 1781*, Paris (1784).
72. C. H. Chen and P. G. P. Ho, "Statistical pattern recognition in remote sensing", *Pattern Recognition*, 41 (9) (2008), p. 2731.
73. L. Xu, A. Krzyzak and C. Y. Suen, "Methods of combining multiple classifiers and their applications to handwriting recognition", *IEEE Trans. Syst. Man Cybern.* 22, 3 (1992), p. 418.
74. F. M. Alkoot and J. Kittler, "Improving the performance of the product fusion strategy", *Proc. ICPR2000, 15th Int. Conf. Pattern Recognition* (2000), p. 164.
75. L. Breiman, J.H. Friedman, R.A. Olson and C.J. Stone, *Classification and Regresssion Trees,* (Wadsworth International Group, Belmont, 1984).
76. I. Bloch, "Information combination operators for data fusion: a comparative review with classification", *IEEE Trans. Syst. Man Cybern.* 26, 1 (1996), p. 52.
77. I. Bloch and H. Maître, "Fusion of Image Information under Imprecision", *Aggregation and Fusion of Imperfect Information*, Ed. B. Bouchon-Meunier, (Springer, Berlin, 1997), p. 189.
78. L. Breiman, "Bagging predictors," *Mach. Learn.* 24, 2 (1996), p. 123.
79. J. Cao, M. Ahmadi and M. Shridhar, "Handwritten numerals with multiple features and multistage classifiers", *IEEE Int. J. Circuits Syst.* 6 (1996), p. 323.
80. K. Chen, L. Wang and H. Chi, "Methods of combining multiple classifiers with different features and their applications to text-independant speaker identification", *Int. J. Pattern Recognition and Artificial Intelligence*, 11, 3 (1997), p. 417.
81. J.M. Hill, E.M Oxley and K.W. Bauer, "Evaluating the fusion of multiple classifier via ROC curves", *Proc. SPIE*, Vol. 5096, (2003), p. 411.
82. M. Saerens and F. Fouss, "Yet another method for combining classifiers outputs: a maximum entropy approach", *L. Notes in Computer Science* 3077 (Springer, Berlin, 2004), p. 82.
83. J. Meynet, Information theoretic combination of classifiers with application to face detection, Ph.D. Thesis at EPFL (EPFL, Lausanne, 2007).
84. T. Denoeux, "Conjunctive and disjunctive combination of belief functions induced by nondistinct bodies of evidence", *Artificial Intelligence*, 172, 2-3 (2008), p. 234.
85. V. Vapnik, The Nature of Statistical Learning Theory (Springer, Berlin, 1995).

CHAPTER 1.3

ON DISSIMILARITY EMBEDDING OF GRAPHS IN VECTOR SPACES

Horst Bunke* and Kaspar Riesen†

Institute of Computer Science and Applied Mathematics, University of Bern
Neubrückstrasse 10, CH-3012 Bern, Switzerland
**bunke@iam.unibe.ch*
†riesen@iam.uniber.ch

Recently, in pattern recognition and related fields an emerging trend of representing objects by graphs can be observed. In fact, graph based object representation offers a versatile alternative to vectorial representations. The domain of graphs, however, contains only little mathematical structure. That is, most of the basic mathematical operations, actually required by many standard pattern recognition algorithms, are not available for graphs. Consequently, we observe a severe lack of algorithmic tools in the domain of graphs. The present chapter aims at introducing a novel approach for graph embedding in vector spaces. The rationale for such an embedding is to bridge the gap between the high representational power and flexibility of graphs and the large amount of algorithms available for object representations in terms of feature vectors. The key-idea of the proposed embedding method is to use the distances of an input graph to a number of training graphs, termed prototypes, as a vectorial description of the graph. In contrast with some other graph embedding methods, there are no restrictions on the type of graphs the proposed embedding framework can deal with. Moreover, the dissimilarity embedding of graphs is particularly able to cope with noisy data. In an experimental evaluation we show that the proposed methodology of first embedding graphs in vector spaces and then applying a statistical classifier has significant potential to outperform classifiers that directly operate in the graph domain.

1. Introduction

Obviously, the question of how to represent the objects from some pattern recognition application in a formal way such that a machine can learn to discriminate between different classes is very important. There are two major ways to tackle this crucial step, viz. the *statistical* and the *structural* approach.

In statistical pattern recognition objects are represented by feature vectors. That is, an object x from some pattern space $\mathcal{X}$ is formally represented as a vector of n measurements, i.e. n numerical features. Hence, an object can be understood as a point in the n-dimensional real space, i.e. $x = (x_1, \ldots, x_n) \in \mathbb{R}^n$. Representing objects or patterns by feature vectors $x \in \mathbb{R}^n$ offers a number of useful properties, in particular, a remarkable wealth of mathematical operations available in a vector space. For example, computing the sum, the product, the mean, or the distance of two entities is well defined in vector spaces and moreover, can be efficiently computed. The convenience and low computational complexity of algorithms that use feature vectors as their input have eventually resulted in a rich repository of algorithmic tools for pattern recognition.[1]

However, the use of feature vectors implicates two quite severe limitations. First, as vectors always represent a predefined set of features, all vectors in a particular application have to preserve the same length regardless of the size or complexity of the corresponding patterns. Furthermore, there is no direct possibility to describe relationships that might exist among different parts of a pattern. These two drawbacks are grave, particularly when the patterns under consideration are characterized by complex structural relationships rather than the statistical distribution of a fixed set of pattern features.

Structural pattern recognition is based on symbolic data structures, such as *strings*, *trees*, or *graphs*, which are better suited for representing complex and interrelated data. As a matter of fact, graphs, which consist of a finite set of nodes connected by edges, are the most general data structure in computer science, and all common data types are special cases of graphs. That is, from an algorithmic perspective both strings and trees are simple instances of graphs. In the present chapter graphs are used as a representation formalism for structural pattern recognition. Although focusing on graphs, the reader should keep in mind that strings and trees are always included as special cases.

Graphs are not only able to describe properties of a pattern but also binary relationships among different parts of the underlying pattern by means of edges. Note that these relationships can be of various nature, viz. spatial, temporal, conceptual, etc. Moreover, graphs are not constrained to a fixed size, i.e. the number of nodes and edges is not limited a priori and can be adapted to the size or the complexity of each individual object under consideration. Hence, the above mentioned drawbacks of feature vectors, namely the size constraint and the lack of representing relationships, can be overcome by graph based representations.[2] Due to these abilities graphs found widespread applications in science and engineering.[2–4] For instance, graphs have been successfully applied in the fields of bioinformatics and chemoinformatics,[5,6] in web content mining,[7] in image classification from various fields,[8,9] or in computer network analysis,[4] to name a few.

One drawback of graphs, when compared to feature vectors, is the significantly increased complexity of many algorithms. For example, the comparison of two feature vectors for identity can be accomplished in linear time with respect to the length of the two vectors. For the analogous operation on general graphs, i.e. testing two graphs for isomorphism, only exponential algorithms are known today. Another serious limitation in the use of graphs for pattern recognition tasks arises from the fact that there is little mathematical structure in the domain of graphs. For example, computing the sum, the weighted sum, or the product of a pair of entities (which are elementary operations required in many classification algorithms) is not possible, or not defined in a standardized way, in the domain of graphs. Due to these general problems in the graph domain, we observe a lack of algorithmic tools for graph based pattern recognition.

A promising approach to overcome this lack is offered through graph embedding in vector spaces. Basically, such an embedding of graphs establishes access to the rich repository of algorithmic tools for pattern recognition originally reserved for vectorial data. A prominent class of graph embedding is based on spectral methods.[11–15] The basic idea is to represent graphs by the eigendecomposition of their structural matrices (i.e. adjacency or Laplacian matrix). The resulting representation exhibits interesting properties for pattern recognition. However, the sensitiveness of the eigendecomposition towards structural errors is a major problem of all spectral methods. That is, spectral methods are not fully able to cope with larger amounts of structural noise, such as missing or spurious nodes or edges. Furthermore, most spectral approaches are only applicable to unlabeled graphs or graphs with severely constrained label alphabets.

The objective of the present chapter is to define a new class of graph embedding procedures which can be applied to both directed and undirected graphs, as well as to graphs with arbitrary labels on their nodes and/or edges. Furthermore, the novel graph embedding framework is distinguished by its ability to handle structural errors. The novel approach to graph embedding is primarily based on the idea proposed in Ref. 16 where the dissimilarity representation for pattern recognition in conjunction with feature vectors was first introduced. The key idea is to use the dissimilarities of an input graph g to a number of training graphs, termed *prototypes*, as a vectorial description of g. In other words, the dissimilarity representation rather than the original graph representation is used for pattern recognition.

The remainder of this chapter is organized as follows. Next, basic concepts and notation are introduced. Particularly, the concept of graph based representation and the graph dissimilarity model of graph edit distance are formally introduced. In Sect. 3 the general graph embedding procedure and some basic properties are discussed. Also the close relationship of the proposed embedding framework to kernel methods as well as the basic problem of prototype selection (which arises with the proposed procedure) are discussed. In Sect. 4 six prototype selection strategies are described in detail. In Sect. 5 an experimental evaluation of the embedding framework is carried out. Finally, in Sect. 6 some conclusions are drawn.

2. Basic Concepts and Notation

2.1. *Graph Based Pattern Representation*

Various definitions for graphs can be found in the literature. The one adopted in this chapter is sufficiently general to cope with a variety of applications.

Definition 1 (Graph). *Let L_V and L_E be a finite or infinite label set for nodes and edges, respectively. A graph g is a four-tuple $g = (V, E, \mu, \nu)$, where V is the finite set of nodes, $E \subseteq V \times V$ is the set of edges, $\mu : V \to L_V$ is the node labeling function, and $\nu : E \to L_E$ is the edge labeling function.*

In the remainder of this chapter, symbol $\mathcal{G}$ will be used to denote the set of all graphs that can be built with labels from L_V and L_E.

Definition 1 allows us to handle arbitrarily structured graphs with unconstrained labeling functions. For example, the labels for both nodes and edges can be given by the set of integers $L = \{1, 2, 3, \ldots\}$, the vector space $L = \mathbb{R}^n$, or a set of symbolic labels $L = \{\alpha, \beta, \gamma, \ldots\}$. Given that the nodes and/or edges are labeled, the graphs are referred to as *labeled graphs*. *Unlabeled graphs* are obtained as a special case by assigning the same label ε to all nodes and edges, i.e. $L_V = L_E = \{\varepsilon\}$.

Edges are given by pairs of nodes (u, v), where $u \in V$ denotes the source node and $v \in V$ the target node of a directed edge. The definition given above corresponds to the case of *directed graphs*. Moreover, the class of *undirected graphs* can be modeled by inserting a reverse edge $(v, u) \in E$ with identical label for each edge $(u, v) \in E$.

Despite adverse mathematical and computational conditions in the graph domain, various procedures for evaluating proximity, i.e. similarity or dissimilarity, of graphs have been proposed in the literature.[2] The process of evaluating the similarity of two graphs is commonly referred to as *graph matching*. The overall aim of graph matching is to find a correspondence between the nodes and edges of two graphs that satisfies some, more or less, stringent constraints. Roughly speaking, there are two categories of algorithms in graph matching, viz. *exact graph matching* and *inexact graph matching*. In the former approach, for a matching to be successful, it is required that a strict correspondence between the two graphs being matched, or at least among their subparts, is found.[17] Based on the exact graph matching paradigms of maximum common subgraph and minimum common supergraph, a few graph similarity measures have been proposed.[18–20] However, the main restriction of exact graph matching and related similarity measures is the requirement that a significant part of the topology together with the corresponding node and edge labels in two graphs have to be identical for obtaining a high degree of similarity. In fact, this is not realistic for many real world applications. Especially if the nodes and/or the edges are continuously labeled, the exact matching paradigm is too restrictive.

In order to make graph matching better applicable to real world problems, several error-tolerant, or inexact, graph matching methods have been proposed.[2] In this approach the requirement of strict correspondences between the nodes of the graphs to be matched is substantially relaxed. That is, inexact matching algorithms are endowed with a certain tolerance to errors, enabling them to detect similarities in a more general way than the exact matching approach. For an extensive review of graph matching methods and applications, the reader is referred to Ref. 2. One of the most flexible methods for error-tolerant graph matching is based on the edit distance of graphs.[21,22] This paradigm is used in the present chapter and described in detail in the next subsection.

2.2. *Graph Edit Distance*

The key idea of graph edit distance is to define the dissimilarity, or distance, of graphs by the minimum amount of distortion that is needed to transform one graph into another. A standard set of distortion operations is given by *insertions*, *deletions*, and *substitutions* of nodes and edges. Other operations, such as *merging* and *splitting* of nodes,[9] can be useful in certain applications but are not considered in this paper.

Given two graphs, the source graph g and the target graph g', the idea of graph edit distance is to delete some nodes and edges from g, relabel (substitute) some of the remaining nodes and edges, and insert some nodes and edges in g', such that g is finally transformed into g'. A sequence of edit operations $e_1, \ldots, e_k$ that transform g into g' is called an *edit path* between g and g'. In Fig. 1 an example of an edit path between two graphs is given. Obviously, for every pair of graphs (g, g'), there exist a number of different edit paths transforming g into g'. Let $\Upsilon(g, g')$ denote the set of all such edit paths. To find the most suitable edit path out of $\Upsilon(g, g')$, one introduces a cost for each edit operation, measuring the strength of the corresponding operation. The idea of such cost functions is to define whether or not an edit operation represents a strong modification of the graph. Hence, between two similar graphs, there should exist an inexpensive edit path, representing low cost operations, while for different graphs an edit path with high costs is needed. Consequently, the *edit distance* of two graphs is defined by the minimum cost edit path between two graphs.

Definition 2 (Graph Edit Distance). *Let $g = (V, E, \mu, \nu)$ be the source graph and $g' = (V', E', \mu', \nu')$ be the target graph. The graph edit distance between g and g' is defined by*

$$d(g, g') = \min_{(e_1, \ldots, e_k) \in \Upsilon(g, g')} \sum_{i=1}^{k} c(e_i),$$

where $\Upsilon(g, g')$ denotes the set of edit paths transforming g into g', and c denotes the edit cost function measuring the strength $c(e_i)$ of edit operation e_i.

Fig. 1. A possible edit path between graph g_1 and g_2 (node labels are represented by different shades of grey).

Exact Computation of GED The computation of the edit distance is usually carried out by means of a tree search algorithm which explores the space of all possible mappings of the nodes and edges of the first graph to the nodes and edges of the second graph. A widely used method is based on the A* algorithm[23] which is a best-first search algorithm. The basic idea is to organize the underlying search space as an ordered tree. The root node of the search tree represents the starting point of our search procedure, inner nodes of the search tree correspond to partial solutions, and leaf nodes represent complete – not necessarily optimal – solutions. Such a search tree is constructed dynamically at runtime by iteratively creating successor nodes linked by edges to the currently considered node in the search tree. In order to determine the most promising node in the current search tree, i.e. the node which will be used for further expansion of the desired mapping in the next iteration, a heuristic function is usually used. Formally, for a node p in the search tree, we use $g(p)$ to denote the cost of the optimal path from the root node to the current node p, i.e. $g(p)$ is set equal to the cost of the partial edit path accumulated so far, and we use $h(p)$ for denoting the estimated cost from p to a leaf node. The sum $g(p) + h(p)$ gives the total cost assigned to an open node in the search tree. One can show that, given that the estimation of the future costs $h(p)$ is lower than, or equal to, the real costs, the algorithm is admissible, i.e. an optimal path from the root node to a leaf node is guaranteed to be found.[23]

Algorithm 1 Graph edit distance algorithm.

Input: Non-empty graphs $g_1 = (V_1, E_1, \mu_1, \nu_1)$ and $g_2 = (V_2, E_2, \mu_2 \nu_2)$, where $V_1 = \{u_1, \ldots, u_{|V_1|}\}$ and $V_2 = \{v_1, \ldots, v_{|V_2|}\}$

Output: A minimum cost edit path from g_1 to g_2 e.g. $p_{min} = \{u_1 \to v_3, u_2 \to \varepsilon, \ldots, \varepsilon \to v_2\}$

initialize $OPEN$ to the empty set $\{\}$
For each node $w \in V_2$, insert the substitution $\{u_1 \to w\}$ into $OPEN$
Insert the deletion $\{u_1 \to \varepsilon\}$ into $OPEN$
loop
  Remove $p_{min} = argmin_{p \in OPEN}\{g(p) + h(p)\}$ from $OPEN$
  if p_{min} is a complete edit path **then**
    Return p_{min} as the solution
  else
    Let $p_{min} = \{u_1 \to v_{i1}, \cdots, u_k \to v_{ik}\}$
    if $k < |V_1|$ **then**
      For each $w \in V_2 \setminus \{v_{i1}, \cdots, v_{ik}\}$, insert $p_{min} \cup \{u_{k+1} \to w\}$ into $OPEN$
      Insert $p_{min} \cup \{u_{k+1} \to \varepsilon\}$ into $OPEN$
    else
      Insert $p_{min} \cup \bigcup_{w \in V_2 \setminus \{v_{i1}, \cdots, v_{ik}\}} \{\varepsilon \to w\}$ into $OPEN$
    end if
  end if
end loop

In Algorithm 1 the A*-based method for optimal graph edit distance computation is given. The nodes of the source graph are processed in the order $(u_1, u_2, \ldots)$. The deletion (line 12) or the substitution of a node (line 11) are considered simultaneously, which produces a number of successor nodes in the search tree. If all nodes of the first graph have been processed, the remaining nodes of the second graph are inserted in a single step (line 14). The set $OPEN$ of partial edit paths contains the search tree nodes to be processed in the next steps. The most promising partial edit path $p \in OPEN$, i.e. the one that minimizes $g(p) + h(p)$, is always chosen first (line 5). This procedure guarantees that the complete edit path found by the algorithm first is always optimal, i.e. has minimal costs among all possible competing paths (line 7).

Note that edit operations on edges are implied by edit operations on their adjacent nodes, i.e. whether an edge is substituted, deleted, or inserted, depends on the edit operations performed on its adjacent nodes. Obviously, the implied edge operations can be derived from every partial or complete edit path during the search procedure given in Algorithm 1. The costs of these implied edge operations are dynamically added to the corresponding paths in $OPEN$.

In order to integrate more knowledge about partial solutions in the search tree, it has been proposed to use heuristics.[23] Basically, such heuristics for a tree search algorithm aim at the estimation of a lower bound $h(p)$ of the future costs. In the simplest scenario this lower bound estimation $h(p)$ for the current node p is set to zero for all p, which is equivalent to using no heuristic information about the present situation at all. The other extreme would be to compute for a partial edit path the actual optimal path to a leaf node, i.e. perform a complete edit distance computation for each node of the search tree. In this case, the function $h(p)$ is not a lower bound, but the exact value of the optimal costs. Of course, the computation of such a perfect heuristic is both unreasonable and intractable. Somewhere in between the two extremes, one can define a function $h(p)$ evaluating how many edit operations have to be performed in a complete edit path at least.[21]

Efficient Approximation Algorithms The method described in the previous section finds an optimal edit path between two graphs. Due to the fact that the error-tolerant nature of edit distance potentially allows every node of a graph g to be mapped to every node of another graph g' (unlike exact graph matching methods such as subgraph isomorphism or maximum common subgraph), the computational complexity of the edit distance algorithm, whether or not a heuristic function $h(p)$ is used to govern the tree traversal process, is exponential in the number of nodes of the involved graphs. Consequently, the edit distance can be computed for graphs of a rather small size only.

In recent years, a number of methods addressing the high computational complexity of graph edit distance computation have been proposed. In some approaches, the basic idea is to perform a local search to solve the graph matching problem, that is, to optimize local criteria instead of global, or optimal ones.[24] In Ref. 25, a linear programming method for computing the edit distance of graphs with unlabeled edges is proposed. The method can be used to derive lower and upper edit distance bounds in polynomial time. Two fast but suboptimal algorithms for graph edit distance computation are proposed in Ref. 26. The authors propose simple variants of an optimal edit distance algorithm that make the computation substantially faster. Another approach to efficient graph edit distance computation has been proposed in Ref. 27. The basic idea is to decompose graphs into sets of subgraphs. These subgraphs consist of a node and its adjacent nodes and edges. The graph matching problem is then reduced to the problem of finding an optimal match between the sets of subgraphs by means of dynamic programming. In Ref. 28 another efficient algorithm for solving the problem of graph edit distance computation is introduced. This approach is somewhat similar to the method described in Ref. 27, i.e. the graph edit distance is approximated by finding an optimal match between nodes of two graphs and their local structure. However, in order to find the optimal match, a bipartite matching procedure is used rather than dynamic programming. Because of its suboptimal nature, this method does not generally return the optimal edit path, but only an approximate one. Nonetheless, it has been shown to work well in various graph classification tasks.[28,40]

A common way to make graph matching more efficient is to restrict considerations to special classes of graphs. Examples include the classes of planar graphs,[29] bounded-valence graphs,[30] trees,[31] and graphs with unique node labels.[10] Recently, a suboptimal edit distance algorithm has been proposed[32] that requires the nodes of graphs to be planarly embedded, which is satisfied in many computer vision applications of graph matching.

3. Graph Embedding by Means of Dissimilarity Representation

3.1. *General Embedding Procedure and Properties*

The idea of the proposed graph embedding framework stems from the seminal work done by Duin and Pekalska[16] where dissimilarities for pattern representation are used for the first time. Later this method was extended so as to map string representations into vector spaces.[33] In the current work we go one step further and generalize and substantially extend the methods described in Refs. 16,33 to the domain of graphs. The key idea of this approach is to use the distances of an input graph to a number of training graphs, termed *prototype graphs*, as a vectorial description of the graph.

Assume we have a set of sample graphs, $\mathcal{T} = \{g, \ldots, g_N\}$ from some graph domain $\mathcal{G}$ and an arbitrary graph dissimilarity measure $d : \mathcal{G} \times \mathcal{G} \rightarrow \mathbb{R}$. Note that $\mathcal{T}$ can be any kind of graph set. However, for the sake of convenience we assume in the following that $\mathcal{T}$ is a training set of existing graphs. After selecting a set of prototypical graphs $\mathcal{P} \subseteq \mathcal{T}$, we compute the dissimilarity of a given input graph

g to each prototype graph $p_i \in \mathcal{P}$. Note that g can be an element of $\mathcal{T}$ or any other graph set $\mathcal{S}$. Given n prototypes, i.e. $\mathcal{P} = \{p_1, \ldots, p_n\}$, this procedure leads to n dissimilarities, $d_1 = d(g, p_1), \ldots, d_n = d(g, p_n)$, which can be arranged in an n-dimensional vector $(d_1, \ldots, d_n)$.

Definition 3 (Graph Embedding). *Let us assume a graph domain $\mathcal{G}$ is given. If $\mathcal{T} = \{g, \ldots, g_N\} \subseteq \mathcal{G}$ is a training set with N graphs and $\mathcal{P} = \{p_1, \ldots, p_n\} \subseteq \mathcal{T}$ is a prototype set with n graphs, the mapping*

$$\varphi_n^{\mathcal{P}} : \mathcal{G} \to \mathbb{R}^n$$

is defined as the function

$$\varphi_n^{\mathcal{P}}(g) = (d(g, p_1), \ldots, d(g, p_n)),$$

where $d(g, p_i)$ is any graph dissimilarity measure between graph g and the i-th prototype graph.

Obviously, by means of this definition we obtain a vector space where each axis corresponds to a prototype graph $p_i \in \mathcal{P}$ and the coordinate values of an embedded graph g are the distances of g to the elements in $\mathcal{P}$. In this way we can transform any graph g from the training set $\mathcal{T}$ as well as any other graph set $\mathcal{S}$ (for instance a validation or a test set of a classification problem), into a vector of real numbers. The embedding procedure is illustrated in Fig. 2.

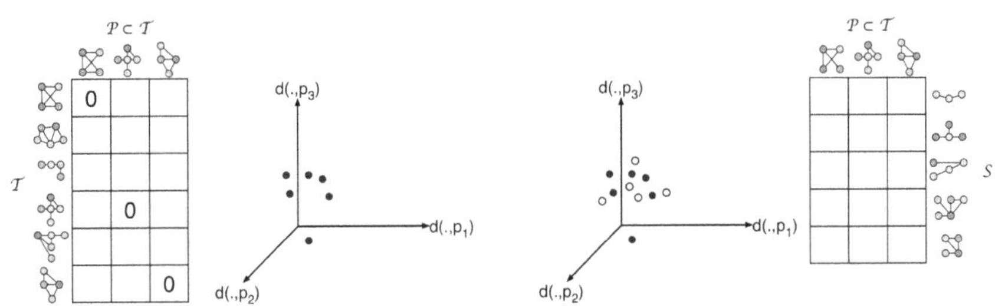

Fig. 2. Based on a prototype set $\mathcal{P} \subseteq \mathcal{T}$ the graphs from the training set $\mathcal{T}$ as well as graphs from some additional data set $\mathcal{S}$ are transformed into a vector of dissimilarities. Black dots represent embedded graphs from $\mathcal{T}$, while white dots are embedded graphs from $\mathcal{S}$. Note that training graphs, which have been selected as prototypes before, have a zero entry in their corresponding vector.

The embedding procedure proposed in this chapter makes use of graph edit distance. Note, however, that any other graph dissimilarity measure can be used as well. Yet, using graph edit distance allows us to deal with a large class of graphs (directed, undirected, unlabeled, node and/or edge labels from any finite or infinite domain). Furthermore, due to the flexibility of graph edit distance, a high degree of robustness against various graph distortions can be expected. Hence, in contrast with other graph embedding techniques (e.g. spectral methods) our approach is characterized by a high degree of flexibility in the graph definition and a high tolerance to structural errors in the graph matching process. Since the computation of graph edit distance is exponential in the number of nodes for general graphs, the complexity of this graph embedding is exponential as well. However, as mentioned in Sect. 2, there exist efficient approximation algorithms for graph edit distance computation with cubic time complexity (e.g. the procedure described in Ref. 28). Consequently, given n predefined prototypes the embedding of one particular graph is established by means of n distance computations with polynomial time.

After two graphs, g and g', have been mapped to the vector space, we observe the following relationship[a]:

$$\begin{aligned}
||\varphi(g) - \varphi(g')|| &= (\langle\varphi(g), \varphi(g)\rangle + \langle\varphi(g'), \varphi(g')\rangle - 2\langle\varphi(g), \varphi(g')\rangle)^{\frac{1}{2}} \\
&= \left(\sum_{i=1}^{n} d(g, p_i)^2 + \sum_{i=1}^{n} d(g', p_i)^2 - 2\sum_{i=1}^{n} d(g, p_i)d(g', p_i)\right)^{\frac{1}{2}} \\
&= \left(\sum_{i=1}^{n} (d(g, p_i) - d(g', p_i))^2\right)^{\frac{1}{2}} \qquad (1) \\
&\leq \left(n \cdot d(g, g')^2\right)^{\frac{1}{2}} \\
&= \sqrt{n} \cdot d(g, g') \qquad (2)
\end{aligned}$$

Hence, the Euclidean distance of a pair of graphs g and g' in the vector space is equal to the square root of the sum of squared differences between the edit distances of g and g' to the prototype graphs (Eq. 1). Given d is a metric, the more similar g and g' are, the smaller will be the terms $(d(g, p_i) - d(g', p_i))^2$ and, consequently, the smaller will be their Euclidean distance in the vector space. Moreover, due to the triangle inequality, the upper bound of the Euclidean distance of a pair of graph maps $\varphi(g)$ and $\varphi(g')$ is given by $\sqrt{n} \cdot d(g, g')$ (Eq. 2). Obviously, defining the embedding procedure given in Def. 3 as

$$\varphi_n^{\mathcal{P}}(g) = (d(g, p_1)/q, \ldots, d(g, p_n)/q) \quad ,$$

with $q = \sqrt{n}$, the upper bound of the Euclidean distance of a pair of graph maps $\varphi(g)$ and $\varphi(g')$ amounts to $d(g, g')$. By means of the normalization by q, the influence of the number of prototypes is reduced. Hence, regardless of the number of prototypes, it is guaranteed that distances in the resulting embedding space are bounded by the original graph edit distance $d(g, g')$.

3.2. *Relation to Kernel Methods*

Dissimilarity embedding is closely related to kernel methods.[34,35] In the kernel approach the patterns are described by means of pairwise kernel functions, and in the dissimilarity approach patterns are described by pairwise dissimilarities. However,

[a]For the purpose of simplicity we write $\varphi(g)$ for $\varphi_n^{\mathcal{P}}(g)$. Furthermore, $\langle\cdot,\cdot\rangle : \mathbb{R}^n \times \mathbb{R}^n \rightarrow \mathbb{R}$ denotes the scalar product of two vectors.

there is one fundamental difference between kernels and dissimilarity embeddings. In the former method, the kernel values are interpreted as dot products in some implicit feature space. By means of kernel machines, the pattern recognition task is eventually carried out in this kernel feature space. In the latter approach, the set of dissimilarities is interpreted as a vectorial description of the pattern under consideration. Hence, no implicit feature space, but an explicit dissimilarity space is obtained.

Although conceptually different, the embedding paradigm established by $\varphi_n^{\mathcal{P}} : \mathcal{G} \to \mathbb{R}^n$ constitutes a foundation for a novel class of graph kernels. One can define a valid graph kernel κ based on the graph embedding $\varphi_n^{\mathcal{P}} : \mathcal{G} \to \mathbb{R}^n$ by computing the standard dot product of two graph maps in the resulting vector space[b]

$$\kappa_{\langle\rangle}(g, g') = \langle \varphi_n^{\mathcal{P}}(g), \varphi_n^{\mathcal{P}}(g') \rangle \quad .$$

Of course, not only the standard dot product can be used but any valid kernel function defined for vectors. For instance an RBF kernel function

$$\kappa_{RBF}(g, g') = \exp\left(-\gamma ||\varphi_n^{\mathcal{P}}(g) - \varphi_n^{\mathcal{P}}(g')||^2\right)$$

with $\gamma > 0$ can thus be applied to graph maps. We refer to this procedure as *graph embedding kernel* using some specific vector kernel function.

This graph embedding kernel is somewhat similar in spirit to the subgraph kernel proposed in Ref. 37. However, rather than using a potentially infinite set of subgraphs, we restrict our considerations to a finite number of prototypes, i.e. dimensions. Furthermore, rather than counting the number of occurrences of a particular subgraph within some host graph, we employ the edit distance in order to capture graph similarity. Obviously, large values of the kernel function occur when both g and g' are similar to the same prototype graphs. Clearly, whenever both g and g' are similar to the same prototype graphs we may conclude that g and g' are similar to each other. Using the graph edit distance rather than counting the number of occurrences of certain subgraphs allows us to flexibly deal with various kinds of distortions that may affect the underlying graphs.

In a recent book, graph kernels were proposed that directly use graph edit distances.[38] This approach turns an existing dissimilarity measure (e.g. graph edit distance) into a similarity measure by mapping low distance values to high similarity values and vice versa. To this end monotonically decreasing transformations are used. Given the edit distance $d(g, g')$ of two graphs g and g', the similarity kernel is defined, for instance, as $\kappa(g, g') = \exp(-\gamma\ d(g, g'))$. Note that this kernel function is not positive definite in general. However, there is theoretical evidence that using kernel machines in conjunction with indefinite kernels may be reasonable if some conditions are fulfilled.[39] We refer to this method as *similarity kernel*, or *sim* for short, in the following.

[b]Note that is approach is very similar to the empirical kernel map described in Ref. 36 where general similarity measures are turned into kernel functions.

Note the fundamental difference between such an approach and our graph embedding procedure. While with the similarity kernel the existing dissimilarity measure is turned into a similarity measure (i.e. a kernel function) and subsequently plugged into a kernel machine, the dissimilarity embedding of graphs uses the distances to n prototypes as features for a new explicit vectorial description of the underlying object.

3.3. *The Problem of Prototype Selection*

The selection of the n prototypes $\mathcal{P} = \{p_1, \ldots, p_n\}$ is a critical issue since not only the prototypes $p_i \in \mathcal{P}$ themselves but also their number n affect the resulting graph mapping $\varphi_n^{\mathcal{P}}(\cdot)$ and thus the performance of the corresponding pattern recognition algorithm. A good selection of n prototypes seems to be crucial to succeed with the classification algorithm in the embedding vector space. A first and very simple solution might be to use all available training graphs from $\mathcal{T}$ as prototypes. Yet, two severe shortcomings arise with such a plain approach. First, the dimensionality of the resulting vector space is equal to the size N of the training set $\mathcal{T}$. Consequently, if the training set is large, the mapping results in (possibly too) high dimensional feature vectors definitely disabling efficient computations. Secondly, the presence of similar prototypes as well as outlier graphs in the training set $\mathcal{T}$ is likely. Therefore, redundant and noisy or irrelevant information will be captured in the graph maps which in turn may harm the performance of the pattern recognition algorithms applied subsequently.

The selection of prototypes for graph embedding has been addressed in various papers.[40–43] In Refs. 40,44, for instance, a number of *prototype selection methods* are discussed. These selection strategies use some heuristics based on the underlying dissimilarities in the original graph domain. The basic idea of these approaches is to select prototypes from $\mathcal{T}$ that best reflect the distribution of the training set $\mathcal{T}$ or cover a predefined region of $\mathcal{T}$. The rationale of this procedure is that capturing distances to significant prototypes from $\mathcal{T}$ leads to meaningful dissimilarity vectors.

A severe shortcoming of prototype selection strategies is that the dimensionality of the embedding space has to be determined by the user. In other words, the number of prototypes to be selected by a certain prototype selection algorithm has to be experimentally defined by minimizing the error rate of the target classifier on a validation set. In order to overcome this limitation, various *prototype reduction schemes*[45] (known from the literature for the task of condensing training sets in nearest-neighbor classification scenarios) are adopted for the task of graph embedding in Ref. 41. In contrast with the heuristic prototype selection strategies, with these procedures the number of prototypes n, i.e. the resulting dimensionality of the vector space, is defined by an algorithmic procedure.

Another solution to the problem of noisy and redundant vectors with too high dimensionality is offered by the following procedure. Rather than selecting the prototypes beforehand, the embedding is carried out first and then the problem of prototype selection is reduced to a feature subset selection problem. That is, for graph embedding all available elements from the training set are used as prototypes, i.e. we define $\mathcal{P} = \mathcal{T}$. Next, a large number of different feature selection strategies[46–49] can be applied to the resulting large scale vectors eliminating redundancies and noise, finding good features, and reducing the dimensionality. In Ref. 42, for instance, principal component analysis (PCA)[48] and Fisher linear discriminant analysis (LDA)[49] are applied to the vector space embedded graphs. Rather than traditional PCA, kernel PCA[50] is used for feature transformation in Ref. 43.

As an example for solving the problem of prototype selection, the procedures from the first category, viz. heuristic prototype selection strategies,[40,44] are discussed in detail in the next Section.

4. Prototype Selection Strategies

The prototypes $\mathcal{P} = \{p_1, \ldots, p_n\}$ serve as reference points in the underlying graph domain in order to transform graphs into real vectors. Hence, the objective of prototype selection is to find reference points which lead to meaningful vectors in the embedding space. Intuitively, prototypes should not be redundant, i.e. we should avoid selecting similar graphs, and simultaneously, prototypes should include as much information as possible, i.e. they should be uniformly distributed over the whole set of patterns. In the present section six prototype selection strategies are discussed, all of them selecting graphs from $\mathcal{T}$ with respect to the underlying dissimilarity information. All of the algorithms take as input the training set $\mathcal{T}$ and the number of prototypes n to be selected. As output the algorithms provide a set of prototypes $\mathcal{P} = \{p_1, \ldots, p_n\} \subseteq \mathcal{T}$.

For five of the proposed prototype selectors, special graphs, termed *set median graph*[51] and *set marginal graph*, will be needed. They are introduced below.

Definition 4 (Set Median and Set Marginal Graph). *Given a set $\mathcal{S}$ of graphs, the set median graph is defined as*

$$median(\mathcal{S}) = \operatorname*{argmin}_{g \in \mathcal{S}} \sum_{g' \in \mathcal{S}} d(g, g')$$

The set median graph is a graph in $\mathcal{S}$ whose sum of edit distances to all other graphs in $\mathcal{S}$ is minimal. Intuitively, the set median graph is located in the center of a given graph set. In contrast to the graph located in the center of a given set, the set marginal graph is located at the border of a given graph set. Formally,

$$marginal(\mathcal{S}) = \operatorname*{argmax}_{g \in \mathcal{S}} \sum_{g' \in \mathcal{S}} d(g, g')$$

The set marginal graph is thus the graph in $\mathcal{S}$ whose sum of distances to all other graphs in $\mathcal{S}$ is maximal.

Assume there are k different classes of graphs $\omega_1, \ldots, \omega_k$ represented in the training set $\mathcal{T}$. We distinguish between classwise and classindependent prototype selection. Classindependent selection means that the selection is executed over the whole training set $\mathcal{T}$ to get n prototypes, while in classwise selection the selection is performed individually for each of the k different classes $\omega_1, \ldots, \omega_k$. In the latter case, n_i prototypes are selected independently for each class ω_i such that $\sum_{i=1}^{k} n_i = n$. Whether the classwise or classindependent method is more convenient depends on a number of factors, including the size of $\mathcal{T}$, the distribution of the graphs in $\mathcal{T}$, whether or not classes are balanced, and the application.

In order to illustrate the behavior of the different prototype selectors we provide illustrative examples in Fig. 3. The data underlying these examples are two-dimensional vectors obtained through multidimensional scaling applied to the original graph edit distances from the Letter data set (this data set is used in the experimental evaluation and is described in detail in Sect. 5).

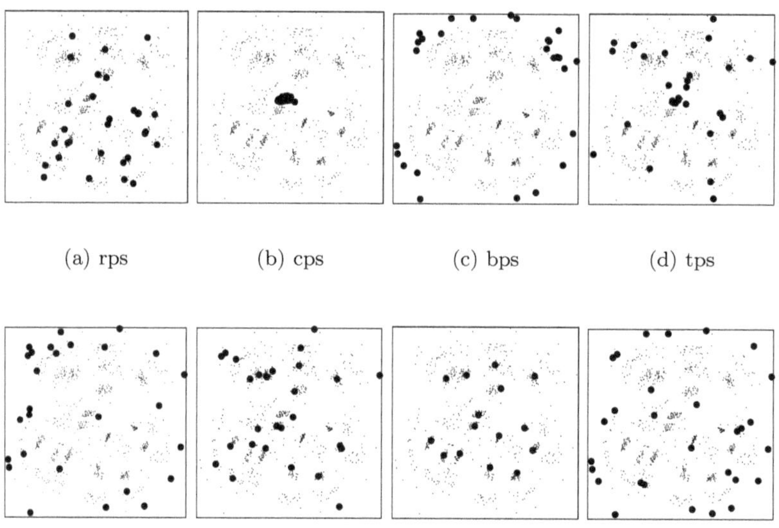

(a) rps (b) cps (c) bps (d) tps

(e) sps (f) k-cps (g) cps-c (h) bps-c

Fig. 3. Illustration of the different prototype selectors applied to the training set $\mathcal{T}$ of the Letter data set. The number of prototypes is defined by $n = 30$. The selected prototypes by the respective selection algorithms are shown with big dots.

Random (rps) A random selection of n unique prototypes from $\mathcal{T}$ is performed (cf. Fig. 3 (a)). Of course, the Random prototype selector can be applied classindependent or classwise (rps-c).

Center (cps) The Center prototype selector selects prototypes situated in, or near, the center of the training set $\mathcal{T}$ (cf. Fig. 3 (b)). The set of prototypes $\mathcal{P}$ is iteratively constructed starting with $\mathcal{P} = \{\}$. In each iteration the set median graph $median(\mathcal{T} \setminus \mathcal{P})$ is selected as a prototype and added to $\mathcal{P}$.

It seems that the Center prototype selector is not very appropriate for selecting n prototypes. Obviously, both requirements, avoiding redundancy and uniform distribution, are not satisfied. Nevertheless, we mention this prototype selector here for the purpose of completeness. To obtain a better distribution, one can apply the Center prototype selector classwise (cps-c). In this case, supposably, the prototypes mirror the given distribution better than the classindependent version. In Fig. 3 (g) the classwise cps is illustrated.

Border (bps) The Border prototype selector selects prototypes situated at the border of the training set $\mathcal{T}$ (cf. Fig. 3 (c)). The set of prototypes $\mathcal{P}$ is iteratively constructed starting with $\mathcal{P} = \{\}$. In each iteration the set marginal graph $marginal(\mathcal{T} \setminus \mathcal{P})$ is moved from $\mathcal{T}$ to the prototype set $\mathcal{P}$.

In contrast to the Center prototype selector, where many prototypes are structurally similar, this kind of selection avoids redundancies much better. However, there are no prototypes selected from the center of the set. Obviously, the requirement of uniform distribution is not satisfied. Furthermore, one has to expect that bps possibly selects outliers. The Border prototype selector can be applied classindependent and classwise (bps-c). In Fig. 3 (h) the classwise bps is illustrated.

Targetsphere (tps) The idea of this prototype selector is to distribute the prototypes from the center to the border as uniformly as possible (cf. Fig. 3 (d)). The Targetsphere prototype selector first determines the set median graph $g_c = median(\mathcal{T})$. After this center graph has been found, the graph furthest away from g_c, i.e. the graph $g_f \in \mathcal{T}$ maximizing the distance to g_c is located. Both graphs, g_c and g_f, are selected as prototypes. The distance from g_c to g_f is referred to as d_{max}, i.e. $d_{max} = d(g_c, g_f)$. The interval $[0, d_{max}]$ is then divided into $n-1$ equidistant subintervals of width $w = \frac{d_{max}}{n-1}$. The $n-2$ graphs for which the corresponding distances to the set median g_c are located nearest to the interval borders in terms of edit distance are selected as remaining prototypes. More formally, the i-th prototype p_i (where $i = 1, \ldots, n-2$) is defined as

$$p_i = \underset{g \in \mathcal{T}}{\text{argmin}} |d(g, g_c) - iw| \quad .$$

The selection of the prototypes considers the distance to the set median graph only. That is, the procedure selects n prototypes uniformly distributed in the interval $[0, d_{max}]$. The individual distances to the prototypes selected before are not considered. The Targetsphere prototype selector can be applied classindependent as well as classwise (tps-c).

Spanning (sps) The Spanning prototype selector considers all distances to the prototypes selected before. The first prototype p_1 is the set median graph $median(\mathcal{T})$. Each additional prototype selected by the spanning prototype selector is the graph furthest away from the already selected prototype graphs. More formally, the i-th prototype p_i (where $i = 2, \ldots, n$) is defined as

$$p_i = \underset{g \in \mathcal{T}}{\text{argmax}} \min_{p \in \mathcal{P}} d(g, p) \quad .$$

In contrast to tps, where only the distances to the center graph are considered during the selection, sps takes all distances to the already selected prototypes into account and tries to cover the whole training set as uniformly as possible (cf. Fig. 3 (e)). The Spanning prototype selector can be applied classindependent or classwise (sps-c).

k-Centers (k-cps) The k-Centers prototype selector tries to choose n graphs from $\mathcal{T}$ so that they are evenly distributed with respect to the dissimilarity information given by d. The algorithm is similar to k-medians clustering.[52] It starts with a prototype set obtained through sps. These prototypes serve as initial cluster centers. Next each of the N training patterns from $\mathcal{T}$ is assigned to the closest cluster center. Based on this initial clustering, the cluster centers (the set medians of each cluster) are recomputed. The two preceding steps, i.e. the assignment of objects to the nearest cluster center and the recomputation of the centers, are repeated until no more changes in the clusters occur. The prototypes $\mathcal{P} = \{p_1, \ldots, p_n\}$ are then given by the set medians of the n disjoint clusters.

The k-Centers prototype selector can be applied classindependent as well as classwise (k-cps-c). In contrast to sps, we note that k-cps avoids selecting prototypes from the border by focusing on graphs that are in the center of densely populated areas (cf. Fig. 3 (e) and (f) for a comparison between sps and k-cps).

5. Experimental Evaluation

In this section we provide results of an experimental evaluation of the proposed embedding procedure. We aim at empirically confirming that the method of prototype based graph embedding and subsequent classification in real vector spaces is applicable to different graph classification problems and matches, or even surpasses, the performance of comparable techniques. For graph edit distance computation the suboptimal algorithm introduced in Ref. 28 has been used. This graph edit distance algorithm shows superior performance in time and accuracy compared to other suboptimal algorithms. For prototype selection we focus on three classwise selectors, viz. bps-c, sps-c, and k-cps-c. The classifier used in the embedding space is the support vector machine (SVM).[53] Of course, any other classifier could be used for this purpose as well. However, we feel that the SVM is particularly suitable because of its theoretical advantages and its superior performance that has been empirically confirmed in many practical classification problems.

5.1. *Graph Data Sets*

For evaluation, six data sets from the IAM graph database repository[c] for graph based pattern recognition and machine learning are used.[54] The first graph data set involves graphs that represent distorted letter drawings (Letter). We consider the 15 capital letters of the Roman alphabet that consist of straight lines only (*A*, *E*, *F*, ..., *Z*). For each class, a prototype line drawing is manually constructed. These prototype drawings are then converted into prototype graphs by representing lines by undirected edges and ending points of lines by nodes. Each node is labeled with a two-dimensional attribute giving its position relative to a reference coordinate system. Edges are unlabeled.

The second graph set (Digit) consists of graphs representing handwritten digits[56] (*0*, *1*, *2*, ..., *9*). During the recording of the digits, the position of the pen was determined with constant time intervals. The resulting sequences of (x, y)-coordinates were converted into graphs by inserting nodes in regular intervals between the starting and ending points of a line. Successive nodes are connected by undirected edges. Each node is labeled with a two-dimensional attribute giving its position relative to a reference coordinate system. The edges are attributed with an angle denoting the orientation of the edge with respect to the horizontal direction.

The third graph data set consists of graphs representing fingerprint images out of the four classes *arch*, *left*, *right*, and *whorl* from the Galton-Henry classification system (Fingerprint). The fingerprint database used in our experiments is based on the NIST reference database of fingerprints.[55] In order to obtain graphs from fingerprint images, the relevant regions are binarized and a noise removal and thinning procedure is applied. This results in a skeletonized representation of the extracted regions. Ending points and bifurcation points of the skeletonized regions are represented by nodes. Additional nodes are inserted in regular intervals between ending points and bifurcation points. Finally, undirected edges are inserted to link nodes that are directly connected through a ridge in the skeleton. Each node is labeled with a two-dimensional attribute giving its position. The edges are attributed with an angle denoting the orientation of the edge with respect to the horizontal direction.

The AIDS data set consists of graphs representing molecular compounds. We construct graphs from the AIDS Antiviral Screen Database of Active Compounds.[57] This data set consists of two classes (*active*, *inactive*), which represent molecules with activity against HIV or not. The molecules are converted into graphs in a straightforward manner by representing atoms as nodes and the covalent bonds as edges. Nodes are labeled with the number of the corresponding chemical symbol and edges by the valence of the linkage.

---

[c] `www.iam.unibe.ch/fki/databases/iam-graph-database`

Table 1. Summary of graph data set characteristics, viz. the size of the training (tr), the validation (va) and the test set (te), the number of classes ($|\Omega|$), the label alphabet of both nodes and edges, the average and maximum number of nodes and edges (∅/max nodes/edges).

Database	size (tr, va, te)	$\|\Omega\|$	node labels	edge labels	∅/max nodes	∅/max edges
Letter	750, 750, 750	15	x, y coordinates	none	4.7/9	4.5/9
Digit	1,000, 500, 2,000	10	x, y coordinates	Angle	11.8/32	13.1/30
Fingerprint	500, 300, 2,000	4	x, y coordinates	Angle	5.4/26	4.4/24
AIDS	250, 250, 1,500	2	Chemical symbol	Valence	15.7/95	16.2/103
Mutagenicity	1,500, 500, 2,337	2	Chemical symbol	Valence	30.3/417	30.8/112
Web	780, 780, 780	20	Word/Frequency	Type	186.1/834	104.6/596

In order to convert molecular compounds of the Mutagenicity data set[58] into attributed graphs the same procedure as for the AIDS data set is applied. The Mutagenicity data set is divided into two classes *mutagen* and *nonmutagen*.

In Ref. 7 several methods for creating graphs from web documents are introduced. For the graphs included in this data set (Web), the following method was applied. First, all words occurring in the web document – except for stop words, which contain only little information – are converted into nodes in the resulting web graph. We attribute each node with the corresponding word and its frequency. Next, different sections of the web document are investigated individually. If a word w_i immediately precedes word w_{i+1}, a directed edge from the node corresponding to word w_i to the node corresponding to the word w_{i+1} is inserted in our web graph. The resulting edge is attributed with the corresponding section label. In our experiments we make use of a data set with documents from 20 categories (*Business*, *Health*, *Politics*, ...).

In Table 1 a summary of the graph data set characteristics is given.

5.2. *Reference Systems and Experimental Setup*

In contrast with the high representational power of graphs, we observe a lack of general classification algorithms that can be applied in the graph domain. One of the few classifiers directly applicable to arbitrary graphs is the k-nearest-neighbor classifier (k-NN). Given a labeled set of training graphs, an unknown graph is assigned to the class that occurs most frequently among the k nearest graphs (in terms of edit distance) from the training set. The decision boundary of this classifier is a piecewise linear function which makes it very flexible. This classifier in the graph domain will serve us as our first reference system.

The second reference system is the similarity kernel described in Sect. 3.2 (sim) in conjunction with an SVM. Comparing the performance of this similarity kernel function with the performance of the dissimilarity embedding procedure offers us the possibility to understand whether the power of our system is primarily due to the embedding process or to the strength of the kernel classifier.

In the validation phase, the costs of the edit operations are determined first. For all considered graph data sets but the web data, node and edge labels are integer numbers, real numbers, or real vectors. Here the substitution cost of a pair of labels is given by a distance measure (e.g. Euclidean distance), and only the deletion and insertion costs have to be determined. For the sake of symmetry, we assume in all experiments identical costs for node deletions and insertions, and for edge deletions and insertions. Hence only two parameters, node deletion/insertion cost τ_n, and edge deletion/insertion cost τ_e, need to be validated. The validation procedure is illustrated in Fig. 4, where the classification results of a k-NN classifier on the validation set as a function of the two edit cost parameters are shown for the Fingerprint data set ($k = 1, 3, 5$). For the web data set we followed the cost model used in Ref. 7, i.e. node substitutions are not admissible and the costs of all other edit operations are set to an arbitrary constant. Consequently, no edit cost validation is needed for this data set.

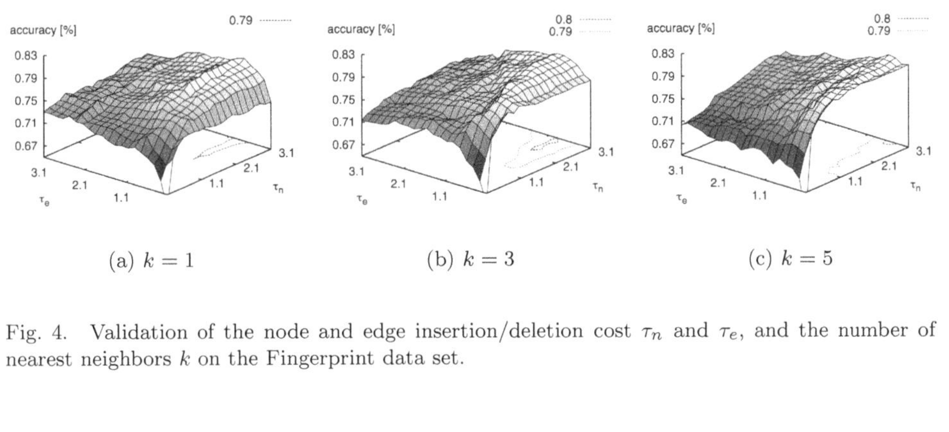

(a) $k = 1$ (b) $k = 3$ (c) $k = 5$

Fig. 4. Validation of the node and edge insertion/deletion cost τ_n and τ_e, and the number of nearest neighbors k on the Fingerprint data set.

For the second reference system (sim) the same edit distances as for the k-NN classifier are used. Hence, the weighting parameter C of the SVM, which controls whether the maximization of the margin or the minimization of the error is more important, and the meta parameter γ in the kernel function $\kappa(g, g') = \exp(-\gamma\ d(g, g'))$ have to be additionally validated.

For the task of graph embedding in real vector spaces one additional meta parameter has to be validated, namely the number of prototypes n, i.e. the dimensionality of the resulting vector space. In order to determine suitable values of n, each graph set is embedded in a vector space with all of the prototype selectors described in Section 4, varying the dimensionality of the target vector space over a certain interval. With the resulting vector sets, which still consist of a validation, a training, and a test set, an SVM is trained. We make use of an SVM with RBF-kernel where besides the weighting parameter C, the meta parameter γ in the kernel

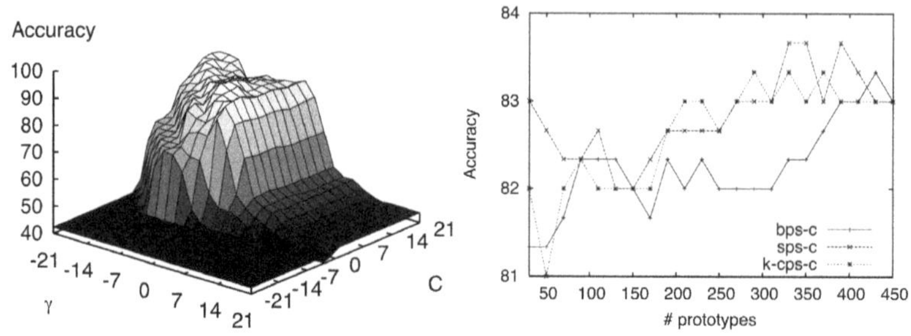

(a) SVM Validation (b) class-wise

Fig. 5. (a) Validation of the meta parameter tuple (C, γ) for a specific prototype selector and a certain number of prototypes (the parameter values are on a logarithmic scale to the basis 2). (b) Validation of three classwise prototype selectors.

function has to be optimized[d]. In Fig. 5 (a) such an SVM parameter validation on the Fingerprint data set is illustrated.

The SVM optimization is performed on a validation set for every possible dimension of the target space and every prototype selection method. Thus, for each embedding procedure the classification accuracy can be regarded as a function of the dimensionality and the prototype selector. This final optimization is illustrated on the Fingerprint data set in Fig. 5 (b) where the accuracies for three classwise prototype selectors (bps-c, sps-c, k-cps-c) and each dimensionality are shown.

By means of this procedure for each prototype selector the number of prototypes n, and the SVM parameters (C, γ) can be determined for every data set on the validation set. The parameter combination that results in the lowest classification error is finally applied to the independent test set.

5.3. *Results and Discussion*

The results obtained on the test set by means of this procedure are reported for three prototype selectors in Table 2. Let us first compare the classification accuracies achieved by our embedding framework with the first reference system (k-NN). It clearly turns out that the novel procedure is much more powerful than the traditional k-NN in the graph domain as on all tested applications the embedding kernel outperforms the first reference system's classification results. Note that 16 out of 17 improvements are statistically significant.

Next we compare the embedding procedure with the similarity kernel (sim). It turns out that the similarity kernel is better on the Letter, Fingerprint, and Web data sets compared to at least one prototype selection method. However,

[d]We make use of the RBF kernel since it is the most widely used kernel and it has been extensively studied in various fields of pattern recognition.[35]

Table 2. Embedding methods vs. reference systems. (①/② Stat. significant improvement over the first/second reference system (k-NN and sim),❷ Stat. significant deterioration compared to the second reference system.)

	ref. systems		embedding methods		
Data Set	k-NN	sim	sps-c	bps-c	k-cps-c
Letter	89.1	92.9	92.3 ①	92.9 ①	92.0 ①
Digit	97.4	98.1	98.6 ①②	98.7 ①②	98.7 ①②
Fingerprint	79.1	82.0	82.0 ①	83.1 ①②	81.7 ①
AIDS	94.9	97.0	98.1 ①②	98.0 ①②	98.1 ①②
Mutagenicity	66.9	68.6	71.6 ①②	70.7 ①	71.8 ①②
Web	80.6	82.9	82.7 ①	80.4 ❷	82.3

the embedding framework outperforms the similarity kernel ten times on the Digit, Fingerprint, AIDS, and Mutagenicity data sets. Note that nine of the improvements, but only one of the deteriorations are statistically significant. From these findings we can conclude that the power of our novel approach primarily results from the embedding process itself and not from to the strength of the kernel classifier.

For a more detailed analysis of the proposed approach including both more thorough descriptions of the achieved results and other tasks than classification (e.g. clustering of vector space embedded graphs) we refer to Refs. 40–42.

6. Conclusions

For objects given in terms of feature vectors a rich repository of algorithmic tools for classification has been developed over the last decades. Graphs are a versatile alternative to feature vectors, and are known to be a powerful and flexible representation formalism. The representational power of graphs is due to their ability to represent not only feature values but also relationships among different parts of an object, and their flexibility comes from the fact there are no size or labeling restrictions that constrain the representation of a given object. However, graph based object representation suffers from the fact that there is only little mathematical structure in the graph domain. In contrast to vectors, most of the basic mathematical operations required for classification do not exist for graphs.

In the present chapter a novel approach to graph embedding, using prototypes and dissimilarities, is proposed. Our embedding procedure explicitly makes use of graph edit distance and can therefore deal with various kinds of graphs (labeled, unlabeled, directed, undirected, etc.). The basic idea of the embedding method is to describe a graph by means of n dissimilarities to a predefined set of graphs termed prototypes. That is, a graph g is mapped explicitly to the n-dimensional real space $\mathbb{R}^n$ by arranging the edit distances of g to all of the n prototypes as a vector. By means of this procedure both statistical classifiers and vector kernels can be applied to the resulting graph maps.

The problem of prototype selection is an issue in the proposed graph embedding framework. Three possible solutions are reviewed in the present chapter, viz. heuristic prototype selection,[40] prototype reduction,[41] and feature selection.[42] Six possible algorithms from the first category are described in detail. Note that none of them performs generally the best. That is, the quality of a particular selection strategy depends on the underlying data set. However, in case where no possibility for independent validation exist, one of the classwise prototype selectors could be the method of choice (e.g. sps-c).

In the experimental evaluation, the proposed graph embedding framework is compared to a nearest-neighbor classifier and another kernel-based classifier. In several experiments a high degree of robustness and flexibility of the proposed approach has been empirically verified in this chapter.

References

1. R. Duda, P. Hart, and D. Stork, *Pattern Classification.* (Wiley-Interscience, 2000), 2nd edition.
2. D. Conte, P. Foggia, C. Sansone, and M. Vento, Thirty years of graph matching in pattern recognition, *Int. Journal of Pattern Recognition and Artificial Intelligence.* **18** (3), 265–298, (2004).
3. H. Bunke, P. Dickinson, M. Kraetzl, and W. Wallis, *A Graph-Theoretic Approach to Enterprise Network Dynamics.* vol. 24, *Progress in Computer Science and Applied Logic (PCS)*, (Birkhäuser, 2007).
4. D. Cook and L. Holder, Eds., *Mining Graph Data.* (Wiley-Interscience, 2007).
5. P. Mahé, N. Ueda, and T. Akutsu, Graph kernels for molecular structures – activity relationship analysis with support vector machines, *Journal of Chemical Information and Modeling.* **45**(4), 939–951, (2005).
6. L. Ralaivola, S. Swamidass, H. Saigo, and P. Baldi, Graph kernels for chemical informatics, *Neural Networks.* **18**(8), 1093–1110, (2005).
7. A. Schenker, H. Bunke, M. Last, and A. Kandel, *Graph-Theoretic Techniques for Web Content Mining.* (World Scientific, 2005).
8. Z. Harchaoui and F. Bach. Image classification with segmentation graph kernels. In *IEEE Conference on Computer Vision and Pattern Recognition*, pp. 1–8, (2007).
9. R. Ambauen, S. Fischer, and H. Bunke. Graph edit distance with node splitting and merging and its application to diatom identification. In eds. E. Hancock and M. Vento, *Proc. 4th Int. Workshop on Graph Based Representations in Pattern Recognition*, LNCS 2726, pp. 95–106. Springer, (2003).
10. P. Dickinson, H. Bunke, A. Dadej, and M. Kraetzl, Matching graphs with unique node labels, *Pattern Analysis and Applications.* **7**(3), 243–254, (2004).
11. B. Luo, R. Wilson, and E. Hancock, Spectral embedding of graphs, *Pattern Recognition.* **36**(10), 2213–2223, (2003).
12. T. Caelli and S. Kosinov, Inexact graph matching using eigen-subspace projection clustering, *Int. Journal of Pattern Recognition and Artificial Intelligence.* **18**(3), 329–355, (2004).
13. R. Wilson and E. Hancock. Levenshtein distance for graph spectral features. In eds. J. Kittler, M. Petrou, and M. Nixon, *Proc. 17th Int. Conf. on Pattern Recognition*, vol. 2, pp. 489–492, (2004).

14. A. Shokoufandeh, D. Macrini, S. Dickinson, K. Siddiqi, and S. Zucker, Indexing hierarchical structures using graph spectra, *IEEE Transactions on Pattern Analysis and Machine Intelligence.* **27**(7), 1125–1140, (2005).
15. A. Robles-Kelly and E. Hancock, A Riemannian approach to graph embedding, *Pattern Recognition.* **40**, 1024–1056, (2007).
16. E. Pekalska and R. Duin, *The Dissimilarity Representation for Pattern Recognition: Foundations and Applications.* (World Scientific, 2005).
17. J. Ullmann, An algorithm for subgraph isomorphism, *Journal of the Association for Computing Machinery.* **23**(1), 31–42, (1976).
18. H. Bunke and K. Shearer, A graph distance metric based on the maximal common subgraph, *Pattern Recognition Letters.* **19**(3), 255–259, (1998).
19. W. Wallis, P. Shoubridge, M. Kraetzl, and D. Ray, Graph distances using graph union, *Pattern Recognition Letters.* **22**(6), 701–704, (2001).
20. M.-L. Fernandez and G. Valiente, A graph distance metric combining maximum common subgraph and minimum common supergraph, *Pattern Recognition Letters.* **22** (6–7), 753–758, (2001).
21. H. Bunke and G. Allermann, Inexact graph matching for structural pattern recognition, *Pattern Recognition Letters.* **1**, 245–253, (1983).
22. A. Sanfeliu and K. Fu, A distance measure between attributed relational graphs for pattern recognition, *IEEE Transactions on Systems, Man, and Cybernetics (Part B).* **13**(3), 353–363, (1983).
23. P. Hart, N. Nilsson, and B. Raphael, A formal basis for the heuristic determination of minimum cost paths, *IEEE Transactions of Systems, Science, and Cybernetics.* **4** (2), 100–107, (1968).
24. S. Sorlin and C. Solnon. Reactive tabu search for measuring graph similarity. In eds. L. Brun and M. Vento, *Proc. 5th Int. Workshop on Graph-based Representations in Pattern Recognition*, LNCS 3434, pp. 172–182. Springer, (2005).
25. D. Justice and A. Hero, A binary linear programming formulation of the graph edit distance, *IEEE Trans. on Pattern Analysis ans Machine Intelligence.* **28**(8), 1200–1214, (2006).
26. M. Neuhaus, K. Riesen, and H. Bunke. Fast suboptimal algorithms for the computation of graph edit distance. In eds. D.-Y. Yeung, J. Kwok, A. Fred, F. Roli, and D. de Ridder, *Proc. 11.th int. Workshop on Strucural and Syntactic Pattern Recognition*, LNCS 4109, pp. 163–172. Springer, (2006).
27. M. Eshera and K. Fu. A Graph Distance Measure for Image Analysis. *IEEE Trans. Systems, Man, and Cybernetics*, **14**(3), 353–363, (1984).
28. K. Riesen and H. Bunke, Approximate graph edit distance computation by means of bipartite graph matching, *Image and Vision Computing.* (2008). accepted for publication.
29. J. Hopcroft and J. Wong. Linear time algorithm for isomorphism of planar graphs. In *Proc. 6th Annual ACM Symposium on Theory of Computing*, pp. 172–184, (1974).
30. E. Luks, Isomorphism of graphs of bounded valence can be tested in polynomial time, *Journal of Computer and Systems Sciences.* **25**, 42–65, (1982).
31. A. Torsello, D. Hidovic-Rowe, and M. Pelillo, Polynomial-time metrics for attributed trees, *IEEE Transactions on Pattern Analysis and Machine Intelligence.* **27**(7), 1087–1099, (2005).
32. M. Neuhaus and H. Bunke. An error-tolerant approximate matching algorithm for attributed planar graphs and its application to fingerprint classification. In eds. A. Fred, T. Caelli, R. Duin, A. Campilho, and D. de Ridder, *Proc. 10th Int. Workshop on Structural and Syntactic Pattern Recognition*, LNCS 3138, pp. 180–189. Springer, (2004).

33. B. Spillmann, M. Neuhaus, H. Bunke, E. Pekalska, and R. Duin. Transforming strings to vector spaces using prototype selection. In eds. D.-Y. Yeung, J. Kwok, A. Fred, F. Roli, and D. de Ridder, *Proc. 11.th int. Workshop on Strucural and Syntactic Pattern Recognition*, LNCS 4109, pp. 287–296. Springer, (2006).
34. B. Schölkopf and A. Smola, *Learning with Kernels.* (MIT Press, 2002).
35. J. Shawe-Taylor and N. Cristianini, *Kernel Methods for Pattern Analysis.* (Cambridge University Press, 2004).
36. K. Tsuda. Support vector classification with asymmetric kernel function. In ed. M. Verleysen, *Proc. 7th European Symposium on Artifical Neural Netweorks*, pp. 183–188, (1999).
37. J. Ramon and T. Gärtner. Expressivity versus efficiency of graph kernels. In *Proc. First International Workshop on Mining Graphs, Trees and Sequences*, pp. 65–74, (2003).
38. M. Neuhaus and H. Bunke, *Bridging the Gap Between Graph Edit Distance and Kernel Machines.* (World Scientific, 2007).
39. B. Haasdonk, Feature space interpretation of SVMs with indefinite kernels, *IEEE Transactions on Pattern Analysis and Machine Intelligence.* **27**(4), 482–492, (2005).
40. K. Riesen and H. Bunke, Graph classification based on vector space embedding, *Int. Journal of Pattern Recognition and Artificial Intelligence.* (2008). accepted for publication.
41. K. Riesen and H. Bunke. Dissimilarity based vector space embedding of graphs using prototype reduction schemes. Accepted for publication in Proc. Machine Learning and Data Mining in Pattern Recognition, (2009).
42. K. Riesen and H. Bunke, Reducing the dimensionality of dissimilarity space embedding graph kernels, *Engineering Applications of Artificial Intelligence.* **22**(1), 48–56, (2008).
43. K. Riesen and H. Bunke. Non-linear transformations of vector space embedded graphs. In eds. A. Juan-Ciscar and G. Sanchez-Albaladejo, *Pattern Recognition in Information Systems*, pp. 173–186, (2008).
44. E. Pekalska, R. Duin, and P. Paclik, Prototype selection for dissimilarity-based classifiers, *Pattern Recognition.* **39**(2), 189–208, (2006).
45. J. Bezdek and L. Kuncheva, Nearest prototype classifier designs: An experimental study, *Int. Journal of Intelligent Systems.* **16**(12), 1445–1473, (2001).
46. P. Pudil, J. Novovicova, and J. Kittler, Floating search methods in feature-selection, *Pattern Recognition Letters.* **15**(11), 1119–1125 (1994).
47. A. Jain and D. Zongker, Feature selection: Evaluation, application, and small sample performance, *IEEE Trans. on Pattern Analysis and Machine Intelligence.* **19**(2), 153–158, (1997).
48. I. Jolliffe, *Principal Component Analysis.* (Springer, 1986).
49. R. Fisher. The statistical utilization of multiple measurements. In *Annals of Eugenics*, vol. 8, pp. 376–386, (1938).
50. B. Schölkopf, A. Smola, and K.-R. Müller, Nonlinear component analysis as a kernel eigenvalue problem, *Neural Computation.* **10**, 1299–1319, (1998).
51. X. Jiang, A. Münger, and H. Bunke, On median graphs: Properties, algorithms, and applications, *IEEE Transactions on Pattern Analysis and Machine Intelligence.* **23** (10), 1144–1151, (2001).
52. L. Kaufman and P. Rousseeuw, *Finding Groups in Data: An Introduction to Cluster Analysis.* (John Wiley & Sons, 1990).
53. V. Vapnik, *Statistical Learning Theory.* (John Wiley, 1998).

54. K. Riesen and H. Bunke. IAM graph database repository for graph based pattern recognition and machine learning. In ed. N. da Vitoria Lobo et al., *Structural, Syntactic, and Statistical Pattern Recognition*, LNCS 5342, pp. 287–297, (2008).
55. C. Watson and C. Wilson. NIST Special Database 4, Fingerprint Database. National Institute of Standards and Technology, (1992).
56. E. Alpaydin and F. Alimoglu. *Pen-Based Recognition of Handwritten Digits.* Dept. of Computer Engineering, Bogazici University, (1998).
57. D. T. P. `http://dtp.nci.nih.gov/docs/aids/aids_data.html`, AIDS antiviral screen, (2004).
58. J. Kazius, R. McGuire, and R. Bursi. Derivation and validation of toxicophores for mutagenicity prediction. *Journal of Medicinal Chemistry* **48**(1), 312–320, (2005).

CHAPTER 1.4

MATCH TRACKING STRATEGIES FOR FUZZY ARTMAP NEURAL NETWORKS

Eric Granger[1*], Philippe Henniges[1], Robert Sabourin[1] and Luiz S. Oliveira[2]

[1]*Laboratoire d'imagerie, de vision et d'intelligence artificielle, École de technologie supérieure (ETS), Montreal, Canada*

[2]*Dept. de Informática Aplicada, Pontifícia Universidade Católica do Paraná (PUCPR), Curitiba, Brazil*

Training fuzzy ARTMAP neural networks for classification using data from complex real-world environments may lead to category proliferation, and yield poor performance. This problem is known to occur whenever the training set contains noisy and overlapping data. Moreover, when the training set contains identical input patterns that belong to different recognition classes, fuzzy ARTMAP will fail to converge. To circumvent these problems, some alternatives to the network's original match tracking (MT) process have been proposed in literature, such as using negative MT, and removing MT altogether. In this chapter, the MT parameter of fuzzy ARTMAP is optimized during training using a new Particle Swarm Optimisation (PSO)-based strategy, denoted `PSO(MT)`. The impact on fuzzy ARTMAP performance of training with different MT strategies is assessed empirically, using different synthetic data sets, and the NIST SD19 handwritten character recognition data set. During computer simulations, fuzzy ARTMAP is trained with the original (positive) match tracking (`MT+`), with negative match tracking (`MT-`), without MT algorithm (`WMT`), and with `PSO(MT)`. Through a comprehensive set of simulations, it has been observed that by training with `MT-`, fuzzy ARTMAP expends fewer resources than with other MT strategies, but can achieve a significantly higher generalization error, especially for data with overlapping class distributions. In particular, degradation of error in fuzzy ARTMAP performance due to overtraining is more pronounced for `MT-` than for `MT+`. Generalization error achieved using `WMT` is significantly higher than other strategies on data with complex non-linear decision bounds. Furthermore, the number of internal categories required to represent decision boundaries increases significantly. Optimizing the value of the match tracking parameter using `PSO(MT)` yields the lowest overall generalization error, and requires fewer internal categories than `WMT`, but generally more categories than `MT+` and `MT-`. However, this strategy requires a large number of training epochs to convergence. Based on this empirical results with `PSO(MT)`, the MT process as such can provide a significant increase to fuzzy ARTMAP performance, assuming that the MT parameter is tuned for the specific application in mind.

***Corresponding author:** ETS, 1100 Notre-Dame Ouest, Montreal, Quebec, H3C 1K3, Canada, email: `eric.granger@etsmtl.ca`, phone: 1-514-396-8650, fax: 1-514-396-8595.

1. Introduction

The fuzzy ARTMAP neural network is capable of self-organizing stable recognition categories in response to arbitrary sequences of analog or binary input patterns. It can perform fast, stable, on-line, unsupervised or supervised, incremental learning, classification, and prediction.[6,7] As such, it has been successfully applied in complex real-world pattern recognition tasks such as the recognition of radar signals,[15,32] multi-sensor image fusion, remote sensing and data mining,[9,31,35,40] recognition of handwritten characters,[3,13,23] and signature verification.[29]

A drawback of fuzzy ARTMAP is its ability to learn decision boundaries between class distributions that consistently yield low generalization error for a wide variety of pattern recognition problems. For instance, when trained for automatic classification of handwritten characters, fuzzy ARTMAP cannot achieve a level of performance that is competitive with some other well-known models.[16] Statistical models (*e.g.*, linear and quadratic discriminant function, Gaussian mixture classifier, k-Nearest-Neighbor (kNN) and Support Vector Machines (SVMs)), and neural networks(*e.g.*, Multi-Layer Perceptron (MLP) and Radial Basis Function (RBF) networks) are commonly used for classification due to their learning flexibility and inexpensive computation.[26] Such recognition problems typically exhibit complex decision boundaries, with moderate overlap between character classes.

In the context of batch supervised learning of a finite training set, the main factors affecting fuzzy ARTMAP's capacity to generalize are:

(1) internal dynamics of network: prototype choice and class prediction functions, learning rule, match tracking process, hyper-parameter values, and representation of categories with hyper-rectangles.
(2) learning process: supervised learning strategy (and thus, the number of training epochs), proportion of patterns in the training subset to those in validation and test subsets, user-defined hyper-parameter values, data normalisation technique, sequential gradient-based learning, and data presentation order.
(3) data set structure: overlap and dispersion of patterns, etc., and therefore of the geometry of decision boundaries among patterns belonging to different recognition classes.

Several ARTMAP networks have been proposed to refine the decision boundaries created by fuzzy ARTMAP. For instance, many variants attempt to improve the accuracy of fuzzy ARTMAP predictions by providing for probabilistic (density based) predictions.[10,14,25,36,39,41]

When learning data from complex real-world environments, fuzzy ARTMAP is known to suffer from overtraining, often referred to in literature as the category proliferation problem. It occurs when the training data set contains noisy and overlapping class distributions.[19,22,24] In this case, increasing the amount of training data requires significantly more internal category neurons, and therefore computational complexity, while yielding a higher generalisation error. The category proliferation problem is directly connected to the match tracking (MT) process of fuzzy ARTMAP. During fuzzy ARTMAP training, when a mismatch occurs between predicted and desired output responses, MT allows selecting alternate category neurones.

The match tracking process is parameterized by hyper-parameter ϵ, and was originally introduced as a small positive value.[10] In fuzzy ARTMAP literature, this parameter is commonly set to a value ($\epsilon = 0^+$) that allows to minimize network resources. Such a choice may however contribute to overtraining, and significantly degrade the capacity to generalize. As a result, some authors have studied the impact on performance of removing the MT altogether, and conclude that the usefulness of MT is questionable.[2,27] However, training without MT may lead to a network with a greater number of internal categories, and possibly a higher generalization error.

In an extreme case, a well known convergence problem occurs when learning inconsistent cases – identical training subset patterns that belong to different classes.[10] The consequence is a failure to converge, as identical prototypes linked to these inconsistent cases proliferate. This anomalous situation is a result of the original match tracking process. This convergence problem may be circumvented by using the feature of ARTMAP-IC[10] called *negative* match tracking (*i.e.*, setting $\epsilon = 0^-$ after mismatch reset). This allows fuzzy ARTMAP training to converge and find solutions with fewer internal categories, but may however lead to a higher generalization error.

In this chapter, the impact on fuzzy ARTMAP performance of training with different MT strategies – the original positive MT (`MT+`), negative MT (`MT-`) and without MT (`WMT`) – is assessed empirically. As an alternative, a Particle Swarm Optimization (PSO)-based approach called `PSO(MT)` is used to optimize the value of MT hyper-parameter ϵ during fuzzy ARTMAP training, such that the generalization error is minimized. The architecture, weights, and MT parameter are in effect selected to minimize generalisation error by virtue of ARTMAP training, which allows to grow the network architecture (*i.e.*, the number of category neurons) with the problem's complexity. An experimental protocol has been defined such that the generalization error and resource requirements of fuzzy ARTMAP trained with different MT strategies may be compared using different types of pattern recognition problems. The first two types consist of synthetic data with overlapping class distributions, and with complex decision boundaries but no overlap, respectively. The third type consists of real-world data - handwritten numerical characters extracted from the NIST SD19.

In the next section, the MT strategies for fuzzy ARTMAP training are briefly reviewed. Section III presents the experimental methodology, e.g., protocol, data sets and performance measures employed for proof of concept computer simulations. Section IV presents and discuss experimental results obtained with synthetic and NIST SD19 data.

2. Fuzzy ARTMAP Match Tracking

2.1. *The fuzzy ARTMAP neural network:*

ARTMAP refers to a family of neural network architectures based on Adaptive Resonance Theory (ART)[4] that is capable of fast, stable, on-line, unsupervised or supervised, incremental learning, classification, and prediction[6].[7] ARTMAP is often applied using the simplified version shown in Figure 1. It is obtained by combining an unsupervised ART neural network[4] with a map field.

In order to improve the performance or robustness to noise of ARTMAP architectures, several variants have been proposed in literature. Some networks, such as fuzzy ARTMAP, ART-EMAP, ARTMAPT-PI, ARTMAP-IC, Default ARTMAP, Simplified ARTMAP, and Distributed ARTMAP, represent each class using one or more fuzzy set hyper-rectangle, and perform category activation using an L_1 norm. Other networks, such as PROBART, Probabilistic Fuzzy ARTMAP, MLANS, Gaussian ARTMAP, Ellipsoid ARTMAP, boosted ARTMAP, and μARTMAP, represent each class using one or more probability density functions (pdfs). The class predictions of probabilistic ARTMAP variants consist in estimating the posterior probability that each class generated a given input pattern. Then, the Bayes decision procedure may be applied to assign one-of-L possible classes to the input according to the maximum posterior probability decision rule. This rule defines decision boundaries among classes that yield the minimum probability of misclassification.

Fuzzy ARTMAP[7] is one of the earliest and most popular ARTMAP architecture. It can process both analog and binary-valued input patterns by employing fuzzy ART[5] as the ART network. The fuzzy ART neural network consists of two fully connected layers of nodes: an M node input layer, F_1, and an N node competitive layer, F_2. A set of real-valued weights $\mathbf{W} = \{w_{ij} \in [0,1] : i = 1,2,...,M;\ j = 1,2,...,N\}$ is associated with the F_1-to-F_2 layer connections. Each F_2 node j represents a recognition category that learns a prototype vector $\mathbf{w}_j = (w_{1j}, w_{2j}, ..., w_{Mj})$. The F_2 layer of fuzzy ART is connected, through learned associative links, to an L node map field F^{ab}, where L is the number of classes in the output space. A set of binary weights $\mathbf{W}^{ab} = \{w_{jk}^{ab} \in \{0,1\} : j = 1,2,...,N;\ k = 1,2,...,L\}$ is associated with the F_2-to-F^{ab} connections. The vector $\mathbf{w}_j^{ab} = (w_{j1}^{ab}, w_{j2}^{ab}, ..., w_{jL}^{ab})$ links F_2 node j to one of the L output classes.

2.2. *Algorithm for supervised learning of fuzzy ARTMAP:*

In batch supervised training mode, ARTMAP classifiers learn an arbitrary mapping between training set patterns $\mathbf{a} = (a_1, a_2, ..., a_m)$ and their corresponding binary supervision patterns $\mathbf{t} = (t_1, t_2, ..., t_L)$. These patterns are coded to have unit value $t_K = 1$ if K is the target class label for $\mathbf{a}$, and zero elsewhere. The following steps describe fuzzy ARTMAP learning:

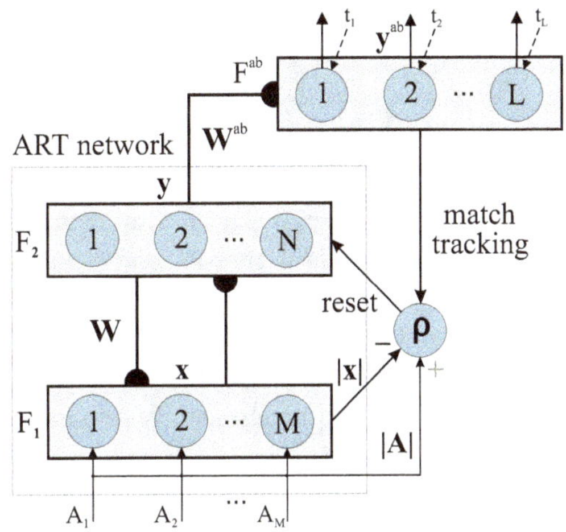

Fig. 1. An ARTMAP neural network architecture for pattern classification.

(1) **Initialisation:** Initially, all the F_2 nodes are uncommitted, all weight values w_{ij} are initialized to 1, and all weight values w_{jk}^{ab} are set to 0. An F_2 node becomes committed when it is selected to code an input vector $\mathbf{a}$, and is then linked to an F^{ab} node. Values of the learning rate $\beta \in [0, 1]$, the choice $\alpha > 0$, the match tracking $0 < \epsilon \ll 1$, and the baseline vigilance $\bar{\rho} \in [0, 1]$ hyper-parameters are set.

(2) **Input pattern coding:** When a training pair $(\mathbf{a}, \mathbf{t})$ is presented to the network, $\mathbf{a}$ undergoes a transformation called complement coding, which doubles its number of components. The complement-coded input pattern has $M = 2m$ dimensions and is defined by $\mathbf{A} = (\mathbf{a}, \mathbf{a}^c) = (a_1, a_2, ..., a_m; a_1^c, a_2^c, ..., a_m^c)$, where $a_i^c = (1 - a_i)$, and $a_i \in [0, 1]$. The vigilance parameter ρ is reset to its baseline value $\bar{\rho}$.

(3) **Prototype selection:** Pattern $\mathbf{A}$ activates layer F_1 and is propagated through weighted connections $\mathbf{W}$ to layer F_2. Activation of each node j in the F_2 layer is determined by the *Weber law choice function*:

$$T_j(\mathbf{A}) = \frac{|\mathbf{A} \wedge \mathbf{w}_j|}{\alpha + |\mathbf{w}_j|} , \tag{1}$$

where $|\cdot|$ is the L^1 norm operator defined by $|\mathbf{w}_j| \equiv \sum_{i=1}^{M} |w_{ij}|$, $\wedge$ is the fuzzy AND operator, $(\mathbf{A} \wedge \mathbf{w}_j)_i \equiv \min(A_i, w_{ij})$, and α is the user-defined *choice parameter*. The F_2 layer produces a binary, winner-take-all pattern of activity $\mathbf{y} = (y_1, y_2, ..., y_N)$ such that only the node $j = J$ with the greatest activation value $J = \arg\max\{T_j : j = 1, 2, ..., N\}$ remains active; thus $y_J = 1$ and $y_j = 0, j \neq J$. If more than one T_j is maximal, the node j with the smallest index is chosen. Node J propagates its top-down expectation, or prototype vector $\mathbf{w}_J$, back onto F_1 and the *vigilance test* is performed. This test compares the degree of match between $\mathbf{w}_J$ and $\mathbf{A}$ against the dimensionless *vigilance parameter* $\rho \in [0, 1]$:

$$\frac{|\mathbf{A} \wedge \mathbf{w}_J|}{|\mathbf{A}|} = \frac{|\mathbf{A} \wedge \mathbf{w}_J|}{M} \geq \rho \,. \tag{2}$$

If the test is passed, then node J remains active and resonance is said to occur. Otherwise, the network inhibits the active F_2 node (*i.e.*, T_J is set to 0 until the network is presented with the next training pair $(\mathbf{a}, \mathbf{t})$) and searches for another node J that passes the vigilance test. If such a node does not exist, an uncommitted F_2 node becomes active and undergoes learning (Step 5). The depth of search attained before an uncommitted node is selected is determined by the choice parameter α.

(4) **Class prediction:** Pattern $\mathbf{t}$ is fed directly to the map field F^{ab}, while the F_2 category $\mathbf{y}$ learns to activate the map field via associative weights $\mathbf{W}^{ab}$. The F^{ab} layer produces a binary pattern of activity $\mathbf{y}^{ab} = (y_1^{ab}, y_2^{ab}, ..., y_L^{ab}) = \mathbf{t} \wedge \mathbf{w}_J^{ab}$ in which the most active F^{ab} node $K = \arg\max\{y_k^{ab} : k = 1, 2, ..., L\}$ yields the class prediction $(K = k(J))$. If node K constitutes an incorrect class prediction, then a *match tracking* (MT) signal raises the vigilance parameter ρ such that:

$$\rho = \frac{|\mathbf{A} \wedge \mathbf{w}_J|}{M} + \epsilon \,, \tag{3}$$

where $\epsilon = 0^+$, to induce another search among F_2 nodes (Step 3). This search continues until either an uncommitted F_2 node becomes active, and learning ensues (Step 5), or a node J that has previously learned the correct class prediction K becomes active.

(5) **Learning:** Learning input $\mathbf{a}$ involves updating prototype vector $\mathbf{w}_J$, and, if J corresponds to a newly-committed node, creating an associative link to F^{ab}. The prototype vector of F_2 node J is updated according to:

$$\mathbf{w}'_J = \beta(\mathbf{A} \wedge \mathbf{w}_J) + (1 - \beta)\mathbf{w}_J \,, \tag{4}$$

where β is a fixed *learning rate parameter*. The algorithm can be set to slow learning with $0 < \beta < 1$, or to fast learning with $\beta = 1$. With complement coding and fast learning, fuzzy ARTMAP represents category j as an m-dimensional hyperrectangle R_j that is just large enough to enclose the cluster of training set patterns $\mathbf{a}$ to which it has been assigned. That is, an M-dimensional prototype vector $\mathbf{w}_j$ records the largest and smallest component values of training subset patterns $\mathbf{a}$ assigned to category j. The vigilance test limits the growth of hyperrectangles – a ρ close to 1 yields small hyperrectangles, while a ρ close to 0 allows large hyperrectangles. A new association between F_2 node J and F^{ab} node K $(k(J) = K)$ is learned by setting $w_{Jk}^{ab} = 1$ for $k = K$, where K is the target class label for $\mathbf{a}$, and 0 otherwise. The next training subset pair $(\mathbf{a}, \mathbf{t})$ is presented to the network in Step 2.

Network training proceeds from one epoch to the next, and is halted for validation after each epoch [a]. Given a finite training data set, batch supervised learning ends after the epoch for which the generalisation error is minimized on an independent validation data set. With the large data sets considered in this chapter, learning through this hold-out validation (HV) is an appropriate validation strategy. If data were limited, k-fold cross-validation would be a more suitable strategy, at the expense of some estimation bias due to crossing.[19,34]

Once the weights $\mathbf{W}$ and $\mathbf{W}^{ab}$ have been found through this process, ARTMAP can predict a class label for an input pattern by performing Steps 2, 3 and 4 without any vigilance or match tests. During testing, a pattern $\mathbf{a}$ that activates node J is predicted to belong to class $K = k(J)$. The time complexity required to process one input pattern, during either a training or testing phase, is $O(MN)$.

2.3. *Match tracking strategies:*

During training, when a mismatch occurs between a predicted response $\mathbf{y}^{ab}$ and a desired response $\mathbf{t}$ for an input pattern $\mathbf{a}$, the original positive MT process (MT+) of fuzzy ARTMAP raises the internal vigilance parameter to $\rho = (|\mathbf{A} \wedge \mathbf{w}_J|)(M)^{-1} + \epsilon$ in order to induce another search among F_2 category nodes. MT+ is parameterized by the MT hyper-parameter ϵ, which was introduced as a small positive value, $0 < \epsilon \ll 1$.[7]

It is well documented that training fuzzy ARTMAP with data from noisy and overlapping class distributions may lead to category proliferation, and that this problem is connected to the MT mechanism. Overlapping between classes, which is responsible for misclassifications during training, requires MT to find a more suitable category for the misclassified pattern. The selected $F2$ node requires a larger prototype vector, and thus a smaller size to pass the vigilance test. Such misclassifications are responsible for the formation of a large number of small categories within the overlapping area, many of which contribute little to the classification process.[24] Category proliferation is intensified with the degree of class overlap.

Category proliferation is an indication of overtraining. Increasing the amount of training data requires significantly more resources (*i.e.*, the number of internal category neurons, thus memory space and computational complexity), yet provides a higher generalisation error.[19,22,24] In addition, the MT parameter is commonly set to the value $\epsilon = +0.001$ in fuzzy ARTMAP literature to minimize network resources.[10] Such a choice may play a significant role in category proliferation, and considerably degrade the capacity to generalize.

Although pruning may help reduce category proliferation, some authors have challenged the need for a MT process.[1,27,37] Training without MT (WMT) implies creating a new category each time that a predictive response $\mathbf{y}^{ab}$ does not match a desired response $\mathbf{t}$. When an node in the $F2$ layer is chosen to represent an input

[a] An *epoch* is defined as one complete presentation of all the patterns of the training set.

pattern, but this node is mapped to the incorrect label, an uncommitted node is activated to represent this pattern. Note that training fuzzy ARTMAP WMT is equivalent to performing MT but setting $\epsilon = 1$. Training WMT may create networks with a greater number of internal categories, and possibly a higher generalization error.

In an extreme case, a convergence problem occurs whenever the training set contains identical patterns that belong recognition classes.[10] The effect is a proliferation of identical prototypes associated with the *inconsistent cases*, and a failure to converge. Consider for example that on the first training epoch, fuzzy ARTMAP learns two completely overlapping, minimum-sized prototypes, $\mathbf{w}_{A.1}$ (linked to class A) and $\mathbf{w}_{B.1}$ (linked to class B), for two identical pulse patterns, $\mathbf{a}_1$ and $\mathbf{a}_2$. In a subsequent epoch, $\mathbf{w}_{A.1}$ is initially selected to learn $\mathbf{a}_2$, since $T_{A.1} = T_{B.1} \simeq 1$, and $\mathbf{w}_{A.1}$ was created prior to $\mathbf{w}_{B.1}$ (index $A.1$ is smaller than $B.1$). Since $\mathbf{w}_{A.1}$ is not linked to class B, mismatch reset raises the vigilance parameter ρ to $(|\mathbf{A}_2 \wedge \mathbf{w}_{A.1}|/M) + \epsilon$, where $|\mathbf{A}_2 \wedge \mathbf{w}_{A.1}| = |\mathbf{A}_2 \wedge \mathbf{w}_{B.1}|$. As a result, $\mathbf{w}_{B.1}$ can no longer pass the vigilance test required to become selected for $\mathbf{a}_2$, and fuzzy ARTMAP must create another minimum-sized prototype $\mathbf{w}_{B.2} = \mathbf{w}_{B.1}$. From epoch to epoch, the same phenomenon repeats itself, yielding ever more prototypes $\mathbf{w}_{B.\text{n}} = \mathbf{w}_{B.1}$ for n $= 3, 4, ..., \infty$.

ARTMAP-IC[10] is an extension of fuzzy ARTMAP that produce a binary winner-take-all pattern $\mathbf{y}$ when training, but use distributed activation of coded $F2$ nodes when testing. ARTMAP-IC is further extended in two ways. First, it biases distributed test set predictions according to the number of times $F2$ nodes are assigned to training set patterns. Second, it uses a *negative MT* process (MT-) to address the problem of inconsistent cases, whereby identical training set patterns correspond to different classes labels.

With negative MT (MT-), ρ is also initially raised after mismatch reset, but is allowed to decay slightly before a different node J is selected. Then, the MT parameter is set to a small negative value, $\epsilon \leq 0$ (typically a value of $\epsilon = -0.001$), which allows for identical inputs that predict different classes to establish distinct recognition categories. In the example above, mismatch reset raises ρ but $\mathbf{w}_{B.1}$ would still pass the vigilance test. This allows to learn fully overlapping prototypes for training set patterns that belong to different classes.

In some applications, incorporation into fuzzy ARTMAP of the MT- feature of ARTMAP-IC may be essential to avoid the convergence problem observed with original MT+. Training fuzzy ARTMAP with MT- would thereby find solutions with fewer internal categories, but may nonetheless lead to a higher generalization error. In other cases, MT reset may also be buffered based on a category's previous predictive success, thereby improving the compression achieved with minimal loss of accuracy.[18] During supervised learning, match tracking search allocates memory based on the degree of similarity between newly encountered and previously encountered inputs, regardless of their prior predictive success.

An alternate approach consists in optimizing the MT hyper-parameter during batch supervised learning of a fuzzy ARTMAP neural network. In effect, both network (weights and architecture) and ϵ values are co-optimized for a given problem, using the same cost function. The next Subsection presents a Particle Swarm Optimization (PSO)-based approach called PSO(MT) that automatically selects a value (magnitude and polarity) of ϵ during fuzzy ARTMAP training such that the generalization error is minimized. This approach is based on the PSO training strategy proposed in,[16] but focused only on a one-dimensional optimization space of $\epsilon \in [-1, 1]$.

2.4. *Particle Swarm Optimisation (PSO) of the match tracking parameter*

PSO is a population-based stochastic optimization technique that was inspired by social behavior of bird flocking or fish schooling.[20] It shares many similarities with evolutionary computation techniques such as genetic algorithms (GAs), yet has no evolution operators such as crossover and mutation. PSO belongs to the class of evolutionary algorithm techniques that does not utilize the "survival of the fittest" concept, nor a direct selection function. A solution with lower fitness values can therefore survive during the optimization and potentially visit any point of the search space.[12] Finally, while GAs were conceived to deal with binary coding, PSO was designed, and proved very effective, in solving real valued global optimization problems, which makes it suitable for this study.

With PSO, each *particle* corresponds to a single solution in the search space, and the population of particles is called a *swarm*. All particles are assigned position values which are evaluated according to the fitness function being optimized, and velocities values which direct their movement. Particles move through the search space by following the particles with the best fitness. Assuming a d-dimensional search space, the position of particle i in an P-particle swarm is represented by a d-dimensional vector $\mathbf{s}_i = (s_{i1}, s_{i2}, \ldots, s_{id})$, for $i = 1, 2, \ldots, P$. The velocity of this particle is denoted by vector $\mathbf{v}_i = (v_{i1}, v_{i2}, \ldots, v_{id})$, while the best previously-visited position of this particle is denoted as $\mathbf{p}_i = (p_{i1}, p_{i2}, \ldots, p_{id})$. For each new iteration $q + 1$, the velocity and position of particle i are updated according to:

$$\mathbf{v}_i^{q+1} = w^q \mathbf{v}_i^q + c_1 r_1 (\mathbf{p}_i^q - \mathbf{s}_i^q) + c_2 r_2 (\mathbf{p}_g^q - \mathbf{s}_i^q) \quad (5)$$

$$\mathbf{s}_i^{q+1} = \mathbf{s}_i^q + \mathbf{v}_i^{q+1} \quad (6)$$

where $\mathbf{p}_g$ represents the global best particle position in the swarm, w^q is the particle inertia weight, c_1 and c_2 are two positive constants called cognitive and social parameters, respectively, and r_1 and r_2 are random numbers uniformly distributed in the range [0,1].

The role of w^q in Equation 5 is to regulate the trade-off between exploration and exploitation. A large inertia weight facilitates global search (exploration), while a small one tends to facilitate fine-tuning the current search area (exploitation).

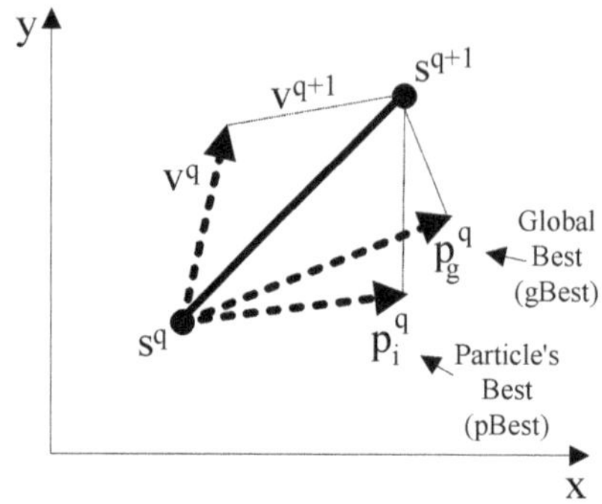

Fig. 2. An ARTMAP neural network architecture specialized for pattern classification.

This is why inertia weight values are defined by some monotonically decreasing function of q. Proper fine-tuning of c_1 and c_2 may result in faster convergence of the algorithm and alleviation of the local minima. Kennedy and Eberhart propose that the cognitive and social scaling parameters be selected such that $c_1 = c_2 = 2$.[21] Finally, the parameters r_1 and r_2 are used to maintain the diversity of the population. Figure 2 depicts the update by PSO of a particle's position from $\mathbf{s}_i^q$ to $\mathbf{s}_i^{q+1}$.

Algorithm 1 shows the pseudo-code of a PSO learning strategy specialized for supervised training of fuzzy ARTMAP neural networks. It essentially seeks to minimize fuzzy ARTMAP generalisation error $E(\mathbf{s}_i^q)$ in the d-dimensional space of hyper-parameter values. For enhanced computational throughput and global search capabilities, Algorithm 1 is inspired by the synchronous parallel version of PSO.[33] It utilizes a basic type of neighborhood called global best or *gbest*, which is based on a sociometric principle that conceptually connects all the members of the swarm to one another. Accordingly, each particle is influenced by the very best performance of any member of the entire swarm. Exchange of information only takes place among the particle's own experience (the location of its personal best $\mathbf{p}_i^q$, lbest), and the experience of the best particle in the swarm (the location of the global best $\mathbf{p}_g^q$, gbest).

The `PSO(MT)` approach is obtained by setting $d = 1$, and particle positions to MT parameter values, $\mathbf{s}_i^q = \epsilon_i^q$. Measurement of any fitness values $E(\mathbf{s}_i^q)$ in this algorithm involves computing the generalisation error on a validation subset for the fuzzy ARTMAP network which has been trained using the MT parameter value at particle position ϵ_i^q. When selecting $\mathbf{p}_i^q$ or $\mathbf{p}_g^q$, if the two fitness values being compared are equal, then the particle/network requiring fewer number of $F2$ category nodes is chosen. The same training and validation sets are used throughout this process. Following the last iteration of Algorithm 1, the overall generalisation error is computed on a test set for the network corresponding to position $\mathbf{p}_g^q$.

Algorithm 1: PSO learning strategy for fuzzy ARTMAP.

A. Initialization:

set the maximum number of iterations $q_{\max}$ *and/or fitness objective* E^*
set PSO parameters P, $\mathbf{v}\max$, w^0, c_1, c_2, r_1 *and* r_1
initialize particle positions at random such that $\mathbf{p}_g^0$, $\mathbf{s}_i^0$ *and* $\mathbf{p}_i^0 \in [-1,1]^d$, *for* $i = 1, 2, \ldots, P$
initialize particle velocities at random such that $0 \leq \mathbf{v}_i^0 \leq \mathbf{v}\max$, *for* $i = 1, 2, \ldots, P$

B. Iterative process:

set iteration counter $q = 0$
while $q \leq q_{\max}$ *or* $E(\mathbf{p}_g^q) \geq E^*$ **do**
  for $i = 1, 2, \ldots, P$ **do**
    train fuzzy ARTMAP using hold-out validation and $\mathbf{s}_i^q$
    compute fitness value $E(\mathbf{s}_i^q)$ *of resulting network*
    if $E(\mathbf{s}_i^q) < E(\mathbf{p}_i^q)$ **then**
      update particle's best personal position: $\mathbf{p}_i^q = \mathbf{s}_i^q$
    end
  end
  select the particle with best global fitness:
  $g = \arg\min\{E(\mathbf{s}_i^q) : i = 1, 2, \ldots, P\}$
  for $i = 1, 2, \ldots, P$ **do**
    update velocity: $\mathbf{v}_i^{q+1} = w^q \mathbf{v}_i^q + c_1 r_1 (\mathbf{p}_i^q - \mathbf{s}_i^q) + c_2 r_2 (\mathbf{p}_g^q - \mathbf{s}_i^q)$
    update position: $\mathbf{s}_i^{q+1} = \mathbf{s}_i^q + \mathbf{v}_i^{q+1}$
  end
  $q = q + 1$
  update particle inertia w^q
end

3. Experimental Methodology

To assess the performance achieved by fuzzy ARTMAP using MT strategies, several data sets were selected for computer simulations. Four synthetic data sets are representative of pattern recognition problems that involve either (1) simple decision boundaries with overlapping class distributions, or (2) complex decision boundaries, were class distributions do not overlap on decision boundaries. A set of handwritten numerical characters from the NIST SD19 database is representative of complex real-world pattern recognition problems. Prior to a simulation trial, these data sets were normalized according to the min-max technique, and partitioned into three parts – training, validation, and test subset.

During each simulation trial, the performance of fuzzy ARTMAP is compared from a perspective of different training subset size, and match tracking strategies. In order to assess the effect on performance of training subset size, the number of training subset patterns used for supervised learning was progressively increased, while corresponding validation and test subsets were held fixed. The performance is compared for fuzzy ARTMAP neural networks trained according to four different

MT strategies: `MT+` ($\epsilon = 0.001$), `MT-` ($\epsilon = -0.001$), `WMT` (equivalent to setting $\epsilon = 1$) and `PSO(MT)`. Training is performed by setting the other three hyper-parameters such that the resources (number of categories, training epochs, etc.) are minimized: $\alpha = 0.001$, $\beta = 1$ and $\overline{\rho} = 0$. In all cases, training is performed using the HV strategy[34] described in Subsection 2.2.

The `PSO(MT)` strategy also uses the hold-out validation technique on fuzzy ARTMAP network to calculate the fitness of each particle, and therefore find the network and ϵ value that minimize generalization error. Other fuzzy ARTMAP hyper-parameters are left unchanged. In all simulations involving PSO, the search space of the MT parameter was set to the following range of $\epsilon \in [-1, 1]$. Each simulation trial was performed with $P = 15$ particles, and ended after a maximum of $q_{\max} = 100$ iterations (although none of our simulations have ever attained that limit). A fitness objective E^* was not considered to end training, but a trial was ended if the global best fitness $E(\mathbf{p}_g^q)$ is constant for 10 consecutive iterations. The initial position $\mathbf{s}_1^0$ of one particle was set according to `MT-` ($\epsilon = -0.001$). All the remaining particle vectors were initialized randomly, according to a uniform distribution in the search space. The PSO parameters were set as follows: $c_1 = c_2 = 2$; r_1 and r_2 were random numbers uniformly distributed in [0,1]; w^q was decreased linearly from 0.9 to 0.4 over the $q_{\max}$ iterations; the maximum velocity $\mathbf{v}_{\max}$ was set to 0.2. At the end of a trial, the fuzzy ARTMAP network with the best global fitness value $\mathbf{p}_g^q$ was retained. Independently trials were repeated 4 times[b] with different initializations of particle vectors, and the network with greatest $\mathbf{p}_g^q$ of the four was retained.

Since fuzzy ARTMAP performance is sensitive to the presentation order of the training data, each simulation trial was repeated 10 times with either 10 different randomly generated data sets (synthetic data), or 10 different randomly selected data presentation orders (NIST SD19 data). The average performance of fuzzy ARTMAP was assessed in terms of resources required during training, and its generalisation error on the test sets. The amount of resources required during training is measured by compression and convergence time. *Compression* refers to the average number of training patterns per category prototype created in the F2 layer. *Convergence time* is the number of epochs required to complete learning for a learning strategy. It does not include presentations of the validation subset used to perform hold-out validation. *Generalisation error* is estimated as the ratio of incorrectly classified test subset patterns over all test set patterns. Given that compression indicates the number of $F2$ nodes, the combination of compression and convergence time provides useful insight into the amount of processing required during training to produce its best asymptotic generalisation error. Average results, with corresponding standard error, are always obtained as a result of the 10 independent

[b]From previous study with our data sets, it was determined that performing 4 independent trials of the PSO learning strategy with only 15 particles leads to better optimization results than performing 1 trial with 60 particles.

simulation trials.

The Quadratic Bayes classifier (CQB) and k-Nearest-Neighbour with Euclidean distance (kNN) classifier were included for reference with generalisation error results. These are classic parametric and non-parametric classification techniques from statistical pattern recognition, which are immune to the effects of overtraining. For each computer simulation, the value of k employed with kNN was selected among $k = 1, 3, 5, 7$, and 9, using hold-out validation. The rest of this section gives some additional details on the synthetic and real data sets employed during computer simulations.

3.1. *Synthetic data sets:*

All four synthetic data sets described below are composed of a total of 30,000 randomly-generated patterns, with 10,000 patterns for the training, validation, and test subsets. They correspond to 2 class problems, with a 2 dimensional input feature space. Each data subset is composed of an equal number of 5,000 patterns per class. In addition, the area occupied by each class is equal. During simulation trials, the number of training subset patterns used for supervised learning was progressively increased from 10 to 10,000 patterns according to a logarithmic rule: 5, 6, 8, 10, 12, 16, 20, 26, 33, 42, 54, 68, 87, 110, 140, 178, 226, 286, 363, 461, 586, 743, 943, 1197, 1519, 1928, 2446, 3105, 3940, 5000 patterns per class. This corresponds to 30 different simulation trials over the entire 10,000 pattern training subset.

These data sets have been selected to facilitate the observation of fuzzy ARTMAP behavior on different tractable problems. Of the four sets, two have simple linear decision boundaries with overlapping class distributions, $D_\mu(\xi_{tot})$ and $D_\sigma(\xi_{tot})$, and two have complex non-linear decision boundaries without overlap, D_{CIS} and D_{P2}. The total theoretical probability of error associated with D_μ and D_σ is denoted by ξ_{tot}. Note that with D_{CIS} and D_{P2}, the length of decision boundaries between class distributions is longer, and fewer training patterns are available in the neighborhood of these boundaries than with $D_\mu(\xi_{tot})$ and $D_\sigma(\xi_{tot})$. In addition, note that the total theoretical probability of error with D_{CIS} and D_{P2} is 0, since class distributions do not overlap on decision boundaries. The four synthetic data sets are now described.

$D_\mu(\xi_{tot})$. As represented in Figure 3(a), this data consists of two classes, each one defined by a multivariate normal distribution in a two dimensional input feature space. It is assumed that data is randomly generated by sources with the same Gaussian noise. Both sources are described by variables that are independent and have equal variance σ^2, therefore distributions are hyperspherical. In fact, $D_\mu(\xi_{tot})$ refers to 13 data sets, where the degree of overlap, and thus the total probability of error between classes differs for each set. The degree of overlap is varied from a total probability of error, $\xi_{tot} = 1\%$ to $\xi_{tot} = 25\%$, with 2% increments, by adjusting the mean vector μ_2 of class 2.

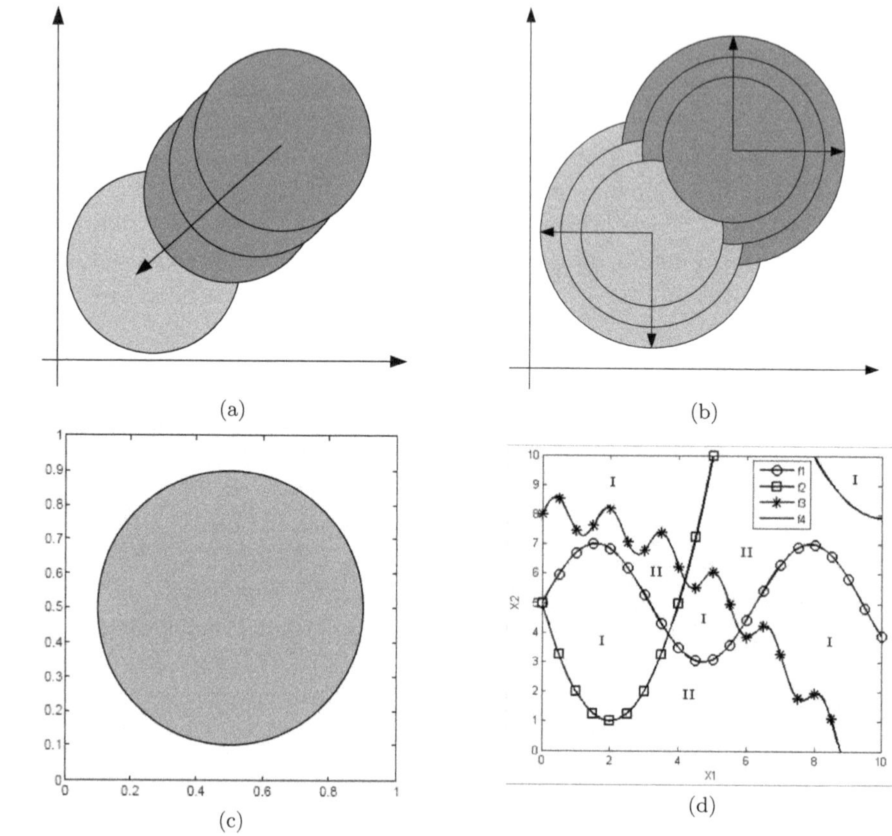

Fig. 3. Representation of the synthetic data sets used for computer simulations: (a) $D_\mu(\xi_{tot})$, (b) $D_\sigma(\xi_{tot})$, (c) D_{CIS} and (d) D_{P2}.

$D_\sigma(\xi_{tot})$. As represented in Figure 3(b), this data is identical to $D_\mu(\xi_{tot})$, except that the degree of overlap between classes is varied by adjusting the variance σ_2^2 of both classes. Note that for a same degree of overlap, $D_\sigma(\xi_{tot})$ data sets have a larger overlap boundary than $D_\mu(\xi_{tot})$ yet they are not as dense.

D_{CIS}. As represented in Figure 3(c), the Circle-in-Square problem[6] requires a classifier to identify the points of a square that lie inside a circle, and those that lie outside a circle. The circle's area equals half of the square. It consists of one non-linear decision boundary where classes do not overlap.

D_{P2}. As represented in Figure 3(d), each decision region of the D_{P2} problem is delimited by one or more of the four following polynomial and trigonometric functions:

$$f_1(x) = 2\sin(x) + 5 \tag{7}$$

$$f_2(x) = (x-2)^2 + 1 \tag{8}$$

$$f_3(x) = -0.1x^2 + 0.6\sin(4x) + 8 \tag{9}$$

$$f_4(x) = \frac{(x-10)^2}{2} + 7.902 \tag{10}$$

and belongs to one of the two classes, indicated by the Roman numbers I and II.[38] It consists of four non-linear boundaries, and class definitions do not overlap. Note that equation $f_4(x)$ was slightly modified from the original equation such that the area occupied by each class is approximately equal.

3.2. *NIST Special Database 19 (SD19):*

Automatic reading of numerical fields has been attempted in several domains of application such as bank cheque processing, postal code recognition, and form processing. Such applications have been very popular in handwriting recognition research, due to the availability of relatively inexpensive CPU power, and to the possibility of considerably reducing the manual effort involved in these tasks.[30]

The NIST SD19[17] data set has been selected due to the great variability and difficulty of such handwriting recognition problems (see Figure 4). It consists of images of *handwritten sample forms* (hsf) organized into eight series, hsf-{0,1,2,3,4,6,7,8}.

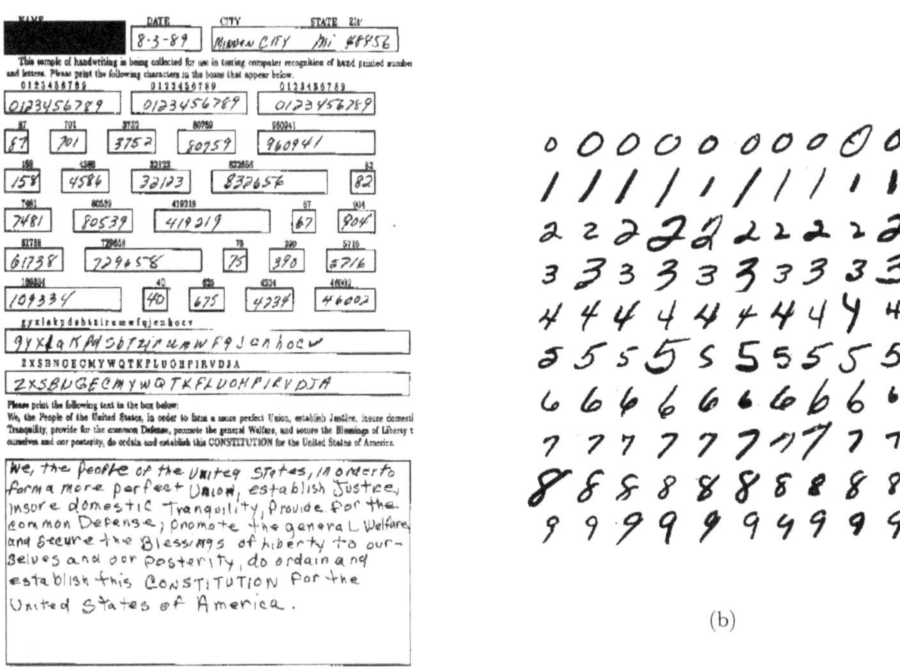

(a) (b)

Fig. 4. Examples in the NIST SD19 data of: (a) a handwriting sample form, and (b) some images of handwritten digits extracted from the forms.

SD19 is divided in 3 sections which contains samples representing isolated handwritten digits ('0', '1', ..., '9') extracted from hsf-{0123}, hsf-7 and hsf-4.

For our simulations, the data in hsf-{0123} has been further divided into training subset (150,000 samples), validation subset 1 (15,000 samples), validation subset 2 (15,000 samples) and validation subset 3 (15,000 samples). The training and validation subsets contain an equal number of samples per class. All 60,089 samples in hsf-7 has been used as a standard test subset. The distribution of samples per class in test sets is approximately equal.

The set features extracted for samples is a mixture of concavity, contour, and surface characteristics.[30] Accordingly, 78 features are used to describe concavity, 48 features are used to describe contour, and 6 features are used to describe surface. Each sample is therefore composed of 132 features that are normalized between 0 and 1 by summing up their respective feature values, and then dividing each one by its summation. With this feature set, the NIST SD19 data base exhibits complex decision boundaries, with moderate overlap between digit classes. Some experimental results obtained with Multi-Layer Perceptron (MLP), Support Vector Machine (SVM), and k-NN classifiers are reported in.[16]

During simulations, the number of training subset patterns used for supervised learning was progressively increased as from 100 to 150,000 patterns, according to a logarithmic rule. The 16 different training subset consist of the first 10, 16, 28, 47, 80, 136, 229, 387, 652, 1100, 1856, 3129, 5276, 8896, and all 15000 patterns per class.

4. Simulation Results

4.1. *Synthetic data with overlapping class distributions:*

Figure 5 presents the average performance obtained when fuzzy ARTMAP is trained with the four MT strategies – `MT-`, `MT+`, `WMT` and `PSO(MT)` – on $D_\mu(13\%)$. The generalisation errors for the Quadratic Bayes classifier (CQB), as well as the theoretical probability of error (ξ_{tot}), are also shown for reference.

As shown in Figure 5(a), `PSO(MT)` generally yields the lowest generalisation error over training set sizes, followed by `WMT`, `MT+`, and then `MT-`. With more than 20 training patterns per class, the error of both `MT-` and `MT+` algorithms tends to increase in a manner that is indicative of fuzzy ARTMAP overtraining.[19] However, with more than about 500 training patterns per class, the generalization error for `MT-` grows more rapidly with the training set size than for `MT+`, `WMT` and `PSO(MT)`. With a training set of 5000 patterns per class, a generalization error of about 21.22% is obtained with `MT+`, 26.17% with `MT-`, 16.22% with `WMT`, and 15.26% with `PSO(MT)`. The degradation in performance of `MT-` is accompanied by a notably higher compression and a lower convergence time than other MT strategies. `MT-` produces networks with fewer but larger categories than other MT strategies because of the

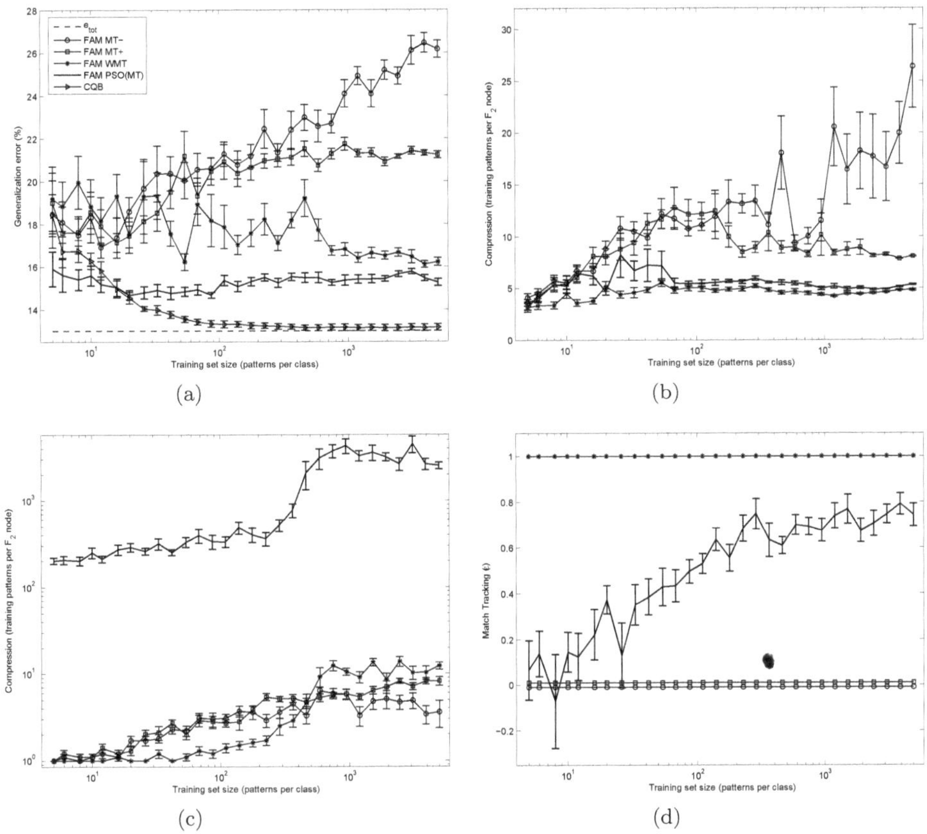

Fig. 5. Average performance of fuzzy ARTMAP (with MT+, MT-, WMT and PSO(MT)) versus training subset size for $D_\mu(\xi_{tot} = 13\%)$: (a) generalisation error, (b) compression, (c) convergence time, and (d) MT parameter for PSO(MT). Error bars are standard error of the sample mean.

MT polarity. Those large categories contribute to a lower resolution of the decision boundary, and thus a greater generalization error.

By training with WMT, the generalization error is significantly lower than both MT- and MT+ especially with a large amount of training patterns, but the compression is the lowest of all training strategies. Based on the error alone, the effectiveness of the MT algorithm is debateable with overlapping data when compared with MT- and MT+, especially for application in which resource requirements are not an issue.

By training with PSO(MT), fuzzy ARTMAP yields a significantly lower generalization error than all other strategies, and a compression that falls between that of WMT and MT- or MT+. With a training set of 5000 patterns per class, a compression of about 8.0 is obtained with MT+, 26.4 with MT-, 4.8 with WMT, and 5.3 with PSO(MT). The convergence time is generally comparable with WMT, MT- and MT+. However, PSO(MT) requires a considerable number of training epochs to complete the optimization process. With a training set of 5000 patterns per class, a convergence time of about 8.2 epochs is obtained with MT+, 3.6 with MT-, 12.3 with WMT, and 2534 with PSO(MT).

Empirical results indicate that the MT process of fuzzy ARTMAP has a considerable impact on performance obtained with overlapping data, especially when ϵ is optimized. As shown in Figure 5(d), when $\alpha = 0.001$, $\beta = 1$ and $\overline{\rho} = 0$, and class distributions overlap, the values of ϵ that minimize error tends from about 0 towards 0.8 as the training set size grows. Higher ϵ settings tend to create a growing number of category hyperrectangles close to the bourdary between classes. Generalisation error of PSO(MT) tends toward that of to WMT on this data set. Furthermore, PSO(MT) and WMT do not show the performance degradation due to overtraining as with MT+ and MT-.

Very similar tendencies are found in simulation results where fuzzy ARTMAP is trained using the other $D_\mu(\xi_{tot})$ and $D_\sigma(\xi_{tot})$ data sets. However, as ξ_{tot} increases,

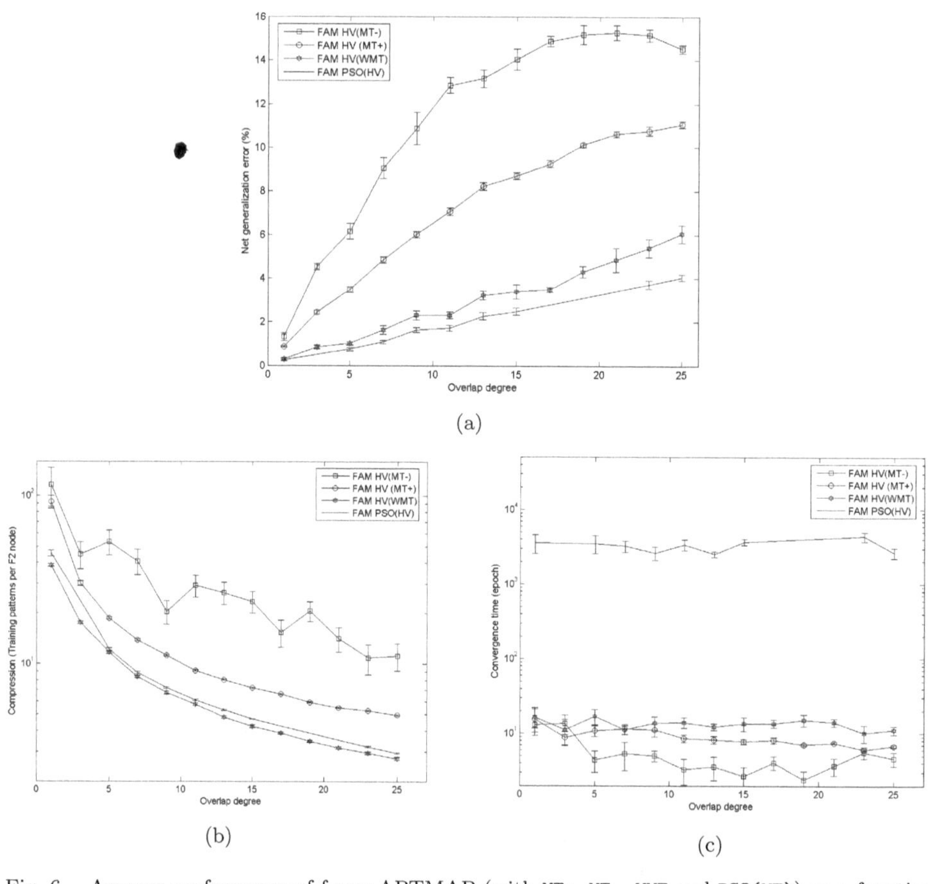

(a)

(b)

(c)

Fig. 6. Average performance of fuzzy ARTMAP (with MT+, MT-, WMT and PSO(MT)) as a function of ξ_{tot} for all $D_\mu(\xi_{tot})$ data sets: (a) net generalisation error, (b) compression, and (c) convergence time.

the performance degradation due to training subset size tends to become more pronounced, and occurs for fewer training set patterns. Let us define the *net error* as the difference between the generalization error obtained by using all the training data (5,000 patterns per class) and the theoretical probability of error ξ_{tot} of the database. Figure 6 shows the performance of fuzzy ARTMAP as a function of ξ_{tot} for all $D_\mu(\xi_{tot})$ data sets. As shown, using PSO(HV) always provides the lowest net error over ξ_{tot} values for overlapping data, followed by `WMT`, `MT+` and `MT-`. Again, `MT-` obtains the highest compression, whereas `PSO(MT)` obtain a compression between `WMT` and `MT+`. The convergence time of PSO(HV) is orders of magnitude longer that the other strategies.

Figure 7 presents an example of decision boundaries obtained for $D_\mu(\xi_{tot} = 13\%)$ when fuzzy ARTMAP is trained with 5,000 patterns per class and different MT strategies. For overlapping class distribution, `MT-` tends to create much fewer $F2$

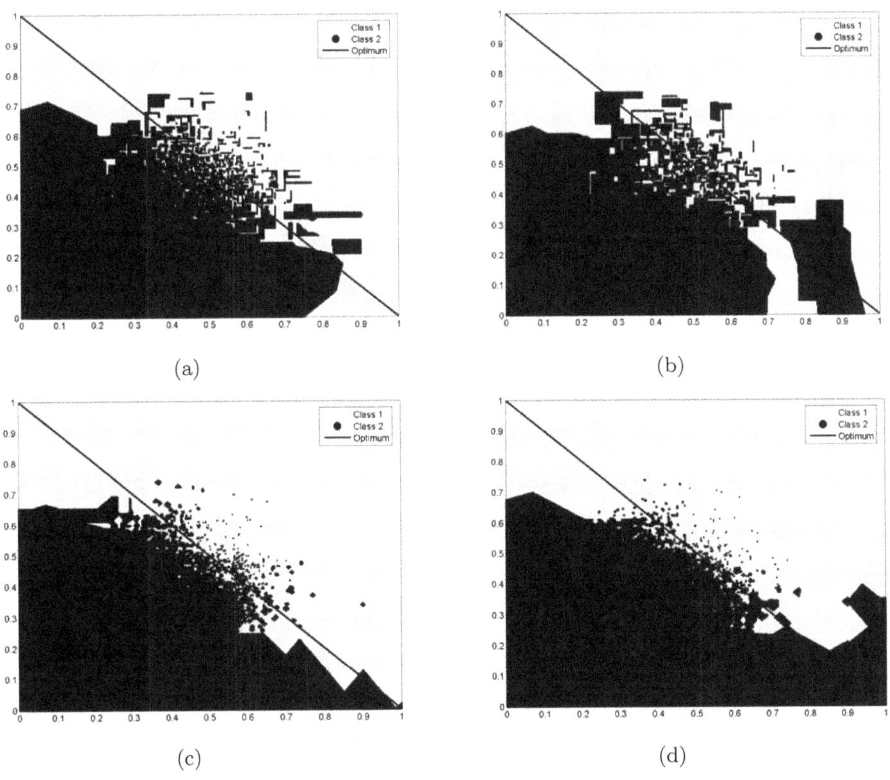

Fig. 7. An Example of decision boundaries formed by fuzzy ARTMAP in the input space for $D_\mu(\xi_{tot} = 13\%)$. Training is performed (a) with `MT+`, (b) with `MT-`, (c) `WMT`, and (d) `PSO(MT)` on 5,000 training patterns per class. The optimal decision boundary for $D_\mu(\xi_{tot} = 13\%)$ is also shown for reference. Note that virtually no training, validation or test subset patterns are located in the upper-left and lower-right corners of these figures.

nodes (908 categories with 5000 patterns per class) than the other MT strategies because of the polarity of ϵ. Although it leads to a higher compression, and can resolve inconsistent cases, the larger categories produce coarse granulation of the decision boundary, and thus a higher generalization error. With `PSO(MT)` and `WMT`, the lower error is a consequence of the finer resolution on overlap regions of the decision boundary between classes.

4.2. *Synthetic data with complex decision boundaries:*

Figure 8 presents the average performance obtained when fuzzy ARTMAP is trained on D_{CIS} using the four MT strategies – `MT-`, `MT+`, `WMT` and `PSO(MT)`. The generalisation error for the k-NN classifier, as well as the theoretical probability of error, ξ_{tot}, are also shown for reference.

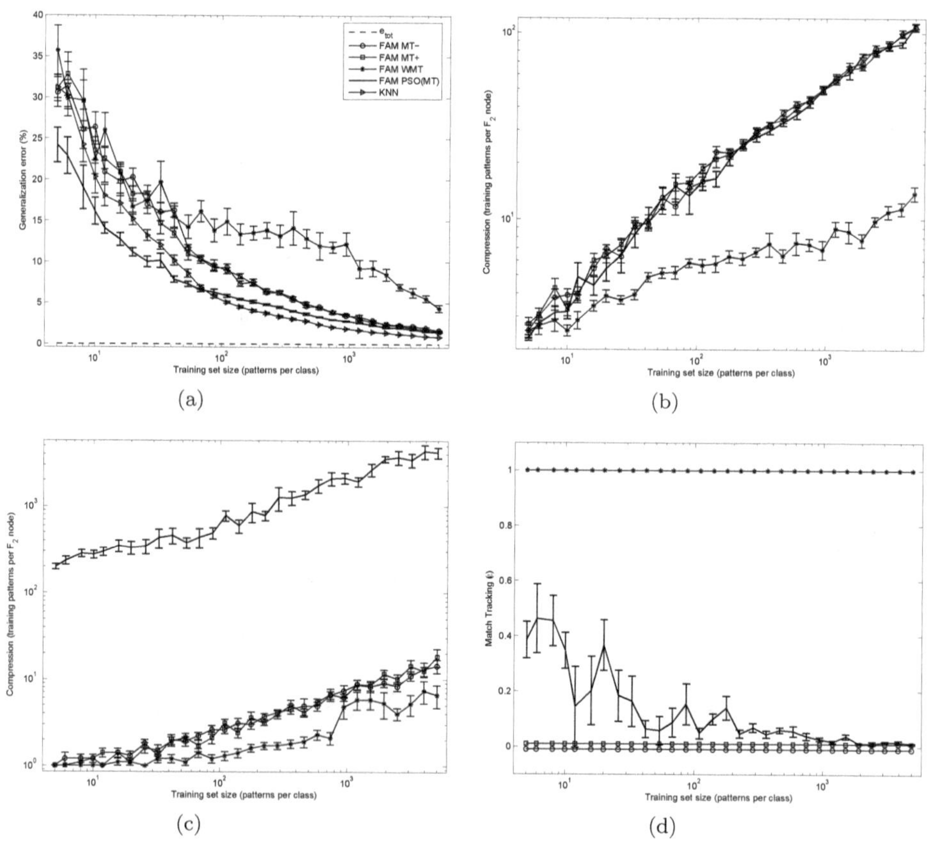

Fig. 8. Average performance of fuzzy ARTMAP (with `MT+`, `MT-`, `WMT` and `PSO(MT)`) versus training subset size for D_{CIS}: (a) generalisation error, (b) compression, (c) convergence time, and (d) MT parameter for `PSO(MT)`. Error bars are standard error of the sample mean.

In this case, MT+, MT- and PSO(MT) obtain a similar generalization error across training set sizes, while WMT yields an error that is significantly higher than the others strategies for larger training set sizes. For example, with a training set of 5000 patterns per class, a generalization error of about 1.51% is obtain with MT+, 1.64% with MT-, 4.36% with WMT, and 1.47% with PSO(MT). Compression of fuzzy ARTMAP as a functions of training set size in a grows in a similar way for MT-, MT+ and PSO(MT). With a training set of 5000 patterns per class, a compression of 107 is obtained with MT+, 108 with MT-, 14 with WMT, and 109 with PSO(MT). WMT does not allow to create a network with higher compression because data structure leads to the creation many small categories that overlap on the decision boundary between classes. However, WMT requires the fewest number of training epochs to converge, while PSO(MT) requires a considerable number of epochs. With a training set of 5000 patterns per class, a convergence time of about 18.4 epochs is required with MT+, 14.4 with MT-, 6.6 with WMT, and 4186 with PSO(MT).

Empirical results indicate that the MT process of fuzzy ARTMAP also has a considerable impact on performance obtained on data with complex decision boundaries, especially when ϵ is optimized. As shown in Figure 8(d), when $\alpha = 0.001$, $\beta = 1$ and $\overline{\rho} = 0$, and decision boundaries are complex, the values of ϵ that minimize error tends from about 0.4 towards 0 as the training set size grows. Lower ϵ settings tend to create fewer category hyperrectangles close to the bourdary between classes. Generalisation error of PSO(MT) tends toward that of to MT+ and MT- on this data.

Similar tendencies are found in simulation results where fuzzy ARTMAP is trained using the D_{P2} data set. However, since the decision boundaries are more complex with D_{P2}, a greater number of training patterns are required for fuzzy ARTMAP to asymptotically start reaching its minimum generalisation error. Moreover, all MT strategies tested on data with non linear decision boundaries generate no overtraining.[19]

Figure 9 presents an example of decision boundaries obtained for D_{CIS} when fuzzy ARTMAP is trained with 5,000 patterns per class and different MT strategies. For data with complex decision boundaries, training fuzzy ARTMAP WMT yields higher generalization error since it initially tends to create some large categories, and then compensates by creating many small categories. This leads to coarse granulation of the decision boundary, and thus a higher generalization error.

Table 1 shows the average generalisation error obtained with the reference classifiers and the fuzzy ARTMAP neural network using different MT strategies on $D_{\mu}(\xi_{tot})$, D_{CIS} and D_{P2}. Training was performed on 5,000 patterns per class. When using PSO(MT), the generalisation error of fuzzy ARTMAP is always lower than when using MT+, MT- and WMT, but is always significantly higher than that of the Quadratic Bayes and k-NN classifiers. When data contains overlapping class distributions, the values of ϵ that minimize error tends towards +1. In contrast, when decision boundaries are complex, these ϵ values tend towards 0.

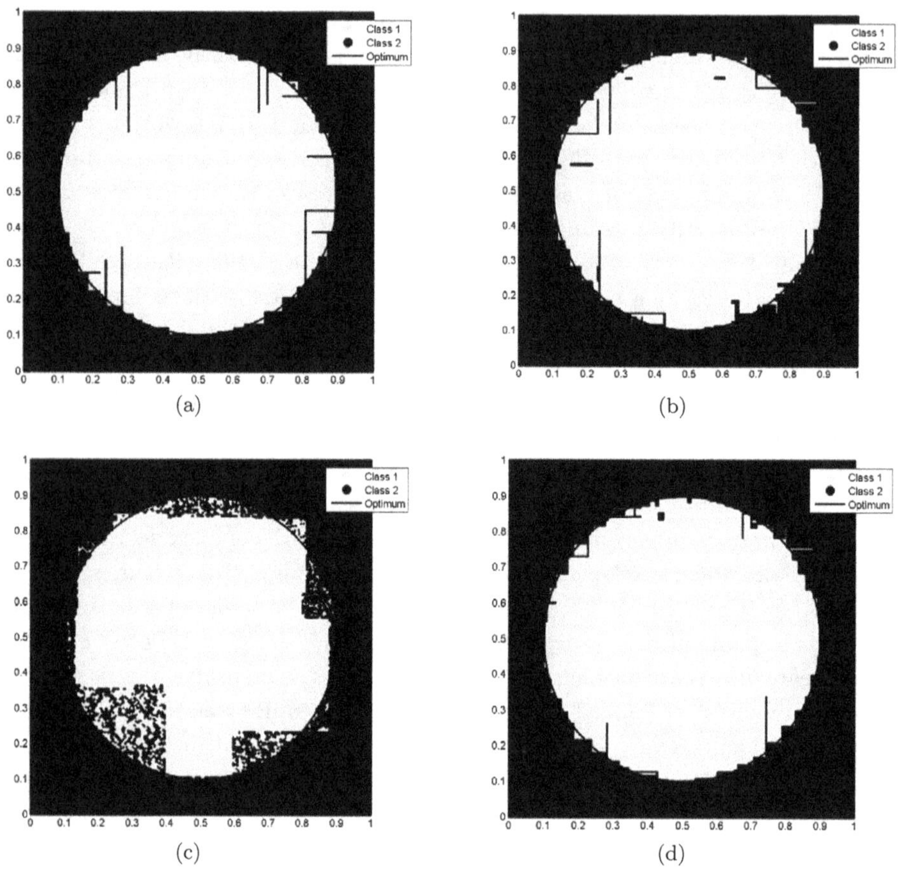

Fig. 9. An Example of decision boundaries formed by fuzzy ARTMAP in the input space for D_{CIS}. Training is performed (a) with `MT+`, (b) with `MT-`, (c) `WMT`, and (d) `PSO(MT)` on 5,000 training patterns per class. The optimal decision boundary for D_{CIS} is also shown for reference.

4.3. *NIST SD19 data:*

Figure 10 presents the average performance obtained when fuzzy ARTMAP is trained on the NIST SD19 data using the four MT strategies – `MT-`, `MT+`, `WMT` and `PSO(MT)`. The generalisation error for the k-NN classifier are also shown for reference.

As shown in this figure, `MT-` and `MT+` obtain similar average generalization error across training set sizes. Using a training set of 52760 patterns, a generalization error of about 5.81% is obtained with `MT+`, 6.02% with `MT-`, 32.84% with `WMT`, and 5.57% with `PSO(MT)`. When optimizing the MT parameter with `PSO(MT)`, generalization error is lower then other MT strategies with a small number of training pattern, and similar to `MT-` and `MT+` with greater number of training pattern. `WMT` is unable to create fuzzy ARTMAP network with low generalization error on NIST SD19.

Table 1. Average generalisation error of reference and fuzzy ARTMAP classifiers using different MT strategies on synthetic data sets. Values in parenthesis are standard error of the sample mean.

Data set	**Average generalisation error (%)**					
	CQB	**k-NN**	**FAM w/** `MT+`	**FAM w/** `MT-`	**FAM w/** `WMT`	**FAM w/** `PSO(MT)` $\rightarrow \epsilon$
$D_\mu(1\%)$	1,00(0,04)	1,08(0,03)	1,87(0,04)	2,31(0,19)	1,30(0,03)	1,24(0,04) $\rightarrow$ 0,61(0,06)
$D_\mu(3\%)$	3,08(0,05)	3,31(0,06)	5,44(0,09)	7,52(0,16)	3,84(0,09)	3,66(0,06) $\rightarrow$ 0,75(0,05)
$D_\mu(5\%)$	4,87(0,07)	5,26(0,08)	8,48(0,13)	11,15(0,36)	6,01(0,07)	5,75(0,08) $\rightarrow$ 0,79(0,04)
$D_\mu(7\%)$	7,00(0,10)	7,48(0,11)	11,85(0,15)	16,05(0,47)	8,63(0,20)	8,07(0,08) $\rightarrow$ 0,73(0,04)
$D_\mu(9\%)$	9,12(0,08)	9,88(0,08)	15,01(0,14)	19,88(0,74)	11,30(0,21)	10,62(0,11) $\rightarrow$ 0,72(0,02)
$D_\mu(11\%)$	11,00(0,08)	11,81(0,13)	18,06(0,18)	23,85(0,37)	13,29(0,15)	12,72(0,12) $\rightarrow$ 0,77(0,05)
$D_\mu(13\%)$	13,16(0,15)	14,27(0,18)	21,22(0,17)	26,17(0,41)	16,22(0,19)	15,26(0,16) $\rightarrow$ 0,74(0,05)
$D_\mu(15\%)$	15,11(0,15)	16,13(0,13)	23,69(0,16)	29,05(0,48)	18,40(0,32)	17,42(0,15) $\rightarrow$ 0,74(0,04)
$D_\mu(17\%)$	16,96(0,10)	18,39(0,09)	26,25(0,16)	31,87(0,25)	20,49(0,13)	19,79(0,33) $\rightarrow$ 0,71(0,08)
$D_\mu(19\%)$	19,25(0,16)	20,71(0,16)	29,13(0,09)	34,19(0,44)	23,30(0,26)	22,24(0,11) $\rightarrow$ 0,79(0,05)
$D_\mu(21\%)$	20,97(0,13)	22,70(0,16)	31,63(0,14)	36,28(0,34)	25,86(0,54)	24,35(0,12) $\rightarrow$ 0,79(0,05)
$D_\mu(23\%)$	22,99(0,12)	25,04(0,13)	33,77(0,21)	38,15(0,28)	28,40(0,41)	26,72(0,19) $\rightarrow$ 0,71(0,03)
$D_\mu(25\%)$	25,11(0,10)	27,23(0,12)	36,08(0,14)	39,52(0,18)	31,05(0,40)	29,05(0,14) $\rightarrow$ 0,72(0,04)
D_{CIS}	N/A	0,86(0,03)	1,51(0,04)	1,64(0,04)	4,36(0,43)	1,47(0,04) $\rightarrow$ 0,01(0,00)
D_{P2}	N/A	1,65(0,04)	3,45(0,19)	4,33(0,22)	7,13 0,48)	3,44(0,06) $\rightarrow$ 0,01(0,00)

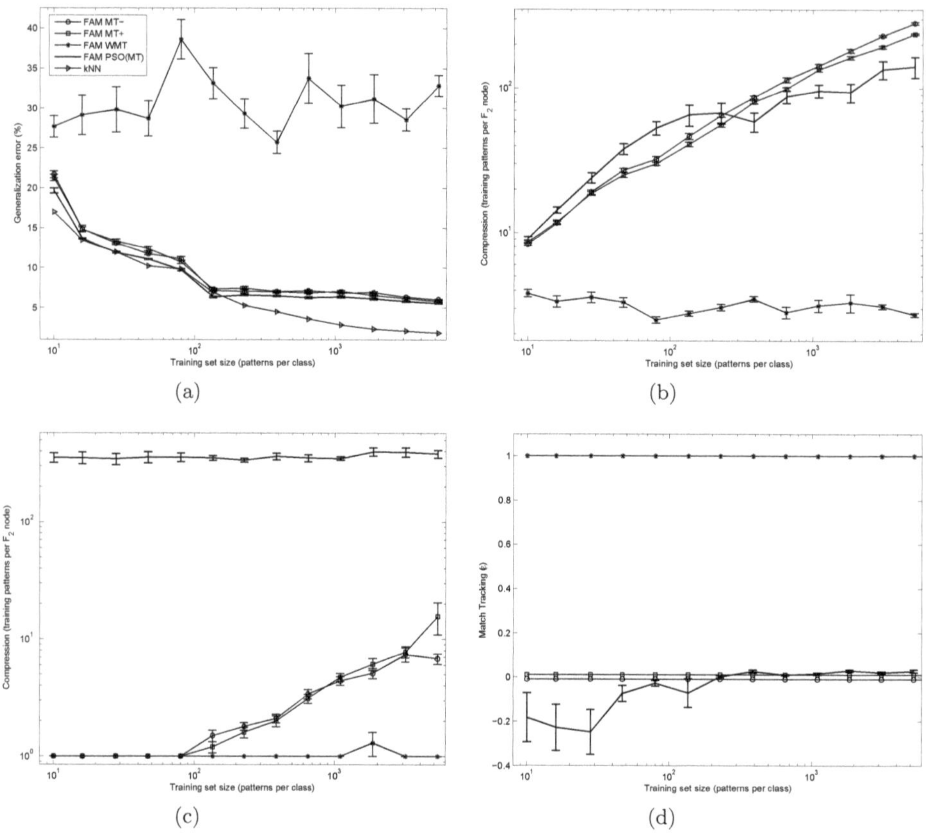

Fig. 10. Average performance of fuzzy ARTMAP (with `MT+`, `MT-`, `WMT` and `PSO(MT)`) versus training subset size for NIST SD19 data set: (a) generalisation error, (b) compression, (c) convergence time, and (d) MT parameter for `PSO(MT)`. Error bars are standard error of the sample mean.

Since NIST database possesses complex decision boundaries with a small degree of overlap, `WMT` cannot generate a good representation of the decision boundaries because it generates too many categories that overlap between classes.

Using all the training data, `MT-` acheives the highest compression, followed by `MT+`, `PSO(MT)` and `WMT`. However, with small amount of training patterns, `PSO(MT)` generates the highest compression. For example, with a training set of 52760 patterns, a compression rate of about 237.4 is obtained with `MT+`, 281.9 with `MT-`, 2.7 with `WMT`, and 141.6 with `PSO(MT)`. `WMT` obtains the lowest compression rate because if creates many very small categories to define the decision boundaries. With a training set of 52760 patterns, a convergence time of about 15.7 epochs is obtain with `MT+`, 6.8 with `MT-`, 1 with `WMT`, and 381 with `PSO(MT)`. `WMT` still possesses the fastest convergence time. The low generalization error of `PSO(MT)` requires a high convergence time (about 24.3 time higher than `MT+` with all training pattern).

As shown in Figure 10(d), when $\alpha = 0.001$, $\beta = 1$ and $\overline{\rho} = 0$, and decision boundaries are complex, the values of ϵ that minimize error tends from about -0.2 towards 0 as the training set size grows. As with D_{CIS} and D_{P2}, Generalisation error of `PSO(MT)` tends toward that of to `MT+` and `MT-` on this data set. Despite promising results training fuzzy ARTMAP with `PSO(MT)`, other pattern classifiers (such as SVM) have achieved significantly lower generalization error.[28,30]

5. Conclusions

A fuzzy ARTMAP neural network applied to complex real-world problems such as handwritten character recognition may achieve poor performance, and encounter a convergence problem, whenever the training set contains noisy and overlapping patterns that belong to different classes. In this chapter, a PSO-based strategy called `PSO(MT)` is used to optimize the MT parameter during training. The impact on fuzzy ARTMAP performance of adopting different MT strategies – the original positive MT (`MT+`), negative MT (`MT-`), without MT (`WMT`), and `PSO(MT)` – is assessed. An experimental protocol has been defined such that the generalization error and resource requirements of fuzzy ARTMAP trained with different MT strategies may be assessed on different types of synthetic problems and on the NIST SD19 handwritten character recognition data sets.

Overall, empirical results indicate that using the MT process for batch supervised learning has a significant impact on fuzzy ARTMAP performance. When data is defined by overlapping class distributions, training with `MT-` tends to produce fewer categories than the other MT strategies, although this advantage coincides with a higher generalization error. The need for `MT+` or `MT-` is debateable as `WMT` yields a significantly lower generalization error. However, `PSO(MT)` has been shown to create fuzzy ARTMAP networks with a finer resolution on decision bounds, and an even lower error than `WMT`. In addition, it has been shown to eliminate the degradation of error due to overtraining. To represent overlapping class distributions with `PSO(MT)`, the lowest errors are obtained for MT parameter values that tend toward the maximum value ($\epsilon = 1$) as the training set size grows. `PSO(MT)` thereby favors the creation of new internal categories to define decision boundaries.

When data is defined by complex decision boundaries, training with `PSO(MT)` creates the decision boundaries that yield the lowest generalization error, followed most closely by `MT-` and then `MT+`. Training with `WMT` yields a considerably higher generalization error and lower compression than the other MT strategies, specially when for larger training set sizes. To represent complex decision boundaries with `PSO(MT)`, the lowest errors are obtained for MT parameter values that tend toward 0 as the training set size grows.

Finally, with the NIST SD19 data set, when using all training pattern the generalization error obtain with `PSO(MT)` is about 0.84% lower than `MT-`, but comes at the expense of lower compression and a convergence time that can be two order of magnitude greater than other strategies. Training with a Multi-Objective PSO (MOPSO) based strategy, where the cost function accounts for both generalization error and compression would provide solutions that require fewer internal categories. In addition light weight versions of PSO may reduce the convergence time.

In this chapter, training fuzzy ARTMAP with `PSO(MT)` has been shown to produce a significantly lower generalization error than with other MT strategies. These results are always produced at the expense of a significantly higher number of training epochs. Nonetheless, results obtained with `PSO(MT)` underline the importance of optimizing the MT parameter during training, for different problems. The MT parameter values found using this strategy vary significantly according to training set size and data set structure, and differ considerably from the popular choice ($\epsilon = 0+$), specially when data has overlapping class distributions.

Acknowledgements

This research was supported in part by the Natural Sciences and Engineering Research Council of Canada, and le Fonds québécois de la recherche sur la nature et les technologies.

References

1. Anagnostopoulos, G. C., Georgiopoulos, M., Verzi, S. J., and Heileman, G. L., "Boosted Ellipsoid ARTMAP," *Proc. SPIE – Applications and Science of Computational Intelligence V*, **4739**, 74-85, 2002.
2. Anagnostopoulos, G. C., and Georgiopoulos, "Putting the Utility of Match Tracking in Fuzzy ARTMAP Training to the Test," *Lecture Notes in Computer Science*, **2774**, 1-6, 2003.
3. Bote-Lorenzo, M. L., Dimitriadis, Y., Gómez-Sánchez, E., "Automatic extraction of human-recognizable shape and execution prototypes of handwritten characters," *Pattern Recognition*, **36:7**, 1605-1617, 2003.
4. Carpenter, G. A., and Grossberg, S., "A Massively Parallel Architecture for a Self-Organizing Neural Pattern Recognition Machine," *Computer, Vision, Graphics and Image Processing*, **37**, 54-115, 1987.
5. Carpenter, G. A., Grossberg, S., and Rosen, D. B., "Fuzzy ART: Fast Stable Learning and Categorisation of Analog Patterns by an Adaptive Resonance System," *Neural Networks*, **4:6**, 759-771, 1991.
6. Carpenter, G. A., Grossberg, S., and Reynolds, J. H., "ARTMAP: Supervised Real-Time Learning and Classification of Nonstationary Data by a Self-Organizing Neural Network," *Neural Networks*, **4**, 565-588, 1991.
7. Carpenter, G. A., Grossberg, S., Markuzon, N., Reynolds, J. H., and Rosen, D. B., "Fuzzy ARTMAP: A Neural Network Architecture for Incremental Supervised Learning of Analog Multidimensional Maps," *IEEE Trans. on Neural Nets*, **3:5**, 698-713, 1992.
8. Carpenter, G. A., and Ross, W. D, "ART-EMAP: A Neural Network Architecture for Object Recognition by Evidence Accumulation," *IEEE Trans. on Neural Networks*, 6:4, 805-818, 1995

9. Carpenter, G.A., Gjaja, M.N., Gopal, S., and Woodcock, C.E., "ART Neural Networks for Remote Sesing: Vegetation Classification from Landsat TM and Terrain Data," *IEEE Trans. on Geosciences and Remote Sensing*, **35:2**, 1997.
10. Carpenter, G. A., and Markuzon, N., "ARTMAP-IC and Medical Diagnosis: Instance Counting and Inconsistent Cases," *Neural Networks*, **11:2**, 323-336, 1998.
11. Carpenter, G. A., Milenova, B. L., and Noeskeand, B. W., "Distributed ARTMAP: a neural network for fast distributed supervised learning," *Neural Nets*, **11**, 793813, 1998.
12. Eberhart, R. C. and Shi, Y., "Comparison Between Genetic Algorihms and Particle Swarm Intelligence," in *Evolutionary Programming VII*, V. W. Porto et al, eds., Springer, 611-616, 1998
13. Gómez-Sánchez, E., Gago-Gonzalez, J. A., Dimitriadis, Y. A., Cano-Izquierdo, J. M., Lopez Coronado, J., "Experimental Study of a Novel Neuro-Fuzzy System for On-Line Handwritten UNIPEN Digit Recognition," *Pattern Recognition Let.*, **19**, 357-364, 1998.
14. Gómez-Sánchez, E., Dimitriadis, Y. A., Cano-Izquierdo, J. M., Lopez-Coronado, J., "μARTMAP: Use of Mutual Information for Category Reduction in Fuzzy ARTMAP," *IEEE Trans. on Neural Networks*, **13:1**, 58-69, 2002.
15. Granger, E., Rubin, M., Grossberg, S., and Lavoie, P., "A What-and-Where Fusion Neural Network for Recognition and Tracking of Multiple Radar Emitters," *Neural Networks*, **14**, 325-344, 2001.
16. Granger, E., Henniges, P., Sabourin, R., and Oliveira, L. S., "Supervised Learning of Fuzzy ARTMAP Neural Networks Through Particle Swarm Optimization," *Journal of Pattern Recognition Research*, **2:1**, 27-60, 2007.
17. Grother, P. J., "NIST Special Database 19 - Handprinted forms and characters database," *National Institute of Standards and Technology (NIST)*, 1995.
18. Grossberg, S., Rubin, M. A., and Streilein, W. W., "Buffered reset leads to improved compression in fuzzy ARTMAP classification of radar range profiles," *Intelligent Engineering Systems Through Artificial Neural Networks*, **6**, 419-424, 1996.
19. Henniges, P., Granger, E., and Sabourin, R., "Factors of Overtraining with Fuzzy ARTMAP Neural Networks," *International Joint Conference on Neural Networks 2005*, 1075-1080, Montreal, Canada, August 1-4, 2005.
20. Kennedy, J., and Eberhart, R. C., "Particle Swarm Intelligence," *Proc. Int'l Conference on Neural Network*, 1942-1948, 1995.
21. Kennedy, J., and Eberhart, R. C., *Swarm Intelligence*, Morgan Kaufmann, 2001.
22. Koufakou, A., Georgiopoulos, M., Anagnostopoulos, G., and Kasparis, T., "Cross-Validation in Fuzzy ARTMAP for Large Databases," *Neural Nets*, **14**, 1279-1291, 2001.
23. Lee, S.-J., and Tsai, H.-L., "Pattern Fusion in Feature Recognition Neural Networks for Handwritten Character Recognition", *IEEE Transactions on Systems, Man, and Cybernetics Part B: Cybernetics*, **28:4**, 612-617, 1998.
24. Lerner B., and Guterman, H., "Advanced Developments and Applications of the Fuzzy ARTMAP Neural Network in Pattern Classification," In *Advanced Computational Intelligence Techniques in Data Analysis and Applications*, Springer-Verlag, **SCI 137**. 77-107, 2008.
25. Lim C. P., and Harrison, R. F., "Modified Fuzzy ARTMAP Approaches for Bayes Optimal Classification Rates: An Empirical Demonstration," *Neural Network*, **10:4**, 755-774, 1997.

26. Liu, C.-L., Sako, H., and Fujisawa, H., "Performance Evaluation of Pattern Classifiers for Handwritten Character Recognition," *Int'l J. on Document Analysis and Recognition*, **4**, 191-204, 2002.
27. Marriott, S., and Harrison, R. F., "A modified fuzzy ARTMAP architecture for the approximation of noisy mappings", *Neural Networks*, **8:4**, 619-41, 1995.
28. Milgram, J., Chriet, M. and Sabourin, R., "Estimating Accurate Multi-class Probabilities with Support Vector Machines," *International Joint Conference on Neural Networks 2005*, 1906-1911, Montral, Canada, August 1-4, 2005.
29. Murshed, N. A., Bortolozzi, F., and Sabourin, R., "A Cognitive Approach to Signature Verification," *International Journal of Pattern Recognition and Artificial Intelligence (Special issue on Bank Cheques Processing*, **11:7**, 801-825, 1997.
30. Oliveira, L. S., Sabourin, R., Bortolozzi, F., and Suen, C. Y., "Automatic Recognition of Handwritten Numerical Strings: A Recognition and Verification Strategy," *IEEE Transactions on Pattern Analysis and Machine Intelligence*, **24:11**, 1438-1454, 2002.
31. Parsons, O., and Carpenter, G. A., "ARTMAP neural network for information fusion and data mining: map production and target recognition methodologies," *Neural Networks*, **16**, 10751089, 2003.
32. Rubin, M.A., "Application of Fuzzy ARTMAP and ART-EMAP to Automatic Target Recognition Using Radar Range Profiles," *Neural Networks*, **8:7**, 1109-1116, 1995.
33. Schutte, J. F., Reinbolt, J. A., Fregly, B. J., Haftka, R. T., and George, A. D., "Parallel Global Optimization with Particle Swarm Algorithm," *International J. of Numerical Methods in Engineering*, **61**, 2296-2315, 2004.
34. Stone, M., "Cross-Validatory Choice and Assessment of Statistical Predictions," *Journal of the Royal Statistical Society*, 111-147, 1974.
35. Sumathi, S., Sivanandam, S. N., and Jagadeeswari, R., "Design of Soft Computing Models for Data Mining Applications," *Indian J. of Engineering and Materials Sciences*, **7:3**, 107-21, 2000.
36. Srinivasa, N., "Learning and Generalization of Noisy Mappings Using a Modified PROBART Neural Net," *IEEE Trans. on Signal Processing*, **45:10**, 2533-2550, 1997.
37. Taghi, M., Baghmisheh, V., and Pavesic, N., "A fast simplified Fuzzy ARTMAP network," *Neural Processing Letters*, **17**, 273316, 2003.
38. Valentini, G., "An Experimental Bias-Variance Analysis of SVM Ensembles Based on Resampling Techniques," *IEEE Trans. Systems, Man, and Cybernetics - Part B: Cybernetics*, **35:6**, 1252-1271, 2005.
39. Verzi, S. J., Heileman, G. L., Georgiopoulos, M., and Healy, M. J., "Boosting the Performance of ARTMAP", *IEEE International Joint Conference on Neural Networks Proceedings 1998*, Anchorage, USA, 396-401, 1998.
40. Waxman, A. M., Verly, J. G., Fay, D. A., Liu, F., Braun, M. I., Pugliese, B., Ross, W., Streilein, W., "A Prototype System for 3D Color Fusion and Mining of Multisensor/Spectral Imagery," *Proc. of the 4th International Conference on Information Fusion*, Vol. 1, pp. WeC1-(3-10), Montreal, Canada, August 7-10, 2001.
41. Williamson, J. R., "A Constructive, Incremental-Learning Neural Network for Mixture Modeling and Classification," *Neural Computation*, **9:7**, 1517-1543, 1997.

CHAPTER 1.5

SHAPE AND COLOR FEATURES FOR OBJECT RECOGNITION SEARCH

Tuan A. Duong[1,*], Vu A. Duong[1] and Allen R. Stubberud[2]

[1]*Bio-Inspired Technologies and Systems*
Jet Propulsion Laboratory
4800 Oak Grove Dr., Pasadena, CA 91109
[*]*taduong@mail.jpl.nasa.gov*

[2]*Department of Electrical Engineering and Computer Science*
University of California, Irvine
Irvine, CA 92697

In an object recognition search, in which objects are examined one at a time, a bio-inspired shape feature is introduced which emulates the integration of the saccadic eye movement and the horizontal layer in the retina of a vertebrate. An optimal computational model for shape extraction, which is based on Principal Component Analysis (PCA) and which reduces processing time and enables real time adaptive capability, is also developed. A color feature of the object, using color segmentation, is used to aid shape feature recognition to better solve object recognition problems in heterogeneous environments where a single technique, using either shape or color, may not be effective. To enable an effective recognition system, an adaptive architecture with an autonomous mechanism is introduced to recognize and adapt the shape and color features of moving objects. These preliminary results set a corner stone for further study of practical and effective discrete modeling of a 3-D object. The adaptive architecture and mechanism provide a major step toward enabling an optimal and practical bio-inspired visual system for the future. Additional work is planned to incorporate a bipolar layer which addresses the scaling issue.

1. Introduction

In spite of many years of research on, and many reports describing approaches to, object recognition [1-10], object recognition continues to be a challenging problem. Recently, several new approaches, including bio-related methods [1,2,4,5], shape and color features based approaches [11,12], and image retrieval based shape and color feature methods [13,14] have been presented. Since the shape and color features of a given object can uniquely define its characteristics and they also agree with biological models, these approaches are strongly related [15-17].

From a physical perspective, an electronic image of an object is a collection of photon reflections from the surface of that object with respect to a fixed camera position. If the object moves and the motion/rotation of the surface of the object are sufficiently small, a new collection of photon reflections may not be correlated with the previous reflections from a *pixel-wise view point*; however, they still maintain a *strong correlation* with a *global view* of the object. To obtain a global view of an object, the shape feature is the logical building block to be used and the color feature, if available, provides additional information. Moreover, for different lighting conditions and different light absorption properties of the various materials in the object, different parts of the object reflect and absorb differently, so that the object color distribution adds a unique characteristic to the uniqueness of the object's signature. Furthermore, when the object moves in an evolving light environment the color response changes locally with respect to its previous response, adding to its temporal signature.

From this perspective, an *adaptative capability* for shape and color based on local color and shape correlation is a logical tool to employ for the recognition of a moving object in an evolving environment. From a practical viewpoint, the major challenge for a color and shape feature based recognition technique is how to obtain the effective features in real time, or near real time, which enable an adaptive capability for dynamically changing objects in a 3-D environment. When the operating environment is evolving, the change of shape and color features are not guaranteed to be local or 'smooth' in all dimensions all the time; however, we will assume that at least one of them changes 'smoothly'. The feedback architecture of the recognition system uses two variables, the color and shape feature vectors, as input parameters to drive the system to optimally converge to the correct attractor i.e., to the correct object recognition.

When assumptions of smoothness of shape and color features in time are violated, the feature database that has been adapted and stored of shape and color features, needs to be employed to search the match features from scratch as a new start every time. This aspect will be discussed elsewhere.

In this chapter, section 2 is a discussion of shape and color feature extraction techniques and an associated adaptive architecture and capabilities. As support for our approach, simulation studies were conducted to validate each building block technique. Section 3 presents the integration of shape and color feature techniques and an associated simulation study; section 4 discusses the techniques and simulation results, and section 5 presents conclusions.

2. Shape and Color Feature Extraction Techniques

In this section, we discuss mathematical models of color and shape features of an object and an adaptive architecture for effectively using the features. Simulation results are presented to confirm that our approach is effective. When the object is exposed to a heterogeneous environment, unexpected changes of the color or shape features may occur, then a combination of color and shape features can provide an effective method for enhancing object recognition capability.

a. Bio-Inspired Shape Feature Extraction (BSFE)

From a biological view point, the saccadic eye movement is well understood [18]; however, its use in visual models has not gone beyond recognition of the phenomenon [19]. In our approach, one of the key technologies is that we map the saccadic eye movements into the statistical form, from which Principal Component Analysis (PCA) can be used to extract the features from an image of a single (or multiple) objects. One can view this BSFE approach as a bio-inspired technique, which emulates the saccadic eye movement and the horizontal layer in the retina of a vertebrate [26]. From this modeling, the significant and salient shape features of the object of interest can be extracted.

1. PCA Feature Extraction

For problems with rich data sources such as image recognition and computer vision, the dimension of the input vector is typically larger than the number of input samples, which leads to overfitting of the data if no care is taken to achieve useful generalization. Furthermore, computing power rapidly becomes insufficient to process the data within a reasonable time. To overcome these two obstacles, a preprocessing step can be effective in the following two ways:

- non-useful and redundant data are eliminated, thus enhancing the operational processing step;
- the salient feature can be selected, thus improving the recognition capability

The combination of these two items enables real time adaptive capability which can mimic the bio-inspired visual system.

Support vector machines [20], learning by hints [21], Principal Component Analysis (PCA) [22] and Cellular Neural Networks (CNN) [23] all reduce the dimension of the input vector set with little or no significant loss of information. Such data extraction methods also have the advantage of eliminating some irrelevant data, such as small amplitude noise, and speeding up the classification step. Both PCA [24,25] and CNN [23] have been used in bio-inspired model to extract features from data. We employ sequential PCA (developed at JPL), which is an efficient tool requiring less computation and being more hardware friendly to enable real time feature extraction for real time adaptative capability.

Principal Component Analysis (PCA) is a second order statistical approach, which has been used to extract the features of a data set [27] or to perform data compression [28-30]. Especially, when the data set is Gaussian, redundant and overwhelmingly large, PCA is a very effective preprocessing step to extract data features for classification and/or to cluster data in the most compact energy vectors for data compression. Unfortunately, PCA requires that the basis vectors be orthogonal, which is typically an artificial assumption.

The PCA procedure is complicated and computationally intensive ($O(N^3)$ [31], where N is the dimension of the vector input), thereby making it impossible to use for rich data sources. To get over the hurdles from the traditional PCA technique, simple sequential PCA techniques have been developed [32]. These techniques are based on a learning approach that sequentially obtains principal component vectors. Implementations of PCA are reported using Hebbian or anti-Hebbian learning [33,34] and gradient-based learning [30,35] or even the more elegant technique of natural gradient descent [36].

For sequential PCA, the gradient descent (GED) technique [30,35] is a more attractive approach for hardware implementation as a straight forward technique compared to others e.g. steepest decent, conjugate gradient, or Newton's second order method, but it exposes some difficulties in learning convergence when other principal component vectors are corresponding to smaller eigen values. In addition, this technique still requires some complicated hardware.

At JPL, we have developed a Dominant-element-based gradient descent and dynamic initial learning rate technique for sequential PCA [38,39] which serves two purposes: 1) simplified hardware implementation, especially in VLSI as a System-On-A-Chip; 2) fast and reliable convergence as compared with counterpart gradient descent. Furthermore, our new technique requires much less computation and its optimized architecture is more suitable for implementing as a real time adaptive learning system in hardware.

The objective function for the learning technique is defined as:

$$J(w) = \sum_{i=1}^{m} J_i(w_i) = \sum_{i=1}^{m}\sum_{t=1}^{k} \left| x_t - w_i w_i^T x_t \right|^2 \tag{1}$$

where m is the number of principal components, k is the number of measurement vectors, x_t, measured at time t and w_i is the ith principal vector (or eigen vector). We also define

$$J_i(w_i) = \sum_{t=1}^{k} \left| y_i^t - w_i w_i^T y_i^t \right|^2 \qquad y_i^t = x_t - \sum_{j=1}^{i-1} w_j w_j^T x_t \tag{2}$$

PCA learning approach

From equation (2), the learning algorithm can be processed sequentially for each principal vector that is based on the gradient descent as follows:

$$\Delta w_{ij} = -\frac{\partial J_i}{\partial w_{ij}} = -\frac{\partial(\left|y_i^t - w_i w_i^T y_i^t\right|^2)}{\partial w_{ij}} \tag{3}$$

From equation (3), only the dominant element [28] is used; the weight update can be obtained as follows:

$$w_{ij}^{new} = w_{ij}^{old} + \zeta \Delta w_{ij} = w_{ij}^{old} + \zeta \varepsilon_{ij} (w_i^T y_i^t + w_{ij} y_{ij}^t) \tag{4}$$

where $\zeta = \frac{E_0}{E_{i-1}}$ and $\hat{y}_i^t = w_i w_i^T y_i^t$

E_0 is the initial energy when the network starts learning and E_{i-1} is the energy of the *(i-1)*th extracted principal.

2. Simulation Results

In this subsection, we present results of two experiments based on bio-inspired shape feature approach, described above.

Static Object Recognition

Here we are addressing an image retrieval application based on object recognition, where the image database and searched image are available. For example, search a specific person in a scene whose appearance is known from DMV's driver license image database.

In this application, only a single image is obtained and used as an image input to prompt to database for the best match.

We used the database of 1520 images (286x384 pixel resolutions) of 23 different persons available at http://www.humanscan.de/support/downloads/facedb.php and we used only a single image of one random person at a time for bio-inspired feature extraction. Three bio-inspired features of an image of each person are used. The processing architecture is provided in Figure 1.

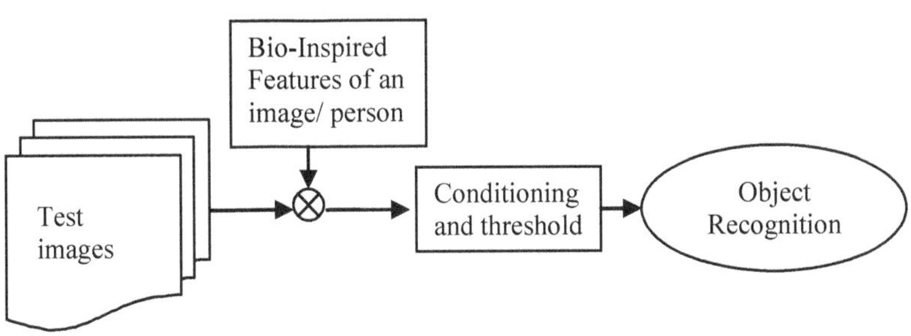

Figure 1: System configuration

We picked 3 out of 23 persons and used a selected image per person shown in Figures 2a, 2b and 2c. We resized the image to 50% of the original in order to improve processing time. These images are extracted to obtain three features (eigen vectors) for each selected image shown in Figures 2a, 2b and 2c.

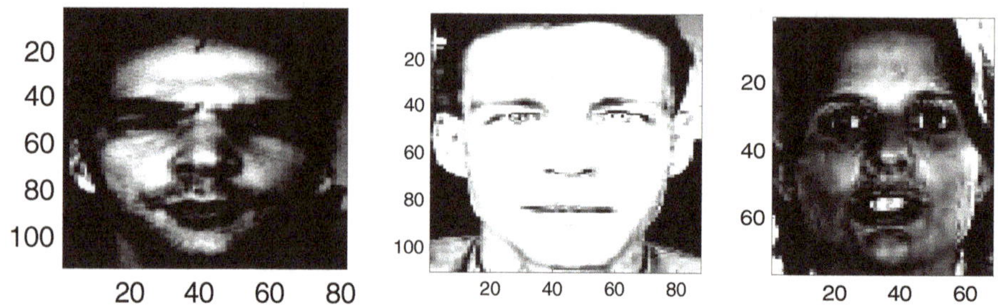

2(a) 2(b) 2(c)

Figure 2(a, b, c): Training images

When done, these feature vectors are used to correlate through a test image to find the maximum correlation within it. The maximum correlation values for test images (1520) are threshold to determine whether it belongs to the same person with having used features.

(3a) (3b) (3c)

Figure 3: Test results; 3a) and 3b) correct recognition; and 3c) missed one

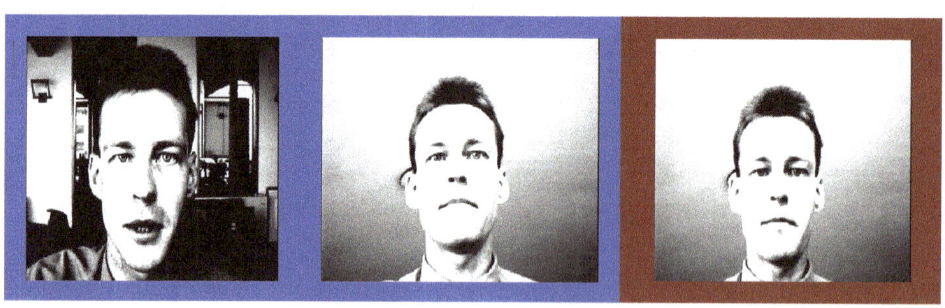

(4a) (4b) (4c)

Figure 4: Test results; 4a) and 4b) correct recognition; and 4c) missed one

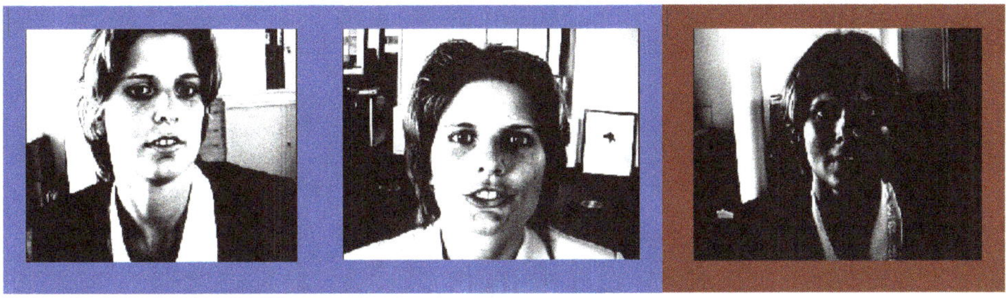

(5a) (5b) (5c)

Figure 5: Test results; 5a) and 5b) correct recognition; and 5c) missed one

The Figures 3, 4 and 5 have shown that the first two images are correctly recognized and the third image is a missed one.

We decided to use one sample image, because it is practical and user friendly and that not many image of the object are available in some of real world applications. Moreover, the search results will be depended on the users' desire by controlling the threshold value to obtain a pool of candidate matches and its size.

In this study, our constraints are not only to maximize the percentage of correct recognition, but also maintain a good generalization capability. Per each sample image/person, it is about 2.4% of the total image of person to be inputted to the system and on average we were able to obtain about 55% correct recognition (generalization) of that person appearance in the entire database. To do so, we have to relax the threshold controlled values, hence introduced more false positive and the simulation results are shown in Table I.

Table I: The simulation results

Features	F1	F1 & F2	F1-F3	Comments
Percentage of correct recognition with *positive false alarm*	86%	92%	94.2%	With one sample input per person (2.4%) and 55% correct generalization (20 different images of the same person)
Percentage of correct recognition with no *positive false alarm*	96.7%	...	...	15% correct generalization (7 different images of the same person) with no positive false alarm

To relax on the generalization to 15% (instead of 55%) correct recognition of a person appearance in the entire database (with 2.4% input image or one image/per person), we were able to 96.7% correct recognition with *no positive false alarm* using only one feature (F1). For example, the first feature (F1) of the person in Figure 2a was used to recognize the same person in all pictures shown in Figure 6.

Figure 6: Search results of the same object as (2a) without positive false alarm

From the experiment described above, we clearly see that 1) increasing the number of features per image improve recognition; 2) with less generalization capability, recognition is improved further with no positive false alarms. The simulation results are excellent in spite of the algorithmic and architectural simplicity of the model.

Dynamic Object Recognition

Here we focus on the dynamic object recognition, attempting to provide a practical technique to be used in the real world environment e.g., intelligent video search etc.

Adaptive Shape Feature Scheme

We specifically use the BSFE technique to detect the object in the new scene based on their correlations. The correlation value of each shape feature corresponding to the similar raw image must be closest to eigen value of each corresponding feature vector (eigen vector).

Whenever a new object is detected in a scene, it is automatically fed into BSFE as a newly found object for extracting new features.

Simulation Results

In this experiment, we focus on the moving object recognition using grayscale images. Due to the grayscale image, only the adaptive shape feature approach is employed. The purpose of this study was to demonstrate the adaptive shape feature mechanism and moving object recognition, to build the foundation for the final object recognition based on shape and color feature signature.

We used moving car object that was acquired from the USC-SIPI Image Data Base in http://sipi.usc.edu/database/ as a benchmark. This sequencing image is carefully designed to show that the car has moved in the orthogonal direction to avoid a trivial movement.

We sampled an object (Figure 6b) (100x100) from the scene and we only used one feature (eigen vector) for this application.

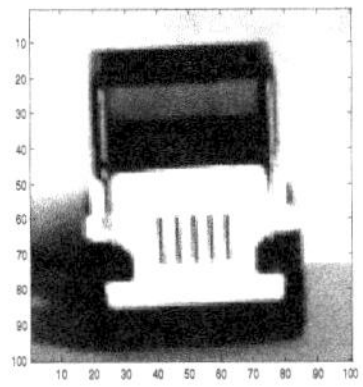

Figure 6b: Initial object for object of interest for recognition

(7a) (7b)

Figure 7: The test image 7a) and the object 7b) is found as the same as the initial object

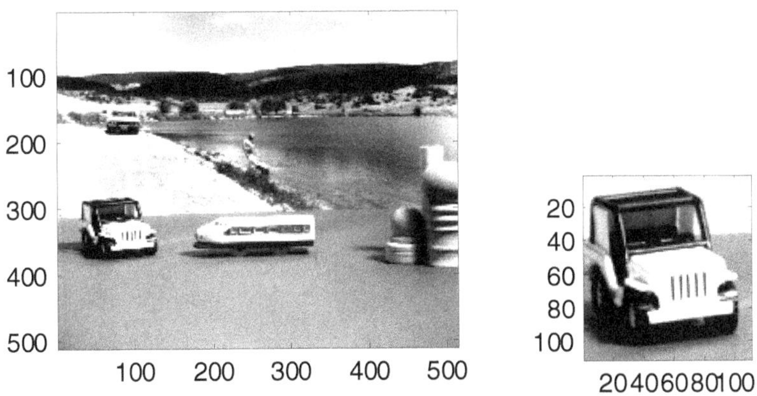

(8a) (8b)

Figure 8: The test image 8a) and the detected object 8b)

(9a) (9b)

Figure 9: The test image 9a) and the detected object 9b)

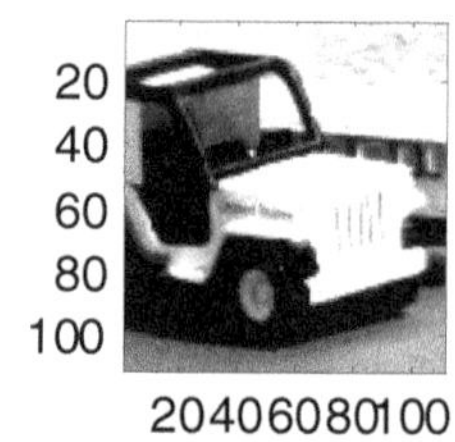

(10a) (10b)

Figure 10: The test image 10a) and the detected object 10b)

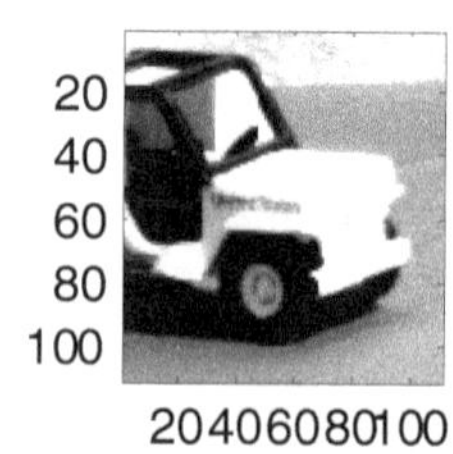

(11a) (11b)

Figure 11: The test image 11a) and the detected object 11b)

(12a) (12b)

Figure 12: The test image 12a) and the detected object 12b)

Our BSFE technique has successfully detected the moving car based on adaptive shape feature in the dynamic scene. However, we have used a fixed window size in this simulation, so the objects detected are not fully exposed in the window as shown in Figures 9b, 10b, 11b and 12b.

The most significant technologies in this section are: powerful bio-inspired feature extraction, fast PCA processing technique [28] and adaptive shape feature architecture and autonomous feedback mechanism. The combination of them enables the real time adaptive capability, which can be an effective technique to solve the object recognition in changing environment.

b. Adaptive Color Feature Extraction

In the heterogeneous environment, the visible shape of an object itself may not be sufficient for unambiguous recognition when a) the object change violates our assumption of "smooth" transition, making the adaptive approach ineffective; and b) ambiguous environment where several similar type of objects exist. To deal with this issue, the color feature of the object can be used to extend a new dimension of search. Perhaps, the color segmentation may be an optimal technique to be used to assist the shape feature for object recognition.

To use the object color feature for helping in object recognition in a friendly and correlated environment, color segmentation is a suitable approach to narrow down the search space. Several color segmentation algorithms have been proposed in the literature [37-46]. The majority of existing color segmentation techniques is based on Red-Green-Blue (RGB) classification in combination with complex data processing. This processing includes spatial clustering to separate targets from the background, multi-histogram analysis [47], Bayesian methods [48], and various neural network approaches [49]. Algorithms have been proposed based on edge detection in color space [40] and normalized color space [41].

In real time applications, algorithms with fast learning and adaptive capabilities are needed. In the previous report [52], we have shown that the Cascade Error Projection (CEP) algorithm provides an excellent tool in fast and simple learning.

1. CEP Algorithm approach

Here we employ a modified CEP [52] feed-forward neural network architecture with inherent learning capability. The CEP neural network architecture is illustrated in Figure 13. Shaded squares and circles indicate frozen weights; a square indicates calculated weights, and a circle indicates learned weights.

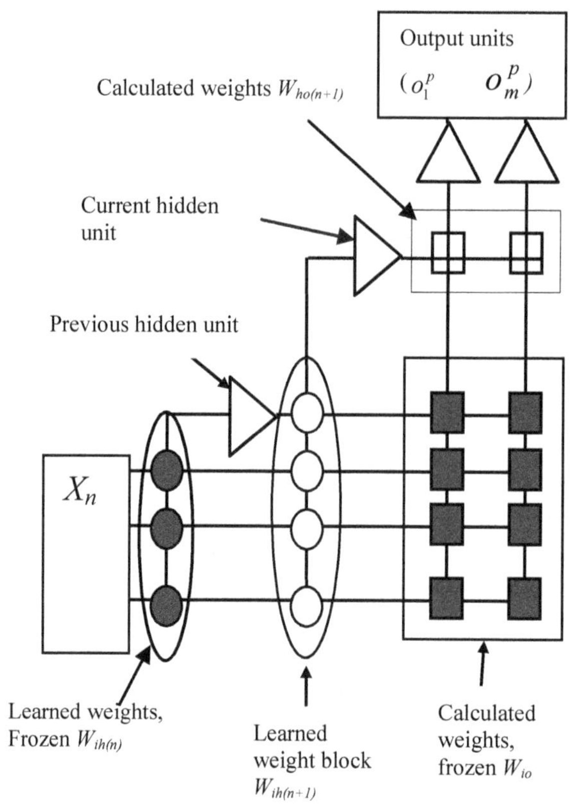

Figure 13: The architecture of CEP includes inputs, hidden units, and output units. The shaded circles or squares indicate the learned or calculated weight set that is computed and frozen. A circle indicates that perceptron learning is applied to obtain the weight set, and a square indicates that the weight set is deterministically calculated.

Mathematical background:
The energy function is defined as:

$$E(n+1) = \sum_{p=1}^{P} \{f_h^p(n+1) - \frac{1}{m}\sum_{o=1}^{m}(t_o^p - o_o^p)\}^2 \tag{5}$$

The weight update between the inputs (including previously added hidden units) and the newly added hidden unit is calculated as follows:

$$\Delta w_{ih}^p(n+1) = -\eta \frac{\partial \mathrm{E}(n+1)}{\partial w_{ih}^p(n+1)} \tag{6}$$

and the weight update between hidden unit *n+1* and the output unit o is

$$w_{ho}(n+1) = \frac{\sum_{p=1}^{P} \varepsilon_o^p f'^p_o f_h^p(n+1)}{\sum_{p=1}^{P} [f'^p_o f_h^p(n+1)]^2} \tag{7}$$

$$\text{with } f(x) = \frac{1-e^{-x}}{1+e^{-x}}$$

The notations used are defined as follows:
m is the number of outputs, P is the number of training patterns.
Error $\varepsilon_o^p = t_o^p - o_o^p(n)$; where $o_o^p(n)$ is the output element o of the actual output $o(n)$ for training pattern p, *and* t_o^p is the target element $\boldsymbol{o}$ for training pattern p. n indicates the number of previously added hidden units .
$f'^p_o(n) = f'^p_o$ denotes the output transfer function derivative with respect to its input.
$f_h^p(n+1)$ denotes the transfer function of hidden unit $n+1$.
The CEP algorithm is processed in two steps:

Single Perceptron learning which is governed by equation (6) to update the weight vector $W_{ih}(n+1)$

When the single Perceptron learning is completed, the weight set $W_{ho}(n+1)$ can be obtained by the calculation governed by equation (7).

Details of the CEP algorithm and convergence analysis can be found in [51].

2. Adaptive Architecture approach

The most challenging aspect in color segmentation is when the light intensity and resolution are dynamically changing. It is easily recognized that the initial knowledge used to train the network will have very little effect at a new location and therefore will need to be updated through learning of newly extracted data.

When the network that has acquired current color knowledge at time t0 is used to test the subsequent image at time t0+Δt, segmentation results from the image at t0+Δt need to be used to extract the training set to update the previous knowledge of the network at time t0. This process of repeatedly segmenting and updating is performed as long as the operation is going on.

While the process of segmenting and updating are desired characteristics of an adaptive processing, the issue that needs addressing is how often such updates are necessary.

The problem with infrequent updates, however, is that the network may not interpolate easily based upon new images, resulting in the newly segmented data may be insufficient for training. To find the optimal sampling rate, we realize Δt must be "sufficiently small" and will depend upon the moving velocity and other environmental changes. This issue will become significant in the actual design and development of a realistic system.

3. Simulation results

For this experiment, we selected a series of color images (Fig. 14) from a JPL-recreated terrestrial "MARS-YARD". To segment an image into multiple color-defined classes, we need to classify each pixel. To that effect, we used a pixel to be classified and its immediate neighbors to form a 3x3 sub-window as the input training pattern (thus each input pattern has 27 elements from 3x3 of RGB pixel). Based on a previous study, the 3x3 RGB input pattern was found to be the optimal size in comparison to using a single RGB input, a 5x5 RGB sub-window, or a 7x7 RGB sub-window [45]. In this study, our objective was to segment the image into three groups: "Rock1", "Rock2", and "Sand". The topology of our network is a 27x5x3 cascading architecture neural network, having 27 inputs, 5 cascaded hidden units, and 3 output units.

In Figure 14a of 3:00PM image, we sampled and collected 408 patterns for training data, 588 patterns for cross validation, and 1200 patterns for testing. With these sample sets, the learning is completed with 91% correct in training, 90% correct in validation, and 91% correct in testing.

After training was performed, the segmented output of the original image in Figure 14a is shown in Figure 14b.

Figure 14a: Original image that was collected at 3:00PM.

Figure 14b: The segmented image after training

With the knowledge acquired in Figure 14a, the network is tested with the image input shown in Figure 15a collected at 4:00PM. The output result is shown in Figure 15b (no intermediate adaptation step was performed). Figure 15c is output result with the network acquired from the intermediate knowledge through adaptive learning, is clearly better than the results shown in Fig. 15b.

Figure 15a: Original image that was collected at 4:00PM

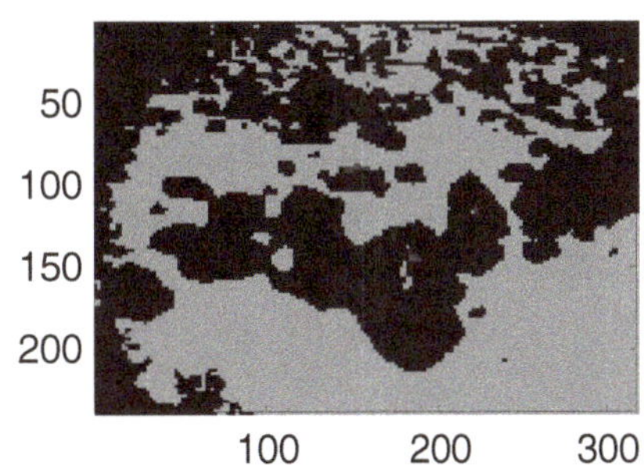

Figure 15b: The segmented image based on the previous training data at 3:00PM (no intermediate adaptive step)

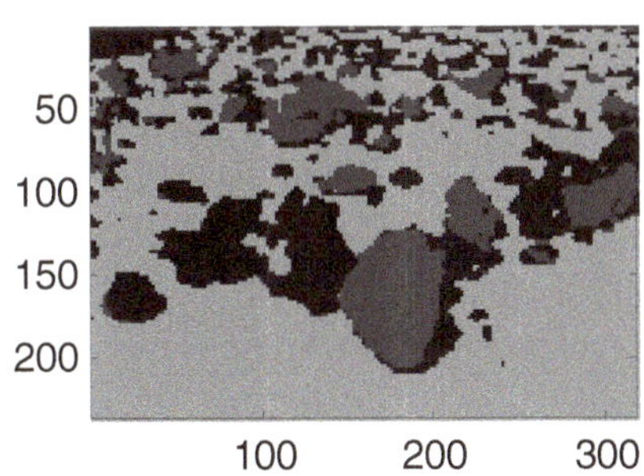

Figure 15c: The segmented image based on the intermediate adaptive step

In a similar manner, the original image shown in Figure 16a was collected at 5PM. Figure 16b is the segmented image with the previous training set at 4PM and Figure 16c is the segmented image with intermediate adaptive step. Once again, the classification results in Fig. 16c are significantly better than those in Fig. 16b.

Figure 16a: Original image that based was collected at 5:00PM

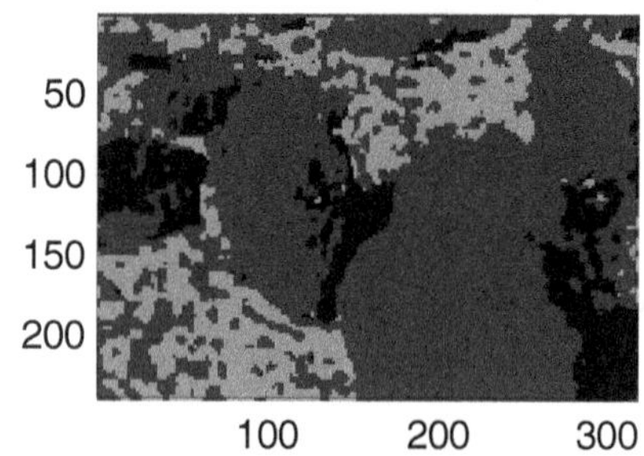

Figure 16b: The segmented image on the previous training data at 4:00PM (no intermediate adaptive step)

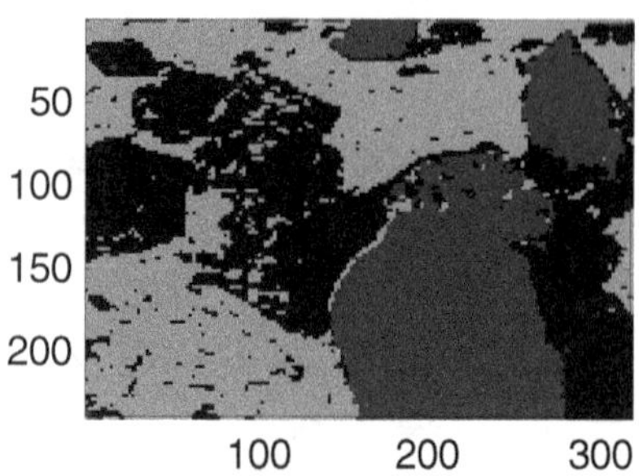

Figure 16c: The segmented image based on the intermediate adaptive step

Based on the aforementioned results, we conclude that the adaptive technique is needed to obtain better segmentation when the environment is changing rapidly.

3. Integration of Color and Shape Feature Based Approach

As discussed earlier, the role of shape feature itself carries substantial information for recognition while the color feature may help to ease the unexpected event which may often occur in the heterogeneous environment. The integration of shape and color feature based approach strengthens our technique to meet the object recognition challenges of the real world applications.

1. Architecture

As shown in Figure 17, the initial knowledge of shape and color feature is obtained in (6) and (7) from the statistical image collection (single or multiple)

and extraction stored in box (2) and box (4) respectively. At time t, when tested and found the objects (5) based on the initial shape and color feature, the object or region of object located (5) is extracted to get new features (6) and is segmented to new color feature (7) at time t+Δt. Depending on the discrepancy of the previous and new features are found whether the feature space (color and shape) will be updated and stored in box 2 and 4.

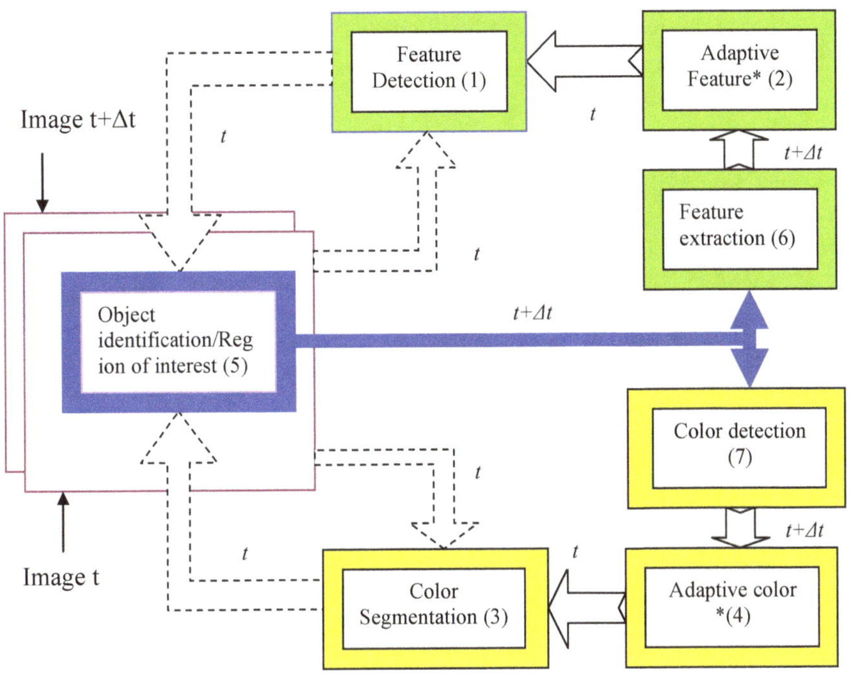

Figure 17: Color and Shape Feature Feedback Adaptive architecture
*Adaptive feature and color box (2 and 4) can serve as feature database for interested objects. From the previous shape adaptive features (2) and color adaptive features (4), the object identification and region of interest (5) are found from the image at time t via feature detection (1) and color segmentation (3). The region of interest will be used as sample data to adapt a new shape and color features for the image at time t+Δt.

2. *Application*

To demonstrate its effectiveness in autonomous object recognition system, the simulation of object recognition in a real world scenario is presented. In this simulation study, we only focus on the object recognition of the moving target. This work can be distinguished from the conventional tracking technique due to the fact that our technique is completely based autonomous object detection. Generally tracking an object in a scene may not require object detection, but it is tremendously advantageous making the tracking more robust.

3. Simulation Results

In this study, we recorded a video at the shopping mall in Southern California. We used this application to demonstrate that our bio-inspired object recognition based on shape and color can be effective to recognize an interested person in the heterogeneous environment where the single technique e.g., shape or color itself exposed its difficulties to perform effective recognition. Moreover, this demo also demonstrates the mechanism and architecture of the autonomous adaptive system to enable the realistic system for the practical use in the future.

For the first frame shown in Figure 18a, we select a window to include an object shown in Figure 18b.

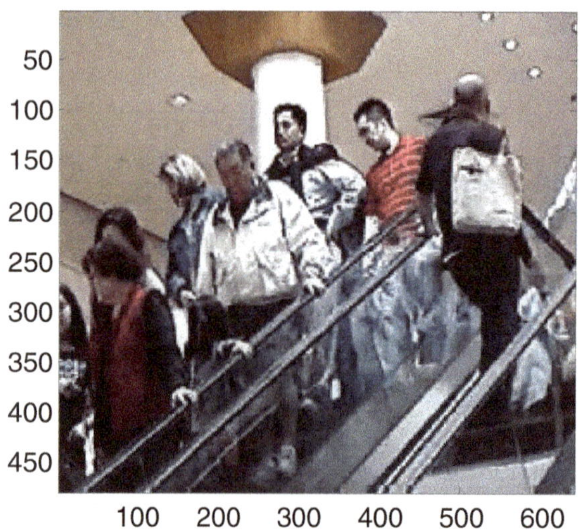

(18a)

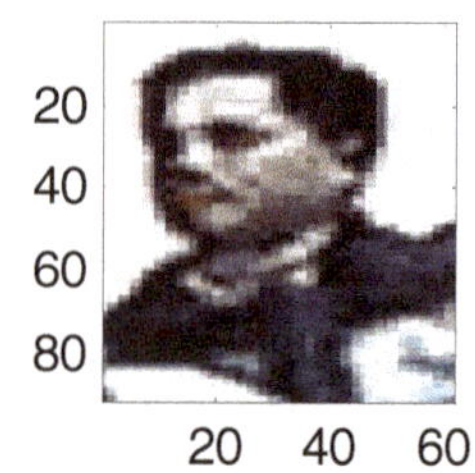

(18b)

Figure 18: a) Initial scene; and b) selected object

When the object is selected, its image is processed to extract the shape and color features of it. For this application, we only used one shape feature and three color features: face skin, hair and background.

The initial knowledge such as shape, face skin color, hair color, and background of the object are obtained and used to validate the new scene. When done, the architecture and feedback mechanism are applied and described in the architecture section. The sequences of the video and its recognition are shown below:

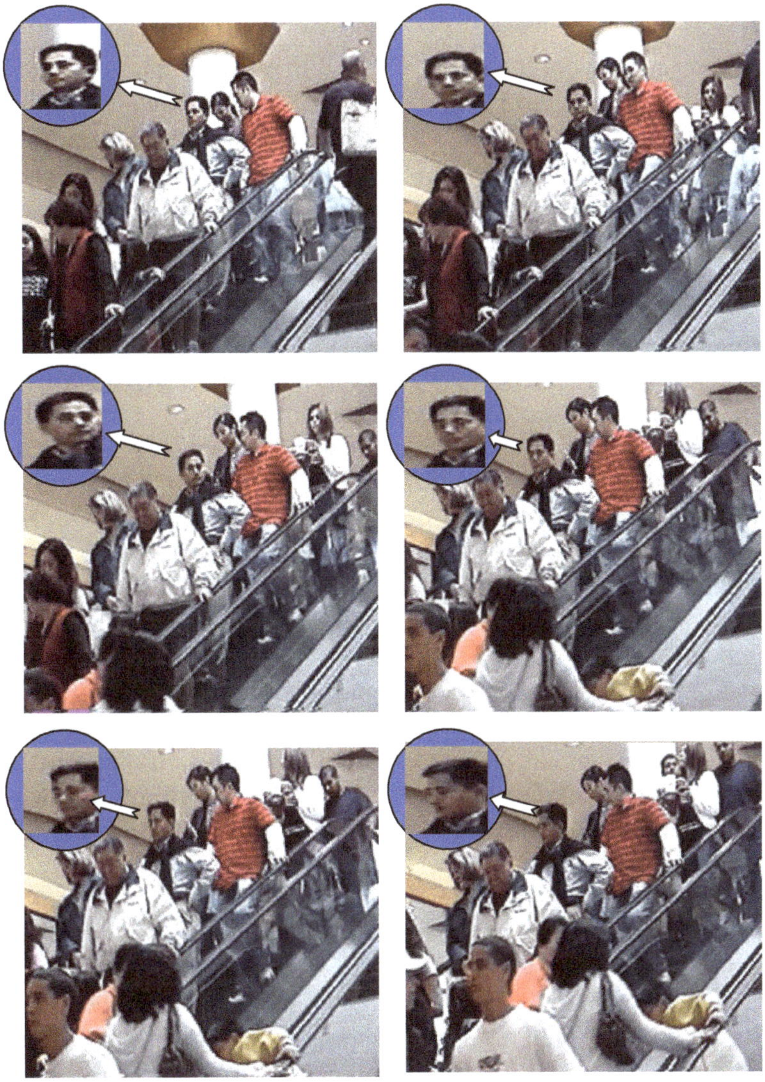

Figure 19: Video sequences and its object recognition (in the upper left corner) in the dynamic environment

This series of images contains the variation of the dynamics in heterogeneous environment and the person of interest has moved to show the rotation and scaling of his face in a changing background. This represents a good test for our approach.

Clearly, our simulation results have shown (Fig. 19) that object recognition in the dynamic scene is indeed successful in our representative experiment.

4. Discussion

In a static object recognition situation, we have attempted to modify a generalized constraint so as to maximize the recognition capability of the same object from different perspectives. The expectation of having very limited information (one input object image) but with a large recognition capability is very attractive, but it is based on a very minimal mathematical foundation and modeling, except for bio-related modeling of which we have insufficient knowledge at this point. Practically and theoretically, a local extrapolation of the input object under local rotation and image scaling should be expected, but not much more. If so, our simulation results suggest that our approach is very effective as shown in Table I (second row result) with no false alarms, full recognition, and a correct *locally* rotated image of the object. This result can be viewed as sampling the finite resolution space of a 3-D modeling approach. This 3-D modeling can then be used to solve the 3-D rotation of an object.

In pattern recognition, scaling of an object is also an issue of great concern and difficult to deal with optimally. From a bio-inspired perspective, the bipolar layer can be modeled to obtain the scaling ratio between the reference object and the potential object. The bio-inspired bipolar layer as an edge feature of the object is used in Cellular Neural Network (CNN) by Chua and Werblin [53].

A Visual Cortex-like approach [1] has employed multiple window resolution and rotation to deal with it effectively; however, this approach may not be an optimal one. In our study of static objects, we have only relied on the bio-inspired shape feature extraction BSFE technique based on saccadic eye movements and vertebrate retina (horizontal layer), but no scaling and rotations are employed; however, the scaling and rotation tolerances are observed in our simulation shown in Figures 4,5,6 and 19.

We plan to extend our study [54] to determine what resolution of rotation can be tolerated. From this study, one can effectively employ the multiple rotation scheme to cover 3-D rotation. For example, if a 30° rotation can be tolerated, it will require 144 (12x12) rotation positions to cover a complete 360° view of a 3-D rotation.

For applications related to dynamic image processing e.g., object recognition search or visual surveillance and monitoring, processing time is always a crucial parameter. It is hoped that obtained results are within a valid time frame; otherwise, they may not be useful. Specially, our approach is not based on local spatio-temporal information, but it is rather global spatio-temporal information of an object. Hence, required processing time is more extensive. JPL-SPCA has demonstrated that it required ~ 2 min. vs. 6 hours (using MATLAB) to extract the feature of an array, containing 100 (saccadic) positions each with a view of 100x100 pixel window. For the static object search, the time consumed is ~ 0.6 sec to completely examine whether the object (112x83 pixel) is in an image and

~3.4 sec for an object (90x60 pixel) search in an image for a dynamic search. These timing numbers above are upper-bound values and can be reduced by parallel processing, skip frame or heuristic reduced search window etc. to meet the timing budget for each application. Moreover, custom-hardware approaches might improve the speed performance tremendously.

In the dynamic object recognition, the local rotation and scaling can be tolerated and adaptively learned from our adaptive architecture and autonomous feedback mechanism.

5. Conclusions

In this study, we have focused on a bio-inspired features approach, an adaptive architecture and autonomous mechanism, and an effective processing technique to enable a real time adaptive visual system. The preliminary simulation results are encouraging and exceed benchmark comparisons for the same data set [55]. The system-level architecture is proposed that may lay a foundation for image based real-time search and intelligent video applications.

This study clearly invites more work on the BSFE technique in conjunction with color segmentation, not only based on the simulation, but also a careful modeling and theoretical analysis.

From a physiological and psychological view point, there is evidence [56,57] that the brain can process a sub-shape feature as a part-based shape [58] e.g., mouth, eyes etc in the face, for object recognition [59,60]. This is also evident in the saccadic eye movement for a part-based object [48]. We plan to pursue this direction of research in the future.

We will also extend our study to compare it with the state of the art e.g., visual cortex-like [1]. We also believe that the combination of our bio-inspired model and the visual cortex-like method can be an exciting next step in moving closer to modeling the bio-visual system.

Acknowledgements

The research described herein was performed by the Jet Propulsion Laboratory, California Institute of Technology under contract with the National Aeronautics and Space Administration (NASA). The authors would like to thank Drs. Anil Thakoor and T. Daud for their discussions and suggestions.

References

1. Thomas Serre, Lior Wolf, Stanley Bileschi, Maximilian Riesenhuber, and Tomaso Poggio, "Robust object recognition with Cortex-like mechanisms", *IEEE Transactions On Pattern Analysis And Machine Intelligence*, Vol. 29, No. 3, 411–426, March 2007.

2. K. Fukushima, Neocognitron: a self-organizing neural network model for a mechanism of pattern recognition unaffected by shift in position. *Biol. Cybern.*, 36: 193–202, 1980.
3. Y. LeCun, L. Bottou, Y. Bengio, and P. Haffner, 'Gradient-based learning applied to document recognition'. *Proc. IEEE*, 86(11): 2278–2324, 1998.
4. M. Riesenhuber and T. Poggio,' Hierarchical models of object recognition in cortex'. *Nat. Neurosci.*, 2: 1019–1025, 1999.
5. D.I. Perrett and M. Oram, Neurophysiology of shape processing. *Image Vis. Comput.*, 11: 317–333, 1993.
6. G. Wallis and E.T. Rolls, A model of invariant object recognition in the visual system. *Prog. Neurobiol.*, 51:167–194, 1997.
7. S.J. Thorpe, "Ultra-rapid scene categorization with a wave of spikes", In: *Proc. Biologically Motivated Comput. Vis.*, Tubingen, Germany, 2002.
8. Y. Amit, Y. and M. Mascaro, "An integrated network for invariant visual detection and recognition". *Vision Res.*, 43(19): 2073–2088, 2003.
9. M. Weber, M. Welling, P. Perona, Unsupervised learning of models for recognition. *IEEE ECCV* 2000 18–32, 2000.
10. D. Arathorn, Map-seeking circuits in visual cognition: A computational mechanism for biological and machine vision. Stanford (California): *Stanford University Press.* 240 p., 2002.
11. B. W. Mel, "SEEMORE: Combining color, shape, and texture histogramming in a neurally inspired approach to visual object recognition', *Neural Computation*, 9, pp. 777-804, 1997.
12. Eitan Sharon, Meirav Galun, Dahlia Sharon, Ronen Basri and Achi Brandt, *Nature*, Vol 442.17, 810-813, August 2006.
13. Theo Gevers and Arnold W. M. Smeulders, "PicToSeek: Combining color and shape invariant features for image retrieval" *IEEE Transactions On Image Processing*, Vol. 9, No. 1, pp. 102-119, January 2000.
14. A.K. Jain and A. Vailaya, "Image retrieval using color and shape" *Pattern Recognition*, vol. 29, pp. 1233–1244, 1996.
15. G. Dyer and L. Chittka, "Biological significance of distinguishing between similar colours in spectrally variable illumination: bumblebees (*Bombus terrestris*) as a case study" *Journal of Comparative Physiology A: Neuroethology, Sensory, Neural, and Behavioral Physiology based on distinguishable color.* pp. 105-114, Volume 190, Number 2 / February, 2004.
16. E. Warrant, D.-E. Nilsson *Invertebrate Vision,* Cambridge University Press, 2006.
17. C. Kock and J.L. Davis, editors, *Large-Scale Neuronal Theories of The Brian* The MIT Press, Second Edition, 1995.
18. A.L. Yarbus, *Eye Movements and Vision.* Plenum Press, New York, 1967.
19. J.K. O'Regan, Eye movemnets and reading. *In: Eye Movements and Their Role in Visual and Cognitive Processes,* Kowler, E. (ed.). Elsevier, Amterdam, 1990.
20. V.N. Vapnik, *Statistical Learning Theory*, Wiley New York, 1998.
21. Y.S. Abu-Mostafa, "Hints and the VC Dimension" *Neural Computation*, 5:278–288, MIT Press, March 1993.
22. R.O. Duda and P.E. Hart, Pattern Classification and Scene Analysis, New York, Wisley, 1973.

23. L.O. Chua, L. Yang, "Cellular neural networks," *IEEE International Symposium on Circuits and Systems*, pp. 985 –988, vol.27-9, June 1988.
24. R. Baddeley and P.J.B. Hancock, *Proc. R. Soc*, B246, 219-223, 1991.
25. R. Baddeley, *Nature*, Vol. 381. 13, 560-561, 1996.
26. F. Werblin and A. Jacobs, "Using CNN to unravel space-time processing in the vertebrate retina," *CNNA-94 Third IEEE Intl' Workshop on Cellular Neural Networks and Their Applications*, Rome Italy, Dec. 18-21, 1994.
27. A.K. Jain, R.P.W. Duin, and J. Mao "Statistical pattern recognition: A Review", *IEEE Transactions on Pattern Analysis and Machine Intelligence*, Vol. 22, No. 1, January 2000.
28. T.A. Duong "Real time on-chip sequential adaptive principal component analysis for data feature extraction and image compression", *GOMAC Tech-03*, Vol. I, pp. 460-464, Tampa, Florida, 31 March - 3 April, 2003.
29. T.A. Duong and V.A. Duong, "Sequential principal component analysis -an optimal and hardware-implementable transform for image compression", *To appear in The 3rd IEEE Space Mission Challenges for Information Technology in* Pasadena, California, 19 - 23 July, 2009.
30. S. Bannour and M.R. Azimi-Sadjadi, "Principal component extraction using recursive least squares learning," *IEEE Trans. On Neural Networks*, Vol. 6, No. 2, March 1995.
31. A.K. Jain, *Fundamentals of Digital Image Processing.* Prentice Hall Information and System Sciences Series, 1989.
32. E. Oja and J. Karhunen, "On stochastic approximation of the eigenvectors and eigenvalues of the expectation of a random matrix." *J. Math. Anal. Appl.*, vol. 106, pp. 69-84, 1985.
33. S. Haykin, Neural Network-A Comprehensive Foundation. New York: Macmillan, 1994.
34. P. Baldi and K. Hornik, "Learning in linear neural networks: A survey," *IEEE Trans. Neural Networks*, Vol. 6, pp. 837-857, 1995.
35. L. Xu, "Least mean square error reconstruction principal for self-organizing neural-nets," *Neural Networks*, 6, pp. 627-648, 1993.
36. S. I. Amari. Natural gradient works efficiently in learning. *Neural Computation*, 1998.
37. M. Celenk. "A color clustering technique for image segmentation." *Computer Vision Graphics Image Process. Graphical Models Image Process.* 52. pp. 145-170, 1990.
38. J. Lui and Y.H. Yang, "Multiresolution color image segmentation," *IEEE Trans. Patt. Anal. Mach. Intel.* 16, 689-700, 1994.
39. E. Littman and H. Ritter. "Adaptive Color Segmentation – A Comparison of Neural and Statistical Methods." *IEEE Trans. Neural Net.* Vol. 8, No. 1, pp. 175-185, 1997
40. F. Perez and C. Kock, "Toward color image segmentation in Analog VLSI: algorithm and hardware," *Int. J. Comp. Vision* 12:1,17-24,1994.
41. G. Healy. "Segmenting images using normalized color." *IEEE Trans. Syst. Man. Cyber.* 22. 1. pp. 64-73, 1992.
42. H. Okii, et al. "Aotumatic color segmentation method using a neural network model for stained images," *IEICE Trans. Inf. Syst.* (Japan) Vol. E770D No. 3, pp. 343-350, March 1994.

43. T. Nakamura and T. Ogasawara, "On-line visual learning method for color image segmentation and object yracking," *Proc. of The 1999 IEEE/RSJ Intelligent Robots and Systems*, pp. 222-228, 1999.
44. P.H. Batavia and S. Singh, "Obstacle detection using adaptive color segmentation and color stereo homography," *Proc. Of the 2001 IEEE International Conference on Robotic and Automation*, Seoul, Korea, May 21-26, 2001.
45. E. Fiesler, S. Campbell, L. Kempen, and T. Duong, "Color sensor and neural processor on one chip." International Symposium on Optical Science, Engineering, and Instrumentation, *Proc. of the SPIE, vol. 3455 'Applications and Science of Neural Networks, Fuzzy Systems, and Evolutionary Computation'*, pp. 214-221, 1998.
46. T.A. Duong, "Real time adaptive color segmentation for Mars landing site identification," *Journal of Advanced Computational Intelligence and Intelligent Informatics* (Japan), pp. 289-293, Vol.7 No.3, 2003.
47. M. Celenk. "A color clustering technique for image segmentation." *Computer Vision Graphics Image Process. Graphical Models Image Process*. 52. pp. 145-170, 1990.
48. J. Lui and Y.H. Yang, "Multiresolution color image segmentation," *IEEE Trans. Patt. Anal. Mach. Intel.* 16, 689-700, 1994.
49. E. Littman and H. Ritter. "Adaptive color segmentation – a comparison of neural and statistical methods." *IEEE Trans. Neural Net.* Vol. 8, No. 1, pp. 175-185, 1997.
50. F. Perez and C. Kock, "Toward color image segmentation in analog VLSI: algorithm and hardware," *Int. J. Comp. Vision* 12:1, 17-24,1994.
51. H. Okii, et al. "Aotumatic color segmentation method using a neural network model for stained images," *IEICE Trans. Inf. Syst.* (Japan) Vol. E770D No. 3, pp. 343-350, March 1994.
52. T. Duong and A. Stubberud, "Convergence analysis of cascade error projection-an efficient learning algorithm for hardware implementation", *IJNS*, Vol. 10, No. 3, pp. 199-210, June 2000.
53. F. Werblin, T. Roska, and L.O. Chua, "The analogic cellular neural network as a Bionic Eye", *International J. Circuit Theory and Applications*, Vol. 23, pp.541-569, 1995.
54. Tuan A. Duong, "3-D object modeling based on JPL bio-inspired approach," In preparation.
55. O. Jesorsky, K. Kirchberg, R. Frischholz. "Robust face detection using the Hausdorff Distance" In J. Bigun and F. Smeraldi, editors, *AVBPA 2001*, pp. 90-95. Springer, 2001.
56. S.E. Palmer, "Hierarchical structure in perceptual representation", *Cogn. Psychol.* **9**, 441–474 1977.
57. E. Wachsmuth, M.W. Oram, and D.L. Perrett, "Recognition of objects and their component parts: responses of single units in the temporal cortex of the macaque". *Cereb. Cortex* **4,** 509–522 1994.
58. D.D. Lee and H.S. Seung, "Learning the parts of objects by non-negative matrix factorization". *Nature* **401**, 788–791, 1999.

59. I. Biederman, "Recognition-by-components: a theory of human image understanding". *Psychol. Rev.* **94**, 115–147, 1987.
60. S. Ullman, High-Level Vision: Object Recognition and Visual Cognition, *MIT Press*, Cambridge, MA, 1996.

Part 2

Basic Methods in Computer Vision and Image Processing

CHAPTER 2.1

PROBABILISTIC GRAPHICAL MODELS AND THEIR APPLICATIONS IN COMPUTER VISION

Lei Zhang*, Yan Tong† and Qiang Ji‡

*3662 Moore St.
Los Angeles, CA 90066
leizhang2009@gmail.com

Rensselaer Polytechnic Institute,
110 8th St., Troy, NY 12180
†tongyan@ge.com
‡qji@ecse.rpi.edu

Probabilistic graphical models (PGMs) have become increasingly popular statistical modeling tools for effectively addressing many real-world problems. Both directed graphical models such as Bayesian Networks (BNs), Dynamic Bayesian Networks (DBNs), and undirected graphical models such as Markov Random Fields (MRFs) and Conditional Random Fields (CRFs) have been widely used to solve computer vision problems. In this chapter, we briefly introduce the basic concepts and theories of different types of PGMs, including BNs, DBNs, MRFs, and CRFs. We further show their applications in two important computer vision problems: facial expression recognition and image segmentation. Specifically, we first illustrate how to employ a DBN to capture the static and dynamic relationships between action units for facial expression recognition. We then introduce a hybrid PGM to model the heterogeneous relationships between different image entities (regions, edges, junctions, etc.) for effective image segmentation. These applications demonstrate the powerful capability of PGMs for flexibly modeling the real-world problems in a rigorous statistical framework and for solving the computer vision problems systematically through a probabilistic inference.

1. Introduction

Probabilistic Graphical Models (PGMs) are graphical representation of the probabilistic relationships among random variables. They represent the marriage between the graph theory and the probability theory. PGMs provide a natural tool for dealing with two problems that occur throughout applied mathematics and engineering: uncertainty and complexity. As a modeling and inference tool, PGMs use an intuitive, powerful and flexible graphical structure to represent the joint probability distributions of multiple random variables. The nodes in the graph correspond to the random variables. The links between the nodes represent the probabilistic relationships among the random variables. The graph structure encodes the conditional dependence and independence relationships among the random variables, which follow the Markov property. PGMs also systematically account for the uncertainties and provide means to propagate uncertainties in a principled manner.

Given the PGM model, the joint (or conditional) probability distributions can be factored as the product of a set of local functions defined on a subset of random variables. Such factorization greatly simplifies the modeling of the multivariate joint probability distributions. Based on the probability theory, many principled algorithms have been developed to exploit the graph structure to efficiently calculate the marginal probability or the joint probability of interest. Probabilistic inference can be then performed to estimate the optimal states of the unobserved random variables.

2. Basic Types of Probabilistic Graphical Models

Different types of PGMs can be roughly grouped into three classes: the undirected PGMs such as MRFs,[1,2] CRFs[3,4] and Factor Graph (FG);[5] the directed acyclic PGMs such as Bayesian Networks (BNs)[6,7] and Dynamic Bayesian Networks (DBNs);[8,9] and the hybrid PGMs such as chain graph,[10,11] which consist of both directed links and undirected links. These PGMs represent and encode different conditional independence relationships on the random variables. The undirected PGMs are often used to capture the spatial correlations or mutual dependencies between random variables, while the directed PGMs often capture the casual relationships among random variables.

Each PGM consists of two parts: the qualitative part and the quantitative part. The qualitative part means the graph representation of a PGM. Each PGM is represented by a graph $G = (V, E)$, where V denotes all the vertices (nodes) in the graph and E denotes all the edges (links) connecting these nodes. The nodes correspond to the random variables and the links represent the probabilistic relationships among the random variables. For undirected PGMs, E only consists of undirected links. In contrast, for directed PGMs, E only consists of directed links. A hybrid PGM, on the other hand, can consist of both directed links and undirected links.

The quantitative part in a PGM means the joint probability distribution represented by such a model. Due to the local conditional independence of the random variables, the joint probability distribution can usually be factored as the product of local potential functions or local conditional probabilities, which simplifies the parametrization and estimation of the joint probability distribution for performing probabilistic inference.

2.1. *Bayesian Networks (BNs)*

BN models the joint probability distribution of a set of random variables through a directed acyclic graph (DAG) $G = (V, E)$. The directed links E explicitly model the probabilistic relationships including causality among these random variables. BN is a powerful tool to model humans' knowledge and has been widely used in medical diagnostic system, bioinformatics, medicine, information retrieval, computer vision, decision support system, data fusion, and law.[6,7]

Figure 1 shows a simple BN model. Through its local Markov property, the graph structure of a BN encodes the conditional independence relationships among the nodes. Basically, a node is conditionally independent of its non-descendants given its parent values. These conditional independence can be easily read from the directed graph structure using the so-called d-separation rules.[6] For example, X_1 and X_5 are conditionally independent given X_2 and X_4 since all paths from X_1 to X_5 are blocked by X_2 and X_4.

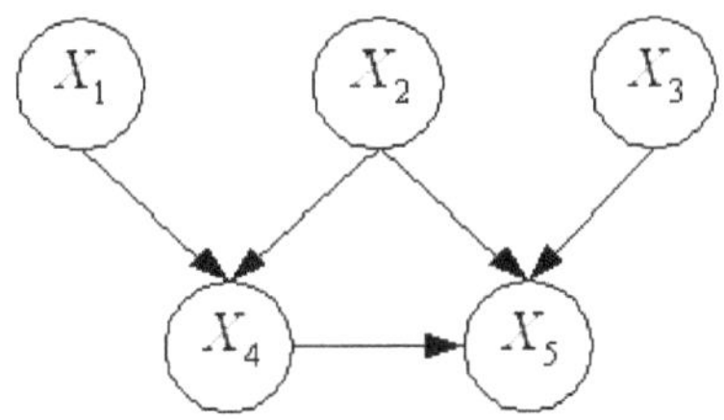

Fig. 1. A simple Bayesian Network with five nodes.

Based on these conditional independence relationships, the joint probability of all random variables in a BN can be factored as a product of the conditional probability of each node given its parent configuration, i.e.

$$P(\{X_i\}) = \prod_i P(X_i|pa(X_i)) \tag{1}$$

where $pa(X_i)$ denotes the parents of X_i. This factorization simplifies the parametrization, parameter learning and probabilistic inference in the BN.

2.2. *Dynamic Bayesian Networks (DBNs)*

DBN is an extension of both BN and the traditional dynamic systems such as Hidden Markov Models (HMM) and the Kalman filter. It is often used to model the dynamic relationships among the random variables at different time slices. Therefore, DBN is a very powerful tool to model the temporal evolution of a dynamic system. Different from the static BNs, DBN includes both intra-slice and inter-slice links. While the intra-slice links capture the static relationships among the random variables, the inter-slice links capture their temporal evolution and interactions. DBN has been applied to many problems requiring a dynamic system, including object tracking,

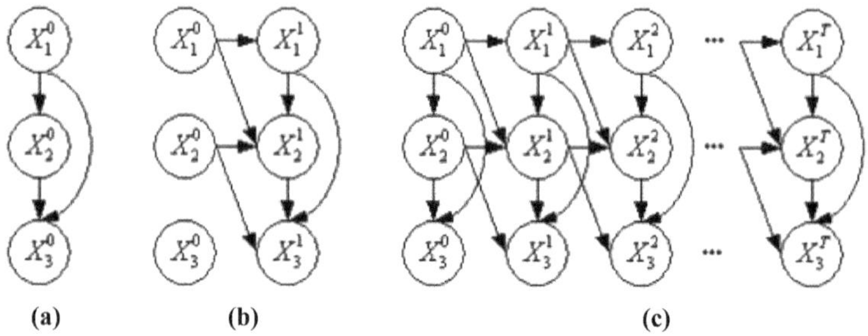

Fig. 2. A simple DBN[12] consists of (a) static network G_0 and (b) transition network $G_\rightarrow$ that defines the dynamic relationships of three random variables X_1, X_2, X_3. (c) The corresponding unrolled DBN for T time slices.

activity recognition, speech recognition, bioinformatics, decision support system, etc.

In a discrete example, let X^t represents a set of random variables at the time slice t. A DBN is defined as a model $DBN = (G, \Theta)$, where G represents the graphical model structure, and Θ represents the model parameters, i.e., the conditional probability tables (CPTs) for all nodes. Assuming the dynamic system follows the first-order stationary Markovian property, the DBN model can be defined by a pair of networks $(G_0, G_\rightarrow)$: (1) the static network G_0, as shown in Figure 2(a), captures the static distribution over all variables X^0; and (2) the transition network $G_\rightarrow$, as shown in Figure 2(b), specifies the transition probability $P(X^{t+1}|X^t)$ for all time slices.[12]

The conditional independence relationships among random variables in a DBN can be easily read from the DBN graph structure, following the same d-separation rules used in static BNs. Given the DBN model and using the Markov property, the joint probability over all variables $X^0, ..., X^T$ can be factored by "unrolling" the DBN into an extended static BN, as shown in Figure 2(c). Based on the conditional independence, the joint probability represented by a DBN is computed as follows:

$$P(\mathbf{X}^0, \mathbf{X}^1, \cdots, \mathbf{X}^T) = P_{G_0}(\mathbf{X}^0) \prod_{t=0}^{T-1} P_{G_\rightarrow}(\mathbf{X}^{t+1}|\mathbf{X}^t) \tag{2}$$

where $P_{G_0}(X_0)$ represents the joint probability of all variables in the static network G_0. In addition, $P_{G_\rightarrow}(X^{t+1}|X^t)$ represents the transition probability and can be further decomposed as follows:

$$P_{G_\rightarrow}(\mathbf{X}^{t+1}|\mathbf{X}^t) = \prod_{i=1}^{M} P_{G_\rightarrow}(X_i^{t+1}|pa(X_i^{t+1})) \tag{3}$$

where M is the number of random variables in each slice and $pa(X_i^{t+1})$ represents the parent values of the ith variable X_i^{t+1} in the transition network $G_\rightarrow$.

2.3. *Markov Random Fields (MRFs)*

MRFs are generative models that use undirected links to capture the mutual dependence among a set of random variables. MRFs model the interactions among the random variables through the local interactions within a selected neighborhood system N. In the undirected graph representation of a MRF, a node X_j is a neighbor of the node X_i if and only if there is an undirected link between them. This type of interaction is modeled as a Markovian prior probability distribution on the random variables. The Markov property in MRFs means a random variable X_i is conditionally independent of other random variables, given the random variables within its neighborhood, i.e.

$$P(X_i|X\backslash X_i) = P(X_i|X_{N_i}) \tag{4}$$

where N_i denotes the neighborhood of X_i.

If a MRF models the joint probability distribution (JPD) of some label variables X and some observations Y, the JPD in the MRF can be formulated as a Gibbs distribution based on the Hammersley-Clifford theorem,[13] i.e.

$$P(X,Y) = \frac{1}{Z}\exp(-\sum_{c\in C} V_c(X_c,Y_c)) \tag{5}$$

where c denotes a clique that is a fully connected sub-graph. $V_c(X_c, Y_c)$ is an energy function defined on the subset of random variables in the clique c. Z is a normalization constant (i.e. the partition function), which can be calculated by marginalizing over all random variables in the MRF.

The size of the clique has crucial influence on the computational complexity in a MRF model. In practice, researchers usually use low-order neighborhood systems. In a MRF model with up to the pairwise cliques, the joint probability distribution can be modeled as

$$P(X,Y) = \frac{1}{Z}\prod_i \phi(X_i,Y_i) \prod_{<i,j>,j\in N_i} \psi(X_i,X_j) \tag{6}$$

where $\phi(X_i, Y_i)$ is the unary potential since it is associated with only one label variable X_i. Similarly, $\psi(X_i, X_j)$ is called the pairwise potential.

MRFs have been widely used in computer vision, pattern recognition, and image processing domains to model the correlations between image entities, object parts, etc. Figure 3 shows a MRF example for image labeling. Given the observations $Y = \{y_i\}$, the MRF model can be used to infer the image labels $X = \{x_i\}$ by probabilistic inference.

2.4. *Conditional Random Fields (CRFs)*

CRF is a discriminative model that directly models the posteriori probability distribution of a set of random variables X given the observations Y. It is first proposed for natural language processing[3] and then extended to other domains such as object

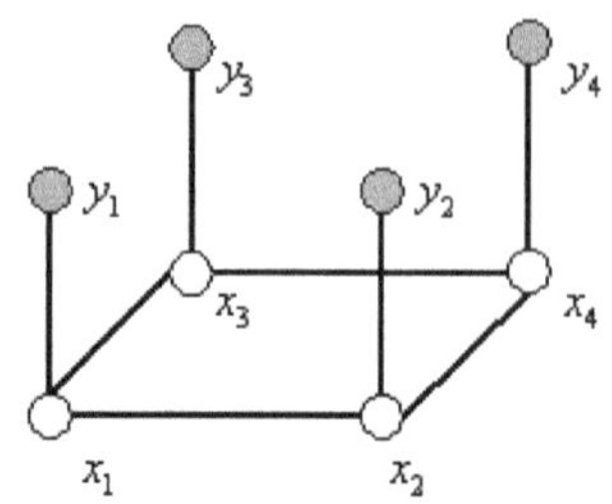

Fig. 3. A simple MRF example for image labeling.

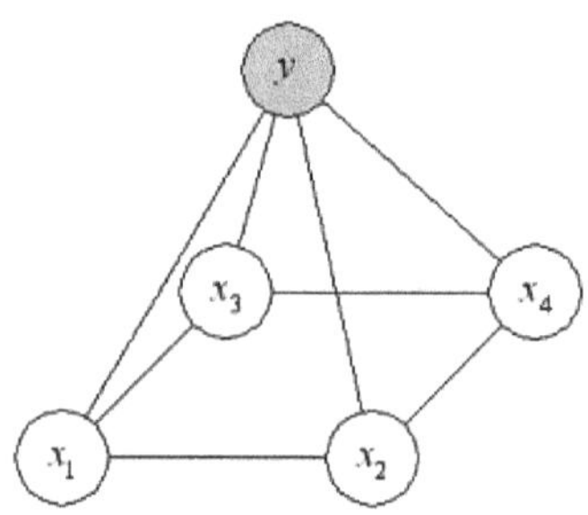

Fig. 4. A simple CRF example for image labeling.

detection and recognition, image segmentation,[4,14–16] etc. CRF assumes the posteriori probability distribution follows the Markov property and can be represented by the product of potential functions, i.e.

$$P(X|Y) = \frac{1}{Z(Y)} \prod_c \exp(-A_c(X_c|Y)) \tag{7}$$

where $A_c(X_c|Y)$ is an energy function defined on a set of random variables X_c in the clique c, conditioned on the observations Y. Note that the energy function also depends on the observations Y. The partition function Z is a function of Y and can be calculated by marginalizing over X.

Figure 4 shows a simple CRF for image labeling. Compared to the MRF model in Figure 3, the potential functions are all related to the observations. Compared to the MRF, a CRF directly models the posteriori probability distribution that will be used for the classification or labeling purpose. It does not model the probability distribution of the observations and focuses directly on the classification problem given the observations. In addition, CRF naturally considers the discontinuity of the labels since the interactions between the labels can be automatically adjusted according to the observations. For example, if the observations at two adjacent pixels are significantly different, it is less likely that their labels should be similar to each other. The pairwise potentials can therefore discourage both pixels are labeled as the same. These characteristics make CRF quickly extended to many domains.

2.5. *Two Fundamental Issues in using PGMs*

Before using a PGM for an application, two fundamental issues have to be addressed: parameter learning and probabilistic inference.

Parameter learning aims at automatically learning the model parameters in PGMs. Given a set of complete training data D (i.e. all random variables are labeled), we can use Maximum Likelihood Estimation (MLE) to learn the model parameters Θ, i.e.

$$\Theta^* = \arg\max_{\Theta} P(D|\Theta) \tag{8}$$

If we also know the prior probability distribution of the parameters Θ, we can use Bayesian estimation to learn the model parameters, i.e.

$$\Theta^* = \arg\max_{\Theta} P(\Theta|D) = \arg\max_{\Theta} P(D|\Theta)P(\Theta) \tag{9}$$

where $P(\Theta)$ is the prior probability distribution of the parameters.

Parameter learning in a directed graphical model with complete training data can usually be reduced to fitting a local conditional probability distribution, which often results in closed-form solutions. For discrete Bayesian Networks, parameter learning can be further simplified as a counting problem, i.e. estimating the frequency of a joint parent-child configuration.

We can also perform parameter learning in undirected PGMs based on the Maximum Likelihood Estimation. However, parameter learning in undirected PGMs is usually more difficult because the partition function Z couples all parameters. It is generally difficult to estimate the partition function or its derivatives w.r.t the model parameters, which makes the parameter learning difficult.

Nevertheless, several approximate parameter learning approaches have been proposed before to deal with learning in undirected PGMs.[17–22] These approximate learning methods include:

(1) Iterative Proportional Fitting (IPF)[23]
(2) Pseudo-likelihood approximation[17]
(3) Saddle point approximation[18]
(4) Piecewise training[19]
(5) Approximate ML estimation by pseudo-moment matching[20]
(6) Parameter learning with Gibbs/MCMC sampling[2]
(7) Learning with contrastive divergence[21]
(8) Max-Margin parameter learning[22]

The key idea behind these approaches is to avoid exactly calculating the log-partition function or its derivative. For example, the pseudo-likelihood method[17] approximates the likelihood using the product of local pseudo-likelihood, which can be calculated by assuming the neighboring variables are known. Given this assumption, the pseudo-likelihood can be locally normalized, which is easy to compute. This property makes it a very efficient parameter learning method with some loss of accuracy. We refer the readers to the original references for detailed discussions of these approximate parameter learning approaches.

Probabilistic inference in a PGM aims at calculating the marginal probability, inferring the maximum a posterior (MAP) solution, or inferring the most probable explanation (MPE) given the evidence. For example, in a MRF model, the MAP inference estimates the optimal states of labels X given the observations Y as follows

$$X^* = \arg\max_X P(X|Y) = \arg\max_X \frac{P(X,Y)}{P(Y)} = \arg\max_X \frac{P(Y|X)P(X)}{P(Y)} \tag{10}$$

where $P(Y|X)$ is the likelihood model and $P(X)$ is the prior probability distribution of labels.

The inference algorithms for PGMs can be roughly divided into the exact inference methods and the approximate inference methods. For the singly connected tree structured PGMs, the Pearl's belief propagation (BP),[6] the sum-product algorithm and max-product algorithm[5] can find the exact solution. All these methods are operating based on the local message-passing principle. They iteratively update the messages between adjacent nodes until convergence and then infer the optimal states for a single node or a set of nodes. However, these methods cannot guarantee finding the exact solution for a PGM with undirected loops. For the PGM with undirected loops, the Junction Tree algorithm[24,25] provides a general way to find the exact solution. Its basic idea is converting the PGM with loops into a simple tree structured PGM by grouping a set of nodes. Exact inference can then be performed on this converted PGM using the belief propagation method.[6] However, the conversion of a large graphical model into a Junction Tree is difficult and time-consuming. Researchers therefore often resort to approximate inference methods to deal with complex graphical models.

The approximate inference methods include loopy belief propagation (LBP), the Mean Field method, variational algorithms (VA),[26,27] etc. The loopy belief propagation method directly applies the belief propagation method[6] to a graphical model with undirected loops. It can usually produce an approximate inference result but sometimes might not guarantee the convergence of the message-passing process. In contrast to the Junction Tree method, LBP is computationally more efficient. In some sense, LBP treats the inference accuracy for the computational efficiency.

The variational algorithms solve the inference problem by changing it into an optimization problem. In VA, we confine ourselves to a family of tractable distributions, and approximate the original complex probability distribution to the one that minimizes the Kullback-Leibler (KL) divergence between them. The Mean Field algorithm, a special case of VA, takes a family of fully factorized distributions. One can then update the approximate marginal probability at each node based only on the neighboring nodes. It assumes the original probability distribution is peaked on several modes. The inference is then simplified as finding these static modes, which sometimes has analytic solutions. Besides the Mean Field approximation, more elaborate approximations of the true probability distribution have been developed.[28,29]

3. Facial Expression Recognition using A Dynamic Bayesian Network

In this section, we will present an application of a Dynamic Bayesian Network (DBN) in facial expression recognition. We are particularly interested in recognizing facial actions from spontaneous facial displays. Spontaneous facial action is characterized by rigid head movements and nonrigid facial muscular movements. More importantly, it is the spatiotemporal interactions among the rigid and nonrigid facial motions that produce a meaningful and natural facial display. Recognizing this fact, we employ a DBN to simultaneously and coherently capture rigid and nonrigid facial motions, their spatiotemporal dependencies, and their image measurements. More details on facial action modeling can be found in.[12,30,31] Facial action recognition is accomplished through DBN inference by systematically integrating measurements of facial motions with the facial action model.

3.1. *Facial Action Modeling*

3.1.1. *Overview of the Facial Action Model*

A spontaneous facial action consists of rigid head motion, nonrigid facial deformations, and their spatiotemporal interactions. In the view of facial action analysis from 2D images, 2D facial shape encodes the information of both rigid and nonrigid facial motions since it is generated from three hidden causes: head pose, 3D facial shape, and nonrigid facial muscular movements. 3D facial shape characterizes an intrinsic property of a subject. Nonrigid facial muscular movements represented by a set of facial Action Units (AUs) as defined in[32] cause 3D shape deformation of the facial surface. Head pose produces the changes in the position and shape of the 2D face on the projected 2D image. In addition, through various computer vision techniques, we can obtain measurements for these hidden causes.

Based on the causal relationships shown in Fig. 5a, we propose to use a Bayesian network (BN) to model the statistical dependencies among rigid motion, nonrigid motions, and their interactions through the 2D facial shape. Furthermore, the nodes in a BN can be grouped into hidden nodes and measurement nodes. 3D facial shape denoted by S_{3D}, facial muscular movements represented by a set of AUs, 3D head pose denoted by $Pose$, and 2D facial shape denoted by S_{2D} are modeled as hidden nodes, and their true states can be inferred from their measurements through the model. And so, we associate each hidden node with a measurement node representing the observation of the corresponding hidden node, as shown in Fig. 5a.

Given the model, facial action recognition is to find the optimal states of head pose and AUs by maximizing the joint probability of pose and AUs given the measurements as follows:

$$Pose^*, \mathbf{AU}^* = \arg \max_{Pose, \mathbf{AU}} p(Pose, \mathbf{AU} | O_{pose}, O_{\mathbf{AU}}, O_{S_{3D}}, O_{S_{2D}}) \tag{11}$$

where **AU** is the set of all target AUs; O_{pose}, $O_{\mathbf{AU}}$, $O_{S_{3D}}$, and $O_{S_{2D}}$ denote the measurements of head pose, AUs, 3D facial shape, and 2D facial shape, respectively. In the next several sections, we gradually show how the relationships in Fig. 5a can be expanded and enriched.

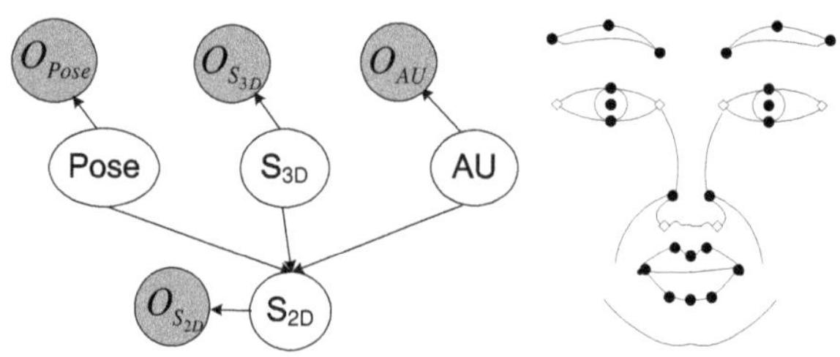

Fig. 5. (a) A graphical model to represent the causal relationships among elements of a facial action, where the shaded nodes represent the measurements of the connected hidden nodes. (b) Facial feature points on a frontal view face: the black dots represent the local feature points, whereas the white dots represent the global feature points.

3.1.2. *Modeling the Interactions Between Rigid and Nonrigid Facial Motions*

In this research, the shape of a 3D face can be represented by a vector of 28 facial feature points, which are located around each facial component (e.g., mouth, eye, nose, and eyebrow), as shown in Fig. 5b. Given a 3D face, the deformation of a 2D facial shape reflects the action of both head pose and facial muscular movements. Specifically, head pose and facial muscular movements may affect different sets of facial feature points. As a result, the facial feature points are further divided into global feature points, which are relatively invariant to facial muscular movements, and local feature points, which are affected by both head pose and facial muscular movements.

(1) Modeling Rigid Motion with 2D Shape

The 2D global shape denoted by $\mathbf{S}_g$ is the projection of the 3D global feature points on the image plane. Therefore, the 3D facial shape governs the shape of the 2D global shape, whereas the 3D head pose controls both the position and shape of $\mathbf{S}_g$. This causal dependency can be represented by a directed link from head pose/3D facial shape to $\mathbf{S}_g$, as shown in Fig. 6a.

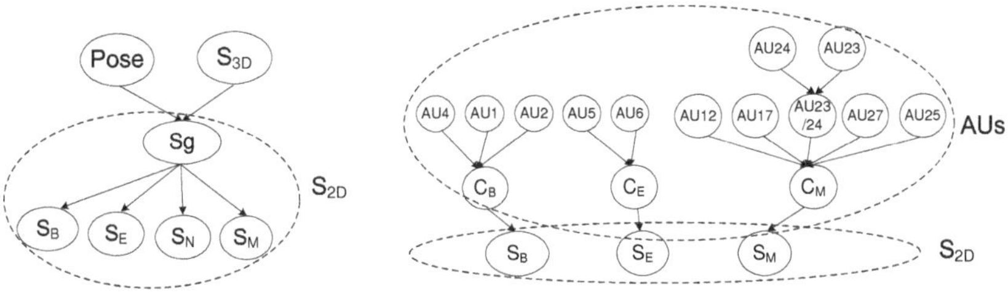

Fig. 6. (a) The head pose and 3D facial shape directly affect the 2D global shape $\mathbf{S}_g$; whereas the 2D global shape $\mathbf{S}_g$ controls the variation of each 2D local facial component shape. (b) The relationships between the facial muscular movements (AUs) and the 2D local facial component shapes through the intermediate nodes (C_B, C_E, and C_M).

Furthermore, the 3D local facial shape can be partitioned into four components: eyebrows, eyes, nose, and mouth. The corresponding 2D local facial component shape is indirectly affected by the head movement through the 2D global shape $\mathbf{S}_g$. Given $\mathbf{S}_g$, the center of each 2D local facial component can be roughly estimated, independent of the head pose. For example, the center of eye can be determined, given the eye corners, which are part of the global shape. Hence, this causal relationship can be represented by a directed link from $\mathbf{S}_g$ to each 2D local facial component shape, as shown in Fig. 6a.

(2) Modeling the Relationships between Nonrigid Motion and 2D Local Facial Component Shapes

The nonrigid facial muscular movements produce significant changes in the 3D shape of the facial component. These 3D facial muscular movements can be systematically represented by AUs as defined in.[32] For example, activating AU27 (mouth stretch) will produce a widely open mouth; and activating AU4 (brow lowerer) makes the eyebrows lower and pushed together.

Since the 3D shape of each facial component is determined by the related AUs, the 2D local facial component shape is controlled by the AUs, besides rigid head movement. We can model such causal relationship by directly connecting the related AUs to the corresponding facial component. For instance, AU1 (inner brow raiser), AU2 (outer brow raiser), and AU4 (brow lowerer) control eyebrow movements, and can be connected to the eyebrow. However, directly connecting all related AUs to one facial component would result in too many AU combinations, most of which rarely occur in the daily life. Thus, only a set of common AUs or AU combinations is sufficient to control the shape variations of the facial component. As a result, a set of intermediate nodes (i.e., "C_B", "C_E", and "C_M" for eyebrow, eye, and mouth, respectively) are explicitly introduced to model the correlations among AUs and to reduce the number of AU combinations. For example, "C_M" has 8 states, each of which represents a common AU or AU combination controlling mouth movement. Fig. 6b shows the modeling of the relationships between nonrigid facial motions (AUs) and the local facial component shapes.

3.1.3. *Modeling Semantic and Dynamic Relationships among AUs*

So far, we have modeled relationships between rigid head motion and nonrigid facial motions through the 2D facial shapes, but have not discussed the modeling of spatial and temporal dependencies among AUs, which are crucial to create a meaningful and natural facial display. Tong et al[12] demonstrated that there are two important spatial and semantic relationships among the AUs: co-occurrence and mutually exclusive relationships. Furthermore, AUs also show strong temporal dependencies to represent different naturalistic facial behaviors. Nishio et al.[33] have shown that when the mouth moves prior to the eye movement, a smile expression is mostly interpreted as a smile of enjoyment. On the contrary, when the eyes move prior to the mouth movement, it is mostly interpreted as a miserable smile.

Generally speaking, there are two types of temporal dependencies among AUs: intra-dependency and inter-dependency. Intra-dependency characterizes the self evolution of an AU, while inter-dependency captures temporal dependencies among different AUs, i.e., an AU will be activated following the activation of another AU. For example, in a spontaneous smile, AU6 (cheek raiser) is activated in an average of 0.4 second after the activation of AU12 (lip corner puller);[34] after both the actions reach their apexes simultaneously, AU6 is relaxed before AU12 is released. Furthermore, due to the variability among individuals and different contexts, the dynamic relationships among AUs are stochastic. Therefore, systematically capturing the spatiotemporal dependencies among AUs and incorporating them into facial action recognition process is especially important for spontaneous facial behaviors. In this work, we propose to use a DBN to model and learn both the spatial and dynamic dependencies among AUs.

As discussed in Section 2.2, a DBN can be defined by a pair $(G_0, G_{\rightarrow})$ representing the static and transition networks. In this work, G_0 is used to capture the spatial dependencies among AUs, while $G_{\rightarrow}$ is used to capture and model the intra-dependency and inter-dependency for AUs in two adjacent time slices. The intra-dependency is modeled as an arc linking AU_i at time $t-1$ (AU_i^{t-1}) to that at time t (AU_i^t) and depicts how a single AU develops over time. The inter-dependency is modeled as an arc from AU_i^{t-1} to AU_j^t and represents the pairwise dynamic dependency between two different AUs. Following our previous work,[12,30] we can learn the structures of both (G_0 and $G_{\rightarrow}$) (Fig. 7), given observed data D consisting of a set of image sequences.

3.1.4. *Modeling Facial Motion Measurements*

In the facial action model, head pose, AUs, 3D facial shape, 2D global facial shape, 2D local facial component shapes, and the intermediate nodes are hidden nodes. To acquire the measurements for the hidden nodes except for the intermediate nodes, we first perform face and eye detection and detect the 28 feature points on neutral face with frontal view. The measurement of 3D facial shape is obtained by personal-

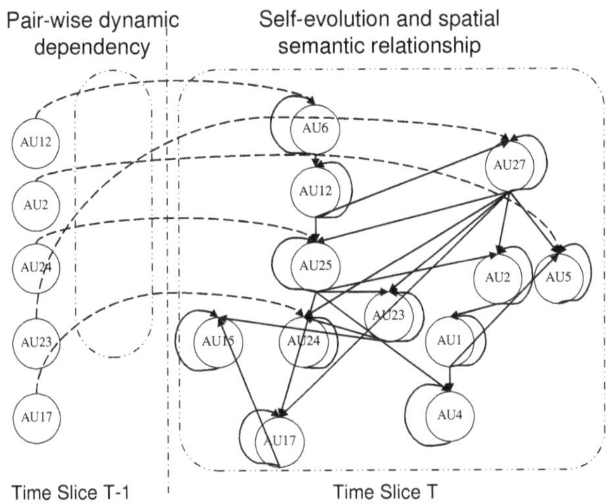

Fig. 7. The learned DBN by the proposed algorithm for AU modeling. The self-arrow at each AU node indicates the intra-dependency. The dashed line with arrow across two time slices indicates the pairwise inter-dependency between different AUs. The links among AUs in the same time slice represent the spatial and semantic relationship among AUs.

izing a trained generic 3D shape model. Afterward, the measurements of 2D global shape and local facial component shapes are obtained by tracking the 28 feature points in each frame. Furthermore, based on the personalized 3D facial shape and the tracked global feature points, three face pose angles (i.e., pan, tilt and roll) are estimated by using a technique similar to.[35] The continuous pan angle is further discretized into frontal, left, and right face pose measurement to model the left-right head rotation. Given the normalized face image, we also extract the measurement for each AU based on Gabor wavelet-based feature representation and AdaBoost classification similar to the work in.[36] To incorporate these measurements into the facial action model, a node (shaded) is introduced to represent each measurement. In addition, a link between a hidden node and its measurement is introduced to model the measurement accuracy.

3.2. *A Complete Facial Action Model and Its Parametrization*

3.2.1. *A Comprehensive Facial Action Model*

Now we have extended and enriched the causal relationships in Fig. 5a to a complete DBN model for facial action understanding, as shown in Fig. 8. Specifically, the interactions among nonrigid facial muscular movements are characterized by the static links among AUs in the same time slice and the temporal links among AUs across consecutive time slices. The 2D shape deformations of the facial components are controlled by both the head pose through the 2D global shape and the related AUs through the intermediate nodes. In this way, the interactions between head pose (rigid motion) and the AUs (nonrigid motions) are indirectly modeled through their relationships with 2D global and local facial component shapes. Finally, the facial motion measurements are systematically incorporated into the model through the shaded nodes. This model, therefore, completely characterizes the spatiotemporal dependencies between rigid and nonrigid facial motions and accounts for the uncertainties in facial motion measurements.

3.2.2. *Model Learning And Parametrization*

Given the model structure shown in Fig. 8, we need to define the states for each node and, then, learn the model parameters associated with each node. For each node without parents, it is parameterized by its prior probability. For the continuous node X with discrete/continuous parents, it is characterized by Conditional Probabilistic Distribution (CPD) and defined as $p(X|pa(X))$; whereas for the discrete node with discrete parents, it is characterized by the CPT defined as $p(X|pa(X))$ similar to CPD.

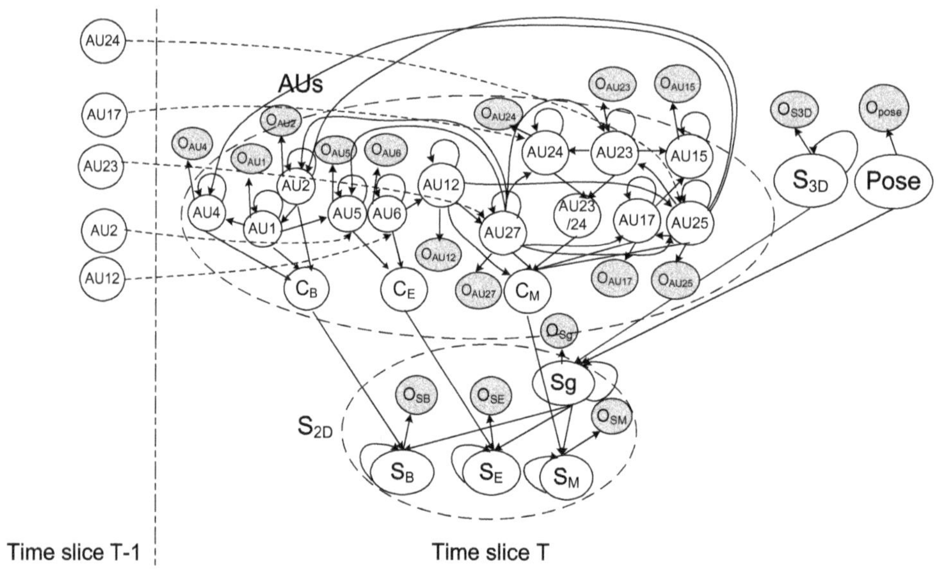

Fig. 8. The DBN for spontaneous facial action understanding. The shaded node indicates the observation for the connected hidden node. The self-arrow at the hidden node represents its temporal evolution from time $t-1$ to time t. The link from AU_i at time $t-1$ to AU_j ($j \neq i$) at time t indicates the dynamic dependency between different AUs.

Specifically, head pose is represented by three states: left, frontal, and right in the proposed system. The prior information of the pose $p(Pose)$ can be learned from training data. 3D facial shape S_{3D} is characterized by a continuous 3D shape vector consisting of 28 feature points from neutral faces. 2D global shape $\mathbf{S}_g$ is represented by a continuous shape vector consisting of global feature points, whereas 2D local shape of the jth facial component is represented by a continuous shape vector $\mathbf{S}_{l_j}$ containing the corresponding local feature points. Each AU has two discrete states representing the presence/absence state of the AU. The intermediate nodes (i.e., C_B, C_E, and C_M) are discrete nodes, each state of which represents a specific AU/AU combination related to a facial component.

Given $Pose = k$ and $\mathbf{S}_{3D} = s_{3D}$, the CPD of $\mathbf{S}_g$ can be defined as follows:[37]

$$p(\mathbf{S}_g = s_g | Pose = k, \mathbf{S}_{3D} = s_{3D}) = (2\pi)^{-\frac{d_g}{2}} |\Sigma_{gk}|^{-\frac{1}{2}} \\ exp(-\frac{(s_g - \mathbf{W}_{gk} * s_{3D} - \mu_{gk})^T \Sigma_{gk}^{-1} (s_g - \mathbf{W}_{gk} * s_{3D} - \mu_{gk})}{2}) \tag{12}$$

where d_g is the dimension of $\mathbf{S}_g$; μ_{gk}, $\mathbf{W}_{gk}$, and Σ_{gk} are the mean shape vector, regression matrix, and covariance matrix, respectively. Based on the conditional independence in the BN, we can learn μ_{gk}, $\mathbf{W}_{gk}$ and Σ_{gk} locally from the training data consisting of $\mathbf{S}_g$, S_{3D}, and head pose.

For each local shape component node (i.e., S_B, S_E, S_N, and S_M), its CPD is parameterized as a Gaussian distribution. Given the training data of each 2D local shape, $\mathbf{S}_g$, and the related AUs, we can learn its CPD locally. The CPT $p(C_i|pa(C_i))$ for each intermediate node is manually specified based on the data analysis. The CPD/CPT of each measurement node is learned to reflect the measurement accuracy of the computer vision technique. The CPTs for all the AUs including the static and dynamic links among AUs are learned simultaneously in the local DBN model, as shown in the Fig. 7. Finally, we learn the transition probability $p_{B_\rightarrow}(X_i^{t+1}|pa(X_i^{t+1}))$ for the other temporal links of the DBN.

3.2.3. *Facial Action Inference*

Once the measurement nodes are observed, we can infer the facial action by maximizing the joint probability of pose and AUs given the measurements. Let $Pose^t$ and $AU^t_{1\cdots N}$ represent the nodes for $Pose$ and N target AUs at time t. Given the available evidence until time t: $\mathbf{O}_{S_{3D}}$, $\mathbf{O}^{1:t}_{Pose}$, $\mathbf{O}^{1:t}_{S_g}$, $\mathbf{O}^{1:t}_{S_{l_{1\cdots L}}}$, $\mathbf{O}^{1:t}_{AU_{1\cdots N}}$ for the measurements of the 3D facial shape, the head pose, the 2D global shape, the 2D local facial component shapes, and the AUs, respectively, where N is the number of target AUs and L is the number of local facial component shapes, the probability $p(Pose^t, AU^t_{1\cdots N}|\mathbf{O}_{S_{3D}}, \mathbf{O}^{1:t}_{Pose}, \mathbf{O}^{1:t}_{S_g}, \mathbf{O}^{1:t}_{S_{l_{1\cdots L}}}, \mathbf{O}^{1:t}_{AU_{1\cdots N}})$ can be factorized and computed via the facial action model by performing the DBN updating process as described in.[38] Then, the true joint states of $Pose$ and the AUs are inferred simultaneously over time by maximizing $p(Pose^t, AU^t_{1\cdots N}|\mathbf{O}_{S_{3D}}, \mathbf{O}^{1:t}_{Pose}, \mathbf{O}^{1:t}_{S_g}, \mathbf{O}^{1:t}_{S_{l_{1\cdots L}}}, \mathbf{O}^{1:t}_{AU_{1\cdots N}})$, given the measurements of the 3D face, head pose, the 2D global shape, the 2D local shapes, and the AUs through probabilistic inference.

3.3. *Experiments on Spontaneous Facial Action Recognition*

To demonstrate the system robustness for recognizing spontaneous facial action, the system is trained and tested on a spontaneous facial expression database, which consists of image videos collected through three sources: (1) Multiple Aspects of Discourse research lab at the University of Memphis;[39] (2) Belfast natural facial expression database;[40] and (3) videos obtained from the website. In these image sequences, the subjects are displaying various spontaneous facial expressions with natural head movements. For this study, all the image sequences are coded into AUs frame by frame. For each AU, the positive samples are chosen as the images containing the target AU at different intensity levels, while the negative samples are defined as the images without the target AU. For training the 2D facial shapes, we also manually marked the 28 feature points on some images in the database.

Currently, the database contains 74 image sequences from 13 subjects, where 63 image sequences are used for training, and 11 image sequences for testing. In this work, we intend to recognize 12 target AUs, as shown in Fig. 8, which frequently occur in the database. Since we intend to recognize AUs under varying head pose, the AdaBoost classifiers are trained on frontal, left, and right face view, respectively, for each AU. Assuming that face pose varies smoothly over time, the AdaBoost classifier corresponding to the face pose estimated in the previous frame is selected to obtain the AU measurement for the current frame.

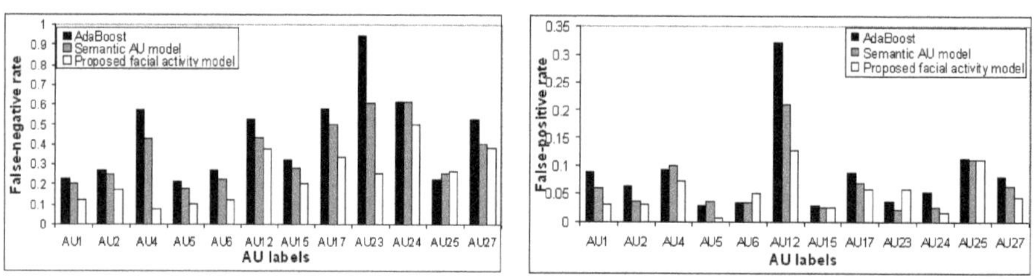

(a) (b)

Fig. 9. AU recognition results on spontaneous facial expression database: (a) average false negative rates, and (b) average false positive rates. In each figure, the black bar denotes the result using the AdaBoost classifier, the grey bar represents the result using the semantic AU model,[12] and the white bar represents the result using the proposed facial action model.

Fig. 9 shows the average AU recognition performance on the spontaneous facial expression database of using the AdaBoost classifiers alone, using the semantic AU model that focuses on modeling the static relationships among AUs and self-development of AUs as in,[12] and using the proposed facial action model, respectively. The AdaBoost classifiers achieve an average false negative rate (the ratio of the misclassified positive samples to the total positive samples) of 44% and an average false positive rate (the ratio of the misclassified negative samples to the total negative samples) of 8.58% for the 12 target AUs. By employing the semantic relationships among AUs, the semantic AU model decreases the average false negative rate to 36.5% and decreases the average false positive rate to 6.6%. With the semantic and dynamic relationships among AUs and the interactions between the rigid and nonrigid motions, the average false negative rate further decreases to 24.3%, and the false positive rate decreases to 5.3%.

Compared to the recognition results using the AdaBoost classifiers, the system performance is greatly improved by using the proposed facial action model for some AUs. For example, the false negative rate of AU23 (lip tighten) decreases from 94.4% to 25.9% with a moderate increasing of false positive rate (from 3.6% to 5.8%); the false negative rate of AU12 (lip corner puller) decreases from 53% to 37.8% with a significant drop of false positive rate (from 32% to 12.8%); and the false negative rate of AU2 (outer brow raiser) decreases from 26.9% to 16.9% with a decreasing of false positive rate from 6.2% to 3.1%. Furthermore, since the spontaneous facial action is often accompanied by natural head movements, only employing the relationships among AUs are not sufficient to deal with the facial appearance variations due to varying head pose. By incorporating the relationships between head pose and AUs through the 2D facial shapes, the complete facial action model further improves the system performance compared to the recognition results by using the semantic AU model.[12] That demonstrates the effectiveness and importance to model the interactions among the rigid motion and nonrigid motions and to model the spatiotemporal relationships among AUs for understanding spontaneous facial action.

4. Image Segmentation using A Hybrid Graphical Model

In this section, we introduce a hybrid probabilistic graphical model for figure/ground image segmentation. The hybrid graphical model combines a CRF part and a BN part to capture the heterogeneous relationships among multiple image entities, including regions, edges, and vertices, for effective image segmentation.[41,42]

4.1. *A Hybrid Graphical Model for Image Segmentation*

Our approach starts from an oversegmentation of the original image. In the oversegmentation, the image is split into smaller regions (also referred to as superpixels) with different labels. Several types of image entities are involved in the figure/ground image labeling problem: regions, edges, and vertices. There are hierarchical causal relationships among these entities. First, two adjacent regions with different labels intersect to form an edge segment between them. Second, multiple edges intersect to form a vertex (or a junction). In addition, adjacent regions have spatial correlations because they tend to have similar labels (i.e. local homogeneity in natural images). These relationships can be captured by different links between image entities, as illustrated by Figure 10. Specifically, we use directed links to capture the causality between different types of image entities and use undirected

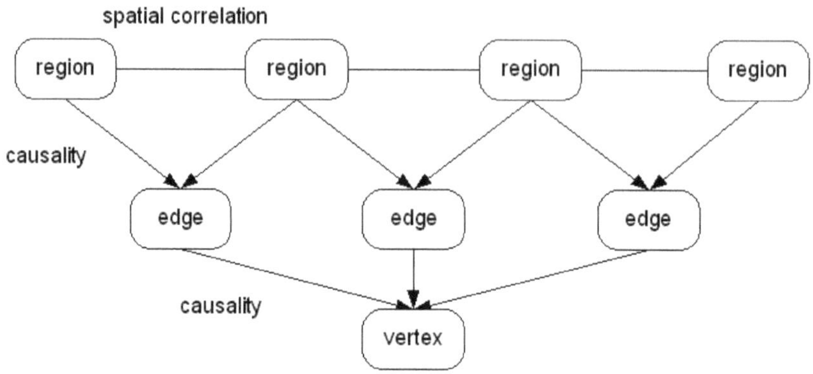

Fig. 10. The heterogeneous relationships among multiple image entities.

links to capture the spatial correlations between adjacent regions. We therefore formulate a hybrid graphical model for image segmentation.

The structure of the proposed graphical model is illustrated in Figure 11 for a synthetic example. It consists of two parts: the Conditional Random Field (CRF) part and the Bayesian Network (BN) part. The CRF part captures the spatial correlations of adjacent regions and performs region-based image segmentation, while the BN part models the hierarchical causal relationships among multiple image entities and performs edge-based image segmentation. The two parts are connected through the region nodes $\{y_i\}$ and form a hybrid probabilistic graphical model.

Given an image, it is first oversegmented to produce an edge map. We can use any suitable segmenter to produce this oversegmentation. Specifically, we use the Edgeflow-based Anisotropic diffusion method[43] for this purpose. The region node y_i correspond to each superpixel in the oversegmentation. We assume that the region labels $\mathbf{y} = \{y_i\}$ and the region observations $\mathbf{x} = \{x_i\}$ form a CRF. They are used to construct the CRF part of the model. Conditioned on the image observation $\mathbf{x}$, the region labels $\mathbf{y}$ follow the Markovian property, i.e. $P(y_i|\{y_j\}_{j\neq i}, x) = P(y_i|N(y_i), x)$, where $N(y_i)$ represents the spatial neighborhood of y_i.

The CRF image segmentation can be thought of as a labeling problem, i.e., to assign a label to the ith superpixel. In the figure/ground image segmentation problem, the node y_i has two possible values, i.e. +1 (the foreground) and -1 (the background). Let $\{y_i\}_{i=1}^n$ be the label variables corresponding to all superpixels, where n is the total number of superpixels in an oversegmentation. $\{x_i\}_{i=1}^n$ are the corresponding image measurements. x_i is represented as a local feature vector extracted from the image. Different types of cues such as intensity, color, and textures can be included in this feature vector for image segmentation. Using the CRF model, we can infer the label for each superpixel from the image measurements and decide whether adjacent superpixels should be merged.

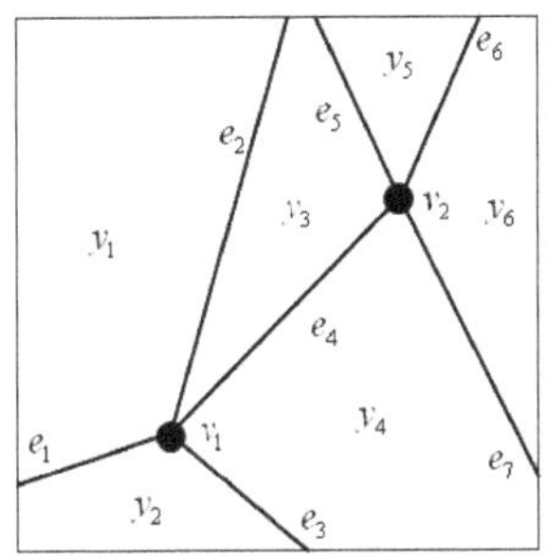

(a)

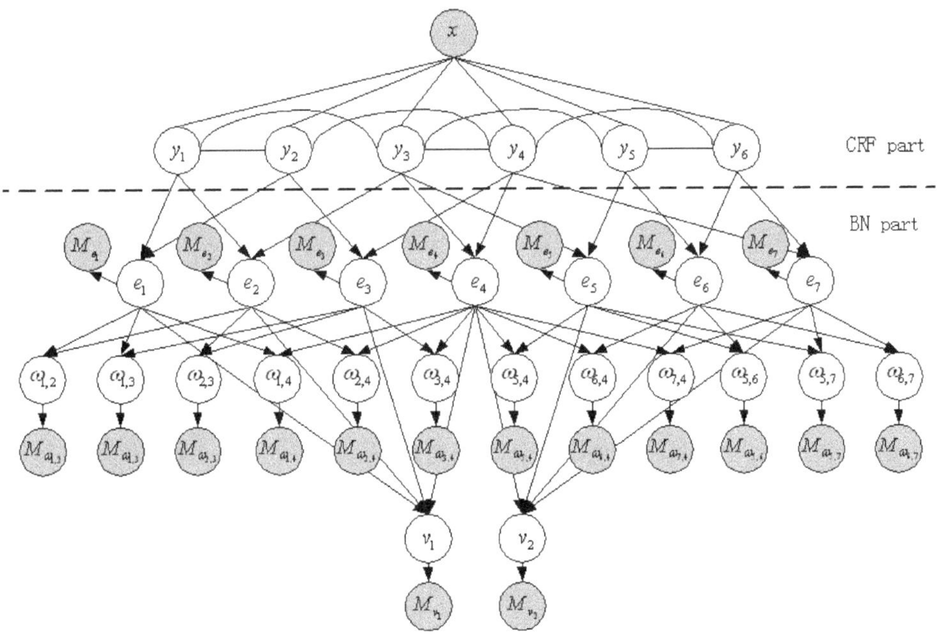

(b)

Fig. 11. A synthetic image oversegmentation (a) and its corresponding hybrid probabilistic graphical model (b).

Although the CRF model can effectively capture the spatial relationships among image entities, they cannot capture other causal relationships that naturally exist among different image entities. Due to this reason, a multi-layer BN is constructed to capture the causalities among the regions, edges, vertices, and their measurements. In Figure 11(a), the **e** nodes represent all the edge nodes in the constructed BN model. These edge nodes correspond to the edge segments in the oversegmentation edge map. An edge node has two values: +1 (on the boundary) or -1 (otherwise). The **v** nodes represent all the vertex nodes. They are automatically detected from the oversegmentation edge map. A vertex node also has two values:

+1 (a corner on the boundary) or -1 (otherwise). The nodes $\mathbf{M_e}$ and $\mathbf{M_v}$ represent the measurements of edges and vertices, respectively. We use the average gradient magnitude as the edge measurement and use the corner response from Harris corner detector to estimate the vertex measurements.[44] Given the BN model, the goal is to infer the edge labels from various image measurements and to decide if an edge segment belongs to the object boundary. Besides the aforementioned image entities, we also introduce another image entity, i.e. the angles $\omega = \{\omega_{j,s}\}$ between edges e_j and e_s, into the BN model. These angle nodes are used for enforcing the local smoothness constraint on connecting edges.[44] An angle node has two possible values: +1 (a very small angle) or -1 (otherwise). If the angle between two edges e_j and e_s is small, it is less likely that both edges belong to the object boundary. Such a constraint is captured by the relationships between an angle node and its parent edge nodes.

Combining the CRF model with the BN model yields a hybrid probabilistic graphical model as shown in Figure 11(b) that capture both the causal relationships and the spatial correlations for image segmentation. We therefore formulate a hybrid graphical model that can exploit both the region-based information and the edge-based information, and more importantly, the causal and non-causal relationships among random variables for image segmentation. Specifically, in Figure 11(b), the region nodes $\{y_i\}$ act as the parents of an edge node. The parents of the edge node correspond to the two regions that intersect to form this edge. The links between the parents and the child represent their causal relationships. If the parent region nodes have different labels, it is more likely that there is an object boundary (i.e., $e_j = 1$) between them. In this way, the proposed model systematically combines the CRF model and the BN model in a single probabilistic framework.

4.2. *Joint Probability Distribution*

The hybrid graphical model unifies the CRF part and the BN part through the causal relationships between the region nodes $\mathbf{y}$ and the edge nodes $\mathbf{e}$, as shown in Figure 11. The CRF part collects region-based information to infer the states of the superpixels. The BN part collects edge-based information and the imposed local constraints to infer the states of the edge segments. The hybrid model therefore integrates two kinds of complementary image segmentation principles in a unified framework. More importantly, it exploits both the spatial correlations of the region labels and the hierarchical causal relationships among regions, edges and vertices for image segmentation.

Based on the graphical structure of the hybrid graphical model, we can factor the joint probability distribution of all variables according to the global Markov property (cf. Chapter 3 of[45]). According to the graphical structure, $\mathbf{x}$ are conditionally independent of other random variables given $\mathbf{y}$. We can derive

$$P(\mathbf{y}, \mathbf{e}, \omega, \mathbf{v}, \mathbf{M_e}, \mathbf{M}_\omega, \mathbf{M_v}, \mathbf{x}) \tag{13}$$
$$= \frac{1}{Z} \prod_{j=1}^{m} [P(M_{e_j}|e_j) P(e_j|pa(e_j))] \prod_{j=1, s\in\Omega_j}^{m} [P(M_{\omega_{j,s}}|\omega_{j,s}) P(\omega_{j,s}|e_j, e_s)]$$
$$\prod_{t=1}^{l} [P(M_{v_t}|v_t) P(v_t|pa(v_t))] \prod_{i\in V} \phi(y_i, x_i) \prod_{<i,j>, j\in\mathcal{N}_i} \exp(y_i y_j \lambda^T g_{ij}(\mathbf{x}))$$

where Z is the normalization constant. V denotes the set of region nodes and $pa(\ )$ denotes the parents of the node in the parenthesis. Among these factored functions, $P(M_{e_j}|e_j)$, $P(M_{\omega_{j,s}}|\omega_{j,s})$ and $P(M_{v_t}|v_t)$ are the likelihood models of the measurements of edges, angles and vertices, respectively. $P(e_j|pa(e_j))$, $P(\omega_{j,s}|e_j, e_s)$ and $P(v_t|pa(v_t))$ are the conditional probabilities of edge nodes, angle nodes and vertex nodes, respectively. The remaining factored functions $\phi(y_i, x_i)$ and $\exp(y_i y_j \lambda^T g_{ij}(x))$ are the unary potentials and pairwise potentials defined in the CRF part. $g_{ij}(\mathbf{x}) = [1, |x_i - x_j|]$ is the pairwise feature vector,[4] where $|\ |$ means the component-wise absolute value. For the unary potential, we use a logistic function to convert the output of a three-layer perceptron classifier into a potential function, similar as that used by.[16]

4.3. *The Model Parameterization and Probabilistic Inference*

In the joint probability distribution of Eq.(13), the local conditional probability distributions and the parameters of pairwise potentials λ are the model parameters. We can use Maximum Likelihood Estimation to learn these model parameters, given a set of complete training data. Specifically, we learn the likelihood models of $P(M_{e_j}|e_j)$, $P(M_{\omega_{j,s}}|\omega_{j,s})$ and $P(M_{v_t}|v_t)$, the conditional probability $P(e_j|pa(e_j))$, and the parameters of pairwise potentials λ. In addition, we manually set the conditional probabilities $P(\omega_{j,s}|e_j, e_s)$ and $P(v_t|pa(v_t))$ to encourage the smooth connection of edges (i.e. local smoothness constraint) and the simple connectivity of object boundary (i.e. connectivity constraint). Both parameters are empirically set in a way similar to the parameter setting in.[44]

The hybrid graphical model in Figure 11 consists of both directed links and undirected links. To perform a probabilistic inference, it is convenient to convert the hybrid graphical model into a Factor Graph (FG) representation[46,47] since it is difficult to directly perform inference in such a hybrid model. Based on the factored joint probability distribution in Eq.(13), we convert the hybrid graphical model into a Factor Graph representation. The converted Factor Graph is shown in Figure 12, where each factor node (i.e. the square) corresponds to one factored local function in the joint probability distribution.

Given the FG model in Figure 12, we use the inference package provided by Hutter et al.[48] to perform Most Probable Explanation (MPE) inference using the

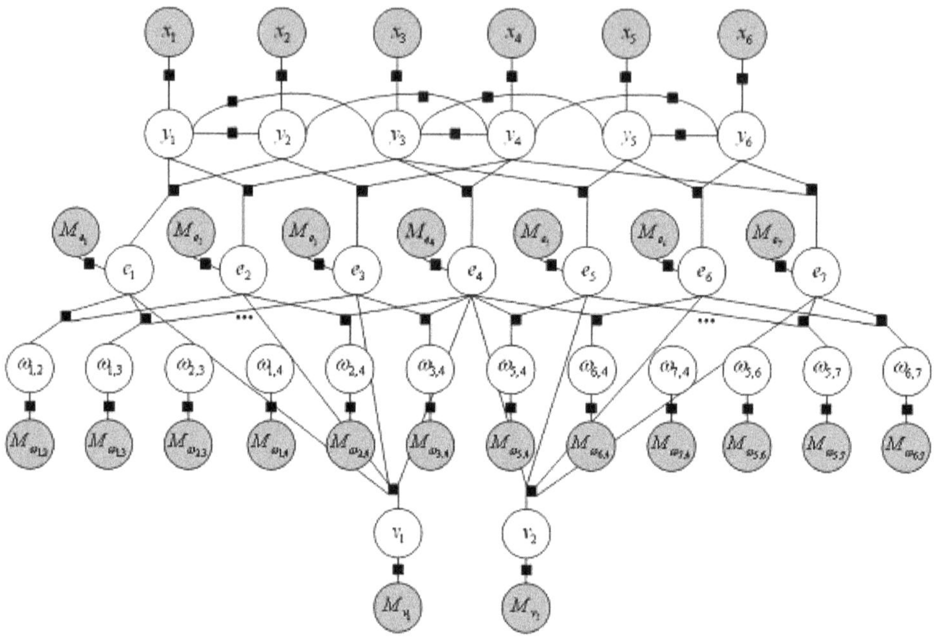

Fig. 12. The converted Factor Graph corresponding to the hybrid graphical model in Figure 11.

stochastic local search (SLS) method, i.e.

$$\mathbf{y}^*, \mathbf{e}^*, \omega^*, \mathbf{v}^* = \arg\max_{y,e,\omega,v} P(\mathbf{y}, \mathbf{e}, \omega, \mathbf{v}, \mathbf{M_e}, \mathbf{M}_\omega, \mathbf{M_v}, \mathbf{x}) \tag{14}$$

where the joint probability distribution is calculated by Eq.(13). In the MPE solution, the region nodes with the foreground labels form the final segmentation.

4.4. *Experiments*

We tested the proposed image segmentation model on the Weizmann horse dataset.[49] This dataset includes the side views of many horses with different appearances and poses in complex backgrounds, which makes it challenging to segment these images. On the other hand, in several related works,[49–53] experiments were conducted on (a subset of) this dataset, too. We can compare our results with the results reported by these state-of-the-art works.

We use 60 horse images as the training data to train the hybrid graphical model. The test images include 120 color images from the Weizmann horse dataset. For the color images, we use the average CIELAB color features and their standard deviations as the local features for each superpixel. These features are simple statistical features that can be extracted from each superpixel. In this case, the length of the feature vector x_i is 6. The three-layer perceptron has a structure with 6 nodes in the input layer, 35 hidden nodes in the hidden layer and 1 node in the output layer.

The number of hidden nodes is selected according to the volume of training data, which are about 4000 superpixels extracted from the training data.

All the training images and the test images are first oversegmented using the Edgeflow-based Anisotropic diffusion software. Given the training images and their ground truth labeling, we automatically train the hybrid graphical model. After training the model, we perform image segmentation on the test images using the inference process described in Section 4.3. Figure 13 shows examples of the color horse images and their segmentation results. We achieved good segmentation results on these images. Most small errors happen on the horse's feet where the appearances of these parts are different from the horse's body. The clutter causes another kind of error. When the background (e.g. the shadow) has a similar appearance as the foreground, this model may not be able to completely separate them.

Fig. 13. Examples of the color image segmentation results produced by the proposed method, arranged in 2 groups of 2 rows. The first row includes the color horse images. The second row includes the segmentation produced by the proposed approach.

In order to quantitatively evaluate our segmentation results and compare them with those aforementioned approaches, we calculate the average percentage of correctly labeled pixels (i.e. labeling accuracy) in all test images. The quantitative results are summarized in Table 1. In this table, we also list the labeling accuracy achieved by other related works.

Table 1. The quantitative comparison of our approach with several related works for segmenting the Weizmann horse images. The average percentage of correctly labeled pixels (i.e. labeling accuracy) is used as the quantitative measurement.

method	image type	train images	test images	labeling accuracy
Cour et al.[51]	color	20	308	94.2%
Winn et al.[52]	color	20	200	93.1%
Zhu et al.[54]	color	12	316	93.3%
Ren et al.[55]	color	172	172	91.0%
Borenstein et al.[50]	grey scale	64	328	93.0%
Winn et al.[52]	grey scale	20	200	93.0%
our hybrid graphical model	grey scale	60	120	94.0%
our hybrid graphical model	color	60	120	95.4%

From the quantitative results in Table 1, we conclude that our results are comparable to (or better than) the results produced by other state-of-the-art approaches. Note that we only use very simple image features (e.g. the color) for segmentation and have not performed any feature selection. Besides, we have not utilized the additional object shape information as some works have done.[50,51]

In addition to the experiments on the Weizmann dataset, we have also performed experiments on the Caltech Cars (1999) dataset,[56] which includes 126 images with cars in complex city backgrounds. In order to test the stability of the image segmentation performance, we have perform a 10-fold experiments, while in each fold we randomly select 66 images for training and then test on the remaining 60 images. Some examples of image segmentation results are shown in Figure 14. We have achieved good segmentation results. We calculate the labeling accuracy for each fold of experiments, as summarized in Table 2. The maximum performance in the 10-fold experiments is 94.6% of labeling accuracy. The average labeling accuracy is 94.1% and the standard deviation is 0.3%, which shows a stable image segmentation performance with respect to the random selections of training and testing data.

Table 2. The labeling accuracies for 10-fold experiments on the Caltech Car (1999) database.

Fold 1	Fold 2	Fold 3	Fold 4	Fold 5	Fold 6	Fold 7	Fold 8	Fold 9	Fold 10
94.5%	94.0%	94.0%	94.0%	93.7%	94.3%	93.7%	94.6%	93.8%	94.1%

5. Summary

In this chapter, we first introduce the concepts of probabilistic graphical models and the basic types of PGMs, including BNs, DBNs, MRFs, and CRFs. These PGMs have been demonstrated to be powerful modeling tools to capture the relationships among random variables and provide principled methods to perform parameter learning and probabilistic inference. We further explain the applications of PGMs into two important computer vision tasks: facial expression recognition and image

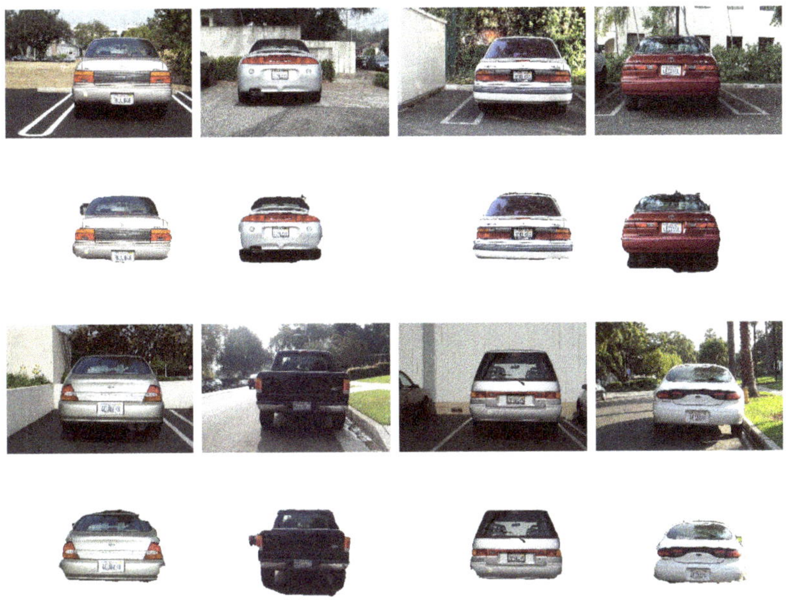

Fig. 14. Examples of the segmentation of car images from the Caltech Cars (1999) database.

segmentation. We demonstrate that the proposed PGMs can successfully capture the useful and informative relationships among facial action units or image entities (regions, edges, vertices) to solve these challenging computer vision problems.

Finally, we should notice that the adoptions of PGMs for solving computer vision problems are on the very edge of its further extension and development. Researchers are continuously studying advanced probabilistic graphical models with more modeling capability and exploring other effective parameter learning methods (e.g. discriminative parameter learning) and more efficient probabilistic inference methods to deal with complex PGMs. In addition, there are still many issues along this direction that are not solved yet, such as the structure learning of PGMs.

References

1. S. Geman and D. Geman, Stochastic relaxation, gibbs distributions, and the bayesian restoration of images, *Pattern Analysis and Machine Intelligence*. **6**(6), 721–741, (1984).
2. S. Z. Li, Markov random field modeling in image analysis, *Springer*. (2001).
3. J. Lafferty, A. McCallum, and F. Pereira. Conditional random fields: Probabilistic models for segmenting and labeling sequence data. In *International Conference on Machine Learning*, pp. 282C–289, (2001).
4. S. Kumar and M. Hebert, Discriminative random fields, *International Journal of Computer Vision*. **68**(2), 179–201, (2006).
5. B. Frey, F. Kschischang, H. Loeliger, and N. Wiberg. Factor graphs and algorithms. In *Proceedings of the 35th Allerton Conference*, (1998).

6. J. Pearl, Probabilistic reasoning in intelligent systems: networks of plausible inference, *Morgan-Kaufmann Publishers Inc. 1988.* (1988).
7. F. V. Jensen, Bayesian networks and decision graphs, *Springer-Verlag, New york.* (2001).
8. T. Dean and K. Kanazawa, A model for reasoning about persistence and causation, *Computational Intelligence.* **5**(3), 142 – 150, (1989).
9. K. P. Murphy, Dynamic bayesian networks: Representation, inference and learning, *UC Berkeley, Computer Science Division, PhD thesis.* (2002).
10. S. L. Lauritzen and N. Wermuth, Graphical models for associations between variables, some of which are qualitative and some quantitative, *Annals of Statistics.* **17**(1), 31–57, (1989).
11. M. Frydenberg, The chain graph markov property, *Scandinavian Journal of Statistics.* **17**, 333–353, (1990).
12. Y. Tong, W. Liao, and Q. Ji, Facial action unit recognition by exploiting their dynamic and semantic relationships, *IEEE Trans. Pattern Analysis and Machine Intelligence.* **29**(10), 1683–1699 (October, 2007).
13. J. Hammersley and P. Clifford, Markov fields on finite graphs and lattices, *unpublished manuscript, Oxford University.* (1971).
14. A. Quattoni, M. Collins, and T. Darrell. Conditional random fields for object recognition. In *Advances in Neural Information Processing Systems*, (2004).
15. A. McCallum, K. Rohanimanesh, and C. Sutton. Dynamic conditional random fields for jointly labeling multiple sequences. In *NIPS 2003 Workshop on Syntax, Semantics, Statistics*, (2003).
16. X. He, R. Zemel, and M. Carreira Perpinan. Multiscale conditional random fields for image labeling. In *Computer Vision and Pattern Recognition*, vol. 2, pp. II–695–II–702, (2004).
17. J. Besag, Efficiency of pseudolikelihood estimation for simple gaussian fields, *Biometrika.* **64**, 616–618, (1977).
18. S. Kumar and M. Hebert. Approximate parameter learning in discriminative fields. In *Snowbird Learning Workshop*, (2004).
19. C. Sutton and A. McCallum. Piecewise training for undirected models. In *Proceedings of the 21th Annual Conference on Uncertainty in Artificial Intelligence (UAI-05)*, p. 568, (2005).
20. M. J. Wainwright, T. S. Jaakkola, and A. S. Willsky. Tree-reweighted belief propagation algorithms and approximate ml estimation by pseudo-moment matching. In *Workshop on Artificial Intelligence and Statistics*, (2002).
21. G. E. Hinton, Training products of experts by minimizing contrastive divergence, *Neural Computation.* **14**, 2002, (2002).
22. B. Taskar, C. Guestrin, and D. Koller. Max-margin markov networks. In *Neural Information Processing Systems Conference (NIPS03)*, (2003).
23. J. N. Darroch and D. Ratcliff, Generalized iterative scaling for log-linear models, *The Annals of Mathematical Statistics.* **43**(5), 1470–1480, (1972).
24. R. D. Shachter, S. K. Andersen, and P. Szolovits. Global conditioning for probabilistic inference in belief networks. In *Uncertainty in Artificial Intelligence*, pp. 514–522, (1994).
25. F. V. Jensen, S. L. Lauritzen, and K. G. Olesen, Bayesian updating in causal probabilistic networks by local computations, *Computational Statistics Quaterly.* **4**, 269–282, (1990).
26. M. I. Jordan, Z. Ghahramani, T. S. Jaakkola, and L. K. Saul, An introduction to variational methods for graphical models, *Learning in Graphical Models.* pp. 105 –

161, (1999).
27. M. Wainwright and M. Jordan. A variational principle for graphical models. In *chapter 11 in: New Directions in Statistical Signal Processing*. MIT Press, (2005).
28. M. A. R. Leisink and H. J. Kappen. General lower bounds based on computer generated higher order expansions. In *Uncertainty in Artificial Intelligence*, pp. 293–300, (2002).
29. M. J. Wainwright and M. I. Jordan. Semidefinite relaxations for approximate inference on grpahs with cycles. In *Advances in Neural Information Processing Systems*, vol. 16, (2004).
30. Y. Tong, J. Chen, and Q. Ji, A unified probabilistic framework for spontaneous facial action modeling and understanding, *IEEE Trans. Pattern Analysis and Machine Intelligence.* in Press.
31. Y. Tong, J. Chen, and Q. Ji. Modeling and exploiting the spatio-temporal facial action dependencies for robust spontaneous facial expression recognition. In *Proc. IEEE Computer Vision and Pattern Recognition Workshop for Human Communicative Behavior Analysis*, (2009).
32. P. Ekman, W. V. Friesen, and J. C. Hager, *Facial Action Coding System: the Manual.* (Research Nexus, Div., Network Information Research Corp., Salt Lake City, UT, 2002).
33. S. Nishio, K. Koyama, and T. Nakamura, Temporal differences in eye and mouth movements classifying facial expressions of smiles, *Proc. Third IEEE Int'l Conf. Automatic Face and Gesture Recognition.* pp. 206–211 (April, 1998).
34. K. Schmidt and J. Cohn, Dynamics of facial expression: Normative characteristics and individual differences, *IEEE Int'l Conf. on Multimedia and Expo.* pp. 728–731, (2001).
35. B. Bascle and A. Blake, Separability of pose and expression in facial tracking and animation, *Proc. Int'l Conf. on Computer Vision.* pp. 323–328, (1998).
36. M. S. Bartlett, G. C. Littlewort, M. G. Frank, C. Lainscsek, I. R. Fasel, and J. R. Movellan, Automatic recognition of facial actions in spontaneous expressions, *J. Multimedia.* **1**(6), 22–35 (September, 2006).
37. K. Murphy, Inference and learning in hybrid bayesian networks, *Technical Report CSD-98-990, Department of Computer Science, U. C. Berkeley.* (1998).
38. K. B. Korb and A. E. Nicholson, *Bayesian Artificial Intelligence.* (Chapman and Hall/CRC, London, UK, 2004).
39. Multiple aspects of discourse research lab http://madresearchlab.org/.
40. E. Douglas-Cowie, R. Cowie, and M. Schroeder, The description of naturally occurring emotional speech, *Fifteenth Int'l Congress of Phonetic Sciences.* (2003).
41. L. Zhang and Q. Ji. A multiscale hybrid model exploiting heterogeneous contextual relationships for image segmentation. In *Proc. IEEE Computer Vision and Pattern Recognition (CVPR)*, (2009).
42. L. Zhang and Q. Ji, A unified graphical model and its application to image segmentation, *accepted by IEEE Transactions on Pattern Analysis and Machine Intelligence (PAMI) subject to minor revision.* (2009).
43. B. Sumengen and B. S. Manjunath. Edgeflow-driven variational image segmentation: Theory and performance evaluation. Technical report, University of California, Santa Barbara, (2005). http://barissumengen.com/seg/.
44. L. Zhang and Q. Ji. Integration of multiple contextual information for image segmentation using a bayesian network. In *IEEE International Workshop on Semantic Learning Applications in Multimedia, in association with CVPR 2008*, (2008).
45. S. L. Lauritzen, Graphical models, *Oxford University Press.* (1996).

46. F. Kschischang, B. J. Frey, and H.-A. Loeliger, Factor graphs and the sum-product algorithm, *IEEE Transactions on Information Theory.* **47**, 498–519, (2001).
47. B. J. Frey. Extending factor graphs so as to unify directed and undirected graphical models. In *Proceedings of Conference on Uncertainty in Artificial Intelligence 19*, pp. 257–26, (2003).
48. F. Hutter, H. H. Hoos, and T. Stutzle. Efficient stochastic local search for mpe solving. In *International Joint Conferences on Artificial Intelligence*, pp. 169–174, (2005).
49. E. Borenstein, E. Sharon, and S. Ullman. Combining top-down and bottom-up segmentation. In *CVPR Workshop on Perceptual Organization in Computer Vision*, pp. 46– 46, (2004).
50. E. Borenstein and J. Malik. Shape guided object segmentation. In *Computer Vision and Pattern Recognition*, pp. 969– 976, (2006).
51. T. Cour and J. Shi. Recognizing objects by piecing together the segmentation puzzle. In *Computer Vision and Pattern Recognition*, pp. 1–8, (2007).
52. J. Winn and N. Jojic. Locus: Learning object classes with unsupervised segmentation. In *International Conference on Computer Vision*, pp. 756–763, (2005).
53. S. Zheng, Z. Tu, and A. Yuille. Detecting object boundaries using low-, mid-, and high-level information. In *Computer Vision and Pattern Recognition*, (2007).
54. L. L. Zhu, C. Lin, H. Huang, Y. Chen, and A. L. Yuille. Unsupervised structure learning: Hierarchical recursive composition, suspicious coincidence and competitive exclusion. In *European Conference on Computer Vision*, pp. 759–773, (2008).
55. X. Ren, C. C. Fowlkes, and J. Malik. Cue integration in figure/ground labeling. In *Advances in Neural Information Processing Systems 18*, (2005).
56. M. Weber, M. Welling, and P. Perona. Towards automatic discovery of object categories. In *IEEE Conference on Computer Vision and Pattern Recognition*, pp. 101–108, (2000).

CHAPTER 2.2

A REGION TREE BASED IMAGE DISCRETE LABELING FRAMEWORK

Cheng Lei and Yee-Hong Yang

Department of Computing Science, University of Alberta
Edmonton, Alberta, Canada
E-mail: {clei, yang}@cs.ualberta.ca

In this chapter, we present a region-tree based framework for general image discrete labeling problems. In particular, the input image is represented as a tree of over-segmented regions and the labeling problem is solved by optimizing an energy function defined on such a region-tree using dynamic programming (DP). By using the over-segmentation based region-tree as image representation, our framework combines the advantages of enabling labeling primitives that contain more information with a large support area compared to that of the pixel based approaches with a small support area and of reducing the risk of propagating segmentation errors into the labeling process compared to the layer based approaches. The region-tree representation further enables the use of fast DP optimization and enhances the performance by avoiding the use of smoothness constraints crossing discontinuities between adjacent labeling primitives. Variants such as coarse-to-fine region-tree and temporal region-tree are also developed for different applications such as binocular stereo matching, optical flow estimation and multi-view space-time consistent video depth recovery. The corresponding experimental evaluations have shown that our proposed framework is very effective and produces promising results.

1. Introduction

Many important tasks in low-level vision, image analysis and pattern recognition such as image restoration, stereo matching and texture modeling can be formulated as discrete labeling problems, where the labels represent some task-specific local attributes such as disparity or depth in stereo matching, intensity in image restoration and motion displacement in optical flow estimation. How to solve this class of labeling problems efficiently and accurately has attracted a significant amount of computer vision research over the past few years. So far, the most popular and successful approach addresses such labeling problems using the energy minimization formulation, by which an energy function that measures the quality of a labeling using a data term and a smoothness term is minimized. In particular, the data term evaluates the consistency of a labeling solution with given observations and the smoothness evaluates term the spatial piecewise coherence of the labeling solution [1]. One justification of using such an energy function definition can be traced back to its equivalence to maximum posteriori estimation of Markov random field (MRF) [1]. In recent years, many powerful new algorithms such as graph cuts [2, 3], and loopy belief propagation [4] have been proposed for optimizing such energy functions, with encouraging progress made, especially in stereo matching, and more recently, in motion estimation.

In addition to optimization algorithms, image representation also plays an important role in solving an image labeling problem. So far, two image representations are most often used, in particular, the traditional pixel grid and the layer representations [5]. The former one uses a pixel as a labeling primitive. It is simple to use but suffers from more ambiguous labeling possibilities. The latter one uses a layer as a labeling primitive. Its main challenge is in the difficulties of correct layer segmentation and layer parameterization without prior knowledge. As a trade-off between enabling labeling primitives to contain enough information with a large support area and reducing the risk of violating the parameterization assumption with a small support area, representations using over-segmented regions have shown great potentials [6,7]. Such representations can reduce the computational complexity compared to pixel-based representations due to much fewer number of regions. Also given the smaller size of the regions, the chance that color segmentation errors propagate into the labeling process using regions is reduced compared to that using layers.

In this chapter, we present a general region-tree based framework for image labeling problems. In this framework, a labeling target image is first over-segmented into a set of regions using the mean-shift algorithm. From the segmentation result, a spanning region tree is extracted from the corresponding region adjacency graph. The labeling problem is solved by optimizing an energy function defined on such a region tree. By using different labeling definitions and energy functions, the same framework is applied to solving the classic binocular stereo, the space-time consistent multi-view stereo matching and the optical flow estimation problems.

This chapter is organized as follows. In section 2, we briefly review related work. Then in section 3, we elaborate the details of the proposed framework. In section 4, a couple of its variants and applications are shown, with the corresponding experimental results reported in section 5. Finally, we conclude in section 6.

2. Related Work

As a step of fundamental importance in many computer vision applications, image labeling is in general NP-hard [1] and usually only approximate solutions can be found. So most efforts have been focused on designing better optimization algorithms. With powerful new optimization algorithms such as graph cuts [2,3], and loopy belief propagation [4] proposed and popularized in the last few years, the performance improvement of various image labeling methods has been encouraging, which in turn further enables many novel applications such as interactive photo editing [8]. Therefore, most proposed labeling frameworks are usually categorized according to the chosen optimization techniques. For a detailed survey of such labeling frameworks, please refer to [1].

On the other hand, image labeling frameworks can also be categorized according to the image representations. In the literature, the most often used image representation is the pixel grid, which can be applied to any rectangular shaped image. However, because it loses some useful scene-based smoothness constraints, which can be implicitly and easily enforced in the region-based or layer-based approaches, its performance in ambiguity discrimination suffers.

Region-based image representation has shown unique potentials in various image labeling problems, especially in the classic binocular stereo matching [9, 10]. Specifically, according to the de facto stereo benchmarking [11], the current top two ranked stereo methods [9, 10] are both region-based. In particular, in [9], an inter-regional cooperative optimization scheme is proposed using color-segmented regions as labeling primitives. Local window-based pixel matching is done first to fit a disparity plane for each region. And then the parameters of all such disparity planes are iteratively optimized to obtain the final disparity map. A similar idea is also used in [10], in which the scene structure is modeled by a set of planar surface patches through color segmentation. Disparity planes are extracted by applying local window-based pixel matching and segmentation-based clustering. The disparity plane label assignment for each region is optimized using belief propagation. One drawback of such disparity plane fitting based approaches is their reliance on accurate segmentation. Error in region/layer segmentation can result in violation of the assumption that regions do not cross label discontinuities. Also for non-structured scenes in which there are no dominant planes, their performance could be degraded. To address these issues, researchers proposed over-segmentation based approaches [5, 6, 7].

Over-segmented region based representation can be regarded as a trade-off between enabling labeling primitives to contain enough information with a large support area and reducing the risk of violating the parameterization assumption with a small support area. On one hand, it can reduce the computational complexity compared to pixel-based representations due to much fewer number of regions. On the other hand, given the smaller size of the regions, the chance that color segmentation errors propagate into the matching process is reduced compared to layer based representation. In particular, in [6], the image is first over-partitioned into a grid of equal sized square patches and then K-means clustering is applied to refine the over-segmentation. Depths of regions are computed using loopy belief propagation over the corresponding region graph. In [7], over-segmentation is also used for simultaneous stereo reconstruction and alpha matting of mixed region boundary pixels. Specifically the segmentation is done w.r.t. both images for consistency and iteratively updated based on region color, depth and shape constraints. For a given segmentation, the depths of the regions are estimated using belief propagation, which is in turn used for updating the segmentation until convergence.

All the above successes of using region-based image representations in stereo matching motivate our work. In this chapter, we introduce a general region-based framework for image labeling problems using a novel image representation form – a spanning tree of over-segmented regions. Using a tree structure instead of a more general graph structure enables the use of fast optimization techniques such as Dynamic Programming (DP). The most related work in this regard is [12, 13]. In particular, [12] extends the traditional DP based stereo matching by introducing a novel pixel-tree structure. The pixel-tree is formed by keeping only the "most important" edges of 4 connected neighbor pixels based on color similarity. Applying DP to such a 2D pixel-tree structure enables vertical consistency enforcement between scanlines, resulting in truly global optimization since the label (disparity) assignment for each pixel depends on assignments of all other pixels. In [13], scan-lines are partitioned into segments and the resulting line segments are used as tree vertices instead of single pixels. In this way, the connections between adjacent pixels within the same segment are preserved as completely as possible, resulting in better stereo matching performance.

3. Proposed Framework

3.1. *Problem Formulation*

In our framework, similar to other methods, we also formulate the image labeling problem as a discrete energy minimization problem. In particular, the labeling target image $\boldsymbol{I}$ is represented as a spatial structure Ω of a set of image primitives $\boldsymbol{\mathcal{P}}$ spanning the whole image. Then the labeling problem is solved by finding a label assignment collection $\boldsymbol{L}(\Omega)$ which assigns each image primitive $p \in \boldsymbol{\mathcal{P}}$ an optimal label ℓ_p from a label set $\boldsymbol{\mathcal{L}}$ so that an energy function $E(\boldsymbol{L}(\Omega))$ is minimized. $E(\boldsymbol{L}(\Omega))$ is equivalent to the log likelihood of posterior distribution of an MRF and consists of a data term $E_{data}(\boldsymbol{L}(\Omega))$ and a smoothness term $E_{smooth}(\boldsymbol{L}(\Omega))$ in the general form of

$$E(\boldsymbol{L}(\Omega)) = E_{data}(\boldsymbol{L}(\Omega)) + \lambda \cdot E_{smooth}(\boldsymbol{L}(\Omega)) \tag{1}$$

where the positive constant λ represents the relative weight.

The data term $E_{data}(\boldsymbol{L}(\Omega))$ sums the data energy over all the labeling primitives as

$$E_{data}(\boldsymbol{L}(\Omega)) = \textstyle\sum_{p \in \Omega} \mathcal{D}(p, \ell_p) \tag{2}$$

where the data energy function $\mathcal{D}(p, \ell_p)$ measures the penalty of assigning label ℓ_p to a primitive p according to some known observations.

The smoothness term $E_{smooth}(\boldsymbol{L}(\Omega))$ sums the smoothness energy over all neighboring image primitive pairs as

$$E_{smooth}(\boldsymbol{L}(\Omega)) = \textstyle\sum_{\{p,q\} \in \mathcal{N}} \mathcal{S}(p, q, \ell_p, \ell_q) \tag{3}$$

where the smoothness energy function $\mathcal{S}(p, q, \ell_p, \ell_q)$ measures the penalty of assigning each neighboring primitive pair $\{p, q\} \in \boldsymbol{\mathcal{N}}$ with labels $\{\ell_p, \ell_q\}$ respectively based on the smoothness prior.

3.2. *Region-tree based Image Representation*

In this chapter, the labeling primitive set $\boldsymbol{\mathcal{P}}$ is a set of over-segmented regions $\boldsymbol{\mathcal{R}}$. The spatial structure Ω is a spanning tree $\boldsymbol{\mathcal{T}}$ over $\boldsymbol{\mathcal{R}}$, which is extracted from the region adjacency graph (RAG) created as the segmentation result. All the edges in $\boldsymbol{\mathcal{T}}$ form the corresponding neighborhood system $\boldsymbol{\mathcal{N}}$ in $\boldsymbol{\mathcal{R}}$.

3.2.1. *Mean-shift based Image Over-segmentation*

To build the corresponding region-tree $\mathcal{T}$, the labeling target image is first over-segmented using an improved variant [14] of the mean-shift method [15].

In particular, the segmentation consists of two steps, namely, mean-shift filtering and pixel/region bottom-up fusion. In the filtering step, in addition to transforming its intensity/spatial information into feature space parameters, a weight is also assigned to each pixel. Such weight is defined using its normalized rank of the gradient magnitude and edge confidence. It is used as the edge-likelihood measure to enhance the contributions of edge-like pixels and to suppress the contribution of non-edge-like pixels, when calculating the mean-shift moves in the feature space. In this way, the discontinuity preserving property of the mean-shift filtering is further enhanced.

After the filtering step is done, the fusion step iteratively fuses smaller regions (pixels in the first iteration) into larger ones by performing transitive closure operations to the corresponding RAG. Specifically, for each edge in the RAG, a boundary strength measure is computed as a "fusibility" measure between regions by averaging the above mentioned weights of pixels on the boundary shared by the two regions. By thresholding the maximum boundary strength between any two to-be-fused regions, discontinuity-preserving region fusion can be achieved. Furthermore, the segmentation granularity is adjusted by controlling the fusion iterations and the minimum region size.

An illustration of a typical over-segmentation result is shown in Figure 1 (a).

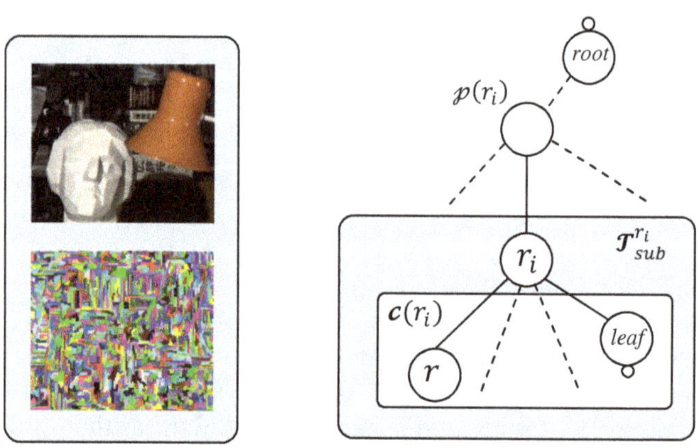

(a) Over-Segmentation Example (b) Region-tree illustration

Figure 1. Illustrations of over-segmentation and region tree.

3.2.2. *Region-tree Generation*

With the results of mean-shift based over-segmentation, the labeling target image is partitioned into a set of regions $\mathcal{R}$ and the corresponding RAG $\mathcal{G}(\mathcal{R}, \mathcal{E}')$ is created. Each region $r_i \in \mathcal{R}$ has N_{r_i} pixels $(x, y) \in r_i$ and corresponds to a mode m_{r_i}, which is a mean-shift clustering point in the multiple dimensional feature space (5 for color images and 3 for grey-level images). Each vertex in $\mathcal{G}(\mathcal{R}, \mathcal{E}')$ corresponds to a region r_i and each edge $e_{(i,j)} \in \mathcal{E}'$ corresponds to a link between two adjacent regions r_i and r_j.

From the RAG $\mathcal{G}(\mathcal{R}, \mathcal{E}')$, we further extract a region tree by keeping only the "most important" edges. Then an obvious question is on how to evaluate the importance of a given edge. An intuitive criterion is that, the more likely that two regions have the same label, the more desirable is to keep the edge between them. Therefore, each edge $e_{(i,j)}$ is assigned with a positive edge weight $w_{e_{(i,j)}}$ encoding the *un*-likelihood for regions r_i and r_j to share the same label. The larger the edge weight $w_{e_{(i,j)}}$, the less likely regions r_i and r_j to have the same label so that the less important is the edge $e_{(i,j)}$. Then we construct a minimum spanning tree (MST) $\mathcal{T}(\mathcal{R}, \mathcal{E})$ by traversing all region vertices in $\mathcal{G}$ and by minimizing the sum of weights of the remaining edges $\sum_{e_{(i,j)} \in \mathcal{E}} w_{e_{(i,j)}}$ using the Prim greedy MST method [16]. In this way, each specific region is only linked to the regions which are most likely to have the same label so that the adaptive piecewise labeling smoothness can be enforced.

Furthermore, it can be easily seen that different edge-weighting schemes may result in different region tree structures and in turn different labeling performance.

3.3. *Energy Minimization over Region-tree*

The optimization of energy function (1) defined over a region-tree structure $\Omega = \mathcal{T}(\mathcal{R}, \mathcal{E})$ can be done by any discrete optimization methods supporting non-grid neighbour system such as graph cuts and belief propagation. However, the unique tree structure makes it possible to use the more efficient dynamic programming (DP) optimization. As we know, to apply DP, the problem in question must exhibit an optimal structure. Our region-tree based labeling optimization does have such an optimal structure. Indeed, the optimal solution (partial labeling) for any sub-region-tree must also be optimal for the solution for the whole region-tree.

Specifically, taking the sub-region-tree $\mathcal{T}_{sub}^{r_i}$ rooted at vertex r_i shown in Figure 1 (b) as an example. Suppose vertex r_i is not the global root of the whole region tree $\mathcal{T}$, then the minimum value of the energy as defined in (1) on the sub-tree $\mathcal{T}_{sub}^{r_i}$ plus the edge between vertex r_i and its parent vertex $\wp(r_i)$ can be recursively written as a function of the label $\ell_{\wp(r_i)}$ assigned to $\wp(r_i)$:

$$E_{r_i}(\ell_{\wp(r_i)}) = \min_{\ell_{r_i}\in\mathcal{L}}\left(\mathcal{D}(r_i,\ell_{r_i}) + \lambda\cdot\mathcal{S}(r_i,\wp(r_i),\ell_{r_i},\ell_{\wp(r_i)}) + \sum_{r\in c(r_i)} E_r(\ell_{r_i})\right) \tag{4}$$

where $c(r_i)$ is the set of the children vertices of vertex r_i.

Then let's further investigate two types of special vertices in the whole region tree: the leaf and the root.

For a leaf vertex, since it has no child vertex, then (4) can be directly evaluated for each possible label assignment for its parent vertex, that is,

$$E_{leaf}(\ell_{\wp(leaf)}) = \min_{\ell_{leaf}\in\mathcal{L}}\left(\mathcal{D}(leaf,\ell_{leaf}) + \lambda\cdot\mathcal{S}(r_{leaf},\wp(leaf),\ell_{leaf},\ell_{\wp(leaf)})\right) \tag{5}$$

Therefore, for the leaf vertex r_{leaf}, if the label of its parent vertex $\wp(leaf)$ has been assigned somehow, its optimal label ℓ_{leaf} can be easily determined.

While for the root vertex, it has no parent vertex so that (4) can be simplified as

$$E_{root} = \min_{\ell_{root}\in\mathcal{L}}\left(\mathcal{D}(root,\ell_{root}) + \sum_{r\in c(root)} E_r(\ell_{root})\right) \tag{6}$$

That is, with E_{root} known for each possible root vertex label assignment and for each child vertex of the root vertex, the best label ℓ_{root} can be directly recovered. Furthermore, it can be seen that the minimum energy E_{root} of the root vertex is just the minimum value of the energy (1) of the whole region tree. So the labeling assignment optimizing E_{root} is just the one optimizing (1).

Based on the above observed properties, the energy function (1) can be optimized using DP through two recursive region-tree traversals. In the first bottom-up traversal starting from the leaf vertices, the optimal label that minimizes the energy (5) and the corresponding minimal energy are determined for each possible parent vertex label. Then upon reaching the root vertex, the optimal label of the root vertex is trivially determined and the region-tree is breadth-first traversed again starting from the root vertex. In this turn, for each visited non-root vertex, since the parent's optimal label has been set, its optimal label can be easily determined accordingly. Therefore, in this way, all the optimal labels of the vertices can be recovered so that the image labeling problem is solved.

4. Applications and Variants

In this section, we apply the above region-tree based framework to solving three different image labeling problems: the classic binocular stereo matching, optical flow estimation and multi-view space-time consistent depth recovery. For the latter two, variants of the original region-tree are developed for problem-specific requirements. In this chapter, we elaborate only the most important part in each application. For more details, please refer to the corresponding papers or technical reports [17, 18].

4.1. *Classic Binocular Stereo Matching*

Applying our framework to binocular stereo matching between two stereo images $\boldsymbol{I}_{ref}$ and $\boldsymbol{I}_{peer}$ is rather straightforward. The label set $\boldsymbol{\mathcal{L}}$ directly maps labels to a set of hypothesis disparities in a known range. Then for each region r_i in the region-tree $\boldsymbol{\mathcal{T}}(\boldsymbol{\mathcal{R}},\boldsymbol{\mathcal{E}})$of the labeling target image $\boldsymbol{I}_{ref}$, the data energy function $\mathcal{D}(r_i,\ell_{r_i})$ measures the matching penalty for assigning the corresponding region r_i with disparity $d_{\ell_{r_i}}$ corresponding to a label ℓ_{r_i}. The choice of function $\mathcal{D}(r_i,\ell_{r_i})$ is arbitrary as long as it is non-decreasing in the absolute (average) intensity difference between the matched regions of the reference image and the matching image $\boldsymbol{I}_{peer}$. We define $\mathcal{D}(r_i,\ell_{r_i})$ based on the zero-mean normalized cross-correlation (ZNCC) measure η [21] as

$$\mathcal{D}(r_i,\ell_{r_i}) = \frac{\sum_{(x,y)\in r_i}\left(1.0-\eta\left(\boldsymbol{I}_{ref}(x,y),\boldsymbol{I}_{peer}\left(x+d_{\ell_{r_i}},y\right)\right)\right)}{N_{r_i}} \tag{7}$$

In our implementation, the matching penalty evaluation of all regions for a specific disparity is done by first calculating the corresponding disparity space image [11] and then aggregating/averaging the penalties of all the pixels in each region.

For each edge $e_{(i,j)}\in\boldsymbol{\mathcal{E}}$, the smoothness energy function $\mathcal{S}\left(r_i,r_j,\ell_{r_i},\ell_{r_j}\right)$ measures the assignment of two adjacent regions r_i and r_j with disparities $d_{\ell_{r_i}}$ and $d_{\ell_{r_j}}$, respectively. It can be any arbitrary non-decreasing function in absolute disparity difference $\left|d_{\ell_{r_i}}-d_{\ell_{r_j}}\right|$. We define it as

$$\mathcal{S}\left(r_i,r_j,\ell_{r_i},\ell_{r_j}\right) = \frac{\left|d_{\ell_{r_i}}-d_{\ell_{r_j}}\right|}{\left(w_{e_{(i,j)}}+\epsilon\right)} \tag{8}$$

where ϵ is a small positive value for avoiding division-by-zero.

Then optimizing the total energy function over the whole region-tree gives us the disparity map for image $\boldsymbol{I}_{ref}$. To correct possible errors due to occlusions, the symmetric cross-checking technique is further done as post processing. Specifically, the above process is also applied to image $\boldsymbol{I}_{peer}$ to estimate its disparity map. In particular, for two pixels matched in two images, if the corresponding disparity d_{ref} is larger than d_{peer}, then the pixel in $\boldsymbol{I}_{ref}$ is marked as occluded due to the violation of the weak consistency constraint defined in [17]. The detected occlusions are corrected by hole-filling [16] or by directly replacing d_{ref} with d_{peer}.

4.2. *Coarse-to-fine Region-tree based Optical Flow Estimation*

Although essentially a continuous problem, in recent years, optical flow estimation can also be solved as a discrete image labeling optimization, as attempted in [19, 20]. Their promising performance is due to the better ability in finding global minima and in applying non-convex smoothness regularization compared to the traditional continuous gradient-descent based optimization methods. Since it bears commonalities with the binocular stereo matching problem and also there is no previous attempt to use image over-segmentation to handle it, in the following we also reformulate optical flow estimation into a labeling problem and solve it using a variant of our region-tree framework.

4.2.1. *Coarse-to-fine Region-tree*

To estimate continuous motion between two images using our discrete labeling framework, the continuous 2D displacement solution space has to be quantized and mapped to a discrete set of labels. However, brute-force discretization usually suffers from the so-called "*discretization bottleneck*" problem, which means that the number of labels required for sampling the search ranges with fine enough precision could be too large for efficient optimization.

To address this problem, we develop a new C2F variant of our original region-tree representation. In particular, during the mean-shift based image segmentation process, by using different region fusion size thresholds, multi-level of image over-segmentations with different segmentation granularities are obtained. As shown in Figure 2, the bottom-up fusion based segmentation process as explained in section 3.2.1 makes it possible to obtain such multiple level over-segmentations efficiently in one single pass and guarantees that each larger region in the coarser level consists of exactly the same smaller regions that are in the larger region in its corresponding finer level segmentation. That is, each region $r \in \mathcal{R}_j$ in a coarser level j consists of a group of adjacent regions $\in \mathcal{R}_k$ of a finer segmentation level $k > j$, such that region r is the "container" region of these adjacent regions. Then for each segmentation level, a corresponding region-tree is built as explained in section 3.2.2. Hence, all the levels together forms a multiple-layer C2F region-tree. Its main difference from the original single-layer region-tree is that there exist implicit edges between layers linking coarse-level container region to its finer-level "child" regions.

Then for label discretization, at each level j, each region $r^{(j)} \in \boldsymbol{\mathcal{R}}_j$ maintains a displacement Look-Up-Table (LUT) $\mathcal{M}_{r^{(j)}}$ which defines a bijection mapping of each label $\ell \in \boldsymbol{\mathcal{L}}$ to the corresponding 2D displacement vector (u_ℓ, v_ℓ). It is mainly used to evaluate the energy terms of the candidate label during optimization and induces the optical flow field $(\boldsymbol{U}, \boldsymbol{V})$ for a given labeling $\boldsymbol{L}$. Such LUT is built by uniformly sampling the corresponding displacement search ranges $\left[u_{min}^{r^{(j)}}, u_{max}^{r^{(j)}}\right]$ and $\left[v_{min}^{r^{(j)}}, v_{max}^{r^{(j)}}\right]$ in both the horizontal and vertical directions with a sampling interval δ_j. At the coarsest level, the displacement search ranges of all regions are initialized globally with

$$u_{max}^{r^{(0)}} = v_{max}^{r^{(0)}} = -u_{min}^{r^{(0)}} = -v_{min}^{r^{(0)}} = \beta \cdot \max(w, h) \tag{9}$$

where w and h are the width and height of the input images, respectively, and β is a positive constant (set as 0.04 in the experiments). That is, we assume that the horizontal and vertical displacements each have upper bounds related to the image dimension. Then with the result of the last coarser level j known, the search range of a region $r^{(k)} \in \boldsymbol{\mathcal{R}}_k$ at the finer level $k > j$ is setup based on its container region $r^{(j)} \in \boldsymbol{\mathcal{R}}_j$ at level j. That is, suppose that the recovered displacement vector for region $r^{(j)}$ is $(u^{r^{(j)}}, v^{r^{(j)}})$, the displacement search ranges for region $r^{(k)} \in \boldsymbol{\mathcal{R}}_k$ will be defined in a neighborhood around $\left(u^{r^{(j)}}, v^{r^{(j)}}\right)$ as $\left[u^{r^{(j)}} - \Delta_k, u^{r^{(j)}} + \Delta_k\right]$ and $\left[v^{r^{(j)}} - \Delta_k, v^{r^{(j)}} + \Delta_k\right]$, each of which is sampled with δ_k to setup the corresponding displacement LUT

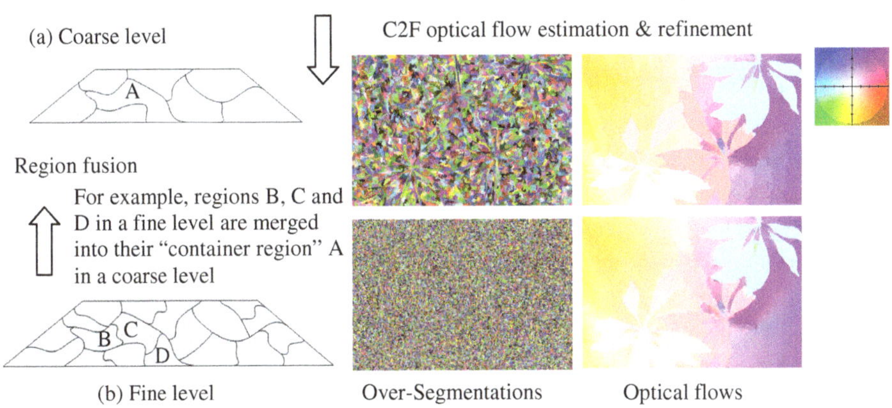

Figure 2. Two-level coarse-2-fine over-segmentations of dataset "Schefflera" and its optical flow recovery procedure.

$\mathcal{M}_{r^{(k)}}$ for region $r^{(k)} \in \mathcal{R}_k$. By decreasing Δ_k and δ_k level by level, an incremental displacement refinement can be achieved.

Hence, with region-dependent displacement search ranges and level by level search range refinement, a small number of labels can achieve a similar quality of sampling to continuous methods, resulting in much better efficiency and speed of convergence.

4.2.2. *Optical Flow Estimation*

Then to recover the optical flow field $(\boldsymbol{U}, \boldsymbol{V})$ between two temporal consecutive input images $\boldsymbol{I}_t$ and $\boldsymbol{I}_{t+1}$, where $(\boldsymbol{U}, \boldsymbol{V})$ denote the horizontal and vertical components of the displacement vector field, respectively, a C2F region-tree is built for $\boldsymbol{I}_t$ as above and multiple energy optimizations over region-tree of each segmentation level are done starting from the coarsest level.

Similar to the above discussed binocular stereo matching problem, the data energy term $E_{data}(\boldsymbol{L}(\Omega))$ used for optical flow estimation measures the brightness matching error between two images correlated by a warping induced from the optical flow $(\boldsymbol{U}, \boldsymbol{V})$ corresponding to the labeling $\boldsymbol{L}$. And the smoothness term enforces the piecewise smoothness regularity of the optical flow by penalizing spatial variance in the flow field $(\boldsymbol{U}, \boldsymbol{V})$.

We also define data energy function $\mathcal{D}(r, \ell_r)$ based on ZNCC as in (7)

$$\mathcal{D}(r_i, \ell_{r_i}) = \frac{\sum_{(x,y)\in r_i}\left(1.0 - \eta\left(\boldsymbol{I}_t(x,y), \boldsymbol{I}_{t+1}\left(x + u_{\ell_{r_i}}, y + v_{\ell_{r_i}}\right)\right)\right)}{N_{r_i}} \tag{10}$$

and define the smoothness energy function differently as

$$\mathcal{S}\left(r_i, r_j, \ell_{r_i}, \ell_{r_j}\right) = \left|u_{\ell_{r_i}} - u_{\ell_{r_j}}\right| + \left|v_{\ell_{r_i}} - v_{\ell_{r_j}}\right| \tag{11}$$

4.3. *Multi-view Space-time Consistent Video Depth Recovery*

In this section, we will extend and combine the above region-tree based stereo matching and optical flow estimation for multi-view space-time consistent depth recovery, which is very important for many applications such as depth based free viewpoint video (FVV) rendering [22]. Because of the space limitation, only a brief overview is given here. Details can be found in an upcoming paper [18].

Formally speaking, given a set of M video sequences $\widehat{\boldsymbol{S}} = \{\boldsymbol{S}_m | m = 0, \ldots, M-1\}$, in which each $\boldsymbol{S}_m$ has T video frames $\widehat{\boldsymbol{I}}_m = \{I_m^t | t = 0, \ldots, T-1\}$ and is captured by a stationary camera C_m for view V_m with known intrinsic calibration matrix $\boldsymbol{K}_m$ and extrinsic parameters $(\boldsymbol{R}_m, \boldsymbol{T}_m)$, the objective of the space-time consistent video depth recovery problem is to estimate M depth map sequences $\widehat{\boldsymbol{D}}_m = \{\boldsymbol{D}_m | m = 0, \ldots, M-1\}$ by maximizing the temporal consistency between any two consecutive depth maps D_m^t and D_m^{t+1} for view V_m, and the spatial consistency between the synchronized depth maps $\widehat{\boldsymbol{D}}^t = \{D_m^t | m = 0, \dots, M-1\}$ for time instant t.

We follow the well-known plane-sweeping scheme [23]. That is, the depth search space is globally discretized into a set of depth planes $\boldsymbol{\Pi}_n = (0,0,d_n)^T$ $(n = 0, \dots, N-1)$ perpendicular to the optical axis of reference camera C_0. The corresponding depths $\{d_n\}$ tessellate a known depth range $[d_{near}, d_{far}]$ non-linearly. Each depth plane $\boldsymbol{\Pi}_n$ is tagged with a label $l = n$, which together form the label set $\boldsymbol{\mathcal{L}}$. In this way, we can solve the video depth recovery problem using our labeling framework.

In particular, our method includes two passes. In each pass, the temporal consistency is enforced using a temporal variant of the region-tree and the spatial consistency is enforced using inconsistency-map based iterative refinement [18]. After all T frames of $\hat{\boldsymbol{I}}_m$ have been processed for each view V_m in the first pass, we estimate the corresponding background color/depth information and used it in the second pass for background biased optimization to further improve the spacetime consistency between depth maps.

5. Experiments

Extensive experiments are done to evaluate the proposed framework in different applications discussed above. For each class of experiments, the same parameter settings are used. For image segmentation, in addition to the default parameters used in [14], two extra parameters controlling the allowed minimal and maximal region size are added to control the segmentation granularity. The maximal region size is only a soft constraint. That is, some regions are allowed to exceed this upper limit since otherwise some very small regions, which violate the lower size limit, may never be fused with its most similar neighbor. These two parameters are set based on image dimension. As for the most sensitive parameter, the smoothness energy term weight λ is automatically calculated for a specific dataset in a similar way (adapted to region-based) as in [24] based on the Kullback-Leiber divergence $\overline{\text{KL}}$. That is, $\lambda = \alpha\overline{\text{KL}}$. For different applications, the multiplicative constant α can be different. In particular, we use $\alpha = 4/3$ for stereo matching and $\alpha = 0.15$ for optical flow estimation (since different smoothness energy functions are used). In all the experiments, we use the same region-tree edge weight scheme defined based on the Euclidean distance $dis(m_{r_i}, m_{r_j})$ between two region's modes in the feature space of mean-shift segmentation and the in-between boundary strength $\varkappa_{r_i \leftrightarrow r_j}$ as

$$w_{e_{(i,j)}} = dis(m_{r_i}, m_{r_j}) \cdot e^{\varkappa_{r_i \leftrightarrow r_j}} \quad (12)$$

5.1. *Binocular Stereo Matching*

(1) Middlebury dataset benchmarking

Four datasets, Tsukuba, Venus, Teddy and Cones in the new Middlebury stereo evaluation scheme [25] are quantitatively benchmarked. Specifically, the stereo matching quality is measured by the percentages of the number of bad pixels (whose absolute disparity error is larger than 1) to the total number of pixels in the whole image, in un-occluded regions and in discontinuity regions. In all the experiments, we set the segmentation granularity to the region size range of (10, 15) pixels. The ZNCC window size used is 3×3.

Figure 3 shows the dense disparity maps produced by our algorithm and the ground truths for the above mentioned dataset. For comparison, the results of the pixel-tree based algorithm [12] are also included. It can be seen that the incorporation of the region-based approach using the region tree instead of the pixel tree does improve the matching accuracy for most datasets, especially the more challenging Teddy and Cones due to their fairly large disparity range. Shown in Table 1 is the ranking of the overall quality of our results. It can be seen that our algorithm performs quite well and is ranked among state-of-the-art algorithms.

(2) Performance comparison of using different optimization methods

Our region-tree based labeling framework can also use other optimization methods which are capable of handling non-grid structure such as belief propagation or graph cuts. To investigate the performance difference of using different optimization methods, we compare DP with graph cuts. In particular, we use the expansion based graph cuts implementation from [1], which provides a generalized coding interface for defining neighborhood system of labeling primitives. In addition to comparing graph cuts on region-tree based stereo matching to our DP on region-tree based one, we also investigate the performance of using graph cuts on region adjacency graph for comparison.

Specifically, the top two rows of Figure 4 are disparity results using graph cuts on region-tree. For Venus, Teddy and Cones datasets, the results are very close to the ones using DP as shown in Figure 3, with the largest quality measure improvement for the Teddy dataset by 0.02%. The only exception is the Tsukuba dataset, for which the non-occlusion error rate decreased from 1.84% for DP to 1.61% for graph cuts. Therefore, we can see for a tree structure, DP has very similar performance to graph cuts in minimizing the corresponding form of energy functions. However, DP optimization is much faster than graph cuts optimization. For the Teddy or Cones datasets, the speedup can be more than 10 times.

C-SemiGlob [19]	17.0	2.61 40	3.29 35	9.89 38	0.25 16	0.57 15	3.24 20	5.14 7	11.8 12	13.0 6	2.77 2	8.35 6	8.20 7	5.76
SO+borders [29]	17.3	1.29 19	1.71 12	6.83 21	0.25 17	0.53 14	2.26 10	7.02 18	12.2 19	16.3 11	3.90 21	9.85 21	10.2 23	6.03
DistinctSM [27]	19.0	1.21 14	1.75 14	6.39 16	0.35 19	0.69 21	2.63 17	7.45 25	13.0 22	18.1 26	3.91 22	9.91 23	8.32 9	6.14
YOUR METHOD	19.0	1.84 30	2.47 29	9.23 35	0.21 11	0.48 11	2.61 16	6.77 16	11.8 11	17.4 24	3.42 13	8.87 15	9.35 17	6.20
OverSegmBP [26]	19.5	1.69 28	1.97 22	8.47 31	0.50 26	0.68 19	4.69 28	6.74 15	11.9 17	15.8 10	3.19 11	8.81 14	8.89 13	6.11
SegmentSupport [28]	20.2	1.25 16	1.62 10	6.68 18	0.25 15	0.64 18	2.59 15	8.43 32	14.2 29	18.2 27	3.77 19	9.87 22	9.77 21	6.44
CostAggr+occ [39]	20.5	1.38 22	1.96 21	7.14 25	0.44 23	1.13 29	4.87 29	6.80 17	11.9 15	17.3 22	3.60 15	8.57 10	9.36 18	6.20
RegionTreeDP [18]	21.6	1.39 24	1.64 11	6.85 22	0.22 12	0.57 15	1.93 8	7.42 24	11.9 16	16.8 16	6.31 44	11.9 38	11.8 31	6.56
EnhancedBP [24]	22.6	0.94 5	1.74 13	5.05 5	0.35 20	0.86 24	4.34 27	8.11 30	13.3 24	18.5 30	5.09 36	11.1 30	11.0 27	6.69

Table 1. Middlebury benchmark ranking.
The first red colored rows are results of our method using ZNCC based data energy function. The second red colored row is our old results using absolute difference based energy function [26]. Using ZNCC improves our ranking.

Reference Image Ground Truth Pixel-Tree results [12] Our results

Figure 3. Disparity results comparison to [12] on 4 Middlebury datasets.

On the other hand, the lower two rows of Figure 4 show the results of using graph cuts on region adjacency graph. As can be seen from the "bad pixel" maps shown, for all four dataset, the results are worse than the ones using region tree, with the results of the Venus dataset being the worst. One possible reason could be the fact the region-tree minimizes the number of edges crossing disparity discontinuities compared to a region graph and hence, avoids unnecessary and error-incurring smoothness enforcement on them. Therefore, using region-tree does have a unique advantage over using region graph.

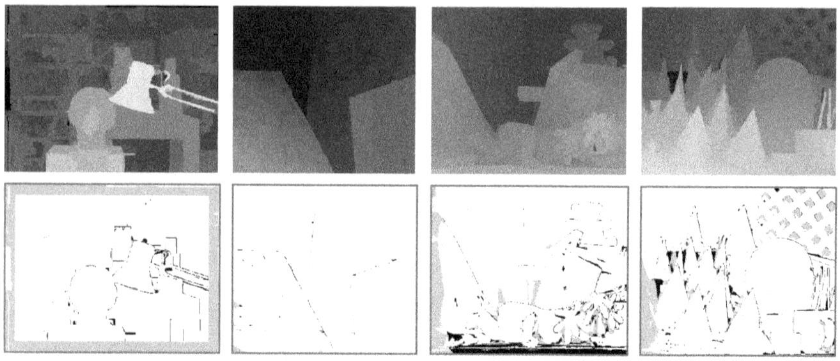

Disparity results and corresponding "bad pixel" maps using graph cuts and region-tree

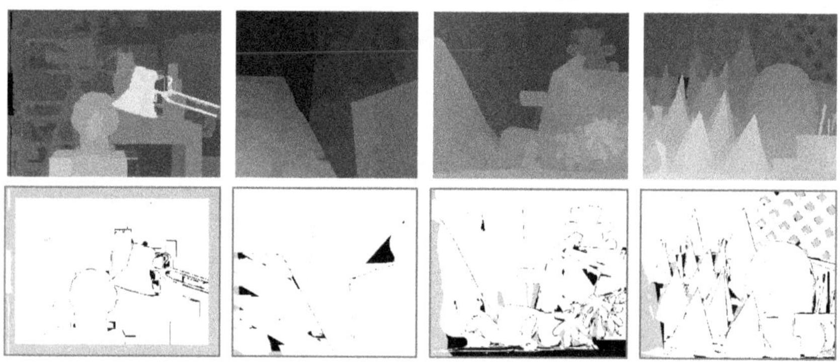

Disparity results and corresponding "bad pixel" maps using graph cuts and region-graph

Figure 4. Comparison of using graph cuts on region-tree and on region-graph.

(3) Performance sensitivity to segmentation granularity

We further investigate the performance sensitivity of our framework to segmentation granularity in binocular stereo matching. In Figure 5, profiling curves of disparity error rate with respect to minimum region size limit, which varies from 5 pixels to 55 pixels, in non-occlusion, all and discontinuity areas for the above 4 Middlebury datasets are shown, respectively. Two examples of the tested segmentation granularities are shown for the Teddy dataset in the bottom row. From the curves, we can see that the general performance is robust to segmentation granularity, though for some datasets with dominant slanted surfaces such the Venus dataset, a larger region size could incur a slightly higher error rate. The main reason is because our current formulation assumes that all pixels in each region have the same label (disparity). Therefore using larger regions will increase errors when disparity changes occur within a region, in which case using smaller regions is better.

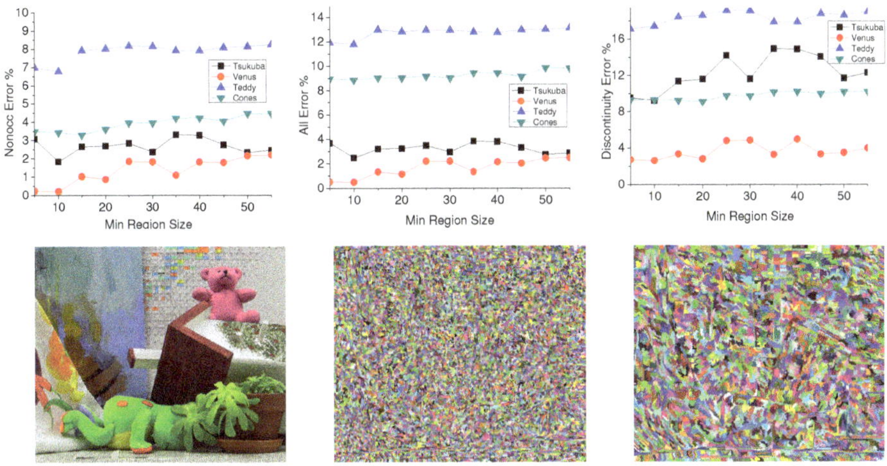

Figure 5. Profiling curves of performance to segmentation granularities on Teddy dataset. Bottom row shows two example segmentation granularities tested.

5.2. *Optical Flow Estimation*

We use the 2-frame color version of the Middlebury optical flow benchmarking datasets [27] to quantitatively evaluate our C2F region-trees based optical flow estimation. In particular, 12 image sequences from hidden fluorescent texture, realistic synthetic, stereo and real video categories are tested. Two segmentation levels are used in all the tests. The granularity in the coarse level is determined by limiting the region number to be around 2500, while for the fine level, the minimal region size is fixed at 5 pixels. The ZNCC window size is 5x5 for the fine level and 3x3 for the coarse level.

In table 2, we show the average angular error (AAE) and average end-point error (AEPE) of the top five algorithms at the time of submission. Our results of 5 datasets for quantitative evaluation are shown in Figures 2 and 6. In this chapter, all optical flow results are color coded using the same scheme as that specified in the Middlebury Stereo Vision site [27]. It can be seen that the overall ranking of our method is quite high (both 4th for AAE and AEPE). In particular, the lowest AAE is obtained for the Teddy dataset and the lowest AEPE is obtained for the Schefflera and Teddy and Grove datasets. Our method obtained superior evaluations around motion discontinuities in most datasets, showing our region-based representation has advantage in preserving motion boundaries. As shown in Figure 7, we compare the performance difference between using a multiple level C2F region tree and a single-level one. Specifically, the optical flow result is also obtained as shown in the second column by using only the finer level regions. However, coarse-to-fine displacement refinement is still performed. As we can see, more errors are incurred along motion boundaries compared to the result using two-level region-trees as shown in the third column. For reference, the result from the state-of-art continuous optimization based method [28] is also included. By comparison, we can see that using over-segmented regions instead of pixels does have a unique advantage in handling sharp motion discontinuities.

Figure 6. Example results (bottom row) on Middlebury datasets along with ground truths (2nd row).

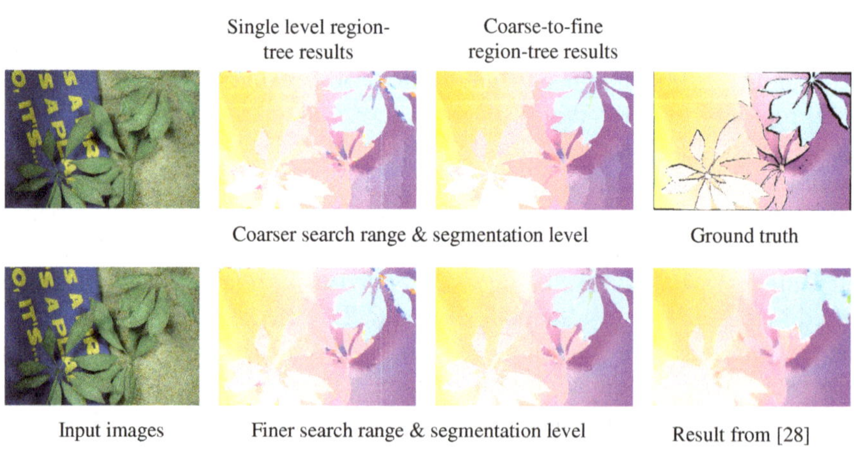

Figure 7. Comparison of using single-level region-tree and C2F region-tree for the "Schefflera" dataset.

Average angle error	avg rank	Army (Hidden texture) all	Army disc	Army untext	Mequon (Hidden texture) all	Mequon disc	Mequon untext	Schefflera (Hidden texture) all	Schefflera disc	Schefflera untext	Wooden (Hidden texture) all	Wooden disc	Wooden untext	Grove (Synthetic) all	Grove disc	Grove untext	Urban (Synthetic) all	Urban disc	Urban untext	Yosemite (Synthetic) all	Yosemite disc	Yosemite untext	Teddy (Stereo) all	Teddy disc	Teddy untext
Spatially variant [22]	4.1	3.73	10.2	3.33	3.02	11.0	2.67	5.36	13.8	2.35	3.67	19.3	1.84	3.81	4.81	3.69	4.48	16.0	3.90	2.11	3.28	2.12	4.66	9.41	4.35
TV-L1-improved [20]	4.8	3.36	9.63	2.62	2.82	10.7	2.23	6.50	15.6	2.73	3.80	21.3	1.76	3.24	4.38	2.39	5.97	16.1	5.67	3.57	4.92	3.43	4.01	9.84	3.44
F-TV-L1 [18]	6.3	5.44	12.5	5.69	5.46	15.0	4.03	7.48	16.3	3.42	5.08	23.3	2.81	3.42	4.34	3.03	4.05	15.1	3.18	2.43	3.92	1.87	3.90	9.35	2.61
Brox et al. [8]	7.8	4.60	14.4	4.29	4.05	13.5	3.71	6.63	16.0	7.26	5.22	22.7	3.22	4.56	6.09	3.40	3.97	17.9	3.41	2.07	3.76	1.18	5.14	11.9	4.28
DPOF [21]	7.8	5.63	10.9	4.16	4.05	12.1	3.31	3.87	8.82	3.17	4.34	16.2	3.13	3.95	4.78	4.17	6.69	15.2	6.27	5.62	6.89	6.60	2.44	4.83	3.74

Average end-point error	avg rank	Army (Hidden texture) all	Army disc	Army untext	Mequon (Hidden texture) all	Mequon disc	Mequon untext	Schefflera (Hidden texture) all	Schefflera disc	Schefflera untext	Wooden (Hidden texture) all	Wooden disc	Wooden untext	Grove (Synthetic) all	Grove disc	Grove untext	Urban (Synthetic) all	Urban disc	Urban untext	Yosemite (Synthetic) all	Yosemite disc	Yosemite untext	Teddy (Stereo) all	Teddy disc	Teddy untext
Spatially variant [22]	4.3	0.10	0.27	0.08	0.22	0.76	0.19	0.43	1.00	0.18	0.19	1.05	0.10	1.05	1.41	1.16	0.59	1.61	0.43	0.13	0.11	0.28	0.96	1.72	1.28
TV-L1-improved [20]	5.0	0.09	0.26	0.07	0.20	0.71	0.16	0.53	1.16	0.22	0.21	1.24	0.11	0.90	1.31	0.72	1.51	1.93	0.84	0.18	0.17	0.31	0.73	1.62	0.87
F-TV-L1 [18]	6.3	0.14	0.35	0.14	0.34	0.98	0.26	0.59	1.19	0.26	0.27	1.36	0.16	0.90	1.30	0.76	0.54	1.62	0.36	0.13	0.15	0.20	0.68	1.56	0.66
DPOF [21]	6.5	0.15	0.30	0.11	0.34	1.01	0.25	0.29	0.59	0.26	0.26	0.94	0.20	0.80	1.13	0.63	0.90	1.85	0.66	0.27	0.22	0.54	0.65	1.20	0.99
Fusion [9]	6.8	0.11	0.34	0.10	0.19	0.69	0.16	0.29	0.66	0.23	0.20	1.19	0.14	1.07	1.42	1.22	1.35	1.49	0.86	0.20	0.20	0.26	1.07	2.07	1.39

Table 2. Middlebury benchmark ranking (average angle & endpoint errors). Red colored rows are results of our method.

5.3. *Multi-view space-time consistent video depth recovery*

We extensively tested the region-tree based multi-view space-time consistent video depth recovery using a variety of multi-view stereo video sequences, including the Microsoft Research Breakdancing and Ballet datasets [22] and sequences captured by our own camera array. All video sequences were captured with synchronized PointGrey 1024×768 cameras at a frame rate of 15 FPS. The Microsoft sequences are particularly challenging due to large indoor textureless surfaces with surface reflections, and fast moving people. Our implementation processes each video in parallel using a cluster of 8 PCs and uses GPUs for match cost evaluation. We use the same depth tessellation scheme as the one used in [22] to tessellate the depth range $[d_{near}, d_{far}]$ non-linearly using $N = 256$ levels. In all the experiments, the same set of empirically tuned parameters are used: $\delta = 0.008|d_{far} - d_{near}|$. Due to the larger image size, we set the segmentation granularity to the region size range of (20, 30) pixels and the ZNCC window size as 5×5 for half-size frame. Furthermore, 3 iterations are found enough for acceptable accuracy.

Shown in Figure 8 is an illustration of the typical steps in our proposed algorithm. In particular, frame 57 of view 7 of the Ballet dataset is used as an example. The second row shows the inconsistency map based iterative optimizations of the depth map in the first pass. After the first pass, the background information is estimated. Then with the background information available, which is shown in the first row, in the second pass, the depth map is iteratively refined. As we can see, in the first iteration of pass one, the depth map has many errors. By iteratively enforcing spatial consistency, many of these errors are corrected and accordingly the number of inconsistent pixels (white pixels in the inconsistency map) decreases during iterations.

Example depth map results for frames 78 and 79 of the Breakdancing dataset are illustrated in Figure 9. By comparison to results of [2], we can see that ours performs better in handling occlusions. Furthermore, in our results, the depth values of static and textureless regions of the scene such as the wall and floor are more temporally consistent than those shown in [2] due to the use of temporal region-tree. Furthermore, the use of background information in the second pass further improves the depth accuracy.

Figure 10 shows an example of the datasets captured using our camera array. Results of 4 of the 8 views are illustrated in the second row, including two boundary ones. In the first row, an example of the optical flow estimation result for two consecutive frames is shown. As we can see, the optical flow result is quite accurate. For low FPS sequences with fast movements like the Breakdancing one [22], even though more errors could be incurred, most static regions (colored as white) are still accurate. Conservatively using only static regions to build temporal region tree can lower the impact of inaccuracies in optical flow estimation while still can improve in enforcing temporal consistency.

In term of efficiency, our un-optimized implementation takes about 6 minutes (two passes) to recover the depth map for each half-size frame using 3 iterations and 256 depth tessellation levels.

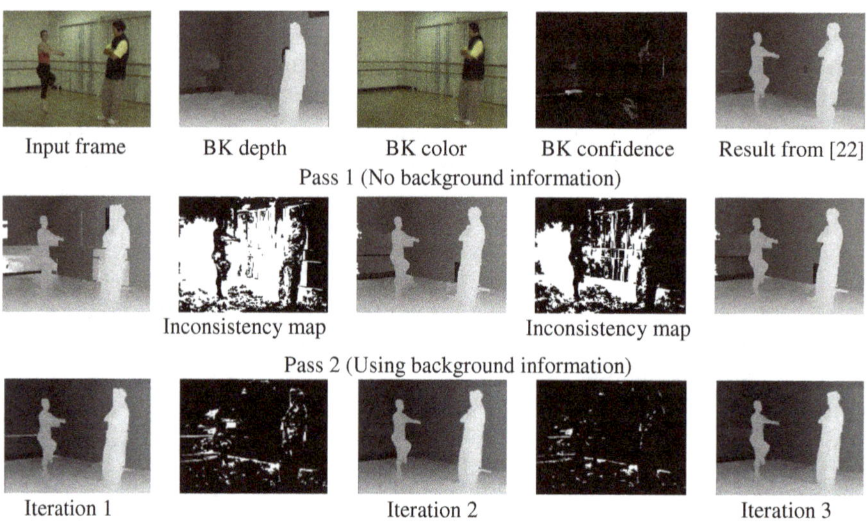

Figure 8. Illustration of the typical steps in our video depth recovery algorithm with frame 57 view 7 of the Ballet dataset.

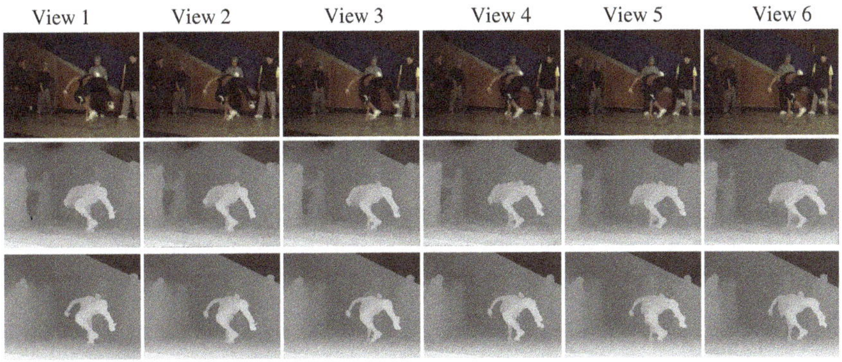

(a) Comparison for frame 78. Top: Input video frames. Middle: Microsoft results [22]. Bottom: Our results

(b) Our results for frame 79

Figure 9. Qualitative comparison of our video depth results to results from [22] on the Breakdancing dataset.

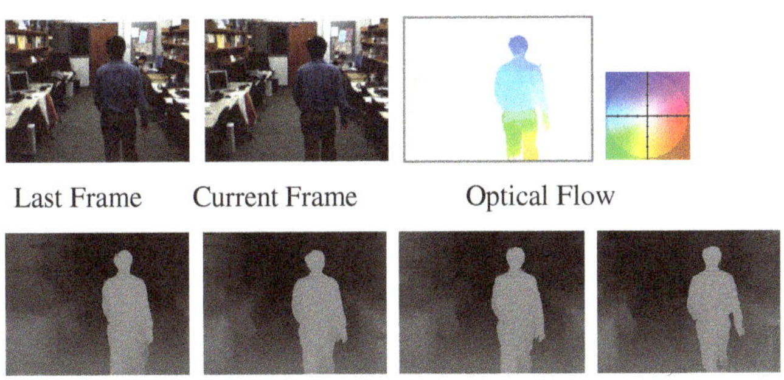

Figure 10. Video depth results of an example lab test dataset.

6. Conclusions and Discussions

In this chapter, we have presented a region-tree based framework for general image discrete labeling problems. By using the over-segmentation based region-tree as image representation, our framework combines the advantages of enabling labeling primitives to contain enough information with a large support area compared to the pixel based approaches and of reducing the risk of propagating segmentation errors into the labeling process compared to the layer based approaches. Compared to the region-graph representation, region-tree enables the use of a fast optimization method such as dynamic programming and enhances the performance by avoiding smoothness constraints between adjacent labeling primitives crossing discontinuities. The presented framework has been applied to a couple of different important computer vision tasks such as stereo matching and optical flow estimation. The extensive experimental evaluations have shown that our proposed framework is very effective and produces promising results. However, our framework does have limitation in handling the class of labeling problems in which the labeling target image itself has to be solved such as in interactive photo editing [8]. Since the labeling target image is unknown before the labeling problem is solved, the corresponding image segmentation cannot be done, for which only pixel based representation can be used.

As for future work, we would like to investigate more applications of our framework and plan to further generalize all the proposed variants.

Acknowledgements

The authors would like to thank NSERC, iCore, and the University of Alberta for their generous financial support. Special thanks are also due to Steve Sutphen who helped to develop the hardware used in this research and the hardware support team, in particular, Rene Leiva, Charles Jobagy, and Rick Hughes for their help in maintaining the hardware in the lab.

References

1. R. Szeliski, R. Zabih, D. Scharstein, O. Veksler, V. Kolmogorov, A. Agarwala, M. F. Tappen, C. Rother. A Comparative Study of Energy Minimization Methods for Markov Random Fields with Smoothness-Based Priors. IEEE Trans. Pattern Anal. Mach. Intell. 30(6): 1068-1080 (2008).
2. Y. Boykov, O. Veksler, and R. Zabih. Fast Approximate Energy Minimization via Graph Cuts. IEEE Trans. Pattern Analysis and Machine Intelligence, vol. 23, no. 11, pp. 1222-1239, Nov. 2001.
3. V. Kolmogorov and R. Zabih. What Energy Functions can be Minimized via Graph Cuts. IEEE Trans. Pattern Analysis and Machine Intelligence, vol. 26, no. 2, pp. 147-159, Feb. 2004.
4. J. Yedidia, W. Freeman, and Y. Weiss. Generalized belief propagation. Advances in Neural Information Processing Systems, pp. 689-695, 2000.
5. M. Bleyer and M. Gelautz. A layered stereo algorithm using image segmentation and global visibility constraints. ICIP 2004.
6. C. L. Zitnick, S. B. Kang. Stereo for image-based rendering using image over-segmentation. International Journal of Computer Vision 75(1): 49-65 (2007).
7. Y. Taguchi, B. Wilburn, C. L. Zitnick. Stereo reconstruction with mixed pixels using adaptive over-segmentation. CVPR 2008
8. A. Agarwala,M. Dontcheva, M. Agrawala, S. Drucker, A. Colburn,B. Curless, D. Salesin, and M. Cohen. Interactive Digital Photomontage. ACM Trans. Graphics, vol. 23, no. 3, pp. 294-302, 2004.
9. Z. Wang and Z. Zheng. A region based stereo matching algorithm using cooperative optimization. CVPR 2008.

10. A. Klaus, M. Sormann and K. Karner. Segment-based stereo matching using belief propagation and a self-adapting dissimilarity measure. ICPR 2006.
11. D. Scharstein and R. Szeliski. A taxonomy and evaluation of dense two-frame stereo correspondence algorithms. in *International Journal of Computer Vision*, volume 47, pp. 7-42, April 2002.
12. O. Veksler. Stereo Correspondence by Dynamic Programming on a Tree. in *Proc. IEEE Conference on Computer Vision and Pattern Recognition*, San Diego, USA, 2005, Vol. 2, pp. 384-390.
13. Y. Deng, X. Y Lin. A Fast Line Segment Based Dense Stereo Algorithm Using Tree Dynamic Programming. ECCV (3) 2006: 201-212
14. C. Christoudias, B. Georgescu, and P. Meer. Synergism in low-level vision. *16th International Conference on Pattern Recognition*, Quebec City, Canada, August 2002, vol. IV, 150-155
15. D. Comaniciu and P. Meer. Mean shift: A robust approach toward feature space analysis. in IEEE *Trans. on Pattern Analysis and Machine Intelligence*, Vol.24:603–619, May 2002.
16. T. H. Cormen, C. E. Lerserson, Ponald L. Rivest, and C. Stein. Introduction to algorithms. The MIT Press, 2nd Edition, 2001
17. C. Lei and Y. H. Yang. Optical Flow Estimation on Coarse-to-Fine Region-Trees using Discrete Optimization. International Conference on Computer Vision, Kyoto, Japan, September 27-October 2, 2009.
18. C. Lei, X. Chen and Y. H. Yang. A New Multiview Spacetime-Consistent Depth Recovery Framework for Free Viewpoint Video Rendering. International Conference on Computer Vision, Kyoto, Japan, September 27-October 2, 2009.
19. V. Lempitsky, S. Roth, and C. Rother. FusionFlow. Discrete-continuous optimization for optical flow estimation. CVPR 2008.
20. B. Glocker, N. Paragios, N. Komodakis. Optical flow estimation with uncertainties through dynamic MRFs. CVPR 2008
21. O. Faugeras, B. Hotz, H. Mathieu, T. Viille, Z. Zhang, P. Fua, E. Theron, L. Moll, G. Berry, J. Vuillemin, P. Bertin and C. Proy. Real time correlation-based stereo: Algorithm, implementations and applications. INRIA, Tech. Report. RR-2013, 1993.
22. C. L. Zitnick, S. B. Kang, M. Uyttendaele, S. A. J. Winder, R. Szeliski. High-quality video view interpolation using a layered representation. ACM Trans. on Graphics. 23(3): 600-608 (2004)
23. R. Yang and M. Pollefeys. Multi-resolution real-time stereo on commodity graphics hardware. in Int. Conf. on Computer Vision and Pattern Recognition," pp. I: 211.217, 2003.
24. J. Sun, Y. Li, S.B. Kang, and H.-Y. Shum. Symmetric stereo matching for occlusion handling. in Proc. *IEEE Conference on Computer Vision and Pattern Recognition*, June 2005, vol. 2, pp. 399-406
25. D. Scharstein and R. Szeliski. A taxonomy and evaluation of dense two-frame stereo correspondence algorithms. in *International Journal of Computer Vision*, volume 47, pp. 7-42, April 2002. http://www.middlebury.edu/stereo/
26. C. Lei, J. M. Selzer, Y. H. Yang. Region-Tree Based Stereo Using Dynamic Programming Optimization. CVPR (2) 2006: 2378-2385
27. S. Baker, D. Scharstein, J. Lewis, S. Roth, M. J. Black, and R. Szeliski. A database and evaluation methodology for optical flow. In Proceedings of ICCV 2007. http://vision.middlebury.edu/flow/
28. M. J. Black and P. Anandan. The robust estimation of multiple motions: parametric and piecewise smooth flow fields. Computer Vision and Image Understanding, 63(1):75–104, January 1996.

Chapter 2.2

binocular stereo matching, 165
dynamic programming, 163
image labeling frameworks, 159
optical flow estimation, 166
over-segmentation, 162
region adjacency graph, 161
region tree, 163

CHAPTER 2.3

STATISTICAL BACKGROUND MODELING FOR FOREGROUND DETECTION: A SURVEY

Thierry Bouwmans, Fida El Baf and Bertrand Vachon

Laboratory MIA, University of La Rochelle, Avenue M. Crepeau, 17000 La Rochelle
E-mail: tbouwman@univ-lr.fr

Background modeling is often used in the context of moving objects detection from static cameras. Numerous methods have been developed over the recent years and the most used are the statistical ones. The purpose of this chapter is to provide a recent survey of these different statistical methods. For this, we have classified them in term of generation following the years of publication and the statistical tools used. We then focus on the first generation methods: Single Gaussian, Mixture of Gaussians, Kernel Density Estimation and Subspace Learning using PCA. These original methods are reminded and then we have classified their different improvements in term of strategies. After analyzing the strategies and identifying their limitations, we conclude with several promising directions for future research.

1. Introduction

Background modeling is used in different applications to model the background and then detect the moving objects in the scene like in video surveillance [1], optical motion capture [2] and multimedia [3]. The simplest way to model the background is to acquire a background image which doesn't include any moving object. In some environments, the background isn't available and can always be changed under critical situations like illumination changes, objects being introduced or removed from the scene. To take into account these problems of robustness and adaptation, many background modeling methods have been developed and the most recent survey can be found in [4]. These background modeling methods can be classified in the following categories: Basic Background Modeling [5, 6, 7], Statistical Background Modeling [8, 9, 10], Fuzzy Background Modeling [11, 12] and Background Estimation [13, 14, 15]. All these modeling approaches are used in background subtraction context which presents the following steps and issues: background modeling, background initialization, background maintenance, foreground detection, choice of the feature size (pixel, a block or a cluster), choice of the feature type (color features, edge features, stereo features, motion features and texture features). Developing a background subtraction method, all these choices determine the robustness of the method to the critical situations met in video sequence [3, 13]: Noise image due to a poor quality image source (NI), Camera jitter (CJ), Camera automatic adjustments (CA), Time of the day (TD), Light switch (LS), Bootstrapping (B), Camouflage (C), Foreground aperture (FA), Moved background objects (MO), Inserted background objects (IBO), Multimodal background (MB), Waking foreground object (WFO), Sleeping foreground object (SFO) and Shadows (S). Different datasets benchmarks [16, 17] are available to evaluate the robustness of the background subtraction methods against these critical situations which have different spatial and temporal characteristics which must be take into account to obtain a good segmentation. This challenge must be made in the context of real-time application which runs on common PC and so two constraints are introduced: less computation time (CT) and less memory requirement (MR) as possible. The performance is evaluated using the ROC analysis [18] or the PDR Analysis [19] or the similarity measure [20].

Reading the literature, two remarks can be made: (1) The most used models are the statistical ones due to their robustness to the critical situations. (2) There are many improvements of these different statistical models as can be seen for the MOG model with the different acronyms found like GMM [21], TLGMM [22], STGMM [23], SKMGM [24], TAPPMOG [25] and S-TAPPMOG [26]. The motivation of this chapter concerns these two remarks. The objective is to categorize these different statistical models in one paper and classify their improvements following the different strategies used. We also discuss them following the challenges met in video sequences.

The rest of this chapter is organized as follows: In the Section 2, we firstly provide a background on the statistical background models and a classification of these models in term of generation. Then, we survey the first generation models and their respective improvements in the Section 3 for the Single Gaussian (SG), in the Section 4 for the Mixture of Gaussians (MOG), in the Section 5 for the Kernel Density Estimation (KDE) and in the Section 6 for the Subspace Learning methods using PCA (SL-PCA). Finally, the conclusion is given in Section 7.

2. Background

The first way to represent statistically the background is to assume that the history over time of intensity values of a pixel can be modeled by a single Gaussian (SG) [8]. However, a unimodal model cannot handle dynamic backgrounds when there are waving trees, water rippling or moving algae. To solve this problem, the Mixture of Gaussians (MOG) has been used to model dynamic backgrounds [9]. This model has some disadvantages. Background having fast variations cannot be accurately modeled with just a few Gaussians (usually 3 to 5), causing problems for sensitive detection. So, a non-parametric technique was developed for estimating background probabilities at each pixel from many recent samples over time using Kernel density estimation (KDE) [10] but it is time consuming. In [4], Subspace Learning using Principal Component Analysis (SL-PCA) is applied on N images to construct a background model, which is represented by the mean image and the projection matrix comprising the first p significant eigenvectors of PCA. In this way, foreground segmentation is accomplished by computing the difference between the input image and its reconstruction. These four models are the first generation ones using basic statistical models. The second generation models use more sophisticated statistical models as Support Vector Machine (SVM) [27], Support Vector Regression (SVR) [28] and Support Vector Data Description (SVDD) [29]. The third generation models generalize the first generation model as the Single General Gaussian (SGG) [30], the Mixture of General Gaussians (MOGG) [31] and subspace learning using Incremental Component Analysis (SL-ICA) [32], Incremental Non-negative Matrix Factorization (SL-INMF) [33] or Incremental Rank-(R_1,R_2,R_3) Tensor (SL-IRT) [34]. The Table 1 shows an overview of the statistical background modeling. The first column indicates the generation and the second column the name of each method. Their corresponding acronym is indicated in the first parenthesis and the number of papers counted for each method in the second parenthesis. The third column gives the name of the authors and the date of the first related publication. As we can see, the MOG (~100 papers) is the most used and improved due to a good compromise between robustness and time/memory requirements.

In the following sections, we survey the first generation methods and we have classify their related improvements in the following way: *intrinsic improvements* which concern the modification made in the initialization, the maintenance and the foreground detection steps, and *extrinsic improvements* which consist in using external tools to perform the results.

Table 1. Statistical Background Modeling: An Overview

Generation	Methods	Authors - Dates
First Generation (1997-2001)	Single Gaussian (SG) (5)	Wren *et al.* (1997) [8]
	Mixture of Gaussians (MOG) (~100)	Stauffer and Grimson (1999) [9]
	Kernel Density Estimation (KDE) (21)	Elgammal *et al.* (2000) [10]
	Subspace Learning using PCA (SL-PCA) (15)	Oliver *et al.* (1999) [89]
Second Generation (2002-2006)	Support Vector Machine (SVM) (3)	Lin *et al.* (2002) [27]
	Support Vector Regression (SVR) (2)	Wang *et al.* (2006) [28]
	Support Vector Data Description (SVDD) (5)	Tavakkoli *et al.* (2006) [29]
Third Generation (Since 2007)	Single General Gaussian (SGG) (3)	Kim *et al.* (2007) [30]
	Mixture of General Gaussian (MOGG) (3)	Allili *et al.* (2007) [31]
	Subspace Learning using ICA (SL-ICA) (1)	Tsai and Lai (2009) [32]
	Subspace Learning using INMF (SL-INMF) (2)	Bucak *et al.* (2007) [33]
	Subspace Learning using IRT (SL-IRT) (1)	Li *et al.* (2008) [34]

3. Single Gaussian (SG)

Principle: Wren *et al.* [8] have proposed to model the background independently at each pixel location (i,j). The model is based on ideally fitting a Gaussian probability density function on the last n pixel's values. In order to adapt the model over time at each new frame time t+1, the mean and the variance are updated using a running average scheme. The foreground detection is made as follows: if $|\mu_{t+1} - X_{t+1}| < k\sigma_{t+1}$, the pixel is classified as background otherwise the pixel is classified as foreground.

Improvements: Medioni *et al.* [35] have operated in the HSV color space. Indeed, the HSV color model is more robust than the RGB one to gradual illumination changes (TD) because it separates the intensity and chromatic information of the pixel. Zhao *et al.* [36] used HSV too but have remarked that the respective distribution of H and S vary naturally a lot and that the distribution of V is the most stable. So, the components H and S are used only when they are stable. Results show better performance in presence of gradual illumination changes (TD) and shadows (S).

4. Mixture of Gaussians (MOG)

Stauffer and Grimson [9] have proposed to model the recent history of the color features of each pixel $\{X_1,...,X_t\}$ by a mixture of K Gaussians. We remind below the algorithm.

Principle: First, each pixel is characterized by its intensity in the RGB color space. Then, the probability of observing the current pixel value is considered given by the following formula in the multidimensional case:

$$P(X_t) = \sum_{i=1}^{K} \omega_{i,t} . \eta(X_t, \mu_{i,t}, \Sigma_{i,t}) \tag{1}$$

where the parameters are K is the number of distributions, $\omega_{i,t}$ is a weight associated to the ith Gaussian at time t with mean $\mu_{i,t}$ and standard deviation $\Sigma_{i,t}$. η is a Gaussian probability density function:

$$\eta(X_t, \mu, \Sigma) = \frac{1}{(2\pi)^{n/2} |\Sigma|^{1/2}} e^{-\frac{1}{2}(X_t - \mu)\Sigma^{-1}(X_t - \mu)} \tag{2}$$

For computational reasons, Stauffer and Grimson [9] assumed that the RGB color components are independent and have the same variances. So, the covariance matrix is of the form:

$$\Sigma_{i,t} = \sigma_{i,t}^2 I \tag{3}$$

So each pixel is characterized by a mixture of K Gaussians. Once the background model is defined, different parameters of the mixture of Gaussians must be initialized. The parameters of the MOG's model are the number of Gaussians K, the weight $\omega_{i,t}$ associated to the ith Gaussian at time t, the mean $\mu_{i,t}$ and the covariance matrix $\Sigma_{i,t}$. K determined the multimodality of the background and by the available memory and computational power. Stauffer and Grimson [9] proposed to set K from 3 to 5. The initialization of the weight, the mean and the covariance matrix is made using an EM algorithm. Stauffer and Grimson [9] used the K-mean algorithm for real time consideration. Once the parameters initialization is made, a first foreground detection can be made and then the parameters are updated. Firstly, Stauffer and Grimson [9] used as criterion the ratio $r_j = \omega_j / \sigma_j$ and ordered the K Gaussians following this ratio. This ordering assumes that a background pixel corresponds to a high weight with a weak variance due to the fact that the background is more present than moving objects and that its value is practically constant. The first B Gaussian distributions which exceed certain threshold T are retained for a background distribution:

$$B = \arg\min_b \left(\sum_{i=1}^{b} \omega_{i,t} > T \right) \tag{4}$$

The other distributions are considered to represent a foreground distribution. Then, when the new frame incomes at times t+1, a match test is made for each pixel. A pixel matches a Gaussian distribution if the Mahalanobis distance $sqrt\left(\left(X_{t+1}-\mu_{i,t}\right)^T.\Sigma_{i,t}^{-1}.\left(X_{t+1}-\mu_{i,t}\right)\right)<k\sigma_{i,t}$ (5) where k is a constant threshold equal to 2.5. Then, two cases can occur:

- Case 1: A match is found with one of the K Gaussians. In this case, if the Gaussian distribution is identified as a background one, the pixel is classified as background else the pixel is classified as foreground.
- Case 2: No match is found with any of the K Gaussians. In this case, the pixel is classified as foreground.

At this step, a binary mask is obtained. Then, to make the next foreground detection, the parameters must be updated. Using the match test (5), two cases can occur like in the foreground detection:

- Case 1: A match is found with one of the K Gaussians. For the matched component, the weight, the mean and the variance are updated using a running average. For the unmatched components, μ and Σ are unchanged, only the weight are decreased.
- Case 2: No match is found with any of the K Gaussians. In this case, the least probable distribution k is replaced with a new one with a low prior weight, a large initial variance and the mean set to the current pixel.

Once the parameters maintenance is made, foreground detection can be made and so on. Complete studies on the signification and the setting of the parameters can be found in [38, 39].

Improvements: The original MOG presents several advantages. Indeed, it can work without having to store an important set of input data in the running process. The multimodality of the model allows dealing with multimodal backgrounds (MB) and gradual illumination changes (TD). Despite it, this model presents some disadvantages: the number of Gaussians must be predetermined, the need for good initializations, the dependence of the results on the true distribution law which can be non-Gaussian and slow recovery from failures. Others limitations are the needs for a series of training frames absent of moving objects and the amount of memory required in this step. To alleviate these limitations, numerous improvements have been proposed over the recent years. All the developed improvements can be classified following the strategies and a complete survey can be found in [40]. We have summarized them in the following. The first strategies called intrinsic improvements (Table 2) consist to be more rigorous in the statistical sense or to introduce spatial and/or temporal constraints in the different step of the model. For example, some authors [41-43] propose to determine automatically and dynamically the number of Gaussians to be more robust to dynamic backgrounds (MB). Other approaches use another algorithm for the initialization [44] or allow presence of foreground objects in the training sequence [45]. For the maintenance, the learning rates can be better set [47, 48] or adapted over time [49]. For the foreground detection, the improvement found in the literature are made using a different measure for the matching test [51], using a Pixel Persistence Map (PPM) [50-52], using the probabilities [53, 54], using a foreground model [55], or using the most dominant background model [56]. Another way to improve the efficiency and robustness of the original GMM consists in using external strategies (Table 3). Some authors used Markov Random Fields [57], hierarchical approaches [58, 59], multi-level approaches [60, 61], multiple backgrounds [62], multi-layer approaches [63], tracking feedback [64] or specific post-processing [65]. All the intrinsic and extrinsic model improvements concern the quality of the foreground detection but there is another manner to improve the original MOG which consists in reducing the computation time. It achieved by using region of interest [66], by using a variable adaption rate [67], by switching the background model [68], by using a space sampling strategies [69] or by using hardware implementation [70]. All the precedent improvements concern directly the original MOG and the

Table 2. MOG - Intrinsic improvements

Parameters Initialization	
Variable K	Zivkovic [41], Cheng *et al.* [42], Shimada *et al.* [43]
Variables μ, σ, ω	Another algorithm: Morellas *et al.* [44]
	Allowing presence of moving objects: Zhang *et al.* [45]
Parameters Maintenance	
Variable K	Zivkovic [41], Cheng *et al.* [42], Shimada *et al.* [43]
Variables μ, σ, ω	Maintenance rules: Han et Lin [46]
	Maintenance mechanisms: Zhang *et al.* [45]
Learning rates α, ρ	Better setting for the learning rates: Zang and Klette [47], White and Shah [48]
	Adaptive learning rates: KaewTraKulPong and Bowden [49]
Foreground Detection	
Different measure for the matching test	Morellas *et al.* [51]
Pixel Persistence Map (PPM)	Pnevmatikakis *et al.* [50], Landabaso *et al.* [52]
Probabilities	Yang and Hsu [53], Lee [54]
Foreground model	Whitagen *et al.* [55]
Most dominant background model	Haque *et al.* [56]

Table 3. MOG - Extrinsic improvements

Markov Random Fields	Schindler and Wang [57]
Hierarchical approaches	Sun and Yuan [58], Chen *et al.* [59]
Multi-level approaches	Zang and Klette [60], Zhong *et al.* [61]
Multiple backgrounds	Su and Hu [62]
Multi-layer approaches	Yang *et al.* [63]
Tracking feedback	Harville [64]
Post-processing	Turdu *et al.* [65]

foreground detection results only from it. Another way to improve this method is to enhance the results of the foreground detection by using cooperation with another segmentation method. It can be achieved by a cooperation with a statistical background disturbance technique [71], with a color segmentation [72] or with a region based motion detection [73].

5. Kernel Density Estimation (KDE)

Principle: Elgammal *et al.* [10] estimated the probability density function for each pixel using the kernel estimator K for N recent sample of intensity values $\{x_1, x_2, ..., x_N\}$ taken consecutively in a time size window W as follows:

$$P(x_t) = \frac{1}{N}\sum_{i=1}^{N} K(x_t - x_i) \quad (1)$$

where K() is the kernel estimator function which is taken as a Normal Gaussian function $N(0, \Sigma)$. So, the probability density function is determined as follows:

$$P(x_t) = \frac{1}{N}\sum_{i=1}^{N} \frac{1}{(2\pi)^{d/2}|\Sigma|^{1/2}} e^{-1/2*(x_t - x_i)^T \Sigma^{-1}(x_t - x_i)} \quad (2)$$

Elgammal *et al.* [15] assumed that the different color channels are independent with a different kernel bandwidths, then the kernel function bandwidth is as follows:

$$\Sigma = \begin{pmatrix} \sigma_1^2 & 0 & 0 \\ 0 & \sigma_2^2 & 0 \\ 0 & 0 & \sigma_3^2 \end{pmatrix}. \quad (3)$$

So, the probability density function can be written as follows:

$$P(x_t) = \frac{1}{N}\sum_{i=1}^{N}\prod_{j=1}^{d}\frac{1}{\sqrt{2\pi\sigma_j^2}}e^{-1/2*(x_{t_j}-x_{i_j})^T/\sigma_j^2} \quad (4)$$

Elgammal *et al.* [10] detected the foreground using the probabilities and a threshold T as follows: if $P(x_t) < T$ then the pixel classified as foreground else the pixel is classified as background. At this step, a binary mask is obtained. Then, to make the next foreground detection, the parameters must be updated. For this, Elgammal *et al.* [10] used two background models: a short term one and a long term one. To combine the advantage of each model and to eliminate their disadvantages, the final foreground detection is obtained by taking the intersection of the two foreground detection coming from each model.

Improvements: The original KDE presents several advantages. The multimodality of the model allows dealing with multimodal backgrounds particularly in fast changes (waving trees, water rippling, etc…). Despite it, this model present some disadvantages: N frames need to be kept in memory during the entire detection process which is costly memory wise when N is large. The algorithm is time consuming too due the complexity in O(N*N). To solve these problems, different improvements have been proposed. The intrinsic improvements consist in changing the kernel function [74-78] as shown in Table 4. For the training, some authors propose to decrease the number of samples by determining a proper size of the frame buffer [75], by using a diversity sampling scheme [79] or by using a sequential Monte Carlo sampling scheme [80].

Table 4. KDE - Intrinsic improvements

Background Model	
Gaussian Kernel Function	Automatic selection of kernel band width: Tavakkoli *et al.* [74]
Rectangular Kernel Function	Constant kernel band width: Ianasi *et al.* [75], Tanaka *et al.* [76] Variable kernel band width: Zivkovic [77]
Cauchy Kernel Function	Ramezani *et al.* [78]
Background Initialization	
Decreasing the number of samples	Adopting the proper size of frame buffer: Ianasi *et al.* [75] Diversity samples scheme: Mao and Shi [79] Sequential Monte Carlo sampling: Tang *et al.* [80]
Background Maintenance	
Recursive Maintenance	Tavakkoli *et al.* [81, 82]
Selective Maintenance	Tavakkoli *et al.* [74], Mao and Shi [79]
Foreground Detection	
Dissimilarity measure	Ianasi *et al.* [75]
Probability	Zivkovic [77]
Foreground model	Tavakkoli *et al.* [81, 82]

Table 5. KDE - Extrinsic improvements

Markov Random Fields	Pahalawatta *et al.* [83]
Hierarchical approaches	Orten *et al.* [84]
Multiples backgrounds	Tanaka *et al.* [85]
Graph cuts	Mahamud [86]

Furthermore, a recursive maintenance [81, 82] can be adopted to reduce the computation time. For the foreground detection, different scheme can be used as proposed by Ianasi *et al.* [75], Zivkovic [77], Tavakkoli *et al.* [81, 82]. For external strategies (Table 5), some authors used Markov Random Fields [83], hierarchical approaches [84], multiple backgrounds [85] and graph cuts [86]. Another way to improve this method is to enhance the results of the foreground detection by using cooperation with another segmentation method. It achieved by a cooperation with the consecutive frame difference [87] or using a subspace learning approach using PCA [88].

6. Subspace Learning using PCA (SL-PCA)

Principle: Oliver *et al.* [89] have proposed to model each background pixel using an eigenbackground model. This model consists in taking a sample of N images $\{I_1, I_2, ..., I_N\}$ and computing the mean background image μ_B and its covariance matrix C_B. This covariance matrix is then diagonalized using an eigenvalue decomposition as follows: $L_B = \Phi_B C_B \Phi_B^T$ (1) where Φ_B is the eigenvector matrix of the covariance of the data and L_B is the corresponding diagonal matrix of its eigenvalues.

In order to reduce, the dimensionality of the space, only M eigenvectors (M<N) are kept in a principal component analysis (PCA). The M largest eigenvalues are contained in the matrix L_M and the M vectors correspond to these M largest eigenvalues in the matrix Φ_M. Once the eigenbackground images stored in the matrix Φ_M are obtained and the mean μ_B too, the input image I_t can be approximated by the mean background and weighted sum of the eigenbackgrounds Φ_M. So, the coordinate in eigenbackground space of input image I_t can be computed as follows:

$$w_t = (I_t - \mu_B)^T \Phi_M \tag{4}$$

When w is back projected onto the image space, a reconstructed background image is created as follows:

$$B_t = \Phi_M w_t^T + \mu_B \tag{5}$$

Then, the foreground object detection is made as follows:

$$|I_t - B_t| > T \quad (6)$$

where T is a constant threshold.

Improvements: The eigenbackground model provides a robust model of the probability distribution function of the background, but not of the moving objects while they do not have a significant contribution to the model. So, the first limitation of this model is that the size of the foreground object must be small and don't appear in the same location during a long period in the training sequence. The second limitation appears for the background maintenance. Indeed, it is computationally intensive to perform model updating using the batch mode PCA. Moreover without a mechanism of robust analysis, the outliers or foreground objects may be absorbed into the background model. The third limitation is that the application of this model is mostly limited to the gray-scale images since the integration of multi-channel data is not straightforward. It involves much higher dimensional space and causes additional difficulty to manage data in general. To solve the first limitation, Xu *et al.* [90] propose to apply recursively an error compensation process to reduce the influence of foreground moving objects on the eigenbackground model. An adaptive threshold method is also introduced for background subtraction, where the threshold is determined by combining a fixed global threshold and a variable local threshold. Results show more robustness in presence of moving objects. Another approach developed by Kawabata *et al.* [91] consists in an iterative optimal projection method to estimate a varied background in real time from a dynamic scene with foreground. Firstly, background images are collected for a while and then the background images are compressed using eigenspace method to form a database. While monitoring the scene, new image is taken by a camera, and the image is projected onto the eigenspace to estimate the background. As the estimated image is much affected by the foreground, the foreground region is calculated by using background subtraction with former estimated background to exclude the region from the projection. Thus the image whose foreground region is replaced by the former background is projected to eigenspace and then the background is updated. Kawabata *et al.* proved that the cycle converges to a correct background image. For the maintenance, some authors propose incremental PCA algorithms. The adaptive incremental PCA proposed by Rymel *et al.* [92] need less computation time than the batch PCA but the background image is contamined by the foreground object. To solve this, Li [93], Skocaj *et al.* [94] and Zhang and Zhuang [95] propose incremental and robust PCA algorithms. Wang *et al.* [96] use a similar approach based on the sequential Karhunen-Loeve algorithm. All these incremental methods avoid the eigen-decomposition of the high dimensional covariance matrix using approximation of it and so a low decomposition is allowed at the maintenance step with less computational load. However, these incremental methods maintain the whole eigenstructure including both the eigenvalues and the exact matrix Φ_M. In this context, Li *et al.* [97] use a fast recursive and robust eigenbackground maintenance avoiding eigen-decomposition. For the third limitation, Han and Jain [98] propose an efficient algorithm using 2-Dimensional Principal Component Analysis (2DPCA). It is shown that the principal components in 2DPCA are computed efficiently by transformation to standard PCA. To perform the computational time, Han and Jain [98] use an incremental algorithm to update eigenvectors to handle temporal variations of background. The proposed algorithm was applied to 3-channel (RGB) and 4-channel (RGB+IR) data. Results show noticeable improvements in presence of multimodal background (MB) and shadows (S).

Table 7. SL-PCA - Intrinsic improvements

Background Maintenance	
Recursive Error Compensation	Xu *et al.* [90]
Iterative Optimal Projection	Kawabata *et al.* [91]
Incremental PCA	Rymel *et al.* [92]
Incremental and robust PCA	Li [93]
Weighted Incremental and Robust PCA	Skocaj *et al.* [94]
Adaptive Weight Selection for Incremental PCA	Zhang and Zhuang [95]
Sequential Karhunen-Loeve algorithm	Wang *et al.* [96]
Fast Recursive Maintenance	Li *et al.* [97]

7. Performance Evaluation

For the performance evaluation, we have compared the original version of the SG, MOG, KDE and SL-PCA using the Wallflower dataset provided by Toyama *et al.* [13]. This dataset is the most used and consists in a set of image sequences where each sequence presents a different type of difficulty that a practical task may meet. The performance is evaluated against hand-segmented ground truth. Three terms are used in evaluation: False Positive (FP) is the number of background pixels that are wrongly marked as foreground; False Negative (FN) is the number of foreground pixels that are wrongly marked as background; Total Error (TE) is the sum of FP and FN. A brief description of the Wallflower image sequences can be made as follows:

- Moved Object (MO) - A person enters into a room, makes a phone call, and leaves. The phone and the chair are left in a different position.
- Time of Day (TOD) - The light in a room gradually changes from dark to bright. Then, a person enters the room and sits down.
- Light Switch (LS) - A room scene begins with the lights on. Then a person enters the room and turns off the lights for a long period. Later, a person walks in the room, switches on the light, and moves the chair, while the door is closed.
- Waving Trees (WT) - A tree is swaying and a person walks in front of the tree.
- Camouflage (C) - A person walks in front of a monitor, which has rolling interference bars on the screen. The bars include similar color to the person's clothing.
- Boostrapping (B) - The image sequence shows a busy cafeteria and each frame contains people.
- Foreground Aperture (FA) - A person with uniformly colored shirt wakes up and begins to move slowly.

Table 8. The top row shows the test image. The second row shows the ground truth. After, each row shows the results of the algorithm cited.

Sequence	MO Frame 985	TD Frame 1850	LS Frame 1865	WT Frame 247	C Frame 251	B Frame 299	FA Frame 489
Test image							
Ground Truth							
SG Wren *et al.* [8]							
MOG Stauffer *et al.* [9]	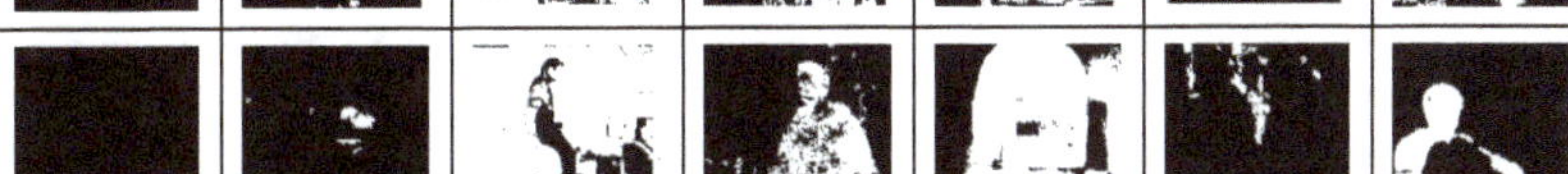						
KDE Elgammal *et al.* [10]							
SL-PCA Oliver *et al.* [89]							

Table 9. Performance of algorithms on the Wallflower sequences

Algorithm	Error	Problem Type							
		MO	TD	LS	WT	C	B	FA	**TE**
SG	FN	0	949	1857	3110	4101	2215	3464	
Wren *et al.* [8]	FP	0	535	15123	357	2040	92	1290	35133
MOG	FN	0	1008	1633	1323	398	1874	2442	
Stauffer and Grimson [9]	FP	0	20	14169	341	3098	217	530	27053
KDE	FN	0	1298	760	103	238	1755	2413	
Elgammal *et al.* [10]	FP	0	125	14153	284	3392	933	624	26078
SL-PCA	FN	0	879	962	1027	350	304	2441	
Oliver *et al.* [89]	FP	1065	16	362	2057	1548	6129	537	17677

The sequences MO, TOD, LS evaluate the adaptation of the model. The sequence WT shows the robustness in the case of dynamic backgrounds. The Table 8 shows the image results obtained for each algorithm on each sequence. The Table 9 shows the performance of each algorithm. Results for SG, MOG and SL-PCA come from [13]. The SL-PCA presents globally the best performance due to its results on the sequence LS but it is not the best algorithm for MO and WT. For dynamic backgrounds (WT), the KDE is the most robust followed by the MOG. This comparison gives not a definitive ranking of these algorithms because the performance is dependent of three factors: the task, the sequence and the application.

8. Conclusion

This chapter attempts to provide a comprehensive survey of research on statistical background modeling. Thus, we have proposed a classification in term of generation: The first generation (SG, MOG, KDE, SL-PCA), the second generation (SVM, SVR, SVDD) and the third generation (SGG, MOGG, SL-ICA, SL-INMF, SL-IRT). Then, we have focus on the different improvements of the first generation methods that we have categorize in two classes intrinsic and extrinsic model improvements. Strategies adding spatial and temporal information in the different steps or in added process have shown their abilities to improve the robustness of the original model to critical situations. Cooperation with another segmentation has shown their interests too. Methods which reduce the computation time permit to deal with the constraints of real-time application. Although significant progress have been made since the original methods, there is still work to be done and we believe that the second and the third generation are very promising methods.

In conclusion, this chapter presents a survey of the recent strategies in statistical background modeling. Our future works concern a recent state-of the-art of background subtraction methods.

References

[1] S. Cheung, C. Kamath. Robust Background Subtraction with Foreground Validation for Urban Traffic Video. Journal of Applied Signal Processing, 14, 2330-2340, 2005.

[2] J. Carranza, C. Theobalt. M. Magnor, H. Seidel. Free-Viewpoint Video of Human Actors. ACM Transactions on Graphics, 22 (3), 569-577, 2003.

[3] F. El Baf, T. Bouwmans, B. Vachon. Comparison of Background Subtraction Methods for a Multimedia Learning Space. SIGMAP 2007, Jul. 2007.

[4] S. Elhabian, K. El-Sayed, S. Ahmed. Moving Object Detection in Spatial Domain using Background Removal Techniques - State-of-Art. Recent Patents on Computer Science, 1 (1), 32-54, Jan. 2008.

[5] B. Lee, M. Hedley. Background Estimation for Video Surveillance. Image and Vision Computing New Zealand, IVCNZ 2002, 315-320, 2002.

[6] N. McFarlane, C. Schofield. Segmentation and tracking of piglets in images. British Machine Vision and Applications, BMVA 1995, 187-193, 1995.

[7] J. Zheng, Y. Wang, N. Nihan, E. Hallenbeck. Extracting Roadway Background Image: A mode based approach. Journal of Transportation Research Report, 1944, 82-88, Mar. 2006.

[8] C. Wren, A. Azarbayejani, T. Darrell, A. Pentland. Pfinder: Real-Time Tracking of the Human Body. IEEE Transactions on PAMI, 19 (7), 780 –785, 1997.

[9] C. Stauffer, W. Grimson, Adaptive background mixture models for real-time tracking. Conference on Computer Vision and Pattern Recognition, CVPR 1999, 246-252, 1999.

[10] A. Elgammal, D. Harwood, L. Davis. Non-parametric Model for Background Subtraction", ECCV 2000. 751-767, Jun. 2000.

[11] M. Sigari, N. Mozayani, H. Pourreza. Fuzzy Running Average and Fuzzy Background Subtraction: Concepts and Application. Journal of Computer Science and Network Security, 8 (2), 138-143, 2008.

[12] F. El Baf, T. Bouwmans, B. Vachon. Type-2 fuzzy mixture of Gaussians model: Application to background modeling. International Symposium on Visual Computing, 772-781, Dec. 2008.

[13] K. Toyama, J. Krumm, B. Brumitt, B. Meyers. Wallflower: Principles and Practice of Background Maintenance. ICCV 1999, 255-261, Sept. 1999.

[14] S. Messelodi, C. Modena, N. Segata, M. Zanin. A Kalman filter based background updating algorithm robust to sharp illumination changes. ICIAP 2005, 3617, 163-170, Sept. 2005.

[15] R. Chang, T. Ghandi, M. Trivedi. Vision modules for a multi sensory bridge monitoring approach. ITSC 2004, 971-976, Oct. 2004.

[16] Wallflower : http://research.microsoft.com/users/jckrumm/WallFlower/TestImages.htm

[17] VSSN Dataset : http://mmc36.informatik.uni-augsburg.de/VSSN06_OSAC/

[18] J. Davis, M. Goadrich. The Relations hip between Precision-Recall and ROC Curves. ICML 2006, 148, 233-240, 2006.

[19] K. Kim, T. H. Chalidabhongse, D. Harwood, L. Davis. PDR: Performance Evaluation Method for Foreground-Background Segmentation Algorithms. Journal on Applied Signal Processing, 2006.

[20] L. Li, W. Huang, I. Gu, Q.Tian. Statistical Modeling of Complex Background for Foreground Object Detection. IEEE Transaction on Image Processing, 13 (11), 1459-1472, Nov. 2004.

[21] B. Han, X. Lin. Update the GMMs via adaptive Kalman filtering. SPIE - The International Society for Optical Engineering, 5960 (3), 1506-1515, 2005.

[22] H. Yang, Y. Tan, J. Tian, J. Liu. Accurate dynamic scene model for moving object detection", International Conference on Image Processing, 6, 157-160, 2007.

[23] W. Zhang, X. Fang, X. Yang, Q. Wu. Spatiotemporal Gaussian mixture model to detect moving objects in dynamic scenes. Journal of Electronic Imaging, 16 (2), Apr. 2007.

[24] P. Tang, L. Gao, Z. Liu. Salient Moving Object Detection Using Stochastic Approach Filtering. ICIG 2007, 530-535, 2007.

[25] M. Harville. A framework for high-level feedback to adaptive, per-pixel, mixture-of-Gaussian background models", 7th European Conference on Computer Vision, 543 -560, May 2002.

[26] M. Cristani, V. Murino. A spatial sampling mechanism for effective background subtraction. VISAPP 2007, 2, 403-410, Mar. 2007

[27] H. Lin, T. Liu, J. Chuang. A probabilistic SVM approach for background scene initialization. International Conference on Image Processing, 3, 893-896, Sept. 2002.

[28] J. Wang, G. Bebis, R. Miller. Robust Video-Based Surveillance by Integrating Target Detection with Tracking. OTCBVS 2006, 137-144, Jun. 2006.

[29] A. Tavakkoli, M. Nicolescu, G. Bebis. A Novelty Detection Approach for Foreground Region Detection in Videos with Quasi-stationary Backgrounds. ISVC 2006, 40-49, Nov. 2006.

[30] H. Kim, R. Sakamoto, I. Kitahara, T. Toriyama, K. Kogure. Robust Foreground Extraction Technique Using Gaussian Family Model and Multiple Thresholds. ACCV2007, Nov. 2007.

[31] M. Allili, N. Bouguila, D. Ziou. A Robust Video Foreground Segmentation by Using Generalized Gaussian Mixture Modeling. CRV 2007, 503-509, 2007.

[32] D. Tsai and C. Lai. Independent component analysis based background subtraction for indoor surveillance. IEEE Transactions on Image Processing, 18(1):158–167, Jan. 2009.

[33] S. Bucak, B. Gunsel, O. Gursoy. Incremental Non-negative Matrix Factorization for Dynamic Background Modelling. International Workshop on Pattern Recognition in Information Systems, Jun. 2007.

[34] X. Li, W. Hu, Z. Zhang, X. Zhang. Robust Foreground Segmentation Based on Two Effective Background Models. MIR 2008, 223-228, Oct. 2008.

[35] A. François, G. Medioni. Adaptive Color Background Modeling for Real-time Segmentation of Video Streams. International Conference on Imaging Science, Systems, and Technology, 227-232, Jun. 1999.

[36] M. Zhao, N. Li, C. Chen. Robust automatic video object segmentation technique. International Conference on Image Processing, Sept. 2002.

[37] N. Friedman, S. Russell. Image Segmentation in Video Sequences: A Probabilistic Approach. Thirteenth Conference on Uncertainty in Artificial Intelligence, 175-181, 1997.

[38] S. Atev, O. Masoud, N. Papanikolopoulos. Practical Mixtures of Gaussians with Brightness Monitoring. Conference on Intelligent Transportation Systems, 423-428, 2004.

[39] Q. Zang, R. Klette. Parameter Analysis for Mixture of Gaussians. CITR Technical Report 188, Auckland University, 2006.

[40] T. Bouwmans, F. El Baf, B. Vachon. Background Modeling using Mixture of Gaussians for Foreground Detection - A survey. Recent Patents on Computer Science, 1 (3), 219-237, Nov. 2008.

[41] Z. Zivkovic. Improved adaptive Gaussian mixture model for background subtraction. International Conference on Pattern Recognition, 2, 28-31, 2004.

[42] J. Cheng , J. Yang, Y. Zhou, Y. Cui. Flexible background mixture models for foreground segmentation. Image and Vision Computing, 24, 473-482, 2006.

[43] A. Shimada, D. Arita, R. Taniguchi. Dynamic Control of Adaptive Mixture-of-Gaussians Background Model. AVSS 2006, 5, Nov. 2006.

[44] V. Morellas, I. Pavlidis, P. Tsiamyrtzis. DETER: detection of events for threat evaluation and recognition. Machine Vision and Applications, 15, 29-45, Jun. 2003.

[45] Y. Zhang, Z. Liang, Z. Hou, H. Wang, M. Tan. An Adaptive Mixture Gaussian Background Model with Online Background Reconstruction and Adjustable Foreground Mergence Time for Motion Segmentation. ICIT 2005, 23-27, Dec. 2005.

[46] B. Han, X. Lin. Update the GMMs via adaptive Kalman filtering. SPIE - International Society for Optical Engineering, 5960 (3), 1506-1515, 2005.

[47] Q. Zang, R. Klette. Evaluation of an Adaptive Composite Gaussian Model in Video Surveillance. CITR Technical Report 114, Auckland University, Aug. 2002.

[48] B. White, M. Shah. Automatically Tuning Background Subtraction Parameters Using Particle Swarm Optimization/ International Conference on Multimedia and Expo, 1826-1829, 2007.

[49] P. KaewTraKulPong, R. Bowden. An Improved Adaptive Background Mixture Model for Real-time Tracking with Shadow Detection. AVBS 2001, Sept. 2001.

[50] A. Pnevmatikakis, L. Polymenakos. 2D Person Tracking Using Kalman Filtering and Adaptive Background Learning in a Feedback Loop. CLEAR Workshop 2006, LNCS 4122, 151-160, 2006.

[51] V. Morellas, I. Pavlidis, P. Tsiamyrtzis. DETER: detection of events for threat evaluation and recognition. Machine Vision and Applications, 15, 29-45, Jun. 2003.

[52] J. Landabaso, M. Pardas, L. Xu. Hierarchical Representation of Scenes using Activity Information. ICASSP 2005, 677 – 680, Mar. 2005.

[53] S. Yang, C. Hsu. Background Modeling from GMM Likelihood Combined with Spatial and Color Coherency. International Conference on Image Processing, 1, 2801-2804, 2006.

[54] D. Lee. Effective Gaussian Mixture Learning for Video Background Subtraction. IEEE Transaction on PAMI, 27, 827-832 2005.

[55] P. Withagen, F. Groen, K. Schutte. EMswitch: a multi-hypothesis approach to EM background modeling. ACIVS 2003, Sept. 2003.

[56] M. Haque, M. Murshed, M. Paul. On Stable Dynamic Background Generation Technique using Gaussian Mixture Models for Robust Object Detection. AVSS 2008, 41-48, Sept. 2008.

[57] K. Schindler, H. Wang. Smooth Foreground-Background Segmentation for Video Processing. Asian Conference on Computer Vision, ACCV 2006, 3852, 581-590, Jan. 2006.

[58] Y. Sun, B. Yuan. Hierarchical GMM to handle sharp changes in moving object detection. Electronic Letters, 40 (13), 801-802, Jun. 2004.

[59] Y. Chen, C. Chen, C. Huang, Y. Hung. Efficient hierarchical method for background subtraction. Pattern Recognition, 40 (10), 2706-2715, Oct. 2007.

[60] Q. Zang, R. Klette. Robust Background Subtraction and Maintenance. International Conference on Pattern Recognition, ICPR 2004, 2, 90-93, 2004.

[61] Q. Zhong, L. Dai, Y. Song, R. Wang. A hierarchical motion detection algorithm with the fusion of the two types of motion information. Pattern Recognition and Artificial Intelligence, 18 (5), 552-557, Oct. 2005.

[62] T. Su, J. Hu. Background Removal in Vision Servo System using Gaussian Mixture Model Framework. ICNSC 2004, 1, 70-75, Mar 2004.

[63] H. Yang, Y. Tan, J. Tian, J. Liu. Accurate dynamic scene model for moving object detection", International Conference on Image Processing, ICIP 2007, 6, 157-160, 2007.

[64] M. Harville. A framework for high-level feedback to adaptive, per-pixel, mixture-of-Gaussian background models. ECCV 2002, 543-560, May 2002.

[65] D. Turdu, H. Erdogan. Improved post-processing for GMM based adaptive background modeling. International Symposium on Computer and Information Science, 1-6, Nov. 2007.

[66] P. Atrey, V. Kumar, A. Kumar, M. Kankanhalli. Experiential sampling based foreground/ background segmentation for video surveillance. ICME 2006. 1809-1812, Jul. 2006.

[67] D. Magee. Tracking Multiple Vehicles using Foreground, Background and Motion Models. Image and Vision Computing, 22, 143-155, 2004.

[68] J. Zuo, Q. Pan, Y. Liang, H. Zhang, Y. Cheng. Model Switching Based Adaptive Background Modeling Approach. Acta Automatica Sinica, 33 (5), 467-473, 2007.

[69] Z. Tang, Z. Miao. Fast Background Subtraction Using Improved GMM and Graph Cut. Congress on Image and Signal Processing, 4, 181-185, May 2008.

[70] H. Jiang, H. Ardo, V. Owall. Hardware accelerator design for video segmentation with multi-modal background modeling. International Symposium on Circuits and Systems, ISCAS 2005, 2, 1142- 1145, May 2005.

[71] A. Al-Mazeed, M. Nixon, S. Gunn. Classifiers Combination for Improved Motion Segmentation. International Conference on Image Analysis and Recognition, 363-371, 2004.

[72] X. Fang, W. Xiong, B. Hu, L. Wang. A Moving Object Detection Algorithm Based on Color Information. Journal of Physics, 48, 384–387, 2006.

[73] J. Zhang, C. Chen. Moving Objects Detection and Segmentation in Dynamic Video Backgrounds. Conference on Technologies for Homeland Security, 64-69, May 2007.

[74] A. Tavakkoli, M. Nicolescu, G. Bebis. Automatic Statistical Object Detection for Visual Surveillance. IEEE Southwest Symposium on Image Analysis and Interpretation, Denver, Colorado, SSIAI 2006, March 2006.

[75] C. Ianasi, V. Gui, C. Toma, D. Pescaru. A Fast Algorithm for Background Tracking in Video Surveillance, Using Nonparametric Kernel Density Estimation. Facta Universitatis, Series: Electronics and Energetics, 18 (1), 127-144, Apr. 2005.

[76] T. Tanaka, A. Shimada, D. Arita, R. Taniguchi. Non-parametric Background and Shadow Modeling for Object Detection. ACCV 2007, 159-168, Nov. 2007.

[77] Z. Zivkovic. Efficient adaptive density estimation per image pixel for the task of background subtraction. Pattern Recognition Letters, 27 (7), 773-780, Jan. 2006.

[78] R. Ramezani, P. Angelov, X. Zhou. A Fast Approach to Novelty Detection in Video Streams using Recursive Density Estimation. International IEEE Symposium on Intelligent Systems, 2, 142-147, Sept. 2008.

[79] Y. Mao, P. Shi. Diversity sampling based kernel density estimation for background modeling. Journal of Shanghai University, 9 (6), 506-509, Dec. 2005.

[80] P. Tang, L. Gao, Z. Liu. Salient Moving Object Detection Using Stochastic Approach Filtering. International Conference on Image and Graphics, ICIG 2007, 530-535, 2007.

[81] A. Tavakkoli, M. Nicolescu, G. Bebis, An Adaptive Recursive Learning Technique for Robust Foreground Object Detection. ECCV 2006, May 2006.

[82] A. Tavakkoli, M. Nicolescu, G. Bebis. Robust Recursive Learning for Foreground Region Detection in Videos with Quasi-Stationary Backgrounds. ICPR 2006, Aug. 2006.

[83] P. Pahalawatta, D. Depalov, T. Pappas, A. Katsaggelos. Detection, Classification, and Collaborative Tracking of Multiple Targets Using Video Sensors. International Workshop on Information Processing for Sensor Networks, 529-544, Apr. 2003.

[84] B. Örten, M. Soysal, A. Alatan. Person Identification In Surveillance Video By Combining Mpeg-7 Experts. WIAMIS 2005, Apr. 2005.

[85] T. Tanaka, A. Shimada, D. Arita, R. Taniguchi. Object Segmentation under Varying Illumination based on Combinational Background Modeling. Proceedings of the 4th Joint Workshop on Machine Perception and Robotics, 2008.

[86] S. Mahamud. Comparing Belief Propagation and Graph Cuts for Novelty Detection", CVPR 2006, 1, 1154-1159, 2006.

[87] J. Hu, J. Jiang, M. Qi. A Motion Detection Approach Based on Context Modeling", CEPS 2006, 78-80, Mar. 2006.

[88] O. Akman. Multi-Camera Video Surveillance Detection, Occlusion Handling, Tracking and Event Recognition. PhD Thesis, Middle East Technical University, 2007.

[89] N. Oliver, B. Rosario, A. Pentland. A Bayesian Computer Vision System for Modeling Human Interactions. International Conference on Vision Systems, 22 (8), 841- 843, Jan. 1999.

[90] Z. Xu, P. Shi, I. Gu. An Eigenbackground Subtraction Method using Recursive Error Compensation. Advances in Multimedia Information Processing, PCM 2006, 4261, 779-787, 2006.

[91] S. Kawabata, S. Hiura, K. Sato. Real-Time Detection of Anomalous Objects in Dynamic Scene. International Conference on Pattern Recognition, 3, 1171 – 1174, Aug. 2006.

[92] J. Rymel, J. Renno, D. Greenhill, J. Orwell, G. Jones. Adaptive Eigen-Backgrounds for Object Detection. International Conference on Image Processing, 3, 1847-1850, Oct. 2004.

[93] Y. Li. On incremental and robust subspace learning. Pattern Recognition, 37 (7), 1509-1518, 2004.

[94] D. Skocaj, A. Leonardis. Weighted and Robust Incremental Method for Subspace Learning. International Conference on Computer Vision, ICCV 2003, 1494-1501, 2003.

[95] J. Zhang Y. Zhuang. Adaptive Weight Selection for Incremental Eigen-Background Modeling. International Conference on Multimedia and Expo, 851-854, Jul. 2007.

[96] L. Wang, L. Wang, M. Wen, Q. Zhuo, W. Wang. Background subtraction using incremental subspace learning. International Conference on Image Processing, 5, 45-48, Oct. 2007.

[97] R. Li, Y. Chen, X. Zhang. Fast Robust Eigen-Background Updating For Foreground Detection. International Conference on Image Processing, ICIP 2006, 1833-1836, Oct. 2006.

[98] B. Han, R. Jain. Real-Time Subspace-Based Background Modeling Using Multi-channel Data. International Symposium on Visual Computing, ISVC 2007, 162-172, Nov. 2007.

CHAPTER 2.4

LATEST ALGORITHMS FOR 3-D RECONSTRUCTION FROM TWO VIEWS

Kenichi Kanatani

Department of Computer Science,
Okayama University, Okayama 700-8530 Japan
kanatani@suri.cs.okayama-u.ac.jp

Yasuyuki Sugaya[1,*] and Yasushi Kanazawa[2,†]

[1]*Department of Information and Computer Sciences,*
[2]*Department of Knowledge-based Information Engineering,*
Toyohashi University of Technology, Toyohashi, Aichi 441-8580 Japan
[*]*sugaya@iim.ics.tut.ac.jp*
[†]*kanazawa@tutkie.tut.ac.jp*

1. Introduction

The problem of computing from point correspondences over two views their 3-D locations, the positions and orientations of the two cameras, and their intrinsic parameters was intensively studied in 1980s and 1990s. Today, it is a well established technology and is used for many applications including robot navigation and virtual reality as a typical example of what is known as *structure from motion* [Kanatani (1996); Hartley and Zisserman (2000)]. The principle is now extended to multiple views and video streams [Hartley and Zisserman (2000); Kanatani (2008)].

Although the mathematical principle is well understood today, ceaseless efforts have still been made to improve component technologies, such as the fundamental matrix computation, with a view to further upgrade the accuracy and computational efficiency. This chapter describes such latest efforts for two view reconstruction. The basic structure of the algorithm was given by Kanatani and Ohta (2003), and we replace its individual components by latest results. Specifically:

(1) Given point correspondences, we must first compute the fundamental matrix, for which Kanatani and Ohta (2003) used a method called *renormalization* [Kanatani (2000)] and then optimally corrected the resulting solution so that it satisfies the rank constraint. This produces a statistically optimal solution up to high order noise terms. Here, we compute the fundamental matrix by

a strictly optimal scheme recently introduced by Kanatani and Sugaya (2009), using the *EFNS* (*Extended Fundamental Numerical Scheme*) [Kanatani and Sugaya (2007b)]. This not only achieves yet higher accuracy but also significantly simplifies the algorithm.

(2) Optimal estimation of the fundamental matrix requires iterations. Kanatani and Ohta (2003) initialized their renormalization by least squares. Here, we use the method of Taubin (1991), which is known to produce a better solution without any iterations [Kanatani and Sugaya (2007a,c)]. This considerably accelerates the convergence of the succeeding optimization of the fundamental matrix.

(3) From the computed fundamental matrix, we next compute the focal lengths of the two cameras. For this, Kanatani and Ohta (2003) used the formula of Bougnoux (1998), which we call the "free focal length method". This method fails when the two camera are in a degenerate configuration [Kanatani et al. (2006)]. However, this configuration occurs very frequently in practical situations. Following Kanatani and Matsunaga (2000) and Kanatani et al. (2006), we also describe the "fixed focal length method", assuming that the two focal lengths are equal.

(4) After the focal lengths are computed, we can determine the relative camera positions [Kanatani (1993)], from which we can triangulate the 3-D locations of the points in the scene. For this, we use the method of Kanatani et al. (2008) for optimally correcting corresponding points and triangulating them relative to a given world coordinate system.

This chapter is organized as follows. Section 2 describes the entire structure of our algorithm. Section 3 focuses on the fundamental matrix computation. We describe the details of the initialization by the Taubin method and the subsequent optimization using the EFNS procedure. In Sec. 4, we discuss in detail the computation of the camera focal lengths. We describe the general theory, the free focal length method, and the fixed focal length method. Section 5 concentrates on computation of 3-D motion and structure from the computed focal lengths and the fundamental matrix. We describe practical procedures for computing the camera translation and rotation and optimally triangulating individual points. We conclude in Sec. 6, summarizing related issues and listing typical situations in which 3-D reconstruction fails if care is not taken. The Appendix describes the details of computational procedures omitted in the body.

2. Entire Algorithm Structure

Figure 1 shows the entire structure of the algorithm to be described here. The input and output are as follows:

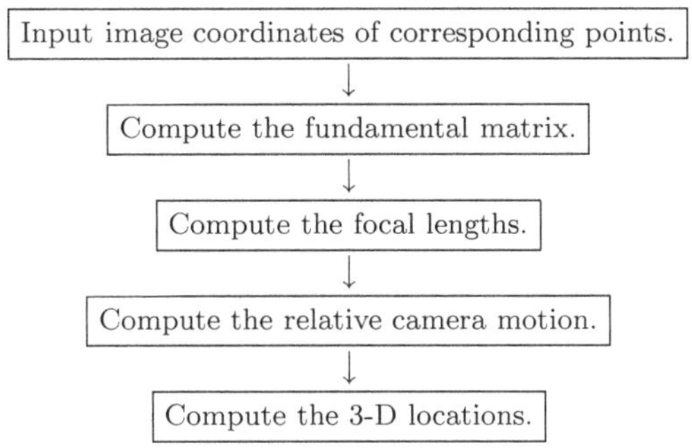

Fig. 1. The entire structure of our algorithm.

Input: The image coordinates (x_α, y_α), (x'_α, y'_α), $\alpha = 1, ..., N$ (≥ 8), of corresponding points.
Output: Their 3-D locations $(X_\alpha, Y_\alpha, Z_\alpha)$, $\alpha = 1, ..., N$.

Our algorithm is based on the following assumptions and conventions:

- The principal point (the intersection of the optical axis and the image plane) is known. We define an xy image coordinate system with the origin $(0,0)$ at the principal point. We take the x-axis *upward* direction and the y-axis *rightward*.
- Image distortion does not exists or already corrected beforehand. Hence, the rows and columns of the pixel array are regarded as physically orthogonal with aspect ratio 1.
- The focal lengths of the two cameras are equal, or equivalently one camera is moved without changing the zoom or the focus.

Given sufficiently many point correspondences over sufficiently many images, we can in principle determine the principal point and also correct the image distortion [Hartley and Zisserman (2000)]. From two views, however, all available information is the epipolar constraint encoded in the 3×3 fundamental matrix (Sec. 3.1). The fundamental matrix has scale indeterminacy, so only five of its elements are independent. From this limited information, 3-D reconstruction is possible only when the principal point is known and no image distortion exists (Sec. 4.1). For most of the digital cameras available on the market today, we can safely assume this. Otherwise, we need prior camera calibration and geometric correction.

The reason we take the x-axis upward and the y-axis rightward is that we view the camera optical axis, which we identify with the z-axis, as extending *away* from the viewer. For humans, this is a natural interpretation of perspective projection. Taking the x-axis rightward and the y-axis upward would define a *left-handed* coordinate system, in which rotation matrices have determinant -1. To avoid this, we could of course take the x-axis downward and the y-axis rightward.

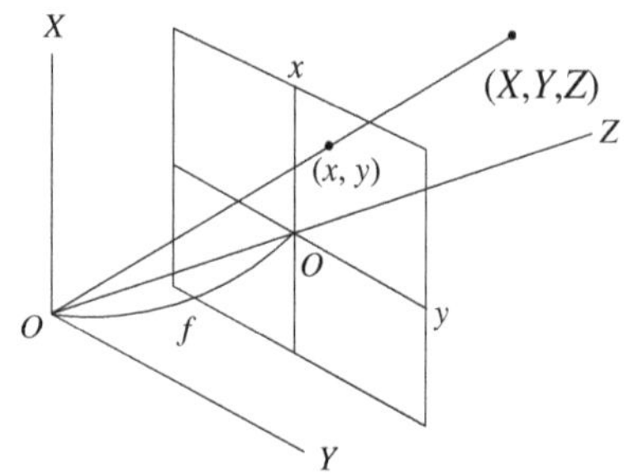

Fig. 2. Interpretation of perspective projection.

Thus, we model the camera imaging geometry as shown in Fig. 2. We take an XYZ coordinate system fixed to the camera in such a way that the origin O is at the center of the lens, which we hereafter call the *viewpoint*, and the Z axis along the optical axis of the lens . We identify the xy image plane with the plane $Z = f$, and call f the *focal length*. The xy image coordinate system is defined so that the image origin $(0,0)$ is at the *principal point* $(0,0,f)$ and the x- and y-axis are parallel to the X- and Y-axis, respectively. A point (X,Y,Z) in the scene is projected to the intersection (x,y) of the image plane with the ray, or the *line of sight*, starting from the viewpoint O and passing through (X,Y,Z). This geometry is known as *perspective projection* or the *pin-hole camera model*.

The assumption of equal focal lengths is not necessary as long as the camera configuration is nondegenerate. However, most practical applications take place in a degenerate configuration, in which 3-D reconstruction is impossible unless the two cameras have the same focal length or their focal lengths are known (Sec. 4.3). The equal focal length assumption is not a strong constraint, since people usually take images without changing the zoom or the focus. The subsequent sections describe the component algorithms according to the flow shown in Fig. 1.

3. Fundamental Matrix Computation

3.1. *Epipolar equation*

The first step is to compute the fundamental matrix $\boldsymbol{F}$ from point correspondences (x_α, y_α) and (x'_α, y'_α), $\alpha = 1, ..., N$. We regard x_α, y_α, x'_α, and y'_α as noisy observations of their true values $\bar{x}_\alpha$, $\bar{y}_\alpha$, $\bar{x}'_\alpha$, and $\bar{y}'_\alpha$. We represent points (x_α, y_α) and (x'_α, y'_α) by 3-D vectors

$$\boldsymbol{x}_\alpha = \begin{pmatrix} x_\alpha/f_0 \\ y_\alpha/f_0 \\ 1 \end{pmatrix}, \qquad \boldsymbol{x}'_\alpha = \begin{pmatrix} x'_\alpha/f_0 \\ y'_\alpha/f_0 \\ 1 \end{pmatrix}, \tag{1}$$

where f_0 is a scaling constant of the order of the image size for stabilizing numerical computation [Hartley (1997)]. Let $\bar{\boldsymbol{x}}_\alpha$ and $\bar{\boldsymbol{x}}'_\alpha$ be the true values of $\boldsymbol{x}_\alpha$ and $\boldsymbol{x}'_\alpha$, respectively. They satisfy the following *epipolar equation* [Hartley and Zisserman (2000)]:

$$(\bar{\boldsymbol{x}}_\alpha, \boldsymbol{F}\bar{\boldsymbol{x}}'_\alpha) = 0. \tag{2}$$

Throughout this chapter, we denote the inner product of vectors $\boldsymbol{a}$ and $\boldsymbol{b}$ by $(\boldsymbol{a}, \boldsymbol{b})$. The matrix $\boldsymbol{F}$ in Eq. (2) is of rank 2 and called the *fundamental matrix*; it depends on the relative positions and orientations of the two cameras and their intrinsic parameters (e.g., their focal lengths) but not on the scene or the choice of the corresponding points. As Eq. (2) implies, the matrix $\boldsymbol{F}$ has scale indeterminacy.

If the noise in (x_α, y_α) and (x'_α, y'_α) is regarded as independent and identical Gaussian variables of mean 0, *maximum likelihood* (*ML*) estimators of $\bar{\boldsymbol{x}}_\alpha$, $\bar{\boldsymbol{x}}'_\alpha$, and $\boldsymbol{F}$ are computed by minimizing the *reprojection error*

$$E = \sum_{\alpha=1}^{N} \Big(\|\boldsymbol{x}_\alpha - \bar{\boldsymbol{x}}_\alpha\|^2 + \|\boldsymbol{x}'_\alpha - \bar{\boldsymbol{x}}'_\alpha\|^2 \Big), \tag{3}$$

subject to Eq. (2). In order to determine the nine elements of $\boldsymbol{F}$ up to scale from Eq. (2), we need to observe at least eight points. If we consider the rank constraint $\det \boldsymbol{F} = 0$, we could theoretically determine $\boldsymbol{F}$ from seven correspondences, but in that case multiple solutions can exist [Hartley and Zisserman (2000)]. Hereafter, we assume that we observe eight or more points.

If the points we observe are in a special configuration in 3-D, however, Eq. (2) can be satisfied by infinitely many $\boldsymbol{F}$. This occurs if all the points and the viewpoints O and O' of the two cameras are on a quadratic surface, known as a *critical surface* [Maybank (1990, 1991), Kanatani (1993, 1996)]. Such a configuration occurs, e.g., when all the points are coplanar in the scene. In the following, we assume that the points are in general position so that Eq. (2) is satisfied by a unique $\boldsymbol{F}$ up to scale. We will return to this issue in Sec. 6.

3.2. *Initializing the fundamental matrix*

All known algorithms for minimizing Eq. (3) subject to Eq. (2) are iterative, so we need an appropriate initial value. The simplest and most popular is the least squares (Appendix A). As pointed out by Kanatani and Sugaya (2007a,c), however, the accuracy of the least squares is rather limited, and a much better solution is obtained by the method of Taubin (1991). The use of the Taubin solution as the initial value considerably accelerates the convergence of the subsequent optimization iterations. For computational convenience, we identify the fundamental matrix $\boldsymbol{F}$ with the 9-D vector

$$\boldsymbol{u} = (F_{11}, F_{12}, F_{13}, F_{21}, F_{22}, F_{23}, F_{31}, F_{32}, F_{33})^\top. \tag{4}$$

The Taubin method goes as follows (see Appendix B for the derivation):

Input: The image coordinates (x_α, y_α), (x'_α, y'_α), $\alpha = 1, \ldots, N$ (≥ 8), of corresponding points.

Output: The 9-D vector encoding the fundamental matrix $\boldsymbol{F}$.

(1) Represent the corresponding points (x_α, y_α) and (x'_α, y'_α) by the following 8-D vectors:

$$\boldsymbol{z}_\alpha = (x_\alpha x'_\alpha,\ x_\alpha y'_\alpha,\ f_0 x_\alpha,\ y_\alpha x'_\alpha,\ y_\alpha y'_\alpha,\ f_0 y_\alpha,\ f_0 x'_\alpha,\ f_0 y'_\alpha)^\top. \tag{5}$$

(2) Compute the following 8×8 matrices $V_0[\boldsymbol{z}_\alpha]$:

$$V_0[\boldsymbol{z}_\alpha] = \begin{pmatrix} x_\alpha^2 + x_\alpha'^2 & x'_\alpha y'_\alpha & f_0 x'_\alpha & x_\alpha y_\alpha & 0 & 0 & f_0 x_\alpha & 0 \\ x'_\alpha y'_\alpha & x_\alpha^2 + y_\alpha'^2 & f_0 y'_\alpha & 0 & x_\alpha y_\alpha & 0 & 0 & f_0 x_\alpha \\ f_0 x'_\alpha & f_0 y'_\alpha & f_0^2 & 0 & 0 & 0 & 0 & 0 \\ x_\alpha y_\alpha & 0 & 0 & y_\alpha^2 + x_\alpha'^2 & x'_\alpha y'_\alpha & f_0 x'_\alpha & f_0 y_\alpha & 0 \\ 0 & x_\alpha y_\alpha & 0 & x'_\alpha y'_\alpha & y_\alpha^2 + y_\alpha'^2 & f_0 y'_\alpha & 0 & f_0 y_\alpha \\ 0 & 0 & 0 & f_0 x'_\alpha & f_0 y'_\alpha & f_0^2 & 0 & 0 \\ f_0 x_\alpha & 0 & 0 & f_0 y_\alpha & 0 & 0 & f_0^2 & 0 \\ 0 & f_0 x_\alpha & 0 & 0 & f_0 y_\alpha & 0 & 0 & f_0^2 \end{pmatrix}. \tag{6}$$

(3) Compute the following 8-D vectors $\bar{\boldsymbol{z}}$ and $\tilde{\boldsymbol{z}}_\alpha$:

$$\bar{\boldsymbol{z}} = \frac{1}{N}\sum_{\alpha=1}^{N} \boldsymbol{z}_\alpha, \qquad \tilde{\boldsymbol{z}}_\alpha = \boldsymbol{z}_\alpha - \bar{\boldsymbol{z}}. \tag{7}$$

(4) Compute the following 8×8 matrices $\boldsymbol{M}_{\rm TB}$ and $\boldsymbol{N}_{\rm TB}$:

$$\boldsymbol{M}_{\rm TB} = \sum_{\alpha=1}^{N} \tilde{\boldsymbol{z}}_\alpha \tilde{\boldsymbol{z}}_\alpha^\top, \qquad \boldsymbol{N}_{\rm TB} = \sum_{\alpha=1}^{N} V_0[\boldsymbol{z}_\alpha]. \tag{8}$$

(5) Solve the following generalized eigenvalue problem (Appendix C) and compute the unit generalized eigenvector $\boldsymbol{v}$ for the smallest generalized eigenvalue:

$$\boldsymbol{M}_{\rm TB}\boldsymbol{v} = \lambda \boldsymbol{N}_{\rm TB}\boldsymbol{v}. \tag{9}$$

(6) Return the 9-D unit vector $\boldsymbol{u}$

$$\boldsymbol{u} = \mathcal{N}[\begin{pmatrix} \boldsymbol{v} \\ -(\bar{\boldsymbol{z}}, \boldsymbol{v})/f_0^2 \end{pmatrix}], \tag{10}$$

where $\mathcal{N}[\,\cdot\,]$ denotes normalization to unit norm ($\mathcal{N}[\boldsymbol{a}] = \boldsymbol{a}/\|\boldsymbol{a}\|$).

The use of f_0 in Eqs. (5) and (6) is for making the vector components and the matrix elements to have the same order of magnitude for stabilizing the numerical computation [Hartley (1997)]. The final solution does not depend on the specific value f_0.

3.3. *Alternating optimization scheme*

Starting from the Taubin solution, we iteratively optimize it so that it minimizes the reprojection error E in Eq. (3) subject to the epipolar equation in Eq. (2). The major obstacle is that the fundamental matrix $\boldsymbol{F}$ is constrained to have rank 2. A popular approach to this is to introduce auxiliary variables so that the rank constraint is automatically satisfied and to do numerical search in the augmented parameter space. Specifically, we first compute from an assumed fundamental matrix a tentative 3-D reconstruction, i.e., the 3-D locations of the observed points, the relative positions and orientations of the two cameras, and their focal lengths. Then, we iteratively adjust the 3-D locations, the camera configuration, and the focal lengths so that the image positions obtained by "reprojecting" the reconstructed 3-D points are as close to the observed points as possible[1], see, e.g., Bartoli and Sturm (2004). This approach is known as *bundle adjustment* [Triggs et al. (2000)].

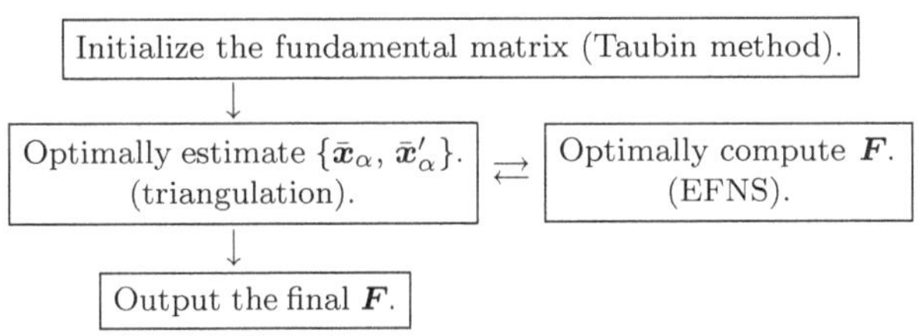

Fig. 3. Fundamental matrix computation.

The search space for bundle adjustment is very high dimensional, since the N locations in 3-D are all taken to be variables as well as the camera configuration and the focal lengths. A well known optimization scheme is the Levernberg-Marquardt method [Press et al. (1992)], but the search is very inefficient unless one incorporates a clever implementation such as preprocessing of the Hessian, which is very high dimensional and very sparse. Here, we describe an efficient method recently proposed by Kanatani and Sugaya (2009). The computation is done only in the 9-D space of the fundamental matrix, yet is equivalent to bundle adjustment. The computation of $\{\bar{\boldsymbol{x}}_\alpha, \bar{\boldsymbol{x}}'_\alpha\}$, and $\boldsymbol{F}$ that minimize Eq. (3) subject to Eq. (2) consists of the following two components, which are alternated until the solution converges (Fig. 3):

- Optimally computing the estimate $\{\hat{\boldsymbol{x}}_\alpha, \hat{\boldsymbol{x}}'_\alpha\}$ of $\{\bar{\boldsymbol{x}}_\alpha, \bar{\boldsymbol{x}}'_\alpha\}$ for a given $\boldsymbol{F}$.
- Optimally computing $\boldsymbol{F}$ for a given estimate $\{\hat{\boldsymbol{x}}_\alpha, \hat{\boldsymbol{x}}'_\alpha\}$ of $\{\bar{\boldsymbol{x}}_\alpha, \bar{\boldsymbol{x}}'_\alpha\}$.

In the first component, we minimize Eq. (3) with respect to $\{\bar{\boldsymbol{x}}_\alpha, \bar{\boldsymbol{x}}'_\alpha\}$ subject to Eq. (2) for a given $\boldsymbol{F}$. This is nothing but what is known as *triangulation* of stereo images. Here, we use the procedure of Kanatani et al. (2008). In the sec-

[1]From this, Eq. (3) comes to be known as the "reprojection error".

ond component, we minimize Eq. (3) with respect to $\boldsymbol{F}$ subject to Eq. (2) and the rank constraint $\det \boldsymbol{F} = 0$ for a given estimate of $\{\bar{\boldsymbol{x}}_\alpha, \bar{\boldsymbol{x}}'_\alpha\}$. For this computation, we use the *EFNS* (*Extended Fundamental Numerical Scheme*) of Kanatani and Sugaya (2007b). This scheme iteratively updates $\boldsymbol{F}$ so that the reprojection error E decreases and $\det \boldsymbol{F}$ approaches 0 at the same time. In other words, the rank constraint $\det \boldsymbol{F} = 0$ is not satisfied in the course of iterations but is satisfied in the end. This type of iterations was introduced by Chojnacki et al. (2004), who called their method *CFNS* (*Constrained Fundamental Numerical Scheme*), which is a modification of their unconstrained optimization scheme called *FNS* (*Fundamental Numerical Scheme*) [Chojnacki et al. (2000)]. However, Kanatani and Sugaya (2007b) pointed out that the CFNS does not necessarily converge to a correct solution while their EFNS is guaranteed to produce an optimal solution.

3.4. *Optimal computation of the fundamental matrix*

Given an initial value of the 9-D vector representation $\boldsymbol{u}$ of the fundamental matrix $\boldsymbol{F}$ in the form of Eq. (4) and observed corresponding points (x_α, y_α), (x'_α, y'_α), $\alpha = 1, \ldots, N$, the following algorithm optimally computes the fundamental matrix $\boldsymbol{F}$ (see Kanatani and Sugaya (2009) and Kanatani (2009) for the derivation).

Input: The image coordinates (x_α, y_α), (x'_α, y'_α), $\alpha = 1, \ldots, N$ (≥ 8), of corresponding points, and the 9-D vector $\boldsymbol{u}$ representing an initial estimate of $\boldsymbol{F}$.
Output: Optimized fundamental matrix $\boldsymbol{F}$.

(1) Let $E^0 = \infty$ (a sufficiently large number), and initialize $\hat{x}_\alpha$, $\hat{y}_\alpha$, $\hat{x}'_\alpha$, $\hat{y}'_\alpha$, $\tilde{x}_\alpha$, $\tilde{y}_\alpha$, $\tilde{x}'_\alpha$, and $\tilde{y}'_\alpha$, $\alpha = 1, \ldots, N$, as follows:

$$\hat{x}_\alpha = x_\alpha, \qquad \hat{y}_\alpha = y_\alpha, \qquad \hat{x}'_\alpha = x'_\alpha, \qquad \hat{y}'_\alpha = y'_\alpha,$$

$$\tilde{x}_\alpha = \tilde{y}_\alpha = \tilde{x}'_\alpha = \tilde{y}'_\alpha = 0. \tag{11}$$

(2) Compute the following 9-D vectors $\boldsymbol{\xi}_\alpha$ and 9×9 matrices $V_0[\boldsymbol{\xi}_\alpha]$, $\alpha = 1, \ldots, N$:

$$\boldsymbol{\xi}_\alpha = \begin{pmatrix} \hat{x}_\alpha \hat{x}'_\alpha + \hat{x}'_\alpha \tilde{x}_\alpha + \hat{x}_\alpha \tilde{x}'_\alpha \\ \hat{x}_\alpha \hat{y}'_\alpha + \hat{y}'_\alpha \tilde{x}_\alpha + \hat{x}_\alpha \tilde{y}'_\alpha \\ f_0(\hat{x}_\alpha + \tilde{x}_\alpha) \\ \hat{y}_\alpha \hat{x}'_\alpha + \hat{x}'_\alpha \tilde{y}_\alpha + \hat{y}_\alpha \tilde{x}'_\alpha \\ \hat{y}_\alpha \hat{y}'_\alpha + \hat{y}'_\alpha \tilde{y}_\alpha + \hat{y}_\alpha \tilde{y}'_\alpha \\ f_0(\hat{y}_\alpha + \tilde{y}_\alpha) \\ f_0(\hat{x}'_\alpha + \tilde{x}'_\alpha) \\ f_0(\hat{y}'_\alpha + \tilde{y}'_\alpha) \\ f_0^2 \end{pmatrix}, \tag{12}$$

$$V_0[\boldsymbol{\xi}_\alpha] = \begin{pmatrix} \hat{x}_\alpha^2 + \hat{x}_\alpha'^2 & \hat{x}'_\alpha \hat{y}'_\alpha & f_0 \hat{x}'_\alpha & \hat{x}_\alpha \hat{y}_\alpha & 0 & 0 & f_0 \hat{x}_\alpha & 0 & 0 \\ \hat{x}'_\alpha \hat{y}'_\alpha & \hat{x}_\alpha^2 + \hat{y}_\alpha'^2 & f_0 \hat{y}'_\alpha & 0 & \hat{x}_\alpha \hat{y}_\alpha & 0 & 0 & f_0 \hat{x}_\alpha & 0 \\ f_0 \hat{x}'_\alpha & f_0 \hat{y}'_\alpha & f_0^2 & 0 & 0 & 0 & 0 & 0 & 0 \\ \hat{x}_\alpha \hat{y}_\alpha & 0 & 0 & \hat{y}_\alpha^2 + \hat{x}_\alpha'^2 & \hat{x}'_\alpha \hat{y}'_\alpha & f_0 \hat{x}'_\alpha & f_0 \hat{y}_\alpha & 0 & 0 \\ 0 & \hat{x}_\alpha \hat{y}_\alpha & 0 & \hat{x}'_\alpha \hat{y}'_\alpha & \hat{y}_\alpha^2 + \hat{y}_\alpha'^2 & f_0 \hat{y}'_\alpha & 0 & f_0 \hat{y}_\alpha & 0 \\ 0 & 0 & 0 & f_0 \hat{x}'_\alpha & f_0 \hat{y}'_\alpha & f_0^2 & 0 & 0 & 0 \\ f_0 \hat{x}_\alpha & 0 & 0 & f_0 \hat{y}_\alpha & 0 & 0 & f_0^2 & 0 & 0 \\ 0 & f_0 \hat{x}_\alpha & 0 & 0 & f_0 \hat{y}_\alpha & 0 & 0 & f_0^2 & 0 \\ 0 & 0 & 0 & 0 & 0 & 0 & 0 & 0 & 0 \end{pmatrix}. \tag{13}$$

(3) Update $\boldsymbol{u}$ by calling the subprogram EFNS in Sec. 3.5.

(4) Update $\tilde{x}_\alpha$, $\tilde{y}_\alpha$, $\tilde{x}'_\alpha$, and $\tilde{y}'_\alpha$, $\alpha = 1, ..., N$, as follows:

$$\begin{pmatrix} \tilde{x}_\alpha \\ \tilde{y}_\alpha \end{pmatrix} \leftarrow \frac{(\boldsymbol{u}, \boldsymbol{\xi}_\alpha)}{(\boldsymbol{u}, V_0[\boldsymbol{\xi}_\alpha]\boldsymbol{u})} \begin{pmatrix} u_1 & u_2 & u_3 \\ u_4 & u_5 & u_6 \end{pmatrix} \begin{pmatrix} \hat{x}'_\alpha \\ \hat{y}'_\alpha \\ f_0 \end{pmatrix},$$

$$\begin{pmatrix} \tilde{x}'_\alpha \\ \tilde{y}'_\alpha \end{pmatrix} \leftarrow \frac{(\boldsymbol{u}, \boldsymbol{\xi}_\alpha)}{(\boldsymbol{u}, V_0[\boldsymbol{\xi}_\alpha]\boldsymbol{u})} \begin{pmatrix} u_1 & u_4 & u_7 \\ u_2 & u_5 & u_8 \end{pmatrix} \begin{pmatrix} \hat{x}_\alpha \\ \hat{y}_\alpha \\ f_0 \end{pmatrix}. \tag{14}$$

(5) Update $\hat{x}_\alpha$, $\hat{y}_\alpha$, $\hat{x}'_\alpha$, and $\hat{y}'_\alpha$, $\alpha = 1, ..., N$, as follows:

$$\hat{x}_\alpha \leftarrow x_\alpha - \tilde{x}_\alpha, \qquad \hat{y}_\alpha \leftarrow y_\alpha - \tilde{y}_\alpha,$$

$$\hat{x}'_\alpha \leftarrow x'_\alpha - \tilde{x}'_\alpha, \qquad \hat{y}'_\alpha \leftarrow y'_\alpha - \tilde{y}'_\alpha. \tag{15}$$

(6) Compute the reprojection error E as follows:

$$E = \sum_{\alpha=1}^{N} (\tilde{x}_\alpha^2 + \tilde{y}_\alpha^2 + \tilde{x}_\alpha'^2 + \tilde{y}_\alpha'^2). \tag{16}$$

(7) If $E \approx E^0$, output

$$\boldsymbol{F} = \begin{pmatrix} u_1 & u_2 & u_3 \\ u_4 & u_5 & u_6 \\ u_7 & u_8 & u_9 \end{pmatrix}. \tag{17}$$

Else, let $E^0 \leftarrow E$ and go back to Step (2).

3.5. *EFNS procedure*

The EFNS procedure called in the Step (3) of the above algorithm is the following computation [Kanatani and Sugaya (2007b)]. It produces the 9-D vector $\boldsymbol{u}$ that represents an optimal estimate of the fundamental matrix $\boldsymbol{F}$ for the current $(\hat{x}_\alpha, \hat{y}_\alpha)$, $(\hat{x}'_\alpha, \hat{y}'_\alpha)$, $\alpha = 1, ..., N$, that satisfies the rank constraint $\det \boldsymbol{F} = 0$.

Input: The 9-D vectors $\boldsymbol{\xi}_\alpha$, $\alpha = 1, ..., N$, that encode the corresponding points and the 9-D vector $\boldsymbol{u}$ that represents the current fundamental matrix.

Output: The 9-D vector $\boldsymbol{u}$ that represents the updated fundamental matrix.

(1) Compute the following 9×9 matrices $\boldsymbol{M}$ and $\boldsymbol{L}$:

$$\boldsymbol{M} = \sum_{\alpha=1}^{N} \frac{\boldsymbol{\xi}_\alpha \boldsymbol{\xi}_\alpha^\top}{(\boldsymbol{u}, V_0[\boldsymbol{\xi}_\alpha]\boldsymbol{u})}, \qquad \boldsymbol{L} = \sum_{\alpha=1}^{N} \frac{(\boldsymbol{u}, \boldsymbol{\xi}_\alpha)^2 V_0[\boldsymbol{\xi}_\alpha]}{(\boldsymbol{u}, V_0[\boldsymbol{\xi}_\alpha]\boldsymbol{u})^2}. \tag{18}$$

(2) Compute the following 9-D vector $\boldsymbol{u}^\dagger$ and 9×9 matrix $\boldsymbol{P}_{\boldsymbol{u}^\dagger}$:

$$\boldsymbol{u}^\dagger = \mathcal{N}[\begin{pmatrix} u_5u_9 - u_8u_6 \\ u_6u_7 - u_9u_4 \\ u_4u_8 - u_7u_5 \\ u_8u_3 - u_2u_9 \\ u_9u_1 - u_3u_7 \\ u_7u_2 - u_1u_8 \\ u_2u_6 - u_5u_3 \\ u_3u_4 - u_6u_1 \\ u_1u_5 - u_4u_2 \end{pmatrix}], \qquad \boldsymbol{P}_{\boldsymbol{u}^\dagger} = \boldsymbol{I} - \boldsymbol{u}^\dagger \boldsymbol{u}^{\dagger\top}. \tag{19}$$

(3) Compute the following 9×9 matrices $\boldsymbol{X}$ and $\boldsymbol{Y}$:

$$\boldsymbol{X} = \boldsymbol{M} - \boldsymbol{L}, \qquad \boldsymbol{Y} = \boldsymbol{P}_{\boldsymbol{u}^\dagger} \boldsymbol{X} \boldsymbol{P}_{\boldsymbol{u}^\dagger}. \tag{20}$$

(4) Compute the 9-D unit eigenvectors $\boldsymbol{v}_1$ and $\boldsymbol{v}_2$ for the smallest two eigenvalues of $\boldsymbol{Y}$.

(5) Compute the following 9-D vectors $\hat{\boldsymbol{u}}$ and $\boldsymbol{u}'$:

$$\hat{\boldsymbol{u}} = (\boldsymbol{u}, \boldsymbol{v}_1)\boldsymbol{v}_1 + (\boldsymbol{u}, \boldsymbol{v}_2)\boldsymbol{v}_2, \qquad \boldsymbol{u}' = \mathcal{N}[\boldsymbol{P}_{\boldsymbol{u}^\dagger} \hat{\boldsymbol{u}}]. \tag{21}$$

(6) If $\boldsymbol{u}' \approx \boldsymbol{u}$ up to sign, return $\boldsymbol{u}'$ as the update of $\boldsymbol{u}$. Else, let $\boldsymbol{u} \leftarrow \mathcal{N}[\boldsymbol{u} + \boldsymbol{u}']$ and go back to Step (1).

The vector $\boldsymbol{u}^\dagger$ in Eqs. (19) encodes the nine elements of the cofactor $\boldsymbol{F}^\dagger$ of the fundamental matrix $\boldsymbol{F}$. The matrix $\boldsymbol{P}_{\boldsymbol{u}^\dagger}$ represents projection of the 9-D space onto the 8-D subspace orthogonal to $\boldsymbol{u}^\dagger$. In Step (4), Kanatani and Sugaya (2009) computed two eigenvectors for the smallest eigenvalues *in absolute values*. Later, it has experimentally been found that simply computing two eigenvectors for the smallest eigenvalues results in better convergence in the presence of large noise. The same phenomenon was observed for the FNS of Chojnacki et al. (2000), for which computing the smallest eigenvalue, rather than the smallest eigenvalue in absolute value, exhibits better convergence [Kanatani and Sugaya (2007a,c)]. In Step (6), we could let $\boldsymbol{u} \leftarrow \boldsymbol{u}'$, but it has been observed that the value $\boldsymbol{u}'$ computed in Step (5) in the next round often reverts to the former value of $\boldsymbol{u}$, falling in infinite looping. To avoid this, the "midpoint" $(\boldsymbol{u}' + \boldsymbol{u})/2$ is normalized to a unit vector $\mathcal{N}[\boldsymbol{u}' + \boldsymbol{u}]$. This greatly improves convergence. In fact, it has been confirmed that this technique also improves the convergence of FNS of Chojnacki et al. (2000), which sometimes oscillates in the presence of very large noise.

4. Focal Length Computation

4.1. *General theory*

From thus computed fundamental matrix $\boldsymbol{F}$, we next compute the focal lengths f and f' (in pixels) of the two cameras and their relative translation $\boldsymbol{t}$ and rotation $\boldsymbol{R}$, which we call the *motion parameters*. The fundamental matrix $\boldsymbol{F}$ has scale indeterminacy, and it should satisfy the rank constraint $\det \boldsymbol{F} = 0$, so it has seven degrees of freedom.

The scale indeterminacy of $\boldsymbol{F}$ corresponds to the scale indeterminacy of the translation $\boldsymbol{t}$, since we would observe the same images after the 3-D scene, the depths, and the camera translation $\boldsymbol{t}$ are simultaneously scaled up/down by the same ratio. In order to fix the scale, we normalize the translation $\boldsymbol{t}$ to unit norm ($\|\boldsymbol{t}\| = 1$), so it has two degrees of freedom. The rotation $\boldsymbol{R}$ has three degrees of freedom. Together with f and f', the unknowns have seven degrees of freedom. Thus, we should be able to compute the focal lengths $\{f,\ f'\}$ and the motion parameters $\{\boldsymbol{t},\ \boldsymbol{R}\}$ from $\boldsymbol{F}$.

If more views were observed, other camera parameters such as the aspect ratio, the skew angle, and the principal point (see Sec. 5.3) could be computed [Hartley and Zisserman (2000)]. As the above counting of the degrees of freedom indicates, however, we cannot reconstruct 3-D from two views unless other parameters are known. Hence, we assume an ideal camera for which the aspect ratio is 1 with no image skew, and the principal point is at the center of the image plane (Fig. 2). If the two cameras are ideal in the above sense, the fundamental matrix $\boldsymbol{F}$ has the following expression in the focal lengths $\{f,\ f'\}$ and the motion parameters $\{\boldsymbol{t},\ \boldsymbol{R}\}$ [Kanatani (1993), Hartley and Zisserman (2000)]:

$$\boldsymbol{F} = \mathrm{diag}(1, 1, \frac{f}{f_0})\,\boldsymbol{t} \times \boldsymbol{R}\,\mathrm{diag}(1, 1, \frac{f'}{f_0}). \tag{22}$$

The symbol $\mathrm{diag}(a, b, c)$ designates the diagonal matrix with diagonal elements a, b, and c in that order. The product $\boldsymbol{a} \times \boldsymbol{V}$ of a vector $\boldsymbol{a}$ and a matrix $\boldsymbol{V}$ is a matrix consisting of columns $\boldsymbol{a} \times \boldsymbol{v}_1$, $\boldsymbol{a} \times \boldsymbol{v}_2$, and $\boldsymbol{a} \times \boldsymbol{v}_3$, where $\boldsymbol{v}_i$ is the ith column of $\boldsymbol{V}$. We define the *essential matrix* $\boldsymbol{E}$ by

$$\boldsymbol{E} = \mathrm{diag}(1, 1, \frac{f_0}{f})\boldsymbol{F}\mathrm{diag}(1, 1, \frac{f_0}{f'})\,(= \boldsymbol{t} \times \boldsymbol{R}). \tag{23}$$

Huang and Faugeras (1988) proved that a matrix $\boldsymbol{E}$ can be written as $\boldsymbol{t} \times \boldsymbol{R}$ for some vector $\boldsymbol{t}$ and some rotation matrix $\boldsymbol{R}$ if and only if $\boldsymbol{E}$ has singular values σ, σ, and 0 for some $\sigma > 0$. Let us define the Frobenius norm of matrix $\boldsymbol{A}$ by $\|\boldsymbol{A}\| = \sqrt{\sum_{i,j=1}^{N} A_{ij}^2}$. Kanatani (1993) wrote this condition as

$$\|\boldsymbol{E}\boldsymbol{E}^\top\|^2 = \frac{1}{2}\|\boldsymbol{E}\|^4 \tag{24}$$

along with $\det \boldsymbol{E} = 0$, which is automatically satisfied by the rank constraint $\det \boldsymbol{F} = 0$ of $\boldsymbol{F}$. If $\sigma_1 \geq \sigma_2 \geq \sigma_3$ (≥ 0) are the singular values of $\boldsymbol{E}$, the rank constraint

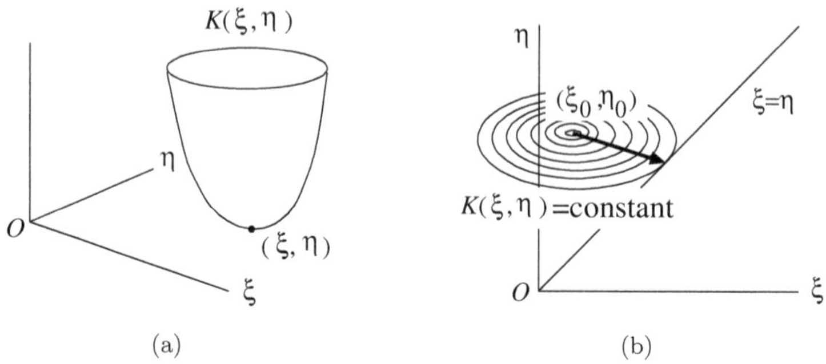

Fig. 4. (a) The focal lengths f and f' are determined by the point (ξ, η) at which the surface of $K(\xi, \eta)$ touches the $\xi\eta$-plane. (b) The point (ξ_0, η_0) is corrected to (ξ, η) on the line $\xi = \eta$ so that the increase of $K(\xi, \eta)$ is minimized.

$\det \boldsymbol{E} = 0$ implies $\sigma_3 = 0$. It can easily be seen that Eq. (24) is then equivalent to $(\sigma_1^2 - \sigma_2^2)^2 = 0$ [Kanatani (1993)]. Thus, if Eq. (23) is substituted, Eq. (24) gives a set of equations for f and f'. Here, we describe the computational theory of Kanatani and Matsunaga (2000). First, we change the variables from f and f' to

$$\xi = \left(\frac{f_0}{f}\right)^2 - 1, \qquad \eta = \left(\frac{f_0}{f'}\right)^2 - 1. \tag{25}$$

Next, we define the function

$$K(\xi, \eta) = \|\boldsymbol{E}\boldsymbol{E}^\top\|^2 - \frac{1}{2}\|\boldsymbol{E}\|^4 \tag{26}$$

of ξ and η obtained by expressing f and f' in Eq. (23) in terms of ξ and η. Equation (24) is equivalent to $K(\xi, \eta) = 0$. Moreover, the following relation holds due to the singular nature of the essential matrix $\boldsymbol{E}$ [Kanatani et al. (2006)]:

$$K = \frac{\partial K}{\partial \xi} = \frac{\partial K}{\partial \eta} = 0. \tag{27}$$

This means that the function $K(\xi, \eta)$ takes its minimum at (ξ, η) *and* its minimum equals 0 (Fig. 4(a)). After some lengthy manipulations, we see that $K(\xi, \eta)$ is a fourth order polynomial in ξ and η, quadratic in each, in the form[2]

$$\begin{aligned} K(\xi, \eta) = {} & (\boldsymbol{k}, \boldsymbol{F}\boldsymbol{k})^4 \xi^2 \eta^2 + 2(\boldsymbol{k}, \boldsymbol{F}\boldsymbol{k})^2 \|\boldsymbol{F}^\top \boldsymbol{k}\|^2 \xi^2 \eta + 2(\boldsymbol{k}, \boldsymbol{F}\boldsymbol{k})^2 \|\boldsymbol{F}\boldsymbol{k}\|^2 \xi \eta^2 \\ & + \|\boldsymbol{F}^\top \boldsymbol{k}\|^4 \xi^2 + \|\boldsymbol{F}\boldsymbol{k}\|^4 \eta^2 + 4(\boldsymbol{k}, \boldsymbol{F}\boldsymbol{k})(\boldsymbol{k}, \boldsymbol{F}\boldsymbol{F}^\top \boldsymbol{F}\boldsymbol{k}) \xi \eta + 2\|\boldsymbol{F}\boldsymbol{F}^\top \boldsymbol{k}\|^2 \xi \\ & + 2\|\boldsymbol{F}^\top \boldsymbol{F}\boldsymbol{k}\|^2 \eta + \|\boldsymbol{F}\boldsymbol{F}^\top\|^2 - \frac{1}{2}\Big((\boldsymbol{k}, \boldsymbol{F}\boldsymbol{k})^2 \xi \eta + \|\boldsymbol{F}^\top \boldsymbol{k}\|^2 \xi \\ & + \|\boldsymbol{F}\boldsymbol{k}\|^2 \eta + \|\boldsymbol{F}\|^2\Big)^2. \end{aligned} \tag{28}$$

Hence, the solution (ξ, η) can be obtained, in principle, by numerically searching for the minimum of this polynomial.

[2]The EFNS procedure in Sec. 3.5 normalizes the vector $\boldsymbol{u}$ representing $\boldsymbol{F}$ to $\|\boldsymbol{u}\| = 1$. This means $\boldsymbol{F}$ is normalized to $\|\boldsymbol{F}\| = 1$.

Here, we need to distinguish two types of algorithms. One is to see f and f' as independent variables and solve for them. Then, if the two cameras have the same focal lengths, the computed f and f' may not be equal, since $\boldsymbol{F}$ is computed from noisy point correspondences. Let us call this the *free focal length method*. The other is to force the equality $f = f'$ from the beginning and solve for them. In this case, Eq. (27) provide redundant constraints, since we have only one unknown. Let us call this the *fixed focal length method*.

4.2. *Free focal length method*

Instead of numerically searching for the minimum of Eq. (28), we can determine f and f' directly from Eq. (22). In fact, the problem of computing the focal lengths from $\boldsymbol{F}$ has attracted many researchers, and various formulas, all mathematically equivalent, have been presented in the past [Hartley (1992), Pan et al. (1995a,b), Newsam et al. (1996), Bougnoux (1998), Kanatani and Matsunaga (2000)]. Here, we describe the modified Bougnoux formula of Kanatani and Matsunaga (2000) (see Appendix D for the derivation). In the following, we define $\boldsymbol{k} = (0,\ 0,\ 1)^\top$.

Input: The fundamental matrix $\boldsymbol{F}$.
Output: The focal lengths f and f' of the two cameras.

(1) Let $\boldsymbol{e}$ and $\boldsymbol{e}'$ be the unit eigenvectors of $\boldsymbol{F}\boldsymbol{F}^\top$ and $\boldsymbol{F}^\top\boldsymbol{F}$, respectively, for the smallest eigenvalue.
(2) Compute the following ξ and η:

$$\xi = \frac{\|\boldsymbol{F}\boldsymbol{k}\|^2 - (\boldsymbol{k}, \boldsymbol{F}\boldsymbol{F}^\top\boldsymbol{F}\boldsymbol{k})\|\boldsymbol{e}' \times \boldsymbol{k}\|^2/(\boldsymbol{k}, \boldsymbol{F}\boldsymbol{k})}{\|\boldsymbol{e}' \times \boldsymbol{k}\|^2\|\boldsymbol{F}^\top\boldsymbol{k}\|^2 - (\boldsymbol{k}, \boldsymbol{F}\boldsymbol{k})^2},$$
$$\eta = \frac{\|\boldsymbol{F}^\top\boldsymbol{k}\|^2 - (\boldsymbol{k}, \boldsymbol{F}\boldsymbol{F}^\top\boldsymbol{F}\boldsymbol{k})\|\boldsymbol{e} \times \boldsymbol{k}\|^2/(\boldsymbol{k}, \boldsymbol{F}\boldsymbol{k})}{\|\boldsymbol{e} \times \boldsymbol{k}\|^2\|\boldsymbol{F}\boldsymbol{k}\|^2 - (\boldsymbol{k}, \boldsymbol{F}\boldsymbol{k})^2}. \tag{29}$$

(3) If $\xi \leq -1$ or $\eta \leq -1$, return the message *failure* and stop. Else return

$$f = \frac{f_0}{\sqrt{1+\xi}}, \qquad f' = \frac{f_0}{\sqrt{1+\eta}}. \tag{30}$$

Equations (30) are simply rewriting of Eqs. (25). The vectors $\boldsymbol{e}$ and $\boldsymbol{e}'$ computed in Step (1) represent the *epipoles* of $\boldsymbol{F}$. Namely, $(f_0e_1/e_3, f_0e_2/e_3)$ is the projection of the viewpoint of the second camera onto the first image, and $(f_0e_1'/e_3', f_0e_2'/e_3')$ is the projection of the viewpoint of the first camera onto the second image (see Eqs. (1)). If $e_3 = 0$ or $e_3' = 0$, the corresponding epipole is located at infinity in the image plane.

If the focal lengths f and f' are known to be equal, we may simply take the average $(f + f')/2$ as their values or replace ξ and η by their average $(\xi + \eta)/2$ in Step (2), but these are too ad hoc. A reasonable approach is to enforce the equality $\xi = \eta$ so that the increase of $K(\xi, \eta)$ is minimized (Fig. 4(b)). Let (ξ_0, η_0) be the solution given by Eqs. (30). The function $K(\xi,\eta)$ is Taylor expanded around (ξ_0,η_0) in the form

$$K(\xi,\eta) = \frac{1}{2}(\begin{pmatrix} \xi-\xi_0 \\ \eta-\eta_0 \end{pmatrix}, \boldsymbol{H}\begin{pmatrix} \xi-\xi_0 \\ \eta-\eta_0 \end{pmatrix}) + \cdots, \tag{31}$$

where $\cdots$ denotes terms of order 3 or higher in $\xi-\xi_0$ and $\eta-\eta_0$. From Eq. (28), the Hessian $\boldsymbol{H} = (H_{ij})$ has the form

$$\begin{aligned}
H_{11} &= 2(\boldsymbol{k},\boldsymbol{F}\boldsymbol{k})^4\eta_0^2 + 4(\boldsymbol{k},\boldsymbol{F}\boldsymbol{k})^2\|\boldsymbol{F}^\top\boldsymbol{k}\|^2\eta_0 + 2\|\boldsymbol{F}^\top\boldsymbol{k}\|^4 - \Big((\boldsymbol{k},\boldsymbol{F}\boldsymbol{k})^2\eta_0 + \|\boldsymbol{F}^\top\boldsymbol{k}\|^2\Big)^2, \\
H_{12} &= 4(\boldsymbol{k},\boldsymbol{F}\boldsymbol{k})^4\xi_0\eta_0 + 4(\boldsymbol{k},\boldsymbol{F}\boldsymbol{k})^2\Big(\|\boldsymbol{F}^\top\boldsymbol{k}\|^2\xi_0 + \|\boldsymbol{F}\boldsymbol{k}\|^2\eta_0\Big) + 4(\boldsymbol{k},\boldsymbol{F}\boldsymbol{k})(\boldsymbol{k},\boldsymbol{F}\boldsymbol{F}^\top\boldsymbol{F}\boldsymbol{k}) \\
&\quad - \Big((\boldsymbol{k},\boldsymbol{F}\boldsymbol{k})^2\xi_0 + \|\boldsymbol{F}\boldsymbol{k}\|^2\Big)\Big((\boldsymbol{k},\boldsymbol{F}\boldsymbol{k})^2\eta_0 + \|\boldsymbol{F}^\top\boldsymbol{k}\|^2\Big) \\
&\quad - (\boldsymbol{k},\boldsymbol{F}\boldsymbol{k})^2\Big((\boldsymbol{k},\boldsymbol{F}\boldsymbol{k})^2\xi_0\eta_0 + \|\boldsymbol{F}^\top\boldsymbol{k}\|^2\xi_0 + \|\boldsymbol{F}\boldsymbol{k}\|^2\eta_0 + \|\boldsymbol{F}\|^2\Big) = H_{21}, \\
H_{22} &= 2(\boldsymbol{k},\boldsymbol{F}\boldsymbol{k})^4\xi_0^2 + 4(\boldsymbol{k},\boldsymbol{F}\boldsymbol{k})^2\|\boldsymbol{F}\boldsymbol{k}\|^2\xi_0 + 2\|\boldsymbol{F}\boldsymbol{k}\|^4 - \Big((\boldsymbol{k},\boldsymbol{F}\boldsymbol{k})^2\xi_0 + \|\boldsymbol{F}\boldsymbol{k}\|^2\Big)^2.
\end{aligned} \tag{32}$$

If we ignore the high order terms $\cdots$, Eq. (31) is minimized subject to $\xi = \eta$ by

$$\xi = \eta = \frac{(H_{11}+H_{12})\xi_0 + (H_{22}+H_{12})\eta_0}{H_{11}+2H_{12}+H_{22}}. \tag{33}$$

(Appendix E.) The corresponding focal lengths $f = f'$ are given by Eqs. (30).

4.3. *Fixed focal length method*

We can immediately see that the computation of Eqs. (29) fails when $K_{33} = (\boldsymbol{k},\boldsymbol{F}\boldsymbol{k}) = 0$. If we recall Eqs. (1), we can view $\boldsymbol{k}$ as representing the image origin $(0,0)$. From Eq. (2), we see that $(\boldsymbol{k},\boldsymbol{F}\boldsymbol{k}) = 0$ implies that the image origin $(0,0)$ of the first camera corresponds to the image origin $(0,0)$ of the second, which means that the optical axes of the two camera intersect in the scene. This camera configuration is said to be *fixating* (Fig. 5(a)). The fact that the focal lengths cannot be determined in fixating camera configurations has been repeatedly pointed out by many researchers [Hartley (1992), Pan et al. (1995a,b), Newsam et al. (1996)]. Kanatani and Matsunaga (2000) showed that for fixating camera configurations the Hessian $\boldsymbol{H}$ in Eq. (31) degenerates to rank 1, so the surface defined by $K(\xi,\eta)$ becomes parabolic ("trough" shape) at (ξ,η) (Fig. 5(b)). Yet, the fixating camera configuration occurs very frequently in real situations.

This difficulty is resolved if the focal lengths f and f' are known to be equal [Brooks et al. (1998), Kanatani and Matsunaga (2000), Sturm (2001)]. Geometrically, the surface of $K(\xi,\eta)$ touches the $\xi\eta$-plane along the bottom of the trough. The equal focal lengths $\xi = \eta$ are obtained by the intersection of the bottom with the line $\xi = \eta$ (Fig. 5(b)). If ξ and η are equated, Eq. (28) reduces to the following

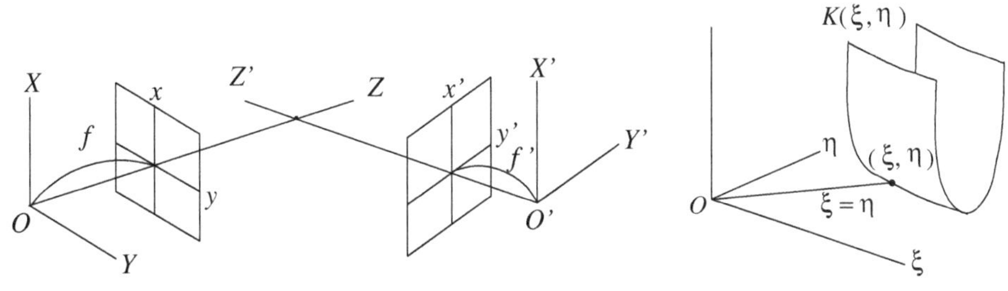

(a) (b)

Fig. 5. (a) Fixating camera configuration. (b) The equal focal lengths $f = f'$ are determined by the intersection of the "bottom' of the surface of $K(\xi, \eta)$ and the line $\xi = \eta$.

fourth-order polynomial in ξ [Kanatani and Matsunaga (2000)]:

$$K(\xi) = a_1\xi^4 + a_2\xi^3 + a_3\xi^2 + a_4\xi + a_5, \tag{34}$$

$$\begin{aligned} a_1 &= \frac{1}{2}(\boldsymbol{k}, \boldsymbol{F}\boldsymbol{k})^4, \\ a_2 &= (\boldsymbol{k}, \boldsymbol{F}\boldsymbol{k})^2(\|\boldsymbol{F}^\top\boldsymbol{k}\|^2 + \|\boldsymbol{F}\boldsymbol{k}\|^2), \\ a_3 &= \frac{1}{2}(\|\boldsymbol{F}^\top\boldsymbol{k}\|^2 - \|\boldsymbol{F}\boldsymbol{k}\|^2)^2 + (\boldsymbol{k}, \boldsymbol{F}\boldsymbol{k})(4(\boldsymbol{k}, \boldsymbol{F}\boldsymbol{F}^\top\boldsymbol{F}\boldsymbol{k}) - (\boldsymbol{k}, \boldsymbol{F}\boldsymbol{k})\|\boldsymbol{F}\|^2), \\ a_4 &= 2(\|\boldsymbol{F}\boldsymbol{F}^\top\boldsymbol{k}\|^2 + \|\boldsymbol{F}^\top\boldsymbol{F}\boldsymbol{k}\|^2) - (\|\boldsymbol{F}^\top\boldsymbol{k}\|^2 + \|\boldsymbol{F}\boldsymbol{k}\|^2)\|\boldsymbol{F}\|^2, \\ a_5 &= \|\boldsymbol{F}\boldsymbol{F}^\top\|^2 - \frac{1}{2}\|\boldsymbol{F}\|^4. \end{aligned} \tag{35}$$

The minimum of $K(\xi)$ is obtained by solving the following cubic equation in ξ:

$$K'(\xi) = 4a_1\xi^3 + 3a_2\xi^2 + 2a_3\xi + a_4 = 0. \tag{36}$$

The corresponding $f = f'$ are given by Eqs. (30). This solution is valid as long as $f = f'$ irrespective of whether the camera configuration is fixating or not. In a strict fixating configuration, we have $(\boldsymbol{k}, \boldsymbol{F}\boldsymbol{k}) = 0$, so $a_1 = a_2 = 0$ and hence Eq. (36) reduces to $2a_3\xi + a_4 = 0$ with the solution

$$\xi = -\frac{2(\|\boldsymbol{F}\boldsymbol{F}^\top\boldsymbol{k}\|^2 + \|\boldsymbol{F}^\top\boldsymbol{F}\boldsymbol{k}\|^2) - (\|\boldsymbol{F}^\top\boldsymbol{k}\|^2 + \|\boldsymbol{F}\boldsymbol{k}\|^2)\|\boldsymbol{F}\|^2}{(\|\boldsymbol{F}^\top\boldsymbol{k}\|^2 - \|\boldsymbol{F}\boldsymbol{k}\|^2)^2}. \tag{37}$$

Thus, we obtain the following algorithm:

Input: The fundamental matrix $\boldsymbol{F}$.
Output: The focal lengths $f = f'$ of the two cameras.

(1) If $(\boldsymbol{k}, \boldsymbol{F}\boldsymbol{k}) \approx 0$, compute the ξ in Eq. (37), and go to Step (3).

(2) Else, solve the cubic equation $K'(\xi) = 0$ in Eq. (36). If it has a single real zero, let it be ξ. If it has three real zeros $\xi_1 \leq \xi_2 \leq \xi_3$, let

$$\xi = \begin{cases} \xi_3 & \xi_2 \leq -1 \ \text{ or } \ K(\xi_1) < 0 \ \text{ or } \ K(\xi_3) \leq K(\xi_1) \\ \xi_1 & 0 \leq K(\xi_1) < K(\xi_3) \\ -1 & \text{otherwise} \end{cases} . \tag{38}$$

(3) If $\xi \leq -1$, return the message *failure* and stop. Else return

$$f = f' = \frac{f_0}{\sqrt{1+\xi}}. \tag{39}$$

4.4. *Issues*

If the camera configuration is strictly fixating, we have no choice but to use Eq. (37). In a practical situation, however, the camera configuration may be nearly fixating but not exactly fixating. In a non-fixating situation, we have two solutions; one obtained using the free focal length method followed by forcing the equality $f = f'$ by Eq. (33); the other using the fixed focal length method. Having two solutions decreases the chance of computational failure. In fact, the inside of the square roots in Eqs. (30) and Eq. (39) frequently becomes negative. This may be due to the inaccurate correspondence point detection but also because the camera model is not accurate enough, in particular, the principal point is not at the center of the image [Hartley and Silpa-Anan (2002)].

There is another source of computational failure, however. It has been pointed out that the focal lengths cannot be computed even for $f = f'$ if the camera configuration is fixating and *symmetric* or the two optical axes and the baseline make an isosceles triangle (including parallel optical axes) [Brooks et al. (1998), Kanatani and Matsunaga (2000), Sturm (2001)] (Fig. 6(a)). In such configurations, the function $K(\xi, \eta)$ in Eq. (28) is symmetric with respect to ξ and η, i.e., $K(\xi, \eta) = K(\eta, \xi)$. Hence, the surface of $K(\xi, \eta)$ touches the $\xi\eta$-plane along the line $\xi = \eta$ [Kanatani and Matsunaga (2000)] (Fig. 6(b)). Unfortunately, this is the most frequently en-

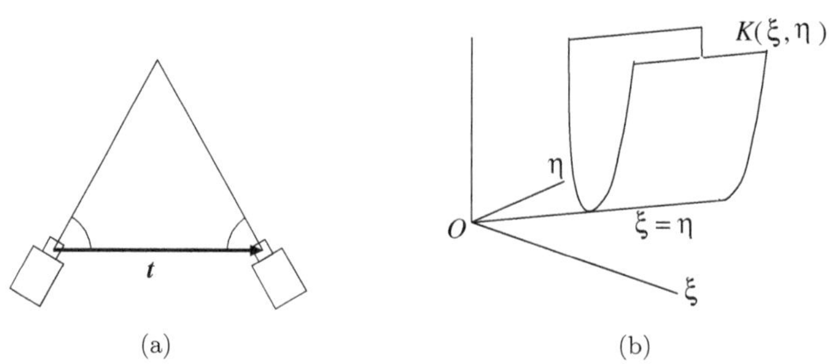

Fig. 6. (a) Symmetric camera configuration. (b) The surface of $K(\xi, \eta)$ touches the $\xi\eta$-plane along $\xi = \eta$.

countered situation in real circumstances; people unconsciously place the cameras in such a configurations.

Thus, in real situations, if one of the two focal length computation methods returns a real solution while the other fails, we can use the real one. If both fail, there is no other way but to use some empirical value for f and f'. If both returns real values, however, which solution should be used? Kanatani et al. (2006) reported that even though images are taken without changing the focus or the zoom, the free focal length method is often more accurate if the camera configuration is non-fixating to some degree. However, setting an appropriate threshold for it is very difficult. A practical solution may be to retain both solutions and compute 3-D reconstruction in two ways. In the end, we choose a better one according to some criterion, e.g., we may check if corners known to be right angles are correctly reconstructed. If no such criterion exists, we may choose the reconstruction for which the reprojection error is smaller (the computation is described in Sec. 5.2).

5. 3-D Reconstruction Procedure

5.1. *Computation of motion parameters*

Once the focal lengths $\{f,\ f'\}$ are computed, it is easy to compute the motion parameters $\{\boldsymbol{t},\ \boldsymbol{R}\}$ that satisfy Eq. (22). The following procedure is well known [Kanatani (1993)]:

Input: The fundamental matrix $\boldsymbol{F}$, the focal lengths f and f', and the image coordinates (x_α, y_α), (x'_α, y'_α), $\alpha = 1, ..., N$, of corresponding points.
Output: The motion parameters $\{\boldsymbol{t},\ \boldsymbol{R}\}$.

(1) Compute from $\boldsymbol{F}$ the essential matrix $\boldsymbol{E}$ in Eq. (23).
(2) Let $\boldsymbol{t}$ be the unit eigenvector of $\boldsymbol{E}\boldsymbol{E}^\top$ for the smallest eigenvalue.
(3) Change the sign of $\boldsymbol{t}$ if

$$\sum_{\alpha=1}^{N} |\boldsymbol{t}, \boldsymbol{x}_\alpha, \boldsymbol{E}\boldsymbol{x}'_\alpha| < 0, \tag{40}$$

where $|\boldsymbol{a}, \boldsymbol{b}, \boldsymbol{c}|$ denotes the scalar triple product of vectors $\boldsymbol{a}$, $\boldsymbol{b}$, and $\boldsymbol{c}$.
(4) Compute the singular value decomposition (SVD) of $-\boldsymbol{t} \times \boldsymbol{E}$ in the form

$$-\boldsymbol{t} \times \boldsymbol{E} = \boldsymbol{U}\mathrm{diag}(\sigma_1, \sigma_2, \sigma_3)\boldsymbol{V}^\top. \tag{41}$$

(5) Compute the rotation $\boldsymbol{R}$ as follows:

$$\boldsymbol{R} = \boldsymbol{U}\mathrm{diag}(1, 1, \det(\boldsymbol{U}\boldsymbol{V}^\top))\boldsymbol{V}^\top. \tag{42}$$

If f and f' are exactly computed so that Eq. (22) is satisfied, we have $\boldsymbol{E} = \boldsymbol{t} \times \boldsymbol{R}$. Hence, $\boldsymbol{E}\boldsymbol{t} = \boldsymbol{0}$, i.e., $\boldsymbol{t}$ is the eigenvector of $\boldsymbol{E}$ for eigenvalue 0. Step (2) ensures a stable computation even when f and f' are not exact. However, eigenvectors have

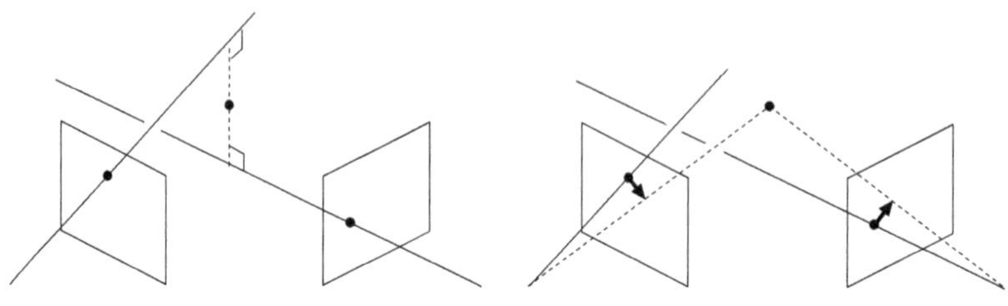

Fig. 7. Triangulation. (a) The mid-point method. (b) Optimal correction.

sign indeterminacy. Step (3) ensures that on the plane passing through $\boldsymbol{x}_\alpha$, $\boldsymbol{t}$, and $\boldsymbol{R}\boldsymbol{x}'_\alpha$, the vectors $\boldsymbol{x}_\alpha$ and $\boldsymbol{R}\boldsymbol{x}'_\alpha$ are on the same side of $\boldsymbol{t}$. This is easily seen from the identity $|\boldsymbol{t}, \boldsymbol{x}_\alpha, \boldsymbol{E}\boldsymbol{x}'_\alpha| = (\boldsymbol{t} \times \boldsymbol{x}_\alpha, \boldsymbol{t} \times \boldsymbol{R}\boldsymbol{x}'_\alpha)$. Step (4) returns the rotation $\boldsymbol{R}$ that minimizes $\|\boldsymbol{t} \times \boldsymbol{R} - \boldsymbol{E}\|$ [Kanatani (1993)].

5.2. *Optimal triangulation*

Having obtained the focal lengths $\{f,\ f'\}$ and the motion parameters $\{\boldsymbol{t},\ \boldsymbol{R}\}$, we can compute the 3-D location of each point. In fact, if we know the intrinsic and extrinsic camera parameters, we can define for each point its line of sight. The 3-D location $(X_\alpha, Y_\alpha, Z_\alpha)$ is the intersection of the lines of sight of (x_α, y_α) and (x'_α, y'_α). Due to noise in observation, however, the lines of sight need not meet in the scene. A naive idea is to regard the "midpoint" of the shortest line segment connecting the two lines of sight as the intersection (Fig. 7(a)). The best way is to displace the corresponding points so that their lines of sight intersect in such a way that the amount of the displacement is minimum (Fig. 7(b)). Mathematically, this amounts to correct (x_α, y_α) and (x'_α, y'_α) to $(\hat{x}_\alpha, \hat{y}_\alpha)$ and $(\hat{x}'_\alpha, \hat{y}'_\alpha)$ in such a way that the reprojection error

$$E_\alpha = (x_\alpha - \hat{x}_\alpha)^2 + (y_\alpha - \hat{y}_\alpha)^2 + (x'_\alpha - \hat{x}'_\alpha)^2 + (y'_\alpha - \hat{y}'_\alpha)^2 \tag{43}$$

is minimized. As noted in Sec. 3.1, this is maximum likelihood (ML) estimation if the noise in (x_α, y_α) and (x'_α, y'_α) is independent and identical Gaussian variables of mean 0. Minimization of Eq. (43) is exactly the same as Sec. 3.4 except that the fundamental matrix is now regarded as given [Kanatani et al. (2008)]:

Input: The focal lengths $\{f,\ f'\}$, the motion parameters $\{\boldsymbol{t},\ \boldsymbol{R}\}$, and the image coordinates (x_α, y_α), (x'_α, y'_α), $\alpha = 1, ..., N$, of corresponding points.

Output: Corrected image coordinates $(\hat{x}_\alpha, \hat{y}_\alpha)$, $(\hat{x}'_\alpha, \hat{y}'_\alpha)$, $\alpha = 1, ..., N$, and the reprojection error E.

(1) Recompute the essential matrix $\boldsymbol{E}$ as follows:

$$\boldsymbol{E} = \boldsymbol{t} \times \boldsymbol{R}. \tag{44}$$

(2) Compute the following 9-D vector $\boldsymbol{u}$:

$$\boldsymbol{u} = (E_{11},\, E_{12},\, E_{13},\, E_{21},\, E_{22},\, E_{23},\, E_{31},\, E_{32},\, E_{33})^\top. \tag{45}$$

(3) Do the following computation for $\alpha = 1, ..., N$:

(a) Let $E_\alpha^0 = \infty$ (a sufficiently large number), and initialize $\hat{x}_\alpha$, $\hat{y}_\alpha$, $\hat{x}'_\alpha$, $\hat{y}'_\alpha$, $\tilde{x}_\alpha$, $\tilde{y}_\alpha$, $\tilde{x}'_\alpha$, and $\tilde{y}'_\alpha$ as follows:

$$\hat{x}_\alpha = x_\alpha, \qquad \hat{y}_\alpha = y_\alpha, \qquad \hat{x}'_\alpha = x'_\alpha, \qquad \hat{y}'_\alpha = y'_\alpha,$$

$$\tilde{x}_\alpha = \tilde{y}_\alpha = \tilde{x}'_\alpha = \tilde{y}'_\alpha = 0. \tag{46}$$

(b) Compute the following 9-D vectors $\boldsymbol{\xi}_\alpha$ and 9×9 matrices $V_0[\boldsymbol{\xi}_\alpha]$:

$$\boldsymbol{\xi}_\alpha = \begin{pmatrix} \hat{x}_\alpha\hat{x}'_\alpha + \hat{x}'_\alpha\tilde{x}_\alpha + \hat{x}_\alpha\tilde{x}'_\alpha \\ \hat{x}_\alpha\hat{y}'_\alpha + \hat{y}'_\alpha\tilde{x}_\alpha + \hat{x}_\alpha\tilde{y}'_\alpha \\ f'(\hat{x}_\alpha + \tilde{x}_\alpha) \\ \hat{y}_\alpha\hat{x}'_\alpha + \hat{x}'_\alpha\tilde{y}_\alpha + \hat{y}_\alpha\tilde{x}'_\alpha \\ \hat{y}_\alpha\hat{y}'_\alpha + \hat{y}'_\alpha\tilde{y}_\alpha + \hat{y}_\alpha\tilde{y}'_\alpha \\ f'(\hat{y}_\alpha + \tilde{y}_\alpha) \\ f(\hat{x}'_\alpha + \tilde{x}'_\alpha) \\ f(\hat{y}'_\alpha + \tilde{y}'_\alpha) \\ ff' \end{pmatrix}, \tag{47}$$

$$V_0[\boldsymbol{\xi}_\alpha] = \begin{pmatrix} \hat{x}_\alpha^2 + \hat{x}'^2_\alpha & \hat{x}'_\alpha\hat{y}'_\alpha & f\hat{x}'_\alpha & \hat{x}_\alpha\hat{y}_\alpha & 0 & 0 & f'\hat{x}_\alpha & 0 & 0 \\ \hat{x}'_\alpha\hat{y}'_\alpha & \hat{x}_\alpha^2 + \hat{y}'^2_\alpha & f\hat{y}'_\alpha & 0 & \hat{x}_\alpha\hat{y}_\alpha & 0 & 0 & f'\hat{x}_\alpha & 0 \\ f\hat{x}'_\alpha & f\hat{y}'_\alpha & ff' & 0 & 0 & 0 & 0 & 0 & 0 \\ \hat{x}_\alpha\hat{y}_\alpha & 0 & 0 & \hat{y}_\alpha^2 + \hat{x}'^2_\alpha & \hat{x}'_\alpha\hat{y}'_\alpha & f\hat{x}'_\alpha & f'\hat{y}_\alpha & 0 & 0 \\ 0 & \hat{x}_\alpha\hat{y}_\alpha & 0 & \hat{x}'_\alpha\hat{y}'_\alpha & \hat{y}_\alpha^2 + \hat{y}'^2_\alpha & f\hat{y}'_\alpha & 0 & f'\hat{y}_\alpha & 0 \\ 0 & 0 & 0 & f\hat{x}'_\alpha & f\hat{y}'_\alpha & ff' & 0 & 0 & 0 \\ f'\hat{x}_\alpha & 0 & 0 & f'\hat{y}_\alpha & 0 & 0 & ff' & 0 & 0 \\ 0 & f'\hat{x}_\alpha & 0 & 0 & f'\hat{y}_\alpha & 0 & 0 & ff' & 0 \\ 0 & 0 & 0 & 0 & 0 & 0 & 0 & 0 & 0 \end{pmatrix}. \tag{48}$$

(c) Update $\tilde{x}_\alpha$, $\tilde{y}_\alpha$, $\tilde{x}'_\alpha$, and $\tilde{y}'_\alpha$ as follows:

$$\begin{pmatrix} \tilde{x}_\alpha \\ \tilde{y}_\alpha \end{pmatrix} \leftarrow \frac{(\boldsymbol{u}, \boldsymbol{\xi}_\alpha)}{(\boldsymbol{u}, V_0[\boldsymbol{\xi}_\alpha]\boldsymbol{u})} \begin{pmatrix} u_1 & u_2 & u_3 \\ u_4 & u_5 & u_6 \end{pmatrix} \begin{pmatrix} \hat{x}'_\alpha \\ \hat{y}'_\alpha \\ f' \end{pmatrix},$$

$$\begin{pmatrix} \tilde{x}'_\alpha \\ \tilde{y}'_\alpha \end{pmatrix} \leftarrow \frac{(\boldsymbol{u}, \boldsymbol{\xi}_\alpha)}{(\boldsymbol{u}, V_0[\boldsymbol{\xi}_\alpha]\boldsymbol{u})} \begin{pmatrix} u_1 & u_4 & u_7 \\ u_2 & u_5 & u_8 \end{pmatrix} \begin{pmatrix} \hat{x}_\alpha \\ \hat{y}_\alpha \\ f \end{pmatrix}. \tag{49}$$

(d) Update $\hat{x}_\alpha$, $\hat{y}_\alpha$, $\hat{x}'_\alpha$, and $\hat{y}'_\alpha$ as follows:

$$\hat{x}_\alpha \leftarrow x_\alpha - \tilde{x}_\alpha, \qquad \hat{y}_\alpha \leftarrow y_\alpha - \tilde{y}_\alpha,$$

$$\hat{x}'_\alpha \leftarrow x'_\alpha - \tilde{x}'_\alpha, \qquad \hat{y}'_\alpha \leftarrow y'_\alpha - \tilde{y}'_\alpha. \tag{50}$$

(e) Compute the reprojection error E_α as follows:

$$E_\alpha = \tilde{x}_\alpha^2 + \tilde{y}_\alpha^2 + \tilde{x}'^{\,2}_\alpha + \tilde{y}'^{\,2}_\alpha. \tag{51}$$

(f) If $E_\alpha \not\approx E_\alpha^0$, let $E_\alpha^0 \leftarrow E_\alpha$ and go back to Step (3b).

(4) Return $(\hat{x}_\alpha, \hat{y}_\alpha)$, $(\hat{x}'_\alpha, \hat{y}'_\alpha)$, $\alpha = 1, ..., N$, and the following (total) reprojection error E:

$$E = \sum_{\alpha=1}^{N} E_\alpha. \tag{52}$$

If we use the focal lengths $\{f,\ f'\}$ computed by the free focal length method (without failure), the resulting values of $(\hat{x}_\alpha, \hat{y}_\alpha)$ and $(\hat{x}'_\alpha, \hat{y}'_\alpha)$ are identical to those computed in Sec. 3.4, because the fundamental matrix $\boldsymbol{F}$ was computed to minimize the reprojection error E in Eq. (52). If the equality $f = f'$ is enforced, however, the results are not necessarily the same, and we need the recomputation as described above. If we have retained two solutions, one by the free focal length method followed by enforcement of $f = f'$, the other by the fixed focal length method, the selection can be done at this stage: we choose the one that has a smaller reprojection error E computed here.

5.3. *3-D location computation*

A 3-D point (X, Y, Z) is projected to a point (x, y) on the image plane by a general perspective camera in the form

$$\begin{pmatrix} x/f_0 \\ y/f_0 \\ 1 \end{pmatrix} \simeq \boldsymbol{P} \begin{pmatrix} X \\ Y \\ Z \\ 1 \end{pmatrix}, \tag{53}$$

where the symbol $\simeq$ means that both sides are equal up to a nonzero constant. Here, $\boldsymbol{P}$ is a 3×4 matrix called the *projection matrix* or the *camera matrix*. If the camera coordinate system is placed at $\boldsymbol{t}$ and rotated by $\boldsymbol{R}$ relative to a fixed world coordinate system, $\boldsymbol{P}$ has form

$$\boldsymbol{P} \simeq \boldsymbol{K} \begin{pmatrix} \boldsymbol{R}^\top & -\boldsymbol{R}^\top \boldsymbol{t} \end{pmatrix}, \tag{54}$$

where $\boldsymbol{K}$ is an upper triangular matrix, called the *matrix of intrinsic parameters*, in the following form [Hartley and Zisserman (2000)]:

$$\boldsymbol{K} = \begin{pmatrix} \gamma f/f_0 & \gamma \tan\theta & u_0/f_0 \\ & f/f_0 & v_0/f_0 \\ & & 1 \end{pmatrix}. \tag{55}$$

Here, f is the focal length, γ is the aspect ratio, θ is the skew angle (defined to be 0 in the skewless case), and (u_0, v_0) is the principal point location. As discussed in Sec. 4.1, we need to assume that $\gamma = 1$, $\theta = 0$, and $(u_0, v_0) = (0, 0)$ for doing 3-D reconstruction from two views. Thus, if the two cameras have focal length f and f' with motion parameters $\{\boldsymbol{t}, \boldsymbol{R}\}$ and if the first camera is identified with the world coordinate system, the projection matrices of the first and the second cameras can be written up scale as follows:

$$\boldsymbol{P} \simeq \mathrm{diag}(1, 1, \frac{f_0}{f})\begin{pmatrix}\boldsymbol{I} & \boldsymbol{0}\end{pmatrix}, \qquad \boldsymbol{P}' \simeq \mathrm{diag}(1, 1, \frac{f_0}{f'})\begin{pmatrix}\boldsymbol{R}^\top & -\boldsymbol{R}^\top\boldsymbol{t}\end{pmatrix}. \tag{56}$$

From Eq. (53), we obtain the following projection relations:

$$\begin{cases} x = f_0\dfrac{P_{11}X + P_{12}Y + P_{13}Z + P_{14}}{P_{31}X + P_{32}Y + P_{33}Z + P_{34}}, \\ y = f_0\dfrac{P_{21}X + P_{22}Y + P_{23}Z + P_{24}}{P_{31}X + P_{32}Y + P_{33}Z + P_{34}}, \end{cases} \quad \begin{cases} x' = f_0\dfrac{P'_{11}X + P'_{12}Y + P'_{13}Z + P'_{14}}{P'_{31}X + P'_{32}Y + P'_{33}Z + P'_{34}}, \\ y' = f_0\dfrac{P'_{21}X + P'_{22}Y + P'_{23}Z + P'_{24}}{P'_{31}X + P'_{32}Y + P'_{33}Z + P'_{34}}. \end{cases} \tag{57}$$

From these, we obtain the following linear equations:

$$\begin{pmatrix} xP_{31} - f_0P_{11} & xP_{32} - f_0P_{12} & xP_{33} - f_0P_{13} \\ yP_{31} - f_0P_{21} & yP_{32} - f_0P_{22} & yP_{33} - f_0P_{23} \\ x'P'_{31} - f_0P'_{11} & x'P'_{32} - f_0P'_{12} & x'P'_{33} - f_0P'_{13} \\ y'P'_{31} - f_0P'_{21} & y'P'_{32} - f_0P'_{22} & y'P'_{33} - f_0P'_{23} \end{pmatrix}\begin{pmatrix} X \\ Y \\ Z \end{pmatrix} = -\begin{pmatrix} xP_{34} - f_0P_{14} \\ yP_{34} - f_0P_{24} \\ x'P'_{34} - f_0P'_{14} \\ y'P'_{34} - f_0P'_{24} \end{pmatrix}. \tag{58}$$

These provide four equations for three unknowns X, Y, and Z. We can choose from among them any three and solve them. Alternatively, we can solve the four redundant equations by least squares (Appendix F). Both give the same solution if we use $(\hat{x}_\alpha, \hat{y}_\alpha)$ and $(\hat{x}'_\alpha, \hat{y}'_\alpha)$ as image coordinates, because the procedure in Sec. 5.2 is done for that purpose in an optimal manner.

However, we need to check the *cheirality* [Hartley and Zisserman (2000)]. The perspective projection described by Eq. (53) stipulates that a 3-D point is projected to the intersection of the ray it defines with the image plane, but no constraint is imposed on the 3-D point to be *in front* of the camera. The condition in Eq. (40) only ensures that the 3-D points are both in front of the two cameras (the correct shape) or both behind them (the mirror image shape). This ambiguity originates from the sign indeterminacy of the fundamental matrix $\boldsymbol{F}$ in Eq. (2); Eq. (40) merely choose the sign of $\boldsymbol{t}$ compatible to $\boldsymbol{F}$. Thus, after computing the 3-D locations $(X_\alpha, Y_\alpha, Z_\alpha)$ of all the points, we reverse the sign of $\boldsymbol{t}$ and all $(X_\alpha, Y_\alpha, Z_\alpha)$ if

$$\sum_{\alpha=1}^{N} \mathrm{sgn}(Z_\alpha) < 0, \tag{59}$$

where $\mathrm{sgn}(x)$ is the signature function that returns 1, -1, and 0 for $x > 0$, $x < 0$, and $x = 0$, respectively. We introduce $\mathrm{sgn}(x)$ because if we require $\sum_{\alpha=1}^{N} Z_\alpha > 0$, the judgment may be reversed when a very large depth $Z_\alpha \approx \infty$ with nearly parallel lines of sight may be computed to be $Z_\alpha \approx -\infty$.

The reconstructed 3-D shape has scale ambiguity, as we discussed in Sec. 4.1, so we introduced the scale normalization $\|\boldsymbol{t}\| = 1$. This ambiguity is resolved if some metric information is available. For example, if the camera translation $\|\boldsymbol{t}\|$ is known, the computed 3-D shape is magnified by $\|\boldsymbol{t}\|$. If the distance between two 3-D points, say (X_1, Y_1, Z_1) and (X_2, Y_2, Z_2), is known to be d_{12}, we can recover the correct scale by multiplying all $(X_\alpha, Y_\alpha, Z_\alpha)$ and $\boldsymbol{t}$ by $d_{12}/\sqrt{(X_2 - X_1)^2 + (Y_2 - Y_1)^2 + (Z_2 - Z_1)^2}$. Instead of taking the first camera as a reference, we can also view the first camera as displaced by $\boldsymbol{t}_c$ and rotated by $\boldsymbol{R}_c$ relative to a fixed world coordinate system. The 3-D locations $(\bar{X}_\alpha, \bar{Y}_\alpha, \bar{Z}_\alpha)$ with respect to this coordinate system are given by

$$\begin{pmatrix} \bar{X}_\alpha \\ \bar{Y}_\alpha \\ \bar{Z}_\alpha \end{pmatrix} = \boldsymbol{R}_c \begin{pmatrix} X_\alpha \\ Y_\alpha \\ Z_\alpha \end{pmatrix} + \boldsymbol{t}_c \quad \left(= \begin{pmatrix} \boldsymbol{R}_c & \boldsymbol{t}_c \end{pmatrix} \begin{pmatrix} X_\alpha \\ Y_\alpha \\ Z_\alpha \\ 1 \end{pmatrix} \right). \tag{60}$$

6. Concluding Remarks

3-D reconstruction from two views has intensively been studied since early 1980s as one of the major problems of computer vision. After nearly 30 years of its history, however, efforts are still being made to improve the computational accuracy and efficiency. This chapter has presented the state-of-the-art technology of this problem.

Since we only need to take two pictures, two view reconstruction is one of the most useful tools in computer vision applications. However, we must also realize that limiting to two views imposes severe restrictions on its applicability. Computer vision practitioners are frequently confounded by the failure of 3-D reconstruction due to the lack of understanding such restrictions. Here, we describe typical *failure modes*:

Infinitely far away scene: If the scene is infinitely far away, the two views are essentially the same (up to camera rotation), so no 3-D information is obtained. In practice, this is the case if the scene is sufficiently far away as compared to the translation $\boldsymbol{t}$,

Planar scene: The fundamental matrix $\boldsymbol{F}$ cannot be computed uniquely if all the points are coplanar in 3-D (Fig. 8(a)). This is a typical instance of the critical surface mentioned in Sec. 3.1; that plane and a plane passing through the two viewpoints O and O' constitute a degenerate quadric surface. However, it is possible to compute 3-D structure and motion by exploiting the geometry of projective transformation, or "homography", instead of the epipolar geometry. [Kanatani (1993, 1996), Hartley and Zisserman (2000)].

Box scene: The fundamental matrix $\boldsymbol{F}$ cannot be computed uniquely if we observe the eight vertices of a cube (Fig. 8(b)). This is also an instance of the critical surface; there always exists a quadric surface that passes through the eight vertices and arbitrarily placed viewpoints O and O'. This configuration

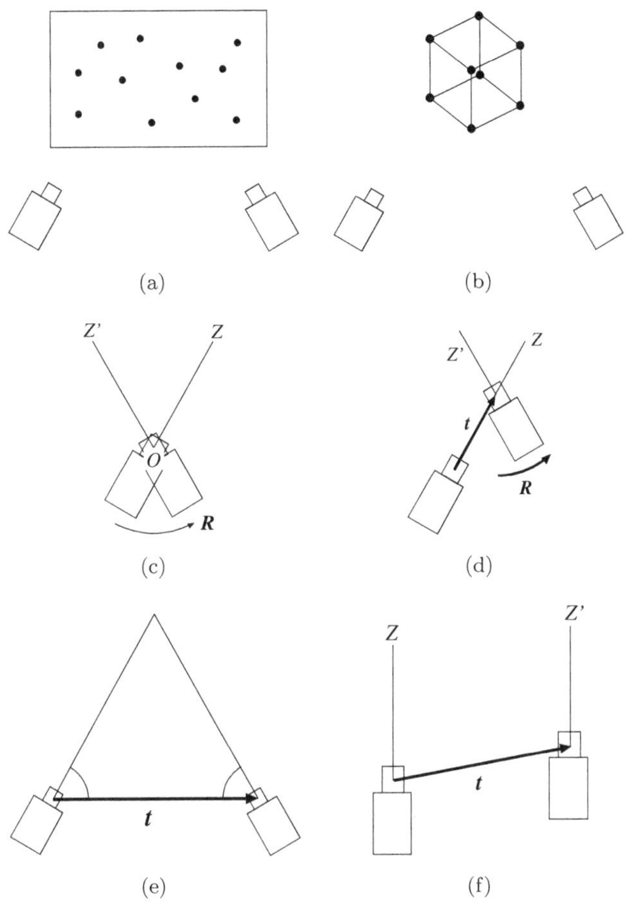

Fig. 8. Typical failure modes of two view 3-D reconstruction. (a) Planar scene. (b) Box scene. (c) Pure rotation. (d) Advancing camera. (e) Symmetric cameras. (f) Pure translation.

usually does not occur in real circumstances, yet this frequently happens in simulation, because people tend to test their algorithms using this configuration. In simulation, we should use general configurations, avoiding too regular or exceptional settings.

Pure rotation: No 3-D information is obtained if we observe the scene by rotating the camera around the lens center (Fig. 8(c)). This is because the lines of sight coming through the lens center are the same before and after the camera rotation. Hence, the fundamental matrix cannot be computed. However, the camera rotation can be determined by computing the homography [Hartley and Zisserman (2000)].

Advancing camera: If one camera is moved along its optical axis and then rotated (Fig. 8(d)), the focal lengths f and f' cannot be computed by the free-focal length method. This is because the two cameras are in a fixating configuration with the fixating point *at the lens center of the second camera*. This fact is easily overlooked in a real situation. In such a case, the focal lengths $f = f'$ can be computed if the fixed focal length method is used as long as the rotation of the second camera is nonzero. However, a care is necessary: the 3-D location of a point at the center of the first image cannot be determined, because it is the epipole of the first camera, i.e., the projection of the viewpoint of the second camera. In fact, a 3-D point P on the line connecting the viewpoints O and O' of the two cameras cannot be "triangulated" if O, O', and P are collinear. In actual computation, the rank of the matrix in Eq. (58) drops to 2, and the solution is indeterminate. Points near the center of the first image can be triangulated, but their computed 3-D locations suffer considerable degradation in accuracy.

Symmetric cameras: As mentioned in Sec. 4.4, the focal lengths f and f' cannot be computed if the camera configuration is symmetric (Fig. 8(e)). We cannot compute f and f' even if we impose the equality $f = f'$. Of course, 3-D reconstruction is possible if the focal lengths f and f' are known. In fact, this is a standard camera configuration of stereo vision.

Pure translation: The focal lengths f and f' cannot be computed if one camera is translated with the optical axis direction fixed (Fig. 8(f)). This is a fixating configuration with the fixation point at infinity. We cannot compute f and f' even if we impose the equality $f = f'$, because the translation $\boldsymbol{t}$ and the two optical axes make an (infinitely elongated) isosceles triangle with the vertex at infinity. Again, 3-D reconstruction is possible if the focal lengths f and f' are known; this is also a standard camera configuration of stereo vision.

Besides avoiding these failure modes, caution is necessary when black box tools are used. For example, the numerical values of the fundamental matrix $\boldsymbol{F}$ depends on the image coordinate system used as well as the scaling constant f_0. For example, if we regard the upper-left corner of the image as the origin $(0, 0)$, take the x-axis direction rightward and the y-axis direction upward, and let $f_0 = 1$, which is a standard convention in image processing applications, the fundamental matrix $\boldsymbol{F}$ has a completely different numerical value. Some authors, e.g., Hartley and Zisserman (2000), define the fundamental matrix $\boldsymbol{F}$ to be the transpose of the one defined in this chapter, so that the epipolar equation is written, instead of Eq. (2), as $(\bar{\boldsymbol{x}}'_\alpha, \boldsymbol{F}\bar{\boldsymbol{x}}_\alpha) = 0$. Thus, users must correctly understand the definitions and conventions used in ready-made tools. Then, the theories and computational techniques described in this chapter will allow users to do two view reconstruction in many practical applications in useful ways.

Finally, it should be stressed that the quality of 3-D reconstruction critically depends on the quality of correspondence detection. The standard procedure is to first detect typical points called *feature points*, *corner points*, *points of interest*, or *keypoints* by image filters such as the Harris operator [Harris and Stephens

(1988)], SUSAN [Smith and Brady (1997)], MSER [Matas, et. al (2002)] and SIFT [Lowe (2004)] in two images independently and then match them using appropriate measures such as intensity correlations, patch similarities, and feature descriptors. Finally, false matches, or *outliers*, that are incompatible with the epipolar geometry are removed by a voting procedure such as RANSAC [Fischer and Bolles (1981)] and LMedS [Rousseeuw and Leroy (1978)], e.g. see Zhang et al. (1995) and Kanazawa and Kanatani (2004). Today, the use of video images is common, and feature points trackers such as the Kanade-Lucas-Tomasi algorithm (KLT) [Tomasi and Kanade (1991)] is widely used. Then, two frames are selected from the sequence, and those points continually tracked between them are used for 3-D reconstruction. However, correct correspondence detection is still a difficult task, and further progress is required for reliable 3-D reconstruction.

Appendix

A. *Least squares*

The simplest scheme for fundamental matrix computation is the following least squares, also known as *algebraic distance minimization* and *Hartley's 8-point algorithm* [Hartley (1997)].

Input: The image coordinates (x_α, y_α), (x'_α, y'_α), $\alpha = 1, ..., N$ (≥ 8), of corresponding points.

Output: The fundamental matrix $\boldsymbol{F}$.

(1) Represent the corresponding points (x_α, y_α) and (x'_α, y'_α), $\alpha = 1, ..., N$, by the following 9-D vectors:

$$\boldsymbol{\xi}_\alpha = (x_\alpha x'_\alpha,\ x_\alpha y'_\alpha,\ f_0 x_\alpha,\ y_\alpha x'_\alpha,\ y_\alpha y'_\alpha,\ f_0 y_\alpha,\ f_0 x'_\alpha,\ f_0 y'_\alpha,\ f_0^2)^\top. \tag{A.1}$$

(2) Let $\boldsymbol{u}$ be the 9-D unit eigenvector of the 9×9 matrix

$$\boldsymbol{M} = \sum_{\alpha=1}^{N} \boldsymbol{\xi}_\alpha \boldsymbol{\xi}_\alpha^\top, \tag{A.2}$$

for the smallest eigenvalue.

(3) Return the following fundamental matrix:

$$\boldsymbol{F} = \begin{pmatrix} u_1 & u_2 & u_3 \\ u_4 & u_5 & u_6 \\ u_7 & u_8 & u_9 \end{pmatrix}. \tag{A.3}$$

If $\boldsymbol{x}_\alpha$ and $\boldsymbol{x}'_\alpha$ are converted in to the 9-D vector in Eq. (A.1) and if the fundamental matrix $\boldsymbol{F}$ is represented by the 9-D vector $\boldsymbol{u}$ in Eq. (4), the epipolar equation $(\boldsymbol{x}_\alpha, \boldsymbol{F}\boldsymbol{x}'_\alpha) = 0$ is equivalently written as

$$(\boldsymbol{u}, \boldsymbol{\xi}_\alpha) = 0. \tag{A.4}$$

Mathematically, the vector $\boldsymbol{\xi}_\alpha$ in Eq. (A.1) is called the *Kronecker product* of $\boldsymbol{x}_\alpha$ and $\boldsymbol{x}'_\alpha$ (multiplied by f_0^2). For noisy data, Eq. (A.4) is not exactly satisfied, so we seek the vector $\boldsymbol{u}$ that minimizes

$$J_{\rm LS} = \sum_{\alpha=1}^{N}(\boldsymbol{u}, \boldsymbol{\xi}_\alpha)^2 = \sum_{\alpha=1}^{N} \boldsymbol{u}^\top \boldsymbol{\xi}_\alpha \boldsymbol{\xi}_\alpha^\top \boldsymbol{u} = (\boldsymbol{u}, \boldsymbol{M}\boldsymbol{u}), \tag{A.5}$$

where $\boldsymbol{M}$ is defined by Eq. (A.2). If we normalized $\boldsymbol{u}$ to a unit vector ($\|\boldsymbol{u}\| = 1$) to remove the scale indeterminacy of $\boldsymbol{F}$, the quadratic form in Eq. (A.5) is minimized by the unit eigenvector of $\boldsymbol{M}$ for the smallest eigenvalue.

The matrix $\boldsymbol{F}$ thus computed does not necessarily satisfy the rank constraint $\det \boldsymbol{F} = 0$. A simple method for remedy is to compute the singular value decomposition (SVD) of $\boldsymbol{F}$ and replace the smallest singular value by 0 [Hartley (1997)]. More sophisticated methods exist for optimally correcting $\boldsymbol{F}$ so that $\det \boldsymbol{F} = 0$ holds based on statistical analysis of noise [Kanatani and Sugaya (2007a)]. However, such sophistication is not necessary in practice, because the least squares solution itself has only limited accuracy [Kanatani and Sugaya (2007a,c)].

B. Taubin method

If the noise in x_α, y_α, x'_α, and y'_α is regarded as independent Gaussian variables of mean 0 and standard deviation σ, the vector $\boldsymbol{\xi}_\alpha$ in Eq. (A.1) has the following covariance matrix [Kanatani and Sugaya (2007a,c)]:

$$V[\boldsymbol{\xi}_\alpha] = \sigma^2 V_0[\boldsymbol{\xi}_\alpha], \tag{B.1}$$

$$V_0[\boldsymbol{\xi}_\alpha] = \begin{pmatrix} x_\alpha^2 + x_\alpha'^2 & x'_\alpha y'_\alpha & f_0 x'_\alpha & x_\alpha y_\alpha & 0 & 0 & f_0 x_\alpha & 0 & 0 \\ x'_\alpha y'_\alpha & x_\alpha^2 + y_\alpha'^2 & f_0 y'_\alpha & 0 & x_\alpha y_\alpha & 0 & 0 & f_0 x_\alpha & 0 \\ f_0 x'_\alpha & f_0 y'_\alpha & f_0^2 & 0 & 0 & 0 & 0 & 0 & 0 \\ x_\alpha y_\alpha & 0 & 0 & y_\alpha^2 + x_\alpha'^2 & x'_\alpha y'_\alpha & f_0 x'_\alpha & f_0 y_\alpha & 0 & 0 \\ 0 & x_\alpha y_\alpha & 0 & x'_\alpha y'_\alpha & y_\alpha^2 + y_\alpha'^2 & f_0 y'_\alpha & 0 & f_0 y_\alpha & 0 \\ 0 & 0 & 0 & f_0 x'_\alpha & f_0 y'_\alpha & f_0^2 & 0 & 0 & 0 \\ f_0 x_\alpha & 0 & 0 & f_0 y_\alpha & 0 & 0 & f_0^2 & 0 & 0 \\ 0 & f_0 x_\alpha & 0 & 0 & f_0 y_\alpha & 0 & 0 & f_0^2 & 0 \\ 0 & 0 & 0 & 0 & 0 & 0 & 0 & 0 & 0 \end{pmatrix}. \tag{B.2}$$

If we regard each $\boldsymbol{\xi}_\alpha$ as perturbed from its true value $\bar{\boldsymbol{\xi}}_\alpha$ by independent Gaussian noise of mean $\boldsymbol{0}$ and covariance matrix $V[\boldsymbol{\xi}_\alpha]$, maximum likelihood (ML) estimation of the fundamental matrix is to minimize the Mahalanobis distance

$$J_{\rm ML} = \sum_{\alpha=1}^{N}(\boldsymbol{\xi}_\alpha - \bar{\boldsymbol{\xi}}_\alpha, V_0[\boldsymbol{\xi}_\alpha](\boldsymbol{\xi}_\alpha - \bar{\boldsymbol{\xi}}_\alpha)), \tag{B.3}$$

subject to

$$(\boldsymbol{u}, \bar{\boldsymbol{\xi}}_\alpha) = 0. \tag{B.4}$$

We can eliminate the constraint of Eq. (B.4) by using Lagrange multipliers and reduce the problem to unconstrained minimization of

$$J_{\rm ML} = \sum_{\alpha=1}^{N} \frac{(\boldsymbol{u}, \boldsymbol{\xi}_\alpha)^2}{(\boldsymbol{u}, V_0[\boldsymbol{\xi}_\alpha]\boldsymbol{u})}. \tag{B.5}$$

This function can be minimized by various means such as *FNS* (*Fundamental Numerical Scheme*) of Chojnacki et al. (2000), *HEIV* (*Heteroscedastic Errors-In-Variables*) of Leedan and Meer (2000), and projective Gauss-Newton iterations of Sugaya and Kanatani (2007a,c), but all requires iterations, which sometimes do not converge in the presence of large noise. If the denominators in Eq. (B.5) are replaced by a constant, the problem reduces to least squares, minimizing Eq. (A.5). The Taubin method replaces the denominators in Eq. (B.5) by their average and minimizes

$$J_{\rm TB} = \frac{\sum_{\alpha=1}^{N}(\boldsymbol{u}, \boldsymbol{\xi}_\alpha)^2}{\sum_{\alpha=1}^{N}(\boldsymbol{u}, V_0[\boldsymbol{\xi}_\alpha]\boldsymbol{u})/N} = \frac{N(\boldsymbol{u}, \boldsymbol{M}\boldsymbol{u})}{(\boldsymbol{u}, \boldsymbol{N}\boldsymbol{u})}, \tag{B.6}$$

where $\boldsymbol{M}$ is the matrix in Eq. (A.2) and the matrix $\boldsymbol{N}$ is defined by

$$\boldsymbol{N} = \sum_{\alpha=1}^{N} V_0[\boldsymbol{\xi}_\alpha]. \tag{B.7}$$

Eq. (B.6) is the ratio of quadratic forms, called the *Rayleigh quotient*, and is minimized by the generalized eigenvector of the generalized eigenvalue problem

$$\boldsymbol{M}\boldsymbol{u} = \lambda \boldsymbol{N}\boldsymbol{u}, \tag{B.8}$$

for the smallest generalized eigenvalue. However, this problem cannot be solved by the standard method (Appendix C), because the matrix $V_0[\boldsymbol{\xi}_\alpha]$ in Eq. (B.2) and hence the matrix $\boldsymbol{N}$ in Eq. (B.7) are singular, the elements in the ninth column and the ninth row all being 0. So, we reduce Eq. (B.8) to a generalized eigenvalue problem of a smaller dimension. We decompose the 9-D vectors $\boldsymbol{u}$, $\boldsymbol{\xi}_\alpha$ and the 9×9 matrix $V_0[\boldsymbol{\xi}_\alpha]$ into their 8-D parts and the 9th elements in the form

$$\boldsymbol{u} = \begin{pmatrix} \boldsymbol{v} \\ F_{33} \end{pmatrix}, \quad \boldsymbol{\xi}_\alpha = \begin{pmatrix} \boldsymbol{z}_\alpha \\ f_0^2 \end{pmatrix}, \quad V_0[\boldsymbol{\xi}_\alpha] = \begin{pmatrix} V_0[\boldsymbol{z}_\alpha] & \boldsymbol{0} \\ \boldsymbol{0}^\top & 0 \end{pmatrix}, \tag{B.9}$$

where $\boldsymbol{z}_\alpha$ is the 8-D vector $\boldsymbol{z}_\alpha$ in Eq. (5) and $V_0[\boldsymbol{z}_\alpha]$ is the 8×8 matrix in Eq. (6). Then, Eq. (B.8) is written as

$$\begin{pmatrix} \sum_{\alpha=1}^{N} \boldsymbol{z}_\alpha \boldsymbol{z}_\alpha^\top & f_0^2 \sum_{\alpha=1}^{N} \boldsymbol{z}_\alpha \\ f_0^2 \sum_{\alpha=1}^{N} \boldsymbol{z}_\alpha^\top & N f_0^4 \end{pmatrix} \begin{pmatrix} \boldsymbol{v} \\ F_{33} \end{pmatrix} = \lambda \begin{pmatrix} \boldsymbol{N}_{\rm TB} & \boldsymbol{0} \\ \boldsymbol{0}^\top & 0 \end{pmatrix} \begin{pmatrix} \boldsymbol{v} \\ F_{33} \end{pmatrix}, \tag{B.10}$$

where the 8×8 matrix $\boldsymbol{N}_{\rm TB}$ is defined by the second of Eqs. (8). If we define $\bar{\boldsymbol{z}}_\alpha$ and o$\tilde{\boldsymbol{z}}_\alpha$ by Eqs. (7) and the 8×8 matrix $\boldsymbol{M}_{\rm TB}$ by the first of Eqs. (8), we have the identity

$$\sum_{\alpha=1}^{N} \boldsymbol{z}_\alpha \boldsymbol{z}_\alpha^\top = \boldsymbol{M}_{\rm TB} + \bar{\boldsymbol{z}}\bar{\boldsymbol{z}}^\top. \tag{B.11}$$

Hence, Eq. (B.10) splits into the following two equations:

$$\boldsymbol{M}_{\mathrm{TB}}\boldsymbol{v} = \lambda \boldsymbol{N}_{\mathrm{TB}}\boldsymbol{v}, \qquad (\bar{\boldsymbol{z}}, \boldsymbol{v}) + f_0^2 F_{33} = 0. \tag{B.12}$$

The second equation determines F_{33} in terms of $\boldsymbol{v}$ by $F_{33} = -(\bar{\boldsymbol{z}}, \boldsymbol{v})/f_0^2$, so if we solve the first equation, which is the same as Eq. (9), the 9-D vector $\boldsymbol{u}$ (normalized to unit norm) is given by Eq. (10). Thus, the procedure in Sec. 3.2 is justified.

C. *Generalized eigenvalue problem*

The generalized eigenvalue problem can be solved by the following procedure:

Input: An $n \times n$ symmetric matrix $\boldsymbol{A}$ and an $n \times n$ positive definite symmetric matrix $\boldsymbol{G}$.

Output: The generalized eigenvalues $\lambda_1, \ldots, \lambda_n$ and the corresponding generalized eigenvectors $\boldsymbol{w}_1, \ldots, \boldsymbol{w}_n$ (normalized to unit norm) that satisfy

$$\boldsymbol{A}\boldsymbol{w}_i = \lambda_i \boldsymbol{G}\boldsymbol{w}_i. \tag{C.1}$$

(1) Compute the eigenvalues $\mu_1, \ldots, \mu_n$ (> 0) of $\boldsymbol{G}$ and the corresponding unit eigenvectors $\boldsymbol{g}_1, \ldots, \boldsymbol{g}_n$.

(2) Define the following $n \times n$ matrices $\boldsymbol{T}$ and $\tilde{\boldsymbol{A}}$:

$$\boldsymbol{T} = \frac{\boldsymbol{g}_1\boldsymbol{g}_1^\top}{\sqrt{\mu_1}} + \cdots + \frac{\boldsymbol{g}_n\boldsymbol{g}_n^\top}{\sqrt{\mu_n}}, \qquad \tilde{\boldsymbol{A}} = \boldsymbol{T}\boldsymbol{A}\boldsymbol{T}. \tag{C.2}$$

(3) Compute the eigenvalues $\lambda_1, \ldots, \lambda_n$ of $\tilde{\boldsymbol{A}}$ and the corresponding unit eigenvectors $\boldsymbol{u}_1, \ldots, \boldsymbol{u}_n$.

(4) Return $\lambda_1, \ldots, \lambda_n$ and the unit vectors $\boldsymbol{w}_1, \ldots, \boldsymbol{w}_n$ defined by

$$\boldsymbol{w}_1 = \mathcal{N}[\boldsymbol{T}\boldsymbol{u}_1], \quad \ldots, \quad \boldsymbol{w}_n = \mathcal{N}[\boldsymbol{T}\boldsymbol{u}_n]. \tag{C.3}$$

This procedure is justified as follows. Since $\{\boldsymbol{g}_i\}$ are an orthonormal system, the first of Eqs. (C.2) implies

$$\boldsymbol{T}^2 = \frac{\boldsymbol{g}_1\boldsymbol{g}_1^\top}{\mu_1} + \cdots + \frac{\boldsymbol{g}_n\boldsymbol{g}_n^\top}{\mu_n} = \boldsymbol{G}^{-1}. \tag{C.4}$$

Multiplied by $\boldsymbol{T}$ from left, Eq. (C.1) is written in the form

$$\boldsymbol{T}\boldsymbol{A}\boldsymbol{T}\boldsymbol{T}^{-1}\boldsymbol{w}_i = \lambda_i \boldsymbol{T}\boldsymbol{G}\boldsymbol{T}\boldsymbol{T}^{-1}\boldsymbol{w}_i. \tag{C.5}$$

However, we see that

$$\boldsymbol{T}\boldsymbol{G}\boldsymbol{T} = \boldsymbol{T}(\boldsymbol{T}^2)^{-1}\boldsymbol{T} = \boldsymbol{T}\boldsymbol{T}^{-1}\boldsymbol{T}^{-1}\boldsymbol{T} = \boldsymbol{I}. \tag{C.6}$$

So, if we define

$$\boldsymbol{u}_i = \boldsymbol{T}^{-1}\boldsymbol{w}_i, \tag{C.7}$$

Eq. (C.5) has the following form of the standard eigenvalue problem:

$$\tilde{\boldsymbol{A}}\boldsymbol{u}_i = \lambda_i \boldsymbol{u}_i. \tag{C.8}$$

Hence, if $\tilde{\boldsymbol{A}}$ has eigenvalues λ_i with corresponding unit eigenvectors $\boldsymbol{u}_i$, then λ_i are the generalized eigenvalues of Eq. (C.1), and from Eq. (C.7) the corresponding generalized eigenvectors $\boldsymbol{w}_i$ are given by $\boldsymbol{w}_i = \mathcal{N}[\boldsymbol{T}\boldsymbol{u}_i]$ after normalization.

D. Free focal length method

From Eq. (22), we can immediately see that the following $\boldsymbol{e}$ and $\boldsymbol{e}'$ are the eigenvectors of $\boldsymbol{F}^\top$ and $\boldsymbol{F}$, respectively, for eigenvalue 0:

$$\boldsymbol{e} \simeq \mathrm{diag}(1, 1, \frac{f_0}{f})\boldsymbol{t}, \qquad \boldsymbol{e}' \simeq \mathrm{diag}(1, 1, \frac{f_0}{f'})\boldsymbol{R}^\top\boldsymbol{t}. \tag{D.1}$$

Hence, we obtain

$$\boldsymbol{t} \simeq \mathrm{diag}(1, 1, \frac{f}{f_0})\boldsymbol{e} \simeq \boldsymbol{R}\mathrm{diag}(1, 1, \frac{f'}{f_0})\boldsymbol{e}'. \tag{D.2}$$

Substituting this into Eq. (22), we obtain

$$\boldsymbol{F} \simeq \boldsymbol{e} \times \mathrm{diag}(1, 1, \frac{f_0}{f})\boldsymbol{R}\mathrm{diag}(1, 1, \frac{f'}{f_0}) \simeq \mathrm{diag}(1, 1, \frac{f}{f_0})\boldsymbol{R}\mathrm{diag}(1, 1, \frac{f_0}{f'}) \times \boldsymbol{e}', \tag{D.3}$$

where we have used the identity $(\boldsymbol{A}\boldsymbol{v}) \times \boldsymbol{I} = (\boldsymbol{A}^{-1})^\top(\boldsymbol{v} \times \boldsymbol{I})\boldsymbol{A}^{-1}$ for an arbitrary vector $\boldsymbol{v}$ and an arbitrary nonsingular matrix $\boldsymbol{A}$. The product $\boldsymbol{A} \times \boldsymbol{v}$ of a 3×3 matrix $\boldsymbol{A}$ and a 3-D vector is defined to be $\boldsymbol{A}(\boldsymbol{v} \times \boldsymbol{I})^\top$. Equation (D.3) implies

$$\begin{aligned} \boldsymbol{F}\mathrm{diag}(1, 1, \frac{f_0}{f'}) &\simeq \boldsymbol{e} \times \mathrm{diag}(1, 1, \frac{f_0}{f})\boldsymbol{R}, \\ \mathrm{diag}(1, 1, \frac{f_0}{f})\boldsymbol{F} &\simeq \boldsymbol{R}\mathrm{diag}(1, 1, \frac{f_0}{f'}) \times \boldsymbol{e}'. \end{aligned} \tag{D.4}$$

Eliminating $\boldsymbol{R}$ by using the orthogonality relation $\boldsymbol{R}^\top\boldsymbol{R} = \boldsymbol{R}\boldsymbol{R}^\top = \boldsymbol{I}$, we obtain the following *Kruppa equations* [Bougnoux (1998), Hartley and Zisserman (2000)]:

$$\begin{aligned} \boldsymbol{F}\mathrm{diag}(1, 1, \frac{f_0^2}{f'^2})\boldsymbol{F}^\top &\simeq \boldsymbol{e} \times \mathrm{diag}(1, 1, \frac{f_0^2}{f^2}) \times \boldsymbol{e}, \\ \boldsymbol{F}^\top\mathrm{diag}(1, 1, \frac{f_0^2}{f^2})\boldsymbol{F} &\simeq \boldsymbol{e}' \times \mathrm{diag}(1, 1, \frac{f_0^2}{f'^2}) \times \boldsymbol{e}'. \end{aligned} \tag{D.5}$$

In terms of $\{\xi, \eta\}$ defined by Eqs. (25), these equations are rewritten as

$$\begin{aligned} \boldsymbol{F}(\boldsymbol{I} + \eta\boldsymbol{k}\boldsymbol{k}^\top)\boldsymbol{F}^\top &\simeq \boldsymbol{e} \times (\boldsymbol{I} + \xi\boldsymbol{k}\boldsymbol{k}^\top) \times \boldsymbol{e}, \\ \boldsymbol{F}^\top(\boldsymbol{I} + \xi\boldsymbol{k}\boldsymbol{k}^\top)\boldsymbol{F} &\simeq \boldsymbol{e}' \times (\boldsymbol{I} + \eta\boldsymbol{k}\boldsymbol{k}^\top) \times \boldsymbol{e}', \end{aligned} \tag{D.6}$$

which reduce to

$$\begin{aligned} \boldsymbol{F}\boldsymbol{F}^\top + \eta(\boldsymbol{F}\boldsymbol{k})(\boldsymbol{F}\boldsymbol{k})^\top &\simeq \boldsymbol{P}_{\mathrm{e}} + \xi(\boldsymbol{e} \times \boldsymbol{k})(\boldsymbol{e} \times \boldsymbol{k})^\top, \\ \boldsymbol{F}^\top\boldsymbol{F} + \xi(\boldsymbol{F}^\top\boldsymbol{k})(\boldsymbol{F}^\top\boldsymbol{k})^\top &\simeq \boldsymbol{P}_{\mathrm{e}'} + \eta(\boldsymbol{e}' \times \boldsymbol{k})(\boldsymbol{e}' \times \boldsymbol{k})^\top, \end{aligned} \tag{D.7}$$

where we define the projection matrices

$$\boldsymbol{P}_{\mathrm{e}} = \boldsymbol{I} - \boldsymbol{e}\boldsymbol{e}^\top, \qquad \boldsymbol{P}_{\mathrm{e}'} = \boldsymbol{I} - \boldsymbol{e}'\boldsymbol{e}'^\top, \tag{D.8}$$

onto the planes perpendicular to $\boldsymbol{e}$ and $\boldsymbol{e}'$, respectively. Multiplying $\boldsymbol{k}$ from the right on both sides of Eqs. (D.7), we obtain

$$\boldsymbol{F}\boldsymbol{F}^\top\boldsymbol{k} + \eta(\boldsymbol{k}, \boldsymbol{F}\boldsymbol{k})\boldsymbol{F}\boldsymbol{k} = c\boldsymbol{P}_{\mathbf{e}}\boldsymbol{k}, \qquad \boldsymbol{F}^\top\boldsymbol{F}\boldsymbol{k} + \xi(\boldsymbol{k}, \boldsymbol{F}\boldsymbol{k})\boldsymbol{F}^\top\boldsymbol{k} = c'\boldsymbol{P}_{\mathbf{e}'}\boldsymbol{k}, \tag{D.9}$$

where c and c' are unknown constants. Computing the inner product of $\boldsymbol{k}$ and the second of Eqs. (D.9) on both sides and the inner product of $\boldsymbol{F}^\top\boldsymbol{k}$ and the second of Eqs. (D.9), on both sides, we obtain

$$\begin{aligned} \|\boldsymbol{F}\boldsymbol{k}\|^2 + (\boldsymbol{k}, \boldsymbol{F}\boldsymbol{k})^2\xi &= c'\|\boldsymbol{e}' \times \boldsymbol{k}\|^2, \\ (\boldsymbol{k}, \boldsymbol{F}\boldsymbol{F}^\top\boldsymbol{F}\boldsymbol{k}) + (\boldsymbol{k}, \boldsymbol{F}\boldsymbol{k})\|\boldsymbol{F}^\top\boldsymbol{k}\|^2\xi &= c'(\boldsymbol{k}, \boldsymbol{F}\boldsymbol{k}), \end{aligned} \tag{D.10}$$

which can be solved for $\{\xi, c'\}$, resulting in the first of Eqs. (29). Similarly, computing the inner product of $\boldsymbol{k}$ and the first of Eqs. (D.9) on both sides and the inner product of $\boldsymbol{F}\boldsymbol{k}$ and the first of Eqs. (D.9) on both sides, we obtain

$$\begin{aligned} \|\boldsymbol{F}^\top\boldsymbol{k}\|^2 + \eta(\boldsymbol{k}, \boldsymbol{F}\boldsymbol{k})^2 &= c\|\boldsymbol{e} \times \boldsymbol{k}\|^2, \\ (\boldsymbol{k}, \boldsymbol{F}^\top\boldsymbol{F}\boldsymbol{F}^\top\boldsymbol{k}) + \eta(\boldsymbol{k}, \boldsymbol{F}\boldsymbol{k})\|\boldsymbol{F}\boldsymbol{k}\|^2 &= c(\boldsymbol{k}, \boldsymbol{F}^\top\boldsymbol{k}), \end{aligned} \tag{D.11}$$

which can be solved for $\{\eta, c\}$, resulting in the second of Eqs. (29) (note the identity $(\boldsymbol{k}, \boldsymbol{F}^\top\boldsymbol{F}\boldsymbol{F}^\top\boldsymbol{k}) = (\boldsymbol{k}, \boldsymbol{F}\boldsymbol{F}^\top\boldsymbol{F}\boldsymbol{k})$). Our derivation is essentially the same as Bougnoux (1998) but has a slightly different appearance.

E. *Equality enforcement*

If we equate ξ with η in Eq. (31) and ignore the higher order terms $\cdots$, we have

$$K(\xi) = \frac{1}{2}\Big(H_{11}(\xi - \xi_0)^2 + 2H_{12}(\xi - \xi_0)(\xi - \eta_0) + H_{22}(\xi - \eta_0)^2\Big). \tag{E.1}$$

Differentiating this with respect to ξ and setting the result to 0, we have

$$H_{11}(\xi - \xi_0) + H_{12}(\xi - \xi_0) + H_{12}(\xi - \eta_0) + H_{22}(\xi - \eta_0) = 0. \tag{E.2}$$

Solving this for ξ, we obtain ξ and η $(= \xi)$ in the form of Eq. (33).

F. *Generalized inverse*

The *singular value decomposition* (*SVD*) of an $m \times n$ matrix $\boldsymbol{A}$ has the form

$$\boldsymbol{A} = \boldsymbol{U}\boldsymbol{\Sigma}\boldsymbol{V}^\top, \qquad \boldsymbol{\Sigma} = \begin{pmatrix} \sigma_1 & & & \\ & \ddots & & \\ & & \sigma_r & \\ & & & 0 \\ & & & & \ddots \end{pmatrix}, \tag{F.1}$$

where $\boldsymbol{U}$ and $\boldsymbol{V}$ are $m \times m$ and $n \times n$ orthogonal matrices, respectively, and $\boldsymbol{\Sigma}$ is an $m \times n$ matrix. The numbers $\sigma_1 \geq \sigma_2 \geq \cdots$ are the *singular values*, and the number

r ($\leq \min(m,n)$) is the *rank* of $\boldsymbol{A}$. Let $\boldsymbol{\Sigma}^-$ be the $n \times m$ matrix

$$\boldsymbol{\Sigma}^- = \begin{pmatrix} 1/\sigma_1 & & & & \\ & \ddots & & & \\ & & 1/\sigma_r & & \\ & & & 0 & \\ & & & & \ddots \end{pmatrix}. \tag{F.2}$$

The (*Moore-Penrose*) *generalized inverse* (or *pseudoinverse*) of $\boldsymbol{A}$ is the $n \times m$ matrix

$$\boldsymbol{A}^- = \boldsymbol{V}\boldsymbol{\Sigma}^-\boldsymbol{U}^\top. \tag{F.3}$$

In practice, the rank r need to be specified by the user, since a theoretically zero value is treated as a very small nonzero value in floating-point arithmetic. If $r = \min(n,m)$, in particular, we can easily see that

$$\boldsymbol{A}^- = \begin{cases} (\boldsymbol{A}^\top\boldsymbol{A})^{-1} & m \geq n \\ (\boldsymbol{A}\boldsymbol{A}^\top)^{-1} & m \leq n \end{cases}. \tag{F.4}$$

The n-D vector $\boldsymbol{x}$ that minimizes

$$J = \|\boldsymbol{A}\boldsymbol{x} - \boldsymbol{b}\|^2, \tag{F.5}$$

for an m-D constant vector $\boldsymbol{b}$ is given by

$$\boldsymbol{x} = \boldsymbol{A}^-\boldsymbol{b}. \tag{F.6}$$

This is shown as follows. Since the norm $\|\cdot\|$ is invariant to orthogonal transformations, i.e., $\|\boldsymbol{a}\| = \|\boldsymbol{U}\boldsymbol{a}\|$ for any orthogonal matrix $\boldsymbol{U}$, we see that

$$\begin{aligned} J &= \|\boldsymbol{A}\boldsymbol{x} - \boldsymbol{b}\|^2 = \|\boldsymbol{U}\boldsymbol{\Sigma}\boldsymbol{V}^\top\boldsymbol{x} - \boldsymbol{b}\|^2 = \|\boldsymbol{\Sigma}\boldsymbol{V}^\top\boldsymbol{x} - \boldsymbol{U}^\top\boldsymbol{b}\|^2 \\ &= \|\boldsymbol{\Sigma}\boldsymbol{y} - \boldsymbol{c}\|^2 = \sum_{i=1}^{r}(\sigma_i y_i - c_i)^2 + \sum_{i=r+1}^{m} c_i^2, \end{aligned} \tag{F.7}$$

where we put $\boldsymbol{y} = \boldsymbol{V}^\top\boldsymbol{x}$ and $\boldsymbol{c} = \boldsymbol{U}^\top\boldsymbol{b}$. Thus, J is minimized by $y_1 = c_1/\sigma_1$, ..., $y_r = c_r/\sigma_r$. If $r < n$, then y_{r+1}, ..., y_n are indeterminate, so we let them be all 0, which minimizes $\|\boldsymbol{y}\| = \|\boldsymbol{x}\|$ among all solutions. Thus, we have

$$\boldsymbol{x} = \boldsymbol{V}\boldsymbol{y} = \boldsymbol{V}\boldsymbol{\Sigma}^-\boldsymbol{c} = \boldsymbol{V}\boldsymbol{\Sigma}^-\boldsymbol{U}^\top\boldsymbol{b} = \boldsymbol{A}^-\boldsymbol{b}. \tag{F.8}$$

References

Bartoli, A., Sturm, P. (2004). Nonlinear estimation of fundamental matrix with minimal parameters, *IEEE Trans. Patt. Anal. Mach. Intell.*, **26**, 3, pp. 426–432.

Bougnoux, S. (1998). From projective to Euclidean space under any practical situation, a criticism of self calibration, in *Proc 6th Int. Conf. Comput. Vision (ICCV'98)* (Bombay, India), pp. 790–796.

Brooks, M. J., de Agapito, L., Huynh, D. Q. and Baumela, L. (1998). Towards robust metric reconstruction via a dynamic uncalibrated stereo head. *Image Vision Comput.* **16**, 14, pp. 989–1002.

Chojnacki, W., Brooks, M.J., van den Hengel, A., Gawley, D. (2000). On the fitting of surfaces to data with covariances, *IEEE Trans. Patt. Anal. Mach. Intell.*, **22**, 11, pp. 1294–1303.

Chojnacki, W., Brooks, M.J., van den Hengel, A., Gawley, D. (2004). A new constrained parameter estimator for computer vision applications, *Image Vis. Comput.*, **22**, 2, pp. 85–91.

Fischler, M. A. and Bolles, R. C. (1981). Random sample consensus: A paradigm for model fitting with applications to image analysis and automated cartography, *Commun. ACM*, **24**, 6, pp. 381–395.

Harris, C. and Stephens, M. (1988). A combined corner and edge detector, in *Proc. 4th Alvey Vision Conf.* (Manchester, U.K.), pp. 147–151.

Hartley, R. I. (1992). Estimation of relative camera position for uncalibrated cameras. in *Proc. 2nd Euro. Conf. Comput. Vision (ECCV'92)* (Santa Margherita Ligure, Italy), pp. 579–587.

Hartley, R. I. (1997). In defense of the eight-point algorithm, *IEEE Trans. Patt. Anal. Mach. Intell.*, **19**, 6, pp. 580–593.

Hartley, R., and Silpa-Anan, C. (2002). Reconstruction from two views using approximate calibration, in *Proc. 5th Asian Conf. Comput. Vision (ACCV'02)* (Melbourne, Australia), Vol. 1, pp. 338–343.

Hartley, R. and A. Zisserman, A. (2000). *Multiple View Geometry in Computer Vision* (Cambridge University Press, Cambridge).

Huang, T. S. and Faugeras, O. D. (1989). Some properties of the E matrix in two-view motion estimation, *IEEE Trans. Patt. Anal. Mach. Intell.*, **11**, 12, pp. 1310–132.

Kanatani, K. (1993). *Geometric Computation for Machine Vision* (Oxford University Press, Oxford).

Kanatani, K. (1996). *Statistical Optimization for Geometric Computation: Theory and Practice* (Elsevier, Amsterdam), reprinted, 2005 (Dover, New York).

Kanatani, K. (2000). Optimal fundamental matrix computation: Algorithm and reliability analysis, in *Proc. 6th Symp. on Sensing via Image Information (SSII'00)* (Yokohama, Japan), pp. 291–298.

Kanatani, K. (2008). Latest progress of 3-D reconstruction from multiple camera images, in X. P. Guo (ed.), *Robotics Research Trends* (Nova Science, Hauppauge, NY), pp. 33–75.

Kanatani, K. (2009). Unified computation of strict maximum likelihood for geometric fitting, in *Proc. IS&T/SPIE Electronic Imaging: Three-Dimensional Imaging Metrology* (San Jose, CA), Vol. 7239.

Kanatani K. and Matsunaga, C. (2000). Closed-form expression for focal lengths from the fundamental matrix, in *Proc. 4th Asian Conf. Comput. Vision*

(ACCV'00) (Taipei, Taiwan), Vol. 1, pp. 128–133.

Kanatani, K., Nakatsuji, A. and Sugaya, Y. (2006). Stabilizing the focal length computation for 3-D reconstruction from two uncalibrated views, *Int. J. Comput. Vision*, **66**, 2, pp. 109–122.

Kanatani, K. and Ohta, N. (2003). Comparing optimal three-dimensional reconstruction for finite motion and optical flow, *J. Electronic Imaging*, **12**, 3, pp. 478–488.

Kanatani, K. and Sugaya, Y. (2007a). High accuracy fundamental matrix computation and its performance evaluation, *IEICE Trans. Inf. Syst*, **E90-D**, 2, pp. 579–585.

Kanatani, K. and Y. Sugaya, Y. (2007b). Extended FNS for constrained parameter estimation, in *Proc. Meeting on Image Recognition and Understanding (MIRU'07)* (Hiroshima, Japan), pp. 219–226.

Kanatani, K. and Sugaya, Y. (2007c). Performance evaluation of iterative geometric fitting algorithms, *Comp. Stat. Data Anal.*, **52**, 2, pp. 1208–1222.

Kanatani, K. and Sugaya, Y. (2009). Compact fundamental matrix computation, in *Proc. IEEE Pacific-Rim Symp. on Image and Video Technology (PSIVT'09)* (Tokyo, Japan), pp. 179–190.

Kanatani, K., Sugaya, Y. and Niitsuma, H. (2008). Triangulation from two views revised: Hartley-Sturm vs. optimal correction, in *Proc. 19th British Machine Vision Conf. (BMVC'08)* (Leeds, U.K.), pp. 173–182.

Kanazawa, Y. and Kanatani, K. (2004). Robust image matching preserving global consistency, in *Proc. 6th Asian Conf. Comput. Vision (ACCV'04)* (Jeju, Korea), Vol. 2, pp. 1128–1133.

Leedan, Y. and Meer, P. (2000). Heteroscedastic regression in computer vision: Problems with bilinear constraint, *Int. J. Comput. Vision* **37**, 2, pp.127–150.

Maybank, S. (1990). The projective geometry of ambiguous surface, *Phil. Trans. Roy. Soc. Lond*, A**332**, pp. 1-47.

LoweD. G. (2004). Distinctive image features from scale-invariant keypoints, *Int. J. Comput. Vision.*, **60**, 2, pp. 91–110.

Matas, J., Chum, O., Urban, M. and T Pajdla, T. (2002). Robust wide baseline stereo from maximally stable extremal regions, in *Proc. 13th British Machine Vision Conf. (BMVC'02)* (Cardiff. U.K.), pp. 384–393.

Maybank, S. (1991). Ambiguity in reconstruction from image correspondences *Image Vision Comput.*, **9**, 2, pp. 93-99.

Newsam, G. N., Huynh, D. Q., Brooks, M. J., and Pan, H.-P. (1996), Recovering unknown focal lengths in self-calibration: An essentially linear algorithm and degenerate configurations, *Int. Arch. Photogram. Remote Sensing*, **31**, B3-III, pp. 575–580.

Pan, H.-P., Brooks, M. J., and Newsam, G. (1995a). Image resituation: initial theory. in *Proc. SPIE: Videometrics IV* (Philadelphia, PA), pp 162–173.

Pan, H.-P., Huynh, D. Q., and Hamlyn, G. (1995b). Two-image resituation: Practical algorithm. in *Proc. SPIE: Videometrics IV*, (Philadelphia, PA), pp. 174–190.

Press, W. H, Teukolsky, S. A., Vettering, W. T. Flannery, B. P. (1992). *Numerical Recipes in C: The Art of Scientific Computing*, 2nd ed. (Cambridge University Press, Cambridge).

Rousseeuw, P. J. and Leroy, A. M. (1987). *Robust Regression and Outlier Detection* (Wiley, New York).

Smith, S. M. and Brady, J. M. (1997). SUSAN—A new approach to low level image processing, *Int. J. Comput. Vision*, **23** 1, pp. 45–78.

Sturm, P. (2001). On focal length calibration from two views, in *Proc. IEEE Conf. Comput. Vision Pattern Recog.* (Kauai, HI), Vol. 2, pp. 145–150.

Taubin, G. (1991). Estimation of planar curves, surfaces, and non-planar space curves defined by implicit equations with applications to edge and rage image segmentation, *IEEE Trans. Patt. Anal. Mach. Intell.*, **13**, 11, pp. 1115–1138.

Tomasi, C. and Kanade, K. (1991). Detection and tracking of point features CMU Tech. Rep. CMU-CS-91-132, Carnegie Mellon University, Pittsburgh, PA. http://vision.stanford.edu/~birch/klt/

Triggs, B., McLauchlan, P. F., Hartley, R. I. and Fitzgibbon, A. (2000). Bundle adjustment—A modern synthesis, in B. Triggs, A. Zisserman, and R. Szeliski, (eds.), *Vision Algorithms: Theory and Practice* (Springer, Berlin), pp. 298–375.

Zhang, Z., Deriche, R., Faugeras, O. and Luong, Q.-T. (1995). A robust technique for matching two uncalibrated images through the recovery of the unknown epipolar geometry, *Artif. Intell.*, **78**, 1/2, pp. 87–119.

CHAPTER 2.5

AFFINE INVARIANT FEATURES IN PATTERN RECOGNITION

Esa Rahtu and Janne Heikkilä

Machine Vision Group
Department of Electrical and Information Engineering
P.O. Box 4500, 90014 University of Oulu, Finland
{erahtu, jth}@ee.oulu.fi

Affine invariant feature extraction is an important topic particularly in image recognition applications. This chapter investigates the area of affine invariant feature construction, and introduces a recently proposed multiscale framework for systematically enhancing the performance of the previous methods. The idea in the approach presented is to extend the given pattern to a set of affine covariant versions, each carrying slightly different information, and then to apply known affine invariants to each of them separately. The key part of the framework is the construction of the affine covariant set, and this is done by combining scaled representations of the original pattern. The application of the multiscale framework is demonstrated by constructing several new affine invariant methods, which are compared to the existing approaches in the experiments.

1. Introduction

Feature representation of a pattern is an essential element in many fields, like texture analysis [1], face detection [2], medical imaging [3], structure from motion [4], and object classification [5]. The design of effective features involves challenges due to the great variety in the appearance of patterns within one class. For example in image processing this is commonly encountered because of changes in the view angle, movements of objects, different illuminations, or scene deformations.

The aforementioned difficulties have been addressed by specifically designing features that are independent or tolerant to a certain types of changes in the patterns. This chapter investigates feature extraction techniques that are insensitive to affine transformations of patterns. The goal is to find a method for extracting a descriptor vector v from the pattern f in such a way that the resulting v is the same for f and affine transformed version f' of f.

Although the methods described are not restricted to certain patterns, we consider here only the case where f is a gray-scale image. Hence the words pattern and image are used interchangeably in the following. The affine transformation

Fig. 1. Two images of a planar object, taken from different viewing angles with a pinhole camera, are connected by a projective transformation, also called homography.

between images f and f' may arise in many circumstances, but to motivate the practical importance of this transformation, the most common practical situation encountered is presented in the following.

Let us have a pinhole camera acquiring a picture of a planar object from different viewing angles, as illustrated in Figure 1. Following from the camera geometry, the resulting images are connected by a projective transformation, also called homography [4]. Using this knowledge, we could define that patterns that are the same up to some homography belong to the same equivalence class.

However, the unrestricted projective transformation gives a great deal of freedom to deform the images, and it is unfortunately very difficult to limit the examination only to the physically relevant cases. This means that many patterns that we would like to distinguish can actually be mapped close to each other by some projective transformation. For example, a square may be mapped to be arbitrarily close to a triangle by a homography, although we would usually like to distinguish between these two shapes. Also any closed contour can be mapped to be very close to a circle with a series of homograhies [6]. Another drawback arises from the fact that the projective transformation of image functions is a nonlinear operation, which can be hard to manage in the algorithms.

A common solution is to relax the requirement of the accurate camera model, and approximate the projective transformation by something that is easier to handle. On many occasions, a proper approximation has been provided by the first order Taylor series expansion of the projective transformation, which is precisely the affine transformation. The approximation is adequate for the entire image under certain conditions [4], but typically it is valid at least locally. The affine covariant area detectors [7; 8; 9; 10; 11] provide a method to find such local patches.

2. The Affine Transformation of Patterns

The affine transformation of a coordinate vector $x \in \mathbb{R}^2$ is defined by a matrix T composed with translation by a vector t.

Definition 1. Define the affine transformation $\mathscr{A} = \mathscr{A}\{T, t\}$ by

$$x' = \mathscr{A}(x) = Tx + t, \tag{1}$$

where $t, x \in \mathbb{R}^2$ and T is a 2×2 nonsingular matrix whose elements belong to $\mathbb{R}$. Any such transformation is invertible with inverse $\mathscr{A}^{-1}(x) = T^{-1}x - T^{-1}t$.

Before applying this transformation to the actual images, they must be defined more specifically.

Definition 2. The patterns examined in this chapter will be gray scale images defined by real valued positive functions in $\mathbb{R}^2$. We will further assume that the functions belong to $L^1(\mathbb{R}^2) \cap L^2(\mathbb{R}^2)$, have compact support, and are not identically zero as L^1 functions.

A function f belongs to $L^p(\mathbb{R}^2)$ if the integral $\int_{\mathbb{R}^2} |f|^p$ exists and has a finite value. f is identically zero as an L^1 function if $\int_{\mathbb{R}^2} |f| \equiv 0$. The additional requirements in Definition 2 have been imposed to ensure that the methods introduced later are well defined. This results in no loss of generality, since practical digital images always satisfy these conditions in Definition 2.

We define the action of affine transformations on image functions as follows:

Definition 3. Suppose $f(x)$ is a gray scale image as defined in Definition 2. We may apply an affine transformation $\mathscr{A}$ to f, which gives a new gray scale image defined by function f' in $\mathbb{R}^2$, where

$$f'(x) = f \circ \mathscr{A}^{-1}(x) = f(T^{-1}x - T^{-1}t). \tag{2}$$

We call f' the $\mathscr{A}$ transformed version of f.

3. Segmentation

A typical problem in pattern recognition is that one has large amounts of data, from which only part is relevant for the recognition task at hand. This nonrelevant background "clutter" quickly causes the performance of the methods to deteriorate. For example, often one needs to recognize if particular sample objects are present in a given image. It would be difficult, if not impossible, to generate the feature vectors directly so that recognition could be done only using them.

A common way to approach this problem is to introduce a preprocessing step, where the regions of interest (ROI) are separated from the nonrelevant parts. The feature extraction methods are then applied only to extracted ROI segments. For example, extracting local interest regions and computing features from them has been shown to give substantial improvements in performance [12; 13; 7].

All the affine feature extraction methods discussed in this chapter require segmentation to be done in one way or another. However, this can involve many different operations, and the best choice again depends heavily on the application. Some recent and generally interesting approaches for image segmentation would include texture based methods [14; 15], mean shift [16], normalized cuts [17], and affine covariant regions [7]. Discussion about image segmentation can be also found from [18; 19; 20].

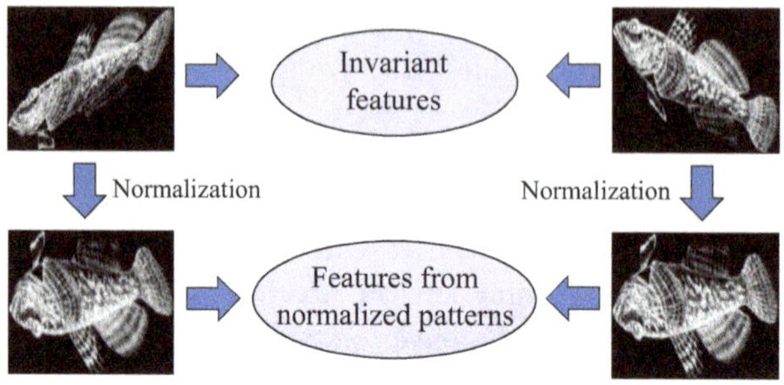

Fig. 2. Illustration of the differences between the invariant and normalizing approaches.

4. Affine Invariance in Feature Extraction

In feature extraction, the idea is to convert the studied pattern to a d-dimensional feature vector describing a point in d-dimensional feature space, which is usually $\mathbb{R}^d$. In this chapter, we consider a special case of general feature extraction, where the idea is to construct a mapping from patterns f to vectors v, which would result in the same v for f and f' if, and only if, the two images are related by an affine transformation. In reality, the known methods require d to be impractically large to satisfy the "only if" part of this condition. In applications, this is not necessarily a problem, if the functions which result in the same feature vector are reasonably close to being the same up to affine transformation.

There are two main approaches to achieving affine invariance, namely normalization and invariant features. Feature extraction methods applying the first approach have a distinguishable normalization part where the pattern is rectified to a canonical form before the actual mapping to the feature vector. The methods in the latter approach map the pattern directly to a feature vector without any normalization phase. Figure 2 illustrates the difference between these two approaches. Another possibility is to use a combination of normalization and invariants. In this kind of approach, one only partially normalizes the image and then applies mapping that is invariant to the deformations left unnormalized. In this chapter, we focus on an invariant approach and consider normalization only briefly in the following. It must be emphasized that in the following discussion it is assumed that the patterns f are segmented and the background is set to be zero.

4.1. *Normalization approach*

The idea in the normalization approach is to rectify the image to a canonical representation before extracting the actual features. The advantage here is that the method producing the features after normalization need not have any invariance. Usually, it can be even better if they do not have much invariance in order to achieve strong discrimination. Some examples of such feature extraction methods, appearing frequently in applications, could be gradient orientation histograms (SIFT) [12] and image moments [21].

The drawback of this approach is, however, that robust affine normalization is not an easy task, and the amount of possible approaches is limited. Perhaps the most important one is based on normalizing the image sequentially, starting from the translation t and continuing to the linear part T. A common example of this procedure, which has been applied in practice, is the following three step algorithm:

(1) Given a pattern f, normalize the translation by moving its centroid, or some other fixed interest point, to the origin.
(2) Normalize the linear part of the affine transformation, excluding final rotation, by transforming f so that its second moment matrix will become the identity matrix.
(3) Normalize the final rotation by turning the pattern in such a way that it has some fixed dominant gradient orientation.

If the above procedure has a unique solution for f, then it should result in the same normalized form also for any $f \circ \mathscr{A}^{-1}$, with arbitrary $\mathscr{A}$. Problems are encountered, however, if no unique solution exists, which often happens, for instance if f has rotational symmetries.

There are many variations of this procedure, where, for example, the second moment matrix is computed from the image gradient magnitudes or substituted with the Hessian matrix, or the final rotation is normalized, based on moments as presented in [22], where also symmetric patterns are considered. The rotation normalization based on gradients is however probably more robust compared to moment approaches and it is applied widely in applications. Affine normalization procedure very similar to the one presented above is also considered with details in [8].

Another possibility for applying normalization is to use it only partially, and then to apply invariant feature extraction to the result. One example, which is used in some moment based features and in some examples of the multiscale framework, is to normalize only the translation part t and leave out the rest of the transformation. Another example could be that we exclude the final step of the presented normalization procedure and apply rotation invariant features. For such features, and other types of invariants, there is a vast number of different approaches. For some examples, one can refer to [23; 24; 25; 13].

4.2. *Invariant features*

Feature extraction methods using the invariant approach do not have a distinguishable normalization part, but the features are created directly from the given pattern. The only exception in this chapter is that also methods that may normalize the translation are categorized as affine invariant features. The advantage compared to the normalizing approach is that one possible error prone step is completely omitted in the process.

Several affine invariant feature extraction methods are known so far. Maybe the most trivial ones are the normalized integral If and histogram Hf

$$If = \frac{1}{\|\chi_f\|_{L^1}} \int_{\mathbb{R}^2} f(x)dx, \qquad Hf = \left[\frac{1}{\|\chi_f\|_{L^1}} \int_{(f)^{-1}(B_j)} dx\right]_{j=1}^{m}, \tag{3}$$

where $B_1, \ldots, B_m$ are disjoint subsets of $\mathbb{R}_+$ (bins), and $\chi_f(x) = 1$ if $f(x) \neq 0$, and $\chi_f(x) = 0$ otherwise. These two descriptors are commonly known as the mean gray value, and the normalized gray value histogram. While they are very easy to evaluate, they have problems as descriptors. The first one of these, which applies to both invariants in (3), is that we ignore the spatial distribution of the gray values in the image. For example, if we discretize the integrals and apply them to digital images, we can permute the pixels into arbitrary order without changing the features. This results in a situation where the set of patterns having the same feature vector is very large. Another problem is the limited number of features. This is especially seen in the case of a normalized integral, which gives only one feature value. It is easy to imagine that this would not be enough to classify complex patterns.

A less trivial solution is provided by the affine invariant moment polynomials, introduced in the 1960's in [26], though later corrected in [27] and [28]. These descriptors make use of the two-dimensional cross products allowing us to incorporate also spatial information to the features. A general formulation, from which all the affine invariant moment polynomials can be constructed, is given in [29] as

$$MPf = \frac{1}{\|f\|_{L^1}^{w+N}} \int_{\mathbb{R}^{2N}} \prod_{1\leq k<l\leq N} C((x_k - \mu(f)), (x_l - \mu(f)))^{n_{kl}} \prod_{i=1}^{N} f(x_i)\, dx_i, \tag{4}$$

where $x_i, x_k, x_l \in \mathbb{R}^2$, $C(x_k, x_l) = \det[x_k x_l]$, $N \geq 2$ and $n_{kl} \geq 0$ are integers, $w = \sum_{k,l} n_{kl}$, and $\mu(f) \in \mathbb{R}^2$ is the centroid of f. Integration is over vectors $x_1, \ldots, x_N$. This formulation can be further simplified to the more familiar form as a polynomial of the image central moments

$$m_{pq}(f) = \int_{\mathbb{R}^2} (x_1 - \mu(f)_1)^p (x_2 - \mu(f)_2)^q f(x)\, dx, \tag{5}$$

where $x = (x_1, x_2)$, $\mu(f) = (\mu(f)_1, \mu(f)_2)$, and the value $p + q$ is called the order of the moment. Only some of these polynomials are independent, and thus worth using as descriptors [29]. As an example, the first two affine invariants would be

$$\begin{aligned} I_1 &= (m_{20}m_{02} - m_{11}^2)/\|f\|_{L^1}^4, \\ I_2 &= (-m_{30}^2 m_{03}^2 + 6m_{30}m_{21}m_{12}m_{03} - 4m_{30}m_{12}^3 \\ &\quad -4m_{21}^3 m_{03} + 3m_{21}^2 m_{12}^2)/\|f\|_{L^1}^{10}. \end{aligned} \tag{6}$$

One can refer to [29] for an extensive list of affine invariant moments. The resulting moment based features are computationally very efficient, and there are number of them. However, in order to produce more features one needs to use higher order moments, and this soon results in poor performance in noisy conditions. The amount of useful features is hence limited and in many cases not enough for recognition.

Another moment based approach was introduced in [30]. There, in addition to the moments computed from image f, they use moments that are computed from the support χ_f of f. In this way, it is possible to construct features that are also insensitive to certain changes in the values of f, which can occur when photographing under varying illuminations. The drawback of these invariants seems to be their sensitivity to the changes in the image. Also the number of features is rather small, which limits the applicability of the method.

One possible solution was presented in [31]. The key part of this approach was the introduction of the cross-weighted moment

$$\mu_{klpq}(f) = \int (x_1 - \mu(f)_1)^k (x_2 - \mu(f)_2)^l (y_1 - \mu(f)_1)^p (y_2 - \mu(f)_2)^q w(x,y) f(x) f(y) dx dy, \quad (7)$$

where x and y are as above, and $w(x,y)$ is a weighting function defined in [31]. Compared to the invariants constructed using traditional moments, this new technique seems to provide clearly more robustness, but at the expense of heavily increased computational cost. One can refer to [31] for explicit expressions for invariants.

Yet another approach to producing affine invariant features was proposed in [32]. The method is called trace transform and it was based on applying a particular combination of functionals to f. The features seem to perform well, but the implementation of the method is nontrivial.

In the methods described above, it was assumed that the patterns f have nonzero support size. In the case of images, this means that there are segments which have a nonzero area. A completely different approach is to construct affine invariants using primitives like points, line segments, or curves with zero support size. In fact, line segments and conics can even be used to construct projective invariants [33]. A number of methods using curves and points to produce affine invariants have been introduced in the literature. For these methods, one can refer to [34; 35; 36; 18].

5. Multiscale Framework

This section defines the multiscale approach for affine invariant feature extraction and discusses some related issues. Several methods based on the framework are also given. Although a multiscale framework is introduced here for feature extraction, it is also applicable to image registration [37; 38] and shape measurements [39], for example. The discussion begins with a short motivation in the following.

5.1. *Motivation*

As already mentioned in Section 4.2, previously proposed affine invariants have some important disadvantages. First of all, the amount of useful descriptors is limited. An extreme example is normalized integration, which produces only one feature. Another example could be the affine invariant moment polynomials [27], where in theory we can construct an infinite amount of features, but in reality only a few invariants which use low order moments are applicable in practice.

The second disadvantage was related to normalized integration and image histogram descriptors, which without any modifications do not include any information about the spatial distribution of gray values. The third problem is related to the computational and implementational difficulties. An example of these could be the trace transform [32], which is able to produce a great number of features with spatial information, but the implementation and evaluation of these features is complicated and time consuming.

To overcome these problems we would ideally like to find a simple way of constructing arbitrarily many descriptors which capture both intensity and the spatial characteristics of the patterns. While this can be difficult, it is possible to modify some existing methods so that they overcome these deficiencies.

5.2. *Definition*

The basic idea in the multiscale approach can be presented as a three step algorithm:

1. Given an image f, represent it in n different scales $f(\alpha_1 x), \ldots, f(\alpha_n x)$.
2. Combine the scaled images to a new image $Gf(x)$. The combination is required to be affine covariant, which means that for any affine transformation $\mathscr{A}$, one has
$$G(f \circ \mathscr{A}^{-1})(x) = (Gf)(\mathscr{A}^{-1}(x)). \tag{8}$$
3. Extract the affine invariant features using f, Gf, and some of the methods presented in Section 4.

Figure 3 illustrates the overall procedure. The advantage of the approach is that by varying the scales α_i and combinations G, we are able to generate a great variety of different descriptors. This is possible regardless of the inherent number of features offered by the descriptor used in step 3. It is also significant that the values of Gf can already carry the spatial information so that this is not necessarily required from the operations in the last step.

The first step, scaling of images, is straightforward. The third step can also be quite simple, depending on which methods are selected, but if we take the mean gray values, gray value histograms, or image centroids, the implementation is not too difficult. The second step is often the most complicated one, and it is the key part of the whole approach. Here, one needs to take the scaled images $f(\alpha_i x)$ and

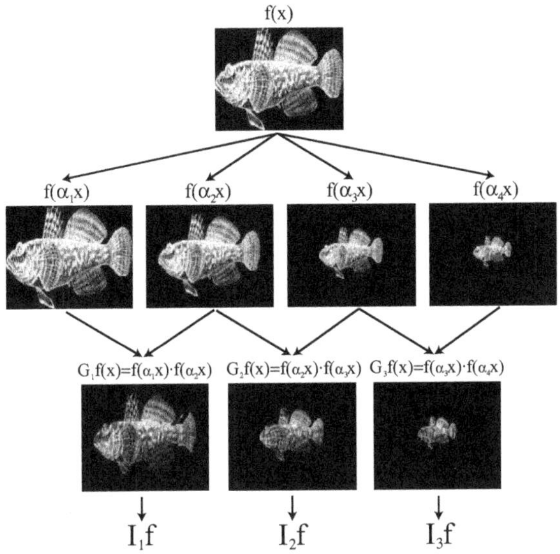

Fig. 3. Example of a multiscale scheme where two scaled representations are combined together with a pointwise product. The final invariants are computed from the resulting images $G_i f$.

combine them to a new image $Gf(x)$ so that Gf and $G(f \circ \mathscr{A}^{-1})$ are related by an affine transformation $\mathscr{A}$. The reason for this requirement is quite obvious, since we need to maintain the affine relationship of the new images $G_i f$ in order to get the methods in step three to work.

The requirement for the affine covariance also provides the motivation for us using scaling instead of other transformations in the construction of Gf. This is because in general only scaling commutes with matrix products, and this makes it easier to find combinations for which this requirement is fulfilled. The translation component of $\mathscr{A}$ can, however, cause some problems. It may either be normalized away by computing the image centroid, or one may choose G more carefully, so that translation invariance is obtained without finding the centroid.

We consider also one possible extension to the basic multiscale framework. We preprocess the image by taking some transform $\mathscr{T}$ of it. The resulting new image $\mathscr{T}(f)$ is then considered as an input to the basic multiscale algorithm. Some extra caution must be paid here, since the relation between these new images is not likely to be the original affine transformation. If the new relation is again affine, then the invariant features can be computed in a straightforward manner.

6. Multiscale Affine Invariants

In this section, we illustrate the multiscale framework by applying it to produce a set of affine invariant feature extraction methods. These techniques have originally been introduced in [40; 41; 42; 43; 44]. They are presented here very briefly and in a slightly different form, representing the viewpoint of the multiscale framework. Many details and further properties can be found in the original publications. We will begin with the simplest construction, and continue to more complex ones.

6.1. *Spatial multiscale affine invariants*

The first example is the Spatial multiscale affine invariants (SMA), originally presented in [40]. Here, we choose Gf to be a product of the original f, and two scaled representations of it, $f(\alpha x)$ and $f(\beta x)$. In this formulation, the translation component must, however, be normalized, and this is done by replacing f with the normalized version $\tilde{f}(x) = f(x + \mu(f))$, where the centroid $\mu(f) = \int_{\mathbb{R}^2} xf(x)dx$. The operator G is given by

$$Gf(x) = \tilde{f}(x)\tilde{f}(\alpha x)\tilde{f}(\beta x), \tag{9}$$

where $\alpha, \beta \in \mathbb{R}$. For the construction of the final invariant, we choose normalized integration. The resulting final SMA invariants are then given by

$$Sf(\alpha, \beta) = \frac{1}{\|f\|_{L^1}} \int_{\mathbb{R}^2} \tilde{f}(x)\tilde{f}(\alpha x)\tilde{f}(\beta x)\, dx. \tag{10}$$

Due to its simplicity, $Sf(\alpha, \beta)$ is very fast to evaluate, and the possibility of varying the scales results in an infinite number of different descriptors. Compared to other more complicated multiscale invariants, SMA is faster, but does not necessarily achieve the same accuracy.

To apply the method to digital images, one needs a discrete form of the invariant, and this is done by discretizing the integral (10) resulting in

$$Sf(\alpha, \beta) = \frac{1}{\sum_x f(x)} \sum_x \tilde{f}(x)\tilde{f}(\alpha x)\tilde{f}(\beta x). \tag{11}$$

Recall that $\tilde{f}(x)$ is a translated version of original image $f(x)$ so that the origin of the image coordinates coincides with the image centroid. To evaluate $\tilde{f}(ax)$ in (11) requires also image resampling, which can be done by linear interpolation.

6.2. *Multiscale autoconvolution*

Another example is multiscale autoconvolution (MSA), originally presented in [45] and later more comprehensively discussed in [41]. In this case, we use a combination of convolutions and products to form Gf. One advantage of this approach is the fact that the translation component does not have to be considered separately. Define

$$Gf(x) = \frac{1}{\|f\|_{L^1}^2} f(x)(f_\alpha * f_\beta * f_\gamma)(x), \tag{12}$$

where $\alpha, \beta \in \mathbb{R}$, $\gamma = 1 - \alpha - \beta$, $f_a(x) = a^{-2}f(x/a)$, and $*$ denotes convolution. The third scaling factor γ is set to $1 - \alpha - \beta$ in order to eliminate the translation component. The actual invariant features are again constructed by normalized integration, which gives the MSA transform

$$Mf(\alpha,\beta) = \frac{1}{\|f\|_{L^1}^3} \int_{\mathbb{R}^2} f(x)(f_\alpha * f_\beta * f_\gamma)(x)\,dx. \tag{13}$$

This formulation is not computationally appealing, but fortunately Mf can be expressed using the Fourier transform $\hat{f}$ as

$$Mf(\alpha,\beta) = \frac{1}{\hat{f}(0)^3} \int_{\mathbb{R}^2} \hat{f}(-\xi)\hat{f}(\alpha\xi)\hat{f}(\beta\xi)\hat{f}(\gamma\xi)\,d\xi. \tag{14}$$

By discretizing the integrals and applying scaling we get

$$Mf(\alpha,\beta) = \frac{1}{N_1 N_2} \frac{1}{\hat{f}(0)^3} \sum_{i=0}^{N_1 N_2 - 1} \hat{f}(-w_i)\hat{f}_\alpha(w_i)\hat{f}_\beta(w_i)\hat{f}_\gamma(w_i), \tag{15}$$

where $f_a(x) = \frac{1}{a^2} f(\frac{x}{a})$ and $\hat{f}(w)$ is now the discrete Fourier transform of $N_1 \times N_2$ digital image f.

6.3. *Multiscale ridgelet invariants*

As mentioned in Section 5.2, it is possible to extend the basic multiscale approach by applying a transform $\mathscr{T}$ to the image function f before application of the multiscale framework. We now give an example of this, presented originally in [43], where we select $\mathscr{T}$ to be the ridgelet transform. We define the ridgelet transform of a function f as

$$Rf(\xi) = \int_{\mathbb{R}^2} f(x)\psi(x \cdot \xi)dx, \tag{16}$$

where $\xi \in \mathbb{R}^2$ and ψ is a wavelet. The function $\psi(x \cdot \xi)$ looks like a ridge, and such functions are called ridgelets in the literature [46; 47]. $Rf(\xi)$ behaves well under the affine transformation, except for the translation part. This must be normalized and we replace f with the normalized version $\tilde{f}$ as in Section 6.1. Further by denoting $f' = f \circ \mathscr{A}^{-1}$ we have the relation

$$R\tilde{f}'(\xi) = |\det(T)| R\tilde{f}(T^t\xi). \tag{17}$$

Now define,

$$Gf(\xi) = \frac{1}{\|f\|_{L^1}^3} R\tilde{f}(\xi) R\tilde{f}(\alpha\xi) R\tilde{f}(\beta\xi), \tag{18}$$

where $\alpha, \beta \in \mathbb{R}$. Then, by applying normalized integration to produce the final invariant, we have

$$RIf(\alpha,\beta) = \frac{1}{\|f\|_{L^1}^2} \int_{\mathbb{R}^2} R\tilde{f}(\xi) R\tilde{f}(\alpha\xi) R\tilde{f}(\beta\xi) d\xi. \tag{19}$$

Implementing RIf is not quite as straightforward as in the case of SMA and MSA. The main idea in the implementation is to apply Radon transform to efficiently evaluate the integrals over the ridgelet transforms. The details of the algorithm are explained in [43].

From the methods presented in this section, multiscale ridgelet invariants are the most complicated and time consuming to evaluate. However it offers the most extensive possibilities for variation, and on some occasions the best performance.

6.4. *Generalized affine moment invariants*

In the previous examples, we used normalized integration, which is perhaps the simplest possible invariant, at the third step of the multiscale approach. Instead, we could use something different, such as affine invariant moment polynomials. This is exactly what is done in the generalized affine moment invariants, presented in [42].

Now let G be one of the operators given in previous sections and as in (4) define

$$If = \frac{1}{\|f\|_{L^1}^{w+N}} \int_{\mathbb{R}^{2N}} \prod_{1 \le k < l \le N} C((x_k - \mu(f)), (x_l - \mu(f)))^{n_{kl}} \prod_{i=1}^{N} Gf(x_i)\, dx_i. \quad (20)$$

With particular choices of N and n_{kl}, If will give all the possible affine invariant moment polynomials of Gf [29]. These invariants result in the same explicit expressions [29] as the traditional affine moment invariants, but using polynomials of modified central moments

$$m_{pq}(G, f) = \int_{\mathbb{R}^2} (x_1 - \mu(f)_1)^p (x_2 - \mu(f)_2)^q Gf(x)\, dx, \quad (21)$$

where $x = (x_1, x_2)$, and $\mu(f) = (\mu(f)_1, \mu(f)_2)$.

The implementation of invariants require evaluation of the polynomials of the modified central moments (21). The implementation of (21) needs a discretization of the integral and the evaluation of Gf, which is done using the same approach described earlier for corresponding normalized integral based multiscale invariants.

6.5. *Multiscale autoconvolution histograms*

In addition to normalized integration and moment polynomials, one may apply a histogramming operation to Gf in order to produce affine invariants. Formally, this is given by

$$If = \left[\frac{1}{\|\chi_{Gf}\|_{L^1}} \int_{(Gf)^{-1}(B_j)} dx \right]_{j=1}^{m}, \quad (22)$$

where $B_1, \ldots, B_m$ are disjoint subsets of $\mathbb{R}_+$ (bins), and $\chi_{Gf}(x) = 1$ if $Gf(x) \neq 0$ and $\chi_{Gf}(x) = 0$ otherwise. These operations have natural discrete analogues. A particular example of this histogramming approach is presented in [44], where

$$Gf(x) = \frac{1}{\|f\|_{L^1}^2} (f_\alpha * f_\beta * f_\gamma)(x). \quad (23)$$

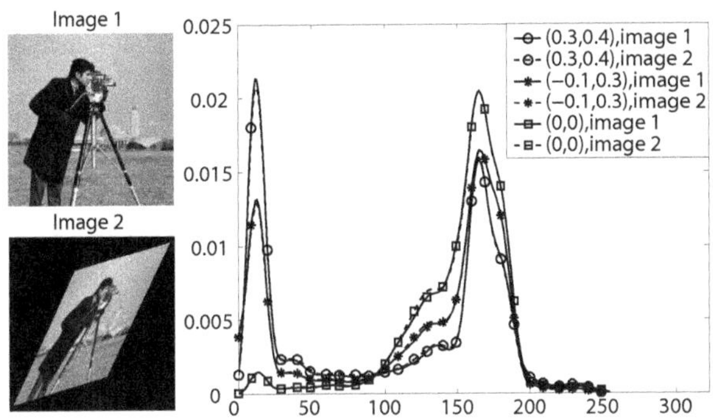

Fig. 4. Examples of the multiscale autoconvolution histograms for an image and affine transformed version of it.

In that paper, the histogram bins are, however, chosen according to the original function f rather than Gf as in (22). For implementing Multiscale autoconvolution histograms, one can use similar approaches to those in Section 6.2. For details, one can refer to [44]. Figure 4 illustrates examples of multiscale autoconvolution histograms with three α, β pairs.

Multiscale autoconvolution histograms seem to result in better performance in the same computational time compared to integral or moment based multiscale methods. However combining the histograms with different (α, β) values may not be as straightforward.

6.6. *Comparison based histograms*

So far, all the examples presented of multiscale invariants construct the combination G, using convolutions or pointwise products. These linear operations can behave robustly in many settings, but they can also compromise the discriminability of the methods to some extent. For this reason, it might be better to utilize nonlinear functionals in the combination of the scaled images.

One example of such an approach, presented in [48], is to use pointwise comparison operations. The motivation for using this method comes from the local binary patterns (LBP) [1], where similar operations were used to construct highly discriminative texture descriptors. In [48], this idea is illustrated in two different cases, based on a formulation similar to SMA and MSA.

In the first of these constructions, define

$$Gf(x) = G_\alpha f(x) = X(\tilde{f}(x), \tilde{f}(\alpha x)), \tag{24}$$

where $\alpha \in \mathbb{R}$, $X(a, b) = 1$ if $a > b$, and $X(a, b) = 0$ otherwise. The G in (24) corresponds to the one defined for SMA in Section 6.1, with some modifications. In (24), the pointwise product is replaced with the comparison operation X, and the affine covariant combination is computed using only one scaled version of f.

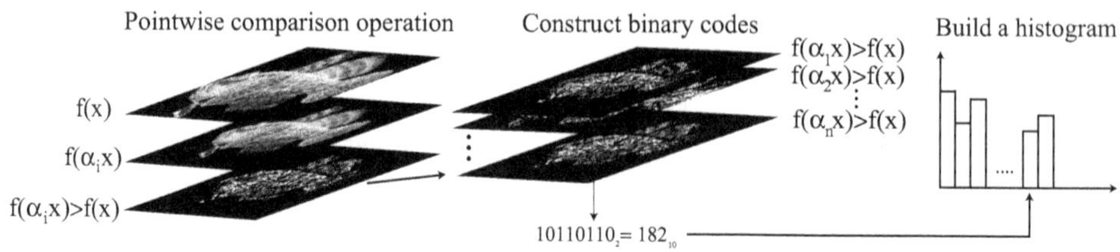

Fig. 5. Illustration of the process for generating the invariant histograms.

The implementation of the method is also similar to SMA, with straightforward modifications. For details see [48].

The second formulation, presented in [48], for G is based on convolution

$$C_\alpha f(x) = \frac{1}{\|f\|_{L^1}}(f_\alpha * f_{1-\alpha})(x), \tag{25}$$

where $\alpha \in \mathbb{R} \setminus \{0\}$ and $f_\alpha(x) = \alpha^{-2} f(x/\alpha)$. Using (25) define

$$Gf(x) = G_\alpha f(x) = X(f(x), C_\alpha f(x)). \tag{26}$$

This formulation is clearly related to MSA, with similar modifications to those in the case of SMA and G defined in (24). Implementation of the method needs slightly more modification of the original MSA implementation than in the previous case with SMA based formulation. These are explained with details in [48].

In the new formulations introduced, the resulting G is a binary image, from which we would have to compute the invariant. This could be done following the same ideas as in the previous case, but since we have such a specific form we can do something better. In [48] it was proposed to compute $Gf(x)$ with a few different α values, say $\alpha_1, \alpha_2, \ldots, \alpha_n$, and then to use a binary code construction as in LBP to combine all the information to a new function

$$Bf(x) = G_{\alpha_1} f(x) + G_{\alpha_2} f(x) \cdot 2 + \ldots + G_{\alpha_n} f(x) \cdot 2^{n-1}. \tag{27}$$

The function Bf has integer values from 0 to $2^n - 1$, and it completely encodes the information in the functions Gf. To compute the final affine invariant, one could just evaluate the normalized integral of Bf. This approach, however, does not make sense, since the different combinations would have an uneven impact on the resulting invariant value. Instead, we construct a histogram $Hf(k)$, with $2^n - 2$ bins, from nonzero values of Bf, and we normalize the histogram so that the sum over all bins is equal to one. Compared to the direct integration of $G_\alpha f$, this makes it possible to preserve the relative spatial arrangements in the functions $G_{\alpha_1} f, \ldots, G_{\alpha_n} f$. The construction of the invariant is illustrated in Figure 5.

7. Experiments

This section presents experiments illustrating the introduced multiscale invariants in classification tasks. The section has been divided into two parts. The first part considers methods where the third stage of the framework (see Section 5.2) applies moment polynomials, and the second part considers methods where the third stage uses normalized histograms.

7.1. *Moment based invariants*

We begin with affine invariant methods where the final invariant is produced using moment polynomials (includes also normalized integral). The methods in this class consist of standard affine moment invariants (AMI) [29], spatial multiscale affine invariants (SMA) [40], multiscale autoconvolution (MSA) [41], and generalized affine invariant moments (GAMI) [42]. The AMI features are generated with 4 and 40 first independent invariant moment polynomials. The SMA features use 36 and the MSA features 37 (α, β) pairs respectively. For GAMI, we apply the combination functions G from SMA and MSA methods, and the 4 first independent moment polynomials.

In the experiment, we classified gray-scale images of 94 different fish. We used these original 94 images to train a nearest neighbor classifier for each method and then classified distorted versions of the same images. The distortions were obtained by first making a random affine transformation and then adding one of the following nonaffine distortions: Gaussian noise, occlusion, and illumination distortion. In addition, we also classified projective transformed versions of the images. Some samples of the original images, and the distorted versions are shown in Figure 6. The classification error rates with different distortions and distortion strengths are illustrated in Figure 7. The results indicate that the performance of the original affine invariants were clearly enhanced by applying a multiscale framework.

7.2. *Histogram based invariants*

This section examines multiscale methods where the final invariant features are generated using normalized histogram. These methods include a Multiscale autoconvolution histogram [44] and a comparison based histogram [48]. For the latter technique, the histograms are constructed using the MSA and SMA like formula-

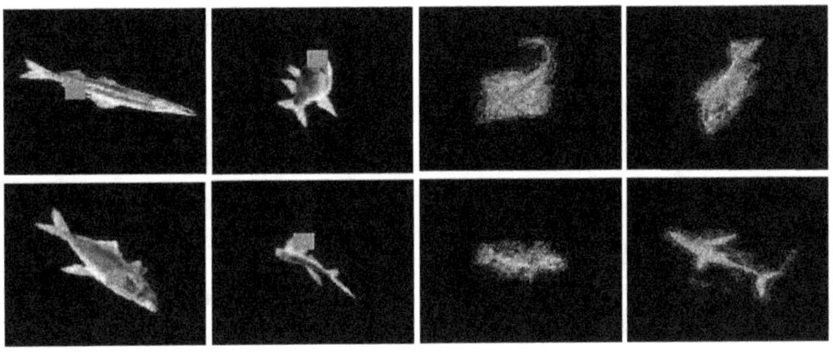

Fig. 6. Examples of the occluded and noise distorted fish images used in the experiments.

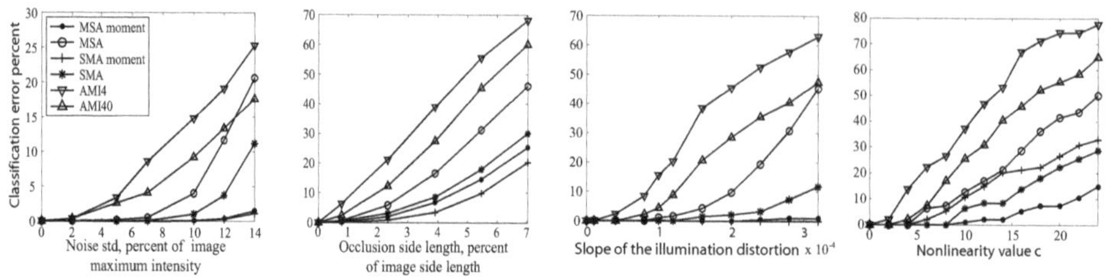

(a) (b) (c) (d)

Fig. 7. Classification error rates for (a) Gaussian noise, (b) occlusion, (c) illumination distortion, and (d) projective transformation. [42] (© 2006 IEEE)

tions for G, and are referred as HM and HS in the experiments, respectively. In addition, we also take the SMA and MSA methods from the previous section for comparison. In these methods, we apply 10 and 36 (α, β) pairs for SMA, and 5 and 19 pairs for MSA.

In the first experiment, we classified 256×256 gray-scale images of postcards, obtained from photographs taken from different viewing angles. The training set included one image from each of the 50 different postcards, and the test set included 175 images of the same postcards taken from various angles in natural light. Some examples of the training and test images are shown in Figure 8. The classification was performed using a simple nearest neighbor classifier, where the distance measure was the histogram intersection for the histogram approaches, and the Euclidean distance for the others. The classification results achieved are given in Table 1, along with the approximate computation times per one 256×256 image using Matlab implementation with a single 2 GHz core. The results indicate that with the same computational load, the histogram based multiscale invariants achieve clearly better performance compared to the similar moment based methods. One can also observe that the comparison based approaches HS and HM result in better discriminability than MSA histograms.

We continued by introducing more distortions to the images. We added illumination distortions by taking a new set of 125 test photos, underexposing them first

Fig. 8. Some examples of training images (first row) and angle distorted images (second row).

Table 1. Classification error percents under real view angle distortions.

	HS	HM	MSA5	MSA19	MSA hist.	SMA10	SMA36
Classification error	3.4 %	5.7 %	22.9 %	3.4 %	27.4 %	36.6 %	26.3 %%
Execution time	1.08 s	1.89 s	2.42 s	11.40 s	0.19 s	2.62 s	5.09 s

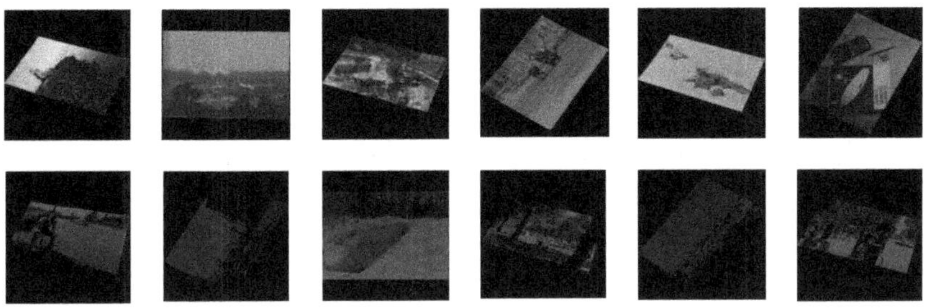

Fig. 9. Examples of illumination distorted images. With the images on first row the underexposure is 1.5 apertures and on the second row 3 apertures.

Table 2. Classification error percents under illumination distortions.

	HS	HM	MSA19	SMA36
Underexposure by 1.5 apertures	2.4 %	2.4 %	1.6 %	10.4 %
Underexposure by 3 apertures	4.0 %	5.6 %	0.0 %	11.2 %
Underexposure by 2 and ISO 3200	24.0 %	30.0 %	2.0 %	4.0 %

by 1.5, and then by 3 aperture steps of the camera. Some examples of the new test sets are shown in Figure 9. In addition, we created one more test set of 50 images, by underexposing by 2 aperture steps and increasing the sensitivity (ISO) value of the camera to 3200. The resulting images were severely contaminated by noise. We omitted the MSA5, MSA histogram, and SMA10 methods from this experiment due to their high error rates. The classification error rates of this experiment are illustrated in Table 2. From the results, we can observe that both comparison based histograms and multiscale autoconvolution seem to provide tolerance against illumination distortions. The last experiment shows also that comparison operations used in HS and HM react more strongly to the substantial changes in gray-scale values than MSA and SMA, which use convolution and product operations.

8. Conclusions and Future Work

This chapter considered affine invariance in feature extraction using normalization and invariant approaches. The main focus was on the invariant features and particularly on the recently proposed multiscale framework. The new framework offers a systematic tool to alleviate some of the basic problems encountered with traditional affine invariant techniques. As the experiments illustrate, using this new framework we can achieve even considerable improvements in the performance of the original methods.

The idea in the multiscale approach is to expand the given pattern to a set of affine covariant versions, each carrying slightly different information, and then to apply known affine invariants to each of them separately. The key part of the framework is the construction of the affine covariant set, and this is done by combining several scaled representations of the original pattern. The usage of the framework was illustrated in seven constructions, where different preprocessing methods, combination schemes, and final feature extraction techniques were applied. The constructions presented are in no way a complete set of possible ways to apply the multiscale approach, and further research may bring new interesting possibilities.

The idea of applying scaled versions of the image in the multiscale framework was introduced particularly due to the properties of the affine transformation. In situations where the interconnections of the patterns are described with some other mappings, similar ideas may be applicable when the scaling is replaced by another suitable operation. For example, in order to construct rotation invariants, one could use combinations of rotated versions of the original pattern.

The multiscale methods and the corresponding experiments were presented using image intensity functions as patterns, since this is the main application where affine relations arise. Nevertheless, the new approach may be additionally applied to other types of patterns where the affine transformation is present, as long as the patterns can be represented by functions. An example of such patterns could be volume data or probability densities.

Acknowledgments

The Academy of Finland is gratefully acknowledged for providing the funding for the research of affine invariant features. Figures 3, 5, 8, 9, and Tables 1 and 2 are taken from [48] with kind permission of Springer Science+Business Media.

References

1. T. Ojala, M. Pietikäinen, and T. Mäenpää, Multiresolution gray-scale and rotation invariant texture classification with local binary patterns, *IEEE Transactions on Pattern Analysis and Machine Intelligence.* **24**(7), 971–987, (2002).
2. P. Viola and M. Jones, Robust real-time face detection, *International Journal of Computer Vision.* **57**, 137154, (2004).
3. T. Lee, D. McLean, and M. Atkins, A new irregularity measure for cutaneous lesions, *Medical Image Analysis.* **7**(1), 47–64, (2003).
4. R. Hartley and A. Zisserman, *Multiple view geometry in computer vision, second edition.* (Cambridge university press, United Kindom, 2003).
5. R. O. Duda, P. E. Hart, and D. G. Stork, *Pattern classification.* (John Wiley & Sons, New York, 2001).
6. K. Åström, Fundamental limitations on projective invariants of planar curves, *IEEE Transactions on Pattern Analysis and Machine Intelligence.* **17**(1), 77–81, (1995).

7. K. Mikolajczyk, T. Tuytelaars, C. Schmid, A. Zisserman, J. Matas, F. Schaffalitzky, T. Kadir, and L. V. Gool, A comparison of affine region detectors, *International Journal of Computer Vision.* **65**(1), 43–72, (2005).
8. K. Mikolajczyk and C. Schmid, Scale & affine invariant interest point detectors, *International Journal of Computer Vision.* **60**(1), 63–86, (2004).
9. J. Matas, O. Chum, M. Urban, and T. Pajdla. Robust wide baseline stereo from maximally stable extremal regions. In *Proc. British Machine Vision Conference*, pp. 384 – 393, (2002).
10. T. Tuytelaars and L. V. Gool, Matching widely separated views based on affine invariant regions, *International Journal of Computer Vision.* **59**(1), 61–85, (2004).
11. T. Kadir, A. Zisserman, and M. Brady. An affine invariant salient region detector. In *Proc. European Conference on Computer Vision*, pp. 404–416, (2004).
12. D. Lowe, Distinctive image features from scale invariant keypoints, *International Journal of Computer Vision.* **2**(60), 91–110, (2004).
13. C. Schmid and R. Mohr, Local grayvalue invariants for image retrieval, *IEEE Transactions on Pattern Analysis and Machine Intelligence.* **19**(5), 530–535, (1997).
14. J.-L. Chen and A. Kundu, Unsupervised texture segmentation using multichannel decompositions and hidden markov models, *IEEE Transactions on Image Processing.* **4**(5), 603–619 (May, 1995).
15. D. Panjwani and G. Healey, Markov random field models for unsupervised segmentation of textured color images, *IEEE Transactions on Pattern Analysis and Machine Intelligence.* **17**(10), 939–954 (October, 1995).
16. D. Comaniciu and P. Meer, Mean shift: a robust approach toward feature space analysis, *IEEE Transactions on Pattern Analysis and Machine Intelligence.* **24**(5), 603–619 (May, 2002).
17. J. Shi and J. Malik, Normalized cuts and image segmentation, *IEEE Transactions on Pattern Analysis and Machine Intelligence.* **22**(8), 888–905 (August, 2000).
18. D. A. Forsyth and J. Ponce, *Computer Vision a Modern Approach.* (Prentice Hall, New Jersey, 2003).
19. S. Umbaugh, *Computer imaging.* (CRC press, Florida, 2005).
20. M. Petrou and P. Bosdogianni, *Image Processing.* (John Wiley & Sons, New York, 1999).
21. R. Gonzalez, R. Woods, and S. Eddins, *Digital Image Processing.* (Prentice Hall, 2002).
22. T. Suk and J. Flusser. Affine normalization of symmetric objects. In *Proc. International Conference on Advanced Concepts for Intelligent Vision Systems*, vol. 3708, pp. 100–107, (2005).
23. H. Burkhardt and S. Siggelkow. Invariant features in pattern recognition - fundamentals and applications. In eds. I. Pitas and C. Kotropoulis, *Nonlinear Model-Based Image/Video Processing and Analysis*, pp. 269–307. John Wiley & Sons, New York, (2001).
24. M. A. Rodrigues, *Invariants for pattern recognition and classification.* (World Scientific, Singapore, 2000).
25. T. H. Reiss, *Recognizing planar objects using invariant image features.* (Springer-Verlag, NJ, 1993).
26. M.-K. Hu, Pattern recognition by moment invariants, *Proceedings of the IRE.* **49**, 1428 (September, 1961).
27. J. Flusser and T. Suk, Pattern recognition by affine moment invariants, *Pattern Recognition.* **26**(1), 167–174 (January, 1993).

28. T. H. Reiss, The revised fundamental theorem of moment invariants, *IEEE Transactions on Pattern Analysis and Machine Intelligence.* **13**(8), 830–834 (August, 1991).
29. T. Suk and J. Flusser. Graph method for generating affine moment invariants. In *Proc. International Conference on Pattern Recognition*, vol. 2, pp. 192–195, (2004).
30. L. Van Gool, T. Moons, and D. Ungureanu. Affine/ photometric invariants for planar intensity patterns. In *Proc. European Conference on Computer Vision*, vol. 1, pp. 642 – 651, (1996).
31. Z. Yang and F. S. Cohen, Cross-weighted moments and affine invariants for image registration and matching, *IEEE Transactions on Pattern Analysis and Machine Intelligence.* **21**(8), 804–814 (August, 1999).
32. M. Petrou and A. Kadyrov, Affine invariant features from the trace transform, *IEEE Transactions on Pattern Analysis and Machine Intelligence.* **26**(1), 30–44 (January, 2004).
33. J. L. Mundy and A. Zisserman, *Geometric Invariance in Computer Vision.* 1992.
34. H. Burkhardt, J. C. Simon, and Y. Neuvo, Eds., *From Pixels to Features: Parallelism in Image Processing: Proceedings of a Workshop, Bonas, France, 27 Aug.-1 Sept., 1990.* (Elsevier Science Inc., New York, 1991).
35. Q. Jin and P. Yan. A new method of extracting invariants under affine transform. In *Proc. International Conference of Pattern Recognition*, vol. 1, pp. 742–745, (1992).
36. M. I. Khalil and M. M. Bayoumi, Affine invariants for object recognition using the wavelet transform., *Pattern Recognition Letters.* **23**, 57–72, (2002).
37. J. Kannala, E. Rahtu, M. Salo, and J. Heikkilä. A new method for affine registration of images and point sets. In *Proc. 14th Scandinavian Conference on Image Analysis*, pp. 224–234, (2005).
38. J. Kannala, E. Rahtu, and J. Heikkilä. Affine registration with multi-scale autoconvolution. In *Proc. International Conference on Image Processing*, vol. 3, pp. 1064–1067, (2005).
39. E. Rahtu, M. Salo, and J. Heikkilä, A new convexity measure based on a probabilistic interpretation of images, *IEEE Transactions on Pattern Analysis and Machine Intelligence.* **28**(9), 1501–1512 (September, 2006).
40. E. Rahtu, M. Salo, and J. Heikkilä. A new efficient method for producing global affine invariants. In *Proc. 13th International Conference on Image Analysis and Processing*, pp. 407–414, (2005).
41. E. Rahtu, M. Salo, and J. Heikkilä, Affine invariant pattern recognition using multiscale autoconvolution, *IEEE Transactions on Pattern Analysis and Machine Intelligence.* **27**(6), 908–918 (June, 2005).
42. E. Rahtu, M. Salo, J. Heikkilä, and J. Flusser. Generalized affine moment invariants for object recognition. In *Proc. 18th International Conference on Pattern Recognition*, vol. 2, pp. 634–637, (2006).
43. E. Rahtu, M. Salo, and J. Heikkilä. A new affine invariant image transform based on ridgelets. In *Proc. 16th British Machine Vision Conference*, vol. 3, pp. 1059–1068, (2006).
44. E. Rahtu, M. Salo, and J. Heikkilä. Multiscale autoconvolution histograms for affine invariant pattern recognition. In *Proc. 16th British Machine Vision Conference*, vol. 3, pp. 1039–1048, (2006).
45. J. Heikkilä. Multi-scale autoconvolution for affine invariant pattern recognition. In *Proc. International Conference of Pattern Recognition*, pp. 119–122, (2002).
46. E. J. Candés, *Ridgelets: Theory and applications, Ph.D. dissertation.* (Dept. Statistics, Stanford Univ., Stanford, USA, 1998).

47. M. N. Do and M. Vetterli, The finite ridgelet transform for image representation, *IEEE Transactions on Image Processing.* **12**(1), 16–28 (January, 2003).
48. E. Rahtu, M. Salo, and J. Heikkilä. Nonlinear functionals in the construction of multiscale affine invariants. In *Proc. 15th Scandinavian Conference on Image Analysis*, pp. 482–491, (2007).

CHAPTER 2.6

STATISTICAL ASPECTS OF DOMINANT PLANE FOR ROBOT NAVIGATION

Naoya Ohnishi* and Atsushi Imiya†

School of Science and Technology, Chiba University, Japan
Institute of Media and Information Technology, Chiba University, Japan
Yayoicho 1-33, Inage-ku, Chiba, 263-8522, Japan
imiya@faculty.chiba-u.jp

The dominant plane is the planar area in the robot workspace corresponding to the largest part of an image or at least half of an image. We present an algorithm for detecting the dominant plane from viewpoint of computational cognitive neuroscience using Independent Component Analysis. Then, using geometrical properties of the dominant plane, we derive a spatial reasoning method for mobile robot navigation. While traditional methods for mobile robot navigation tend to detect obstacles in the environment using various sensors, our method detects the dominant plane as a feasible region for navigation.

1. Introduction

Robot navigation in a featureless environment is an important task for a mobile robot to explore in an unknown area, since any features and landmarks are hardly obtained in general environments. Furthermore, since the robot moves over a planar area without any models of the environment, featureless robot navigation can be operated with a small storage capacity of memory to store a database of environmental maps and with a short processing time for navigation. Therefore, an algorithm of featureless navigation is also suitable for real time robot operation. Since optical flow is a fundamental cue for understanding environment both for animals and robots, we develop a featureless navigation algorithm using optical flow fields from a pair of successive images.[1]

We assume that the robot autonomously moves in a man-made environment without any stored maps in the memory or any markers in the environment. Our robot is required to navigate and avoid collisions with obstacles without using any maps and marker. Since the robot moves in a man-made environment, we can assume that the environment consists of plane segments. Therefore, we develop

*Present Address: Power and Industry Systems Research and Development Centers, Power Systems Company, TOSHIBA, Fuchu, Tokyo, Japan.
†All academic correspondences are addressed to A. Imiya.

a strategy for computing the navigating direction using a hierarchical expression of plane segments in the workspace. Since our robot captures a sequence of images, we propose an algorithm for segmenting multiple planes using an optical flow field computed from successive images captured by an uncalibrated moving camera. Furthermore, we show that the segmented planes enable the robot to autonomously explore in featureless environments. This method also yields local obstacle maps by detecting the dominant plane in images.

In visual perception systems in a human brain, it is said that visual information is decomposed into independent components. Beck *et al.*[2] proposed a method of hierarchical optical flow analysis in a biologically inspired algorithm. Their algorithm uses the optical flow field in multiresolution layers and unifies motion information from layers to detect global motion and local motion in an original image. This algorithm is motivated by the cognitive property that the motion perception in the brain is achieved by the separation and unification of layers based on the multiresolution expression of image. On the other hand, Tsotsos analysed and implemented motion analysis method based on independent channel expression[3] using Gabor filters. His algorithm is based on multichannel data processing in the brain which separates information in to independent properties and unifies local information to derive global information.

Independent component analysis (ICA)[4] extracts statistically independent features from signals and still images. In neuroscience, it is known that the medial superior temporal (MST) area performs visual motion processing. For motion cognition in the MST area of the brain,[5,6] it has been shown that independent components of optical flow are used. Therefore, we develop an algorithm to separate the ground plane from image sequences using ICA. Since the optical flow field on an image can be represented as a linear combination of independent components of optical flow,[5] we can use ICA for the detection of the dominant plane by separating obstacles and the dominant part in an image.

For the concurrent detection of local and global motion, we use independent components of optical flow fields on pyramidal layers.[7] It is known that animals, insects, and human beings use the independent components of optical flow fields for visual behaviour.[5,8,9] For human object recognition, hierarchical model has been proposed.[10,11] Furthermore, to compute optical flow, the pyramid transform of an image sequence is used for the analysis of global motion and local motion.[12,13] The pyramid transform generates multiresolution images as layered images. These layered images are used for computation of optical flow in its original images from the image in the lowest layer. This idea is based on the assertion that global motion is described as the collection of local motion.

Spatial reasoning is an important problem in robot exploration, localisation, mapping, and navigation.[14,15] We use a monocular camera system mounted on a mobile robot to capture the geometrical structures in the robot workspace. If a planar surface of a polyhedral obstacle is observed in the right half of the image captured by the robot, the robot will turn to the left. Therefore, the configuration of planar areas on the ground is a fundamental cue for robot navigation. The configuration of planar areas is estimated by extracting plane segments from the image. For the extraction of plane segments, we use a sequence of images captured by the camera mounted on the robot. There are many methods for the detection of planar areas in vision systems.[16–18] When a mobile robot captures an image sequence of moving objects, the optical flow,[19–21] which is the flow of movement in the scene, is obtained as the fundamental features to generate environmental information around the robot. Additionally, optical flow is considered to be fundamental information for obstacle detection in the context of biological data processing.[22] Therefore, the use of optical flow is an appropriate method for the construction of environmental maps from the viewpoint of the affinity between robots and human beings.

Plane segmentation in an image captured by a camera is a long-studied problem,[23,24] since planes in an image are fundamental features for the modelling and 3D reconstruction of a scene.[25] Furthermore, planes in an image are used for camera calibration[26] and mobile robot navigation.[1,27] There are many methods for detecting planar areas using vision systems.[17] For example, edge detection in images observed by omnidirectional and monocular camera systems,[28] and the observation of predefined landmarks,[29] are classical methods. However, since these methods are dependent on the environment around a robot, it is not easy to use them in general environments. Dominant-plane detection is used as the main method of robot navigation.[1] We show that our dominant-plane-detection strategy allows us to segment planar areas in an image by applying dominant-plane detection successively to the image. Therefore, using the dominant-plane-detection algorithm as the main method of visual control for an autonomous mobile robot, the robot can navigate in general environments and determine the three-dimensional configuration of its workspace. On the other hand, optical flow,[19–21] which is the apparent motion in the local environment, is a fundamental feature for the construction of environment information in the context of biological data processing.[9] Therefore, the use of the optical-flow field is an appropriate method of image processing for autonomous mobile robots.

Planar-area detection and segmentation methods using optical flow have also been proposed.[30–34] Enkelmann[30] proposed a plane-detection method using the model vectors from motion parameters. Santos-Victor and Sandini[33] also proposed a plane-detection algorithm for a mobile robot using the inverse projection of optical flow to the ground, assuming that the motion of the camera system mounted on a robot can be modelled as a pure translation with uniform velocity. However, even if a camera is mounted on a wheel-driven robot, the vision system does not move with uniform velocity due to mechanical errors of the robot and the unevenness of the floor.

For the purpose of the layer separation of an image sequence, Weiss presented the EM algorithm for segmenting an image sequence by fitting optical flow,[35] and Wang

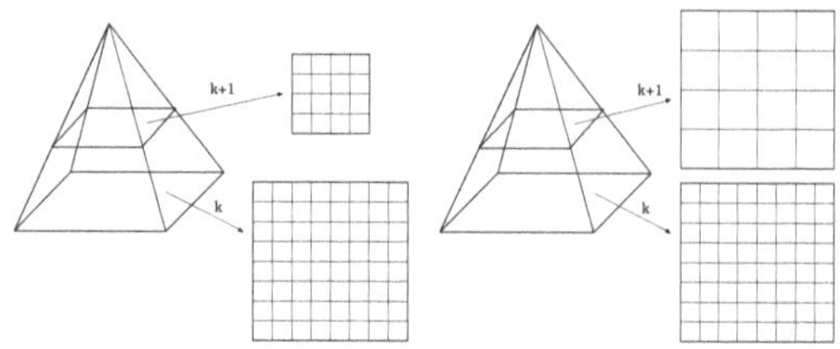

(a) (b)

Fig. 1. Two expressions of the images in the pyramid transform. (a) shows the pyramid transform with equipixel size in each layer. (b) shows the pyramid transform with equilandscape size in each layer.

and Adelson represented an image sequence with sets of layers.[36] In video sequence analysis for the detection of 3D spatial information, depth-layer separation is a fundamental problem. It is computed from the mathematical relation between depth and optical flow for a stationary moving camera. These approaches are categorised by object-based methods. Feasible region separation is the detection of a layer that corresponds to the ground plane. Algorithms for motion segmentation and image segmentation by ICA have also been proposed.[37,38] However, the difference between video image processing and the robot navigation task is in the motion orbit of the camera mounted on a mobile robot. In the robot navigation task, the motion orbit of the camera is a spatial curve determined by the robot trajectory. In video image processing, on the contrary, the motion orbit of the camera is predetected or calibrated using a sequence of images in the video sequence. Optical flow is also used in robot navigation to detect the ground plane.[39,40] Ke and Kanade assumed that the camera is calibrated and proposed an algorithm for ground-plane detection and ego-motion estimation.[39]

2. Dominant Plane and Optical Flow

2.1. *Pyramid-Based Multiresolution Analysis*

Setting $I^0(x,y,t) = I(x,y,t)$ is the original image and $I^l(x,y,t)$ is the pyramid transformation of the image $I(x,y,t)$ at level l, the pyramid representation is expressed as

$$I^{l+1}(x,y,t) = \sum_{i,j=-1}^{1} a_{ij} I^l(2x-\alpha, 2y-\beta, t), \tag{1}$$

where $a_{\pm i \pm j} = \frac{1}{4}\left(1 - \frac{1}{2}|i|\right)\left(1 - \frac{1}{2}|j|\right)$ for $|i| \leq 1$ and $|j| \leq 1$.

Traditionally, the pyramid transform yields a reduced image sequence. This interpretation comes from the fundamental property that the pyramid transform in each step reduces the size of an image using downsampling process, if we use the same pixel size in each layer to express digital images. However, if we use the same landscape size on each layer, the transform in each step yields the lower resolution image. We adopt this equisized-landscape constraint to express the image on each layer. Therefore, image features extracted from the higher layer images of the pyramid transform describe global properties comparing to those extracted in the lower layer from the hierarchical expression. as shown in Fig. 1.

2.2. *Optical Flow Computation*

Setting $\boldsymbol{u}^l = (\dot{x}^l, \dot{y}^l)^\top = (u^l, v^l)^\top$ to be the optical flow of $I^l(x, y, t)$, $\boldsymbol{u}^l$ satisfies $I^l_x u^l + I^l_y v^l + I^l_t = 0$. This equation can be solved by assuming that the optical flow vector of pixels is constant in the neighbourhood of each pixel. We set the window size to be 5×5. Then, we have a system of linear equations,

$$I^l_{x(\alpha\beta)} u^l + I^l_{y(\alpha\beta)} v^l + I^l_{t(\alpha\beta)} = 0, \quad |\alpha| \leq 2, |\beta| \leq 2 \tag{2}$$

where $I^l_{x(\alpha\beta)} = I^l_x(x + \alpha, y + \beta, t)$, $I^l_{y(\alpha\beta)} = I^l_y(x + \alpha, y + \beta, t)$, and $I^l_{t(\alpha\beta)} = I^l_t(x + \alpha, y + \beta, t)$. Therefore, if the vectors

$$\boldsymbol{a}^l = (I^l_{x(-2\,-2)}, I^l_{x(-2\,-1)}, \cdots, I^l_{x(22)})^\top, \ \boldsymbol{b}^l = (I^l_{y(-2\,-2)}, I^l_{y(-2\,-1)}, \cdots, I^l_{y(22)})^\top \tag{3}$$

are independent, we have a unique solution.

$$\boldsymbol{u}^l = -(\boldsymbol{D}_l^\top \boldsymbol{D}_l)^{-1} \boldsymbol{D}_l \boldsymbol{c}^l \tag{4}$$

where $\boldsymbol{D}^l = (\boldsymbol{a}^l, \boldsymbol{b}^l)$ and $\boldsymbol{c}^l = (I^l_{t(-2\,-2)}, I^l_{t(-2\,-1)}, \cdots, I^l_{t(22)})^\top$ for the centre point of the 5×5 window on each layer.

The pyramidal optical flow fields drive global and local motion in the image sequence, that is, $\boldsymbol{u} := \boldsymbol{u}^0$ and $\boldsymbol{u}^l$ for $l \geq 1$ are the motion of pixels on the original image and pixels on the images in the pyramid hierarchy, respectively. Optical flow computation by the Lucas-Kanade method using pyramids is summarised in Algorithm 1. In this algorithm

$$w(I^{l-1}(t+1), \boldsymbol{u}^l) = f(\boldsymbol{x} + E(\boldsymbol{u}^l), t+1), \ \ E(\boldsymbol{u}^l)_{mn} = \frac{1}{4} \sum_{i,j=-2}^{2} a_{ij} \boldsymbol{u}_{\frac{m-i}{2}\ \frac{n-j}{2}} \tag{5}$$

where the summation is achieved when both $\frac{m-i}{2}$ and $\frac{n-j}{2}$ are integers.

Algorithm 1. Optical Flow Computation by the Lucas-Kanade Method Using Pyramids

> **Data**$I^l(t)$, $I^l(t+1)$, $\quad 0 \leq n \leq$ *maximum of number of layers*;
> **Results** $\boldsymbol{u}$;
> $l :=$ *maximum of number of layers*
> **while** $n \neq 0$; $\quad \boldsymbol{u}^l := u(I^l(t), I^l(t+1))$
> $\quad I^{l-1}(t+1) := w(I^{l-1}(t+1), E(\boldsymbol{u}^l))$
> $\quad l := l - 1$
> **end**

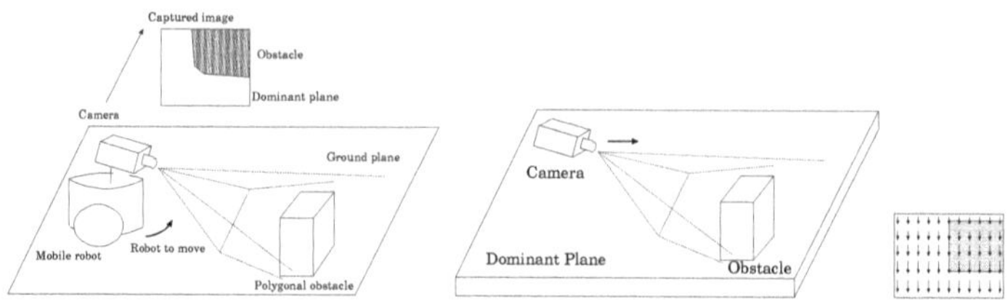

(a) (b) (c)

Fig. 2. Camera motion and its optical flow field. (a) Robot navigation using planar areas. If a planar surface of a polyhedral obstacle in front of the robot is observed in the right half of the image captured by the camera mounted on the robot, the robot will turn to the left. The obstacle area is detected by removing the dominant plane from the image. The dominant plane is a planar area in the robot workspace, and it corresponds to the largest part in the image. (b) Camera moves toward obstacle. (c) Dominant plane of captured image. The white area is the dominant plane. Since the obstacle is placed in front of the camera, the shaded area is not the dominant plane.

2.3. *Definition of the Dominant Plane*

We define the dominant plane as the planar area in the environment corresponding to the largest part in an image. Assuming that the dominant plane in an image corresponds to the ground plane on which the robot moves, the detection of the dominant plane enables the robot to detect the feasible region for navigation in its workspace. An example of this definition is shown in Fig. 2. In Fig. 2, an obstacle is placed in front of a camera and the camera is moved toward the obstacle. Therefore, the top-right area of the image is not detected as the dominant plane.

If the following four assumptions are accepted, the ground plane can be assumed to be the dominant plane.

1. The ground plane is the planar area.
2. The camera mounted on a mobile robot is pointed downward.
3. The robot observes the world using the camera mounted on itself.
4. The robot moves on the ground plane without collision with obstacles.

Therefore, if there are no obstacles around the robot, since the robot does not touch the obstacles, the ground plane corresponds to the dominant plane in the image observed through the camera mounted on the mobile robot. These assumptions are illustrated in Fig. 2(a).

Assuming that the camera displacement is small, on the image plane the corresponding points $\boldsymbol{x} = (x, y)^\top$ and $\boldsymbol{x}' = (x', y')^\top$ in the dominant planes between a pair of successive two images are connected with an affine transform such that

$$\boldsymbol{x}' = \boldsymbol{A}\boldsymbol{x} + \boldsymbol{b}, \tag{6}$$

where $\boldsymbol{A}_l$ and $\boldsymbol{b}_l$ are a 2×2 affine-coefficient matrix and a 2-dimensional vector, respectively. We can estimate the affine coefficients using the RANSAC-based algorithm.[41]

Using the estimated affine coefficients, we can estimate the optical flow on the dominant plane $\hat{\boldsymbol{x}} = (\hat{x}, \hat{y})^\top$,

$$\hat{\boldsymbol{x}} = \boldsymbol{A}\boldsymbol{x} + \boldsymbol{b} - \boldsymbol{x}, \tag{7}$$

for all points $\boldsymbol{x}$ in the image. We call $\hat{\boldsymbol{x}}$ the *planar flow* and $\hat{\boldsymbol{x}}(x, y, t)$ the *planar flow field* at time t, which is a set of planar flows $\hat{\boldsymbol{x}}$ computed for all pixels in an image. Algorithm 2 computes the affine coefficients to describe the planar flow.
Algorithm 2. Planar Flow Estimation

repeat
Randomly select three points from $\{\boldsymbol{x}\}$
Estimate $\boldsymbol{A}$ *and* $\boldsymbol{b}$ *in Eq. (6) from* $\boldsymbol{x} = \boldsymbol{x}' - \boldsymbol{x}$
Compute planar flow field $\hat{\boldsymbol{x}}$ *using Eq. (7)*
until$\dot{\boldsymbol{x}} = \hat{\boldsymbol{x}}$ *for more than half of image*

If an obstacle exists in front of the robot, the planar flow on the image plane differs from the optical flow on the image plane. Since the planar flow $\hat{\boldsymbol{x}}$ is equal to the optical flow vector $\dot{\boldsymbol{x}}$ on the dominant plane, we use the difference between these two flows to detect the dominant plane. We set ε to be the tolerance of the difference between the optical flow vector and the planar flow vector. Therefore, if the inequality

$$|\dot{\boldsymbol{x}} - \hat{\boldsymbol{x}}| < \varepsilon, \text{ for } \hat{\boldsymbol{x}} = (\boldsymbol{A}\boldsymbol{x} + \boldsymbol{b}) - \boldsymbol{x}, \quad \nabla I^\top \dot{\boldsymbol{x}} + \partial_t I = 0 \tag{8}$$

is satisfied, we accept point $\boldsymbol{x}$ as a point on the dominant plane.[1]

3. Dominant Plane Detection using Independent Components of Optical Flow Field

3.1. *ICA of Optical Flow Field*

Similar to ICA which separates mixture signals into independent components, the MST area in the brain separates the motion fields obtained from visual perception into independent components.[5,6] We apply the idea of ICA separation to the understanding of an environment for robot navigation. Therefore, we accept the assumption that optical flow fields observed by a moving camera are linear combinations of optical flow fields of the dominant plane and the obstacles. That is, setting $\dot{\boldsymbol{u}}_{\text{dominant}}$ and $\dot{\boldsymbol{u}}_{\text{obstacle}}$ to be the optical flow fields of the dominant plane and the obstacles, respectively, the observed optical flow field $\dot{\boldsymbol{u}}$ is approximately expressed by a linear combination of $\dot{\boldsymbol{u}}_{\text{dominant}}$ and $\dot{\boldsymbol{u}}_{\text{obstacle}}$ as

$$\dot{\boldsymbol{u}} = a_1 \dot{\boldsymbol{u}}_{\text{dominant}} + a_2 \dot{\boldsymbol{u}}_{\text{obstacle}}, \tag{9}$$

where a_1 and a_2 are the mixture coefficients, as shown in Fig. 3. This assumption is numerically and geometrically acceptable if motion displacement is small compared with the size of obstacles. Therefore, ICA is suitable for the separation of optical

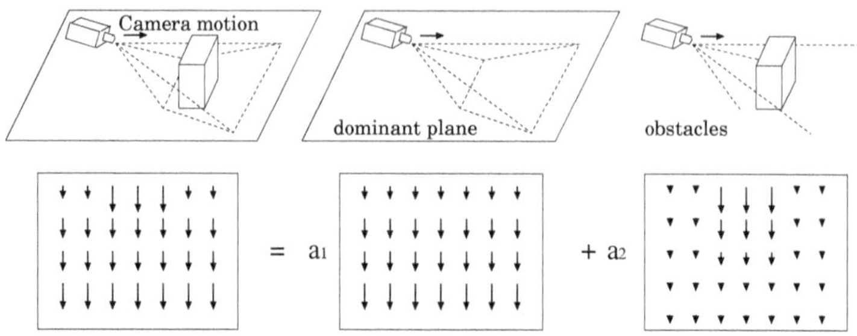

Fig. 3. Linear combination of optical flow field in the scene. Top: example of camera displacement and the environment. Bottom left: optical flow field observed through the moving camera. Bottom middle: optical flow field of the dominant plane. Bottom right: optical flow field of the other objects. The optical flow field (bottom left) is expressed as a linear combination of those shown in the bottom middle and the bottom right. a_1 and a_2 are the mixture coefficients.

flow into the independent flow components. For each image in a sequence, we consider that optical flow vectors in the dominant plane correspond to independent components.

The first reason why we use ICA is that ICA is robust against noises and errors. The following noises and errors exist in our strategy.

1. The noise of images captured by a moving camera.
2. The computational error in estimating the optical flow.
3. The noise produced during the extraction of the dominant plane from noisy optical flow.

ICA is robust against these noises and errors because it has a scale ambiguity for the output vector, and our method is independent of the scale of vectors. We conclude that ICA is suitable for developing an algorithm that is robust against noise. The second reason for using ICA is to examine the statistical properties of optical flow fields from the viewpoint of neuroscience. This kind of study is actively discussed as a collaboration between computer vision and neuroscience.[42,43]

However, it is known that the ICA output signals have ambiguity in the order of the independent components.[4] That is, ICA cannot distinguish whether components have the optical flow of the dominant plane or of obstacle areas. Therefore, we must decide the output order. Since the optical flow of the dominant plane should be smooth compared with the optical flow of obstacle areas, we can use the variance of the output signals for the determination of the output order. Some methods have been proposed ordering of independent components obtained by ICA.[44,45]

3.2. *Dominant Plane Detection by ICA*

ICA requires at least two input signals for separation into two independent components. Thus, we use the optical flow field $\dot{\boldsymbol{u}} = \{(\dot{u}, \dot{v})_{ij}^{\top}\}_{i=1,j=1}^{h,w}$ and planar flow field $\hat{\boldsymbol{u}} = \{(\hat{u}, \hat{v})_{ij}^{\top}\}_{i=1,j=1}^{h,w}$ as the input vectors of ICA for the detection of the dominant plane, where w and h are the width and height of an image, respectively. Since planar flow is the motion of the dominant plane relative to the robot motion, the use of a planar flow is suitable for separation into the dominant plane and obstacles.

For the ICA of optical flow fields, we rewrite the matrix of two-dimensional vectors as a one-dimensional array. The sorted vector field

$$\dot{\boldsymbol{u}} = \begin{bmatrix} (\dot{u},\dot{v})_{11}^\top & (\dot{u},\dot{v})_{12}^\top & \cdots & (\dot{u},\dot{v})_{1w}^\top \\ (\dot{u},\dot{v})_{21}^\top & (\dot{u},\dot{v})_{22}^\top & \cdots & (\dot{u},\dot{v})_{2w}^\top \\ & & \vdots & \\ (\dot{u},\dot{v})_{h1}^\top & (\dot{u},\dot{v})_{h2}^\top & \cdots & (\dot{u},\dot{v})_{hw}^\top \end{bmatrix} \tag{10}$$

is transformed to the vector

$$\dot{\boldsymbol{u}} \to ((\dot{u},\dot{v})_1 \cdots (\dot{u},\dot{v})_k \cdots (\dot{u},\dot{v})_n)^\top \to (\dot{u}_1 \cdots \dot{u}_n \; \dot{v}_1 \cdots \dot{v}_n)^\top = \mathrm{vec}\dot{\boldsymbol{u}}, \tag{11}$$

where the index is $k = j + iw$ and the size of the vector is $n = hw$. Since the relation

$$\dot{\boldsymbol{w}} = \alpha\dot{\boldsymbol{u}} + \beta\dot{\boldsymbol{v}} \tag{12}$$

leads to the relation

$$\mathrm{vec}\dot{\boldsymbol{w}} = \alpha\mathrm{vec}\dot{\boldsymbol{u}} + \beta\mathrm{vec}\dot{\boldsymbol{v}}, \tag{13}$$

this transform does not influence Eq. (9). These steps are invertible. Therefore, it is possible to extract regions corresponding to $\dot{\boldsymbol{u}}$ and $\dot{\boldsymbol{v}}$ if the observation $\dot{\boldsymbol{w}}$ is decomposed into two independent components $\dot{\boldsymbol{u}}$ and $\dot{\boldsymbol{v}}$. We use this vector, derived from a vector-valued image, as input to ICA.

Setting $\boldsymbol{v}_\alpha$ and $\boldsymbol{v}_\beta$ to be the output optical flow fields of ICA, Eq. (9) corresponds to

$$\begin{cases} \dot{\boldsymbol{u}} = a_{11}\boldsymbol{v}_\alpha + a_{12}\boldsymbol{v}_\beta \\ \hat{\boldsymbol{u}} = a_{21}\boldsymbol{v}_\alpha + a_{22}\boldsymbol{v}_\beta \end{cases}, \tag{14}$$

where a_{ij} are the mixture coefficients. ICA estimates the independent optical flow fields $\boldsymbol{v}_\alpha$ and $\boldsymbol{v}_\beta$ from optical flow fields $\dot{\boldsymbol{u}}$ and $\hat{\boldsymbol{u}}$. The output one-dimensional array is realigned to the matrix of two-dimensional vectors $\boldsymbol{v} = \{(\dot{u},\dot{v})_{ij}\}_{i=1,j=1}^{h,w}$.

The outputs $\boldsymbol{v}_\alpha$ and $\boldsymbol{v}_\beta$ have ambiguity in the order of each component, since the results of ICA have ambiguity in the order of the independent components.[4] We are required to determine whether components have the optical flow of the dominant plane or of the obstacle areas. We solve this problem using the difference between the variances of the norms of $\boldsymbol{v}_\alpha$ and $\boldsymbol{v}_\beta$.

Setting $\boldsymbol{l}_{\alpha,\beta} = \{l_{ij}\}_{i=1,j=1}^{h,w}$ to be the norm of $\boldsymbol{v}_{\alpha,\beta} = \{(\dot{u},\dot{v})_{ij}\}_{i=1,j=1}^{h,w}$, such that $l_{ij} = |(\dot{u},\dot{v})_{ij}|$, the variance σ^2 is computed as

$$\sigma^2 = \frac{1}{hw}\sum_{i=1,j=1}^{h,w}(l_{ij} - \bar{l})^2, \tag{15}$$

for $\bar{l} = \frac{1}{hw}\sum_{i=1,j=1}^{h,w} l_{ij}$.

The motions of the dominant plane and obstacles in the images are different, and the dominant-plane motion in the images is smooth compared with the obstacle motion, as shown in Fig. 4(a). Consequently, the output signal of the obstacle motion has a larger variance than the output signal of the dominant-plane motion. Therefore, if $\sigma_\alpha^1 > \sigma_\beta^2$, we use the norm $\boldsymbol{l}_\alpha$ of output optical flow field $\boldsymbol{v}_\alpha$ for dominant-plane detection; otherwise, we use the norm $\boldsymbol{l}_\beta$ of output optical flow field $\boldsymbol{v}_\beta$.

Since the planar flow field is subtracted from the optical flow field including obstacle motion, $\boldsymbol{l}$ is constant on the dominant plane. However, the length of $\boldsymbol{l}$ is ambiguous, because of the form of Eq. (14). Thus, we use the median value of $\boldsymbol{l}$ for the detection of the dominant plane. Since the dominant plane occupies the largest domain in the image, we compute the distance between $\boldsymbol{l}$ and the median of $\boldsymbol{l}$, as shown in Fig. 4(b). The area that has the median value of the component is detected as the dominant plane. Setting m to be the median value of the elements in $\boldsymbol{l}$, the distance $\boldsymbol{d} = \{d_{ij}\}_{i=1,j=1}^{h,w}$ is

$$d_{ij} = |l_{ij} - m|. \tag{16}$$

We detect the area in which $d_{ij} \approx 0$ as the dominant plane.

The procedure for dominant-plane detection by ICA is summarised as follows.
ICA-based Dominant Plane Detection

(1) Compute optical flow $(\dot{u}, \dot{v})^\top$ from two successive images.
(2) Compute affine coefficients in Eq. (6) by random selection of three points.
(3) Estimate planar flow $(\hat{u}, \hat{v})^\top$ from affine coefficients.
(4) Input optical flow field $\dot{\boldsymbol{u}}$ and planar flow field $\hat{\boldsymbol{u}}$ to ICA, and output optical flow fields $\boldsymbol{v}_\alpha$ and $\boldsymbol{v}_\beta$.
(5) Compute norms $\boldsymbol{l}_\alpha$ and $\boldsymbol{l}_\beta$ from $\boldsymbol{v}_\alpha$ and $\boldsymbol{v}_\beta$, respectively.
(6) Compute variances σ_α^2 and σ_β^2 from $\boldsymbol{l}_\alpha$ and $\boldsymbol{l}_\beta$, respectively.
(7) If $\sigma_\alpha^2 > \sigma_\beta^2$, then $\boldsymbol{l} = \boldsymbol{l}_\alpha$, else $\boldsymbol{l} = \boldsymbol{l}_\beta$.
(8) Compute distance $\mathbf{d}$ between $\boldsymbol{l}$ and median of $\boldsymbol{l}$.
(9) Detect area in which $d_{ij} \approx 0$ as dominant plane.

Figure 5 shows the procedure of dominant-plane detection from the image sequence using ICA.

4. Application of Dominant Plane Detection

4.1. *Iterative Multiple Plane Segmentation*

Using the dominant-plane-detection algorithm iteratively, we develop an algorithm for multiple-plane segmentation in an image. After removing the region corresponding to the dominant plane from an image, we can extract the second dominant planar region from the image. Then, it is possible to extract the third dominant plane by

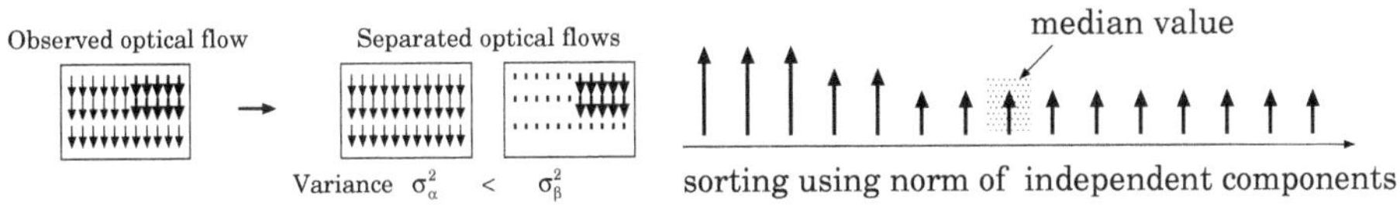

Fig. 4. Difference in the motion of the dominant plane and obstacles, and sorting using the norm of independent components for determination of output order. (a) The dominant plane motion is smooth on the images compared with obstacle motion. Therefore, the order of the components can be determined by using variances σ_α^2 and σ_β^2. (b) Sorting using the norm of independent components for determination of output order. The area that has the median value of the components is detected as the dominant plane, since the dominant plane occupies the largest domain in the image.

removing the second dominant planar region. This process is expressed as

$$D_k = \begin{cases} \mathbf{A}(R \setminus D_{k-1}), & k \geq 2, \\ \mathbf{A}(R), & k = 1, \end{cases} \tag{17}$$

where $\mathbf{A}$, R, and D_k stand for the dominant-plane-extraction algorithm, the region of interest observed by a camera, and the kth dominant planar region, respectively. The algorithm is stopped a predetermined iteration time or the size of kth dominant plane is smaller than predetermined size.

Setting R to be the root of the tree, this process derives a binary tree such that

$$R\langle D_1, R \setminus D_1 \langle D_2, R_2 \setminus D_2 \langle \cdots, \rangle \rangle. \tag{18}$$

Assuming that D_1 s the ground plane on which the robot moves, D_k for $k \geq 2$ are the planar areas on the obstacles. Therefore, this tree expresses the hierarchical structure of planar areas on the obstacles. We call this tree the binary tree of planes. Using this tree constructed by the dominant-plane detection algorithm, we obtain the geometrical properties of planes in a scene.

For an image R^2, setting D and $\overline{D}$ to be the dominant plane area in R^2 and its conjugate $\overline{D} = R^2 \setminus D$, respectively, we have the tree structure $R^2\langle D, \overline{D}\rangle$. Applying the dominant-plane-detection algorithm to $\overline{D}$, we have the tree structure $R^2\langle D, \overline{D}\langle D_1, \overline{D}_1\rangle\rangle$. Here, we affix labels L, R, and M to this tree, where L, R, and M express the position of the dominant plane in the hierarchy in the image plane. We have the following possible trees:

$$T_L = R^2\langle D_L, D_R\langle D_{RL}, D_{RR}\rangle\rangle, \tag{19}$$

$$T_M = R^2\langle D_M, \overline{D_M}\rangle, \tag{20}$$

$$T_R = R^2\langle D_L\langle D_{LL}, D_{LR}\rangle, D_R\rangle. \tag{21}$$

These three trees correspond to the spatial configurations of the planes in front of the robot, as shown in Fig. 6. The correspondence between the trees and the spatial configurations indicates the directions of collision-free paths for the mobile robot. In accordance with these trees, which express the configurations of the free space

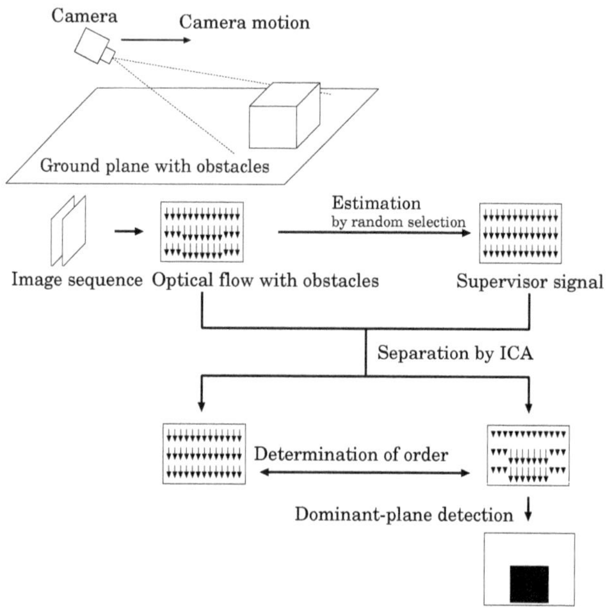

Fig. 5. Procedure for dominant-plane detection using ICA. First, we compute the optical flow $(\dot{u}, \dot{v})^\top$ from two successive images. Then we compute the affine coefficients in Eq. (6) by RANSAC[40] and estimate the planar flow field $(\hat{u}, \hat{v})^\top$ from the affine coefficients. We input optical flow field $\dot{\boldsymbol{u}}$ and planar flow field $\hat{\boldsymbol{u}}$ to ICA, and output the optical flow fields $\boldsymbol{v}_\alpha$ and $\boldsymbol{v}_\beta$. Using the norms and variances of the output optical flow fields, we detect the dominant plane.

and the obstacles in front of the robot, the mobile robot decides which direction to move.

While the robot is moving, it obtains a sequence of trees $T(t)$. When the robot detects the transition of trees, that is, $T(t+1) \neq T(t)$, the robot is required to control its direction for collision-free navigation. This property is used to derive the control rule as follows.

Algorithm 3. Direction Control Rule

if $T(t) = T(t-1)$ **then** *the robot moves in direction of label of* $T_*(t-1)$
  else *the robot moves in direction of label of* $T_*(t)$
  $t := t + 1$

4.2. *Hierarchy of Dominant Plane*

The dominant-plane-detection algorithm is separately processed at layers $l = 0, \cdots, L$ in the pyramid transform. Using the optical flow field $\boldsymbol{u}^l(x, y, t)$ at layer l, we detect obstacles in an image sequence $I^l(x, y, t)$ at time t.

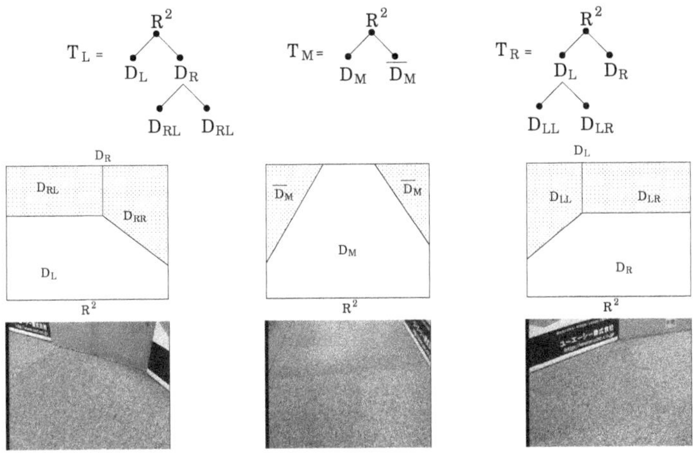

Fig. 6. Binary trees and corresponding images of hierarchical structure of dominant planes. R^2 is an image plane and D_* are hierarchical dominant planes. $\overline{D_M} = R^2 \setminus D_M$.

On each layer the corresponding points $\boldsymbol{x}_l = (x_l, y_l)^\top$ and $\boldsymbol{x}'_l = (x'_l, y'_l)^\top$ in the dominant planes between a pair of successive images are connected with an affine transform such that $\boldsymbol{x}'_l = \boldsymbol{A}_l\boldsymbol{x}_l + \boldsymbol{b}_l$, where $\boldsymbol{A}_l$ and $\boldsymbol{b}_l$ are a 2×2 affine-coefficient matrix and a 2-dimensional vector, respectively. On each layer, we can estimate the optical flow on the dominant plane $\hat{\boldsymbol{x}}_l = (\hat{x}_l, \hat{y}_l)^\top$, $\hat{\boldsymbol{x}}_l = \boldsymbol{A}_l\boldsymbol{x}_l + \boldsymbol{b}_l - \boldsymbol{x}_l$, for all points $\boldsymbol{x}_l$ in the image.

We call $\hat{\boldsymbol{x}}_l$ the *planar flow on the lth layer* and $\hat{\boldsymbol{x}}_l(x_l, y_l, t)$ the *planar flow field on the lth layer* at time t, which is a set of planar flows $\hat{\boldsymbol{x}}_l$ computed for all pixels in an image. Therefore, on each layer, if the inequality

$$|\dot{\boldsymbol{x}}_l - \hat{\boldsymbol{x}}_l| < \varepsilon, \text{ for } \hat{\boldsymbol{x}}_l = (\boldsymbol{A}_l\boldsymbol{x}_l + \boldsymbol{b}_l) - \boldsymbol{x}_l, \quad \nabla I^{l\top}\dot{\boldsymbol{x}}_l + \partial_t I^l = 0 \tag{22}$$

is satisfied, we accept point $\boldsymbol{x}_l$ as a point on the dominant plane[1] on the lth layer. We detect obstacles from optical flow fields on each layer in the pyramid using ICA as shown in Fig. 7. Therefore, our algorithm allows us to achieve both global perception and local perception for the robot vision.

We set O^l to be the obstacle region on the lth layer. From the geometrical properties of the pyramid transform, as shown in Fig. 8, for O^l we have the inclusive relation

$$O^0 \supset O^1 \supset \cdots \supset O^L \tag{23}$$

as the hierarchical expression of obstacle regions on layers. This relation implies the inclusive relation between the dominant planes

$$D^0 \subset D^1 \subset \cdots \subset D^L \tag{24}$$

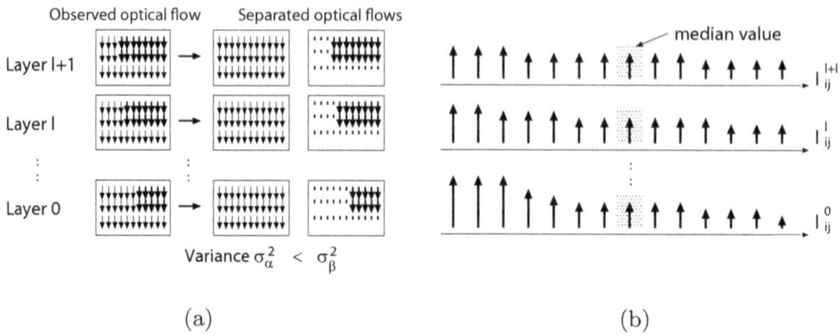

Fig. 7. Dominant plane detection on each layer using its variance. (a) Difference in the motion of the dominant plane and obstacles. The dominant plane motion is smooth on the images compared with obstacle motion. Therefore, the order of the components can be determined using variances σ_α^2 and σ_β^2. (b) Sorting using the norm $\boldsymbol{l}$ to determine output order. $\boldsymbol{l}_{ij}^l$ means the norm $\boldsymbol{l}$ at l-th layer. The area which has the median value of the component is detected as the dominant plane, since the dominant plane occupies the largest domain in the image.

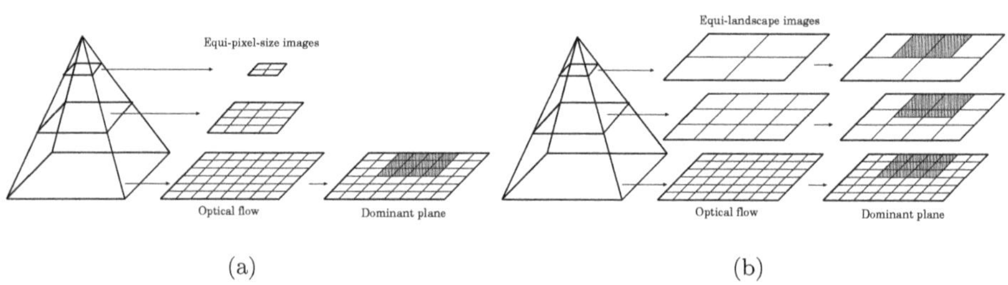

Fig. 8. Hierarchical extraction of the dominant planes. (a) Dominant plane detection using the pyramid transform with equipixel size in each layer. (b) Hierarchical dominant plane detection using the pyramid transform with equilandscape size in each layer. The upper layer images and lower layer images in this hierarchy of the dominant planes provide global features and precise features for robot navigation, respectively. The global features define the feasible region for controlling and topological configuration of the obstacles, and precise features derive geometrical configuration for real-time map generation from visual information captured by a camera mounted on the robot.

for the dominant plane D^l in the pyramid hierarchy since $D^l \cap O^l = \phi$ and $O^l \cup D^l = R$, where R is the total region of the image. These relations imply that a pair $C^l = (D^l, O^l)$ exhibits global and local configurations in the workspace for a larger and a smaller l, respectively. This hierarchical relation is automatically detected from the pyramid-based hierarchical expression of images used for optical flow computation. The system selectively uses C^l for navigation and spatial perception.

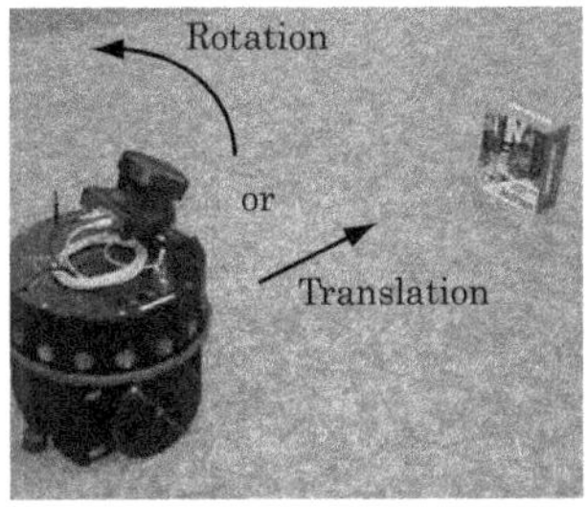

Fig. 9. Motion of mobile robot. The motion of the mobile robot is translational or rotational.

5. Experiments

5.1. *Dominant Plane Detection*

We evaluated the performance of the dominant-plane detection procedure described in the previous section. The motion of a wheel-driven mobile robot is separated into translational motion and rotational motion, as shown in Fig. 9. Therefore, in the experiments, we prepared image sequences obtained for translational motion and rotational motion without loss of generality.

For the visual representation of the results of dominant-plane detection, the value of d_{ij} in Eq. (16) is normalised in the range of 0-255. The image of the detected dominant plane $D(u,v)$ is defined as $D(i,j) = \frac{d_{ij}\times 255}{\max(d_{ij})}$. For $d_{ij} = |l_{ij} - m|$ in Eq. (16), the area where $d_{ij} \approx 0$ corresponds to the dominant plane.

We evaluate the accuracy of the results of dominant-plane detection by comparing with the ground truth of the dominant plane and the results. Setting $G(u,v)$ to be the ground truth of the dominant plane,

$$G(u,v) = \begin{cases} 0 \text{ if } (u,v)^\top \text{ is a point in the dominant plane,} \\ 255 \text{ if } (u,v)^\top \text{ is a point in the obstacle area.} \end{cases} \tag{25}$$

Then, we calculate the detection ratio R:

$$R = \frac{||\{(u,v)^\top|(D(u,v) < 50) \wedge (G(u,v) = 0)\}||}{||\{u,v|G(u,v) = 0\}||} \times 100, \tag{26}$$

where $||A||$ is the number of elements in A and the value of 50 is the threshold for the detection of the dominant plane. If the gray value of the normalised image $D(i,j)$ is 50 or less, the area is detected as the dominant plane. The ratio R of Eq. (26) indicates how accurately the dominant plane can be detected compared with the ground truth.

We use a real image sequence obtained from a camera mounted on a mobile robot in the evaluation experiments. The image sequences and optical flow fields are shown in Figs. 10 and 14 for translational motion and rotational motion, respectively. Figures 11 - 13 show the experimental results for $D(u,v)$, the norm l_{ij}, and d_{ij}, respectively, for translational motion. The corresponding results for rotational motion are shown in Figs. 15 - 17.

Fig. 10. Input optical flow fields to our ICA-based algorithm for translational motion in a real environment. The first, second, and third rows show the image sequences for translational motion, the computed optical flow field $\dot{\boldsymbol{u}}$ for the first input signal, and the estimated planar flow field $\hat{\boldsymbol{u}}$ for the second input signal, respectively.

Table 1. Detection ratio R of dominant plane using a real image sequence observed during translational motion.

	Fig. 11(a)	Fig. 11(b)	Fig. 11(c)	Fig. 11(d)
Detection ratio [%]	89.9	93.1	95.7	85.0

Table 2. Detection ratio R of dominant plane using a real image sequence observed during rotational motion.

	Fig. 15(a)	Fig. 15(b)
Detection ratio [%]	96.7	94.7

In Figs. 11 and 15, the first, second, third, and fourth rows show the output signal $\boldsymbol{v}_\alpha$, the output signal $\boldsymbol{v}_\beta$, the image of the dominant plane, and the results overlaid on the observed images, respectively. Tables 1 and 2 show the detection ratio R, defined in Eq. (26), for these experiments. The ground truth of the dominant plane is manually extracted from the captured images.

For a real image sequence, our algorithm can detect the dominant plane correctly. The dominant plane in the result corresponds to the ground plane in the robot workspace. Therefore, our algorithm enables the detection of a corridor path in a robot workspace for mobile robot navigation.

5.2. *Multiple Plane Detection*

The image in Fig. 18 is captured by a camera mounted on a mobile robot in a real environment. The robot moves forward and two obstacles are placed in front of the robot. These results also show that our algorithm accurately detects multiple

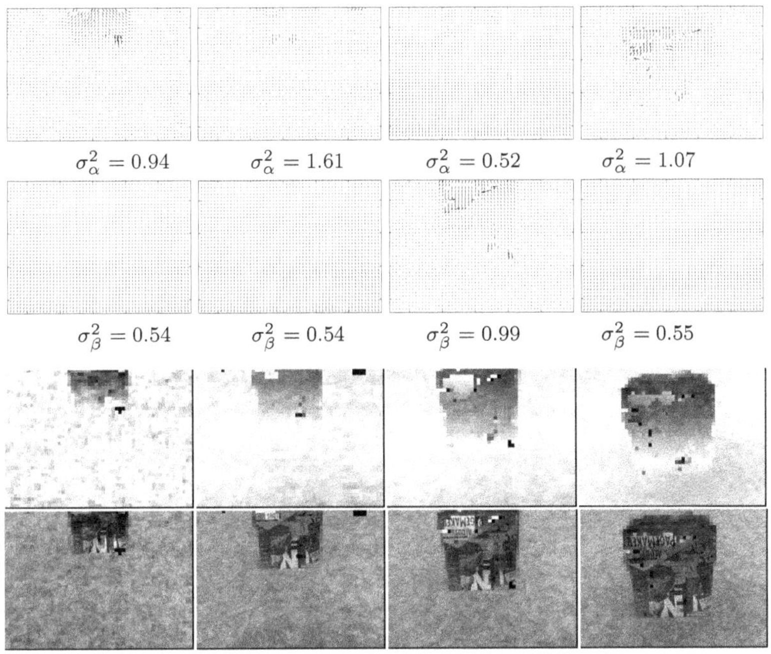

(a) (b) (c) (d)

Fig. 11. Output optical flow fields and detected dominant plane for the input optical flow fields Fig. 10. The first, second, third, and fourth rows show the output signal $\boldsymbol{v}_\alpha$, the output signal $\boldsymbol{v}_\beta$, images of the dominant plane, and the obstacle area in the experimental results overlapped with the images, respectively.

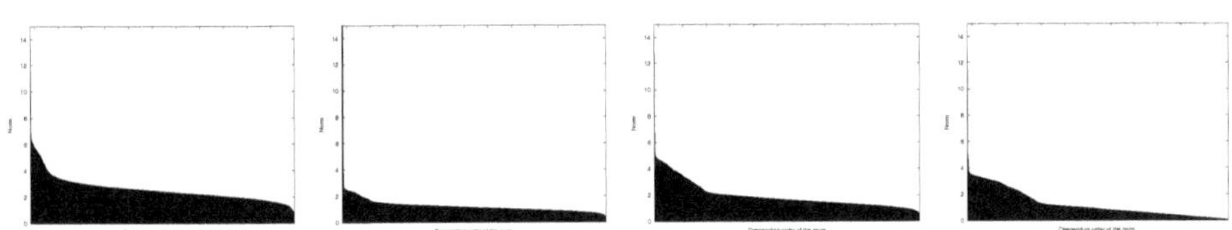

(a) (b) (c) (d)

Fig. 12. Sorted norm $\boldsymbol{l}$ of output $\boldsymbol{v}_\alpha$. (a) - (d) correspond to Figs. 11(a) - (d), respectively. (a) Median value $m = 2.41$. (b) Median value $m = 1.13$. (c) Median value $m = 1.63$. (d) Median value $m = 0.89$.

planes and layers for real image sequences. These noises are caused by the error of the optical flow computation.

5.3. *Hierarchy of Dominant Plane*

We set the maximum layer $L = 3$. For the visual representation of the results of obstacle detection, the value of d_{ij} in Eq. (16) is normalised in the range from 0

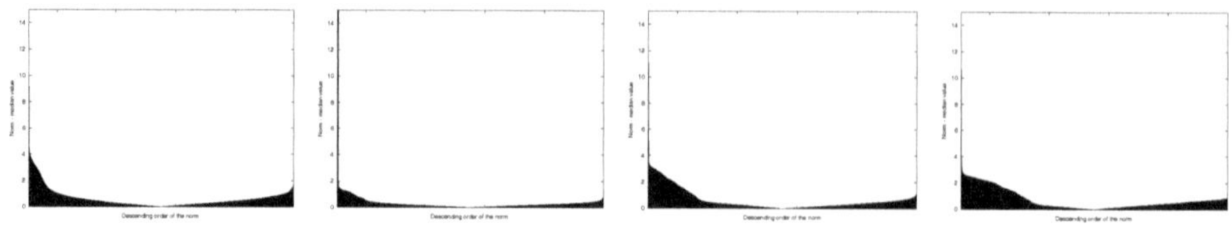

(a) (b) (c) (d)

Fig. 13. Distance d_{ij} between norm $\boldsymbol{l}_{ij}$ and median value m in Eq. (16). The area where $d_{ij} \approx 0$ corresponds to the dominant plane.

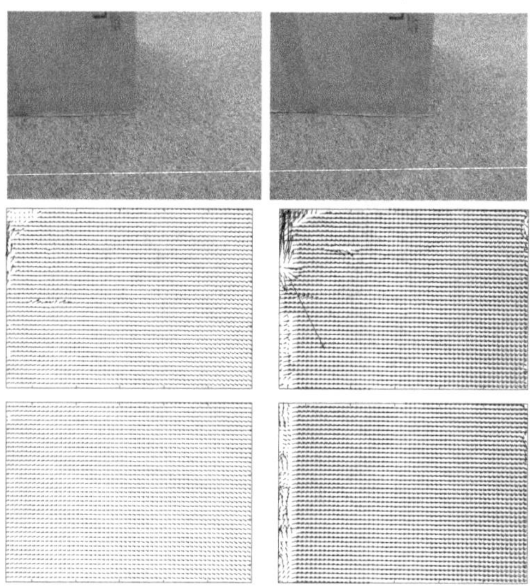

Fig. 14. Optical flow fields input to our ICA-based algorithm for rotational motion in a real environment. The first, second, and third rows show the image sequences for rotational motion, the computed optical flow field $\dot{\boldsymbol{u}}$ for the first input signal, and the estimated planar flow field $\hat{\boldsymbol{u}}$ for the second input signal, respectively.

to 255. The image of the detected obstacle $D^l(u, v)$ on the lth layer is defined as

$$D^l(i, j) = \frac{d_{ij} \times 255}{\max(d^l_{ij})}, \tag{27}$$

where d^l_{ij} is d_{ij} on the lth layer.

Figure 19 shows the captured images in a real environment using a mobile robot. The mobile robot moves toward the obstacle in front of the robot. Figures 20 and 21 show computed optical flow fields and detected obstacles on each layer, respectively. These examples show that on each layer the obstacle regions are detected. Therefore, the algorithm detects the global configuration of obstacles from higher layer images, although the lower layer images allow us to detect the detailed configuration of obstacles.

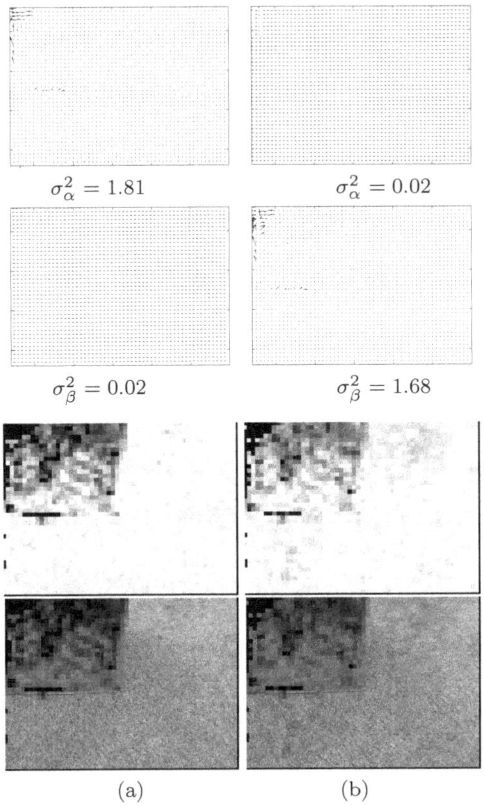

Fig. 15. Output optical flow fields and detected dominant plane for the input optical flow fields of Fig. 14. The first, second, third, and fourth rows show the output signal $\boldsymbol{v}_\alpha$, output signal $\boldsymbol{v}_\beta$, the images of the dominant plane, and the obstacle area of the results overlapped with the images, respectively.

For the performance evaluation, we define a probabilistic inclusive relation between regions in Euclidean space as $P(B \subset A) = \frac{|A \bigcap B|}{|B|}$. $P(B \subset A)$ satisfies the relations

$$P(B \subset A) = \begin{cases} 0, & \text{if } A \cap B = \emptyset, \\ 1, & \text{if } B \subset A, \\ a(0 < a < 1), & \text{otherwise.} \end{cases} \tag{28}$$

for a pair of sets A and B such that $|A| \geq |B|$, where $|A|$ is the measure of the set A.

Definition 1. If $1 \geq P(B \subset A) \geq (1 - \varepsilon)$, we say that B is included in A with the dissimilarity ε. Furthermore, if A is included in B with dissimilarity ε we express $B \sqsubset_\varepsilon A$.

Figure 22 shows geometric relations between A and B.

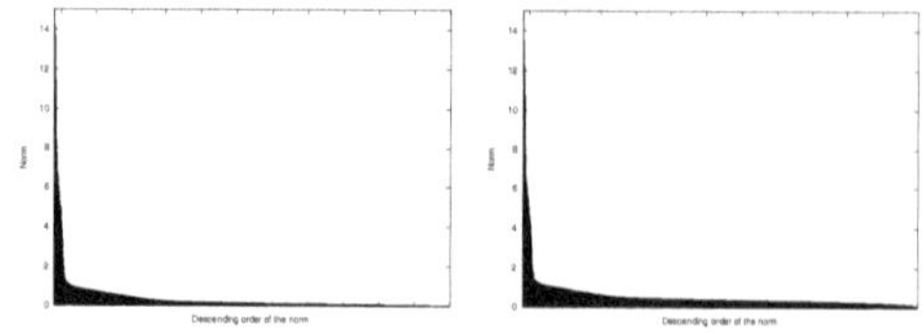

(a) (b)

Fig. 16. Sorted norm $\boldsymbol{l}$ of output $\boldsymbol{v}_\alpha$. (a) and (b) correspond to Figs. 15(a) and (b), respectively. (a) Median value $m = 0.14$. (b) Median value $m = 0.39$.

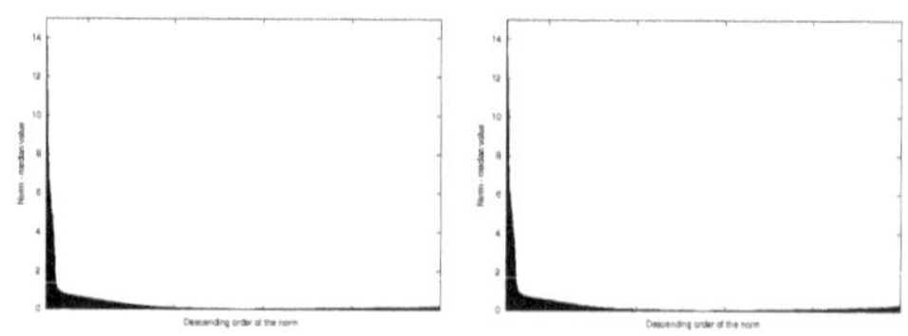

(a) (b)

Fig. 17. Distance d_{ij} between norm $\boldsymbol{l}_{ij}$ and median value m in Eq. (16). The area where $d_{ij} \approx 0$ corresponds to the dominant plane.

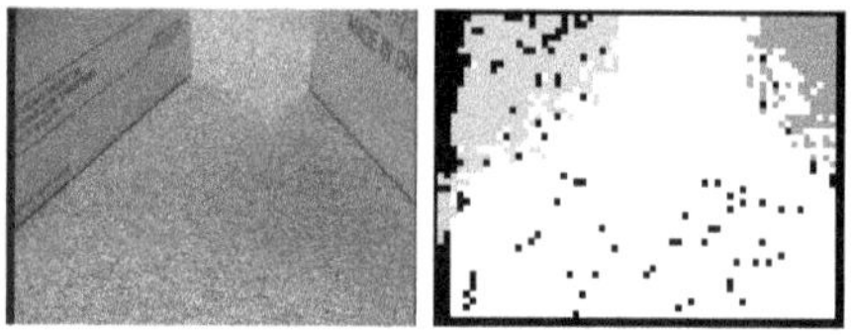

Fig. 18. Image and result for multiple plane detection in a real environment. The white area is the first dominant plane. The light-gray and dark-gray areas are the second and third dominant planes.

Using this probabilistic inclusive relation, we examine whether the dominant plane in the lower layer is included in the higher level for most of the image. Table 3 shows inclusive relation for dominant planes of Fig. 21 detected from the image sequence of Fig. 19. This example shows that in each layer the obstacle regions are detected. Therefore, the algorithm detects the global configuration of obstacles from higher layer images, although the lower layer images allow us to detect the detailed configuration of obstacles.

Table 3. Experimental results of probabilistic inclusive relation for the image sequence of Fig. 19.

Layer	$P(D^0 \subset D^1)$	$P(D^1 \subset D^2)$	$P(D^2 \subset D^3)$
Mobile robot	0.917	0.704	0.802

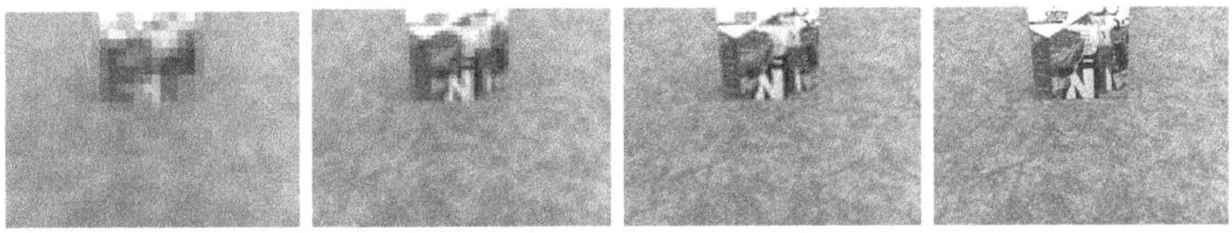

(a) I^3 (b) I^2 (c) I^1 (d) Image I

Fig. 19. Pyramidal representation of captured images in a real environment.

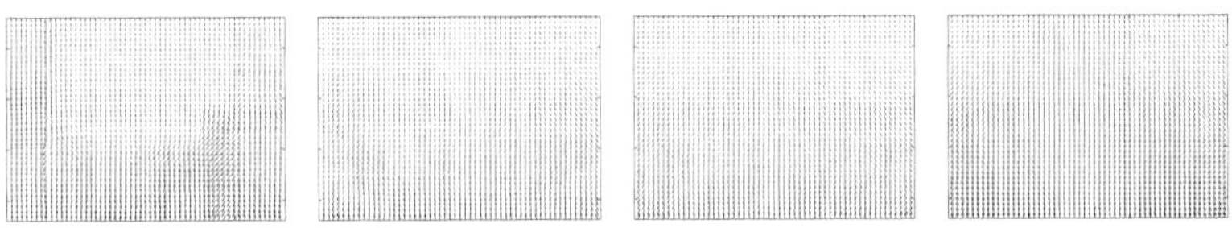

(a) $\boldsymbol{u}^3$ (b) $\boldsymbol{u}^2$ (c) $\boldsymbol{u}^1$ (d) $\boldsymbol{u}^0$

Fig. 20. Layer optical flow fields. $\boldsymbol{u}^l$ is the optical flow field from I^l in Fig. 19.

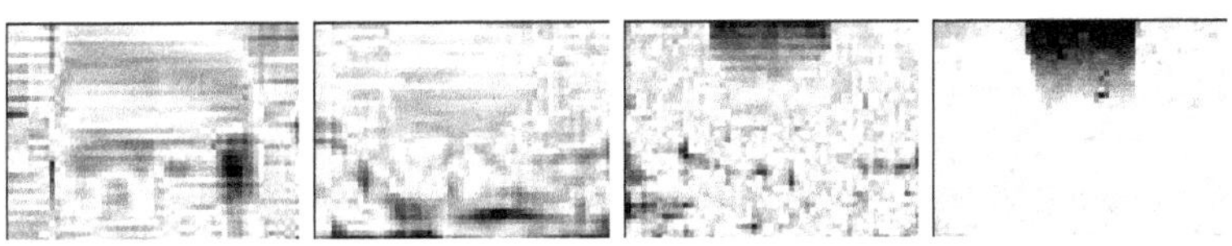

(a) D^3 (b) D^2 (c) D^1 (d) D^0

Fig. 21. Detected obstacle on each layer.

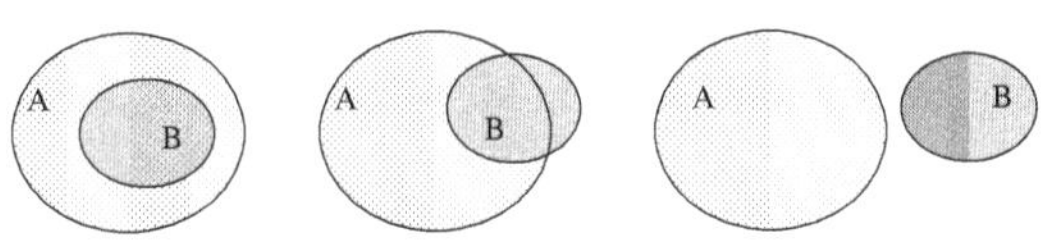

(a) (b) (c)

Fig. 22. Geometric expressions of inclusive relations. (a) $A \supset B$. (b) $(A \cap B) \subset A$. (c) $A \cap B = \phi$.

6. Summary

Using a statistical model, we developed an algorithm for dominant-plane detection from a sequence of images observed through an uncalibrated moving camera. The use of optical flow enables a robot to detect a feasible region in which it can move using simple algorithms based on ICA and RANSAC. Experimental results indicate that the proposed algorithm can detect the dominant plane from real image sequences. These experimental results support the application of our method to the navigation and path planning of a mobile robot mounting a vision system. The assumption that the optical flow field is a linear combination of the optical flow fields on a dominant plane and obstacle region enables to separate these two regions using ICA.

The two-dimensional version of ICA for signal separation involves the extraction of independent features from images. Our algorithm is an application of ICA for vector-valued images. This is an extension of ICA-based image analysis to vector-valued images. In the application, we proposed an ordering technique for independent components of vector-valued images. The ordering allowed us to separate obstacles and ground planes from a sequence of images captured by a camera mounted on an autonomous robot. For each image in a sequence, it was shown that the dominant plane corresponds to an independent component. This relationship provides us with a statistical definition of the dominant plane.

The combination of our ICA based image analysis and multiresolution image representation provides a method which allows us to detect hierarchical configuration of obstacles. This hierarchical processing is achieved concurrently for the simultaneous detection of the local and global obstacle configurations in the work space. We conclude that the process emulates biological perception mechanism for visual behaviour.

The dimension of the optical flow field of an $N \times N$ image is $2(N \times N)$. Therefore, $2(N \times N)$ is the maximum number of independent components needed to express the optical flow field of an image. Our results show that as the first approximation of the full decomposition into independent components, this optical flow field is separated into two components corresponding to the ground plane and obstacles. This geometrical property supports our assumption that the optical flow field is a linear combination of the optical flow fields on a dominant plane and obstacle region for the optical flow field used for robot navigation. Furthermore, the experimental results show that the first two independent components are dominant among all the components that express the flow field in a screen. Moreover, by successive decomposition of the component corresponding to the obstacles, it is possible to classify the flow field derived from obstacles.[46] The robot moves on the dominant plane while avoiding obstacles. Therefore, the robot does not move too close to the obstacles. We have proposed a motion strategy for a mobile robot in references.[1,47]

Whereas Jabri et al. discussed the relationships between optical flow and the MST area in the visual cortex,[8] we presented an algorithm for mobile robot navigation using optical flow and ICA. The algorithm separates optical flow fields by MST-like processing. We demonstrated that the concept of biology-inspired vision is applicable for a vision system in robot navigation.

References

1. N. Ohnishi and A. Imiya, Featureless robot navigation using optical flow, *Connection Science.* **17**, 23–46, (2005).
2. C. Beck, T. Gottbehuet, and H. Neumann. Integration of multiple temporal and spatial scales for robust optic flow estimation in a biologically inspired algorithm. In *International Conference on Compuer Analysis of Image and Pattern*, (2007).
3. J. Tsotsos. On attention, saliency, recognition and feature binding. In *International Conference on Computer Vision Systems*, (2007).
4. A. Hyvarinen and E. Oja, Independent component analysis: algorithms and application, *Neural Networks.* **13**, 411–430, (2000).
5. K.-Y. Park, M. Jabri, S.-Y. Lee, and T. J. Sejnowski. Independent components of optical flows have MSTd-like receptive fields. In *2nd International Workshop on ICA and Blind Signal Separation*, pp. 597–601, (2000).
6. R. S. Zemel and T. J. Sejnowski, A model for encoding multiple object motions and self-motion in area MST of primate visual cortex, *Neuroscience.* **18**, 531–547, (1998).
7. N. Ohnishi and A. Imiya. Independent component analysis of layer optical flow and its application. In *2nd International Symposium on Brain, Vision and Artificial Intelligence*, LNCS 4729, pp. 171–180, (2007).
8. M. A. Jabri, K.-Y. Park, S.-Y. Lee, and T. J. Sejnowski. Properties of independent components of self-motion optical flow. In *IEEE International Symposium on Multiple-Valued Logic*, pp. 355–362, (2000).
9. L. M. Vaina, S. A. Beardsley, and S. K. Rushton, *Optic flow and beyond.* (Kluwer Academic Publishers, 2004).
10. I. Biederman, Human image understanding: Recent research and a theory, *Computer Vision, Graphics, and Image Processing.* **32**, 29–73, (1985).
11. R. G. Domenella and A. Plebe. A neural model of human object recognition development. In *1st International Symposium on Brain, Vision and Artificial Intelligence*, pp. 116–125, (2005).
12. J. Y. Bouguet. Pyramidal implementation of the Lucas Kanade feature tracker description of the algorithm. In *Intel Corporation, Microprocessor Research Labs, OpenCV Documents*, (1999).
13. M. R. Mahzoun, J. Kim, S. Sawazaki, K. Okazaki, and S. Tamura, A scaled multigrid optical flow algorithm based on the least rms error between real and estimated second images, *Pattern Recognition.* **32**, 657–670, (1999).
14. B. Kuipers and Y.-T. Byun, A robot exploration and mapping strategy based on a semantic hierarchy of spatial representations, *Robotics and Autonomous Systems.* **8**, 47–63, (1991).
15. T. Wagner, U. Visser, and O. Herzog, Egocentric qualitative spatial knowledge representation for physical robots, *Robotics and Autonomous Systems.* **49**, 25–42, (2004).
16. S. Baker and I. Matthews, Lucas-kanade 20 years on: a unifying framework, *International Journal of Computer Vision.* **56**, 221–255, (2004).
17. N. D. Guilherme and C. K. Avinash, Vision for mobile robot navigation: A survey, *IEEE Transaction on Pattern Analysis and Machine Intelligence.* **24**, 237–267, (2002).
18. N. Winters, J. Gaspar, G. Lacey, and J. Santos-Victor. Omni-directional vision for robot navigation. In *IEEE Workshop on Omnidirectional Vision*, pp. 21–28, (2000).
19. J. L. Barron, D. J. Fleet, and S. S. Beauchemin, Performance of optical flow techniques, *International Journal of Computer Vision.* **12**, 43–77, (1994).
20. B. K. P. Horn and B. G. Schunck, Determining optical flow, *Artificial Intelligence.* **17**, 185–203, (1981).

21. B. Lucas and T. Kanade. An iterative image registration technique with an application to stereo vision. In *7th International Joint Conference of Artificial Intelligence*, pp. 674–679, (1981).
22. H. A. Mallot, H. H. Bulthoff, J. J. Little, and S. Bohrer, Inverse perspective mapping simplifies optical flow computation and obstacle detection, *Biological Cybernetics*. **64**, 177–185, (1991).
23. J. Kaufhold, R. Collins, A. Hoogs, and P. Rondot. Recognition and segmentation of scene content using region-based classification. In *International Conference on Pattern Recognition*, pp. 755–760, (2006).
24. M. I. A. Lourakis, A. A. Argyros, and S. C. Orphanoudakis. Plane detection in an uncalibrated image pair. In *British Machine Vision Conference*, vol. 2, pp. 587–596, (2002).
25. T. S. Huang and A. N. Netravali, Motion and structure from feature correspondences: A review, *Proceedings of the IEEE*. **82**, 252–268, (1994).
26. P. F. Sturm and S. J. Maybank. On plane-based camera calibration: A general algorithm, singularities, applications. In *International Conference on Computer Vision and Pattern Recognition*, pp. 432–437, (1999).
27. N. Pears and B. Liang. Ground plane segmentation for mobile robot visual navigation. In *IROS*, pp. 1513–1518, (2001).
28. S. B. Kang and R. Szeliski, *3D environment modeling from multiple cylindrical panoramic images*, In eds. R. Benosman and S. B. Kang, *Panoramic Vision: Sensors, Theory, Applications*, pp. 329–358. Springer-Verlag, (2001).
29. F. Fraundorfer. A map for mobile robots consisting of a 3D model with augmented salient image features. In *26th Workshop of the Austrian Association for Pattern Recognition,*, pp. 249–256, (2002).
30. W. Enkelmann, Obstacle detection by evaluation of optical flow fields from image sequences, *Image and Vision Computing*. **9**, 160–168, (1991).
31. K. Kawamoto, N. Ohnishi, A. Imiya, R. Klette, and K. Hirota, Motion-based template matching for obstacle detection, *Journal of Advanced Computational Intelligence and Intelligent Informatics*. **8**(5), 469–477, (2004).
32. E. Mémin and P. Pérez, Dense estimation and object-based segmentation of the optical flow with robust techniques, *IEEE Transactions on Image Processing*. **7**, 703–719, (1998).
33. J. Santos-Victor and G. Sandini, Uncalibrated obstacle detection using normal flow, *Machine Vision and Applications*. **9**, 130–137, (1996).
34. M. Zucchelli, J. Santos-Victor, and H. I. Christensen. Multiple plane segmentation using optical flow. In *British Machine Vision Conference*, (2002).
35. Y. Weiss. Smoothness in layers: Motion segmentation using nonparametric mixture estimation. In *International Conference on Computer Vision and Pattern Recognition*, pp. 520–527, (1997).
36. J. Y. A. Wang and E. H. Adelson, Representing moving images with layers, *IEEE Transactions on Image Processing*. **3**, 625–638, (1994).
37. D. Amanatidis, D. Tsaptsinos, P. Giaccone, and G. Jones, Optimizing motion and colour segmented images with neural networks, *Neurocomputing*. **62**, 197–223, (2004).
38. H.-J. Park and T.-W. Lee, Capturing nonlinear dependencies in natural images using ICA and mixture of Laplacian distribution, *Neurocomputing*. **69**, 1513–1528, (2006).
39. Q. Ke and T. Kanade. Transforming camera geometry to a virtual downward-looking camera: Robust ego-motion estimation and ground-layer detection. In *IEEE International Conference on Computer Vision and Pattern Recognition*, pp. 390–397, (2003).

40. N. Ohnishi and A. Imiya, Dominant plane detection from optical flow for robot navigation, *Pattern Recognition Letters.* **27**, 1009–1021, (2006).
41. M. A. Fischler and R. C. Bolles, Random sample consensus: A paradigm for model fitting with applications to image analysis and automated cartography, *Communications of the ACM.* **24**, 381–395, (1981).
42. C. Beck, P. Bayerl, and H. Neumann. Neural mechanisms of motion integration and disambiguation: from neuroscience to real-world applications. In *ICVS Workshop on Computational Cognitive Neuroscience to Computer Vision*, (2007).
43. N. Bruce and J. Tsotsos. Attention based on information maximization. In *ICVS Workshop on Computational Cognitive Neuroscience to Computer Vision*, (2007).
44. A. D. Back and T. P. Trappenberg, Selecting inputs for modeling using normalized higher order statistics and independent component analysis, *IEEE Transactions on Neural Networks.* **12**, 612–617, (2001).
45. A. Hyvarinen, Survey on independent component analysis, *Neural Computing Surveys.* **2**, 94–128, (1999).
46. N. Ohnishi and A. Imiya. Model-based plane-segmentation using optical flow and dominant plane. In *3rd International Conference on MIRAGE*, LNCS 4418, pp. 295–306, (2007).
47. N. Ohnishi and A. Imiya. Navigation of nonholonomic mobile robot using visual potential field. In *International Conference on Computer Vision Systems*, (2007).

CHAPTER 2.7

EXTRACTING PARTS OF 2D SHAPES USING LOCAL AND GLOBAL INTERACTIONS SIMULTANEOUSLY

Sibel Tari*

Middle East Technical University, Department of Computer Engineering, Ankara, Turkey 06531

stari@metu.edu.tr

Perception research provides strong evidence in favor of part based representation of shapes in human visual system. Despite considerable differences among different theories in terms of how part boundaries are found, there is substantial agreement on that the process depends on many local and global geometric factors. This poses an important challenge from the computational point of view. In the first part of the chapter, I present a novel decomposition method by taking both local and global interactions within the shape domain into account. At the top of the partitioning hierarchy, the shape gets split into two parts capturing, respectively, the gross structure and the peripheral structure. The gross structure may be conceived as the least deformable part of the shape which remains stable under visual transformations. The peripheral structure includes limbs, protrusions, and boundary texture. Such a separation is in accord with the behavior of the artists who start with a gross shape and enrich it with details. The method is particularly interesting from the computational point of view as it does not resort to any geometric notions (e.g. curvature, convexity) explicitly. In the second part of the chapter, I relate the new method to PDE based shape representation schemes.

1. Introduction

Perception research provides strong evidence in favor of part based representation of shapes in human visual system.[1] Recent work using single cell recordings in area V4 in the primate visual cortex supports part based coding at intermediate levels.[2] Many influential shape representation theories either explicitly or implicitly assume an organization in terms of constituent components. In Binford;[3] Marr and Nishiara;[4] and Biederman,[5] shape is represented via pre-defined simple shapes (primitives) and their spatial layout. Medial axis or local symmetry set (one of the influential ideas in shape representation) is closely connected with the notion of parts. Large number of computational methods utilize medial axis to infer part

*Work is supported in part by TUBITAK 108E015.

structure.[6–11] In a recent work, Super[12] presents a quite successful part based recognition scheme.

There are powerful theories accompanied by computational implementations on what constitutes a good partition without resorting to predefined shapes or categorical units, e.g.[13–19] Despite considerable differences among different theories in terms of how part boundaries are found, there is substantial agreement on that the process depends on many geometric factors both at the global and the local levels.[1,13,18–21] Indeed, part of the difficulty in devising computational mechanisms for shape decomposition lies in the difficulty in integrating local boundary concavities with non-local shape descriptions. Recent works by Mi and DeCarlo[7] and Zeng et al.[6] address this challenging issue using contour curvature and local symmetry axis simultaneously. In another recent work, Xu, Liu and Tang[22] combine effectiveness of both local and global features for matching shapes. One of the successful recognition schemes, the shape context by Belongie, Malik and Puzicha,[23] is based on quantifying non-local interactions among boundary points. Moreover, some of the successful methods for partitioning images into pixel groups[24,25] are based on non-local relations.

The rest of the chapter is organized as follows. In 2, the new decomposition method is presented. The new method takes both local and global interactions within the shape domain into account. At the top of the partitioning hierarchy, the shape gets split into two parts capturing, respectively, the gross structure and the peripheral structure. The gross structure may be conceived as the least deformable part of the shape which remains stable under visual transformations. The peripheral structure includes limbs, protrusions, and boundary texture. Such a separation is in accord with the experimental studies suggesting that the global shape is processed before the details;[20,21] and with the behavior of the artists who start with a gross shape and enrich it with details.[26]

In 3, the new method which is formulated in a discrete setting is related to PDE based shape representation approach with particular emphasis given to a recent skeleton based scheme which I had previously developed with my student Cagri Aslan.

2. The New Method

The basic idea is to create a field within the shape domain with emergent structures capturing the parts automatically. This field is computed by minimizing an energy which captures both local and global; and both region and boundary-based interactions among shape points.

Let us define a function ω defined on a discrete shape domain Ω as the minimizer of an energy $E(\omega)$. Let this energy be a sum of a region-based energy $E_{Reg}(\omega)$ and a boundary-based energy $E_{Bdy}(\omega)$; and the region-based energy $E_{Reg}(\omega)$ is a sum of two energies which model the global (G) and the local (L) interactions within the shape domain Ω:

$$\begin{aligned} E(\omega) &= E_{Reg}(\omega) + w_{Bdy} E_{Bdy}(\omega) \\ &= E^G_{Reg}(\omega) + E^L_{Reg}(\omega) + w_{Bdy} E_{Bdy}(\omega) \end{aligned} \tag{1}$$

Assuming that each of the three terms in (1) can be expressed as a sum of energies defined at each pixel i, j, the following form is obtained:

$$E(\omega) = \sum_{i,j \in \Omega} E^G_{Reg}(\omega_{i,j}) + E^L_{Reg}(\omega_{i,j}) + w_{Bdy} E_{Bdy}(\omega_{i,j}) \tag{2}$$

In the absence of any specific purpose or bias, equal importance can be given to both the region and the boundary by setting $w_{Bdy} = 2$. For computational reasons, it is preferable to choose a quadratic form for each energy. Let $E^G_{Reg}(\omega_{i,j})$ be:

$$E^G_{Reg}(\omega_{i,j}) = \frac{1}{|\Omega|} \left(\sum_{k,l \in \Omega} \omega_{k,l} \right)^2 \tag{3}$$

The minimizer $\omega_{i,j}$ of (3) satisfies:

$$\frac{1}{|\Omega|} \sum_{k,l \in \Omega} \omega_{k,l} = 0 \tag{4}$$

The condition satisfied by the minimizer of $E^G_{Reg}(\omega_{i,j})$ is independent of the pixel location (i, j) and it explicitly states that the global average over the shape domain should be zero. It forces ω to attain both positive and negative values within the shape domain Ω. This behavior, when complemented with the behavior induced by the other terms, will be shown to be quite instrumental in obtaining robust and parameter free separation of the gross structure and the peripheral structure.

The second term of the region-based energy, E^L_{Reg}, has the following form:

$$E^L_{Reg}(\omega_i, j) = -0.5\,(\omega_{i+1,j} \cdot \omega_{i-1,j} + \omega_{i,j+1} \cdot \omega_{i,j-1}) \tag{5}$$

Clearly, $E^L_{Reg}(\omega_i)$ is minimized when the values of the neighboring pixels are similar. Thus the second term of the energy imposes smoothness on ω. The condition for the minimizer of E^L_{Reg} is not straightforward to calculate as that of E^G_{Reg}. First, a local continuous approximation at location i, j is considered with the help of Taylor series. Second, the Gateaux derivative is calculated. Third, the Gateaux derivative is discretized and set to *zero*, to obtain the condition for the minimizer:

$$(-2 + 4) * \omega_{i,j} - \omega_{i-1,j} - \omega_{i+1,j} - \omega_{i,j-1} - \omega_{i,j+1} = 0 \tag{6}$$

The boundary based energy $E_{Bdy}(\omega)$ is chosen as a measure of pairwise interaction between two properly chosen boundary points such that the pairs indicate parts. One motivation to consider pairwise interaction between two boundary points comes from the well-established minima rule[13,14] which is used in many computational procedures for shape decomposition. However, computationally, it is not an easy task to model pairwise interactions among boundary points. These interactions are neither local nor global. A simple alternative is constructed by exploiting the connection between the concept of pairwise interaction among boundary points and the shape skeleton.[27] With the help of this connection, E_{Bdy} is expressed as an energy defined over the entire shape domain.

The connection can be explained with the help of the grass-fire model by Blum.[27] Assume that one initiates fire fronts at time $t = 0$ along all the points of the shape boundary, and lets these fronts propagate toward the center of the shape at a uniform speed. The locus of points where these fronts meet and extinguish, defines the shape skeleton. Each skeleton point is formed as a result of interaction between two or more boundary points. During the course of the propagation, the time t may be thought of as a function $t_{i,j}$ defined over the shape domain by setting the value to the time when the propagating fronts passes through the pixel (i, j). The value of $t_{i,j}$ will be proportional to the minimum distance from (i, j) to the nearest boundary points. Skeleton points are the ones which are equidistant from at least two boundary points[27] (Fig. 1.)

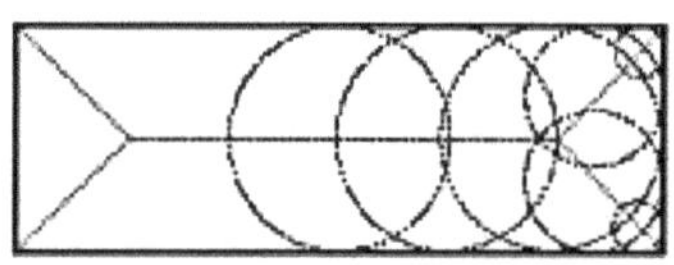

Fig. 1. Each skeleton point is formed as a result of interaction between two or more boundary points. Skeleton points are the ones which are equidistant from at least two boundary points.[27]

This insight enables the expression of $E_{Bdy}(\omega)$ as a quadratic energy defined over the entire shape region Ω, as follows:

$$E_{Bdy}(\omega_{i,j}) = (\omega_{i,j} - t_{i,j})^2 \tag{7}$$

It is straightforward to calculate the condition for the minimizer of the $E_{Bdy}(\omega_{i,j})$ given in (7) as:

$$(\omega_{i,j} - t_{i,j}) = 0 \tag{8}$$

In the absence of other terms, (8) states that the field ω should be equal to the distance transform defined over Ω. Putting together all three competing terms, (4,6,8), the first order derivative w.r.t. each unknown $\omega_{i,j}$ takes the following form:

$$\begin{aligned}\frac{\partial E}{\partial \omega_{i,j}} = \frac{1}{|\Omega|}\left(\sum_{k,l\in\Omega} \omega_{k,l}\right) + (w_{Bdy} - 2 + 4)\omega_{i,j} - w_{Bdy} t_{i,j}\\ - \omega_{i-1,j} - \omega_{i+1,j} - \omega_{i,j-1} - \omega_{i,j+1}\end{aligned} \tag{9}$$

Setting the derivative equal to *zero* yields that the minimizer of $E(\omega)$ satisfies the following condition, at each pixel (i, j):

$$w_{Bdy}t_{i,j} = (w_{Bdy} - 2 + 4)\omega_{i,j} - \omega_{i-1,j} - \omega_{i+1,j} - \omega_{i,j-1} - \omega_{i,j+1} + \frac{1}{|\Omega|}\left(\sum_{k,l\in\Omega} \omega_{k,l}\right) \tag{10}$$

Recall that, in the absence of any specific purpose or bias, the weight w_{Bdy} is set to 2 to give equal importance to both the region and the boundary. Thus, $\omega_{i,j}$ is computed by solving (11) given below, at all the pixels simultaneously, assuming that the values are zero at the boundary pixels.

$$t_{i,j} = 4\omega_{i,j} - \omega_{i-1,j} - \omega_{i+1,j} - \omega_{i,j-1} - \omega_{i,j+1} + \frac{1}{|\Omega|}\left(\sum_{k,l\in\Omega} \omega_{k,l}\right) \tag{11}$$

The field ω, computed by solving (11), is depicted in Fig. 2 for two sample shapes. It attains both negative and positive values. This behavior is dictated by the global region energy E^G_{Reg} which explicitly states that the global average over the shape domain should be small.

In Fig. 2 (a), the restriction of ω where it is negative is displayed. This set is denoted by Ω^-. It captures the peripheral structure (protrusions, limbs, boundary texture). The darkest blue denotes *zero*; the darkest red denotes the lowest negative value. The removed inner part is the part on which ω is positive; and is denoted by Ω^+. This blob-like part captures the gross structure. The gross structure is the least deformable part of the shape which remains stable under a variety of visual changes as demonstrated in Fig. 3 using eight different instances of a hand silhouette.

In Fig. 2 (b), the absolute value of ω is displayed on the entire shape domain. For visualization purposes, the negative values and the positive values are normalized, separately, to the $[0, 1]$ interval. The darkest blue denotes *zero*; the darkest red denotes *one*. Notice that various local maxima capture intuitive parts such as the body, the head, the tail, and the legs of the horse. These parts can be easily extracted by considering a growth starting from each local maxima. Separate growths from each pair of neighboring maxima meet at a saddle point.

2.1. *Experimental Results and Discussion*

The method is is discussed via a set of highly illustrative examples. These examples are silhouettes collected from various sources.[6,10,28,29] Some of the original silhouettes are modified by the author to obtain shapes with holes, missing and/or extra parts.

In Figs. (4-7), some decomposition results are provided. These results are obtained by applying Matlab's *watershed* command to Ω^-. (This is equivalent to considering a growth starting from each local minima of ω.)

In Fig. 4, the first column depicts the given shape. The second column depicts the normalized absolute value of ω. The third column depicts the parts. Bright

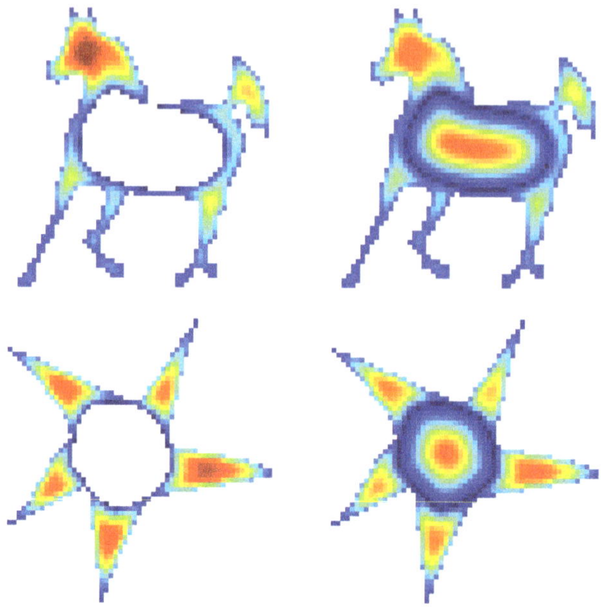

(a) (b)

Fig. 2. The field ω computed by solving (11). For visualization purposes, the values are normalized. (a) The restriction of ω to areas where its values are negative. This part denotes the peripheral structure *i.e.* protrusions, limbs, and boundary texture. The removed inner part on which the values are positive is the gross structure. (b) The absolute value of ω.

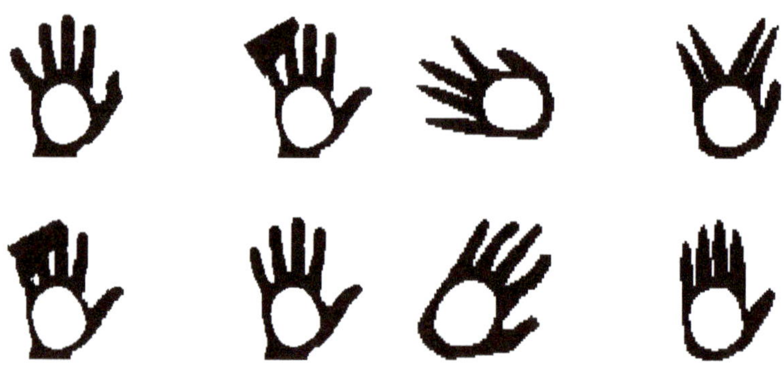

Fig. 3. The areas where $\omega < 0$ is shown in black. The gross structure (inner white blob) may be conceived as the least deformable part of the shape. It remains stable under a variety of changes.

colors are used for parts on which ω is negative; and gray is used for the part on which ω is positive.

The silhouette shown in the first row is a sampled down version of a human silhouette[29] from its original resolution of 414×459 to 60×60. The silhouettes on the remaining rows are drawn by the author to introduce holes, missing portions, occlusions. In each case, semantically meaningful parts (the torso, the legs, the

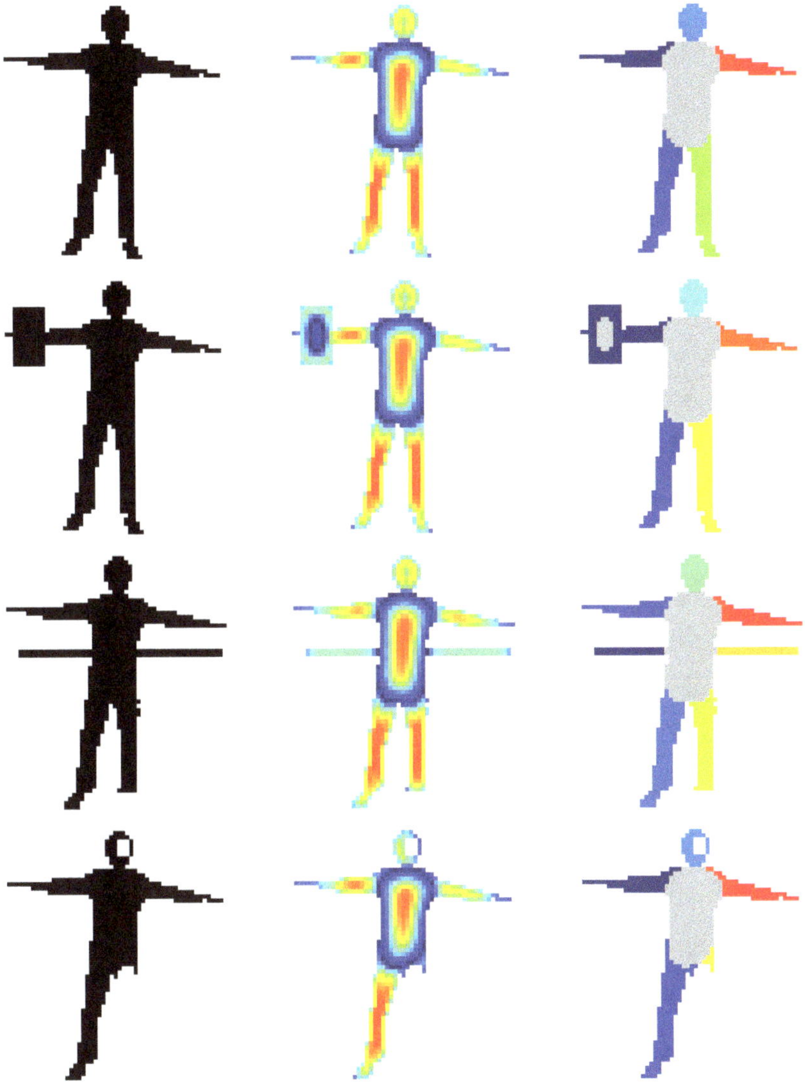

Fig. 4. The method is robust with respect to occlusion, missing data, extra objects, and holes. (a) Input silhouettes. (b) Absolute value of w. (c) Parts extracted by applying Matlab's *watershed* command to w.

arms, the head) are captured. Even though the formulation includes a global term, the local changes, no matter how significant they are, do not affect the detected parts.

In Fig. 5, the applicability of the method when the input consists of disconnected sets (multiple objects in a scene) is demonstrated.

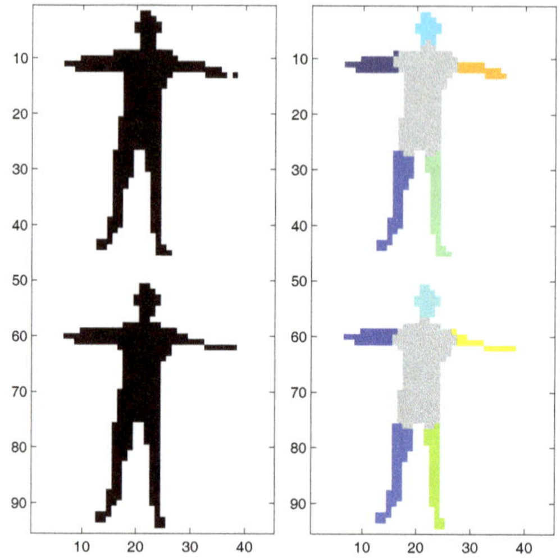

Fig. 5. A scene with two silhouettes. The method is applicable to disconnected sets.

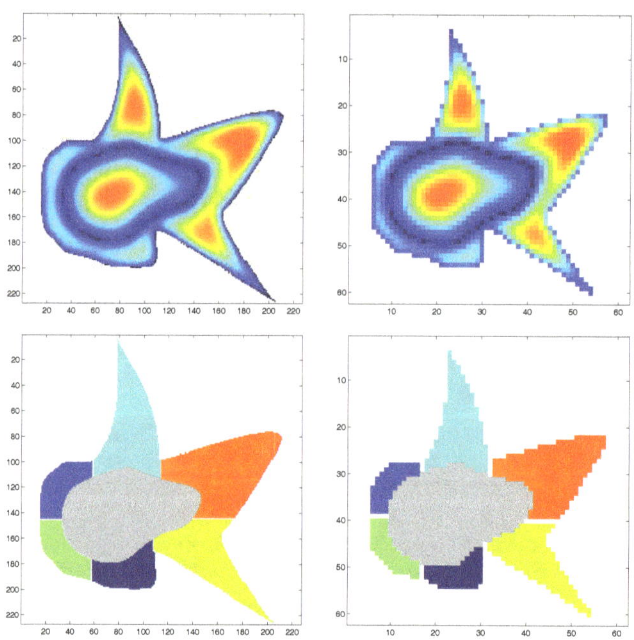

(a) (b)

Fig. 6. The method is robust with respect to resolution changes. An artificial shape[10] on (a) 220×220, (b) 60×60 lattices.

In Fig. 6, the robustness of the method with respect to changes in resolution is demonstrated. An artificial shape from Aslan and Tari[10,30] is used in its original resolution in (a), and in a reduced resolution in (b).

In Fig. 7, the decomposition results for a variety of shapes are provided. The decompositions are consistent; similar shapes are partitioned similarly and the captured parts are compatible with our intuition.

In some cases, as in Fig. 8, the decomposition process starting from each and every local minima (*i.e.* ignoring saliency) may create un-intuitive parts. Normalized absolute value of ω for three sample turtle shapes are depicted in the first row. In all of the three cases, one can easily spot five local maxima in the peripheral part, and one local maxima in the central part. However, the peripheral structure of the first turtle shape is decomposed into seven pieces. For the second turtle, there are six pieces corresponding to the head, the two legs, the two arms, and the tail. For the third turtle, there are five pieces corresponding to the head, the two legs, the two arms. In the third turtle, due to smoother transition between the two legs, the tail part is missed. This produces inconsistencies among silhouettes from the same category. However, inconsistent parts are not as salient as the consistent parts. A

Fig. 7. Sample decompositions. Similar shapes are partitioned similarly; and the parts are compatible with intuition.

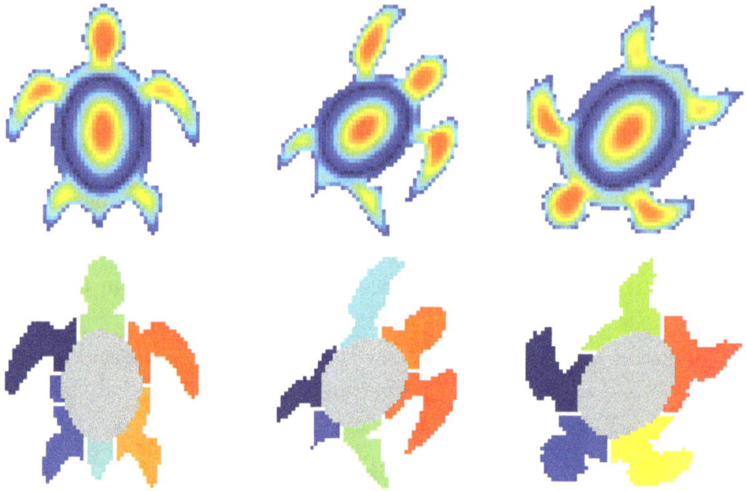

Fig. 8. Un-intuitive parts. See text.

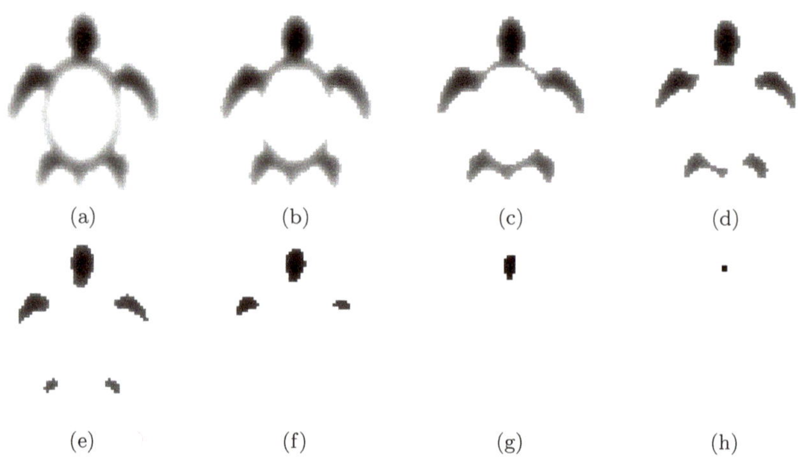

Fig. 9. Saliency of a part. In each figure, the restriction of ω to locations, where its value is less than a given threshold, is depicted. The thresold is increased gradually. (a)-(h) $\omega < -4, -9, -14, -19, -24, -29, -34, -37$, respectively.

clarifying illustration is given in Fig. 9 with the help of the first turtle which is decomposed into eight pieces including the torso. A growth process is simulated starting from Fig. 9 (h) and ending at Fig. 9 (a). In each sub figure, the restriction of ω to the locations, where its value is less than a given threshold is depicted. The threshold is increased gradually. At the first threshold level in (h), only the head appears. At the third threshold level in (f), the arms appear. The head still continues as an individual blob. At the fourth threshold in (e) level the two legs appear. The five pieces remain separate. At the next threshold level in (d) there are still five pieces. However, notice that in somewhere between (e) and (d), the tail piece comes to existence and then gets combined with the rightmost leg. It is appropriate to say that the saliency of the tail is quite low compared to the saliency of the other five pieces, due to its short life span as an individual entity. At the next threshold level in (c), the two legs combine to form a single piece corresponding to the lower body; and the two arms and the head combine to form the upper body. The separation of the peripheral structure into upper and lower body divides the elliptical gross structure from the minor axis. Finally in (a), the upper and lower parts combine to form a single peripheral structure.

In all of the previous examples, the gross structure is shown in gray, as a single part. There may be shapes such that the gross structure is composed of multiple parts, i.e. shapes with strong necks. The parts of the butterfly shape in Fig. 10 (a) are shown in (b) and (c). In (b), the parts in the peripheral structure are shown in bright colors; the gross structure is in gray. In (c) the parts in the gross structure are shown in bright colors; the peripheral structure is in gray. For reference, the restriction of ω to the peripheral structure, i.e. where it is negative, is shown in (d). As the neck which connects the two lobes of the butterfly gets thinner, the gross structure may be split into two disjoint sets. (See Fig. 11.)

2.2. *Comparison to Recent Decomposition Methods by Mi and DeCarlo[7] and Zeng et al.[6]*

In two recent papers, Mi and DeCarlo,[7] and Zeng *et al.*[6] present decomposition methods which exploit both the skeleton and the boundary curvature information.

(a) (b) (c) (d)

Fig. 10. A shape whose gross structure is composed of two blobs. (a) A butterfly shape on a 60×60 lattice. (b-c) The parts. (d) The restriction of ω to areas where the values are negative.

(a) (b) (c) (d)

Fig. 11. When the neck gets thinner, the gross structure may split into two disjoint sets.

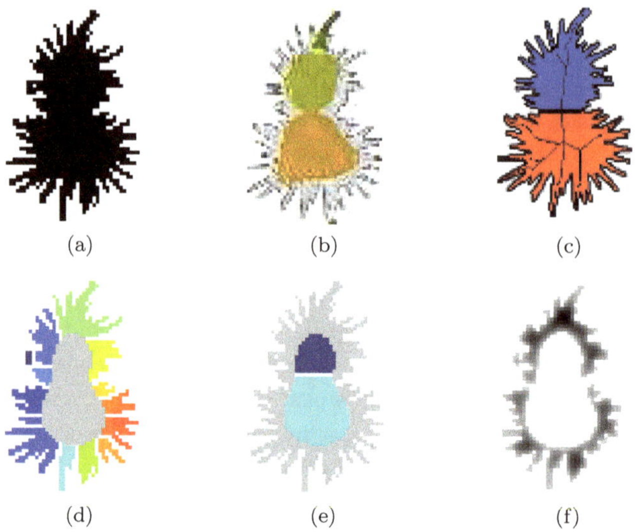

Fig. 12. A noisy shape composed of two blobs. (a) The prickly pear on a 80×80 lattice. (b-c) The decompositions presented in[7] and,[6] respectively. (d-e) The decomposition by the new method. (f) The restriction of ω to where $\omega < -3$. See text. [(b-c) taken from the original sources[7] and[6]]

Neither of the methods consider a fully global context. It is worth comparing the three methods using an illustrative example: *the prickly pear*, which is shown in Fig. 12(a).

The decompositions obtained in Mi and DeCarlo[7] and Zeng *et al.*[6] using local symmetry axis and contour curvature are shown in (b) and (c), respectively. The new decomposition (using a reduced 80×80 resolution) is shown in (d) and (e). All of the three methods find two blobs. Similar to the new decomposition, the decomposition by Mi and DeCarlo[7] separates the boundary texture (shown in gray in (b)) from the main structure. On the other hand, the decomposition by Zeng *et al.*[6] does not separate boundary texture from the main structure leading to the interpretation of the shape as two prickly balls glued together.

As experimental studies on human subjects demonstrate,[16] multiple (and mutually exclusive) parses of a given shape are possible. Thus, in a purely bottom-up process without considering a specific application or a context, one can not decide which partitioning scheme is the best.

I remark that the advantages of the new method are purely from the computational point of view. It does not involve any parameters or thresholds. It can work at very low resolutions as opposed to other methods which involve the computation of curvature or local symmetry axes, since their computation requires certain resolution.

Notice that the restriction of the field ω to where the values are less than a given threshold -3, in (f), indicates the first partitioning of the peripheral structure along the minor axis, similar to the turtle case in Fig. 8. This indication is consistent with the result of the method of Zeng *et al.*[6] which computes the partition line by sequentially eliminating the boundary detail using Discrete Curve Evolution.[15]

The new method is essentially a parameter free method with the assumption that equal importance should be given to local and global terms. However, the other two methods do not use global features (note that non-local is not necessarily global). Thus, for a comparative evaluation purpose, it is worth trying to reduce the effect of the global term by imagining a constant weight $c < 1$ in front of E^G_{Reg} in 2. In Fig. 13 (a-b), $c = 0.5$. One can notice the slight reduction of the peripheral region. The global term is responsible for the balance between the negative values and the positive values of ω. As its importance decreases, more pixels tend to attain positive values. In (c), $c = 0.125$. As c decreases, the peripheral structure shrinks further and the implied decomposition approaches to the one shown in Fig. 12 (c). In Fig. 14, the effect of reducing the importance of the global term is demonstrated using butterfly shapes.

(a) (b) (c)

Fig. 13. The effect of reducing the importance of the global term. (a-b) $c = 0.5$. (c) $c = 0.125$.

Fig. 14. The effect of reducing the importance of the global term. $c = 0.25$.

In Fig. 15, the importance of the global term is significantly reduced by setting $c = 0.025$. The decomposition result using the new method is shown in (a). For reference, the decomposition results by the previous methods are shown in (b-c).

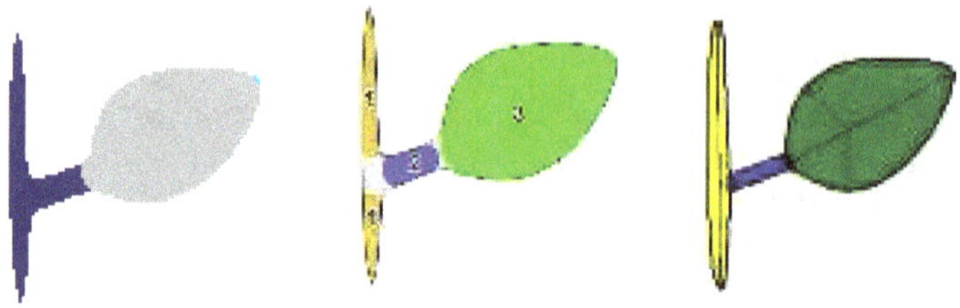

Fig. 15. The effect of significantly reducing the importance of the global term. (a) The decomposition with the new method, $c = 0.025$. (b-c) The decompositions presented by Mi and DeCarlo[7] and Zeng *et al.*,[6] respectively. [(b-c) taken from the original sources[6,7]]

3. Connection to the methods of Tari, Shah and Pien,[31–33] Aslan and Tari,[10,30,34] and Gorelick *et al.*

Recall that the field ω is the minimizer of the following energy :

$$E(\omega) = \sum_{i,j\in\Omega} E^G_{Reg}(\omega_{i,j}) + E^L_{Reg}(\omega_{i,j}) + w_{Bdy}E_{Bdy}(\omega_{i,j})$$

Let us omit the term E^G_{Reg} that models the global interaction among the shape pixels to obtain a reduced energy:

$$\sum_{i,j\in\Omega} -0.5\left(\omega_{i+1,j}\cdot\omega_{i-1,j} + \omega_{i,j+1}\cdot\omega_{i,j-1}\right) + w_{Bdy}\left(\omega_{i,j} - t_{i,j}\right)^2 \tag{12}$$

By setting the first derivative of (12) with respect to $\omega_{i,j}$ equal to *zero*, the condition satisfied by the minimizer is obtained as:

$$(w_{Bdy} - 2)\omega_{i,j} - w_{Bdy}t_{i,j} + 4\omega_{i,j} - \omega_{i-1,j} - \omega_{i+1,j} - \omega_{i,j-1} - \omega_{i,j+1} = 0 \tag{13}$$

Letting $(w_{Bdy} - 2) = \alpha > 0$ gives

$$(4\omega_{i,j} - \omega_{i-1,j} - \omega_{i+1,j} - \omega_{i,j-1} - \omega_{i,j+1}) - \alpha\omega_{i,j} = (\alpha + 2)\,t_{i,j} \tag{14}$$

(14) is clearly the discretization using central difference approximation, of the PDE (15) given below:

$$\begin{aligned} (\triangle - \alpha)\,w(x,y) &= f(x,y) \\ \text{with } w(\mathbf{x}) = 0 \ \text{ for } \mathbf{x} &= (x,y) \in \partial\Omega \end{aligned} \tag{15}$$

where $\triangle$ denotes the Laplace operator, and the right hand side inhomogeneity $f(x,y)$ is a scaled distance transform. (15) is defined on a planar shape domain Ω which is a connected, bounded, open domain of $\mathbf{R}^2$.

Interestingly, when the right hand side inhomogeneity is replaced with a constant function $f(x,y) = 1$ and α is set to *zero* (i.e. $w_B dy = 2$ as in the new method) one obtains the Poisson equation which has been recently proposed by Gorelick *et al.*[29] as a shape representation tool. On the other hand, when $\alpha > 0$ and $f(x,y) = -\alpha$, one obtains the PDE:

$$(\triangle - \alpha)\, v = -\alpha \qquad (16)$$
$$\text{with } v(\mathbf{x}) = 0 \ \text{ for } \mathbf{x} = (x, y) \in \partial\Omega$$

which has been proposed earlier by Tari, Shah and Pien.[31–33] The qualitative behavior of the v function for small α is essentially the same with that of the function obtained by solving the Poisson[29] equation. Both of them are essentially weighted distance transforms[35,36] where the local steps between neighboring points are given different costs. Tari, Shah and Pien[31] have initially proposed v function as a linear and a computationally efficient alternative to the curve evolution scheme by Kimia, Tanenbaum and Zucker[37] by showing that the successive level curves of v mimic the motion of curves with a curvature dependent speed in the direction of the inward normal.[31]

Furthermore, they demonstrated that the gradient of v along a level curve approximates the curvature of level curves, thus, suggesting a robust method for skeleton extraction by locating the extrema of the gradient of v along the level curves. The skeleton computation method of Tari, Shah and Pien[31–33] exploits the connection among morphology, distance transforms and fronts propagating with curvature dependent speeds. Such connections have stimulated many interesting approaches in solving shape related problems.[36] The importance of Tari, Shah and Pien approach is that it is the first attempt to unify segmentation and local symmetry computation into a single formulation by exploiting the connection between (16) and the Mumford and Shah[38] segmentation functional (via its Ambrosio and Tortorelli[39] approximation). It naturally extends to shapes in arbitrary dimension.[33] (In a related publication,[40] the author introduces an additive normalization term to (16) which forces the solution to oscillate, yielding the same boundary texture and gross structure separation. The proposed method is connected to a variety of morphological ideas as well as to the method of Tari, Shah and Pien in the variational calculus and PDE setting.)

In Fig. 16, the basic method of Tari, Shah and Pien is illustrated using an example by C. Aslan.[30,34] At the top row, the level curves of v function, mimicking the behavior of fronts propagating with curvature dependent speed, are shown. In each column, a different α value is used when computing v via (16). As α decreases from left to right, the relative speed of the high curvature points increases. Consequently, the inner level curves lose their concavities and become smoother earlier. Thus the value of α determines the level of smoothing (or *diffusion*). The arrows indicate the maxima and the saddle points. Topological interpretation of the shape varies with α. When, in the first column, $\alpha = 1/\sqrt{4}$, the evolving shape boundary gets split into three curves, each of which shrinks into three distinct maxima separated by the two saddle points, indicating three parts. In (b) and (c) α is reduced to $1/\sqrt{8}$ and $1/\sqrt{16}$, respectively. The level curves lose their concavities (which imply parts) earlier and evolving curves split into two parts instead of three.

Skeleton branches computed from the respective v functions (displayed at the bottom row) typically track the evolution of the indentations and protrusions of

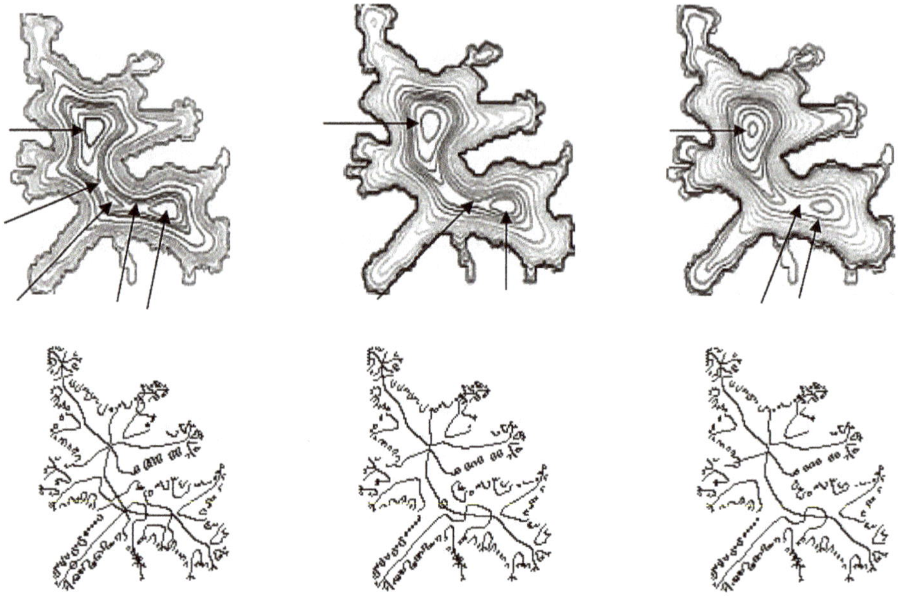

(a) (b) (c)

Fig. 16. Tari, Shah and Pien method using three different α values. (a) $\alpha = 1/\sqrt{4}$ (b) $\alpha = 1/\sqrt{8}$ (c) $\alpha = 1/\sqrt{16}$. The top row (level curves of v) illustrates different topological interpretations by varying α. The arrows indicate the maxima and the saddle points. The bottom row displays the skeletons computed from the respective v functions. [Figures by C. Aslan[30,34]].

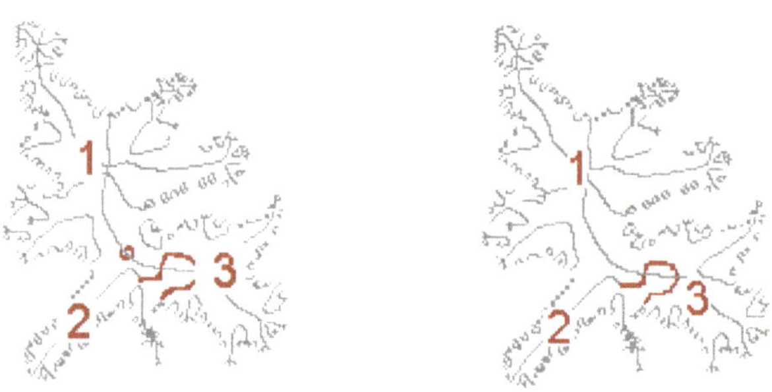

(a) (b)

Fig. 17. Un-intuitive skeleton branches (red color) in Tari, Shah and Pien.[31,32] (a) $\alpha = 1/\sqrt{8}$ (b) $\alpha = 1/\sqrt{16}$. See text for discussion. [Figures by C. Aslan[30,34]].

the shape. However, some of them exhibit a pathological behavior that frequently occurs in the Tari, Shah and Pien method when a limb is close to a neck. Notice that the skeletons contain branches that do not correspond to any protrusion or indentation of the shape. Such branches are marked with red color in Fig. 17. Aslan and Tari[10,30,34] claim that the reason of this pathology is *insufficient diffusion.*

As seen in Fig. 17 increasing the amount of diffusion by decreasing α makes such branches disappear. In (a), the computation stopped while the shape was transforming from a shape with three major blobs to a shape with two major blobs. The circular branch colored with red is due to the interaction of the center of parts two and the neck between parts one and two. As shown in (b), increasing the amount of diffusion by decreasing α makes this branch disappear since the topological change is complete. This time, the shape is between the state with two blobs (parts one and two together and part three) and the state with one blob. The red branch is due to the interaction of the center point of part three and the neck between parts two and three.

Thus, as a remedy, they propose to increase the diffusion by gradually decreasing α so that almost every shape is forcefully interpreted as a single blob ignoring the part structure. Following this ad-hoc modification, the new v function has been successfully applied in shape matching applications.[30,34,41,42]

However, the method can not be applied to shapes which can not be reduced to a single blob. Such cases include:

- shapes with holes;
- thin and long shapes with constant width;
- shapes with more than one equally prominent parts.

The strategy adopted[10,30,34] for shapes with two equally prominent parts (dumbbell-like shapes) is to retain their dumbbell-like character. This ad-hoc solution introduces a representational instability as the width of the neck that separates two prominent parts change.

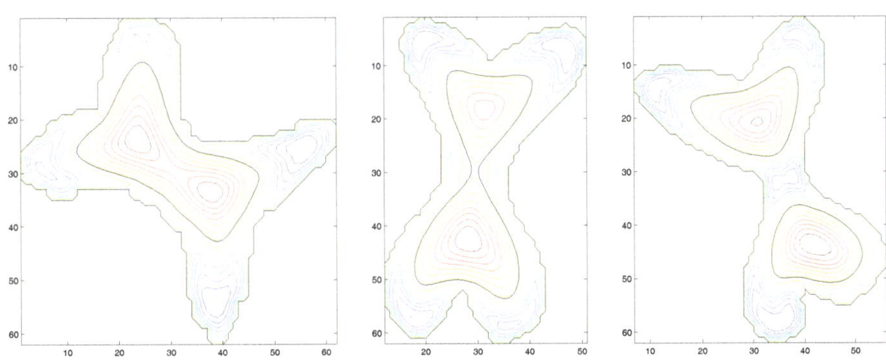

Fig. 18. The level curves of ω. The inner black level curve is the zero-level curve. In all of the three cases, Ω^+ denotes the gross structure.

Fig. 19. Parts of the peripheral structure for dumbbell-like shapes of varying neck thickness.

In Fig. 18-19, the level curves of w are shown for three dumbbell-like shapes with varying neck thickness. In all of the three cases, Ω^+ denotes the gross structure. Instead of relying on a single point to be the shape center as in the method of Aslan and Tari,[10,30,34] the new method takes a different attitude. Robustness is obtained by replacing a point estimate for the center with an interval estimate. Parts of the peripheral structure are depicted in Fig. 19.

The method of Aslan and Tari[30,34] implicitly codes the part structure via disconnected skeleton branches. Fig. 20 demonstrates the possibility of extracting parts from these disconnected branches. In (a) skeleton points detected with the method of Tari, Shah and Pien[32] using the modified function[10,30,34] is shown. In (b) the result of pruning and grouping procedure is shown. In (c) final representation (called the disconnected Aslan skeleton[10,30,34] to distinguish it from the Tari, Shah and Pien skeleton[31,32]) is depicted. In (d) parts are extracted, for each dis-

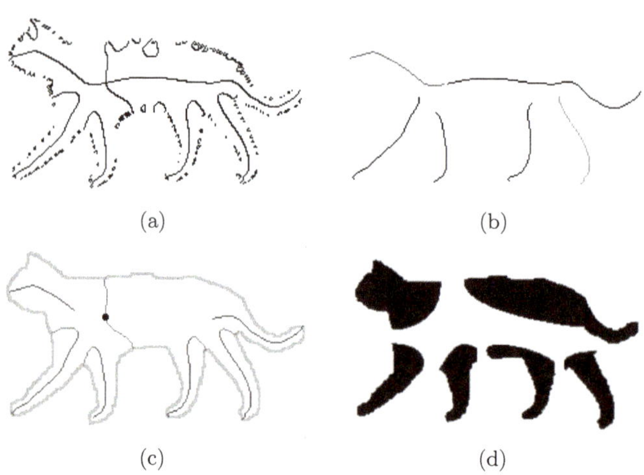

Fig. 20. (a) Skeleton points detected with the method of Tari, Shah and Pien[32] using the modified function.[10,30,34] (b) After pruning and grouping skeleton points. (c) Disconnecting the major skeleton branch.[34] (d) Parts obtained from disconnected branches. [Unpublished result by E. Baseski.]

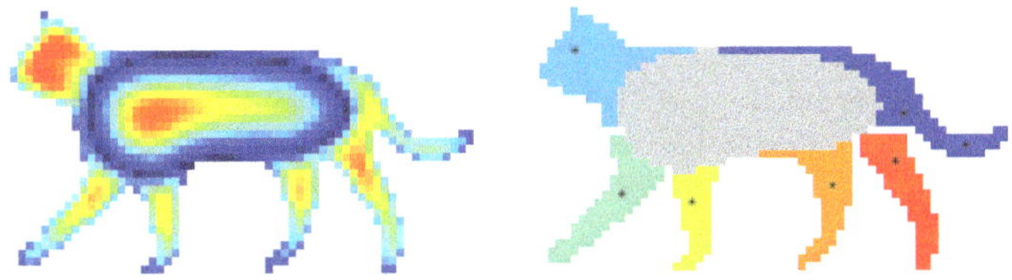

Fig. 21. (a) $|\omega|$. (b) Parts.

connected branch, by fitting a spline that passes through the disconnection point and the nearest indentations. The parts captured by the new method is presented in Fig. 21 for the same cat shape (reduced resolution) for comparison.

References

1. E. Barenholtz and J. Feldman, Visual comparisons within and between object-parts: evidence for a single- part superiority effect., *Vision Research.* **43**, 1655–1666, (2003).
2. A. Pasupathy and C. Connor, Population coding of shape in area V4, *Nature Neuroscience.* **5**(2), 1332–1338, (2002).
3. T. Binford. Visual perception by computer. In *IEEE Conference on Systems and Control*, (1971).
4. D. Marr and H. K. Nishiara. Representation and recognition of spatial organization of three dimensional shapes. In *Proceedings of the Royal Society of London, Series B, Biological Sciences, 200*, pp. 269–294, (1978).
5. I. Biederman, Recognition-by-components: A theory of human image understanding., *Psychological Review.* **94**(2), 115–117, (1987).
6. J. T. Zeng, R. Lakaemper, X. Wei, and X. Li. 2d shape decomposition based on combined skeleton-boundary features. In *Advances in Visual Computing, 4th International Symposium, ISVC*, pp. 682–691, (2008).
7. X. Mi and D. DeCarlo. Separating parts from 2d shapes using relatability. In *ICCV*, pp. 1–8, (2007).
8. S. Svensson and G. S. di Baja, Using distance transforms to decompose 3d discrete objects, *Image Vision Comput.* **20**(8), 529–540, (2002).
9. C. Arcelli and L. Serino, From discs to parts of visual form, *Image Vision Comput.* **15**(1), 1–10, (1997).
10. C. Aslan and S. Tari. An axis-based representation for recognition. In *ICCV*, pp. 1339–1346, (2005).
11. H. Rom and G. Medioni, Hierarchical decomposition and axial shape description, *IEEE Trans. Pattern Anal. Mach. Intell.* **15**(10), 973–981, (1993).
12. B. J. Super, Knowledge-based part correspondence, *Pattern Recognition.* **40**(10), 2818–2825, (2007).
13. E. H. Cohen and M. Singh, Geometric determinants of shape segmentation: Tests using segment identification., *Vision Research.* **47**, 2825–2840, (2007).
14. D. Hofmann and W. Richards, Parts of recognition, *Cognition.* **18**, 65–96, (1984).
15. L. J. Latecki and R. Lakaemper, Convexity rule for shape decomposition based on discrete contour evolution, *Comput. Vis. Image Underst.* **73**(3), 441–454, (1999).
16. L. W. Renninger. *Parts, Objects and Scenes: Psychophysics and Computational Models.* PhD thesis, UC Berkeley, USA, (2003).

17. K. Siddiqi and B. Kimia, Parts of visual form: Computational aspects, *IEEE Trans. Pattern Anal. Mach. Intell.* **17**, 239–251, (1995).
18. K. Siddiqi, K. J. Tresness, and B. Kimia, Parts of visual form: Ecological and psychophysical aspects, *Perception.* **25**, 399–424, (1996).
19. J. de Winter and J. Wagemans, Segmentation of object outlines into parts: A large-scale integrative study., *Cognition.* **99**, 275–325, (1999).
20. C. A. Burbeck and S. M. Pizer, Object representation by cores: Identifying and representing primitive spatial regions., *Vision Research.* **35**, 1917–1930, (1995).
21. D. Navon, Forest before trees: The precedence of global features in visual perception., *Cognitive Psychology.* **9**, 355–383, (1977).
22. C. Xu, J. Liu, and X. Tang, 2d shape matching by contour flexibility, *IEEE Trans. Pattern Anal. Mach. Intell.* **31**(1), 180–186, (2009).
23. S. Belongie, J. Malik, and J. Puzicha. Shape context: A new descriptor for shape matching and object recognition. In *NIPS*, pp. 831–837, (2000).
24. J. Shi and J. Malik. Normalized cuts and image segmentation. In *CVPR*, pp. 731–737, (1997).
25. A. Buades, B. Coll, and J. M. Morel. A non-local algorithm for image denoising. In *CVPR*, pp. 60–65, (2005).
26. J. J. Koenderink and A. J. van Doorn, The shape of smooth objects and the way contours end., *Perception.* **11**, 129–137, (1982).
27. H. Blum, Biological shape and visual science., *Journal of Theoretical Biology.* **38**, 205–287, (1973).
28. http://www.lems.brown.edu/vision/.
29. L. Gorelick, M. Galun, E. Sharon, R. Basri, and A. Brandt, Shape representation and classification using the poisson equation, *IEEE Trans. Pattern Anal. Mach. Intell.* **28** (12), 1991–2005, (2006).
30. C. Aslan. Disconnected skeletons for shape recognition. Master's thesis, Department of Computer Engineering, Middle East Technical University (May, 2005).
31. S. Tari, J. Shah, and H. Pien. A computationally efficient shape analysis via level sets. In *MMBIA '96: Proceedings of the 1996 Workshop on Mathematical Methods in Biomedical Image Analysis*, pp. 234–243, (1996).
32. S. Tari, J. Shah, and H. Pien, Extraction of shape skeletons from grayscale images., *CVIU.* **66**(2), 133–146, (1997).
33. S. Tari and J. Shah. Local symmetries of shapes in arbitrary dimension. In *ICCV*, pp. 1123–1128, (1998).
34. C. Aslan, A. Erdem, E. Erdem, and S. Tari, Disconnected skeleton: Shape at its absolute scale, *IEEE Trans. Pattern Anal. Mach. Intell.* **30**(12), 2188–2203, (2008).
35. G. Borgefors, Distance transforms in digital images., *Comput. Vision, Graphics and Image Processing.* **34**, 344–371, (1986).
36. P. Maragos and M. A. Butt, Curve evolution, differential morphology and distance transforms as applied to multiscale and eikonal problems., *Fundamentae Informatica.* **41**, 91–129, (2000).
37. B. B. Kimia, A. R. Tannenbaum, and S. W. Zucker, Shapes, shocks and deformations, *Int. J. Comput. Vision.* **15**(3), 189–224, (1995).
38. D. Mumford and J. Shah, Optimal approximations by piecewise smooth functions and associated variational problems., *Commun. Pure Appl. Math.* **42**(5), 577–685, (1989).
39. L. Ambrosio and V. Tortorelli, On the approximation of functionals depending on jumps by elliptic functionals via Γ-convergence., *Commun. Pure Appl. Math.* **43**(8), 999–1036, (1990).

40. S. Tari. Hierarchical shape decomposition via level sets. In *International Symposium on Mathematical Morphology*, pages 217–228, (2009).
41. E. Baseski, A. Erdem, and S. Tari, Dissimilarity between two skeletal trees in a context, *Pattern Recogn.* **42**(3), (2009).
42. A. Erdem and S. Tari. Coarse-to-fine matching of shapes using disconnected skeletons by learning class-specific boundary deformations. In *Workshop on Graph-based Representations in Pattern Recognition*, pp. 21–30, (2009).

Part 3

Recognition Applications

CHAPTER 3.1

CONTEXTUAL CLASSIFICATION OF MULTISENSOR OPTICAL-SAR REMOTE SENSING IMAGES

Gabriele Moser[1], Raffaele Gaetano[2], Gianni Poggi[2],
Giuseppe Scarpa[2], and Sebastiano B. Serpico[1]

[1] *University of Genoa, Dept. of Biophysical and Electronic Eng., Via Opera Pia 11a, I-16145 Genoa (Italy), sebastiano.serpico@unige.it*

[2] *University of Naples "Federico II," Dept. of Electronic and Telecommunications Eng., Via Claudio 21, I-80125 Naples (Italy), poggi@unina.it.*

Multispectral and synthetic aperture radar (SAR) images are known to exhibit complementary properties: unlike optical sensors, SAR provides information about soil roughness and moisture and acquires useful data despite clouds and sun-illumination conditions. However, the analysis of the resulting images is more difficult relative to the use of optical imagery due to the noise-like speckle phenomenon. In order to exploit this complementarity for classification purposes, the challenge lies in the definition of accurate joint optical-SAR statistical models due to the different physical natures of these data types and to the corresponding differences in the related parametric models.

In this chapter, the problem of multisensor optical-SAR data classification is addressed and a novel region-based semiparametric classification technique is proposed. The method combines the tree-structured Markov random field approach to segmentation with the dependence tree approach to probability density estimation and with case-specific bivariate models for the distributions of optical and SAR data. A Bayesian decision rule is formulated at the segment level in order to incorporate spatial-contextual information and to gain robustness against noise. Experimental results are presented on a real data set that includes optical multispectral bands and SAR multipolarization and multifrequency channels.

1. Introduction

In the remote-sensing field, significant advances in sensor design have occurred in recent years thanks to the systematic and fast evolution of related technologies. As a consequence, a larger and larger amount of different types of remotely sensed data is currently available. In this context, an effective synergistic use of the data acquired by multiple and (possibly) heterogeneous data sources (e.g., optical sensors and passive and active microwave sensors) by means of data-fusion techniques allows for improvement of the capacity to analyze land covers and their states and temporal evolution by exploiting the complementary natures of the information provided by such sensors.[1] Such complementarity can be observed when dealing with optical and SAR sensors: unlike optical sensors, SAR provides information about the roughness and the moisture content of soil and acquires useful data despite the possible presence of clouds and/or the absence of solar illumination. However, the analysis of the resulting images is more difficult relative to the use of optical imagery due to the presence of the noise-like speckle phenomenon.[2] In order to jointly exploit the information conveyed by optical and SAR data, multisensor analysis strategies are required that take into account the specific problems involved with such heterogeneous sources.

In the specific context of image classification, a critical difficulty lies in the definition of accurate joint optical-SAR statistical models due to the different physical natures of these data types. Several parametric families have been proposed for optical channels[1] or for a SAR amplitude or intensity,[2] but no accurate parametric model is available so far for the joint statistics of optical-SAR data. In addition, both the optical and the SAR channels convey a huge amount of spatial contextual information. Taking advantage of this information is crucial in order to regularize the classification results and minimize the impact of noise (especially speckle).

In this chapter, the problem of multisensor optical-SAR image classification is addressed and a novel region-based semiparametric technique is proposed to this end. The method is based on two key ideas. First, a semiparametric probability density function (PDF) estimation approach, combining the nonparametric dependence-tree method[3] with a set of parametric bivariate distributions, is used to model the joint statistics of optical-SAR data. Then, a region-based classification approach incorporating a segmentation technique[4,5] is adopted to integrate the contextual information in the classification process. The method is tested on a multisensor data set acquired over an agricultural area and consisting of optical channels and of multipolarization and multifrequency SAR amplitudes.

The chapter is organized as follows. Sec. 2 summarizes the main approaches proposed in the literature for optical-SAR image classification and region-based contextual classification. Secs. 3 and 4 describe the proposed method and the related experimental results, respectively. Conclusions are drawn in Sec. 5.

2. Previous work on optical-SAR and region-based classification

Due to the aforementioned difficulty in defining a common optical-SAR parametric statistical model, either semiparametric/nonparametric techniques are adopted or the data acquired by each sensor are separately processed and the resulting extracted information merged at a higher level (i.e., at the level of spatial patterns or at the decision level).[1] More generally, many methods have been proposed to classify multisensor data; most are not specifically for SAR and optical images, but can usually also be applied to this data type. Specifically, the most widespread approaches are based on statistical methods,[6] neural networks,[7] support vector machines (SVMs),[8] symbolic processing[9] or hybrid techniques.[10]

Among the statistical methods, the stacked-vector[1] approach turns out to be the simplest and one of the most commonly used: each pixel is assigned a feature vector generated by column juxtaposition of the measures (features) acquired by all available sources. The resulting vector is classified by adopting a statistical model in the joint feature space. Adopting a parametric approach would lead to the assumption of a single analytical model for each class-conditional probability density function (PDF) of the multisensor feature vector.[1] However, due the abovementioned heterogeneity, the application of this approach to optical-SAR data is difficult. The typical statistical model for optical data is the multivariate Gaussian,[1] whereas the most widely used models for SAR are the Rayleigh, Nakagami-Gamma, K, lognormal and Weibull distributions.[2] Among the multisensor fusion studies that explicitly take into account such statistics, we recall the technique proposed by Lombardo *et al.*[11] for image segmentation and classification.

On the other hand, a nonparametric estimator (e.g., "K nearest neighbors" – K-NN, or Parzen algorithm) does not require the definition of an analytical model for the unknown PDF, but usually presents parameters to be set prior to the application of the method.[1] For instance, the algorithm proposed by Datcu *et al.*[3] combines the Parzen method with the dependence-tree approach in order to estimate the PDF of optical-SAR data. Cell algorithms with data inflation have been proposed to minimize the computational burden in K-NN classification of optical-SAR images.[12]

Other statistical approaches are the stratification,[13] relaxation labeling[1,14] and Markov random field (MRF) methods, which also integrate the spatial-contextual information in the classification process.[15] Benediktsson and Swain[16] developed a modified version of the relaxation labeling method by introducing reliability factors associated with the available information sources. Solberg *et al.*[15] proposed an MRF-based approach to merge optical data, SAR images and prior information derived from a geographical information system. A statistical optical-SAR image classifier was proposed by combining the approach based on reliability factors with a multiplicative autoregressive random field model for the SAR image.[6]

As an alternative to statistical techniques, several fusion methodologies have been developed based on neural networks[7,17–20] to better deal with the heterogeneous properties of the joint feature statistics, thanks to the completely nonparametric characteristics of the neural approach.[1] However, the architecture of a neural network requires time-consuming optimization procedures involving many parameters. In addition, a neural network is a black box, giving the user no information to control and understand the data analysis procedures. For instance, Gamba *et al.*[21,22] proposed a method that integrated a simplified adaptive resonance theory neural network in the classification chain in order to simplify the optimization of the network parameters.

Several approaches to multisensor data classification have recently been proposed based on SVM and kernel-based methods, thanks to both their fully nonparametric nature and to their robustness to the dimensionality issues possibly involved in stacking together many optical and SAR channels. The "support vector domain description" approach was applied to the classification of optical-SAR data of an urban area.[23] Camps-Valls *et al.*[24] developed a general framework to integrate multisource and multitemporal data in support-vector techniques by combining features from multiple sources through the definition of specific kernel functions. The SVM and decision-fusion approaches were combined by separately generating an SVM classification from each subset of single-sensor features and then by training a further SVM to fuse the single-sensor results.[8] Multilevel segmentation and the tree-based random-forest classifier were also integrated in this scheme.[25]

Several hybrid approaches, which combine different methodologies, have been proposed to overcome the main drawbacks of the classical multisensor techniques. For instance, Benediktsson *et al.*[10] applied a statistical model to each individual source and used neural networks to state a consensus decision. Bruzzone *et al.*[17] proposed a nonparametric approach combining multilayer perceptron neural networks and the expectation-maximization method for PDF estimation.

The symbolic methods based on the Dempster-Shafer evidence theory[26] provide a further solution to the problem of multisensor fusion. Usually, the Dempster-Shafer theory is applied to classify multisource data by taking into account the uncertainties related to the different data sources.[9,14,27] Unsupervised approaches to multisource classification based on the Dempster-Shafer evidence theory were also reported.[9] A thorough comparison among several different approaches (fuzzy, genetic, neural, statistical) to the fusion of the information extracted from multisource data has been published by Benediktsson and Sveinsson.[28]

Both in the specific framework of optical-SAR data classification and in the more general context of remote-sensing image classification, the use of spatial information plays a basic role in optimizing the classification accuracy and improving the robustness to noise. For instance, the aforementioned MRFs constitute a prominent approach to introduce contextual information in Bayesian image analysis schemes.

Spatial-contextual information can also be exploited by region-based and object-based methods, usually formulated by integrating classification and segmentation algorithms in order to assign a thematic-class label not to a single pixel but to a homogeneous region.[29] A segmentation method aims at partitioning the image data into a set of homogeneous regions (or segments), typically corresponding to objects or portions of objects. Several object-based and region-based approaches to remote-sensing image classification have been proposed, that involve fuzzy techniques,[30] edge-based methods and probabilistic relaxation,[31] SVM,[32] hierarchical MRFs for joint segmentation and classification,[33] greylevel co-occurrence and variogram texture features,[34] topological models of the relationships between segments,[29] hierarchical multiscale segmentation and probabilistic latent semantic analysis.[35]

3. Methodology

3.1. *Overview of the proposed method*

Let $x_1, x_2, \ldots, x_N$ be the feature vectors of the pixels in a multisensor optical-SAR image of a geographical area of interest (N being the number of pixels), and let $\omega_1, \omega_2, \ldots, \omega_M$ be the thematic classes that are present in the imaged area. These classes are assumed to be endowed with training samples. Let $x_k = (x_{k,1}, x_{k,2}, \ldots, x_{k,n})$ be composed of n features, where $x_{k,1}, x_{k,2}, \ldots, x_{k,m}$ $(0 < m < n)$ are multispectral channels and $x_{k,m+1}, x_{k,m+2}, \ldots, x_{k,n}$ are SAR amplitudes $(k = 1, 2, \ldots, N)$. The extension to the case of SAR intensities is straightforward. The proposed method exploits the spatial contextual information in the multisensor image by adopting a region-based approach. Specifically, in order to perform a region-based classification, a segmentation map is first generated and a classification rule is defined to assign a thematic class label to each resulting segment $\sigma_1, \sigma_2, \ldots, \sigma_L$ (L being the total number of segments in the segmentation map) and not pointwise to each single pixel. A tree-structured Markov random field (TS-MRF) model-based segmentation algorithm[4] is used in the first step (details on this step are given in Sec. 3.2).

Denoting by $\xi_\ell = (x_k : k \in \sigma_\ell)$ the juxtaposition of the feature vectors of all pixels lying in segment σ_ℓ, and assuming the feature vectors of distinct pixels to be independent and identically distributed (i.i.d.), when conditioned to the related class labels, the PDF of ξ_ℓ, given ω_i, turns out to be $(\ell = 1, 2, \ldots, L)$:

$$p(\xi_\ell|\omega_i) = \prod_{k\in\sigma_\ell} p(x_k|\omega_i), \tag{1}$$

where $p(x_k|\omega_i)$ is the PDF of x_k conditioned to ω_i $(k = 1, 2, \ldots, N; i = 1, 2, \ldots, M)$. By estimating this PDF by the semiparametric method described in Sec. 3.3 and the prior probability $P(\omega_i)$ of each class ω_i as the corresponding relative frequency in the training set, this formulation of $p(\xi_\ell|\omega_i)$ allows the application of the Bayesian maximum-*a-posteriori* (MAP) rule to each segment σ_ℓ $(\ell = 1, 2, \ldots, L)$, thus performing a thematic classification at the segment level. Accordingly, segment σ_ℓ is assigned the class label ω_i that maximizes the following loglikelihood functional $(i = 1, 2, \ldots, M; \ell = 1, 2, \ldots, L)$:

$$\mathcal{L}(\omega_i|\sigma_\ell) = \ln P(\omega_i) + \sum_{k\in\sigma_\ell} \ln p(x_k|\omega_i). \tag{2}$$

We note that the aforementioned conditional independence assumption is usually violated by real-world images. Here, it is accepted as a simplifying working hypothesis to ensure the analytical tractability of the problem.

3.2. *Multispectral TS-MRF segmentation*

The goal of segmentation is to associate a map of class[a] labels $y = \{y_1, y_2, \ldots, y_N\}$, $y_k \in \{\sigma_1, \sigma_2, \ldots, \sigma_L\}$, with the multisensor image $x = \{x_1, x_2, \ldots, x_N\}$, $x_k \in \mathbb{R}^n$. In the statistical framework, the problem is approached by choosing suitable probabilistic models for both the data x and the unknown segmentation map y and by estimating the map according to some useful statistical criteria. A popular choice is the MAP criterion, where y is selected as the map that maximizes the joint probability distribution $p(x|y)p(y)$. For the likelihood term we assume, as usually done, pixel-wise conditional independence, that is, $p(x|y) = \prod_{k=1}^{N} p(x_k|y_k)$, where each feature vector x_k has a suitable distribution whose parameters depend on the class label. Under such an assumption, the prior term $p(y)$ has the key role of taking into account the spatial interactions.

The MRF model[36] is a relatively simple, yet effective, tool to encompass prior knowledge in the segmentation process, and in fact the interest in MRFs has been steadily growing in recent years. By modeling the segmentation map as an MRF, one assumes that each pixel depends statistically on the rest of the image only through a selected group of neighbors. This greatly simplifies the problem of assigning a meaningful prior, since only local characteristics of the image need be specified.

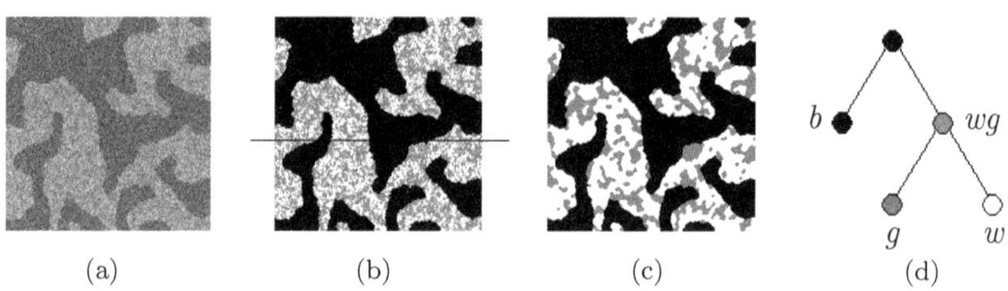

Fig. 1. Synthetic experiment: (a) noisy data; (b) hidden segmentation map; (c) segmentation by Potts model; (d) tree structure of the data: b, black class; g, grey class; w, white class; wg, white-grey merging class.

Many MRF models have been proposed[36] to describe the different features encountered in the images of interest. One of the most challenging (and largely unsolved) modeling tasks is to account for the spatial variations of image statistics. Let us consider for example the image in Fig. 1, where the data (a) were generated by adding Gaussian noise to a synthetic segmentation map (b). As can be seen, following for example the traced line, the class random process is clearly non-stationary. Conventional models provide poor results in the presence of such structured data, like the segmentation map shown in (c), obtained with a second order Potts model.[36] Here, one class is well detected while the other two are mixed almost everywhere because of an over-regularization. Even more sophisticated hierarchical MRFs will

[a]Note that segmentation comes before classification and is totally unsupervised. Hence, the *classes* considered here do not coincide, typically, with those of interest for the final classification.

not cope well with such data unless their parameters depend on the classes. In fact, the correct model for this example is the binary tree of classes shown in Fig. 1(d), where the "black" class is first separated from the other two based on a dedicated binary MRF model, while the "white" and "gray" classes are further separated by means of a different binary MRF model.

This simple example makes clear the potential advantage of using a structured model, like the tree-structured MRF (TS-MRF)[4] described below, since many real images present a similar hierarchical structure although not always so easy to detect. To this end, the TS-MRF model aims at describing the hidden structure of the data by a sequence of binary MRFs, each corresponding to a node in the tree. In such a model, parameters can be defined locally to each node, so as to correctly split the classes depending only on their statistical characteristics. In addition, thanks to its hierarchical structure, the TS-MRF model allows for recursive optimization procedures, which help to reduce the complexity.

3.2.1. *MRF model-based segmentation*

A random field y defined on a lattice $\mathcal{S}$ is said to be an MRF with respect to a given neighborhood system if the Markovian property holds for each site s. Quite often, in order to limit complexity, only the 4 or 8 closest pixels (system η^1 and η^2, respectively) are included in a pixel's neighborhood. Nonetheless, the MRF model proves quite powerful because most of the dependencies can be captured through local interactions. The distribution of a positive MRF can be proved to have a Gibbs form,[36] i.e.:

$$p(y) = \frac{1}{Z}\exp[-U(y)] = \frac{1}{Z}\exp\left[-\sum_{c\in\mathcal{C}} V_c(y_c)\right] \tag{3}$$

where the $V_c(\cdot)$ functions are called potentials, $U(y)$ denotes the energy, Z is a normalizing constant, c indicates a clique of the image, which is a small set of neighboring sites, and $\mathcal{C}$ is the set of admissible cliques. Note that each potential V_c depends only on the values taken on the clique sites $y_c = \{y_s, s \in c\}$ and, therefore, accounts only for local interactions. As a consequence, local dependencies in y can easily be modeled by defining suitable potentials $V_c(\cdot)$. In the following developments we will focus on the Potts model where only two-site cliques are taken into account, with potential defined as

$$V_c(y_c) = \begin{cases} \beta \text{ if } y_p \neq y_q,\ p, q \in c \\ 0 \text{ otherwise} \end{cases} \tag{4}$$

where $\beta > 0$ is the edge-penalty parameter. Such a choice allows some simplifications but is by no means necessary and the TS-MRF approach works with any basic MRF model, no matter how complex.

3.2.2. TS-MRF model

In the Potts model, as well as other commonly used MRFs, the potential function of a two-site clique depends only on the presence or absence of an edge (a class transition), without regard to the involved classes. As a consequence one parameter is sufficient to define the model. If we remove such a constraint and use a different potential function for each different pair of classes, many more parameters become necessary, $\frac{1}{2}(L-1)L$ where L is the number of classes. In this generalized Potts model, the clique potentials will become

$$V_c(y_c) = \begin{cases} \beta_{kh} \text{ if } y_p = k \neq y_q = h,\, p, q \in c \\ 0 \text{ otherwise} \end{cases} \tag{5}$$

where $\beta_{kh} = \beta_{hk} > 0$, $k \neq h$, is the edge-penalty parameter for a transition h-k.

Starting from this very general model, it is possible to reduce the number of parameters by taking into account some class properties. Looking at the example of Fig. 1, it is easy to realize that the white (w) and gray (g) classes have the same relationship with the black class (b), that is, a b-w edge has the same statistical properties as a b-g edge. Therefore, only two independent parameters should be used instead of three. Such class properties can easily be represented by means of the hidden binary structure of Fig. 1(d). In the example considered, the two relevant parameters are associated with the two internal nodes of the tree. The former, associated with the root, penalizes b-w and b-g edges (assumed to be equivalent), while the latter, associated with the other internal node, penalizes w-g edges. Therefore, at the root level, where b is split from w and g, it does not matter if a site is labeled as w or g, but only whether it falls into the macro-class wg. In general, given a binary structure with L terminal nodes, the number of internal nodes, and hence the number of parameters, will be $L-1$, rather than $\frac{1}{2}(L-1)L$. This small number of parameters can be reliably estimated from the available data but, at the same time, allows one to accurately describe class-dependent image nonstationarities.

We now proceed to a more formal description of the model. Let us consider a binary tree T, identified by its nodes and their mutual relationships, and composed of internal nodes $\overline{T}$ and terminal nodes $\widetilde{T}$. Also, let $\wedge$ be a binary operator that gives the nearest common ancestor of two nodes of T. We can define a tree-structured MRF through its clique potentials, still expressed by (5), but with the additional $\frac{1}{2}L(L-3)+1$ constraints that $\beta_{kh} = \beta_{pq} = \beta_t$ if $k \wedge h = p \wedge q = t$

$$V_c(y_c) = \begin{cases} \beta_{y_p \wedge y_q} & \text{if } y_p \neq y_q,\, p, q \in c \\ 0 & \text{otherwise.} \end{cases} \tag{6}$$

Then, the distribution of the TS-MRF becomes simply

$$p(y) = \frac{1}{Z} \exp\left[-\sum_{t \in \overline{T}} \beta_t \mathcal{N}_t\right] = \prod_{t \in \overline{T}} \frac{1}{Z_t} \exp[-\beta_t \mathcal{N}_t]. \tag{7}$$

where $\mathcal{N}_t = \mathcal{N}_t(y)$ is the number of cliques with edge-penalty β_t.

The complexity of this model could still seem prohibitive for a practical implementation because of the dimensionality of the parameter space, dependent on the number of classes. However, thanks to its structural constraints, a recursive optimization procedure can be used that, although sub-optimal, involves only one edge-penalty at a time.

Given a binary tree structure T, which relates the classes hierarchically, the label field y can be completely expressed through the set of the binary fields y^t associated with the internal nodes of T. Considering the example of Fig. 1, only two binary MRF components are needed to build the three-class global TS-MRF field. One associated with the root node separates the class b from the (merging) class wg, the other, associated with node wg, will split the classes w and g. This way, with the exception of the root, each binary field is defined on an irregular lattice resulting from the realization of the field associated with the parent node. Therefore, each binary field y^t depends only on its ancestor fields $y^{a(t)} = \{y^t\}_{t \in a(t)}$, where $a(t)$ denotes the set of ancestors of t. On the other hand, the number $\mathcal{N}_t$ of cliques with edge-penalty β_t depends only on y^t and, from the considerations above, on $y^{a(t)}$, while it is independent of other component binary fields, that is $\mathcal{N}_t = \mathcal{N}_t(y^t, y^{a(t)})$.

Hence, in order to find a MAP estimate of a segmentation with TS-MRF prior, we can maximize recursively the terms in (7), together with the corresponding likelihoods, starting from the root and descending the tree until all leaves are reached. Each term depends only on a single binary field once its ancestor fields are given and can be maximized by itself with conventional techniques like simulated annealing, iterated conditional mode, etc.[15] Note, again, that in the step corresponding to node t, only the parameter β_t must be estimated, and that $\mathcal{N}_t$ is a sufficient statistic for β_t.

Of course, the tree structure is itself an item to be estimated during the segmentation process. To this aim a proper criterion has been proposed[4] that drives the tree evolution. A *split gain* parameter is computed locally for each node, weighting the split and non-split hypotheses: a gain larger than 1 indicates that the data can be better described by two or more classes, and hence that the node must be split.

The general optimization scheme of the TS-MRF segmentation algorithm is summarized in the flow chart of Fig. 2, where $\widetilde{T} \subseteq T$ is the current set of terminal nodes, G^τ is the split gain of node τ, whose computation is performed by the *Check Node* procedure, and finally $l(\tau)$ and $r(\tau)$ are the children nodes provided by the split of node τ through the *Split Tree* function which performs a binary segmentation of the region corresponding to node τ.

3.2.3. *Splitting disconnected regions*

Unlike other segmentation tools, MRF-based methods are oriented more towards a global classification than a local detection of segments because of the importance given to the likelihood component $p(x|y)$. This property is useful in many instances but is not of interest here because the segmenter is only asked to provide an initial

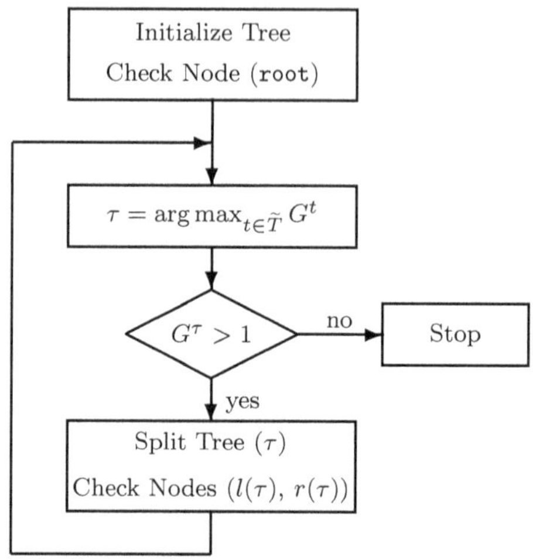

Fig. 2. High-level flow chart of the classic TS-MRF algorithm.

fragmentation of the image, while region-based classification is left to the subsequent step (detailed in Sec.3.3). Therefore, it represents only an undesired constraint. For this reason, we have used here a particular evolution[5] of the TS-MRF algorithm that separates all unconnected elements after each binary split, treating them as independent regions from that point on. This way, local accuracy is increased at the price of a reduced global consistency, which was a useless property anyway.

It has been observed experimentally that the splitting procedure used in TS-MRF quite often originates children regions composed of several spatially distinct areas. An example is shown in Fig. 3, where the original remote-sensing image (a), is separated by a binary split in two regions (b), loosely identified as "dark" and "light," which are further segmented in subsequent steps leading to the three- and four-class maps (c) and (d). It is clear that the two regions of map (b) are both composed of multiple non-connected areas that share similar statistical characteristics. Splitting such areas from one another, no matter how similar they are, is necessarily associated with a positive gain since no new edges (and edge penalties) are introduced in the process. Actually, assuming that all parameters, including the split gain, are correctly estimated, such kind of splits should occur anyway, sooner or later, while the segmentation tree grows.

On the basis of the considerations above, a different version of the TS-MRF algorithm is obtained[5] by forcing, after each binary split, the separation of non-connected areas to form new nodes. With reference to the maps of Fig. 3(b) and (c), the associated tree structure looks as shown in Fig. 4(left). The dark region, associated with a temporary node (empty circle), is automatically split in four subregions associated with actual nodes (solid circles), while the light region is likewise split in three subregions. As a consequence, a single binary split produces 7 non-connected regions. Each of these regions are then tested for possible further

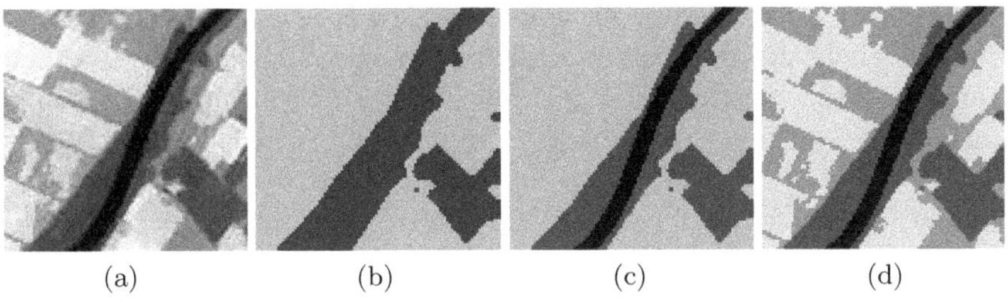

Fig. 3. Example tree-structured segmentation.

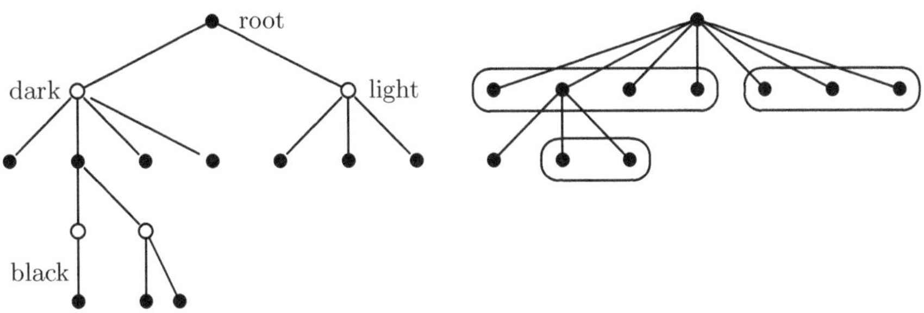

Fig. 4. Complete tree structure generated by the proposed algorithm with separation of non-connected regions (left), and its compact representation (right).

split: in the example, the split of the region including the river leads eventually to three new nodes. Fig. 4(right) shows a more compact representation of the resulting tree, where the ovals group together nodes that share similar statistics.

Thanks to its increased local accuracy, this particular version of the TS-MRF algorithm has been selected here for carrying out the segmentation required by the subsequent region-wise joint optical-SAR classifier.

3.3. *Semiparametric probability density estimation*

In order to address the PDF estimation task required in the application of Eq. (1), the dependence tree approach is adopted, which nonparametrically approximates each class-conditional PDF by a product of bivariate PDFs of couples of features,[3] i.e., $(k = 1, 2, \ldots, N)$:

$$p(x_k|\omega_i) \simeq \frac{\prod_{j=1}^{n-1} p(x_{k,\pi_i(j)}, x_{k,\pi_i(j+1)}|\omega_i)}{\prod_{j=2}^{n-1} p(x_{k,\pi_i(j)}|\omega_i)}, \tag{8}$$

where $\pi_i(\cdot)$ is a permutation of the index set $\{1, 2, \ldots, n\}$ of the features $(i = 1, 2, \ldots, M)$. This choice is explained by the fact that, even though no parametric models have been developed so far for the joint statistics of an arbitrary set of SAR and optical channels, a few models have been proposed for pairs of SAR amplitudes or optical bands.[1,37–39] The optimal permutation $\pi_i(\cdot)$ (or, equivalently, the optimal choices of the couples of features) related to each class ω_i ($i = 1, 2, \ldots, M$) is identified by minimizing the Kullback-Leibler divergence between the true unknown PDF and its DTC approximation given by Eq. (8).[40] This minimization task is formalized as a shortest-spanning-tree problem, addressed by the well-known Kruskal's algorithm.[41] Details on the computation of the optimal permutations can be found in Datcu *et al.*[3]

The application of the DTC approximation in the segment-level MAP decision function of Eq. (2) requires estimating the joint bivariate PDFs $p(x_{k,\pi_i(j)}, x_{k,\pi_i(j+1)}|\omega_i)$ and the related marginal PDFs $p(x_{k,\pi_i(j)}|\omega_i)$ from the available training data ($i = 1, 2, \ldots, M; j = 2, 3, \ldots, n-1$). The key idea of the proposed approach to multisensor classification is to exploit the available prior information on the optical-SAR nature of the available features and to consequently integrate the DTC nonparametric approximation with data-specific parametric models for bivariate PDFs. For ease of notation, we focus on a single class and a single couple of features. Accordingly, we set $r = x_{k,\pi_i(j)}$, $s = x_{k,\pi_i(j+1)}$ and we denote simply by $p(r, s)$, $p(r)$ and $\mathcal{D}$ the unknown PDFs $p(x_{k,\pi_i(j)}, x_{k,\pi_i(j+1)}|\omega_i)$ and $p(x_{k,\pi_i(j)}|\omega_i)$ and the set of training samples assigned to ω_i, thus dropping both the indices i and j and the explicit indication of the considered class ω_i ($i = 1, 2, \ldots, M; j = 2, 3, \ldots, n-1$). Thanks to the aforementioned class-conditional i.i.d. assumption, the dependence on the pixel position $k \in \{1, 2, \ldots, N\}$ can be neglected as well. Then, the proposed method separately iterates for each class ω_i and each permutation index j ($i = 1, 2, \ldots, M; j = 2, 3, \ldots, n-1$) the estimation procedure described below in terms of the random variables r and s.

In particular, each couple (r, s) of features can be made of two SAR features, or of an optical feature and a SAR one, or of two optical features. For couples of optical features we adopt a bivariate Gaussian model, usually accepted for this data type.[1] The parameters of the model, i.e., the means and variances of r and s and the correlation coefficient between them are estimated by using the classical sample-mean and sample-covariance estimators computed on the set $\mathcal{D}$.

In the case of two SAR features, a parametric model taking into account the specific SAR natures of the channels is used. Parametric models for the joint statistics of an arbitrary number of SAR amplitudes have not been proposed in literature, but a few models have been developed for the case of two amplitudes. Here, we adopt the model developed by Lee *et al.*[37] for the joint PDF of two multilook SAR intensities[2] and applied here in the domain of SAR amplitudes (this model will be denoted by "bivariate Nakagami"), or the bivariate Weibull distribution,[38] generalizing to the two-dimensional case the univariate Weibull PDF,[2] frequently adopted as an empirical model for SAR features. Details on these two bivariate models and on the related parameter-estimation procedures are given in Secs. 3.3.1 and 3.3.2

For "mixed" couples (a SAR feature and an optical one) a characterization of the joint PDF is not known. Hence, we propose to use the approach by Storvik *et al.*,[39] which defines an estimate for the joint PDF of a feature vector in terms of parametric estimates of the marginal PDFs of the single features and of estimates of the correlations among them computed according to a Gaussian model in a suitable nonlinearly transformed feature space. Details are given in Sec. 3.3.3.

3.3.1. *The bivariate Nakagami model*

Assuming r and s to be SAR amplitudes, the bivariate Nakagami model for the related unknown PDFs is expressed by $(r, s > 0)$[b]:

$$p(r,s) = A(rs)^{\nu} \exp\left[-\frac{\nu}{1-\zeta}\left(\frac{r^2}{\lambda}+\frac{s^2}{\mu}\right)\right] I_{\nu-1}\left(\frac{2\nu rs}{1-\zeta}\sqrt{\frac{\zeta}{\lambda\mu}}\right)$$
$$p(r) = \frac{2}{\Gamma(\nu)}\left(\frac{\nu}{\lambda}\right)^{\nu} r^{2\nu-1}\exp\left(-\frac{\nu r^2}{\lambda}\right) \tag{9}$$

where ν $(\nu \geq 1)$ is a shape parameter representing the equivalent number of looks (ENL) of the SAR data, λ and μ $(\lambda, \mu > 0)$ are scale parameters characterizing the marginal distributions of r and s, respectively,[2] ζ is the correlation coefficient between r^2 and s^2, $\Gamma(\cdot)$ and $I_q(\cdot)$ are the well-known Gamma function and q-th order $(q \geq 0)$ modified Bessel function of the first kind,[42] and A is a normalization constant given by:[37]

$$A = \frac{4\nu^{\nu+1}}{(\lambda\mu)^{(\nu+1)/2}\Gamma(\nu)(1-\zeta)\zeta^{(\nu-1)/2}}. \tag{10}$$

More precisely, the bivariate PDF in Eq. (9) has been developed by assuming r^2 and s^2 to be ν-look SAR intensities, ν being a positive integer number.[37] Here, the model is further generalized by letting ν take on arbitrary values in $[1, +\infty)$, thus interpreting this parameter as the joint ENL of the SAR channels r and s. For the estimation of this parameter, as well as of λ and μ, several estimation strategies have been proposed.[2,43] Here, they are estimated by the method of log-cumulants (MoLC), a parametric modeling technique recently introduced in the SAR literature and based on the generalization of the concepts of characteristic function and moment-generating function[44] by the use of Mellin transforms instead of the usual Fourier and Laplace transforms.[43,45,46] Specifically, MoLC exploits the Mellin-transform theory to state a set of equations relating the unknown parameters to logarithmic moments or logarithmic cumulants ("log-cumulants" for short) of the graylevel distribution, thus expressing the parameter-estimation problem as the solution of a set of (typically nonlinear) equations. The method has been developed in the context of SAR-oriented parametric models and has proven to be numerically feasible and fast for many SAR-specific models (e.g., the Gamma, Nakagami, K, Weibull and log-normal distributions,[43] the symmetric-α-stable generalized Rayleigh model[47] and the generalized Gaussian-Rayleigh model[46]) and to

[b]SAR amplitudes take values in $(0, +\infty)$, so we explicitly define the related density and distribution functions only for $r, s > 0$ and implicitly set $p(r, s) = 0$ for $r \leq 0$ or $s \leq 0$.

exhibit good theoretical estimation properties.[43,46] When applied to the Nakagami model in Eq. (9), MoLC provides the following nonlinear system[c]:[43]

$$\begin{cases} \ln \lambda + \Psi(\nu) - \ln \nu = 2E\{\ln r\} \\ \ln \mu + \Psi(\nu) - \ln \nu = 2E\{\ln s\} \\ \Psi(1, \nu) = 4\mathrm{Var}\{\ln r\}, \end{cases} \tag{11}$$

where $\Psi(\cdot)$ and $\Psi(1, \cdot)$ are the digamma and first-order polygamma functions, respectively (i.e., the first and second logarithmic derivatives of $\Gamma(\cdot)$).[42] Therefore, after estimating the logarithmic means and variances of r and s as sample-log-moments on $\mathcal{D}$, an estimate of ν is first derived by numerically solving the third equation. Thanks to the strict monotonicity and smoothness of $\Psi(1, \cdot)$, this numerical task is not critical and can be addressed by the simple Newton's method.[42] Then, plugging the resulting value of ν in the first and second equation, estimates for λ and μ are analytically derived.

Finally, since $\zeta = \mathrm{Corr}\{r^2, s^2\}$, an estimate for ζ can be simply obtained as the sample-correlation coefficient between r^2 and s^2, computed again with the training samples in $\mathcal{D}$.

3.3.2. *The bivariate Weibull model*

When adopting the bivariate Weibull model for the couple (r, s) of SAR amplitudes, the unknown PDFs $p(r, s)$ and $p(r)$ are formulated as $(r, s > 0)$:[38]

$$\begin{aligned} p(r, s) &= \frac{\alpha\beta}{rs} G\left(\left(\frac{r}{\theta}\right)^{\alpha/\delta}, \left(\frac{s}{\phi}\right)^{\beta/\delta}, \frac{1}{\delta}\right) \\ p(r) &= \frac{\alpha}{\theta^\alpha} r^{\alpha-1} \exp\left[-\left(\frac{r}{\theta}\right)^\alpha\right] \end{aligned} \tag{12}$$

where δ $(0 < \delta \leq 1)$ is a parameter characterizing the statistical dependence between r and s, α and β $(\alpha, \beta > 0)$ are shape parameters of the marginal distributions of r and s, respectively, θ and ϕ $(\theta, \phi > 0)$ are scale parameters of those distributions, and the function $G(\cdot)$ is given by $(a, b, c > 0)$:

$$G(a, b, c) = \frac{ab[(a+b)^{1/c} + c - 1]}{(a+b)^{2-1/c}} \exp[-(a+b)^{1/c}]. \tag{13}$$

In particular, $\delta = 1$ implies independence between r and s, whereas smaller values of δ in the interval $(0, 1)$ indicate correlated features. We note that the bivariate Weibull model allows, in general, distinct values for the shape parameters α and β of the marginal distributions of r and s, whereas the bivariate Nakagami PDF is characterized by a common shape parameter ν. As in the case of bivariate Nakagami, the marginal PDF parameters of the bivariate Weibull distribution are

[c]Given a random variable v, we denote by $E\{v\}$ and $\mathrm{Var}\{v\}$ the mean and variance of v. Given a second random variable u, we denote by $\mathrm{Corr}\{u, v\}$ the correlation coefficient between u and v.

estimated by MoLC. When applied to this parametric model, MoLC provides the following nonlinear system:[43]

$$\begin{cases} \ln\theta + \Psi(1)\alpha^{-1} = E\{\ln r\} \\ \Psi(1,1)\alpha^{-2} = \mathrm{Var}\{\ln r\} \\ \ln\phi + \Psi(1)\beta^{-1} = E\{\ln s\} \\ \Psi(1,1)\beta^{-2} = \mathrm{Var}\{\ln s\} \end{cases} \tag{14}$$

Therefore, after estimating the logarithmic means and variances of r and s as sample-log-moments on $\mathcal{D}$, the solution of this system of equations can be easily carried out in closed form and estimates for α, β, θ, and ϕ are consequently derived. Moreover, the following relationship can be proved to hold[38] between the PDF parameters and the correlation coefficient between r and s:

$$H\left(\frac{\delta}{\alpha}+1, \frac{\delta}{\beta}+1, \frac{1}{\alpha}+1, \frac{1}{\beta}+1\right) = \mathrm{Corr}\{r, s\}, \tag{15}$$

where the function $H(\cdot)$ is given by $(a, b, c, d > 1)$:

$$H(a,b,c,d) = \frac{\Gamma(a)\Gamma(b)\Gamma(c+d-1) - \Gamma(c)\Gamma(d)\Gamma(a+b-1)}{\Gamma(a+b-1)\sqrt{[\Gamma(2c-1)-\Gamma(c)^2][\Gamma(2d-1)-\Gamma(d)^2]}}. \tag{16}$$

This allows the parameter δ to be estimated by first plugging into Eq. (15) both a sample-correlation-coefficient estimate for $\mathrm{Corr}\{r, s\}$ (computed again with the samples in $\mathcal{D}$) and the estimates obtained by MoLC for the other PDF parameters, and then by numerically solving Eq. (15) for δ. Since the mapping $\delta \mapsto \mathrm{Corr}\{r, s\}$ has been proved to be strictly monotonic (under mild assumptions[38]), this numerical task is not critical; here, it is addressed by the *regula-falsi* method.[42]

3.3.3. *Joint model for an optical and a SAR channel*

In the case of a mixed couple (r, s) of features, we adopt the approach developed by Storvik *et al.* to model the joint statistics of correlated radar channels.[39] Specifically, modeling the marginal PDFs of r and s represents no additional task since (either Gaussian, Nakagami, or Weibull) single-variate estimates are already available as byproducts of the analyses carried out in the previous sections. Thus, focusing on the problem of modeling the joint PDF $p(r, s)$, let r and s be a SAR amplitude and an optical channel, respectively, and let us assume that the marginal PDFs $p(s)$ and $p(r)$ have already been estimated by adopting a Gaussian model with mean η and variance σ^2 and a Nakagami or Weibull model, respectively. Let $F(r)$ be the cumulative distribution function (CDF) of r, i.e., $(r > 0)$:

$$F(r) = 1 - \exp\left[-\left(\frac{r}{\theta}\right)^{\alpha}\right], \tag{17}$$

if r is Weibull-distributed with parameters (α, θ), and

$$F(r) = \mathcal{P}\left(\nu, \frac{\nu r^2}{\lambda}\right), \tag{18}$$

if r is Nakagami-distributed with parameters (ν, λ) and if $\mathcal{P}(\cdot, \cdot)$ indicates the incomplete Gamma function.[42] Denoting by $\Phi(\cdot)$ the CDF of a normalized (i.e., zero-mean and unit-variance) Gaussian random variable and by $\Phi^{-1}(\cdot)$ the corresponding inverse function, the transformed variables $r' = \Phi^{-1}[F(r)]$ and $s' = (s - \eta)/\sigma$ are easily proved to be marginally normalized Gaussian.[44] The key idea of the approach by Storvik *et al.* is to assume r' and s' to be also jointly Gaussian.[39] Under this hypothesis, straightforward probability-theory calculations allows the joint PDF of r and s to be expressed as $(r > 0, s \in \mathbb{R})$:

$$p(r, s) = p(r)p(s)K\left(\Phi^{-1}[F(r)], \frac{s - \eta}{\sigma}, \rho\right), \tag{19}$$

where $\rho = \text{Corr}\{r', s'\}$ and the function $K(\cdot)$ is given by $(a, b \in \mathbb{R}, |\rho| < 1)$:

$$K(a, b, \rho) = \frac{1}{\sqrt{1 - \rho^2}} \exp\left[-\frac{\rho^2 a^2 + \rho^2 b^2 - 2\rho ab}{2(1 - \rho^2)}\right]. \tag{20}$$

This essentially formulates the unknown joint PDF of r and s in terms of the related marginal PDFs and of a multiplicative correction factor related to the correlation among r and s, measured in the nonlinearly transformed space of the variables r' and s'. The parameters of the two marginal PDFs can be estimated as discussed in the previous sections. An estimate for ρ is obtained by transforming through the mapping $(r, s) \mapsto (r', s')$ all training data in $\mathcal{D}$ and by computing the sample correlation coefficient between the resulting samples of r' and s'.

4. Experimental results

4.1. *Data set for experiments*

The proposed method was validated by experiments on a multisensor data set acquired over an agricultural area near Feltwell (UK) and consisting of six optical Daedalus 1268 Airborne Thematic Mapper (ATM) channels and of nine multipolarization (HH, HV, VV) and multifrequency (C, P, and L bands) SAR amplitudes recorded by the fully polarimetric NASA-JPL airborne sensor (Figs. 5(a)-(c)). Five classes were present, i.e., "carrots," "potatoes," "bare soil," "stubble," and "sugar." Ground truth data were available for such classes; spatially disjoint training and test fields were defined from this ground truth. The corresponding training and test sample sizes are reported in Table 1. Further details on the data set can be found in Serpico *et al.*[20]

4.2. *Region-based classification results*

The segmentation map obtained by TS-MRF was made of 2880 segments (Fig. 6(a)). Table 1 shows the classification accuracies of the proposed method, applied with this segmentation and with the bivariate Nakagami and bivariate Weibull models for couples of SAR channels. Similar and quite high overall (OAs) and average accuracies (AAs) were obtained when using both PDFs, even though no denoising was applied and most classes were vegetated land covers, strongly overlapping with one another in the multisensor feature space. This is interpreted as a consequence of the proposed region-level decision rule, which effectively allows spatial contextual information to be integrated in the classification process and the impact of speckle and noise on the classification result to be consequently minimized.

Table 1. Classification accuracies obtained by the proposed region-based (RB) method applied with either bivariate Nakagami or Weibull models for couples of SAR amplitudes, by the non-contextual variant applied with (NCF) or without (NC) noise filtering, and by the DTC method.

class	training samples	test samples	bivariate Nakagami RB	bivariate Nakagami NC	bivariate Nakagami NCF	bivariate Weibull RB	bivariate Weibull NC	bivariate Weibull NCF	DTC (NC)
carrots	814	967	75.6%	76.4%	92.9%	78.7%	76.9%	96.1%	83.5%
potatoes	1411	884	90.8%	81.0%	83.8%	90.2%	68.2%	88.0%	76.4%
bare soil	341	555	84.5%	82.9%	77.3%	66.3%	45.0%	65.0%	66.5%
stubble	1070	1371	85.3%	81.3%	83.7%	93.2%	90.0%	88.1%	87.0%
sugar	1488	2043	94.1%	89.9%	92.8%	93.9%	88.9%	78.0%	92.4%
	overall accuracy		87.6%	83.6%	87.8%	88.0%	79.8%	87.0%	84.7%
	average accuracy		86.1%	82.3%	86.1%	84.5%	73.8%	83.1%	81.1%

In order to more specifically focus on the role of contextual information in the method, two experimental comparisons were carried out. First, a noncontextual variant of the technique was considered by applying the MAP rule at the pixel level after estimating the class-conditional PDFs of the multisensor feature vector by the algorithm in Sec. 3.3. Due to class overlapping in the feature space and especially to the strong presence of speckle (Fig. 5(c)), significantly less accurate results were obtained in this case (Table 1). As compared to this noncontextual variant, incorporating spatial information through segmentation yields an almost 4% increase in both OA and AA in the case of the bivariate Nakagami PDF, an 8.2% increase in OA and a 10.6% increase in AA in the case of the bivariate Weibull model. A visual analysis of the related classification maps confirms these comments and points out a strong impact of speckle on the noncontextual results (Fig. 6(d)), whereas the maps generated by the proposed region-based method were not significantly affected (Figs. 6(b) and (c)).

Then, the noncontextual variant was applied after noise filtering (a simple moving-average filter was used with a 5×5 and 9×9 window for the multispectral and SAR channels, respectively). Similar accuracies were obtained in this case and by the proposed region-based method (Table 1). However, a visual analysis points out a significant oversmoothing of the edges between adjacent regions in the map obtained after noise filtering (Fig. 6(e)). On the contrary, no such effect was remarked in the maps generated by the proposed region-based technique (Fig. 6(c)). This further confirms the effectiveness of this approach in incorporating spatial-contextual information in optical-SAR classification. We note that edge effects did

Fig. 5. Data set for experiments: ATM-3 (a), ATM-7 (b), and C-HH (c) channels after histogram equalization.

not affect the numerical accuracies in Table 1 because the test samples were located inside homogeneous areas and not at the borders between distinct regions (this choice is usual practice in remote sensing and is explained by the need to prevent the inclusion of mixed pixels in ground-truth data).

A further comparison was performed between the proposed technique and the nonparametric dependence-tree classifier (DTC), which combines the dependence-tree approximation with Parzen estimates for the related bivariate PDFs.[3] Here, the Parzen algorithm was applied with Gaussian kernels and with the reference-density approach to kernel-width optimization, applied with a Rayleigh reference density (details can be found in Moser *et al.*[45]). The resulting accuracies turned out to be lower than the ones achieved by the proposed method and quite similar to those given by the noncontextual variant. This is consistent with the fact that DTC is a noncontextual approach whose application does not allow the spatial information to be exploited for classification and further underscores the importance of contextual techniques for multisensor image classification.

5. Conclusions

In the present chapter the problem of supervised classification of multisensor optical-SAR images was addressed by first revising the main approaches proposed in the literature and then by proposing a novel optical-SAR image contextual classification method. This technique combines the dependence-tree approach to PDF estimation with Markovian segmentation and with parametric bivariate models for the PDFs of optical and SAR channels. This allows the information related both to spatial context and to the optical or SAR nature of each available channel to be exploited in the classification process.

Accurate results were obtained on a challenging 15-band optical-SAR data set, including both multispectral observations and multipolarization and multifrequency

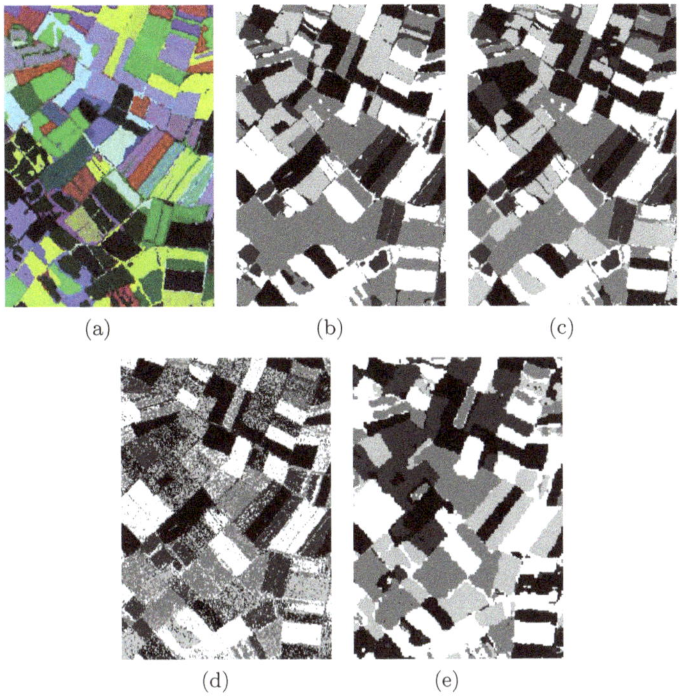

Fig. 6. Experimental results: TS-MRF segmentation map (a) and classification maps generated by the proposed method with bivariate Nakagami (b) and Weibull (c) models, and by the noncontextual variant applied with bivariate Weibull without (d) and after denoising (e). Legend for the classification maps: white = "sugar," light grey = "bare soil," middle grey = "potatoes," dark grey = "carrots," black = "stubble."

SAR channels and related to an agricultural scene with strongly overlapping vegetated classes. This suggests the effectiveness of the method in exploiting multisensor information for classification purposes. In particular, the method exhibited a sharply improved robustness to speckle, as compared to a noncontextual variant that classifies each pixel *per se* without using segmentation information. The application of this variant after noise filtering also yielded undesired edge-oversmoothing effects, whereas the use of segmentation prevented such phenomena. This further confirms the importance of spatial context in the classification of multisensor optical-SAR data and suggests the effectiveness of the proposed approach in exploiting this information. In this perspective, a future extension of the proposed technique could be the integration of edge information in the method (e.g., by line processes[15]) in order to further optimize the classification accuracy in image areas near the spatial borders between different classes. In addition, more sophisticated Bayesian approaches could be integrated in the algorithm in order to allow a region-level decision rule to be formulated without involving class-conditional independence assumptions, thus yielding a more general classifier and further maximizing the classification accuracy.

Acknowledgments

This research was carried out within the framework of the PRIN-2005 project entitled "Development and validation of innovative data-fusion techniques for environmental remote sensing," which was funded by the Italian Ministry of Education, University, and Research. The support is gratefully acknowledged.

References

1. J. Richards and X. Jia, *Remote sensing digital image analysis.* (Springer, 2005).
2. C. Oliver and S. Quegan, *Understanding synthetic aperture radar images.* (SciTech Publishing, 2004).
3. M. Datcu, F. Melgani, A. Piardi, and S. B. Serpico, Multisource data classification with dependence trees, *IEEE Trans. Geosci. Remote Sensing.* **40**(3), 609–617, (2002).
4. C. D'Elia, G. Poggi, and G. Scarpa, A tree-structured Markov random field model for Bayesian image segmentation, *IEEE Trans. Image Process.* **12**(10), 1259–1273, (2003).
5. C. D'Elia, G. Poggi, and G. Scarpa. Improved tree-structured segmentation of remote sensing images. In *Proc. of IGARSS-2003, Toulouse, France*, vol. 3, pp. 1805–1807, (2003).
6. A. H. S. Solberg, T. Taxt, and A. K. Jain, Multisource classification of remotely sensed data: fusion of Landsat TM and SAR images, *IEEE Trans. Geosci. Remote Sensing.* **32**(3), 768–778, (1994).
7. J. A. Benediktsson, P. H. Swain, and O. K. Ersoy, Neural network approaches versus statistical methods in classification of multisource remote sensing data, *IEEE Trans. Geosci. Remote Sensing.* **28**, 540–552, (1990).
8. B. Waske and J. A. Benediktsson, Fusion of support vector machines for classification of multisensor data, *IEEE Trans. Geosci. Remote Sensing.* **45**, 781–790, (2007).
9. S. L. Hegarat-Mascle, I. Bloch, and D. Vidal-Madjar, Application of Dempster-Shafer evidence theory to unsupervised classification in multisource remote sensing, *IEEE Trans. Geosci. Remote Sensing.* **35**, 1018–1021, (1997).
10. J. A. Benediktsson, J. R. Sveinsson, and P. H. Swain, Hybrid consensus theoretic classification, *IEEE Trans. Geosci. Remote Sensing.* **35**, 833–843, (1997).
11. P. Lombardo, J. Oliver, T. Macrì-Pellizzeri, and M. Meloni, A new maximum-likelihood joint segmentation technique for multitemporal SAR and multiband optical images, *IEEE Trans. Geosci. Remote Sensing.* **41**, 2500–2518, (2003).
12. A. Palau, F. Melgani, and S. B. Serpico, Cell algorithms with data inflation for non-parametric classification, *Patt. Rec. Letters.* **27**, 781–790, (2006).
13. C. F. Hutchinson, Techniques for combining landsat and ancillary data for digital classification improvement, *Photogram. Eng. and Remote Sensing.* **48**, 123–130, (1982).
14. T. Lee, J. A. Richards, and P. H. Swain, Probabilistic and evidential approaches for multisource data analysis, *IEEE Trans. Geosci. Remote Sensing.* **GE-25**, 283–293, (1987).
15. A. H. S. Solberg, T. Taxt, and A. K. Jain, A Markov random field model for classification of multisource satellite imagery, *IEEE Trans. Geosci. Remote Sensing.* **34**(1), 100–113, (1996).
16. J. A. Benediktsson and P. H. Swain. A method of statistical multisource classification with a mechanism to weight the influence of the data sources. In *Proc. of IGARSS-2089, Vancouver, Canada*, pp. 517–520, (1989).

17. L. Bruzzone, D. F. Prieto, and S. B. Serpico, A neural-statistical approach to multitemporal and multisource remote-sensing image classification, *IEEE Trans. Geosci. Remote Sensing.* **37**, 1350–1359, (1999).
18. L. Bruzzone, C. Conese, F. Maselli, and F. Roli, Multisource classification of complex rural areas by statistical and neural network approaches, *Photogram. Eng. and Remote Sensing.* **63**(5), 523–533, (1997).
19. S. B. Serpico and F. Roli, Classification of multisensor remote-sensing images by structured neural networks, *IEEE Trans. Geosci. Remote Sensing.* **33**, 562–578, (1995).
20. S. B. Serpico, L. Bruzzone, and F. Roli, An experimental comparison of neural and statistical nonparametric algorithms for supervised classification of remote-sensing images, *Pattern Recognit. Lett.* **17**(3), 1331–1341, (1996).
21. P. Gamba and B. Houshmand, An efficient neural classification chain of SAR and optical urban images, *Int. J. Remote Sensing.* **22**(8), 1535–1553, (2001).
22. P. Gamba and F. Dell'Acqua, Increased accuracy multiband urban classification using a neuro-fuzzy classifier, *Int. J. Remote Sensing.* **24**(4), 827–834, (2003).
23. J. Munoz-Mari, L. Bruzzone, and G. Camps-Valls, A support vector domain description approach to supervised classification of remote sensing images, *IEEE Trans. Geosci. Remote Sensing.* **45**, 2683–2692, (2007).
24. G. Camps-Valls, L. Gomez-Chova, J. Munoz-Mari, J. L. Rojo-Alvarez, and M. Martinez-Ramon, Kernel-based framework for multitemporal and multisource remote sensing data classification and change detection, *IEEE Trans. Geosci. Remote Sensing.* **46**, 1822–1835, (2008).
25. B. Waske and S. van der Linden, Classifying multilevel imagery from SAR and optical sensors by decision fusion, *IEEE Trans. Geosci. Remote Sensing.* **46**, 1457–1446, (2008).
26. G. Shafer, *A mathematical theory of evidence.* (Princeton Univ. Press, 1979).
27. G. Chust, D. Ducrot, and J. L. L. Pretus, Land cover discrimination potential of radar multitemporal optical multispectral images in a mediterranean cultural landscape, *Int. J. Remote Sensing.* **25**(17), 3513–3528, (2004).
28. J.A.Benediktsson and J. Sveinsson, Multisource remote sensing data classification based on consensus and pruning, *IEEE Trans. Geosci. Remote Sensing.* **41**(4), 932–936, (2003).
29. Y. Liua, Q. Guob, and M. Kelly, A framework of region-based spatial relations for non-overlapping features and its application in object based image analysis, *ISPRS J. Photogramm. Rem. Sens.* **63**, 461–475, (2008).
30. A. K. Shackelford and C. H. Davis, A combined fuzzy pixel-based and object-based approach for classification of high-resolution multispectral data over urban areas, *IEEE Trans. Geosci. Remote Sensing.* **41**, 2354–2363, (2003).
31. W. Sun, V. Heidt, P. Gong, and G. Xu, Information fusion for rural land-use classification with high-resolution satellite imagery, *IEEE Trans. Geosci. Remote Sensing.* **41**, 883–890, (2003).
32. J. Inglada, Automatic recognition of man-made objects in high resolution optical remote sensing images by SVM classification of geometric image features, *ISPRS J. Photogramm. Rem. Sens.* **62**, 236–248, (2007).
33. A. Katartzis, I. Vanhamel, and H. Sahli, A hierarchical Markovian model for nultiscale region-based classification of vector valued-image, *IEEE Trans. Geosci. Remote Sensing.* **43**, 548–558, (2005).
34. K. Johansen, N. C. Coops, S. E. Gergel, and Y. Stange, Application of high spatial resolution satellite imagery for riparian and forest ecosystem classification, *Rem. Sens. Environ.* **110**, 29–44, (2007).

35. H. G. Akcay and S. Aksoy, Automatic detection of geospatial objects using multiple hierarchical segmentations, *IEEE Trans. Geosci. Remote Sensing.* **46**, 2097–2111, (2008).
36. S. Z. Li, *Markov random field modeling in image analysis.* (Springer-Verlag, 2001).
37. J.-S. Lee, K. W. Hoppel, S. A. Mango, and A. R. Miller, Intensity and phase statistics of multilook polarimetric and interferometric SAR imagery, *IEEE Trans. Geosci. Remote Sensing.* **32**(5), 1017–1028, (1994).
38. R. A. Johnson and W. Lu, Proof load designs for estimation of dependence in a bivariate Weibull model, *Stat. Probabil. Lett.* **77**, 1061–1069, (2007).
39. B. Storvik, G. Storvik, and R. Fjortoft. Joint distributions for correlated radar images. In *Proc. of IGARSS-2003, Toulouse, France*, vol. 3, pp. 2011–2013, (2003).
40. C. K. Chow and C. N. Liu, Approximating discrete probability distributions with dependence trees, *IEEE Trans. Inform. Theory.* **14**, 462–467, (1968).
41. J. B. Kruskal, On the shortest spanning subtree of a graph and the traveling salesman problem, *Proc. Amer. Math. Soc.* **7**, 48–50, (1956).
42. W. H. Press, S. A. Teukolsky, W. T. Wetterling, and B. P. Flannery, *Numerical recipes in C.* (Cambridge University Press, Cambridge (UK), 2002).
43. J.-M. Nicolas, Introduction aux statistiques de deuxième espéce: applications des log-moments et des log-cumulants a l'analyse des lois d'images radar, *Traitement du Signal.* **19**, 139–167, (2002).
44. A. Papoulis and S. U. Pillai, *Probability, random variables, and stochastic processes.* (McGraw-Hill, 2002).
45. G. Moser, J. Zerubia, and S. B. Serpico. Dictionary-based Stochastic Expectation-Maximization for SAR amplitude probability density function estimation. Research Report 5154, INRIA (Mar., 2004). URL `http://www.inria.fr/rrrt/rr-5154.html`.
46. G. Moser, J. Zerubia, and S. B. Serpico, SAR amplitude probability density function estimation based on a generalized Gaussian model, *IEEE Trans. Image Process.* **15** (6), 1429–1442, (2005).
47. E. E. Kuruoglu and J. Zerubia, Modeling SAR images with a generalization of the Rayleigh distribution, *IEEE Trans. Image Process.* **13**(4), 527–533, (2004).

CHAPTER 3.2

APPROACHES BASED ON SUPPORT VECTOR MACHINE TO CLASSIFICATION OF REMOTE SENSING DATA

Lorenzo Bruzzone and Claudio Persello

Dept. of Information Engineering and Computer Science, University of Trento, Via Sommarive, 14, I-38123 Trento, Italy; E-mail: lorenzo.bruzzone@ing.unitn.it; claudio.persello@disi.unitn.it

This chapter presents an extensive and critical review on the use of kernel methods and in particular of support vector machines (SVMs) in the classification of remote-sensing (RS) data. The chapter recalls the mathematical formulation and the main theoretical concepts related to SVMs, and discusses the motivations at the basis of the use of SVMs in remote sensing. A review on the main applications of SVMs in classification of remote sensing is given, presenting a literature survey on the use of SVMs for the analysis of different kinds of RS images. In addition, the most recent methodological developments related to SVM-based classification techniques in RS are illustrated by focusing on semisupervised, domain adaptation, and context sensitive approaches. Finally, the most promising research directions on SVM in RS are identified and discussed.

1. Introduction

In the last two decades there have been significant improvements both in the technology associated with the development of the sensors used in remote sensing (RS) to acquire signals and images for earth observation and in the analysis techniques adopted for extracting information from these data useful for operational applications. The modern technology resulted in the definition of different kinds of sensors for Earth observation based on different principles and with different properties. We have observed the development of passive multispectral and hyperspectral scanners, and active instruments like Synthetic Aperture Radar (SAR) and Lidar, as well as other instruments devoted to specific applications. Looking at this scenario from an historical viewpoint, we are passed from multispectral data having relatively low spatial and spectral resolution (like the MSS of the Landsat satellites) to a new generation of sensors characterized by very high geometrical resolution (VHR); (e.g., Ikonos, Quickbird, Geoeye-1, etc.), which can acquire images with a metric or sub-metric resolution. Hyperspectral sensors (on airborne or satellite platforms) can acquire images characterized by very high spectral resolution, with hundreds of channels having less than 2 nm of bandwith. The acquisition of very high resolution SAR images from satellite platforms has also become possible thanks to the recent TerraSar-X and Cosmo-Skymed missions. In this context, the challenging properties of new generation of sensors require the definition of novel data analysis methods.

In this chapter we focus our attention on RS image classification methodologies, which are devoted to translate the features that represent the information present in the data in thematic maps representing land cover types according to the solution of a pattern recognition problem. In particular, we concentrate our attention on supervised classification algorithms, which require the availability of labeled samples for the training of the classification model. In this context, the availability of last generation RS images allowed the development of new applications that require the mapping of the Earth surface with high geometric precision and a high level of thematic details. However, the huge amount of data associated with these images requires the development of sophisticated automatic classification techniques capable to obtain accurate land-cover maps in a reasonable processing time.

In the last decades, a great effort has been devoted to exploit machine learning methods for classification of remote sensing images. This has been done by introducing the use of neural networks (NN) in remote sensing (with the pioneering work presented in [1]) for solving many different classification tasks. Several different paradigms and models of NN have been used in recent years for addressing remote sensing image classification problems, ranging from standard Multilayer Perceptron (MLP) network[1-3], to Radial Basis Functions (RBF) neural network[4,5], structured neural networks[6] and hybrid architectures. Also more complex and structured architecture have been exploited for solving specific problems, like compound classification of multitemporal data[7], multiple classification systems made up of neural algorithms[8,9], etc. All these methods share as common property the idea to perform the learning of the classification algorithm according to the minimization of the empirical risk, defined in different ways. However, the last frontiers of machine learning classifiers in RS are represented by methods based on the structural risk minimization principle (which allows one to effectively tune the tradeoff between empirical risk and generalization capability) rather than on the empirical risk minimization. The related statistical learning theory (formulated from Vapnik[10]) is at the basis of the support vector machine (SVM) classification approach. SVM is a classification technique based on kernel methods that has been proved very effective in solving complex classification problems in many different application domains. In the last few years, SVM gained a significant credit also in remote sensing applications. The pioneering work of Gualtieri in 1998[11] related to the use of SVM for classification of hyperspectral images has been followed from several different experiences of other researchers that analyzed the theoretical properties and the empirical performances of SVM applied to different kinds of classification problems[12-28]. The investigations include classification of hyperspectral data[11-18], multispectral images[19-26], VHR images[27], as well as multisource and multisensor classification scenarios[28-30]. SVMs revealed to be very effective classifiers and currently they are among the most adequate techniques for the analysis of last generation of RS data.

In all these cases the success of SVMs is due to the important properties of this approach, which integrated with the effectiveness of the classification procedure and the elegance of the theoretical developments, result in a very solid classification methodology in many different RS data classification domains. As it will be explained in the following section, this mainly depends on the fact that SVMs implement a classification strategy that exploits a margin-based "geometrical" criterion rather than a purely "statistical" criterion. In other words, SVMs do not require an estimation of the statistical distributions of classes to carry out the classification task, but they define the classification model by exploiting the concept of margin maximization.

The main properties that make SVM particularly attractive in RS applications can be summarized as follows[31-33]:

- their intrinsic effectiveness with respect to traditional classifiers thanks to the structural risk minimization principle, which results in high classification accuracies and very good generalization capabilities (especially in classification problems defined in high dimensional feature spaces and with few training samples, which it is a typical situation in the classification of last generation of RS images);
- the possibility to exploit the kernel trick to solve non-linear separable classification problems by projecting the data into a high dimensional feature space and separating the data with a simple linear function;
- the convexity of the objective function used in the learning of the classifier, which results in the possibility to solve the learning process according to linearly constrained quadratic programming (QP) characterized from a unique solution (i.e., the system cannot fall into sub-optimal solutions associated with local minima);

- the possibility of representing the convex optimization problem in a dual formulation, where only non-zero Lagrange multipliers are necessary for defining the separation hyperplane (which is a very important advantage in the case of large datasets). This is related to property of sparseness of the solution;

Moreover, SVMs exhibit important advantages with respect to NN approaches. Among the others we recall: i) higher generalization capability and robustness to the Hughes phenomenon; ii) lower effort required for the model selection in the learning phase (i.e., they involve less control parameters and thus computational time for their optimum values selection) and the implicit automatic architecture definition; iii) optimality of the solution obtained by the learning algorithm.

The objective of this chapter is to review the state of the art of SVM for the classification of RS data. In particular, Section II recalls the basic principles of SVM for pattern classification. Section III presents a literature survey about the most relevant papers that report studies about the application of SVM to the classification of different kinds of RS images and papers that propose advanced systems based on the SVM approach for the analysis of RS data. Along with this state-of-the-art review, we discuss about the operative adoption of SVM for the analysis of RS images and the direction of the future research on this topic. Finally, section IV draws the conclusion of the chapter.

2. Support Vector Machines Classifiers

Let us consider the problem of supervised classification of a generic d-dimensional image Υ of size $I \times J$ pixel. Let us assume that a training set $\mathcal{D}=\{\mathcal{X},\mathcal{Y}\}$ made up of N pairs $(\mathbf{x}_i, y_i)_{i=1}^N$ is available, where $\mathcal{X}=\{\mathbf{x}_i \mid \mathbf{x}_i \in \mathbb{R}^d\}_{i=1}^N \subset \Upsilon$ is a subset of Υ and $\mathcal{Y}=\{y_i\}_{i=1}^N$ is the corresponding set of labels. For the sake of simplicity, since SVMs are binary classifiers, we first focus the attention on the two-class case (the general multiclass case will be addressed later). Accordingly, let us assume that $y_i \in \{+1;-1\}$ is the binary label of the pattern $\mathbf{x}_i$. The goal of the binary SVM is to divide the d-dimensional feature space in two subspaces, one for each class, through a separating hyperplane $H: y=\langle \mathbf{w} \cdot \mathbf{x} \rangle + b = 0$. The final decision rule used to find the membership of a test sample is based on the sign of the discrimination function $f(\mathbf{x})=\langle \mathbf{w} \cdot \mathbf{x} \rangle + b$ associated to the hyperplane. Therefore, a generic pattern $\mathbf{x}$ will be labeled according to the following rule:

$$\begin{array}{lll} f(\mathbf{x})>0 & \Rightarrow & \mathbf{x}=+1 \\ f(\mathbf{x})\leq 0 & \Rightarrow & \mathbf{x}=-1 \end{array} \tag{1}$$

The training of an SVM consists in finding the position of the hyperplane H, estimating the values of the vector $\mathbf{w}$ and the scalar b, according to the solution of an optimization problem. From a geometrical point of view, $\mathbf{w}$ is a vector perpendicular to the hyperplane H and thus defines its orientation. The distance of the H to the origin is $b/\|\mathbf{w}\|$, while the distance of a sample $\mathbf{x}$ to the hyperplane is $f(\mathbf{x})/\|\mathbf{w}\|$. Let us define the *functional margin* $F=\min\{y_i f(\mathbf{x}_i)\}$, $i=1,\dots,N$ and the *geometric margin* $G=F/\|\mathbf{w}\|$. The geometric margin represents the minimum Euclidean distance between the available training samples and the hyperplane.

A. Training of Linear SVM - Maximal Margin algorithm

In the case of a linearly separable problems, the learning of an SVM can be performed with the maximal margin algorithm, which consists in finding the hyperplane H that maximizes the geometric margin G. Rescaling the hyperplane parameters $\mathbf{w}$ and b such that the functional margin $F=1$, it turns out that the optimal hyperplane can be determined as the solution of the following convex quadratic programming problem:

$$\begin{cases} \min\limits_{\mathbf{w},b}: \frac{1}{2}\|\mathbf{w}\|^2 \\ y_i \cdot [\langle \mathbf{w}\cdot\mathbf{x}_i\rangle + b] \geq 1, \quad \forall i=1,\dots,N \end{cases} \tag{2}$$

Let H_1 and H_2 be two hyperplane parallel to the separating hyperplane H and equidistant from it:

$$\begin{array}{l} H_1: f(\mathbf{x})=\langle \mathbf{w}\cdot\mathbf{x}\rangle + b = +1 \\ H_2: f(\mathbf{x})=\langle \mathbf{w}\cdot\mathbf{x}\rangle + b = -1 \end{array} \tag{3}$$

The goal of the training phase is to find the values of $\mathbf{w}$ and b such that the geometric distance between H_1 and H_2 is maximized with the condition that there is no sample between them. Since direct handling of inequality constraints is difficult, Lagrange theory is usually exploited by introducing Lagrange multipliers $\alpha_{i=1}^{N}$ for the quadratic optimization problem. This leads to an alternative dual representation:

$$\begin{cases} \max\limits_{\alpha}: \left\{ \sum_{i=1}^{N} \alpha_i - \frac{1}{2} \sum_{i=1}^{N}\sum_{j=1}^{N} y_i y_j \alpha_i \alpha_j \langle \mathbf{x}_i, \mathbf{x}_j \rangle \right\} \\ \sum_{i=1}^{N} y_i \alpha_i = 0, \quad \alpha_i \geq 0, \quad 1\leq i \leq N \end{cases} \tag{4}$$

The Karush–Kuhn–Tucker (KKT) complementarity conditions provide useful information about the structure of the solution. They state that the optimal solution $\boldsymbol{\alpha}^*$, $(\mathbf{w}^*, b^*)$ should satisfy:

$$\alpha_i^*\left[y_i(\langle \mathbf{w}^* \cdot \mathbf{x}_i \rangle + b^*) - 1\right] = 0, \qquad i = 1, \ldots, N \tag{5}$$

This implies that only input samples $\mathbf{x}_i$ for which the functional margin is one (and that therefore lie closest to the hyperplane, i.e., lie on H_1 or H_2) are associated to Lagrange multipliers $\alpha_i > 0$. All the other multipliers α_i^* are zero. Hence, only these samples are involved in the expression for the weight vector. It is for this reason that they are called *support vectors* (SV). Thus we can write that $\mathbf{w}^* = \sum_{i=1}^{N} y_i \alpha_i^* \mathbf{x}_i = \sum_{i \in SV} y_i \alpha_i^* \mathbf{x}_i$. It is worth noting that the term b does not appear in the dual problem, and should be calculated making use of the primal constraints:

$$b = -\frac{\max_{y_i=-1}(\langle \mathbf{w} \cdot \mathbf{x}_i \rangle) + \min_{y_i=+1}(\langle \mathbf{w} \cdot \mathbf{x}_i \rangle)}{2} \tag{6}$$

Once the values for $\mathbf{w}$ and b are determined by solving the optimization problem, one generic test sample is classified on the basis of the sign of the discriminant function, that can be expressed as:

$$f(\mathbf{x}) = \langle \mathbf{w} \cdot \mathbf{x} \rangle + b = \left(\sum_{i \in SV} y_i \alpha_i^* \mathbf{x}_i \right) \cdot \mathbf{x} + b = \sum_{i \in SV} y_i \alpha_i^* \langle \mathbf{x}_i \cdot \mathbf{x} \rangle + b. \tag{7}$$

Note that the training samples appear only in the form of dot product. This property of the dual form will be exploited later to extend the formulation to nonlinear problems.

B. Training of Linear SVM – Soft Margin algorithm

The maximum margin training algorithm can not be used in many real world problems where the available training samples are not linearly separable because of noisy samples and outliers (this is very common in real RS classification problems). In these cases, the soft margin algorithm is used in order to handle nonlinear separable data. This is done by defining the so called slack variables as:

$$\xi[(\mathbf{x}_i, y_i), (\mathbf{w}, b)] = \xi_i = \max[0, 1 - y_i(\langle \mathbf{w} \cdot \mathbf{x}_i \rangle + b)] \tag{8}$$

Slack variables allow one to control the penalty associated with misclassified samples (Fig. 1). In this way the learning algorithm is robust to both noise and outliers present in the training set, thus resulting in high generalization capability. The optimization problem can be formulated as follows:

$$\begin{cases} \min_{\mathbf{w},b} \left\{ \frac{1}{2}\|\mathbf{w}\|^2 + C\sum_{i=1}^{N}\xi_i \right\} \\ y_i(\langle \mathbf{w}\cdot\mathbf{x}_i\rangle + b) \geq 1-\xi_i,\ \xi_i \geq 0,\ \forall i=1,\ldots,N \end{cases} \tag{9}$$

where $C \geq 0$ is the regularization parameter that allows one to control the penalty associated to errors (if $C=\infty$ we come back to the maximal margin algorithm), and thus to control the tradeoff between the number of allowed mislabeled training samples and the width of the margin. If the value of *C* is too small, many errors are permitted and the resulting discriminant function will poorly fit with the data; on the opposite, if *C* is too large, the classifier may overfit the data instances, thus resulting in low generalization ability. A precise definition of the value of the *C* parameter is crucial for the accuracy that can be obtained in the classification step and should be derived through an accurate model selection phase.

Similarly to the case of the maximal margin algorithm, the optimization problem (9) can be rewritten in an equivalent dual form:

$$\begin{cases} \max_{\alpha}: \left\{ \sum_{i=1}^{N}\alpha_i - \frac{1}{2}\sum_{i=1}^{N}\sum_{j=1}^{N} y_i y_j \alpha_i \alpha_j \langle \mathbf{x}_i, \mathbf{x}_j \rangle \right\} \\ \sum_{i=1}^{N} y_i\alpha_i = 0,\ 0 \leq \alpha_i \leq C,\ 1 \leq i \leq N \end{cases} \tag{10}$$

Note that the only difference between (10) and (4) is in the constraint on the multipliers $\{\alpha_i\}_{i=1}^{N}$ that for the soft margin algorithm are bounded by the parameter *C*. For this reason this problem is also known as box constrained problem. The KKT conditions become in this case:

$$\begin{cases} \alpha_i[y_i(\langle \mathbf{w}\cdot\mathbf{x}_i\rangle + b) - 1 + \xi_i] = 0, & i=1,\ldots,l \\ \xi_i(\alpha_i - C) = 0, & i=1,\ldots,l \end{cases} \tag{11}$$

Varying the values of the multipliers $\{\alpha_i\}_{i=1}^{N}$ three cases can be distinguished:

(i) if $\alpha_i = 0 \Rightarrow \xi_i = 0$ and $y_i(\langle \mathbf{w}\cdot\mathbf{x}_i\rangle + b) \geq 1$;

(ii) if $0 < \alpha_i < C$, we have that $y_i(\langle \mathbf{w}\cdot\mathbf{x}_i\rangle + b) + \xi_i = 1$, but given that $\xi_i = 0$ we have that $y_i(\langle \mathbf{w}\cdot\mathbf{x}_i\rangle + b) = 1$;

(iii) if $\alpha_i = C$, $\Rightarrow$ $y_i(\langle \mathbf{w}\cdot\mathbf{x}_i\rangle + b) + \xi_i = 1$, but given that $\xi_i \geq 0$ we have that $y_i(\langle \mathbf{w}\cdot\mathbf{x}_i\rangle + b) \leq 1$.

The KKT conditions can therefore be rewritten as:

$$\begin{cases} \alpha_i = 0 \quad \Rightarrow y_i f(\mathbf{x}_i) \geq 1 \\ 0 < \alpha_i < C_i \Rightarrow y_i f(\mathbf{x}_i) = 1 \\ \alpha_i = C_i \quad \Rightarrow y_i f(\mathbf{x}_i) \leq 1 \end{cases} \tag{12}$$

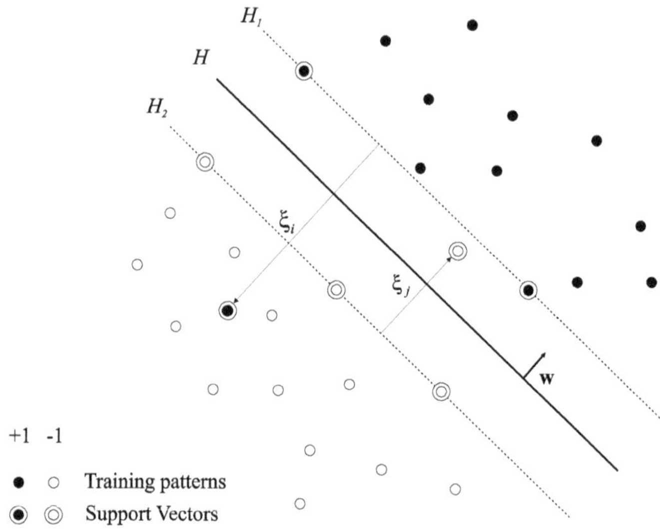

Fig. 1. Qualitative example of a separating hyperplane in the case of a non linear separable classification problem.

The *support vectors* with multiplier $\alpha_i = C$ are called *bound support vectors* (BSV) and are associated to slack variables $\xi_i \geq 0$; the ones with $0 < \alpha_i < C_i$ are called *non bound support vectors* (NBSV) and lie on the margin hyperplane H_1 or H_2 ($y_i f(\mathbf{x}_i) = 1$).

C. Training of Non Linear SVM – Kernel Trick

An important improvement to the above-described methods consists in considering non linear discriminant functions for separating the two information classes. This can be obtained by transforming the input data into a high dimension (Hilbert) feature space $\Phi(\mathbf{x}) \in \Re^{d'}$ ($d' > d$) where the transformed samples can be better separated by a hyperplane (Fig. 2). The main problem is to explicitly choose and calculate the function $\Phi(\mathbf{x}) \in \Re^{d'}$ for each training samples. But given that the input points in dual formulation [see (10)] appear in the form of inner products, we can do this mapping in an implicit way by exploiting the so called kernel trick. Kernel methods provide an elegant and effective way of dealing with this problem by replacing the inner product in the input space with a kernel function such that:

$$K(x_i, x_j) = \langle(\phi(\mathbf{x}_i) \cdot \phi(\mathbf{x}_j)\rangle \qquad i, j = 1, ..., N \tag{13}$$

implicitly calculating the inner product in the transformed space.

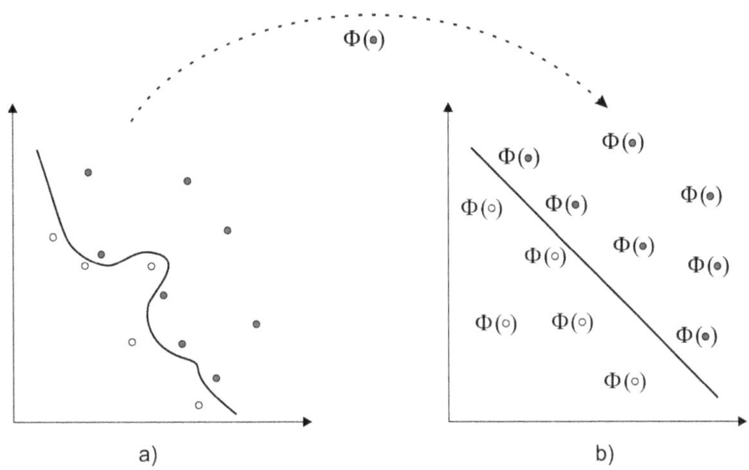

Fig. 2. Transformation of the input data by means of a kernel function into a high dimension feature space. a) Input feature space; b) kernel induced high dimensional feature space.

The soft margin algorithm for nonlinear function can be represented by the following optimization problem:

$$\begin{cases} \max\limits_{\alpha}\left\{\sum\limits_{i=1}^{N}\alpha_i-\dfrac{1}{2}\sum\limits_{i=1}^{N}\sum\limits_{j=1}^{N}y_iy_j\alpha_i\alpha_jk(\mathbf{x}_i,\mathbf{x}_j)\right\} \\ \sum_{i=1}^{N}y_i\alpha_i=0,\ 0\le\alpha_i\le C \text{ and } 1\le i\le N \end{cases} \tag{14}$$

And the discrimination function becomes:

$$f(\mathbf{x})=\sum_{i\in SV}y_i\alpha_i^{*}k(\mathbf{x}_i\cdot\mathbf{x})+b \tag{15}$$

The condition for a function to be a valid kernel is given by the Mercer's theorem[32]. The most widely used non-linear kernel functions are the following[31]:

- homogeneous polynomial function: $k(\mathbf{x}_i,\mathbf{x}_j)=(\mathbf{x}_i\cdot\mathbf{x}_j)^p,\quad p\in Z$
- inhomogeneous polynomial function: $k(\mathbf{x}_i,\mathbf{x}_j)=(c+(\mathbf{x}_i\cdot\mathbf{x}_j))^p,\quad p\in Z, c\ge 0$
- Gaussian function: $k(\mathbf{x}_i,\mathbf{x}_j)=e^{\frac{\|\mathbf{x}_i-\mathbf{x}_j\|^2}{2\sigma^2}},\quad \sigma\in\Re$

D. Multiclass architectures

As stated in the previous section, SVMs are binary classifiers. However, several strategies have been proposed to address multiclass problems with SVMs. Let $\Omega=\{\omega_1,\ldots,\omega_L\}$ be the set of L information classes associated with the different land cover types present in the study area. In order to define a multiclass architecture based on different binary classifiers, the general approach consists of: i) defining an ensemble of binary classifiers; and ii) combining them according to some decision rules. The definition of the ensemble of binary classifiers involves the definition of a set of two-class problems, each modeled with two groups Ω_A and Ω_B of classes. The selection of these subsets depends on the kind of approach adopted to combine the ensemble. In the following, we describe the two most widely adopted (parallel) multiclass strategies, i.e., the *One-Agains-All* (OAA) and *One-Against-One* (OAO) strategies.

1) *One-Against-All:* the one-against-all (OAA) strategy represents the earliest and one of the most common multiclass approach used for SVMs. It involves a parallel architecture made up of L SVMs, one for each class (Fig. 3). Each SVM solves a two-class problem defined by one information class against all the others, i.e.,

$$\begin{cases} \Omega_A = \omega_i \\ \Omega_B = \Omega - \omega_i \end{cases} \tag{16}$$

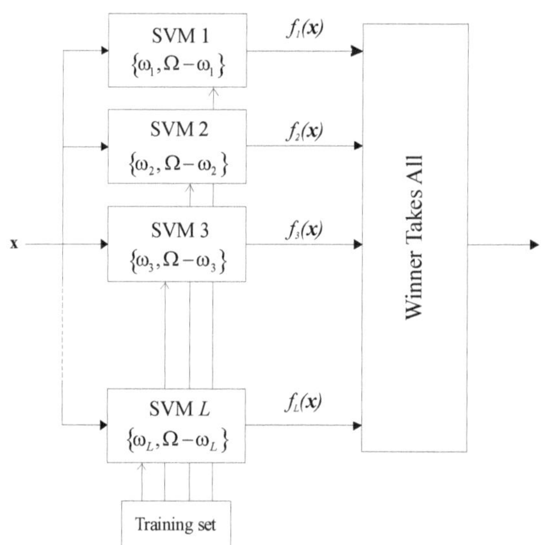

Fig. 3. Block diagram of the *One-Against-All* multiclass architecture.

The *winner-takes-all* rule is used for the final decision, i.e., the winning class is the one corresponding to the SVM with the highest output (discriminant function value).

2) *One-Against-One:* the main problem of the OAA strategy is that the discrimination between an information class and all the others often leads to the

estimation of complex discriminant functions. In addition, a problem with strongly unbalanced prior probabilities should be solved by each SVM. The idea behind the *one-against-one* (OAO) strategy is that of a different reasoning, in which simple classification tasks are made possible thanks to a parallel architecture made up of a large number of SVMs. The OAO strategy involves $L(L-1)/2$ SVMs, which model all possible pairwise classifications. In this case, each SVM carries out a binary classification in which two information classes ω_i and ω_j are analyzed against each other by means of a discriminant function $f_{ij}(\mathbf{x})$. Consequently, the grouping becomes:

$$\begin{cases} \Omega_A = \omega_i \\ \Omega_B = \omega_j \end{cases} \tag{17}$$

Before the decision process, it is necessary to compute for each class $\omega_i \in \Omega$ a score function $S_i(\mathbf{x})$, which sums the favorable and unfavorable votes expressed for the considered class

$$S_i(\mathbf{x}) = \sum_{\substack{j=1 \\ j \neq i}}^{L} \operatorname{sgn}[f_{ij}(\mathbf{x})] \tag{18}$$

The final decision in the OAO strategy is taken on the basis of the *winner-takes-all* rule, which corresponds to the following maximization (Fig. 4):

$$\mathbf{x} \in \omega \Leftrightarrow \omega = \arg\max_{i=1,\dots,L} \{S_i(\mathbf{x})\} \tag{19}$$

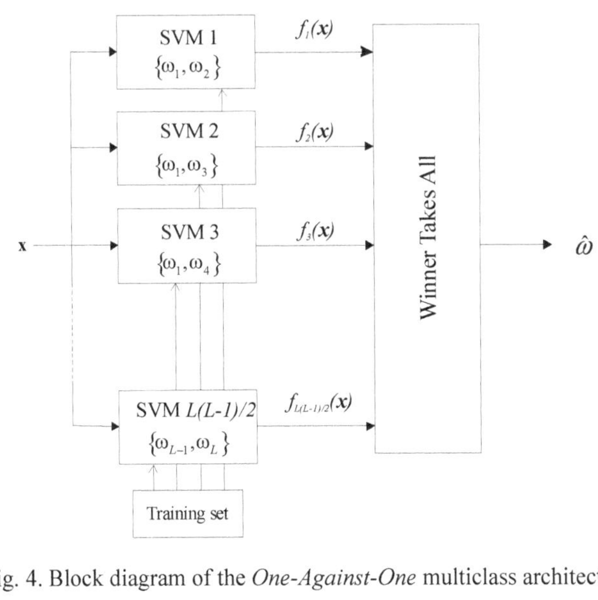

Fig. 4. Block diagram of the *One-Against-One* multiclass architecture.

Other multiclass architectures proposed in the literature are the Directed Acyclic Graph SVM (DAGSVM)[34] and different approaches based on binary hierarchical trees (BHT)[16,35].

3. SVM for the classification of RS Data

In the last decade many studies have been published in the RS literature on the application of SVM classifiers to the analysis of RS data.

Table 1 (which is not exhaustive) presents some relevant papers about the applications of SVM to the classification of RS data, providing a short description of the study and the kind of data used for the experimental analysis. The SVM approach has been first applied to the classification of hyperspectral data[11], which require the classifier to operate in large dimensional feature spaces. Supervised classification of hyperspectral images is a very complex methodological problem due to many different issues, among which we recall the typical small value of the ratio between the number of training samples and the number of available spectral channels, which results in the so-called course of dimensionality (Hughes phenomenon)[36]. Thanks to the structural risk minimization principle and the margin-based approach, SVMs represent an effective choice for the classification of this specific kind of data. Several papers[11-18] confirm the effectiveness of SVMs in the classification of hyperspectral images, which outperform other classification algorithms both in terms of classification accuracy and generalization ability. In particular, in [16] it is found that SVMs are much more effective than other conventional nonparametric classifiers (i.e., the RBF neural networks and the *K*-NN classifier) in terms of classification accuracy, computational time, stability to parameter setting, and generalization ability. In [15], the SVM approach was compared with neural networks and fuzzy methods on six hyperspectral images acquired with the 128-band HyMap spectrometer. The authors of the study concluded that SVMs yield better outcomes than neural networks regarding accuracy, simplicity, and robustness. In [17], SVMs were compared with other kernel-based methods, i.e., with regularized radial basis function NN, kernel Fisher discriminant analysis, and regularized AdaBoost. The results obtained on an AVIRIS dataset show that SVMs are more beneficial, yielding better results than other kernel-based methods, ensuring sparsity and lower computational cost.

Nevertheless, SVMs revealed adequate for the analysis of many different kinds of RS data, i.e., multispectral imagery and SAR imagery (with different resolutions) and LIDAR data. Several papers present a comparison between SVM and other supervised algorithms applied to the classification of different kinds of RS images[20,23,25,30]. In [20], for instance, the authors compared the accuracies obtained by the classification of a Landsat Thematic mapper (TM) scene with four different supervised classifiers, i.e., SVM, maximum likelihood (ML), MLP neural networks (NN), and decision tree classifier (DTC). The obtained results show that SVM was in general sharply more accurate than ML and DTC, and more accurate than NN in most of the cases. In [21], the SVM algorithm was applied to the classification of ASTER data acquired in an urban area of Beer Sheva, Israel. Field validations show that the classification is reliable for urban studies with high classification accuracy. In [23], the SVM classifier, as well as the well-known ML classifier and a context-based classifier based on Markov random fields, were applied to the automatic land cover classification of a Landsat TM image taken on the Tenerife Island. The authors found that SVM was more accurate than the other classification algorithms, but the classification map was not completely satisfying when investigated visually. In the experimental analysis conducted in [25], it is observed that SVM leaded to slightly higher classification accuracies than (MLP) NN. For both classifiers, the accuracy depends on factors such as the number of hidden nodes in the case of NN, and kernel parameters in the case of SVM. Thus, the model selection phase is fundamental for obtaining good results, but the training time required by the SVM is less than the one taken by NN.

SVM can be particularly effective also in the analysis of very high resolution (VHR) images. The typical poor spectral resolution of VHR images requires the extraction of additional features (e.g., texture and geometric measures) to characterize the objects present in the scene under investigation and to discriminate different land-cover classes. Different features modeling objects at different scales are generally necessary for an adequate characterization of the information classes[27], thus resulting in classification problems characterized by large dimensional feature spaces (with some analogies with the problems related to the classification of hyperspectral image). The study proposed in [27] points out that SVM can be effectively applied to the classification of VHR images using a feature extraction block that aims at adaptively modeling the spatial context of each pixel according to a hierarchical multilevel segmentation of the scene. A similar approach can also be adopted for the joint classification of SAR and optical data with SVM, as presented in [29]. In [30], an analysis is proposed on the joint use of hyperspectral and LIDAR data for the classification of complex forest areas. The experimental results obtained in [29-30] show that SVMs are effective for combining multisensor data in complex classification problems and outperforms other more traditional classifiers.

Table 1 – Selected papers related to the application of SVM to classification of different kinds of RS data.

Authors	Description	RS data
J. A. Gualtieri and S. Chettri[13]	The authors introduce SVM for the classification of RS data. In particular they applied SVM to hyperspectral data acquired by NASA's AVIRIS sensor and the commercially available AISA sensor. The authors discuss the robustness of SVM to the course of dimensionality (Hughes phenomenon).	AVIRIS (224 spectral bands) and AISA (20-40 bands)
F. Melgani, L. Bruzzone[16]	The authors propose a theoretical discussion and experimental analysis aimed at understanding and assessing the potentialities of SVM classifiers in hyperdimensional feature spaces. Then, they assess the effectiveness of SVMs with respect to conventional feature-reduction-based approaches and their performances in hypersubspaces of various dimensionalities. To sustain such an analysis, the performances of SVMs are compared with those of two other nonparametric classifiers (i.e., radial basis function neural networks and the K-nearest neighbor classifier). Four different multiclass strategies are analyzed and compared: the one-against-all, the one-against-one, and two hierarchical tree-based strategies.	AVIRIS (224 spectral bands)
G. Camps-Valls, L. Bruzzone[17]	This paper presents the framework of kernel-based methods in the context of hyperspectral image classification, illustrating from a general viewpoint the main characteristics of different kernel-based approaches and analyzing their properties in the hyperspectral domain. In particular, we assess performance of regularized radial basis function neural networks (Reg-RBFNN), standard support vector machines (SVMs), kernel Fisher discriminant (KFD) analysis, and regularized AdaBoost (Reg-AB).	AVIRIS (224 spectral bands)
G. M. Foody, A. Mathur[19]	A approach to multiclass classification of airborne sensor data by a single SVM analysis is evaluated against a series of classifiers that are widely used in remote sensing, with particular regard to the effect of training set size on classification accuracy. In addition to the SVM, the same datasets were classified using a discriminant analysis, decision tree, and multilayer perceptron neural network. For each classification technique, accuracy was positively related with the size of the training set. In general, the most accurate classifications were derived from the SVM approach.	Airborne Thematic Mapper (ATM) (11 spectral bands)
C. Huang, L.S. Davis, J.R.G. Townshend[20]	This paper introduces the theory of SVM and provides an experimental evaluation of its accuracy, stability, and training speed in deriving land cover classifications from satellite images. SVM algorithm is compared with other supervised algorithms: maximimum likelihood (ML) classifier, neural network classifier, and decision tree classifier.	(Spatially degraded) Landsat Thematic Mapper (TM)
G. Zhu, D. G. Blumberg[21]	This paper presents a study on the mapping of urban environments using ASTER data and SVM-based classification algorithms. A case study of the classification of the area of Beer Sheva, Israel is presented. Field validation shows that the classification is reliable and precise.	Advanced Spaceborne Thermal Emission and Reflectance Radiometer (ASTER)

Table 1 – Continued

Authors	Description	RS data
L. Su, M. J. Chopping, A. Rango, J. V. Martonchik, D. P. C. Peters[22]	This paper present a study on mapping and monitoring the desert environment using SVM for the analysis of Multi-angle Imaging Spectro-Radiometer (MISR) RS data. Many classification experiments have been implemented to find the optimal combination of MISR multi-angle data for maximizing the classification accuracy.	Multi-angle Imaging Spectro-Radiometer (MISR)
J. Keuchel, S. Naumann, M. Heiler, A. Siegmund[23]	This paper presents three different approaches to the classification of satellites images: maximum likelihood classifier, SVM, and iterated conditional model (ICM) to perform contextual classification using Markov random field model. The classification algorithms were applied to a Landsat 5 TM image of Tenerife, the largest of the canary Island.	Landsat 5 TM
B. Dixon, N. Candade[25]	This paper presents a study on the comparison between SVM and NN for the classification of RS data. An experimental analysis is carried on Landsat 5 TM data, acquired in the South West of Florida. The obtained results confirmed that SVM and NN outperform the traditional ML classifier. SVM classification resulted slightly more accurate than NN, but SVM required much less computational effort in the training phase.	Landsat 5 TM
L. Bruzzone, L. Carlin[27]	This paper proposes a system for the classification of VHR images. The proposed system is made up of two main blocks: 1) a feature-extraction block that aims at adaptively model the spatial context of each pixel according to a hierarchical multilevel segmentation of the scene and 2) a classification block based on SVM. Experimental results on VHR images confirm the effectiveness of the proposed system.	Quickbird
B. Waske, S. Van der Linden[29]	This paper presents a strategy for the joint classification of multiple segmentation levels from multisensor imagery, using SAR and optical data. The two datasets are separately segmented at different scale levels and independently classified by two SVM-based classifiers. The fusion strategy is based on the application of an additional classifier, which takes in input the soft output of the pre-classified results of the two datasets. The obtained experimental results show that the useful combination of multilevel-multisensor data is feasible with machine learning techniques like SVM and Random forest.	Multitemporal SAR data and Landsat 5 TM
M. Dalponte, L. Bruzzone, and D. Gianelle[30]	The authors propose an analysis on the joint use of hyperspectral and light detection and ranging (LIDAR) data for the classification of complex forest areas. In greater detail, they present: 1) an advanced system for the joint use of hyperspectral and LIDAR data in complex classification problems; 2) an investigation on the effectiveness of the very promising SVM and Gaussian ML with leave-one-out covariance algorithm for the analysis of forest areas characterized from a high number of species; and 3) an analysis of the effectiveness of different LIDAR returns and channels for increasing the classification accuracy obtained with hyperspectral images.	Hyperspectral (126 spectral bands) and LIDAR

The RS literature related to SVM is not limited to the use of this approach on different data and different application domains. Recently, more advanced SVM-based classifiers have been developed for facing complex problems related to the properties of remote sensing images. A list of relevant papers that introduced advanced techniques based on SVM for the classification of RS data is reported in Table 2. These papers represent the most recent (and in some cases on-going) research activities in this field and give insight about the research direction for the next years.

In this context, it is worth mentioning the semi-supervised SVM classifiers[37-43], which are devised for addressing ill-posed problems characterized by a very small ratio between the number of available training samples and the number of features by reinforcing the learning procedure with the use of unlabeled samples. It is worth noting, that even if SVMs have very good generalization capability, they cannot model the classification problem when very few training samples are available ("strongly" ill-posed problems). In these cases, the exploitation of the unlabeled samples to enrich the information of the training samples can result in a significant improvement in the model estimation. The first work on Semisupervised SVM in RS was presented in [37,38]. The presented semisupervised SVM (S^3VM) is based on transductive inference that exploits a specific iterative algorithm which gradually searches a reliable separating hyperplane in the kernel space with a process that incorporates both labeled and unlabeled samples in the training phase. In [39], an S^3VM classification technique is proposed, where the learning phase is performed by optimizing the objective function directly in the primal formulation (without exploiting the dual representation that can be obtained with Lagrange multipliers). In [40], the Laplacian SVM technique[41] is introduced in the RS community. This technique adopts an additional regularization term on the geometry of both labeled and unlabeled samples by using the graph Laplacian. This method follows a non-iterative optimization procedure in contrast to most transductive learning methods and provides out-of-sample predictions in contrast to graph-based approaches. Experimental results confirm the effectiveness of S^3VM techniques for solving ill-posed remote-sensing classification problems. In general S^3VM provides higher accuracy and better generalization ability than standard supervised SVM. In this respect, a more detailed picture of the status on the research on the application of S^3VM to hypedimensional problems can be found in [43].

Other studies address the inclusion of the spatial-context information of the single pixel in the SVM classification process. To this end,[44] proposes a framework for applying the maximum a posteriori (MAP) estimation principle in remote sensing image segmentation, which incorporates contextual and geometrical information in the SVM classification process by means of Markov random field (MRF). In [45], the use of composite kernels is introduced in remote sensing to adopt different kernel functions for different subsets of features to combine spatial and spectral information in an effective way. In [47], a context-sensitive semisupervised SVM is proposed, which exploits the contextual information of the pixels during the learning phase, in order to improve the robustness to possible mislabeled training patterns (which are not unlikely to be present in the reference data due to different kinds of errors that may occur in the collection of labeled samples). This is achieved according to both the design of a semisupervised procedure and the definition of a contextual term in the cost function associated with the learning of the classifier. In the experimental analysis, the authors also studied the robustness to mislabeled training patterns of some widely used supervised and semisupervised classification algorithms (i.e., conventional SVM, progressive semisupervised SVM, Maximum Likelihood, and *k*-Nearest Neighbor). Thanks to their high generalization capability, SVM-based approaches resulted more robust than other classification approaches (e.g., statistical approaches) to the presence of mislabeled training patterns.

The study in [48] addresses the problem of automatic updating the land-cover maps by using RS images periodically acquired over the same investigated area under the hypothesis that a reliable ground truth is not available for all the considered acquisitions. The problem is modeled under the domain-adaptation framework by introducing a novel method designed for land-cover map updating, which is based on a domain-adaptation SVM (DASVM) technique. Given two RS images I_1 and I_2 acquired over the same area at different times (t_1 and t_2, respectively), the goal of the DASVM is to obtain an accurate classification of I_2 by exploiting the labeled training samples from reference image I_1 and the unlabeled samples from the new image I_2. The DASVM algorithm is based on an iterative process, which starts by training an SVM classifier with the original training samples of I_1 and gradually introduces semilabeled samples of I_2 and erases the original training samples. At convergence a final classification function ruled only by semilabeled samples at time t_2 is obtained. In addition, the authors propose a circular accuracy assessment strategy for the validation of the results obtained by domain-adaptation classifiers when no reference data for the considered image I_2 are available.

Another recent and promising approach to the analysis RS data is associated with active learning[49-50], which allows an interactive classification of RS images. The active learning approach is based on the iteration on three different

Table 2 – Relevant papers about advanced techniques based on SVM for the classification of RS data.

Authors	Description
L. Bruzzone, M. Chi, M. Marconcini[38]	This paper introduces a semisupervised classification method that exploits both labeled and unlabeled samples for addressing ill-posed problems with SVMs. The proposed method exploit an iterative algorithms which gradually search a reliable separating hyperplane in the kernel space with a process that incorporates both labeled and unlabeled samples in the training phase. The authors propose a novel modified transductive SVM classifier designed for addressing ill-posed remote-sensing problems, which has the following properties: 1) it exploits a weighting strategy for unlabeled patterns based on a time-dependent criterion; 2) it is able to mitigate the effects of suboptimal model selection; and 3) can address multiclass cases.
M. Chi, L. Bruzzone[39]	This paper addresses classification of hyperspectral remote-sensing images with kernel-based methods defined in the framework of semisupervised SVM (S^3VMs). In particular, the authors analyzed the critical problem of the nonconvexity of the cost function associated with the learning phase of S^3VMs by considering different (S^3VMs) techniques that solve optimization directly in the primal formulation of the objective function. The presented techniques are compared with S^3VMs implemented in the dual formulation in the context of classification of real hyperspectral remote sensing images.
L. Gomez-Chova, G. Camps-Valls, J. Munoz-Mari, J. Calpe[40]	This letter presents a semisupervised method based on kernel machines and graph theory for remote sensing image classification. The support vector machine (SVM) is regularized with the unnormalized graph Laplacian, thus leading to the Laplacian SVM (LapSVM). The method is tested in the challenging problems of urban monitoring and cloud screening, in which an adequate exploitation of the wealth of unlabeled samples is critical.
A. A. Farag, R. M. Mohamed, A. El-Baz[44]	This paper proposes a complete framework for applying the maximum a posteriori (MAP) estimation principle in remote sensing image segmentation. The MAP principle provides an estimate for the segmented image by maximizing the posterior probabilities of the classes defined in the image. The posterior probability can be represented as the product of the class conditional probability (CCP) and the class prior probability (CPP). For the CCP, a supervised algorithm which uses the SVM density estimation approach is proposed. For the CPP estimation, Markov random field (MRF) is a common choice which incorporates contextual and geometrical information in the estimation process.
G. Camp-Valls, L. Gomez-Chova, J. Muñoz-Marí, J. Vila-Francés, and J. Calpe-Maravilla[45]	This letter presents a framework of composite kernel machines for enhanced classification of hyperspectral images. This novel method exploits the properties of Mercer's kernels to construct a family of composite kernels that easily combine spatial and spectral information. This framework of composite kernels demonstrates: 1) enhanced classification accuracy as compared to traditional approaches that take into account the spectral information only: 2) flexibility to balance between the spatial and spectral information in the classifier; and 3) computational efficiency.
M. Marconcini, G. Camps-Valls, L. Bruzzone[46]	This letter presents a novel composite semisupervised SVM for the spectral–spatial classification of hyperspectral images. In particular, the proposed technique exploits the following: 1) unlabeled data for increasing the reliability of the training phase when few training samples are available and 2) composite kernel functions for simultaneously taking into account spectral and spatial information included in the considered image. Experiments carried out on a hyperspectral image pointed out the effectiveness of the presented technique.

Table 2 – Continued

Authors	Description
L. Bruzzone, C. Persello[47]	This paper presents a novel context-sensitive semisupervised SVM (CS^4VM) classifier, which is aimed at addressing classification problems where the available training set is not fully reliable, i.e., some labeled samples may be associated to the wrong information class. Unlike standard context-sensitive methods, the proposed CS^4VM classifier exploits the contextual information in the learning phase to improve the robustness to possible mislabeled training patterns. This is achieved according to both the design of a semisupervised procedure and the definition of a novel contextual term in the cost function associated with the learning of the classifier. In the experimental analysis, the authors studied the robustness to mislabeled training patterns of the proposed CS^4VM and of some widely used supervised and semisupervised classification algorithms (i.e., conventional SVM, progressive semisupervised SVM, Maximum Likelihood, and *k-NN*)
L. Bruzzone, M. Marconcini[48]	The authors address automatic updating of land-cover maps by using remote-sensing images periodically acquired over the same investigated area under the hypothesis that a reliable ground truth is not available for all the considered acquisitions. The problem is modeled in the domain-adaptation framework by introducing a novel method designed for land-cover map updating, which is based on a domain-adaptation SVM technique. In addition, a novel circular accuracy assessment strategy is proposed for the validation of the results obtained by domain-adaptation classifiers when no ground-truth labels for the considered image are available.
D. Tuia, F. Ratle, F. Pacifici, A. Pozdnoukhov, M. Kanevski, F. Del Frate, D. Solimini, W. J. Emery[50]	In this paper, an active learning method is proposed for the semi-automatic selection of training sets in remote sensing image classification. The method adds iteratively to the current training set the unlabeled pixels for which the prediction of an ensemble of classifiers based on bagged training sets show maximum entropy. This way, the algorithm selects the pixels that are the most uncertain and that will improve the model if added in the training set.. Experiments were carried out using SVM.

conceptual steps. In the first step the learning process queries unlabeled samples to select the most informative ones; in the second step the supervisor (e.g., the user) provides a label to the selected samples interacting with the system; and in the third step the learner updates the classification rule by retraining with the updated training set. In [49], it is noted that SVMs are particularly suited to active learning since they are characterized by a small set of support vectors (SVs) which can be easily updated over successive learning iterations. Moreover, one of the most efficient query functions is based on the selection of the sample closest to the separating hyperplane defined at the considered iteration. For additional information about recent developments in kernel methods for the analysis of RS images, we refer the reader to [51].

4. Discussion and Conclusion

In this chapter we presented a review on SVMs in the classification of remote-sensing data, recalling their theoretical formulation, and discussing the motivations at the basis of their use in remote sensing. We presented a literature survey about the adoption of SVMs for the analysis of different kinds of RS images. We observed a large variety of studies published on the use of SVMs for the analysis of different kinds of RS data, which confirm that SVMs represent a valuable and effective tool for the analysis of RS data and can be used in many different applications in the context of RS. We observed that one of the most appealing properties of SVM for the classification of RS data is its high generalization capability and robustness to the Hughes effect, which allow SVMs to operate in large dimensional feature spaces with few training samples. For this reason, SVMs represent an effective choice for the classification of hyperspectral data. Nevertheless, the SVM approach turned out to be particularly effective also in the classification of very high resolution (VHR) images, which typically require the extraction of several additional features to characterize and discriminate the different land-cover classes. Thus, both the classification of VHR and hyperspectral images typically result in classification problems characterized by large dimensional feature spaces. Moreover, thanks to its distribution-free approach and the capability to cope with strongly non-linear problems by means of the kernel function, SVMs are a valuable tool also for the classification of data acquired by different information sources.

In addition, we pointed out the most recent works about the development of advanced SVM-based techniques for the analysis of RS data. Among these developments, we recall semisupervised and domain-adaptation SVM, techniques based on SVM that exploit the spatial-context information, and active learning methods. Semisupervised SVMs have shown to be effective in exploiting both labeled and unlabeled samples for the learning of the classification algorithm, further augmenting the generalization capability and the robustness to the Hughes phenomenon with respect to standard supervised SVM. Domain-adaptation SVM resulted effective for addressing the problem of automatic updating land-cover maps by using RS images periodically acquired over the same investigated area. Context-sensitive techniques based on SVM have been proposed for both regularizing the classification map (exploiting the context information in the classification phase) or for improving the robustness to mislabeled training samples (using the context information in the learning phase of the algorithm). Another promising approach is active learning, which allows one an interactive analysis of RS data, by driving the user to label unlabeled samples that are selected by a query function as most informative.

We can conclude that the SVM approach showed to be very promising for the classification of RS data and recent works demonstrate that SVM can be used as basis for the development of advanced techniques for solving specific RS problems or for exploiting particular properties of the RS data. However, still effort should be devoted to the development of advanced techniques that can effectively extract useful information from the rich and complex data acquired by the last generation of RS sensors. Moreover, effort is required also for applying the SVM-based approaches developed in the research activities in real-world RS problems. Indeed, at the present, the most of the real problems related to RS image classification are still solved with standard classifiers (like maximum likelihood or *k*-NN) that, even if simple, cannot guarantee the accuracy and generalization capabilities of SVMs in complex problems.

References

1. J.A. Benediktsson, P.H. Swain, and O.K. Ersoy, "Neural Network Approaches Versus Statistical Methods in Classification of Multisource Remote Sensing Data," *IEEE Transactions on Geoscience and Remote Sensing*, vol. GE-28, no. 4, pp. 540-552, July 1990.
2. P.D. Heermann, N. Khazenie, "Classification of multispectral remote sensing data using a back-propagation neural network", *IEEE Transactions on Geoscience and Remote Sensing*, vol. 30, no. 1, pp. 81 - 88, Jan. 1992.
3. H. Bischof, W. Schneider, A.J. Pinz, "Multispectral classification of Landsat-images using neural networks", *IEEE Transactions on Geoscience and Remote Sensing*, vol. 30, no. 3, pp. 482 - 490, May 1992.
4. L. Bruzzone, D.F. Prieto, "A technique for the selection of kernel-function parameters in RBF neural networks for classification of remote-sensing images", *IEEE Transactions on Geoscience and Remote Sensing*, vol. 37, no. 2, Part 2, pp. 1179 - 1184, March 1999.

5. L. Bruzzone, M. Marconcini, U. Wegmüller, and A. Wiesmann, "An Advanced System for the Automatic Classification of Multitemporal SAR Images", *IEEE Transactions on Geoscience and Remote Sensing*, vol. 42, no. 6, pp. 1351 - 1362, June 2004.
6. S.B. Serpico, and F. Roli, "Classification of multisensory remote-sensing images by structured neural networks", *IEEE Transactions on Geoscience and Remote Sensing,* Vol. 33, No. 3, pp. 562-578, May 1995.
7. L. Bruzzone, D.F. Prieto, S.B. Serpico, "A neural-statistical approach to multitemporal and multisource remote-sensing image classification", *IEEE Transactions on Geoscience and Remote Sensing,* Vol. 37, No. 3, Part 1, pp. 1350 – 1359, May 1999.
8. J.A. Benediktsson, I. Kanellopoulos, "Classification of multisource and hyperspectral data based on decision fusion", *IEEE Transactions on Geoscience and Remote Sensing,* Vol. 37, No. 3, Part 1, pp. 1367 - 1377, May 1999.
9. L.O. Jimenez, A. Morales-Morell, A. Creus, "Classification of hyperdimensional data based on feature and decision fusion approaches using projection pursuit, majority voting, and neural networks", *IEEE Transactions on Geoscience and Remote Sensing,* Vol. 37, No. 3, Part 1, pp. 1360 - 1366, May 1999.
10. V.N. Vapnik, *Statistical Learning Theory*. New York: Wiley, 1998.
11. J.A. Gualtieri and R.F. Cromp, "Support Vector Machines for Hyperspectral Remote Sensing Classification", 27'th AIPR Workshop, *Proc. of the SPIE,* vol. 3584, pp. 221-232, 1998.
12. J.A. Gualtieri, S.R. Chettri, R.F. Cromp, and L.F. Johnson, "Support Vector Machine Classifiers as Applied to AVIRIS Data", Summaries of the Eighth JPL Airborne Earth Science Workshop, JPL Publication 99-17, pp. 217-227, 1999, ftp://popo.jpl.nasa.gov/pub/docs/workshops/99_docs/toc.html.
13. J.A. Gualtieri and S. Chettri, "Support vector machines for classification of hyperspectral data", *Proc. of IEEE-IGARSS 2000,* Hawaii, pp. 813-815, 2000.
14. F. Melgani, L. Bruzzone, "Support vector machines for classification of hyperspectral remote-sensing images", *IEEE International Geoscience and Remote Sensing Symposium (IGARSS 2002)*, Toronto, Canada, pp. 506-508, June 2002.
15. G. Camps-Valls, L. Gomez-Chova, J. Calpe-Maravilla, J.D. Martin-Guerrero, E. Soria-Olivas, L. Alonso-Chorda, J. Moreno, "Robust support vector method for hyperspectral data classification and knowledge discovery", *IEEE Transactions on Geoscience and Remote Sensing*, Vol. 42, No. 7, pp. 1530 - 1542 , July 2004.
16. F. Melgani, L. Bruzzone, "Classification of hyperspectral remote-sensing images with support vector machines", *IEEE Transactions on Geoscience and Remote Sensing*, Vol. 42, No. 8, pp. 1778-1790, August 2004.
17. G. Camps-Valls, L. Bruzzone, "Kernel-based methods for hyperspectral images classification", *IEEE Transactions on Geoscience and Remote Sensing*, Vol. 43, No. 6, pp. 1351-1362, June 2005.
18. M. Fauvel, J.A. Benediktsson, J. Chanussot, J.R. Sveinsson, "Spectral and Spatial Classification of Hyperspectral Data Using SVMs and Morphological Profiles", *IEEE Transactions on Geoscience and Remote Sensing*, Vol. 46, No. 11, pp. 3804-3814, November 2008.
19. G.M. Foody, A. Mathur, "A relative evaluation of multiclass image classification by support vector machines", *IEEE Transactions on Geoscience and Remote Sensing*, Vol. 42, No. 6, pp. 1335 - 1343, June 2004.
20. C. Huang, L.S. Davis, J.R.G. Townshend, "An assessment of support vector machines for land cover classification", *Int. journal of Remote Sensing*, Vol. 23, No. 4, pp. 725-749, 2002.
21. G. Zhu, D.G. Blumberg, "Classification using ASTER data and SVM algorithms; The case study of Beer Sheva, Israel", *Remote Sensing of Environment*, Vol. 80, No. 2, pp. 233-240, May 2002.

22. L. Su, M.J. Chopping, A. Rango, J.V. Martonchik, D.P.C. Peters, "Support vector machines for recognition of semi-arid vegetation types using MISR multi-angle imagery", *Remote Sensing of Environment*, Vol. 107, No. 1-2, pp. 299-311, March 2007.
23. J. Keuchel, S. Naumann, M. Heiler, A. Siegmund, "Automatic land cover analysis for Tenerife by supervised classification using remotely sensed data", *Remote Sensing of Environment*, Vol. 86, No. 4, pp. 530-541, 2003.
24. Lizarazo, "SVM-based segmentation and classification of remotely sensed data", *int. journal of Remote Sensing*, Vol. 29, No. 24, pp. 7277–7283, December 2008.
25. B. Dixon, N. Candade, "Multispectral landuse classification using neural networks and support vector machines: one or the other, or both?", *Int. journal of Remote Sensing*, Vol. 29, No. 4, pp. 1185–1206, February 2008.
26. H. Carrao, P. Goncalves, M. Caetano, "Contribution of multispectral and multitemporal information from MODIS images to land cover classification", *Remote Sensing of Environment*, Vol. 112, No. 3, pp. 986–997, March 2008.
27. L. Bruzzone, L. Carlin, "A Multilevel Context-Based System for Classification of Very High Spatial Resolution Images", *IEEE Transactions on Geoscience and Remote Sensing*, Vol. 44, No. 9, 2006, 2587-2600.
28. B. Waske, J.A. Benediktsson, "Fusion of Support Vector Machines for Classification of Multisensor Data", *IEEE Transactions on Geoscience and Remote Sensing*, Vol. 45, No. 12, December 2007, pp. 3858-3866.
29. B. Waske, S. Van der Linden, "Classifying Multilevel Imagery From SAR and Optical Sensors by Decision Fusion ", *IEEE Transactions on Geoscience and Remote Sensing*, Vol. 46, No. 5, pp. 1457-1466, May 2008.
30. M. Dalponte, L. Bruzzone, and D. Gianelle, "Fusion of Hyperspectral and LIDAR Remote Sensing Data for Classification of Complex Forest Areas", *IEEE Transactions on Geoscience and Remote Sensing*, Vol. 46, No. 5, pp. 1416-1427, May 2008.
31. C. Burges, "A tutorial on support vector machines for pattern recognition," *Data Mining and Knowledge Discovery*, vol. 2, no. 2, pp. 121–167, 1998.
32. N. Cristianini and J. Shawe-Taylor, An introduction to support vector machines and other kernel-based learning methods, Cambridge, U.K.: University press, 1995.
33. B. Schölkopf and A. Smola, *Learning With Kernels*, Cambridge, MA: MIT Press, online: http://www.learning-with-kernels.org, 2002.
34. Chih-Wei Hsu; Chih-Jen Lin, "A comparison of methods for multiclass support vector machines", *IEEE Transactions on Neural Networks*, Vol. 13, No. 2, pp. 415 - 425, March 2002.
35. Lili Cheng, Jianpei Zhang, Jing Yang, Jun Ma, "An Improved Hierarchical Multi-Class Support Vector Machine with Binary Tree Architecture", *International Conference on Internet Computing in Science and Engineering*, 2008.
36. G.F. Hughes, "On the mean accuracy of statistical pattern recognition," *IEEE Transactions on Information Theory*, vol. 14, no. 1, pp. 55–63, January 1968.
37. L. Bruzzone, M. Chi, M. Marconcini, "Transductive SVM for Semisupervised Classification of Hyperspectral Data," *IEEE International Geoscience and Remote Sensing Symposium (IGARSS 2005)*, Seoul, Korea, July 2005.
38. L. Bruzzone, M. Chi, M. Marconcini, "A Novel Transductive SVM for the Semisupervised Classification of Remote-Sensing Images", *IEEE Transactions on Geoscience and Remote Sensing*, Vol. 44, No. 11, pp. 3363-3373, 2006.
39. M. Chi, L. Bruzzone, "Semi-supervised Classification of Hyperspectral Images by SVMs Optimized in the Primal", *IEEE Transactions on Geoscience and Remote Sensing*, Vol. 45, No. 6, Part 2, pp. 1870-1880, 2007.

40. L. Gomez-Chova, G. Camps-Valls, J. Munoz-Mari, J. Calpe, "Semisupervised Image Classification With Laplacian Support Vector Machines", *IEEE Geoscience and Remote Sensing Letters,* Vol. 5, No. 3, pp. 336-340, July 2008.
41. M. Belkin, P. Niyogi, and V. Sindhwani, "Manifold regularization: A geometric framework for learning from labeled and unlabeled examples," J. Mach. Learn. Res., vol. 7, pp. 2399–2434, Nov. 2006.
42. G. Camps-Valls, T. Bandos Marsheva, D. Zhou, "Semi-Supervised Graph-Based Hyperspectral Image Classification", *IEEE Transactions on Geoscience and Remote Sensing,* Vol. 45, No. 10, pp. 3044-3054, October 2007.
43. L. Bruzzone, M. Chi, M. Marconcini, "Semisupervised Support Vector Machines for Classification of Hyperspectral Remote Sensing Images", in *Hyperspectral Data Exploitation: Theory and Applications,* Ed: C-I. Chang, *John Wiley & Sons, Inc.*, 2007, chapter 11, pp. 275-311.
44. A.A. Farag, R.M. Mohamed, A. El-Baz, "A unified framework for MAP estimation in remote sensing image segmentation," *IEEE Transactions on Geoscience and Remote Sensing*, vol. 43, no. 7, pp. 1617-1634, July 2005.
45. G. Camp-Valls, L. Gomez-Chova, J. Muñoz-Marí, J. Vila-Francés, and J Calpe-Maravilla, "Composite Kernels for Hyperspectral Image Classification" *IEEE Geoscience and Remote Sensing Letters,* Vol. 3, No. 1, January 2006.
46. M. Marconcini, G. Camps-Valls, L. Bruzzone, "A Composite Semisupervised SVM for Classification of Hyperspectral Images", *IEEE Geoscience and Remote Sensing Letters*, Vol. 6, No. 2, pp. 234-238, 2009.
47. L. Bruzzone, C. Persello, "A Novel Context-Sensitive Semisupervised SVM Classifier Robust to Mislabeled Training Samples", *IEEE Transactions on Geoscience and Remote Sensing,* 2009, in press.
48. L. Bruzzone, M. Marconcini, "Toward an Automatic Updating of Land-Cover Maps by a Domain Adapatation SVM Classifier and a Circular Validation Strategy", *IEEE Transactions on Geoscience and Remote Sensing*, Vol. 47, No. 4, pp. 1108-1122, 2009.
49. P. Mitra, B.U. Shankar, S.K. Pal, "Segmentation of multispectral remote sensing images using active support vector machines", *Pattern Recognition Letters*, vol.25, pp. 1067-1074, 2004.
50. D. Tuia, F. Ratle, F. Pacifici, A. Pozdnoukhov, M. Kanevski, F. Del Frate, D. Solimini, W.J. Emery, "Active Learning of Very-High Resolution Optical Imagery with SVM: Entropy vs Margin Sampling", in Proc. *IEEE International Geoscience and Remote Sensing Symposium (IGARSS 2008)*, Boston, USA, pp. IV-73 - IV-76, July 2008.
51. G. Camps-Valls and L. Bruzzone, "Kernel Methods for Remote Sensing Data Analysis", John Wiley & Sons, Inc., 2009, in press.

CHAPTER 3.3

WORD SPOTTING TECHNIQUES IN DOCUMENT ANALYSIS AND RETRIEVAL — A COMPREHENSIVE SURVEY

Muhammad Ismail Shah Ching Y. Suen

CENPARMI (Center for Pattern Recognition and Machine Intelligence), Computer Science and Software Engineering Department, Concordia University, Montreal, QC, H3G 1M8, Canada

E-mails: {mu_shah, suen}@cenparmi.concordia.ca

In this chapter, we present a comprehensive review of word spotting techniques and concepts. Word spotting has been adopted and used by various researchers as a complementary technique to Optical Character Recognition for document analysis and retrieval. The various applications of word spotting include document indexing, retrieval and information filtering. Word spotting techniques are based on matching the visual similarity between two images. Unlike OCR techniques, conversion of documents into machine readable codes and machine recognition is not required in word spotting techniques. Proper estimation of bounding boxes, selection and use of proper features as well as robust image matching methods are considered to be the most important aspects of a word spotting system. Here we include the important aspects of word spotting techniques, such as pre-processing, features, matching algorithms and evaluation methods.

1. Introduction

There are millions of very useful books, research papers, theses, journals, and other important historical documents archived in world wide libraries. These documents are highly valuable due to their historical background as well as the information and knowledge they contain. Most of these documents have been degraded to a great extent due to reading and aging processes. Nowadays, there is a great concern about how to preserve and maintain these valuable knowledge assets for future uses. Moreover, it would be desirable to make these sources available for global access over the Internet in the form of future digital libraries.

The demand for an efficient online system to access these important records and valuable sources of knowledge increases every day. With recent developments in information technology, it has become easily possible to scan and store these valuable documents. However, the actual problem is the retrieval of these documents, that is, after digitization, how to reliably and efficiently access these documents for some specific information. There is a need for a robust system which can effectively manipulate and classify the contents of these documents.

For the last several decades the Information Retrieval community has been trying to find a well-organized way for indexing these huge amounts of documents and for retrieval of only certain relevant documents based on each user's query[53,54]. Optical Character Recognition (OCR) techniques have been applied by some researchers[39,59,60] to retrieve specific documents from document databases. However, these techniques are not effective because most of the documents within libraries and offices are either in a much degraded condition or written in languages for which presently no OCR technologies exist.

Traditional OCR technology strictly depends on careful analysis of page layouts, segmentation of words into isolated characters and conversion of those characters into machine-readable texts followed by a recognition process. OCR techniques have been applied to limited areas, e.g., bank cheque recognition[61], and postal code/address recognition[62]. These applications require complex classifiers trained with a small size of domain-specific lexicons. For good quality typewritten documents, the existing OCR technologies can perform satisfactorily. However, these traditional OCR techniques can not be used to retrieve information from degraded documents with unlimited/unseen vocabulary and complex layout of pages. Moreover, for oriental languages such as Indo-Iranian languages written in cursive scripts, the traditional OCR technology can not perform well, even in the printed documents.

The Word Spotting approach, also called "word searching", is used to detect and locate instances of the given template/query image in the document image databases. When documents in which the instances of the template are detected, they are retrieved and presented to the user. The basic idea of word spotting is that a template image is selected from a set of predefined keywords, i.e., words of interest, and then a search is initiated to find out its other instances in the target set of digitized documents. The process is totally based on image matching and conversion of whole documents to machine readable texts, i.e., ASCII codes, and machine recognition are not involved at all. This factor makes the approach more flexible and suitable for indexing and retrieval of degraded and historical documents written in multiple languages.

A more robust and effective approach is required for managing these multilingual typewritten and handwritten documents, i.e., indexing and retrieving. Word spotting, originally adopted by the speech recognition community[51,52], has been practiced as an alternative technique to OCR technology by many researchers for the last one and half a decades. This approach has been used for experiments on indexing handwritten documents[1,13,19,29] as well as for information retrieval from printed[12,21,42,47] and handwritten[8,9,41] document image databases. Word spotting techniques have been shown to be more effective than OCR, especially for degraded and handwritten documents. Decurtins[16] has compared a keyword spotting system with OCR and has shown that word spotting system is more robust. Yue, et al.[42] have also compared their word spotting method with an OCR tool and have reported that their word spotting system is more effective and efficient.

The rest of the chapter has been organized as follows. Section 2 gives a brief background review, i.e., various scripts addressed by the word spotting community and some Word Spotting Systems. Section 3 is the major section of this chapter, which describes the basic components of the word spotting techniques. Its four subsections include the various word spotting approaches that have been used for pre-processing, features extraction, image matching and results evaluation, respectively. In Section 4, we describe system evaluation techniques in word spotting, i.e., precision and accuracy. In Section 5, we discuss some basic word spotting issues and the future possible work in this regard. Section 6 provides conclusive comments based on the entire study presented in this survey. We hope that this work will prove to be a rich source of useful information for future researchers interested in word spotting

2. Background

2.1. *Scripts Addressed So Far*

In this section, we present an overview of various scripts that have been addressed by the research community in the domain of word spotting. So far, word spotting techniques have been applied to various scripts, i.e., Latin, Arabic, Greek, Chinese, Devanagari, and Amharic, as shown in Table 1.

These scripts differ from each other, based on various factors such as: number of characters, writing direction and cursiveness. For example, languages in Latin scripts are written from left to right in horizontal direction only, either cursively or non-cursively. However, languages in Arabic Scripts are written from right to left, in a horizontal direction, and are totally cursive. On the other hand the Chinese scripts, e.g., Chinese and Korean, have thousands of characters and are written in two dimensions, i.e., left-to-right horizontally and top-to-bottom vertically. Languages written in Devanagari scripts are written horizontally, from left-to-right, in a complex cursive way. Greek and Amharic scripts are written from left-to-right and are non-cursive. Furthermore, each isolated character of the Chinese scripts has certain meanings/semantics whereas the isolated characters of other scripts, e.g., Latin, Arabic, Greek and Devanagari, etc., have no meanings.

Table 1: The scripts used for word spotting.

Publications	Language	Scripts	Domain	
			Type	Print
Rath et al.[1,13], Manmatha et al.[19], Cao et al.[3], Rothfeder et al.[35], Tomai et al.[39], Adamek et al.[4], Srihari et al.[9], Zhang et al. [41]	English	Latin	Off-line	Handwritten
Keaton et al.[18], Kotcz et al.[15]	Spanish			
Leydier et al.[6,10], Christophe et al.[49]	French			
Ntzios et al.[5]	Greek	Greek		
Srihari et al.[8,9], Al-Khatib et al.[34], Shahab et al.[38], Saabni et al.[64]	Arabic	Arabic		
Jain et al.[33], Burl et al.[17]	English	Latin	On-line	
Chen et al.[12,21,28], Kuo et al.[20,24], Linlin et al.[7], Spitz[26], Jeffrey et al.[25], Jawahar et al.[47] Balasubramanian et al.[48], Shijian et al.[37], Yue et al.[42,43]	English	Latin	Offline	Type-written
Josep et al.[50]	Spanish			
Yue et al.[14]	Chinese	Chinese		
Soo et al.[45], Park et al.[46], Cho et al.[44]	Korean			
Gatos et al.[11], Konidaris et al.[2]	Greek	Greek		
Srihari et al.[9]	Sanskrit	Devanagari		
Jawahar et al.[47], Balasubramanian et al.[48]	Hindi			
Jawahar et al.[47]	Amharic	Amharic		
Saabni et al.[64]	Arabic	Arabic		

The proposed techniques for word spotting in one language can be easily applied to other languages written in the corresponding scripts. Most of the researchers have focused on only one type of script, e.g., Latin[1,13,19,35], and they have not tested their methods on other scripts. However, some authors have tried to test their methods on more than one type of scripts. For example, Leydier et al.[6,10] have applied their proposed method to two different types of scripts i.e., Latin and Arabic. Srihari et al.[9] have applied their method to three types of scripts, i.e., Arabic, Latin and Devanagari. Jawahar et al.[47] have also used the same method for three different scripts, i.e., Latin, Devanagari and Amharic.

2.2. *Word Spotting Systems*

So far, various software systems used for word spotting have been reported by the document retrieval research community. Here, we give a brief summary of some word spotting systems:

Srihari et al.[8] presented an interactive software system called CEDARABIC designed for word spotting in handwritten Arabic documents. The proposed system provides a Graphical User Interface for inputting a query and displaying the results of the query. The working mechanism is like a dictionary look-up in which the user enters a query word, typed in English text, and the system finds its Arabic equivalent words in the database. The database contains pre-segmented handwritten Arabic documents accompanied with well-organized ground truth information such as the alphabet sequence, meaning and pronunciation of each word. In the first step, the word image is retrieved from the reference data set and used as a template. In the second step, the documents which contain instances of the given template are retrieved and ranked. The ranking is based on the value of similarity between the template and the instances found in each corresponding document.

Srihari et al.[9] made some modifications to the system presented by Srihari et al.[8]. They extended the system's functionalities to perform word spotting in three languages, i.e., handwritten Arabic and English as well as typewritten Sanskrit (Devanagari) documents. Furthermore, the first version of the system[8] allows the users to enter the query only as an English text whereas in the modified version[9], the users can also enter word images as queries.

Al-Khatib et al.[34] presented an interactive word spotting system for indexing and retrieving historical Arabic manuscripts. The system is based on the framework of a digital library described by Shahab et al.[38]. It has several important functionalities, such as pre-processing, feature extraction, image matching and a well-designed Graphical User Interface for input and output applications. The system uses the user's feedback to index the manuscript pages, which enables efficient retrieval of the relevant documents in future queries.

DeCurtins et al.[23] presented a software system, called SCRIBBLE, for word spotting in machine printed English documents. The system requires some sort of pre-training in the form of an *a priori model*, which specifies the usage of patterns for various words. The word shape codes along with a voting technique are used to compute the likelihood of a given word within a text line. This method is robust due to its voting scheme and can be applied to noisy documents, where the possibility of missing or fake features is higher. DeCurtins[23] compared SCRIBBLE with an OCR system and reported that this word spotting system has a higher performance.

Jawahar et al.[47] proposed a prototype of a word spotting system to retrieve relevant documents from a large collection of printed documents. The documents were initially indexed by clustering the similar words, whereby each cluster was assigned a weight and relevance to the corresponding documents. The system has been applied to documents printed in three types of scripts, i.e., Latin, Devanagari, and Amharic. The system works at two levels: first, feature vectors are aligned using dynamic programming and then structural features are used to compare the shapes of the images. The same framework has been implemented by Balasubramanian et al.[48], however, they applied it to documents printed in Latin and Devanagari scripts only.

3. Word Spotting Methodologies

3.1. *Pre-processing*

The robustness of any word spotting system is directly related to and dependent upon the pre-processing phase. This phase usually involves noise removal, skew correction, bounding box estimation and feature extraction. Based on these pre-processing methods, the word spotting techniques can be broadly divided into two groups. In the first group, the entire document is considered as a single image and no page segmentation is required[6,10]. The given document images are scanned in the direction of writing to find out the instances of a given template based on shape similarities. In the second group, the document image is considered as a collection of sub-images, i.e., words and/or characters. Before keyword searching is initiated, the given document image is segmented into its sub-images.

Approaches for estimation of bounding boxes within a document image can be either Top-down or Bottom-up[14]. In the Top-down methods, the layout of the documents is carefully analyzed and used to segment the document images into various sections based on the type of contents they contain such as text, graphics, tables, charts, etc. The search for a given template image is initiated in the text portion which is usually divided into headlines and body text. This approach is effective in the domain of printed documents with a constant specific format[14]. In the Bottom-up approaches, the connected components within a given document are initially detected and analyzed. The documents are segmented by estimating the bounding boxes of the connected components. Estimation of the bounding boxes within a text document image is a very important step as the overall accuracy and throughput of the system depends on it. For printed documents with stable and uniform layouts, bounding box estimation is not a big problem, however, for handwritten documents it is a real challenge. In the word spotting literature, the bounding boxes have been extracted at three levels, as shown in Table 2, i.e., Lines, Words or Characters.

Table 2: Bounding box estimation at three levels.

Publications	Bounding Box Estimation
Rath et al.[1,13,29], Rothfeder et al.[35], Adamek et al.[4], Manmatha et al.[19,36], Al-Khatib et al.[34], Shahab et al.[38], Chen et al.[21,28], Tomai et al.[39], Jain et al.[33], Jawahar et al.[47], Balasubramanian et al.[48], Srihari et al.[8,9]	Word Level
Cao et al.[3], Yue et al.[14], Park et al.[46], Soo et al.[45]	Character Level
Kotcz et al.[15], Chen et al.[12]	Line Level

In the line oriented approaches, the page is first segmented into lines. Each line is then scanned along the direction of writing to spot a word similar in shape to the given word model/template[15]. The widely used approach is based on estimation of word level bounding boxes. The document images are segmented into words using connected component analysis. The template/keyword is either compared with every word image in the document image[33] or with only those images which have aspect ratios equal/close to the keyword[1]. In the latter case, a method called pruning is used according to which unlikely instances of the keywords are not considered for matching, i.e., words having aspect ratios which are either very large or very small as compared to the given keywords[1]. In the character level approaches, the bounding boxes are estimated for each character. This approach has been widely used for word spotting in printed Chinese[14] as well as Korean[44,45,46] documents. Furthermore, this approach has also been employed in the word spotting approaches where shape codes are derived by using the character level shape information[26].

In addition, Keaton et al.[18] have used a method in which a set of candidate locations are identified and extracted, i.e., text blocks that may contain the instances of the keywords. Correlation strength is used to rank different candidate locations and those with high correlations are selected for pre-processing. The instances of keywords are then searched by scanning these locations.

Page segmentation for word spotting is a rich area of research and so far, many useful approaches have been reported. For example, Manmatha et al.[58] have proposed a scale space technique for automatically segmenting historical handwritten documents. Mahadevan et al.[56] have used Gap metrics for segmenting words within handwritten text lines. Laurence et al.[55] have surveyed various segmentation approaches suitable for segmentation of historical documents. In this paper, we are not explaining the segmentation methods as it is a vast field of research in its own.

3.2. *Features Used*

In this section, we present an overview of the various important features used by different researchers for word spotting. In addition to feature selection, other important factors such as image format and representation must be carefully considered, as they have a great impact on the system's overall performance. Furthermore, the extraction of features also depends upon the image format and image representation used. As shown in Table 3, most of the researchers have used the features extracted from Binary Images. However, most of the word spotting work done in the domain of historical handwritten documents is based on Gray-Scale Images[1,6,35]. The reason for this is that binarization of word images results in the loss of important information which can lead to unsatisfactory results.

There are three types of well-known image representations, i.e., Pixels, Skeletons and Contours. From Table 3, we can find that about 85% of researchers have adopted Pixel representation for word spotting. Similar to image format, image representation can also affect the system's performance in terms of computational cost and information losses. For example, using Skeletons can significantly reduce the computational cost but it can cause a loss of useful information during binarization and thinning operations. In Pixel representation, extraction of most of the features depends upon correct estimation of each word's baseline. On the other hand, baseline estimation is not required in Contour representation. The selection of representation largely depends upon the features used in the process. For example, for extraction of point features, i.e., end points, intersections, etc., the images are required to be skeletonized. Similarly, for extraction of projection profiles the Pixel representation is necessary.

Table 3: Features, image formats and representations used in word spotting.

<table>
<tr><th>Publications</th><th>Features Used</th><th>Image Format</th><th>Representation</th></tr>
<tr><td>Rath et al.[1]</td><td>Projection profiles, Upper/lower word profiles, Background-to-ink transitions</td><td rowspan="5">Gray Scale</td><td rowspan="23">Pixels</td></tr>
<tr><td>Rothfeder et al.[35]</td><td>Corner features</td></tr>
<tr><td>Cao et al.[3]</td><td>Gabor features</td></tr>
<tr><td>Leydier et al.[6]</td><td>Gray level, Gradient orientation</td></tr>
<tr><td>Chen et al.[12]</td><td>Column pixel values</td></tr>
<tr><td>Kotcz et al.[15]</td><td>Upper/lower word profiles, Background-to-ink transitions</td><td rowspan="18">Binary</td></tr>
<tr><td>Gatos et al.[11], Konidaris et al.[2]</td><td>Mesh features, Upper/lower word profiles</td></tr>
<tr><td>Srihari et al.[8,9], Zhang et al.[41]</td><td>Gradient-based binary features (GSC Binary Features)</td></tr>
<tr><td>Al-Khatib et al.[34], Shahab et al.[38]</td><td>Angular line, Concentric circle, 15-point DFT transform of vertical/horizontal profiles, Height, Width, Area and Aspect Ratios</td></tr>
<tr><td>Jawahar et al.[47], Balasubramanian et al.[48]</td><td>Projection profiles, Upper/lower word profiles, Background-to-ink transitions, Moments, Mean, Black pixel distribution, Strokes' curvature</td></tr>
<tr><td>Kuo et al.[20,24]</td><td>Pixel's value, Transactional information, Pixel's relevant position in row and column</td></tr>
<tr><td>Yue et al.[42,43]</td><td>Straight stroke lines, Traversal features</td></tr>
<tr><td>Yue et al.[14]</td><td>Strokes' density</td></tr>
<tr><td>Saabni et al.[64]</td><td>Geometric Features</td></tr>
<tr><td>Soo et al.[45]</td><td>Mesh features</td></tr>
<tr><td>Spitz[26]</td><td>Ascenders and Descenders, East concavity feature</td></tr>
<tr><td>Jeffrey et al.[25]</td><td>Spatial Moments: means, variances</td></tr>
<tr><td>Shijian et al.[37]</td><td>Holes, water reservoir, ascenders/descenders</td></tr>
<tr><td>Linlin et al.[7]</td><td>Straight vertical/non-vertical strokes, Ascenders and Descenders</td></tr>
<tr><td>Park et al.[46]</td><td>Mesh features, Upper/lower and right/left word profiles, Wavelet coefficients</td></tr>
<tr><td>Christophe, et al.[49]</td><td>Image column height, Neighbor hood size</td></tr>
<tr><td>Keaton et al.[18]</td><td>Projection profiles, Cavity features</td><td rowspan="3">Binary</td><td rowspan="3">Skeletons</td></tr>
<tr><td>Josep et al.[50]</td><td>Skeleton points</td></tr>
<tr><td>Ntzios et al.[5]</td><td>Open/closed concavity features</td></tr>
<tr><td>Adamek et al.[4]</td><td>Multi-scale convexities/concavities features</td><td>Binary</td><td rowspan="2">Contours</td></tr>
<tr><td>Chen et al.[21,28]</td><td>Upper/lower contours, Auto-correlations</td><td>Gray Scale</td></tr>
</table>

Similar to character recognition, in word spotting the selection of proper and effective features plays a vital role, especially in historical and degraded handwritten documents which contain various types of noises. Rath et al.[13] have explained some of the features used for word spotting in historical manuscripts.

In addition to the selection of proper features, the pattern of their usage is also important for getting satisfactory results. Using various features in a combined way can give a high performance as compared to the corresponding features individually/separately. Rath et al.[1] and Keaton et al.[18] have shown that using a combination of features can produce a high performance and more satisfactory results. In addition, the best choice of features to be used also depends on the matching method. For example, the profile features have been proved to be more effective when using the Dynamic Time Warping method because these features can be easily represented as a time series.

3.3. *Image Matching*

Image matching is not only the most important, but also the most difficult phase in word spotting. In the following sub-sections, we describe image matching concepts and various approaches adopted by researchers for word spotting.

3.3.1. *Similarity/Dissimilarity Measurement*

Two images are usually matched based on either similarity or dissimilarity functions. An image I_{target} is considered to be an instance of a given image $I_{template}$ if the output of their similarity function is greater than some predefined threshold, i.e.,

$$[I_{template} \equiv I_{target}] \rightarrow [Similarity\ (I_{template}, I_{target}) > Threshold]$$

On the other hand, an image I_{target} is considered to be an instance of a given image $I_{template}$ if the output of their distance/dissimilarity function is lower than some predefined threshold, i.e.,

$$[I_{template} \equiv I_{target}] \rightarrow [Dissimilarity\ (I_{template}, I_{target}) < Threshold]$$

Some of the well-known similarity measurement algorithms are: Dynamic Time Warping (DTW)[1], Shape Context (SC), and Correlation Similarity Measure (CORR)[31]. In addition to CORR, Tubbs[31] has also defined seven other types of similarity measures for matching two binary patterns.

The well-known Minkowski distance, i.e., City Block Distance (*1-norm*) and Euclidean Distance (*2-norm*), have been widely used by researchers for matching dissimilarity between two images. City Block Distance (CBD) and Euclidean

Distance Measure (EDM) are very useful distance measurement techniques. These techniques are not only used as stand-alone matching methods but other image matching methods, such as DTW and K-NN, also depend on some distance function defined by either City Block Distance (CBD)[15,18] or Euclidean Distance Measure (EDM)[1,33,47,48].

3.3.2. *Matching Algorithms/Methods*

In the word spotting literature, there are various types of algorithms and methods used for image matching. These matching methods can be represented hierarchically, as depicted in Figure 1.

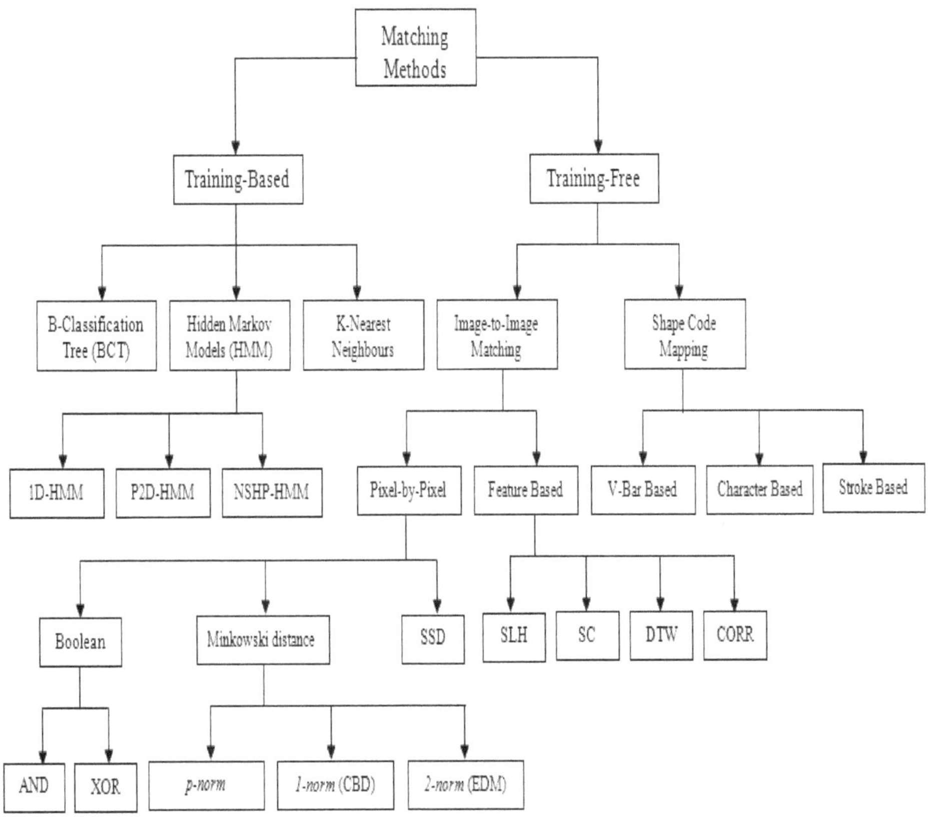

Figure 1: Hierarchical representation of various matching methods.

We have categorized the various image matching techniques into three groups. The first two groups represent those matching methods which are training-free, i.e., Image-to-Image Matching and Shape Code Mapping. The third group represents those matching methods which require a so-called training step, i.e., Hidden Markov Model (HMM), K-nearest Neighbor (K-NN). Here we give a brief summary of various matching methods used in word spotting:

Group 1: Image-to-Image Matching: This type of matching is also called template matching, where a pre-defined template/keyword is matched with the target images within a given set of documents. The matching can be either pixel-by-pixel or feature-based, where initially some features are extracted from the two images and then aligned and compared to find similarities between the template and the target image. The image may be a word, phrase or character depending on the required applications.

In the pixel-by-pixel matching methods, the template and target images are raster scanned and the distance between the corresponding pixel values is calculated. The Minkowski distance has been widely used for this type of matching, i.e., City Block Distance[45,46,11,2] and Euclidean Distance[14,34,38]. In addition to the Minkowski distance measurement techniques, some other pair-wise image matching methods have been reported. For example, Manmatha et al.[19] have used the XOR method for matching two images, where the two binary images are aligned and a difference image, i.e., XOR-image, is calculated by XORing the pixels of the two images. The matching cost is represented by the number of difference pixels. The XOR method is similar to Euclidean Distance Measure (EDM), however, in EDM the difference pixels in a larger cluster/group result from structural differences between the two matching images, hence their weight is heavier. In the Sum of Squared Differences (SSD) method, the two images are translated relative to each other and the matching cost, i.e., minimum cost, is calculated based on the Sum of Squared Differences[32].

In the feature-based image-to-image matching, a fixed number of features are extracted and represented as feature vectors. Similarity between the template and target images is measured by comparing their corresponding feature vectors. This approach has been adopted by many researchers for word spotting and has proven to be more successful and robust than pixel-by-pixel image matching techniques. Rath et al.[1] and Rothfeder et al.[35] have conducted comparative experiments and have reported that the feature-based techniques are more robust and tolerant to handwriting variations than the pixel-by-pixel matching methods.

Scott et al.[57] have presented an algorithm which matches features extracted from two patterns. This algorithm has been named after its developers, i.e., Scott and Longuet-Higgins as the SLH algorithm. The SLH algorithm has been used by various researchers[19,32] for word spotting in historical documents. In this method, sample points are taken from the template and target images and an affine warping transform is recovered between them. The matching cost is represented by the residual between the template points and the warped candidate points[32].

Belongie et al.[30] introduced a shape similarity measurement algorithm called Shape Context (SC), which is based on taking sample points from the outlines of the two images which are to be matched. Corresponding sample points in the two images are recovered by assigning a shape context histogram to each corresponding point. These correspondences describe the distribution of sample points in the shape with respect to the reference sample point, i.e., the point at which it is generated. The candidate image is warped by iteratively performing the matching operation and the cost associated with the chosen correspondences is used to determine the matching cost[1,32].

Rothfeder et al.[35] have used an approach in which points of interest in the template and candidate images are initially determined, and then similarities between those points are recovered using correlation. The number of such similarities and the relative locations of the corresponding points are used to calculate the matching cost between the two images. To measure similarity between the two images, a voting method is used for counting the number of recovered correspondences. Every pair of features points whose similarity is greater than a given threshold is considered as a vote. The total number of votes received by each image is used to rank the retrieved images.

The Dynamic Time Warping algorithm is used to find the similarity between two images by aligning and comparing the sequences of feature vectors represented as time series. The time series are created by extracting a fixed number of features for every image column where each column represents the time axis in a horizontal direction. For each type of feature, a separate time series is created. These time series are then aligned and compared either jointly as a multivariate feature vector[1] or one by one separately[33]. The Dynamic Time Warping algorithm (DTW) has proven to be more successful for word spotting in handwritten and degraded documents[1,29,32].

Srihari et al.[8,9] and Zhang et al.[41] have used the well-known Correlation Similarity Measure (CORR) defined by Tubbs[31]. However, instead of using CORR directly, they have made some modifications and redefined it as a dissimilarity measure[63].

Group 2: Mapping Shape Codes: In this group, we have included those matching methods in which a word image is encoded into a relatively smaller set of predefined symbols, which is easier to recognize as compared to the original character set. For defining shape codes, the word image is initially divided into three zones: ascenders-zone, x-zone and descenders-zone, by estimating the base-line, x-line, mid-line, top-line and bottom-line, as shown in Figure 2. Generally,

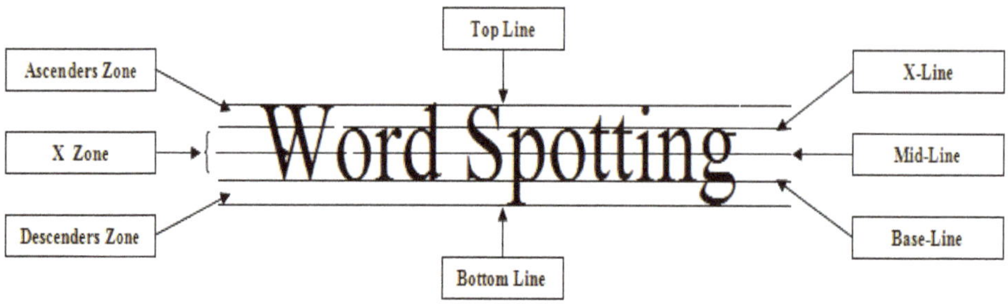

Figure 2: Illustration of important regions and lines used for Shape Codes estimation.

three different schemes for shape coding have been reported, based on vertical bar patterns or character-level shapes or strokes.

The pioneering work for keyword spotting in document images, using character shape codes, was presented by Tanaka et al.[22]. Their proposed system, called Transmedia Machine, identifies and encodes word-level components using two bits per character. The higher order bit encodes the presence of an ascender or descender, whereas the lower order bit encodes the frequency of crossing the mid-line by the corresponding character. The performance of this system improves when ascenders and descenders are treated separately by extending the original code. The Keyword is encoded in the same way and a search is initiated into coded documents for spotting other instances of the given query word.

Spitz et al.[26] have also used a similar but more robust approach, in which document images are transformed into character shape codes. These codes are then aggregated into word shape tokens and the tokenized texts are stored in a look-up table, i.e., an ASCII file. A query word, encoded in a similar way, is mapped to the look-up table using string matching techniques. The use of word shape tokens instead of actual words enables this method to be efficiently used for indexing and retrieval of scanned documents. A total number of six codes have been defined based on the features: ascenders, descenders, number of connected components and deep-eastward concavity. These codes are used to classify the characters. This approach depends on the lexicon and segmentation of character cells for defining the character shape codes.

Tan et al.[27] have proposed a shape coding scheme based on vertical bars. In this approach, local extrema points, i.e., local minimum/maximum pixels, located on the word's contour are paired up and different vertical bar patterns are extracted. These vertical bars are classified into three groups, based on the presence of ascenders and descenders.

Yue et al.[42] have adopted a stroke-based approach for defining 29 word shape codes, which is based on straight strokes and traversal features. The straight strokes include vertical stroke lines, left-down diagonal lines and right-down diagonal lines. These strokes are extracted by using a run-length-based method. Whereas the traversal features represent the background-to-ink transitions, which are computed by scanning the word image column by column. These features are used to extract a code string called Left-to-Right Primitive String (LRPS) as it is sequenced from leftmost-to-right-most of a word, whereby a primitive is described by the ascender/descender features plus the shape of the stroke. Each character is given a standard Primitive String Token (PST). These tokens are stored in a look up table. A query word is converted into a code string in a similar way and the string matching technique is used to spot other instances of the given keyword. In this approach, partial phrases are also located by using a dynamic programming technique called inexact feature matching.

Another stroke-based shape coding method for word spotting has been presented by Linlin et al.[7], where the presence of straight/non-straight vertical strokes, ascenders/descenders and the total number of components are all used for encoding word images into code strings. Word images are decomposed into strokes, and each stroke is assigned a shape code. For decomposition of word images into strokes the pixels lying on the mid-line are considered as decision points. If a pixel at this point, within an image column, is related to the background, then the entire column is converted to the background, otherwise the entire column is retained intact. There are eight codes used in total, which is a small amount compared to Yue et al.[42] where 29 complex codes are used. The instances of a query word are spotted using string matching techniques. This method of word spotting has been reported to be 20-40 times faster than OCR techniques[7].

Group 3: Training-based Matching: The third group, include those matching methods in which a so-called pre-trained model is used for keyword spotting. The well-known matching methods used for training-based word spotting include Hidden Markov Model (HMM)[12,24,49], K-Nearest Neighbor (K-NN)[4,18] and Binary Decision Tree[5].

Chen et al.[21,28] presented an approach for keyword spotting in printed documents which is based on Hidden Markov Model. Appropriate context-dependent character HMMs and sub-character HMMs are used for creating keyword HMMs, and non-keyword HMMs, respectively. Based on the two models, an *HMM network* is created, which is then searched by using Viterbi decoding on it, to spot the given keywords. Chen et al.[12] have modified the approach described by Chen et al.[28] by enabling it to spot phrases/partial words in document images.

Kuo et al.[20,24] have used pseudo 2D Hidden Markov Models, an extension to the standard HMM, which analyzes the word image in both horizontal as well as vertical directions. The model is called pseudo 2D because in order to avoid exponential complexity it is not fully connected[44]. Two models are created, of which one models the keywords while the other models all the extraneous words, i.e., words other than the keyword. The search space is reduced by eliminating the unlikely words, i.e., any word that is not a keyword. Words are spotted by using the doubly embedded Viterbi decoding algorithm on the two HMMs.

Christophe[49] has used a hybrid model called Non-Symmetric Half Plane HMM (NSHP-HMM). This model combines HMM and Markov Random Field (MRF) where MRF is a second type of 2D-HMM with reduced connectivity[44]. At the HMM level, the binary patterns are analyzed column-by-column, whereas at the MRF level they are analyzed pixel-by-pixel.

3.3. *Matching Levels*

In the literature, image matching has been conducted at two levels for word spotting, i.e., whole word matching and partial word matching, as shown in Table 4.

Table 4: Matching levels adopted for word spotting by various researchers.

Publications	Matching Level
Rath et al.[1], Jain et al.[33], Jawahar et al.[47], Balasubramanian et al.[48], Manmatha et al.[19], Rothfeder et al.[35], Cao et al.[3], Keaton et al.[18], Kotcz et al.[15], Adamek et al.[4], Srihari et al.[8,9], Zhang et al.[41], Al-Khatib et al.[34], Shahab et al.[38], Chen et al.[21]	Whole Word
Chen et al.[12,28], Kuo et al.[20,24], Cho et al.[44], Christophe et al.[49], DeCurtins et al.[23]	Partial Word / Phrase

In the first type, the global features of the whole word have to be matched with that of the template. In other words, the entire target word has to be similar to the template, based on some threshold value. Otherwise it will be rejected. In the second type, not only do the exact instances of the given template have to be detected and located, but also the variants if any, have to be located, i.e., partial instances. For example, given a template word '*Science*', the possible search results might be: *Science, Scientific, Scientist, Scientifically, etc.*

4. System Performance and Evaluation

The performance of a word spotting system is evaluated by using Precision-Recall Methods adopted from Information Retrieval literature. *Precision* is a measure of the degree of exactness, i.e., how much the results are exact. It can be calculated as follows:

$$\text{Precision} = \frac{\text{Total number of relevant instances retrieved}}{\text{Total number of instances retrieved}}$$

On the other hand, *Recall* represents the measure of completeness of a word spotting system. It can be calculated as follows:

$$\text{Recall} = \frac{\text{Total number of relevant instances retrieved}}{\text{Total number of instances present}}$$

The overall performance of a word spotting system can be described with the help of a graph called P-R-graph[6,9]. This is drawn between the values of Precision and Recall measures. In addition, a software program called ***trec-eval***, available on the internet, can also be used for evaluation of results[1]. In Table 5, we have presented the good precision rates achieved for each type of scripts in

Table 5: Leading Precision Rates achieved in off-line word spotting of various scripts.

Publications	Average Precision	Characteristics of Data Sets		
		Size	Variability & Quality of Data	Scripts
Rath et al.[1]	65.34 %	2381	Handwriting of a single writer. Extracted from degraded and historical documents. Scanned from microfilms of George Washington's manuscripts.	Latin (Offline)
Rothfeder et al.[35]	62.57 %			
Zhang et al.[41*]	92.18%	9312	Handwriting of 776 individuals. The test database contains only four different words, thus 2304 instances for each word. The quality of data is good.	
Kuo et al.[20]	99.00 %	26000	Printed in a single font size. A synthesized database with optimal quality.	
Kuo et al.[20]	96.00 %	26000	Printed in different font sizes. A synthesized database with optimal quality.	
Linlin et al.[7]	96.22 %	1845	Printed in different font sizes from 10 to 24 points and four types of font styles. Data is of good quality.	

[*] The precision shown is for the top 100 matches only. The authors have presented the precision rates for 12 different top matches.

Table 5: Continued

Publications	Average Precision	Characteristics of Data Sets		
		Size	Variability & Quality of Data	Scripts
Chen et al.[21]	96.00 %	2100	Printed in 8 different fonts including serif and sans-serif. Extracted from a text passage and tables of contents from five journals and conference proceedings. Data is of good quality.	Latin (Offline)
Jawahar et al.[47]	95.89 %	2507	Printed in various font sizes. Extracted from scanned pages of English books stored at the Digital Library of India (DLI). Images are of good quality.	
Jain et al.[33]	92.3 %	6672	Handwritings of 10 different writers. A digital device, i.e., Cross Pad has been used to capture the handwritten data.	Latin (Online)
Jawahar et al.[47]	94.51 %	2547	Printed in various font sizes. Extracted from scanned pages of an Amharic newspaper called Addis Zemen. Images are of good quality.	Amharic (Offline)
Jawahar et al.[47]	92.62 %	3354	Printed in various font sizes. Extracted from scanned Hindi documents archived at the Digital Library of India (DLI). Images are of good quality.	Devanagari (Offline)
Park et al.[46]	89.84 %	1600	Printed in different font sizes ranging from 8 to 10 points. Extracted from 8 document images, printed in Korean. The quality of images is good.	Chinese (Offline)
Yue et al.[14]	84.34 %	128	Printed in different font sizes and font styles. Extracted from various newspapers, printed in both traditional as well as simplified Chinese. Images are of good quality.	
Saabni et al.[64]	85 %**	8000	Extracted from Handwritten documents written by five different writers, Printed documents in different fonts and historical documents. The data is of reasonable quality.	Arabic (Offline)
Srihari et al.[8,9]	70%***	20000	Handwritings of 10 writers. A synthesized database with good quality and accompanied with ground truth information.	
Ntzios et al.[5]	89.06 %	12332	Handwritten characters and ligatures extracted from various historical manuscripts. Data is of poor quality, which is manually labeled and accompanied with ground truth information.	Greek (Offline)

** Average precision for the three categories of data used in experiments.
*** The given precision rate was achieved when 4000 words written by 2 writers were used as the Test Data Set.

which word spotting techniques have been applied. It is important to note that the results shown in Table 5 have been achieved on different data sets, except for Rath et al.[1] and Rothfeder et al.[35] who have used the same data set.

5. Discussions: Issues and Future Work

Important factors which affect the precision rate are variability of data and noises. In handwriting documents, the variability may be due to different writers, whereas in machine printed documents it can be caused by various font sizes and styles. From the work of Kuo et al.[20], it is evident that the precision rate for documents printed in a single font is higher than the rate for documents printed in many different fonts. Noises can be caused by degradation of documents, scanning sources, and most importantly, the page segmentation processes. The accuracy of the segmentation algorithms directly affects the precision rate of a word spotting system. The word spotting techniques which are not based on page segmentation can avoid the problems of improper segmentation. However, in segmentation-free methods accuracy may be affected by the methods used for correctly determining zones of interest and by estimation of bounding boxes of the target images.

There is a need for more research to find efficient ways of image matching, page segmentation as well as to discover more effective features. Furthermore, a generalized model is required, which can be adopted for word spotting in any type of scripts. As discussed in section 2.1, several researchers have applied their proposed methods to different scripts, e.g., Leydier et al.[6,10], Srihari et al.[9], and Jawahar et al.[47], etc. They have shown that the performance of the same word spotting system used for the various scripts is very similar. From their work, we can derive the hypothesis that: "It may be possible to design and develop a single and standard Word Spotting System which can be applied to almost all types of scripts and can achieve nearly equal performance for each type of scripts".

6. Conclusion

Word spotting is an emerging technology and has various useful applications such as document indexing, information retrieval and information filtering. This technology is still in the experimental stages and has been applied to limited sets of documents. It has produced good results for retrieving degraded, multilingual and complex documents as compared to OCR techniques. Word spotting in typewritten documents with good quality has given satisfactory results and can be commercially applied as an alternative to OCR systems. In the domain of degraded and historical documents, printed as well as handwritten, Word Spotting has achieved a higher accuracy rate as compared to OCR techniques. However, it has not achieved a precision rate acceptable for commercial applications. More efforts are needed to achieve a higher accuracy rate, improved performance and reduced computational cost. Robust and reliable page segmentation methods, proper feature selection and efficient image matching algorithms are required to take word spotting in handwriting documents to a commercial level.

Acknowledgement

We are very thankful to Shira Katz, CENPARMI Research Assistant, for her excellent editorial support and to Mr. Nicola Nobile, CENPARMI Research Manager, for his technical support. In addition, we are very grateful to Dr. Wumo Pan and Dr. Louisa Lam for their sincere cooperation.

References

1. Rath, T. M., Manmatha, R.: Word spotting for historical documents. In: International Journal of Document Analysis and Recognition (IJDAR), Vol. 9, No. 2, pp. 139-152, (2007).
2. Konidaris, T., Gatos, B., Ntzios, K., Pratikakis, I., Theodoridis, S., Perantonis, S.J.: Keyword-guided word spotting in historical printed documents using synthetic data and user feedback. In: International Journal of Document Analysis and Recognition (IJDAR), Vol. 9, No. 2, pp. 166-177, (2007).
3. Cao, H., Govindaraju, V.: Template-free word spotting in low-quality manuscripts. In: Proceedings of the 6th International Conference on Advances in Pattern Recognition (ICAPR), pp. 45-53, (2007).
4. Adamek, T., O'Connor, N. E., Smeaton, A. E.: Word matching using single closed contours for indexing handwritten historical documents. In: International Journal of Document Analysis and Recognition (IJDAR), Vol. 9, No. 2, pp. 153-165, (2007).
5. Ntzios, K., Gatos, B., Pratikakis, I., Konidaris, T.: An old Greek handwritten OCR system based on an efficient segmentation-free approach. In: International Journal of Document Analysis and Recognition (IJDAR), Vol. 9, No. 2, pp. 179-192, (2007).
6. Leydier, Y., Lebourgeois, F., Emptoz, H.: Text search for medieval manuscript images. In: Pattern Recognition, Vol. 40, pp. 3552-3567, (2007).
7. Linlin, L., Shijian, L., Chew, L. T.: A fast keyword-spotting technique. In: Proceedings of 9th International Conference on Document Analysis and Recognition (ICDAR'07), pp. 68-72, (2007)
8. Srihari, S. N., Srinivasan, H., Babu, P., Bhole, C.: Handwritten Arabic word spotting using the CEDARABIC document analysis system. In: Proceedings of Symposium on Document Image Understanding Technology (SDIUT-05), College Park, MD, pp. 123-132, (2005).
9. Srihari, S. N., Srinivasan, H., Huang, C., Shetty, S.: Spotting words in Latin, Devanagari and Arabic scripts. In: Indian Journal of Artificial Intelligence, Vol. 16, No. 3, pp. 2-9, (2006).
10. Leydier, Y., Bourgeois, F. L., Emptoz, H.: Omnilingual segmentation-free word spotting for ancient manuscripts indexation. In: Proceedings of the 8th International Conference on Document Analysis and Recognition (ICDAR), Vol. 1, pp. 533-537, (2005).

11. Gatos, B., Konidaris, T., Ntzios, K., Pratikakis, I., Perantonis, S. J.: A segmentation-free approach for keyword search in historical typewritten documents. In: Proceedings of the 8th International Conference on Document Analysis and Recognition (ICDAR'05), pp. 54-56, (2005).
12. Chen, F. R., Bloomberg, D. S., Wilcox, L. D.: Spotting phrases in lines of imaged text. In: Vincent L and Baird H (Eds), Proceedings of SPIE – Document Recognition II. The International Society for Optical Engineering (SPIE) Vol. 2422, pp. 256-269, (1995).
13. Rath, T. M., Manmatha, R.: Features for word spotting in historical manuscripts. In: Proceedings of International Conference on Document Analysis and Recognition (ICDAR), Vol. 1, pp. 218-222, (2003).
14. Yue, L., Tan, C. L.: Chinese word searching in imaged documents. In: International Journal of Pattern Recognition and Artificial Intelligence, Vol. 18, No. 2, pp. 229-246, (2004).
15. Kołcz, A., Alspector, J., Augusteijn, M., Carlson, R., Popescu, G. V.: A line-oriented approach to word spotting in handwritten documents. In: Pattern Analysis & Applications, pp. 153-168, (2000).
16. Decurtins, J.: Comparison of OCR vs. Word shape recognition for keyword spotting. In: Symposium on Document Image understanding Technology, pp. 205-213, (1997).
17. Burl, M., Perona, P.: Using hierarchical shape models to spot keywords in cursive handwriting data. In: Proceedings of IEEE-CS Conference on Computer Vision and Pattern Recognition, pp. 535-540, (1998).
18. Keaton, P., Greenspan, H., Goodman, R.: Keyword spotting for cursive document retrieval. In: Proceedings of Workshop on Document Image Analysis, San Juan, Puerto Rico, pp. 74-81, (1997).
19. Manmatha, R., Han, C., Riseman, E. M.: Word spotting: A new approach to indexing handwriting. In: Proceedings of Computer Vision and Pattern Recognition Conference, pp. 631-637, (1996).
20. Kuo, S., Agazzi, O. E.: Keyword spotting in poorly printed documents using pseudo 2-d Hidden Markov Models. In: IEEE Transactions on Pattern Analysis and Machine Intelligence. Vol. 16, No. 8, pp. 842-848, (1994).
21. Chen, F. R., Vilcox, L. D. Bloomberg, D. S.: Word spotting in scanned images using Hidden Markov Models. In: Proceeding of the International Conference on Acoustics, Speech and Signal Processing, Vol. 5, pp. 1-4, (1993).
22. Tanaka, Y., Torii, H.: Transmedia machine and its keyword search over image texts. In: Recherche d'Information assitée par Ordinateur, Cambridge, MA, pp. 248-258, (1988).
23. DeCurtins, J., Chen, E.: Keyword spotting via word shape recognition. In: Vincent L and Baird H (Eds.), Proceedings of the SPIE – Document Recognition II. The International Society for Optical Engineering (SPIE), Vol. 2422, pp. 270-277, (1995).
24. Kuo, S. S., Agazzi, O. E.: Machine vision for keyword spotting using Pseudo 2-D Hidden Markov Models. In: Proceedings of the International Conference on Acoustics, Speech and Signal Processing. Vol. 5, pp. 11-84, (1993).
25. Jeffrey, C., O'Neill, Alfred O., Hero III., Williams, W. J.: Word spotting via spatial point processes. In: Proceedings of International Conference on Image processing: IEEE, pp. 217-220, (1996).
26. Spitz, A. L.: Using character shape codes for word spotting in document images. In: Dori D and Bruckstein A (Eds.), Shape, Structure, and Pattern Recognition, World Scientific, Singapore, pp. 382-389, (1995).
27. Tan, C. L., Huang, W., Sung, S. Y., Yu, Z., Xu, Y.: Text retrieval from document images based on word shape analysis. In: Applied Intelligence, pp. 257-270, (2003).

28. Chen, F. R., Wilcox, L., Bloomberg, D.: Detecting and locating partially specified keywords in scanned images using Hidden Markov Models. In: Proceedings of the 2nd International Conference on Document Analysis and Recognition. IEEE Computer Society Press, pp. 133-138, (1993).
29. Rath, T. M., Manmatha, R.: Word image matching using Dynamic Time Warping. In: Proceedings of IEEE Computer Society on Computer Vision and Pattern Recognition, Vol. 2, pp. 521-527, (2003).
30. Belongie, S., Malik, J.: Matching with shape contexts. In: Proceedings of IEEE Workshop on Content-based Access of Image and Video Libraries, pp. 20-26, (2000).
31. Tubbs, J. D.: A note on binary template matching. In: Pattern Recognition, Vol. 22, No. 4, pp. 359-356, (19989).
32. Manmatha, R., Rath, T. M.: Indexing of handwritten historical documents—recent progress. In: Proceedings of Symposium on Document Image Understanding Technology (SDIUT), pp. 77-85, (2003).
33. Jain, A. K., Namboodiri A. M.: Indexing and retrieval of on-line handwritten documents. In: Proceedings of the 7th International Conference on Document Analysis and Recognition (ICDAR'03), Vol. 2, pp. 655-659, (2003).
34. Al-Khatib, W. G., Shahab, S. A., Mahmoud, S. A.: Digital library framework for Arabic manuscripts. In: Proceedings of IEEE/ACS International Conference on Computer Systems and Applications, pp. 458-465, (2007).
35. Rothfeder, J. L., Feng, S., Rath, T. M.: Using corner feature correspondences to rank word images by similarity. In: Proceedings of Conference on Computer Vision and Pattern Recognition Workshop, pp. 30-35, (2003).
36. Manmatha, R., Han, C., Riseman, E. M., Croft, W. B.: Indexing handwriting using word matching. In: Proceedings of Digital Libraries'96, 1st ACM International Conference on Digital Libraries, pp. 151–159, (1996).
37. Shijian, L., Tan, C. L.: Keyword spotting and retrieval of document images captured by a digital camera. In: Proceedings of International Conference on Document Analysis and Recognition (ICDAR'07), pp. 23-26, (2007).
38. Shahab, S. A., Al-khatib, W. G., Mahmoud, S. A.: Computer aided indexing of historical manuscripts. In Proceedings of International Conference on Computer Graphics, Imaging and Vision. IEEE Computer Society, pp. 287-295, (2006).
39. Tomai, C. I., Zhang, B., Govindaraju, V.: Transcript mapping for historic handwritten document images. In: Proceedings of 8th International Workshop on Frontiers in Handwriting Recognition (IWFHR'02), pp. 413–418, (2002).
40. Khoubyari, S., Hull, J. J.: Keyword location in noisy document image. In: Proceedings of Second Annual Symposium on Document Analysis and Information Retrieval, UNLV, Las Vegas, pp. 217-231, (1993).
41. Zhang, B., Srihari, S. N., Huang, C.: Word image retrieval using binary features. In: Document Recognition and Retrieval XI, SPIE, Vol. 5296, pp. 45-53, (2004).
42. Yue, L., Tan, C. L.: Information Retrieval in document image databases. In: IEEE Transactions on Knowledge and Data Engineering, Vol. 16, No. 11, pp. 1398-1410, (2004).
43. Yue, L., Tan, C. L.: Word searching in document images using word portion matching. In: Proceedings of 5th IAPR International Workshop on Document Analysis Systems, pp. 19-21, (2002).
44. Cho, B-J., Kim, J. H.: Print keyword spotting with dynamically synthesized pseudo 2D HMMs. In: Pattern Recognition Letters, Vol. 25, pp. 999-1011, (2004).
45. Soo, H. K., Sang, C. P., Chang, B. J., Ji, S. K., Park, H. R., Guee, S. L.: Keyword spotting on Korean document images by matching the keyword image. In: Digital Libraries: Implementing Strategies and Sharing Experiences, Vol. 3815, pp. 158-166, (2005).

46. Park, S. C., Son, H. J., Jeong, C. B., Soo, H. K.: Keyword spotting on Hangul document images using two-level image-to-image matching. In: Proceedings of the 18th International Conference on Innovations in Applied Artificial Intelligence, pp. 79-81, (2005).
47. Jawahar, C. V., Balasubramanian, A., Meshesha, M.: Word-level access to document image datasets. In: Proceedings of the Workshop on Computer Vision, Graphics and Image Processing (WCVGIP), pp. 73-76, (2004).
48. Balasubramanian, A., Meshesha, M., Jawahar, C. V.: Retrieval from document image collections. In: Proceedings of Seventh IAPR Workshop on Document Analysis Systems, (LNCS 3872), pp. 1-12, (2006).
49. Christophe, C.: Dynamic handwritten keyword spotting based on the NSHP-HMM. In: Proceedings of 9th International Conference on Document Analysis and Recognition (ICDAR'07), pp. 242-246, (2007).
50. Josep, L., Gemma, S.: Indexing historical documents by word shape signatures. In: 9th International Conference on Document Analysis and Recognition (ICDAR'07), Vol. 1, pp. 362-366, (2007).
51. Rohlicek, J. R., Russell, W., Roukos, S., Gish, H.: Continuous Hidden Markov Modeling for speaker-independent word spotting. In: Proceedings of IEEE ICASSP, pp. 627-630, (1989).
52. Paul D. B., Rose, R. C.: A Hidden Markov Model based keyword recognition system. In: Proceedings of IEEE ICASSP, pp. 129-132, (1990).
53. Doermann D.: The indexing and retrieval of document images: a survey. In: Computer Vision and Image Understanding, Vol. 70, No. 3, pp. 287-298, (1998).
54. Mitra, M., Chaudhuri, B. B.: Information retrieval from documents: a survey. In: Information Retrieval, Vol. 2, pp. 141-163, (2000).
55. Laurence, L-S., Abderrazak Z., Bruno T.: Text line segmentation of historical documents: a survey. In: International Journal of Document Analysis and Recognition (IJDAR'07), Vol. 9, No. 2, pp. 123-138, (2007).
56. Mahadevan, U., Nagabushnam, R. C.: Gap metrics for word separation in handwritten lines. In: Proceedings of the Third International Conference on Document Analysis and Recognition, Los Alamitos, CA, pp. 124-127, (1995).
57. Scott, G. L., Longuet-Higgins, H. C.: An Algorithm for associating the features of two patterns. In: Proceedings of the Royal Society of London, B224, pp. 21-26, (1991).
58. Manmatha, R., Rothfeder, J.: A scale space approach for automatically segmenting words from historical handwritten documents. In: IEEE Transactions, Pattern Analysis and Machine Intelligence, Vol. 27, No. 8, pp. 1212-1225, (2005).
59. Bokser, M.: Omnidocument technologies. In: Proceedings of IEEE, Vol. 80, No. 7, pp. 1066-1078, (1992).
60. Agrawal, M., Kumar, P. M. N. S. S. K., Jawahar, C. V.: Indexing and retrieval of devanagari text from printed documents. In: Proceedings of NCDAR, pp. 244-251, (2003).
61. Suen, C. Y., Xu, Q., Lam, L.: Automatic recognition of handwritten data on cheques –fact or fiction? In: Pattern Recognition Letters, Vol. 20, pp. 1287-1295, (1999).
62. Kabir, E., Downton, A. C., Birch, R.: Recognition and verification of postcodes in handwritten and hand-printed addresses. In: Proceedings of 10th International Conference on Pattern Recognition, Vol. 1, pp. 469-473, (1990).
63. Zhang, B., Srihari, S. N.: Binary Vector Dissimilarity Measures for Handwriting Identification. In: Proceedings of the SPIE, Document Recognition and Retrieval X, pp. 155-166, (2003).
64. Saabni, R., El-Sana, J.: Keyword Searching for Arabic Handwritten Documents. In: Proceedings of 11th International Conference on Frontiers in Handwritten Recognition, pp. 271-277, (2008).

CHAPTER 3.4

OPTIMIZATION OF REJECTION PARAMETERS TO ENHANCE RELIABILITY IN HANDWRITING RECOGNITION

Chun Lei He Louisa Lam Ching Y. Suen

CENPARMI (Centre for Pattern Recognition and Machine Intelligence)
Computer Science and Software Engineering Department
Concordia University, Montreal, Quebec, Canada
E-mails: {cl_he, llam, suen}@cenparmi.concordia.ca

In pattern recognition, optimizing the parameters in rejection can achieve a high reliability while maintaining a high recognition rate. When rejection is considered to be a two-class problem of accepting the classification result or not, Linear Discriminant Analysis (LDA), which is an effective classification method, can be used to optimize the criterion for rejection. In this work, a Linear Discriminant Analysis Measurement (LDAM) is designed to take into consideration the confidence values of the classifier outputs and the relations between them. In the process, the First Rank Measurement (FRM), First Two Ranks Measurement (FTRM), and other traditional rejection measurements are reviewed. The use of FRM is an absolute rejection strategy that considers only the top choice for rejection, while using FTRM is a relative rejection strategy that considers the top two choices. Experiments are conducted on the CENPARMI Arabic Isolated Numerals Database with outputs that can represent distances or probabilities from different classifiers such as HeroSVM and LibSVM. The results show that the use of LDAM is more optimal than FRM and FTRM in producing reliable recognition results.

1. Introduction

Optical Character Recognition (OCR) is one of the most successful applications in pattern recognition, and it has been under investigation since the mid 1950's. In the early days, research was focused on printed and then handwritten character recognition. When research advanced from printed to handwritten character recognition, many challenges became apparent because of the variations in handwriting styles, ambiguities in handwritten numerals, different kinds of noise that may create breaks in the strokes of characters or change their topology, and so on. Even though the problem is intrinsically complicated, many researchers continued to develop and implement algorithms for recognizing unconstrained handwritten characters, including numerals. This is because the recognition of unconstrained handwritten numerals is an important aspect of OCR that has applications in numerous fields such as automatic postal sorting, automatic bank cheque processing, and financial slip processing [35, 28, 24, 39, 23]. Some current research methods have achieved high recognition rates of over 99% on some Latin numeral databases such as on MNIST [28, 40, 23] and CENPARMI Database [26].

However, even these low error rates can be costly in some applications. For example, in processing financial forms, mistaking an amount or an identification number would cause transaction errors that ultimately must be detected and corrected manually. Therefore, it is the general expectation that OCR machines should achieve a high recognition rate as well as high reliability. In other words, the OCR machine should have a high recognition rate, and the data that cannot be recognized correctly should be rejected since the reject option can be very useful in preventing excessive misclassifications in applications that require high classification reliability [15]. The recognition rate, substitution rate, rejection rate and reliability are defined as follows [35]:

$$Recognition\ rate = \frac{Number\ of\ correctly\ recognized\ samples}{Total\ number\ of\ test\ samples} \times 100\%$$

$$Substitution\ rate = \frac{Number\ of\ incorrectly\ recognized\ samples}{Total\ number\ of\ test\ samples} \times 100\%$$

$$Rejection\ rate = \frac{Number\ of\ rejected\ samples}{Total\ number\ of\ test\ samples} \times 100\%$$

$$Reliability = \frac{Recognition\ rate}{100\%\text{-}Rejection\ rate} \times 100\%$$

Achieving both high recognition and high reliability requires methods capable of assigning generally higher confidences to correct recognition results than to incorrect ones. This confidence scoring method may consist of implementing a simple function with appropriate parameters drawn directly from the recognition process, or it may be a learning task in which a classifier is trained to use an array of parameters to distinguish correct recognitions from misclassifications [30].

It seems that the Bayes' decision rule embodies a rejection rule; namely, the decision can be based on the maximum confidence value (called First Rank Measurement (FRM) in this paper) provided this maximum exceeds a certain threshold value. However, this approach did not perform satisfactorily when experiments were performed on the CENPARMI Arabic Isolated Numerals Database with LibSVM [8, 38] and HeroSVM [13]. LibSVM maps the outputs from a Support Vector Machine (SVM) [38] to posterior probabilities for all classes; and HeroSVM provides distances from an input pattern to the optimal separating hyperplane (OSH) of each class. In LibSVM, the distribution (Figure 1(a)) of incorrectly classified samples is not Gaussian in shape, but remains flat throughout a range of confidence values. This is the case while only the correctly classified samples follow a Gaussian distribution. In HeroSVM, although the distributions (Figure 1(b)) of correctly and incorrectly classified samples are Gaussian in shape, their measurements do overlap for almost half of the range. Therefore, it is difficult to design a rejection strategy based on the measurement of maximum confidence value. The results on the training set are shown in Figure 1.

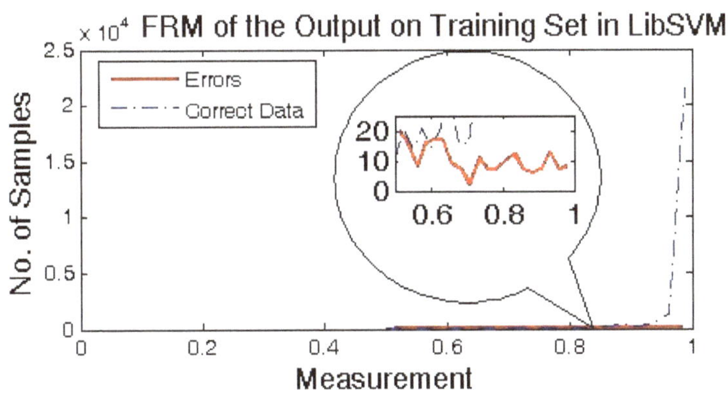

(a)

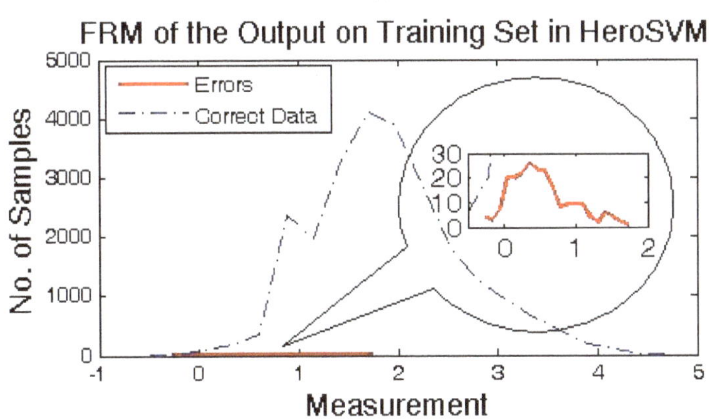

(b)

Figure 1. Distribution of the output on the Training set in LibSVM and HeroSVM

In this work, we design and implement a method that applies Linear Discriminant Analysis (LDA) [14] to the measurement level outputs of a classifier in order to determine an optimal threshold for the rejection option. LDA is a supervised classification method widely used to find an optimal linear combination of features for separating two or more classes. The main idea of LDA is to project high-dimensional data onto a line and perform classification in this one-dimensional space. It provides a linear projection of the data with the outcome of maximum between-class variance and minimum within-class variance. By finding the feature space that can best discriminate an object from others, these discriminative methods have been successfully used in pattern classification applications including Chinese character recognition [17], face recognition [4, 37], image retrieval [36], tracking [25], and marketing [11].

Offline handwritten character recognition in languages such as English, Chinese, and Japanese has been researched extensively for over twenty years. However, Arabic handwriting recognition is a relatively new area of research even though Arabic is one of the most widely used languages in the world [31]. There are a few databases on Arabic handwriting. For instance, the IFN/ENIT databases [12] developed in 2002 consists of 26,549 images of Tunisian town/village names written by 411 writers. Another database is the AHDB database [3] developed in 2002 which contains words frequently used in legal amounts on Arabic checks together with some other frequently used Arabic words. At the Centre for Pattern Recognition and Machine Intelligence (CENPARMI), a number of Arabic Script databases have been developed. Al-Ohali et al. [1] developed an Arabic check database for research on the recognition of Arabic handwritten checks in 2000. The data includes Arabic legal amounts and Arabic sub-words present in checks. In 2006, Solimanpour et al. [33] designed a Farsi database consisting of Farsi isolated digits, numeral strings, letters, legal amounts, and dates. Recently, Alamri et al. [2] developed the CENPARMI Arabic database which contains isolated Indian/Arabic numerals, numeral strings, Arabic isolated letters, and Arabic words. This database was produced by many writers of different genders, ages, educational levels and nationalities, with both left-handed and right-handed writers included. The experiments reported in this paper are conducted on the isolated numerals in this database.

Detecting samples recognized with low confidence for rejection and thus achieving a high reliability in handwritten numeral recognition is our objective in this research. In Section 2, we review related work including the rejection strategies employed by various classifiers including SVM's. We define a novel rejection measurement of LDA Measurement (LDAM), and we compare it to the rejection measurements of First Rank Measurement (FRM) and First Two Ranks Measurement (FTRM) in Section 3. Then we describe the experiments on the CENPARMI Arabic Isolated Numerals Database and compare the results obtained from using these three measurements (Section 4). Finally, in Section 5, we draw conclusions about the use of the measurements.

2. Related Work

In the literature, a number of rejection strategies have been proposed for various underlying classifiers and applications. In this section, we review the state-of-the-art rejection strategies implemented by various offline handwriting recognition systems that make use of different levels of classifier outputs.

2.1. *Outputs from Classifiers*

Generally speaking, classification algorithms supply outputs at three levels [19]:

1) The abstract level: a classifier e outputs a likely unique label/class j; or in some extensions, e outputs a subset $J \subset \Lambda$, where Λ is the set of all classes.

2) The rank level: e ranks all the labels in Λ (or a subset $J \subset \Lambda$) according to the likelihoods that the input sample x has those labels.

3) The measurement level: e attributes to each label in Λ a measurement value. This measurement can be a probability that x has that label, or the distance of x from the class having that label.

Among the three levels, the measurement level provides the most information and the abstract level the least. From the measurement attributed to each label, we could rank all the labels in Λ in ascending or descending order. By choosing the label at the top rank, or directly by choosing the label with the maximal or minimal value at the measurement level, we can assign a unique label to x. In other words, from the measurement level to the abstract level there is an information reduction or abstraction process. On the other hand, when classification provides only abstract level outputs, it would be difficult to design a rejection strategy.

In this work, both support vector classifiers HeroSVM [13] and LibSVM [8] considered do provide measurement level outputs.

In HeroSVM, the outputs represent the distances between the input vector and the margins of each class. HeroSVM[1] is a fast and high-performance SVM software package which introduces a parallel optimization step to quickly remove

[1] Available at: http://www.cenparmi.concordia.ca/~jdong/HeroSvm.html.

most of the nonsupport vectors and applies an effective integration of kernel caching and kernel matrix computation for classification. The strategy applied here in multi-class problems is to consider one class against all others [41]. Taking the training samples with one label as one class and all others as the other class, it reduces to a two-class problem. For k classes of data ($k > 2$), k SVM classifiers are formed and denoted by SVM_i, $i=1,2, \ldots k$. For the test sample x, $d_i(x) = w_i \cdot x + b_i$ can be obtained by using SVM_i, where d_i is the decision function for class i, w_i is a normal vector perpendicular to the hyperplane separating class i from all other classes, and the parameter b_i is the distance from the origin to the hyperplane along the normal vector w_i. The test sample x is considered to belong to the jth class where $d_j(x) = \max_{1-1,2,\ldots,k} d_i(x)$.

Unlike HeroSVM, the classification outputs of LibSVM represent probabilities. LibSVM is an implementation of SVM which applies a one-against-one (or pairwise) strategy in multi-class problems. With the pairwise approach, k^2 support vector machines are trained for a k-class problem. Given k classes of data and any test sample x, the goal is to estimate p_i (the probability that x belongs to class i), which is obtained from r_{ij} ($i, j=1,2, \ldots k$). r_{ij} is a one-against-one class probability obtained from the known training data by solving the following optimization problem:

$$\min_p \tfrac{1}{2} \sum_{i=1}^{k} \sum_{j:j\neq i} (r_{ji} p_i - r_{ij} p_j)^2$$

$$subject\ to \sum_{i=1}^{k} p_i = 1,\ p_i \geq 0, \forall i,$$

where

$$p_i = p(y = i \mid x), \text{ for class label } y \text{ of } x, \text{ and}$$

$$r_{ij} \approx p(y = i \mid y = i\ or\ j, x).$$

2.2. *Rejection Strategies for Offline Handwriting Recognition*

A recognition rule can be considered optimum if for a given recognition rate, it minimizes the error rate (error probability) and places the testing candidates into a reject category when their identities cannot be established with high confidence [10]. When a feature vector has the highest conditional probability for the correct class and low conditional probabilities in all other classes, it should be accepted;

otherwise it should be rejected. Rejections can be applied in single classifiers as well as Multiple Classifier Systems (MCS) in order to increase the reliability of the recognition results. Various handwriting recognition systems for offline handwritten numerals [40, 7], characters [30], words [21], and text lines [5], as well as check processing [18] and address reading [6] systems have explored rejection methodologies in their implementations.

In general, the recognizer estimates posterior probabilities for the various classes, so it is possible to make optimal (Bayesian) decisions by comparing the probabilities of samples to a threshold so that probabilities below the threshold would result in rejections. Generally, rejection strategies can be divided into two categories: absolute and relative rejections.

In absolute rejection, only the top choice among the outputs (called FRM in this paper) is used as criterion for rejection. This has been implemented in handwritten numeral recognition [7] and in character recognition [30]. The latter work evaluated and explored a variety of scoring functions including the "raw" recognition score, and a sample is assigned to the class with the highest score provided this score is large enough.

In relative rejection, the relationship between various confidence measurements is taken into consideration. Examples of such relationships include the likelihood ratio (ratio of the highest and second-highest confidence values) and the estimated posterior probability (ratio of the highest confidence value to the sum of all confidence values) [30]. Other examples include class-dependent and hypothesis-dependent thresholds, since they consider the average of certain confidence measurements or the difference between the top two confidence values, as applied to word recognition in [21]. The distance between the first and second choices is used as a rejection criterion in handwritten numeral recognition [7] and in a German address reading system [6]. This is called First Two Ranks Measurement or FTRM in this paper.

Rejections have been applied to recognition systems with a single classifier or multiple classifiers. In [7], a rejection strategy for convolutional neural network models is proposed. In [30, 18], all confidence measures are used as input to a Multi-Layer Perceptron to finalize the result of recognition. Hidden Markov Models (HMMs) have been used in error-rejection of a word recognition system [21, 5, 6]. In [13], rejection with the FRM is used for a Support Vector Machine (SVM).

In multiple classifiers systems (MCSs), cooperation is a sequential (as opposed to a parallel) architecture [19]. With this topology, classifiers can be applied in succession, with each classifier producing a reduced set of possible classes for each pattern, so that the individual classifiers or experts can become increasingly focused [22]. In handwritten numerals recognition, Zhang *et al.* [40] implemented rejection in a cascade ensemble classifier system which is a sequential combination of multiple classifier ensembles. Rejections from one layer of classification are input to the next layer's classifier. The relationship of the error, rejection, and recognition rates of each Multi-Layer Perceptron classifier is analyzed with the use of Bayesian probability theory. In [16], after linearly combining four types of classifiers with a posteriori probabilities estimation, the absolute rejection strategy with FRM is applied.

3. Rejection Measurements

In considering the outputs of classifiers for the rejection option as a two-class problem, (acceptable or rejected classification), the outputs at the measurement level can be considered as features for the rejection option. In an output vector, whose components may represent distances or probabilities, we expect the confidence value (measure) of the first rank (most likely class) to be far distant from the confidence values or measures of the other classes. In other words, good outputs should be easily separated into two classes: the confidence value of the first rank and others. In the following discussion, we assume that the classifier outputs the probabilities of the patterns for each class, and the discussion would be analogous for the case when the classifier outputs distances.

3.1. *First Rank Measurement (FRM) & First Two Ranks Measurement (FTRM)*

Generally, rejection strategies can be directly applied to the outputs with probability estimation. In an M-class problem, suppose $P(x)=\{p_1(x), p_2(x),\ldots, p_M(x)\}$ is the classification output vector of the given pattern x, with probabilities $p_i(x)$ in descending order. The decision function would be based on $\mathrm{sgn}(\Phi_1(x)-T_1)$, where T_1 is a threshold derived from the training data, and $\Phi_1(x)=p_1(x)$.

If $\Phi_1(x)\le T_1$, the classifier rejects the pattern and does not assign it to a class (it might instead be passed to a human operator). This has the consequence that on the remaining patterns, a lower error rate can be achieved. This method uses the First Rank Measurement (FRM) [5, 7, 13, 30].

Under this method, the frequency distribution according to confidence values of samples in the training set is considered and the threshold T_1 is determined. However, FRM cannot distinguish between reliable and unreliable patterns with the probability distribution of erroneous samples shown in Fig. 1.

To overcome this deficiency of FRM, First Two Ranks Measurement (FTRM) has been designed [6, 7, 21, 35], and it uses the difference between the probabilities $p_1(x)$ and $p_2(x)$ of the first two ranks as a condition of rejection. In FTRM, the measurement function is $\Phi_2(x) = \| p_1(x) - p_2(x) \|$, where $\|.\|$ can be any distance measurement, and the decision function is based on $\mathrm{sgn}(\Phi_2(x) - T_2)$, where T_2 is a threshold derived from the training set.

However, FTRM cannot solve the problem in some cases. For example, if $\| p_1(x) - p_2(x) \|$ is relatively large compared to T_2, but the distance $\| p_2(x) - p_3(x) \|$ is much larger, this pattern may still be accepted, when this pattern should be considered for rejection since the top two classes are close together in terms of relative distance.

3.2. *LDA Measurement (LDAM)*

To consider the relative difference between the measurements in the first two ranks and all other measurements, LDAM is defined and applied. Since rejection in classification can be considered as a two-class problem (acceptance or rejection), we apply LDA [14] for two classes, to implement rejection. LDA approaches the problem by assuming that the conditional probability density functions of the two classes are both normally distributed. There are $n = n_1 + n_2$ observations with d features in the training set, where $\{x_{1i}\}_{i=1}^{n_1}$ arise from class ω_1 and $\{x_{2i}\}_{i=1}^{n_2}$ arise from class ω_2. Gaussian-based discrimination assumes two normal distributions: $(x \mid \omega_1) \sim N(\mu_1, \Sigma_1)$ and $(x \mid \omega_2) \sim N(\mu_2, \Sigma_2)$. In LDA, the projection axis (discriminant vector) w for discriminating between two classes is estimated to maximize the Fisher criterion:

$$J(w) = tr((w^T S_w w)^{-1} (w^T S_B w))$$

where tr(·) denotes the trace of a matrix, $\mathbf{S}_B$ and $\mathbf{S}_w$ denote the between-class scatter matrix and within-class scatter matrix respectively, and w is the optimal discriminant vector. For the two classes ω_1 and ω_2, with *a priori* probabilities p_1 and p_2 (it is often assumed that $p_1 = p_2 = 0.5$), the within-class and between-class scatter matrices can be written as

$$S_w = p_1\Sigma_1 + p_2\Sigma_2 = \Sigma_{12}$$
$$S_B = (\mu_1 - \mu_2)(\mu_1 - \mu_2)^T,$$

where Σ_{12} is the average variance of the two classes. It can be shown that the maximum separation occurs when:

$$w = S_w^{-1}(\mu_1 - \mu_2) = (p_1\Sigma_1 + p_2\Sigma_2)^{-1}(\mu_1 - \mu_2)$$
$$= \Sigma_{12}^{-1}(\mu_1 - \mu_2)$$

To apply this principle to the outputs for the rejection option as a one-dimensional application, we define the two sets $G^{(1)}(x) = \{p_1(x)\}$, and $G^{(2)}(x) = \{p_2(x), p_3(x), ..., p_M(x)\}$.

Then it follows that

$$\mu_1 = p_1(x),$$

$$\mu_2 = \frac{1}{M-1}\sum_{i=2}^{M} p_i(x),$$

$$\Sigma_1 = (p_1(x) - \mu_1)^2 = 0,$$

$$\Sigma_2 = \frac{1}{M-1}\sum_{i=2}^{M}(p_i(x) - \mu_2)^2,$$

$$\text{and} \quad \Sigma_{12} = \tfrac{1}{2}\Sigma_2.$$

Thus, in LDA,

$$w = \Phi_3(x) = \frac{\sum_{i=2}^{M} \| p_1(x) - p_i(x) \|}{(M-1)\cdot\Sigma_{12}}.$$

Then the decision function would be based on $\mathrm{sgn}(\Phi_3(x) - T_3)$, where T_3 is a threshold derived from the training set, and all values are scaled to [0, 1].

In summary, when compared to FRM and FTRM, LDAM should be more reliable and informative since it compares the relative difference of the measures in the first two ranks with all other measures.

4. Experiments & Results

The CENPARMI Arabic Isolated Numerals Database is used for the experiments. It contains 18,585, 6,199, and 6,199 samples in the Training, Validation, and Test sets, respectively. Since validation was not implemented in this experiment, the Training and Validation sets were combined to form the Training set. There are 10 classes (0-9), and five samples of each numeral are shown in Figure 3.

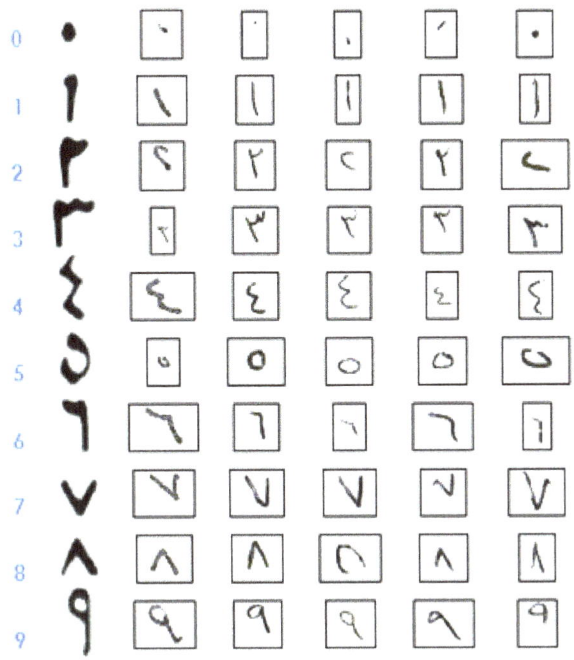

Figure 2. Samples from CENPARMI Arabic Isolated Numerals Database

In the recognition process, the standard procedures of image pre-processing, feature extraction, and classification were implemented. In image pre-processing, we performed noise removal, grayscale normalization, size normalization [21], and binarization of the gray-scale images. Gradient features were extracted from pseudo gray-scale images [32]. The Robert's filter, which uses the masks $\begin{bmatrix} 0 & 1 \\ -1 & 0 \end{bmatrix}$ and $\begin{bmatrix} 1 & 0 \\ 0 & -1 \end{bmatrix}$, was applied to calculate the gradient strengths and directions of pixels. The direction of the gradient was quantized to 32 levels with an interval of $\pi/16$. The normalized character image was divided into 81 (9×9) blocks. After extracting the strengths and directions in each image, the spatial resolution was reduced from 9×9 to 5×5 by down sampling every two horizontal and every two vertical blocks with a 5×5 Gaussian filter. Similarly, the directional resolution was reduced from 32 to 16 levels by down sampling with a weight vector $[1\,4\,6\,4\,1]^T$, to produce a feature vector of size 400 (5×5×16). Moreover, the transformation $y = x^{0.4}$ was applied to make the distribution of the features Gaussian-like. Finally, we scaled the feature vectors by a constant factor so that the values of feature components would range from 0 to 1.

Support Vector Machines [38] was chosen as a classifier. SVMs with different kernel functions can transform a non-linear separable problem into a linear separable one by projecting data into the feature space, and then SVMs can find the optimal separating hyperplane. Radial Basis Function (RBF) was chosen as the kernel in both HeroSVM and LibSVM in this research. The recognition rates on the test set are 98.48% and 97.76% for LibSVM and HeroSVM respectively (Table 1), which are significantly higher than the performance (93.60%) of [2] on the same database. Out of the 6199 samples in the test set, the numbers of misclassifications were 94 (1.52%) and 139 (2.24%) for LibSVM and HeroSVM respectively, and most of these samples cannot be correctly identified even by human beings.

Table 1. Performances on test set with LDAM compared with [2]

Classifier	LibSVM	HeroSVM	[2]
Recognition Rate (%)	98.48	97.76	93.60
Error Rate (%)	1.52	2.24	6.40

In Figures 3 and 4, we show the distributions of samples for the three measurements (FRM, FTRM, and LDAM) obtained from the outputs of LibSVM and HeroSVM respectively on the test set. The solid lines represent the distributions of errors, and the dotted lines represent the distributions of correctly recognized samples.

With LibSVM, the distributions of the training and test data are similar for FRM (distributions of training data are shown in Figure 1(a)). Although the correctly classified samples display a Gaussian distribution, the errors are distributed almost evenly for confidence values (measurements) ranging from 0.4 to 1, which means that correctly and incorrectly classified samples cannot be distinguished based on FRM. When compared to FRM, FTRM is more discriminating, as the range of measurements in FTRM is wider than in FRM (the range is (0, 1) for FTRM). However, the distribution of errors in FTRM is rather even as well. LDAM is the most discriminating of the three measurements, because the errors together with correctly classified samples with low confidence values are assigned small measurements. In LDAM, most incorrectly classified samples (78/94) retain very low measurements (less than 0.05), which results in a high reliability (99.7%) when the threshold is set at this value.

In Figure 4 (showing the results for HeroSVM), it can be observed that for FRM, both the correctly and incorrectly classified samples display Gaussian distributions but with overlapping ranges of measurement values, which means that correctly and incorrectly classified samples cannot be adequately distinguished. When compared to FRM, FTRM is more discriminating, as the

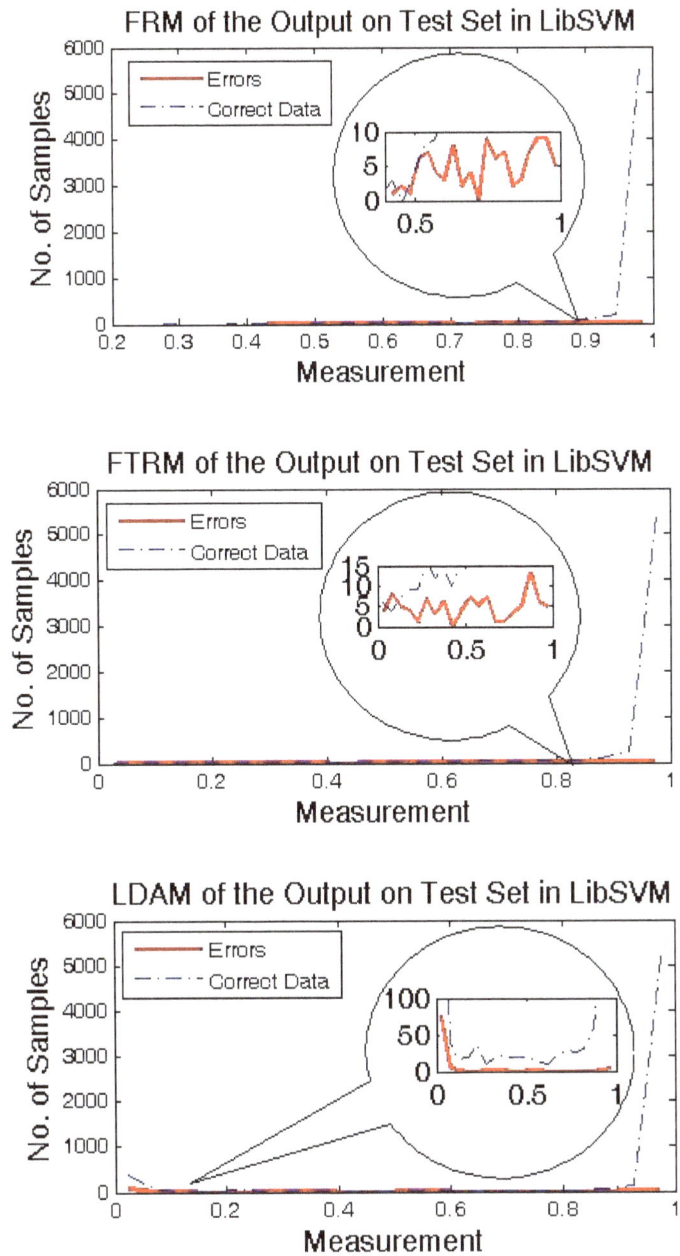

Figure 3. Distributions of the three measurements on the Test Set with LibSVM

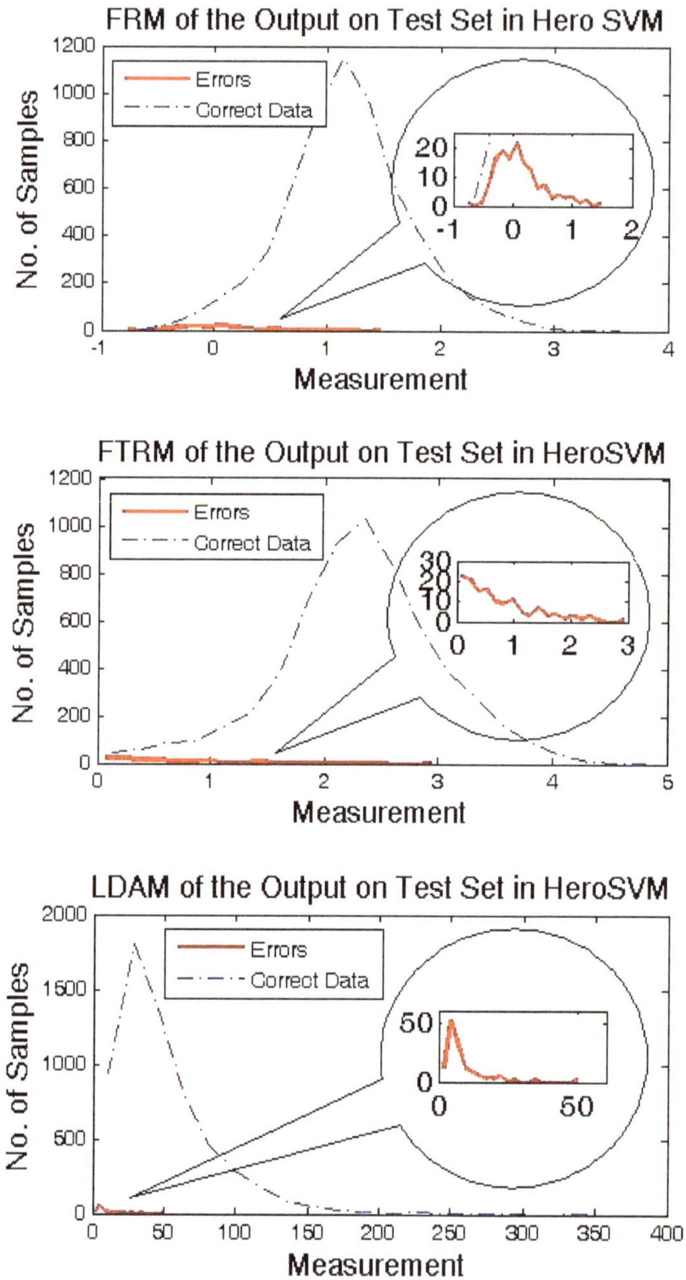

Figure 4. Distributions of the three measurements on the Test Set in with HeroSVM

peaks of the distributions are located farther apart (at 2.35 and 0.08 respectively). However, the distribution of errors in FTRM is even as well. LDAM is more discriminating than FRM and FTRM because, similarly to the distributions in LibSVM, most errors are assigned small measurements, and the distribution of errors decreases sharply from the peak. With LDAM, most incorrectly classified samples (111/139) have very low measurements from 0 to 10 out of a total range from 0 to 351, and this results in a high reliability of 99.51%.

From these experimental results, it can be seen that LDAM enables the rejection of samples classified with low reliability when the thresholds are obtained from the training set for both LibSVM and HeroSVM.

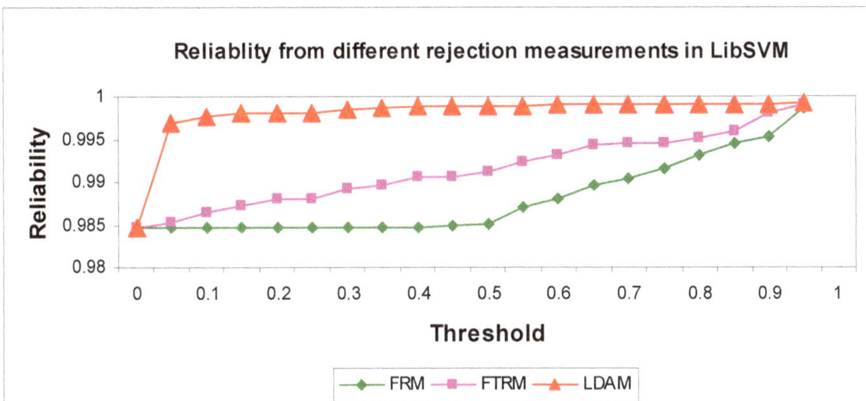

(a)

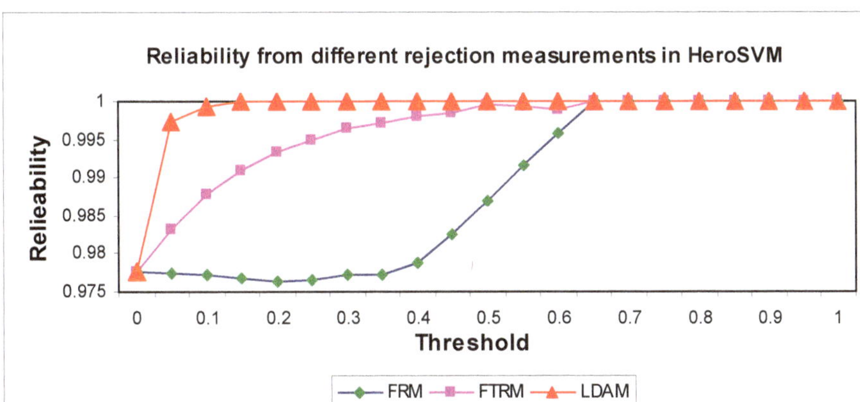

(b)

Figure 5. Reliability with different thresholds used on the three measurements in: (a) LibSVM (b) HeroSVM

The performances using different thresholds with the various measurements are shown in Figure 5. As illustrated, when the threshold T_3 is set to 0.05 in LibSVM, the reliability increases to 99.69% with LDAM, while the reliabilities with FRM and FTRM are 98.48% and 98.52% respectively. Similarly, when the threshold T_3 is set to 0.01 with HeroSVM, the reliability increases to 98.05% with LDAM, while the reliabilities with FRM and FTRM are 97.76% and 97.87%, respectively. These results show that LDAM is the most effective measurement for obtaining reliable results from both LibSVM and HeroSVM.

When the reliabilities of LDAM on LibSVM and HeroSVM are 99.70% and 99.51% respectively, there are 16 and 28 errors (out of 6199 samples) respectively, which are shown in Figure 6. As can be seen, these errors are reasonable since even human beings would have difficulty in recognizing them, or to distinguish between samples of "2" and "3" written in the same styles in Arabic.

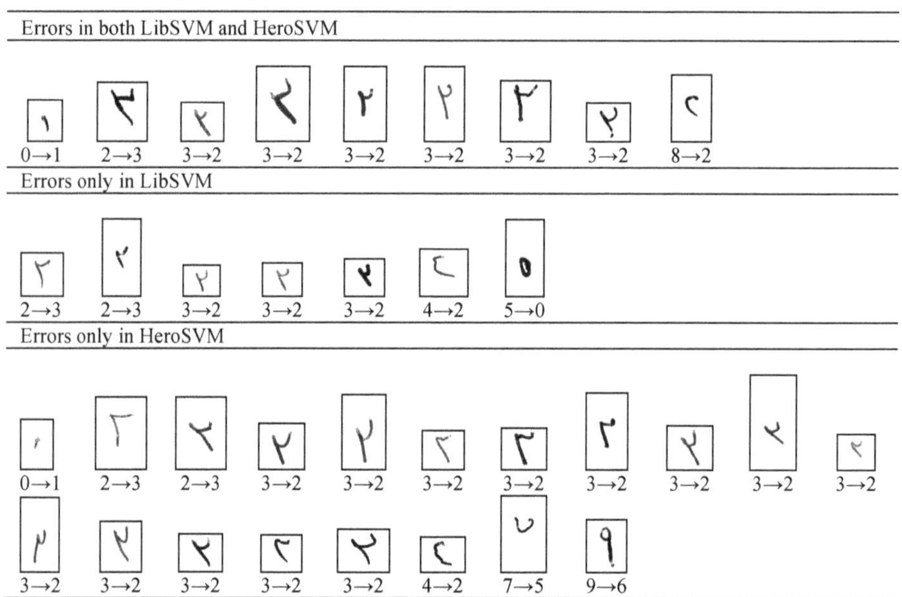

Figure 6. Incorrectly classified images

5. Conclusion

The rejection option is very useful for preventing misclassifications, which is important in applications which require high reliabilities. We designed a novel rejection criterion using the LDA Measurement (LDAM), which relies on the principle of Linear Discriminant Analysis and considers relationships among the probabilities in each output vector. We trained this rejection measurement on the training set in the CENPARMI Arabic Isolated Numerals Database, and tested it on the test set with two classifiers whose outputs are either probabilities or distances. At the same time, we compared LDAM with other measurements such as FRM and FTRM. The results indicate that LDAM achieved a higher reliability than the other measurements.

In the future, we can apply this rejection method to multi-classifier systems such as the combination of LibSVM and HeroSVM. Moreover, we can apply this methodology to semi-supervised learning, in order to reject the data with unreliable classification results produced by supervised learning.

References

[1] Al-Ohali, Y., Cheriet, M., and Suen, C.Y. Database for Recognition of Handwritten Arabic Cheques. *Proc. of the 7th Int. Workshop on Frontiers in Handwriting Recognition (IWFHR 2000)*, The Netherlands, pp. 601–606. (2000).

[2] Alamri, H. Sadri, J., Suen, C. Y., and Nobile, N. A Novel Comprehensive Database for Arabic Off-Line Handwriting Recognition. *Proc. of the 11th Int. Conference on Frontiers in Handwriting Recognition (ICFHR 2008)*, Montreal, Canada, pp. 664–669. (2008).

[3] Alma'adeed, S., Elliman, D., and Higgins, C.A. A Database for Arabic Handwritten Text Recognition Research. *Proc. 8th Int. Workshop of Frontiers in Handwriting Recognition (IWFHR 2002)*, Niagara-on-the Lake, Canada, pp. 485–589. (2002).

[4] Belhumeur, P. N., Hespanda, J., and Kriegman, D. Eigenfaces vs. Fisherfaces: Recognition using class specific linear projection. *IEEE Trans. Pattern Analysis and Machine Intelligence*, 19 (7), pp. 711–720. (1997).

[5] Bertolami, R., Zimmermann, M. and Bunke, H. Rejection strategies for offline handwritten text line recognition. *Pattern Recognition Letters*, 27 (16), pp. 2005–2012. (2006).

[6] Brakensiek, A. and Rigoll, G. Handwritten address recognition using hidden Markov models. *Reading and Learning: adaptive content recognition*, Springer, vol. 2956, pp. 103–122. (2004).

[7] Cecotti, H. and Belaid, A. Rejection strategy for convolutional neural network by adaptive topology applied to handwritten digits recognition. *Proc. of the 8th Int. Conference on Document Analysis and Recognition (ICDAR'05)*, Seoul, Korea, pp. 765–769. (2005).

[8] Chang, C.-C. and Lin, C.-J. LIBSVM: a library for support vector machines. Software available at: http://www.csie.ntu.edu.tw/~cjlin/libsvm. (2001).

[9] Choma, M. A., Sarunic, M. V., Yang, C., and Izatt, J. A. Sensitivity advantage of swept source and Fourier domain optical coherence tomography. *Optics Express*, 11(18), pp. 2183–2189. (2003).

[10] Chow, C.K. On optimum recognition error and reject tradeoff. *IEEE Trans. Inf. Theory*, 26(1), pp. 40–46. (1970).

[11] Dillon, W.R. and Schiffman, L. Appropriateness of Linear Discriminant and Multinomial Classification Analysis in Marketing Research. *Journal of Marketing Research*, vol. 15, pp. 103–112. (1978).

[12] Dimauro, G., Impedovo, S., Modugno, R., and Pirlo, G. A new database for research on bank-check processing. *Proc. of the 8th Int. Workshop on Frontiers in Handwriting Recognition (IWFHR 2002)*, Niagara-on-the Lake, Canada, pp. 524–528. (2002).

[13] Dong, J. X., Krzyzak, A., and Suen, C.Y. Fast SVM training algorithm with decomposition on very large datasets, *IEEE Trans. Pattern Analysis and Machine Intelligence*, 27(4), pp. 603–618. (2005).

[14] Fisher, R.A. The use of multiple measurements in taxonomic problems. *Annals of Eugenics*, 7, pp. 179–188. (1936).

[15] Fumera, G. and Roli, F. Analysis of error-reject trade-off in linearly combined multiple classifiers. *Pattern Recognition*, 37(6), pp. 1245–1265. (2004).

[16] Fumera, G., Roli, F., and Vernazza, G. A method for error *rejection* in multiple classifier systems. *Proc. of the 11th Int. Conference on Image Analysis & Processing*, Palermo, Italy, pp. 454–458. (2001).

[17] Gao, T.-F. and Liu, C.-L. High accuracy handwritten Chinese character recognition using LDA-based compound distances. *Pattern Recognition*, 41, pp. 3442–3451. (2008).

[18] Gorski, N. Optimizing error-reject trade off in recognition systems. *Proc. of the 4th International Conference on Document Analysis and Recognition (ICDAR'97)*, Ulm, Germany, vol. 2, pp. 1092–1096. (1997).

[19] Gunes, V., Menard, M., Loonis, P., and Petit-Renaud, S. Combination, cooperation and selection of classifiers: A state of the art. *Int. Journal of Pattern Recognition and Artificial Intelligence*, vol. 17, No. 8, pp. 1303 – 1324. (2003).

[20] He, C. L., Zhang, P., Dong, J. X., Suen, C. Y., and Bui, T. D. The role of size normalization on the recognition rate of handwritten numerals. *Proc. of IAPR TC3 Workshop of 8th International Conference on Document Analysis and Recognition (ICDAR'05): Neural Networks and Learning in Document Analysis and Recognition*, Seoul, Korea, pp. 8 –12. (2005).

[21] Koerich, A.L. Rejection strategies for handwritten word recognition. *Proc. of the 9th International Workshop on Frontiers in Handwriting Recognition (IWFHR 2004)*, Tokyo, Japan, pp. 479–484. (2004).

[22] Lam, L. Classifier combinations: Implementations and theoretical issues. *Multiple Classifier Systems, First International Workshop*, MCS 2000, Cagliari, Italy, pp. 77–86. (2000).

[23] Lauer, F., Suen, C. Y., and Bloch, G. A trainable feature extractor for handwritten digit recognition. *Pattern Recognition*, 40(6), pp. 1816–1824. (2007).

[24] Le Cun, Y. , Boser, B., Denker, J. S., Howard, R. E., Habbard, W., Jackel, L. D., and Henderson, D. Handwritten digit recognition with a back-propagation network. *Advances in neural information processing systems 2*, Morgan Kaufmann Publishers Inc., San Francisco, CA, pp. 396–404. (1990).

[25] Lin, R.S., Yang, M.H., and Levinson, S.E. Object tracking using incremental Fisher discriminant analysis. *ICPR*, pp. 23–26. (2004).

[26] Liu, C. L., Nakashima, K., Sako, H. and Fujisawa, H. Handwritten digit recognition: Investigation of normalization and feature extraction techniques. *Pattern Recognition*, 37(2), pp. 265–279. (2004).

[27] Liu, C.-L. and Suen, C. Y. A new benchmark on the recognition of handwritten Bangla and Farsi numeral characters, *Proc. of the 11th Int. Conference on Frontiers in Handwriting Recognition (ICFHR 2008)*, Montreal, Canada, pp. 278–283. (2008).

[28] MacQueen, J. B. Some methods for classification and analysis of multivariate observations. *Proc. 5th Berkeley Symposium on Mathematical Statistics and Probability*, 1, pp. 281–297. (1967).

[29] Otsu, N.. A threshold selection method from gray-level histogram, *IEEE Trans. System Man Cybernet.*, vol. 9, pp. 1569–1576. (1979)

[30] Pitrelli, J. F. and Perrone, M. P. Confidence-scoring post-processing for off-line handwritten-character recognition verification. *Proc. of 7th Int. Conference on Document Analysis and Recognition (ICDAR'03)*, Edinburgh, Scotland, vol. I, pp. 278–282. (2003).

[31] Sagheer, A. El., Tsuruta, N., and Taniguchi, R.-I. Arabic lip-reading system: A combination of Hypercolumn Neural Network Model with Hidden Markov Model. *Artificial Intelligence and Soft Computing (ASC 2004)*, Marbella, Spain, 9.1–9.3. (2004).

[32] Shi, M., Fujisawa, Y., Wakabayashi, T., and Kimura, F. Handwritten numeral recognition using gradient and curvature of gray scale image, *Pattern Recognition*, vol. 35, no. 10, pp. 2051–2059. (2002).

[33] Solimanpour, F., Sadri, J., and Suen, C. Y. Standard Databases for Recognition of Handwritten Digits, Numerical Strings, Legal Amounts, Letters and Dates in Farsi Language. *Proc. of the 10th Int. Workshop on Frontiers in Handwriting Recognition (IWFHR 2006)*, La Baule, France, pp. 3–7. (2006).

[34] Suen, C. Y., Nadal, C. P., Mai, T. A., Legault, R., and Lam, L. Recognition of handwritten numerals based on the concept of multiple experts. *Proc. Int. Workshop on Frontiers in Handwriting Recognition (IWFHR 1990)*, Montreal, Canada, pp. 131–144. (1990).

[35] Suen, C. Y., Nadal, C., Legault, R., Mai, T. A., and Lam, L. Computer recognition of unconstrained handwritten numerals. *Proc. IEEE*, 80(7), pp. 1162–1180. (1992).

[36] Swets, D.L. and Weng, J. Hierarchical discriminant analysis for image retrieval. *IEEE Trans. Pattern Analysis and Machine Intelligence*, 21(5), pp. 386–401. (1999).

[37] Tang, F. and Tao, H. Fast linear discriminant analysis using binary bases. *Pattern Recognition Letters*. 28 (16), pp. 2209–2218. (2007).

[38] Vapnik, V. and Lerner, A. Pattern recognition using generalized portrait method. *Automation and Remote Control*, 24, pp. 774–780. (1963).

[39] Wu, T.-F., Lin, C.-J., and Weng, R. C. Probability estimates for multi-class classification by pairwise coupling, *Journal of Machine Learning Research*, vol. 5, pp. 975–1005. (2004).

[40] Zhang, P., Bui, T. D., and Suen, C. Y. A novel cascade ensemble classifier system with a high recognition performance on handwritten digits. *Pattern Recognition*, 40(12), pp. 3415–3429. (2006).

[41] Zhao B., Liu, Y., and Xia, S. W. Support vector machines and its application in handwritten numeral recognition, *Proc. of 15th Int. Conference on Pattern Recognition*, Barcelona, Spain, vol. 2, pp. 720 – 723. (2000).

CHAPTER 3.5

RECOGNITION TECHNIQUES FOR WHITEBOARD NOTES WRITTEN IN ROMAN SCRIPT

Marcus Liwicki and Horst Bunke

German Research Center for Artificial Intelligence,
Trippstadter Str. 122, 67663 Kaiserslautern, Germany
marcus.liwicki@dfki.de

Institute for Computer Science and Applied Mathematics
Neubrückstr. 10, 3012 Bern, Switzerland
bunke@iam.unibe.ch

In this chapter we describe various methods for the automatic recognition of handwritten whiteboard notes. A handwriting recognition system for Roman Script is usually divided into units which iteratively process the handwritten input data to finally obtain the desired ASCII transcription: the preprocessing, where noise in the raw data is reduced; the normalization, where various steps take place to remove writer-specific characteristics of the handwriting; the feature extraction, where the normalized data is transformed into a sequence of feature vectors; the recognition, where a classifier generates a list of word sequence candidates; and the post-processing, where language information is used to improve the results. We review different approaches for all of these stages and describe selected approaches in more detail. Furthermore, we introduce some preprocessing steps which have been developed especially for whiteboard notes. In order to assess the advantages of different methods, we present the results of a broad experimental analysis on a large database of handwritten whiteboard notes.

1. Introduction

Handwriting recognition is the transcription of handwritten data into a digital format. It is a classical pattern recognition problem. The task lies in assigning a pattern to one class out of a number of classes,[1] for example, assigning a handwritten sample to a specific character. The goal of the automated recognition is to process the data with the same or nearly the same accuracy as humans.

The domain of handwriting recognition has traditionally been divided into online and offline recognition. In offline recognition the text to be recognized is captured by a scanner and stored as an image. Therefore only spatial information in terms of an image of the text is available. In the online mode the handwriting is produced by means of an electronic pen or a mouse and acquired as a time ordered sequence of coordinates, representing the movement of the pen-tip. In some cases, also the pressure, the tilt, and the angle of the pen are known for each point. Good progress has been achieved in both online and offline recognition.[2–6] Because of the greater ease of extracting relevant features and the availability of temporal information, online recognition generally yields better results.[7]

Another crucial division is that between the recognition of isolated characters or words, and the recognition of whole lines of text. Unsurprisingly, the latter task is substantially harder than the two former ones, and the excellent results that have been obtained for e.g. digit and character recognition[4,5] have never been matched in case of complete text lines. This is due to the fact that segmentation is difficult for cursive unconstrained text, unless the words have already been recognized. This creates a circular dependency between segmentation and recognition that is often referred to as Sayre's paradox.[8] Therefore the aim of an unconstrained recognition system would be to integrate segmentation into the recognition[9–14] or to allow for several segmentation alternatives.

In this chapter we consider a relatively new task, which is the recognition of Roman script written on a whiteboard. A similar task has been considered in Refs. 15,16. However, while in Refs. 15,16 a video camera was used to capture the handwriting, we use the eBeam interface which is based on infrared sensing. This system is easier to use than a video camera and it is less vulnerable to artifacts arising from poor lighting conditions, self-occlusion and low image resolution. This task is relevant in a number of novel applications, for example, computer assisted learning[17] and smart meeting rooms.[18,19] While the electronic whiteboard has already been a research topic in Refs. 20,21, the task of recognizing whiteboard notes written in Roman script is relatively new.

The rest of this chapter is organized as follows. Section 2 gives an overview of the general processing steps for common handwriting recognition systems. Sections 3–7 review existing methods for the individual steps and present the methods used and evaluated in this chapter. A broad experimental study is described in Section 8, and finally Section 9 draws some conclusions.

2. Processing Steps

A recognition system for unconstrained Roman script is usually divided into consecutive units which iteratively process the handwritten input data to finally obtain the transcription. The main units are illustrated in Fig. 1 and summarized in this section. Certainly, there are differences between offline and online processing, but the principles are the same. Only the methodology for performing the individual steps differs.

First, preprocessing steps are applied to reduce noise in the raw data (see Section 3). Next, the data is normalized, i.e., it is attempted to remove writer-specific characteristics of the handwriting to make writings from different authors looking more similar to each other (see Section 4). Subsequently, features are extracted

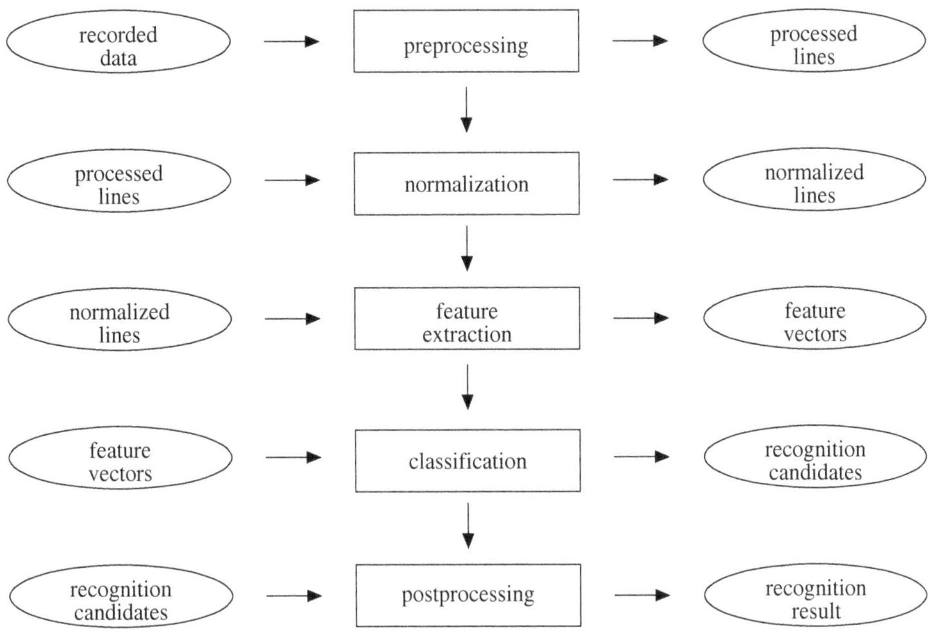

Fig. 1. Recognition system overview

from the normalized data (see Section 5). This particular step is needed because the recognizers need numerical data as their input. In the next phase, a classification system is applied which generates a list of candidates or even a recognition lattice (see Section 6). In the final step some post-processing methods are applied, e.g., integrating language information from a language model (see Section 7).

3. Preprocessing

The first unit is preprocessing, where the input is raw handwritten data and the output usually consists of extracted text lines. The amount of effort that need to be invested into the preprocessing depends on the given data. If the data have been acquired from a system that does not produce any noise and only single words have been recorded, there is nothing to do in this step. But usually the data contains noise which need to be removed to improve the quality of the handwriting, e.g., by means of image enhancement.[22] Another preprocessing step is the extraction of the handwritten text lines,[23–27] which may also include the separation of text and graphics.[28–30] Depending on the recognition system, the preprocessing may also include layout analysis,[31–33] word extraction,[23,34–36] and even character segmentation.[5,23]

In the system presented in this chapter, the whiteboard data have been acquired in the online format. The data usually contain noisy points and gaps within the

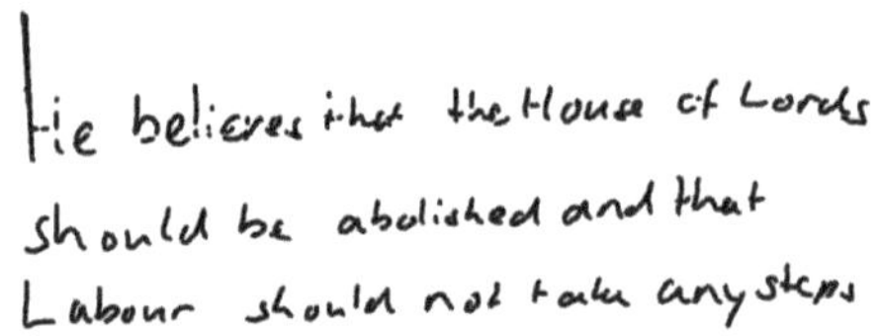

Fig. 2. Recorded text

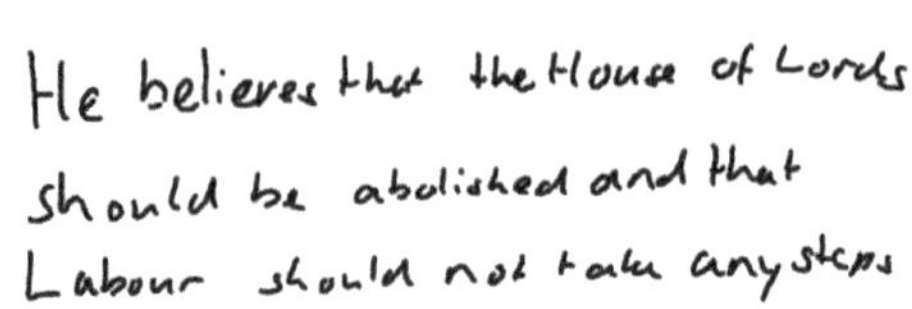

Fig. 3. Text after removing noise

strokes. In Fig. 2 examples of both cases are shown. In the word *He* a spurious point occurs that leads to the introduction of a large artifact, i.e., two long additional strokes. Furthermore, in the first line there are many gaps within the words, caused by loss of data.

Two preprocessing steps are applied to the data to recover from artifacts. For the first type of artifact, spurious points, the following strategy is chosen. Let $p_1, ..., p_n$ be the points of a given stroke and q_1 be the first point of the succeeding stroke, if any. If the distance between two consecutive points p_i, p_{i+1} is larger than a fixed threshold, one of the points is deleted. To decide which point must be deleted, the number of points within a small neighborhood of p_i and p_{i+1} is determined, and the point with the smaller number of neighbors is deleted.

To recover from artifacts of the second type, i.e., gaps within strokes, the distance between the timestamps of p_n and q_1 is taken into account. If it is below a fixed threshold the strokes are merged into one stroke. Note that the thresholds used for preprocessing depend on the recording environment. They can be derived from the frame-rate and the width of the pen strokes, or by scaling the average distance between succeeding timestamps in the recorded data[a]. The result of the preprocessing is shown in Fig. 3. Obviously, the handwriting is now of much better quality.

The text lines on a whiteboard usually have no uniform skew along the whole line and the slant and size of the letters is not the same at the beginning and at the end of a line. This is caused by the fact that people stand, rather than sit, during writing and the arm does not rest on a table. Therefore, the text line is split into subparts and the rest of the preprocessing is done for each subpart separately. To accomplish the splitting, all gaps within a line are determined. The text line is then split at a gap if the size of this gap is larger than the median gap size. Furthermore the size of both parts must be greater than a predefined threshold.

4. Normalization

After preprocessing, the recorded data need to be normalized. This is a very important step in any handwriting recognition system, because the writing styles of the

[a]In the conducted experiments, a threshold of $2 \cdot m$ was used, where m is the average value of the time intervals, or distances, respectively. This is reasonable because at $2 \cdot m$, there are distinct gaps in the histogram of all values.

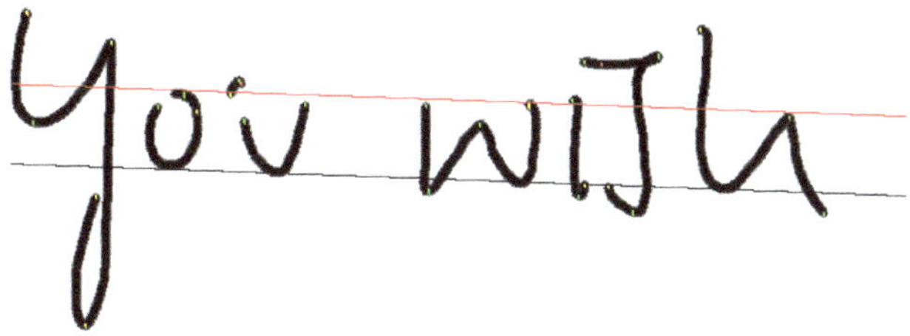

Fig. 4. Baseline and corpus line of a part of an example text line

writers differ with respect to skew, slant, height, and width of the characters. In the literature there is no standard way of normalizing the data, but many systems use similar techniques. First, the text line is corrected in regard to its skew, i.e., it is rotated, so that the baseline is parallel to the x-axis. Then, slant correction is performed so that the slant becomes upright. The next important step is the computation of the baseline and the corpus line (see Fig. 4). These two lines divide the text into three areas: the upper area, which mainly contains the ascenders of the letters; the middle area, where the corpus of the letters is present; and the lower area with the descenders of some letters. These three areas are normalized to predefined heights. Often, some additional normalization steps are performed, depending on the domain. In offline recognition, thinning and binarization may be applied. In online recognition the delayed strokes, e.g., the crossing of a "t" or the dot of an "i", are usually removed, and equidistant resampling is applied.

In this chapter, the following normalization methods are applied to the online data. First, the individual text lines are corrected with respect to their skew using a linear regression through all points. Next, for slant normalization, we compute the histogram over all angles between the lines connecting two successive points of the trajectory and the horizontal line.[37] The histogram ranges from -90° to 90° with a step size of 2°. In the online recognition system described in this chapter, each histogram entry is smoothed using its direct neighbors and a weighting window $(0.25, 0.5, 0.25)$. This is done because in some cases the correct slant is at the border of two angle intervals, and a single peak in one of the adjacent intervals may be slightly higher. This peak will become smaller after smoothing. The histogram values are additionally weighted with a Gaussian whose mean is at the vertical angle and whose variance is chosen empirically. This is beneficial since some words are not properly corrected if a single long straight line is drawn in the horizontal direction, which results in a large histogram value.

After slant normalization, delayed strokes, e.g. the crossing of a "t" or the dot of an "i" are removed, using simple heuristics. To retain the information of delayed strokes for the recognizer, the *hat-feature* is used in the feature set (see Section 5). Next, we perform an equidistant resampling of the point sequence, i.e., the original sequence of points is replaced by a sequence of points on the trajectory where all consecutive points have the same distance to each other. This step is needed because different writers write at a different speed. The distance is set to a fraction of the corpus height.

The next important step is the computation of the baseline and the corpus line. Therefore two linear regressions through the minima and maxima are computed with the constraint that the two resulting lines have to have the same slope. After the regression lines have been determined, the least fitting points are removed and another linear regression is performed. This correction step is done twice, which then results in the estimated baseline (minima) and corpus line (maxima). The baseline is subtracted from all y-coordinates to make it equal to the x-axis and the height of the three main writing areas is normalized. Finally the width of the characters is normalized. This is done by scaling the text horizontally with a fraction of the number of strokes crossing the horizontal line between the baseline and the corpus line. This preprocessing step is needed because the x-coordinates of the points are taken as a feature.

5. Feature Extraction

Feature extraction is essential for any handwriting recognition system, because any classifier needs numeric data as input. However, no standard method for computing the features exists in the literature. One common method in offline recognition of handwritten text lines is the use of a sliding window moving in the writing direction over the text.[12] Features are extracted at every window position, resulting in a sequence of feature vectors. In the case of online recognition the points are already available in a time-ordered sequence, which makes it easier to get a sequence of feature vectors in writing order. If there is a fixed size of the input pattern, such as in character or word recognition, one feature vector of a constant size can be extracted[2,38] for each pattern.

In the system described in chapter state-of-the-art feature extraction methods are applied to extract the features from the preprocessed data. The feature set input to the recognizer consists of 25 features which utilize information from both the real online data stored in XML format, and pseudo offline information automatically generated from the online data. For each (x, y)-coordinate recorded by the acquisition device a set of 25 features are extracted, resulting in a sequence of 25-dimensional vectors for each given text line.

The features can be divided into two classes. The first class consists of features extracted for each point by considering the neighbors with respect to time. The features of this class are the following (the numbers in parentheses indicate the individual number of numerical values extracted by this feature):

- *pen-up/pen-down (1):* a binary feature indicating whether the pen-tip touches the board or not. Consecutive strokes are connected with straight lines for which this feature has the value *false*.

- *hat-feature (1):* this binary feature indicates whether a delayed stroke has been removed at the same horizontal position as the considered point.
- *speed (1):* the velocity is computed before resampling and then interpolated.
- *x-coordinate (1):* the x-position is taken after high-pass filtering, i.e., after subtracting a moving average from the real horizontal position.
- *y-coordinate (1):* this feature represents the vertical position of the point after normalization.
- *writing direction (2):* this pair of features is given by the cosine and sine of the angle between the line segment starting at the point and the x-axis.
- *curvature (2):* similarly to the writing direction, this is a pair of features given by the cosine and sine of the angle between the lines to the previous and the next point.
- *vicinity aspect (1):* this feature is equal to the aspect of the trajectory in the vicinity of a fixed size (see Fig. 5):

$$\frac{\Delta y(t) - \Delta x(t)}{\Delta y(t) + \Delta x(t)}$$

  The vicinity aspect is large for strokes that are nearly straight in horizontal or vertical direction.
- *vicinity slope (2):* this pair of features is given by the cosine and sine of the angle α of the straight line from the first to the last vicinity point (see Fig. 5).
- *vicinity curliness (1):* this feature is defined as the length of the trajectory in the vicinity divided by $\max(\Delta x(t), \Delta y(t))$ (see Fig. 5).
- *vicinity linearity (1):* here the average squared distance d^2 of each point in the vicinity to the straight line from the first to the last vicinity point is used (see Fig. 5).

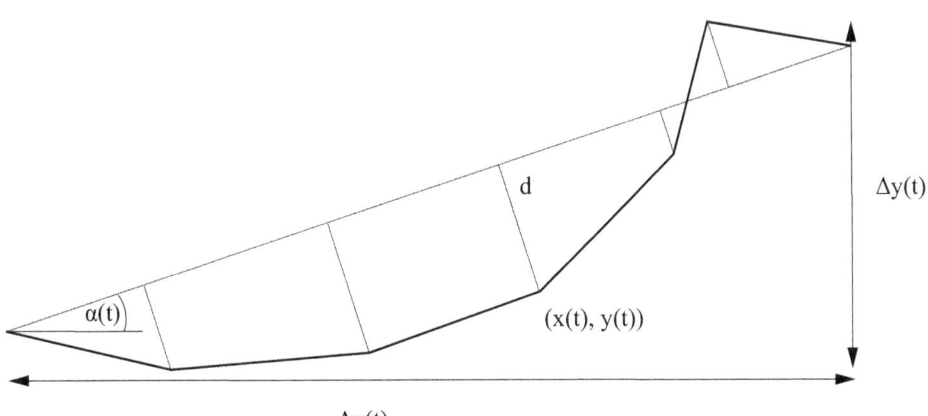

Fig. 5. Features of the vicinity

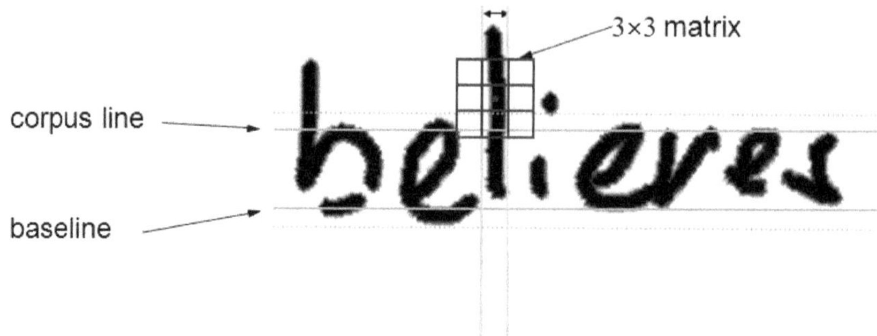

Fig. 6. Features of the offline matrix

The features of the second class take the offline matrix representation into account, i.e., they are based on spatial information. These features are all computed by using a two-dimensional matrix $B = b_{i,j}$ representing the offline version of the data. For each position $b_{i,j}$ the number of points on the trajectory of the strokes is stored. This can be seen as a low-resolution image of the handwritten data. The following features are used:

- *ascenders/descenders (2):* these two features count the number of points above the corpus line (ascenders) and below the baseline (descenders). Only points which have an x-coordinate in the vicinity of the current point are considered. Additionally the points must have a minimal distance to the lines to be considered as part of an ascender or a descender. This can be seen as a tolerance area. The corresponding distances as well as the vicinity area are set to a predefined fraction of the corpus height.
- *context map (9):* the two-dimensional vicinity of the current point is transformed into a 3×3 map. The number of black points in each region is taken as a feature value. Altogether nine features of this type are obtained.

Figure 6 illustrates the extracted features. Note that the strokes are widened for ease of visualization.

For the experiments described in this paper we also investigated an offline recognizer. Therefore we generated offline images generated from the online data, and applied normalization steps as described in Ref. 39. To extract the feature vectors from the normalized images, a sliding window is used. The width of the window is one pixel, and nine geometrical features are computed at each window position. Thus an input text line is converted into a sequence of feature vectors in a 9-dimensional feature space. For more details about the features we refer to Ref. 39.

6. Classification

After feature extraction, the same methods can be used for offline and online recognition. The next unit is the classification of the input features. If there is a fixed length feature vector, neural networks,[40,41] support vector machines,[42] nearest neighbor classifiers,[43] and other techniques can be applied. For sequences of feature vectors, *hidden Markov models* (HMMs)[11,44] or *time delay neural networks* (TDNNs)[37] can be used for classification.

To combine the advantages of neural networks and HMMs, hybrid approaches have been developed. Hybrid systems use HMMs to model long-range dependencies in the data and neural networks to provide localized classifications. Hybrid approaches which use recurrent neural networks (RNNs) combined with HMMs have been proposed in Refs. 45–47. There exist also a number of hybrid approaches based on other forms of neural networks. Classical neural networks have been used for state modeling in Ref. 48. Time delay neural networks combined with HMMs have been applied to online word recognition in Ref. 13. The application of a hybrid approach similar to Ref. 13 has been proposed in Ref. 37 on an advanced feature set. This application has further been improved with pruning techniques, enabling it to be used in real-time with large dictionaries. A more detailed analysis of TDNN/HMM systems is given in Ref. 49.

The output of the classification system is a single recognition result, a list of alternative candidates, or even a recognition lattice, i.e., a directed graph where each node represents a word candidate which is then connected to possible successors via the edges. The latter two formats usually include a probability for each output candidate.

In this chapter two different methods are evaluated for the recognition steps, namely HMMs and bidirectional long short-term memory networks (BLSTM). The HMM-based recognizer uses one HMM for each of the 82 characters in the character set. In all HMMs the linear topology is used. In the emitting states, the observation probability distributions are estimated by mixtures of Gaussian components. In other words, continuous HMMs are used. The number of Gaussian mixtures is optimized on a validation set as suggested in Ref. 50. The character models are concatenated to represent words and sequences of words. For training, the Baum-Welch algorithm[51] is applied. In the recognition phase, the Viterbi algorithm[52] is used to find the most probable word sequence. The output of the recognizer is a sequence of words.

Recurrent neural networks (RNNs) are able to access a wide range of context when transcribing letters, words or word sequences.[53] However, whereas standard RNNs make use of previous context only, bidirectional RNNs (BRNNs) are able to incorporate context on both sides of every position in the input sequence.[54] This is useful in handwriting recognition, since it is often necessary to look at both the context to the right and to the left of a given letter in order to identify it. Figure 7

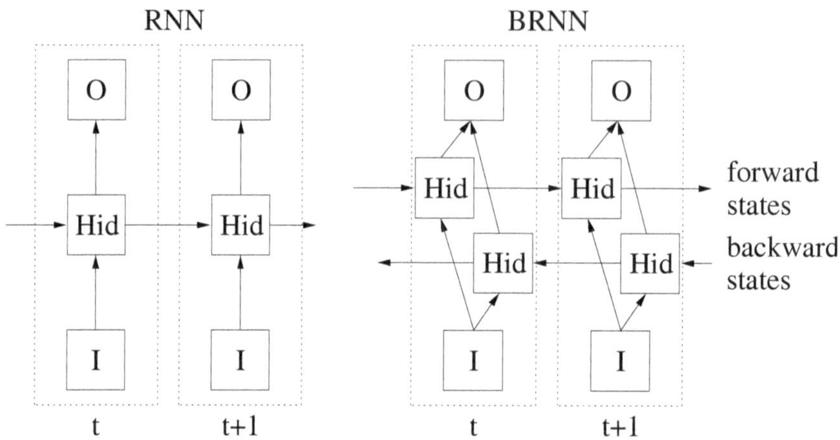

Fig. 7. Recurrent neural network and bidirectional recurrent neural network unfolded in time

shows a comparison of a standard RNN and a BRNN at two different points in time. In the BRNN there exist two hidden layers, i.e. one layer for each direction, and the input/output neurons (marked with I/O) are connected to both.

Long Short-Term Memory (LSTM[55]) is an RNN architecture specifically designed to bridge long time delays between relevant input and target events, making it suitable for problems where long range context is required to disambiguate individual labels as in handwriting recognition. An LSTM layer consists of a set of recurrently connected blocks, known as memory blocks. Each memory block contains one or more recurrently connected memory cells and three multiplicative units, called the input, output and forget gates that provide write, read and reset operations on the cells.

The recently introduced connectionist temporal classification (CTC)[56] is an RNN objective function designed for labeling whole sequences at once. It uses the network to define a probability distribution over a fixed set of labels plus an additional 'blank', or 'no label' unit. It then interprets the sequence of network outputs as a probability distribution over all possible transcriptions for a given input sequence, and trains the network by maximizing the log probabilities of the correct transcriptions on the training set. In the experiments described in this chapter we use a BRNN with LSTM network cells and a CTC output layer to transcribe samples of handwritten text. For more details on the implementation of this recognizer, we refer to.[57]

7. Post-Processing

The last recognition step illustrated in Fig. 1 is the postprocessing. This step is only possible if additional knowledge about the domain is available. For example, in the case of word recognition, the output sequence of characters can be assigned to the most similar word in the dictionary. In the case of word sequence recognition, statistical language models can be applied.[58,59] The motivation of language models is the observation that a person is often able to guess a missing word when reading a given text. Hence the information of the surrounding words can be used to improve the performance of the recognizer. Several kinds of language models exist, including n-grams, decision tree models, linguistically motivated models, exponential models, and adaptive models.

In this chapter bigram language models are used which are a special case of n-gram language models. The *n-gram language models* take advantage of the observation that the probability of a word is highly depend on the previous words. First, the probability of a sequence of words $w_1^k = w_1, \ldots, w_k$ is denoted by:

$$p(w_1^k) = p(w_1)p(w_2|w_1)\cdots p(w_k|w_1^{k-1}) \tag{1}$$

where two word sequence histories are treated as equal if they end in the same $n-1$ words:

$$p(w_k|w_{k-n+1}^{k-1}) = p(w_k|w_1^{k-1}), k \geq n \tag{2}$$

Given these two equations it is easy to derive that for an n-gram language model and a vocabulary size v, there are $v^n - 1$ independent parameters (probabilities).

The estimation of the parameters by examining a given text t is called training. The concept is to count the number of occurrences of each possible word sequence. This can be done iteratively with the sequential maximum likelihood estimation, which, in general, does not lead to *consistent models*[60][b]. However, if the given text t is large enough, the model will be nearly consistent. If the vocabulary is large, it is hard to train a suitable n-gram language model. This is due to the large number of parameters to be estimated. Furthermore, not all possible word sequences that may occur are present in the training text. Thus, for training the language models on smaller resources, several methods for optimizing the training process have been proposed, such as combining classes, interpolating, or smoothing. For more details and an empirical study see.[61]

In this chapter, bigram language models have been chosen. This method approximates the previous text by the last word and models the dependency by the probability $p(w_i|w_{i-1})$, where w_i represents the considered word and w_{i-1} stands for the previous word. The probability $p(W)$ of a text line $W = (w_1, \ldots, w_n)$ can

[b]The order-n conditional probabilities of an n-gram model form the transition matrix of an associated Markov model. The states of this Markov model are sequences of $n-1$ morphs, and the probability of a transition from the state $w_1, w_2 \ldots, w_{n-1}$ to the state $w_2, w_3 \ldots, w_n$ is $P(w_n|w_1, w_2 \ldots, w_{n-1})$. An n-gram language model is called *consistent* if, for each string w_1^{n-1}, the probability that the model assigns to w_1^{n-1} is the steady state probability for the state w_1^{n-1} of the associated Markov model.

then be computed as follows:

$$p(W) = p(w_1)\left(\prod_{i=2}^{n} p(w_i|w_{i-1})\right) \tag{3}$$

Including the output likelihoods of the recognizer, the most likely word sequence $\hat{W} = (w_1, \ldots, w_m)$ for a given observation sequence X is computed in the following way:[62]

$$\hat{W} = \underset{W}{\operatorname{argmax}} \ \log p(X|W) + \alpha \ \log p(W) + m \ \beta \tag{4}$$

According to Eq. (4) the model $p(X|W)$, which results from the decoding step, is combined with the likelihood $p(W)$ obtained from the language model. Two additional parameters α and β are necessary to control the integration of the language model. The parameter α is called *grammar scale factor* (GSF) and weights the influence of the language model against the optical model. The term *word insertion penalty* (WIP) is used for the parameter β which prevents the system from oversegmentation or undersegmentation. That is, the larger the value of β is, the higher is the system's tendency to split a text into many short words.

Figure 8 illustrates the influence of the two parameters GSF and WIP on an example text line. Multiple recognition results are produced for the handwritten text line "their sale being by the". Increasing the parameter α eliminates nonsense word sequences, for example, "being by in E". Furthermore, increasing the parameter β amplifies the average number of words returned by the recognizer (including punctuation marks).

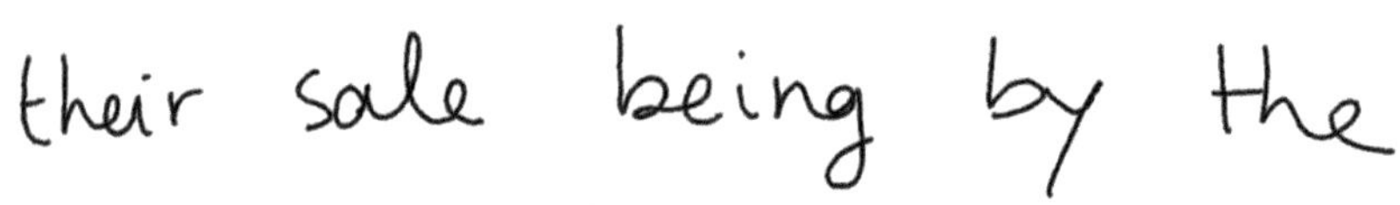

GSF α	WIP β	Recognition result
0	-40	their sale being by by
0	0	their sale being by in E
15	0	their sale being long
15	20	their sale being by the
90	10	their sale being long
90	130	their sale be in a by this ,

Fig. 8. Influence of the parameters GSF and WIP

8. Experiments

All experiments have been performed on the IAM-OnDB,[63] a large on-line handwriting database acquired from a whiteboard[c]. It consists of 13,040 written lines, containing 86,272 instances of 11,050 distinct words. Two benchmark tasks have been defined for the IAM-OnDB. In this chapter, the IAM-OnDB-t2 benchmark task has been used for the experiments. In this task the database is divided into four predefined disjoint sets: a training set containing 5,364 lines; a first validation set containing 1,438 lines; a second validation set containing 1,518 lines, which can be used, for example, to optimize a language model; and a test set containing 3,859 lines. No writer appears in more than one set. Thus a writer-independent recognition task is considered.

The recognition task for all the experiments is to transcribe all the text lines in the test set, given a predefined dictionary. The measure of performance is the *word accuracy*:

$$100 * \left(1 - \frac{\text{insertions} + \text{substitutions} + \text{deletions}}{\text{total length of test set transcriptions}}\right)$$

where the number of word insertions, substitutions and deletions is summed over the whole test set. The IAM-OnDB-t2 benchmark task is based on an open vocabulary that contains the 20,000 most frequent words out of three corpora (LOB Corpus,[64] Brown Corpus,[65] Wellington Corpus[66]). Note that there exist words, called out of vocabulary words (OOV), which are in the transcription of the test set but not in the lexicon. This limits the recognition rate of recognizers that are only able to recognize words included in the lexicon. On the test set the OOV rate is 5.6 %, resulting in an upper limit of 94.4 % for the BLSTM and HMM classifiers.

A total of six classification systems are investigated in the experiments described in this chapter. Three of these systems are based on HMMs while the rest use neural networks. The systems differ in the way of normalization and feature extraction. The first system (denoted by "offline (1)") uses offline features . The second system (denoted by "online (2)") uses online features and applies the standard normalization methods, i.e., methods which have been introduced especially for whiteboard data are not applied (splitting into sub-parts and normalization of the slant histogram). The third systems (denoted by "on-line (3)") also applies the refined preprocessing and normalization methods. Since neural networks are randomly initialized, we have performed ten consecutive runs for each set of features and state the average performance.

A summary of the recognition systems and their accuracies is given in Table 1. While the highest accuracy is obtained by an "online (1)" BLSTM (81.26 %), the average accuracy of the "online (2)" BLSTM (80.1 %) is higher than that of the "online (1)" BLSTM (79.8 %). The offline BLSTM systems perform with an average

[c]The IAM-OnDB is publicly available for download at
http://www.iam.unibe.ch/~fki/iamOnDB

Table 1. Recognition performance on the test set

System		#	Accuracy (average for BLSTM)
BLSTM	offline (1)	10	76.43 % (73.6 %)
	online (2)	10	81.26 % (79.8 %)
	online (3)	10	81.07 % (80.1 %)
HMM	offline (1)	1	57.3 %
	online (2)	1	63.2 %
	online (3)	1	63.9 %
Microsoft©		1	71.3 %
Vision Objects©		1	79.2 %

accuracy of 73.6 %. Note that the standard deviation of the BLSTM systems is less then 0.5 % for each of the three feature sets. The HMM-based recognizer has the lowest accuracy. The recognition rates of the "offline (1)" and "online (2)" system are significantly lower than that of the "online (3)" system.

To put these results in perspective, the Microsoft tablet PC handwriting recognizer[67] and the Vision Objects© recognizer[68,69] have been applied to the same test set. Their results appear at the bottom of Table 1. The Microsoft© recognizer gave a word accuracy score of 71.32% on the online test set and the Vision Objects© recognizer performs with 79.2 %. These results are not directly comparable to our own, since, on the one hand, these systems were trained on different training setsbut, on the other hand, use considerably more sophisticated language modelling than the HMM and BLSTM systems we implemented. However, it indicates that the BLSTM-based recognizer is competitive with the best commercial systems for unconstrained handwriting.

9. Conclusions

In this chapter we presented handwriting recognition systems for whiteboard notes written in Roman Script. We reviewed several methods for the individual processing units of handwriting recognition systems and described selected approaches in more detail. In our experiments we investigated the use of several feature sets and classifiers. Finally, the results were compared to commercial recognizers. The overall best accuracy of 81.26 % was obtained by an "online (1)" BLSTM system.

Acknowledgements

This work was supported by the Swiss National Science Foundation program "Interactive Multimodal Information Management (IM2)" in the Individual Project "Visual/Video Processing". The authors of this chapter would like to thank Alex Graves for providing the neural network toolkit.

References

1. R. O. Duda, P. E. Hart, and D. G. Stork, *Pattern Classification.* (Wiley-Interscience Publication, 2001).
2. H. Bunke. Recognition of cursive Roman handwriting – past present and future. In *Proc. 7th Int. Conf. on Document Analysis and Recognition*, vol. 1, pp. 448–459, (2003).
3. P. M. Lallican, C. Viard-Gaudin, and S. Knerr. From off-line to on-line handwriting recognition. In *Int. Workshop on Frontiers in Handwriting Recognition*, pp. 303–312, (2000).
4. R. Plamondon and S. N. Srihari, On-line and off-line handwriting recognition: a comprehensive survey, *IEEE Trans. Pattern Analysis and Machine Intelligence.* **22** (1), 63–84, (2000).
5. C. C. Tappert, C. Y. Suen, and T. Wakahara, The state of the art in online handwriting recognition, *IEEE Trans. Pattern Analysis and Machine Intelligence.* **12**(8), 787–808, (1990).
6. A. Vinciarelli, A survey on off-line cursive script recognition, *Pattern Recognition.* **35** (7), 1433–1446, (2002).
7. R. Seiler, M. Schenkel, and F. Eggimann. Off-line cursive handwriting recognition compared with on-line recognition. In *ICPR '96: Proceedings of the International Conference on Pattern Recognition (ICPR '96) Volume IV-Volume 7472*, p. 505, Washington, DC, USA, (1996). IEEE Computer Society.
8. K. M. Sayre, Machine recognition of handwritten words: a project report, *Pattern Recognition.* **5**(3), 213–228, (1973).
9. S. Bercu and G. Lorette. On-line handwritten word recognition: an approach based on hidden Markov models. In *Proc. 3rd Int. Workshop on Frontiers in Handwriting Recognition*, pp. 385–390, (1993).
10. T. Starner, J. Makhoul, R. Schwartz, and G. Chou. Online cursive handwriting recognition using speech recognition techniques. In *Int. Conf. on Acoustics, Speech and Signal Processing*, vol. 5, pp. 125–128, (1994).
11. J. Y. Hu, M. K. Brown, and W. Turin, HMM based online handwriting recognition, *IEEE Trans. Pattern Analysis and Machine Intelligence.* **18**(10), 1039–1045, (1996).
12. U.-V. Marti and H. Bunke, Using a statistical language model to improve the performance of an HMM-based cursive handwriting recognition system, *Int. Journal of Pattern Recognition and Artificial Intelligence.* **15**, 65–90, (2001).
13. M. Schenkel, I. Guyon, and D. Henderson, On-line cursive script recognition using time delay neural networks and hidden Markov models, *Machine Vision and Applications.* **8**, 215–223, (1995).
14. A. El-Yacoubi, M. Gilloux, R. Sabourin, and C. Y. Suen, An HMM-based approach for off-line unconstrained handwritten word modeling and recognition, *IEEE Trans. Pattern Analysis and Machine Intelligence.* **21**(8), 752–760, (1999). doi: http://doi.ieeecomputersociety.org/10.1109/34.784288.
15. G. A. Fink, M. Wienecke, and G. Sagerer. Video-based on-line handwriting recognition. In *Proc. 6th Int. Conf. on Document Analysis and Recognition*, pp. 226–230, (2001).
16. M. E. Munich and P. Perona. Visual input for pen-based computers. In *Proc. 3rd Int. Conf. on Pattern Recognition*, pp. 33–37, (1996).
17. N. Iwayama, K. Akiyama, H. Tanaka, H. Tamura, and K. Ishigaki. Handwriting-based learning materials on a tablet pc: a prototype and its practical studies in an elementary school. In *Proc. 9th Int. Workshop on Frontiers in Handwriting Recognition*, pp. 533–

538, (2004).
18. D. Moore. The IDIAP smart meeting room. Technical report, IDIAP-Com, (2002).
19. A. Waibel, T. Schultz, M. Bett, R. Malkin, I. Rogina, R. Stiefelhagen, and J. Yang. SMaRT: the smart meeting room task at ISL. In *Proc. IEEE ICASSP*, vol. 4, pp. 752–755, (2003).
20. H. Bandoh, H. Nemoto, S. ichi Sawada, B. Indurkhya, and M. Nakagawa. Development of educational software for whiteboard environment in a classroom. In *Int. Workshop on Advanced Learning Technologies*, pp. 41–44, (2000).
21. G. Friedland, L. Knipping, J. Schulte, and E. Tapia, E-Chalk: a lecture recording system using the chalkboard metaphor, *Int. Journal of Interactive Technology and Smart Education (ITSE)*. **1**, 9–20, (2004).
22. Z. Shi and V. Govindaraju. Historical document image enhancement using background light intensity normalization. In *Proc. 17th Int. Conf. on Pattern Recognition*, vol. 1, pp. 473–476, (2004).
23. E. Kavallieratou, N. Fakotakis, and G. Kokkinakis, An unconstrained handwriting recognition system, *Int. Journal on Document Analysis and Recognition*. **4**(4), 226–242, (2002).
24. Y. Li, Y. Zheng, D. Doermann, and S. Jger. A new algorithm for detecting text line in handwritten documents. In *Proc. 10th Int. Workshop on Frontiers in Handwriting Recognition*, pp. 35–40, (2006).
25. M. Liwicki, E. Indermühle, and H. Bunke. On-line handwritten text line detection using dynamic programming. In *Proc. 9th Int. Conf. on Document Analysis and Recognition*, vol. 1, pp. 447–451, (2007).
26. R. Manmatha and J. L. Rothfeder, A scale space approach for automatically segmenting words from historical handwritten documents, *IEEE Trans. Pattern Analysis and Machine Intelligence*. **27**, 1212–1225, (2005).
27. B. Yu and A. K. Jain, A robust and fast skew detection of algorithm for generic documents, *Pattern Recognition*. **29**(10), 1599–1629, (1996).
28. S. Rossignol, D. Willems, A. Neumann, and L. Vuurpijl. Mode detection and incremental recognition. In *Proc. 9th Int. Workshop on Frontiers in Handwriting Recognition*, pp. 597–602, (2004).
29. M. Shilman, Z. Wei, S. Raghupathy, P. Simard, and D. Jones. Discerning structure from freeform handwritten notes. In *Proc. 7th Int. Conf. on Document Analysis and Recognition*, vol. 1, pp. 60–65, (2003).
30. L. Wenyin. On-line graphics recognition: state-of-the-art. In *5th Int. Workshop on Graphics Recognition*, pp. 291–304, (2003).
31. D. J. Kennard and W. A. Barrett. Separating lines of text in free-form handwritten historical documents. In *Proc. 1st Int. Workshop on Document Image Analysis for Libraries*, pp. 12–23, (2006).
32. S. N. Srihari and G. Kim. PENMAN: a system for reading unconstrained handwritten page images. In *Symposium on Document Image Understanding Technology*, pp. 142–153, (1997).
33. S. N. Srihari and Z. Shi. Forensic handwritten document retrieval system. In *Proc. 1st Int. Workshop on Document Image Analysis for Libraries*, pp. 188–194, (2004).
34. S.-H. Kim, C. B. Jeong, H. K. Kwag, and C. Y. Suen. Word segmentation of printed text lines based on gap clustering and special symbol detection. In *Proc. 16th Int. Conf. on Pattern Recognition*, vol. 2, pp. 320–323, (2002).
35. U. Mahadevan and S. N. Srihari. Hypothesis generation for word separation in handwritten lines. In *Proc. 5th Int. Workshop on Frontiers in Handwriting Recognition*, pp. 453–456, (1996).

36. T. Varga and H. Bunke. Tree structure for word extraction from handwritten text lines. In *Proc. 8th Int. Conf. on Document Analysis and Recognition*, vol. 1, pp. 352–356, (2005).
37. S. Jäger, S. Manke, J. Reichert, and A. Waibel, Online handwriting recognition: the NPen++ recognizer, *Int. Journal on Document Analysis and Recognition*. **3**(3), 169–180, (2001).
38. C. Y. Suen, K. Kim, Q. Xu, J. Kim, and L. Lam. Handwriting recognition – the last frontiers. In *Proc. 15th Int. Conf. on Pattern Recognition*, vol. 4, pp. 4001–4010, (2000).
39. M. Liwicki and H. Bunke, Handwriting recognition of whiteboard notes – studying the influence of training set size and type, *Int. Journal of Pattern Recognition and Artificial Intelligence*. **21**(1), 83–98, (2007).
40. S.-W. Lee, Off-line recognition of totally unconstrained handwritten numerals using multilayer cluster neural network, *IEEE Trans. Pattern Analysis and Machine Intelligence*. **18**(6), 648–652, (1996).
41. I.-S. Oh and C. Y. Suen, A class-modular feedforward neural network for handwriting recognition, *Pattern Recognition*. **35**(1), 229–244, (2002).
42. C. Bahlmann, B. Haasdonk, and H. Burkhardt. Online handwriting recognition with support vector machines – a kernel approach. In *Proc. 8th Int. Workshop on Frontiers in Handwriting Recognition*, pp. 49–54, (2002).
43. C.-L. Liu and M. Nakagawa, Evaluation of prototype learning algorithms for nearest-neighbor classifier in application to handwritten character recognition, *Pattern Recognition*. **34**(3), 601–615, (2001).
44. H. Bunke, M. Roth, and E. G. Schukat-Talamazzini, Off-line cursive handwriting recognition using hidden Markov models, *Pattern Recognition*. **28**(9), 1399–1413, (1995).
45. A. W. Senior and A. J. Robinson, An off-line cursive handwriting recognition system, *IEEE Trans. Pattern Analysis and Machine Intelligence*. **20**(3), 309–321, (1998).
46. G. Rigoll, A. Kosmala, and D. Willett. A new hybrid approach to large vocabulary cursive handwriting recognition. In *Proc. 14th Int. Conf. on Pattern Recognition*, pp. 1512–1514, (1998).
47. J. Schenk and G. Rigoll. Novel hybrid NN/HMM modelling techniques for on-line handwriting recognition. In *Proc. 10th Int. Workshop on Frontiers in Handwriting Recognition*, pp. 619–623, (2006).
48. S. Marukatat, T. Artières, B. Dorizzi, and P. Gallinari. Sentence recognition through hybrid neuro-Markovian modelling. In *Proc. 6th Int. Conf. on Document Analysis and Recognition*, pp. 731–735, (2001).
49. E. Caillault, C. Viard-Gaudin, and A. R. Ahmad. MS-TDNN with global discriminant trainings. In *Proc. 8th Int. Conf. on Document Analysis and Recognition*, pp. 856–861, (2005).
50. S. Günter and H. Bunke, HMM-based handwritten word recognition: on the optimization of the number of states, training iterations and Gaussian components, *Pattern Recognition*. **37**, 2069–2079, (2004).
51. A. P. Dempster, N. M. Laird, and D. B. Rubin, Maximum likelihood from incomplete data via the EM algorithm, *Journal of Royal Statistical Society B*. **39**(1), 1–38, (1977).
52. G. D. Forney. The Viterbi algorithm. In *Proc. IEEE*, vol. 61, pp. 268–278, (1973).
53. R. Rojas, *Neural Networks – A Systematic Introduction*. (Springer-Verlag, 1996).
54. M. Schuster and K. K. Paliwal, Bidirectional recurrent neural networks, *IEEE Trans. Signal Processing*. **45**, 2673–2681 (November, 1997).
55. S. Hochreiter and J. Schmidhuber, Long short-term memory, *Neural Computing*. **9**(8),

1735–1780, (1997).
56. S. Fernández, A. Graves, and J. Schmidhuber. Sequence labelling in structured domains with hierarchical recurrent neural networks. In *Proc. 20th Int. Joint Conf. on Artificial Intelligence*, pp. 774–779, (2007).
57. M. Liwicki, A. Graves, H. Bunke, and J. Schmidhuber. A novel approach to on-line handwriting recognition based on bidirectional long short-term memory networks. In *Proc. 9th Int. Conf. on Document Analysis and Recognition*, vol. 1, pp. 367–371, (2007).
58. F. Jelinek, *Statistical Methods for Speech Recognition.* (The MIT Press, 1998). ISBN 0262100665.
59. R. Rosenfeld. Two decades of statistical language modeling: Where do we go from here? In *Proc. IEEE*, vol. 88, pp. 1270–1278, (2000).
60. P. F. Brown, P. V. deSouza, R. L. Mercer, V. J. D. Pietra, and J. C. Lai, Class-based n-gram models of natural language, *Computational Linguistics.* **18**(4), 467–479, (1992).
61. S. F. Chen and J. Goodman. An empirical study of smoothing techniques for language modeling. In *Proc. 34th Annual Meeting of the Association for Computational Linguistics*, pp. 310–318, (1996).
62. M. Zimmermann and H. Bunke. Optimizing the integration of a statistical language model in HMM-based offline handwritten text recognition. In *Proc. 17th Int. Conf. on Pattern Recognition*, pp. 541–544, (2004).
63. M. Liwicki and H. Bunke. IAM-OnDB – an on-line English sentence database acquired from handwritten text on a whiteboard. In *Proc. 8th Int. Conf. on Document Analysis and Recognition*, vol. 2, pp. 956–961, (2005).
64. S. Johansson, *The tagged LOB Corpus: User's Manual.* (Norwegian Computing Centre for the Humanities, Norway, 1986).
65. W. N. Francis and H. Kucera, *Manual of Information to Accompany A Standard Corpus of Present-Day Edited American English, for use with Digital Computers.* (Department of Linguistics, Brown University, Providence, USA, 1979).
66. L. Bauer, *Manual of Information to Accompany the Wellington Corpus of Written New Zealand English.* (Department of Linguistics, Victoria University, Wellington, New Zealand, 1993).
67. J. A. Pittman, Handwriting recognition: Tablet pc text input, *Computer.* **40**(9), 49–54, (2007).
68. S. Knerr, V. Anisimov, O. Baret, N. Gorsky, D. Price, and J. Simon, The A2iA INTERCHEQUE system: Courtesy amount and legal amount recognition for French checks, *Int. Journal of Pattern Recognition and Artificial Intelligence.* **11**(4), 505–548, (1997).
69. S. Knerr and E. Augustin. A neural network-hidden Markov model hybrid for cursive word recognition. In *Int. Conf. on Pattern Recognition*, vol. 2, pp. 1518–1520, (1998).

CHAPTER 3.6

DETECTION OF AFFECTIVE PATTERNS IN PHYSIOLOGICAL SIGNALS TOWARDS IMPROVING AUTOMATIC EMOTION RECOGNITION

Jonghwa Kim and Elisabeth André

Institute of Computer Science,
University of Augsburg,
Eichleitnerstr. 30, 86159 Augsburg, Germany
{kim, andre}@informatik.uni-augsburg.de

In this chapter, we investigate the potential of physiological signals as reliable channels for emotion recognition. All essential stages of an automatic recognition system are discussed, from the recording of a physiological dataset to a feature-based multiclass classification. In order to collect a physiological dataset from multiple subjects, we developed a musical induction method, without any deliberate lab setting. Four-channel biosensors were used to measure electromyogram, electrocardiogram, skin conductivity, and respiration changes. A wide range of physiological features from various analysis domains is proposed to find the best emotion-relevant features and correlate them with emotional states. The best features extracted are specified in detail and their effectiveness is proven by classification results. Classification of four musical emotions (positive/high arousal, negative/high arousal, negative/low arousal, positive/low arousal) is performed by using an extended linear discriminant analysis (pLDA). Furthermore, by exploiting a dichotomic property of the 2D emotion model, we develop a novel scheme of emotion-specific multilevel dichotomous classification (EMDC) and compare its performance with direct multiclass classification using the pLDA.

1. Introduction

In advanced human-computer interaction (HCI) today, resolving absence of mutual sympathy (rapport) in interaction between human and machine is one of the most important issues. With exponentially evolving technology, it is no exaggeration to say that any interface that disregards human affective states in the interaction - and thus fails to pertinently react to the states - will never be able to inspire confidence. Instead, users will perceive it as cold, untrustworthy, and socially inept. In human communication, expression and understanding of emotions helps achieve mutual sympathy. To approach this in human-computer interaction, we need to equip machines with the means to interpret and understand human emotions without the input of a user's translated intention. Hence, one of the most important prerequisites for realizing such an advanced user interface is a reliable emotion recognition system which guarantees acceptable recognition accuracy, robustness against any artifacts, and adaptability to practical applications. Developing such a system requires the following stages: to model, analyze, process, train, and classify emotional features measured from the implicit emotion channels of human communication, such as speech, facial expression, gesture, pose, physiological responses, etc. In this chapter we concentrate on identifying emotional cues in various physiological measures.

Recently, numerous studies on engineering approaches to automatic emotion recognition have been published, though research in that field is relatively new compared to the long history of emotion research in psychology and psychophysiology. In particular, many efforts have been deployed to recognize human emotions using audiovisual channels of emotion expression, i.e. facial expressions, speech, and gestures. Little attention, however, has been paid so far to using physiological measures, as opposed to audiovisual emotion channels [1]. This is due to some significant limitations that come with the use of physiological signals for emotion recognition. The main difficulty lies in the fact that it is a very hard task to uniquely map physiological patterns onto specific emotional states. As an emotion is a function of time, context, space, culture, and person, physiological patterns may widely differ from user to user and from situation to situation. Above all, humans use non-discrete labels to describe emotions.

In the next section, we give a brief overview of related research on automatic emotion recognition using physiological signals. Section 3 gives the motivation and rationale for our experimental setting of musical emotion induction and is followed by a detailed explanation of all the biosensors we used. A systematic description of signal analysis methods and classification procedure using extended linear discriminant analysis is given in Section 4. In Section 5, we present the best emotion-relevant ANS features with the recognition results we achieved. In addition, the performance of the novel EMDC scheme is tested and its potential is proven by improved recognition accuracy. In Section 6, we discuss the problems faced during our work including the difficulty in subject-independent recognition. We then conclude with perspectives related to future work.

2. Related Research

A significant amount of work has been conducted by Picard and colleagues at MIT Lab showing that certain affective states may be recognized by using physiological data including heart rate, skin conductivity, temperature, muscle activity and respiration velocity [2]. They used personalized imagery to elicit target emotions from a single subject who had two years' experience in acting, and they achieved an overall recognition accuracy of 81% for eight emotions by using hybrid linear discriminant classification. Nasoz et al. [3] used movie clips based on the study by Gross and Levenson [4] for eliciting target emotions from 29 subjects and achieved an emotion classification accuracy of 83% using the Marquardt Backpropagation algorithm (MBP). In work [5], the IAPS photoset [6] is used to elicit target emotions with positive and negative valence and variable arousal level from a single subject. The arousal and valence dimensions of the emotions were classified separately using a neural network classifier and recognition accuracy rates of 96.6% and 89.9% respectively were achieved.

More recently, an interesting user-independent emotion recognition system was reported by Kim et al. [7]. They developed a set of recording protocols using multimodal stimuli (audio, visual, and cognitive) to evoke targeted emotions (sadness, stress, anger, and surprise) from 175 children aged five to eight. A classification ratio of 78.43% was achieved for three emotions (sadness, stress, and anger) and a ratio of 61.76% for four emotions (sadness, stress, anger, and surprise) by adopting support vector machines as pattern classifier. Most interestingly, analysis steps in the system were fitted to handle relatively short lengths of the input signals (segmented in 50 seconds) compared to previous works that required longer signal lengths of about 2-6 min.

The aforementioned approaches achieved average accuracy rates of over 80% which seem to be acceptable for practical applications. It is true, however, that recognition rates are strongly dependent on the datasets that are used and on the application context. Moreover, the physiological datasets used in most of these works were gathered by using visual elicitation materials in a lab setting. The subjects then "tried and felt" or "acted out" the target emotions while looking at selected photos or watching movie clips that were carefully prearranged to elicit the emotions. In other words, to put it bluntly, the recognition results were achieved for specific users in specific contexts with "forced" emotional states.

Most of the previous works provide evidence of the fact that the accuracy of arousal discrimination is always higher than that of valence differentiation. The reason might be the that the change of the arousal level corresponds directly to the intensity of discharge in ANS activities, such as sweat glands and blood pressure, which is straightforward to measure, while valence differentiation of emotion requires a multifactor analysis of cross-correlated ANS reactions. This finding led us to develop an emotion-specific classification scheme and to calculate a wide range of features in various analysis domains in order to extract valence-relevant features from ECG and RSP signals.

3. Setting of Experiment

3.1. *Musical emotion induction*

To collect a database of physiological signals in which the targeted emotions corresponding to the four quadrants in the 2D emotion model (i.e. EQ1, EQ2, EQ3, and EQ4 in Fig. 1) can be *naturally* reflected without any deliberate expression, we decided to use the musical induction method, i.e. to record physiological signals while the subjects were listening to different pieces of music.

A well established mechanism of emotion induction consists in triggering emotions by resorting to imagination or individual memories. Emotional reaction can be triggered by a specific cue and be evoked by an experimental instruction to imagine certain events. On the other hand, it can spontaneously be resurged in memory. Music is a pervasive element accompanying many highly significant events in human social life and particular pieces of music are often connected to significant personal memories. Following this, music can be a powerful cue in awakening emotional experiences and bringing back memories. Since listening to music is often done by an individual in isolation, the possible artifacts of social masking and social interaction can be minimized in the experiment. Furthermore, like odors, music can be treated at lower levels of the brain that are particularly resistant to modifications by later input, contrary to cortically based episodic memory. This is even the case when the listening occurs at the same time as other activities within a social setting since musical emotion cannot co-occur with social interaction in general.

The subjects were three males (one of the co-authors and two student researchers recruited from the authors' lab) aged 25-38 and who all enjoy listening to music in their everyday life. The subjects were not paid, but allowed to perform the experiments during their regular working hours. They individually handpicked four songs that were intended to spontaneously evoke emotional memories and certain moods corresponding to the four target emotions. Figure 1[a] shows the musical emotion model referred to for the selection of their songs. Generally, emotional responses to music varies greatly from individual to individual depending on their unique past experiences. Moreover, cross-cultural comparisons in literature suggest that emotional responses can be quite differentially emphasized by different musical cultures and training. This is why we advised the subjects to choose themselves the songs they believed would help them recall their individual special memories with respect to the target emotions.

For the experiment, we prepared a quiet listening room in our institute in order to ensure that the subjects could experience the emotions evoked by the music undisturbed. For the recording, the subject had to position the sensors following the instructions posted in the room, to put on the headphones, and select a song from his song list saved in the computer. When clicking on the selected song, the recording and music systems were automatically set up by preset values for each song, such as volume, treble, and bass. Most importantly, before the start of the experiment, the subjects were shown how to prepare the skin by using an antiseptic spray and a skin preparation gel for reducing electrode-impedance, and how to correctly position the sensors. Recording schedules were decided by the subjects themselves and the recordings took place whenever they felt like listening to music. They were also free to choose the songs they wanted to listen to. Thus, in contrast

[a]Metaphoric cues for song selection: song1 (positively exciting, energizing, joyful, exuberant), song2 (noisy, loud, irritating, discord), song3 (melancholic, sad memory), song4 (blissful, pleasurable, slumberous, tender)

to methods used in other studies, the subjects were not forced to participate in a lab setting scenario and to use prespecified stimulation material. We believe that this voluntary participation of the subjects during our experiment might help obtain a high-quality dataset with natural emotions.

During the three months, a total of 360 samples (90 samples for each emotion) from three subjects were collected. The signal length of each sample was between 3-5 minutes depending on the duration of the songs.

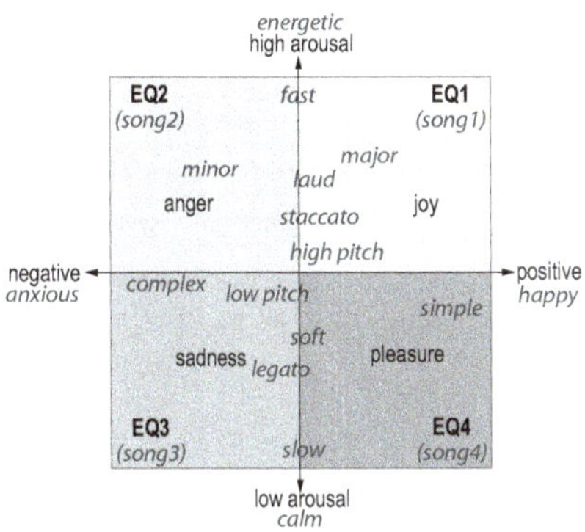

Fig. 1.: Reference emotional cues in music based on the 2D emotion model. EQ1 = positive/high arousal, EQ2 = negative/high arousal, EQ3 = negative/low arousal, EQ4 = positive/low arousal

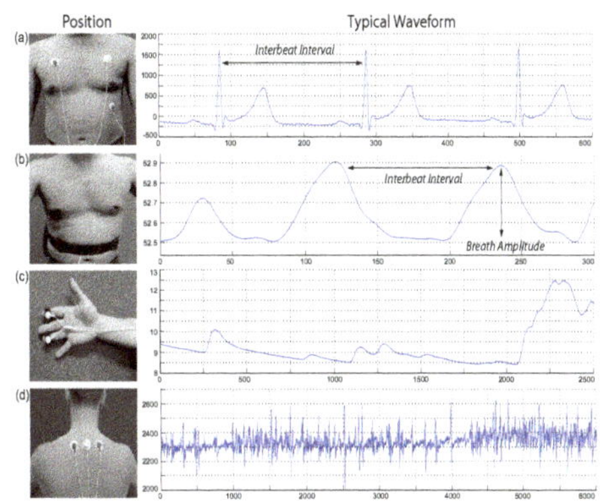

Fig. 2.: Position and typical waveforms of the biosensors: (a) ECG, (b) RSP, (c) SC, (d) EMG.

3.2. *Biosensors*

The physiological signals were acquired using the Procomp[b] Infiniti™ with four biosensors, electromyogram (EMG), skin conductivity (SC), electrocardiogram (ECG), and respiration (RSP). The sampling rates were 32 Hz for EMG, SC, and RSP, and 256 Hz for ECG. The positions and typical waveforms of the biosensors we used are illustrated in Fig. 2.

4. Methodology

The overall structure of our recognition system is illustrated in Figure 3.

After the preprocessing stage for signal segmentation and denoising we calculated 110 features from the 4-channel biosignals and selected the most significant features by using the sequential backward search method. For classification, various machine learning methods (supervised classification in our case) can be used [8].

[b]This is an 8 channel multi-modal Biofeedback system with 14 bit resolution and a fiber optic cable connection to the computer. www.MindMedia.nl

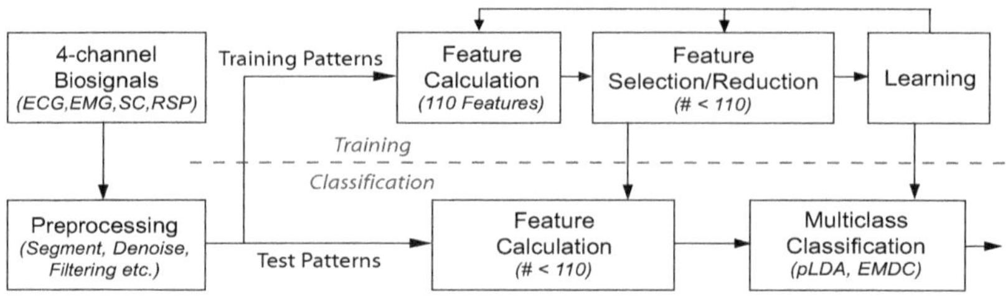

Fig. 3.: Block diagram of supervised statistical classification system for emotion recognition

After having tested some classifiers, such as k-nearest neighbor (k-NN), multilayer perceptron (MLP), and linear discriminant analysis (LDA), we chose the LDA which outperformed with higher recognition accuracy in our case. It should, however, be noted that there is no single best classification algorithm and the choice of the best classification method strongly depends on the characteristics of the dataset to be classified.

4.1. *Preprocessing*

Different types of artifacts were observed in all the four channel signals, such as transient noise due to movement of the subjects during the recording, mostly at the beginning and at the end of each recording. Thus, uniformly for all subjects and channels, we segmented the signals into final samples of a 160 seconds each, obtained by taking the middle part of each signal. It is important to note that the EMG signal generally requires additional pre-processing such as deep smoothing or signal separation, depending on the position of the sensor, because the nature of the signal is such that all the muscle fibers within the recording area of the sensor contract at different rates. In our case, the EMG signal contains artifacts generated by heart beat and respiration, since we positioned the sensor at the upper trapezius muscle. Using an adaptive bandpass filter we removed the artifacts (Fig. 4). For other signals we used pertinent lowpass filters to remove noises without loss of information.

4.2. *Measured features*

From the four channel signals we calculated a total of 110 features from various analysis domains including conventional statistics in time series, frequency domain, geometric analysis, multiscale sample entropy, subband spectra, etc. For the signals with non-periodic characteristics, such as EMG and SC, we focused on capturing the amplitude variance and localizing the occurrences (number of transient changes) in the signals. In the following sections, we describe the feature calculation methods in detail.

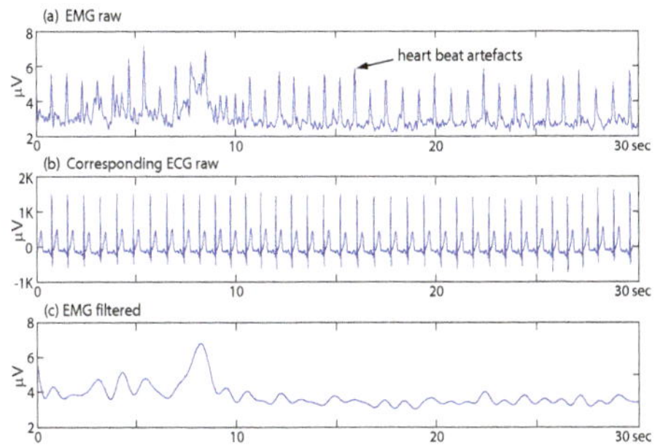

Fig. 4.: Example of EMG signal with heart beat artifacts and denoised signal

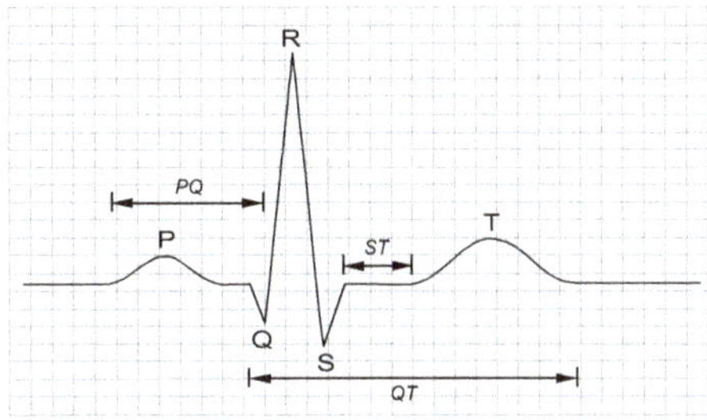

Fig. 5.: QRS waveform in ECG signal. Usual lengths: P-wave (0.08-0.10s), QRS (0.06-0.10s), PR-interval (0.12-0.20s), QT_c-interval ($QT/\sqrt{RR} \leq 0.44$s) [9]

4.2.1. *Electrocardiogram (ECG)*

ECG measures depolarized electrical changes of muscular contraction associated with cardiovascular activity. In general, the ECG is measured at the body surface along the axis of the heart and results from the activation first of the two small heart chambers, the atria, and then of the two larger heart chambers, the ventricles. The contraction of the ventricles produces the specific waveform known as the QRS complex (see Fig. 5).

To obtain the subband spectrum of the ECG signal we used the typical 1024 points fast Fourier transform (FFT) and partitioned the coefficients within the frequency range 0-10 Hz into eight non-overlapping subbands with equal bandwidth. First, as features, power mean values of each subband and fundamental frequency (F0) are calculated by finding maximum magnitude in the spectrum within the range 0-3 Hz. To capture peaks and their locations in subbands, subband spectral entropy (SSE) is computed for each subband. Entropy plays an important role in information theory as a measure of disorganization or uncertainty in a random variable. In pattern recognition it is generally used to measure the degree of a classifier's confidence. To compute the SSE, it is necessary to convert each spectrum into a probability mass function (PMF) like form. Eq. 1 is used for the normalization of the spectrum.

$$x_i = \frac{X_i}{\sum_{i=1}^{N} X_i}, \qquad \text{for } i = 1 \ldots N \tag{1}$$

where X_i is the energy of ith frequency component of the spectrum and $\tilde{\mathbf{x}} = \{x_1 \ldots x_N\}$ is to be considered as the PMF of the spectrum. In each subband the SSE is computed from $\tilde{\mathbf{x}}$ by

$$H_{\text{sub}} = -\sum_{i=1}^{N} x_i \cdot \log_2 x_i \tag{2}$$

By packing the eight subbands into two bands, i.e., subbands 1-3 as the low frequency (LF) band and subbands 4-8 as the high frequency (HF) band, the ratios of the LF/HF bands are calculated from the power mean values and the SSEs.

In biomedical engineering, the analysis of the local morphology of the QRS waveform and its time varying properties has been a standard method for assessing cardiac health. Importantly, heart rate variability (HRV) is one of the most often used measures for ECG analysis. To obtain the HRV from the continuous ECG signal, each QRS complex is detected and the RR intervals (all intervals between adjacent R waves) or the normal-to-normal (NN) intervals (all intervals between adjacent QRS complexes resulting from sinus node depolarization) are determined. We used the QRS detection algorithm of Pan and Tompkins [10] in order to obtain the HRV time series. Figure 6 shows examples of R wave detection and interpolated HRV time series, referring to the increases and decreases over time in the NN intervals.

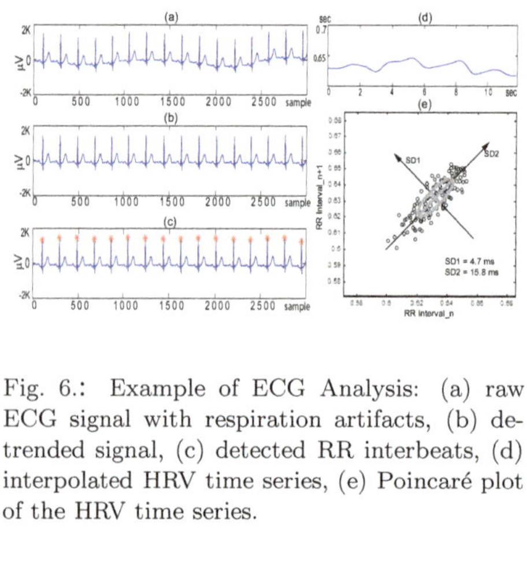

Fig. 6.: Example of ECG Analysis: (a) raw ECG signal with respiration artifacts, (b) detrended signal, (c) detected RR interbeats, (d) interpolated HRV time series, (e) Poincaré plot of the HRV time series.

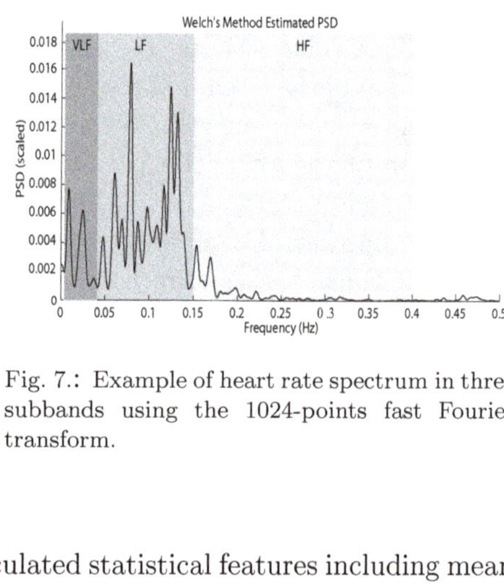

Fig. 7.: Example of heart rate spectrum in three subbands using the 1024-points fast Fourier transform.

In the time-domain of the HRV, we calculated statistical features including mean value, standard deviation of all NN intervals (SDNN), standard deviation of the first difference of the HRV, the number of pairs of successive NN intervals differing by more than 50 ms (NN50), the proportion derived by dividing NN50 by the total number of NN intervals. By calculating the standard deviations in different distances of RR interbeats, we also added Poincaré geometry in the feature set to capture the nature of interbeat (RR) interval fluctuations. Poincaré plot geometry is a graph of each RR interval plotted against the next interval and provides quantitative information of the heart activity by calculating the standard deviations of the distances of $R-R(i)$ to lines $y=x$ and $y=-x+2*R-R_m$, where $R-R_m$ is the mean of all $R-R(i)$. Figure 6 (e) shows an example plot of the Poincaré geometry. The standard deviations SD_1 and SD_2 refer to the fast beat-to-beat variability and longer-term variability of $R-R(i)$ respectively.

Entropy-based features from the HRV time series were also considered. Based on the so-called *approximate entropy* and *sample entropy*, a multiscale sample entropy (MSE) was introduced [11] and successfully applied to physiological data, especially for analysis of short and noisy biosignal. Given a time series $\{X_i\} = \{x_1, x_2, ..., x_N\}$ of length N, the number $(n_i^{(m)})$ of similar m-dimensional vectors $y^{(m)}(j)$ for each sequence vectors $y^{(m)}(i) = \{x_i, x_{i+1}, ..., x_{i+m-1}\}$ is determined by measuring their respective distances. The relative frequency to find the vector $y^{(m)}(j)$ within a tolerance level δ is defined by

$$C_i^{(m)}(\delta) = \frac{n_i^{(m)}}{N - m + 1} \tag{3}$$

The approximate entropy, $h_A(\delta, m)$, and the sample entropy, $h_S(\delta, m)$ are defined as

$$h_A(\delta, m) = \lim_{N \to \infty} [H_N^{(m]}(\delta) - H_N^{(m+1)}(\delta)], \tag{4}$$

$$h_S(\delta, m) = \lim_{N \to \infty} -\ln \frac{C^{(m+1)}(\delta)}{C^{(m)}(\delta)}, \tag{5}$$

where

$$H_N^{(m)}(\delta) = \frac{1}{N - m + 1} \sum_{i=1}^{N-m+1} \ln C_i^{(m)}(\delta), \tag{6}$$

Because it has the advantage of being less dependent on the time series length N, we applied the sample entropy h_S to coarse-grained versions $(y_j^{(\tau)})$ of the original HRV time series $\{X_i\}$,

$$y_j(\tau) = \frac{1}{\tau} \sum_{i=(j-1)\tau+1}^{j\tau} x_i, \quad 1 \leq j \leq N/\tau, \quad \tau = 1, 2, 3, ... \tag{7}$$

The time series $\{X_i\}$ is first divided into N/τ segments by non-overlapped windowing with length of scale factor τ and then the mean value of each segment is calculated. Note that for scale one $y_j(1) = x_j$. From the scaled time series $y_j(\tau)$ we obtain the m-dimensional sequence vectors $y^{(m)}(i, \tau)$. Finally, we calculate the sample entropy h_S for each sequence vector $y_j(\tau)$. In our analysis we used $m = 2$ and fixed $\delta = 0.2\sigma$ for all scales, where σ is the standard deviation of the original time series x_i. Note that using the fixed tolerance level δ as a percentage of the standard deviation corresponds to initial normalizing of the time series and it thus ensures that h_S does not depend on the variance of the original time series, but only on their sequential ordering.

In the frequency-domain of the HRV time series, three frequency bands are of general interest: the very-low frequency (VLF) band (0.003-0.04 Hz), the low frequency (LF) band (0.04-0.15 Hz), and the high frequency (HF) band (0.15-0.4 Hz). From these subband spectra, we computed the dominant frequency and power of each band by integrating the power spectral densities (PSD) obtained by using Welch's algorithm, as well as the ratio of power within the low-frequency band to that within the high-frequency band (LF/HF). Since parasympathetic activity dominates at high frequency, the LF/HF ratio is generally thought to distinguish sympathetic effects from parasympathetic effects. Figure 7 shows the heart rate spectrum from one of the subjects.

4.2.2. *Respiration (RSP)*

RSP signal (breathing rate and intensity) is commonly acquired by measuring physical change of the thoracic expansion with a rubber band around the chest or belly and contains less artifact in general than the other sensors using electrodes, e.g., ECG, EMG, SC etc. Including the typical statistics of the raw RSP signal, we calculated similar types of features, such as the ECG features, the power mean values of three subbands (obtained by dividing the Fourier coefficients within the range 0-0.8 Hz into non-overlapped three subbands with equal bandwidth), and the set of subband spectral entropies (SSE). In order to investigate inherent correlation between respiration rate and heart rate, we considered a novel feature content for the RSP signal. Since an RSP signal exhibits a quasi periodic waveform with sinusoidal properties, it does not seem unreasonable to conduct an HRV-like analysis for the RSP signal, i.e. to estimate breathing rate variability (BRV). After detrending using the mean value of the entire signal and lowpass filtering, we calculated the BRV by detecting the peaks in the signal using the maxima ranks within each zero-crossing (Fig. 8).

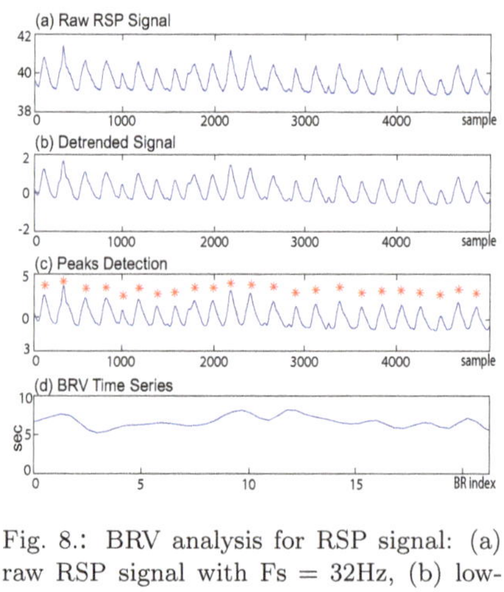

Fig. 8.: BRV analysis for RSP signal: (a) raw RSP signal with Fs = 32Hz, (b) lowpassed and detrended signal of (a), (c) peaks detection, (d) BRV time series.

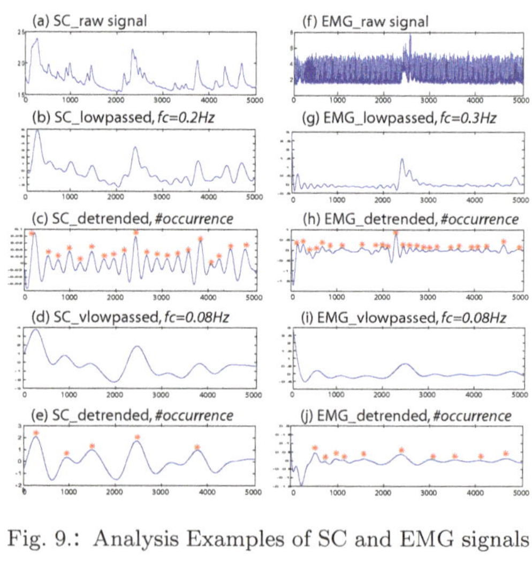

Fig. 9.: Analysis Examples of SC and EMG signals

From the BRV time series, we calculated the mean value, SD, SD of the first difference, MSE, Poincaré analysis, etc. In the spectrum of the BRV, peak frequency, power of the two subbands, the low-frequency band (0-0.03Hz) and the high-frequency band (0.03-0.15 Hz), and the ratio of the power within the two bands (LF/HF) were calculated.

4.2.3. *Skin Conductivity (SC)*

The SC signal includes two types of electrodermal activity, the DC level component and the skin conductance response (SCR). The DC level in the SC signal indicates a general activity of the perspiratory glands influenced by body temperature or external temperature. The SCR is the distinctive short waveform in the SC signal and is considered to be useful for emotion recognition as it is linearly correlated with the intensity of arousal responding to internal/external stimuli. The mean value, standard deviation, and mean of first and second derivations were extracted as features from the normalized SC signal and the low-passed SC signal using a cutoff frequency of 0.2 Hz. To obtain a detrended SCR waveform without DC-level components, we removed the continuous, piecewise linear trend in the two low-passed signals, i.e., the very low-passed (VLP) and the low-passed (LP) signal with a cutoff frequency of 0.08 Hz and 0.2 Hz, respectively (see Fig. 9 (a)-(e)).

The baseline of the SC signal was calculated and subtracted to consider only relative amplitudes. By finding two consecutive zero-crossings and the maximum value between them, we calculated the number of SCR occurrences within 100 seconds from each LP and VLP signal, the mean of the amplitudes of all occurrences, and the ratio of the SCR occurrences within the low-passed signals (VLP/LP).

4.2.4. *Electromyography (EMG)*

For the EMG signal, we calculated types of features similar to those of the SC signal. The mean value of the entire signal, the mean of the first and second derivations, and the standard deviation were extracted as features from the normalized and low-passed signals. The occurrence number of myo-responses and the ratio of that within VLP and LP signals were also added to the feature set and were determined in the same way as the SCR occurrence but using cutoff frequencies with 0.08 Hz (VLP) and 0.3 Hz (LP) (see Fig. 9 (f)-(j)).

In the end, we obtained a total of 110 features from the 4-channel biosignals; 53 (ECG) + 37 (RSP) + 10 (SC) + 10 (EMG).

4.3. *Classification*

4.3.1. *Feature selection*

A large number of algorithms for feature subset selection have been proposed in the literature [12]. Although sequential backward selection (SBS) is computationally more demanding than sequential forward selection (SFS), we decided to use SBS in our recognition system because it outperformed SFS and other methods in the feature space. Nevertheless we note that the performance of all the selection methods proposed is strongly dependent on the given dataset.

We did not consider integrating a dimensionality reduction method in our recognition scheme, such as principle component analysis (PCA) and Fisher projection, which are commonly used in combination with a classifier. Dimensionality reduction amounts to projecting high-dimensional data to a lower dimensional space with a minimal loss of information. This means that new features are created by the transformation of original feature values, rather than by selecting a feature subset from a given feature set. Such feature reduction methods were not suitable for the purpose of our work since we seeked to determine the best emotion-relevant features which preserve their origins of analysis domain and value. We use Fisher projection exclusively to preview the distribution of the features.

4.3.2. *Classifying using extended linear discriminant analysis*

In discriminant analysis, for a given dataset, three scatter matrices, within-class (S_w), between-class (S_b), and mixture scatter matrices (S_m) are defined as follows;

$$S_b = \sum_{i=1}^{c} N_i(\mu_i - \overline{\mathbf{x}})(\mu_i - \overline{\mathbf{x}})^T = \mathbf{\Phi}_b\mathbf{\Phi}_b^T, \tag{8}$$

$$S_w = \sum_{i=1}^{c} \sum_{j \in C_i} (\mathbf{x}_j - \mu_i)(\mathbf{x}_j - \mu_i)^T = \mathbf{\Phi}_w\mathbf{\Phi}_w^T, \tag{9}$$

$$S_m = S_b + S_w = \sum_{i=1}^{N} (\mathbf{x}_i - \overline{\mathbf{x}})(\mathbf{x}_i - \overline{\mathbf{x}})^T = \mathbf{\Phi}_m\mathbf{\Phi}_m^T, \tag{10}$$

where N is the number of all samples, N_i is the number of samples in class $C_i (i = 1, 2, ..., c)$, μ_i is the mean of the samples in class C_i, and $\overline{\mathbf{x}}$ is the mean of all samples, i.e.,

$$\mu_i = \frac{1}{N_i} \sum_{i \in C_i} \mathbf{x}_i, \tag{11}$$

$$\overline{\mathbf{x}} = \frac{1}{N} \sum_{i=1}^{c} \mathbf{x}_i = \frac{1}{N} \sum_{i=1}^{c} N_i \mu_i. \tag{12}$$

Note that the mixture scatter matrix S_m is the covariance matrix of all samples regardless of their class assignments and all the scatter matrices are designed to be invariant under coordinate shifts. The idea in LDA is to find an optimal transformation W which satisfies

$$\mathcal{J}(W) = \underset{W}{\operatorname{argmax}} \frac{|W^T S_b W|}{|W^T S_w W|}, \tag{13}$$

such that the separation between classes is maximized while the variance within a class is minimized (Fisher's criterion). Finding the optimal W is equivalent to finding the generalized eigenvectors satisfying $S_b W = \lambda S_w W$, for $\lambda \neq 0$. Transformation W can be obtained by applying the eigenvalue decomposition to the matrix $S_w^{-1} S_b$ if S_w is nonsingular, or to the matrix $S_b^{-1} S_w$ if S_b is nonsingular, and taking the rows of the transformation matrix to be the eigenvectors corresponding to the $n-1$ largest eigenvalues. Applying the singular value decomposition (SVD) on the scatter matrices of the training set is a stable way to compute the eigenvalue decomposition [13]. Since there are at most $c-1$ nonzero generalized eigenvectors of the scatter matrix, the upper bound of the number of retained dimensions in classical LDA is $c-1$ and the dimensionality can be further reduced, for example, by incorporating in W only those eigenvectors corresponding to the largest singular values determined in the scatter SVD. Given the transformation W, classification can be performed in the transformed space based on some distance measures d, such as Euclidean distance. The new instance, $\mathbf{v}$, is classified to

$$\underset{k}{\operatorname{argmin}}\, d\left(\mathbf{v}W, \overline{\mathbf{x}}_k W\right) \tag{14}$$

where $\overline{\mathbf{x}}_k$ is the centroid of k-th class and $k = 1, 2, ..., c$.

Note that a limitation of conventional LDA is that its objective function requires that one of the scatter matrices be nonsingular. It means that for a given c-class, p-dimensional classification problem, at least $c+p$ samples are required to guarantee that the within-class scatter matrix S_w does not become singular. To deal with the singularity problem, several extended LDA methods are proposed such as PCA+LDA, pseudoinverse LDA, regularized LDA, and LDA using generalized singular value decomposition (GSVD). In our work we used the pseudoinverse LDA (pLDA), a natural extension of classical LDA, applying the eigenvalue decomposition to the matrix $S_b^+ S_w$, $S_w^+ S_b$, or $S_m^+ S_b$. Pseudoinverse matrix is a generalization of the inverse matrix and exists for any $m \times n$ matrix. The computationally simplest way to get the pseudoinverse is using SVD; if $A = U\Sigma V^T$ is the singular value decomposition of A, then the pseudoinverse $A^+ = V\Sigma^+ U^T$. For a diagonal matrix such as Σ, we get the pseudoinverse by taking the reciprocal of each nonzero element on the diagonal.

5. Results

5.1. *Classification using SBS + pLDA*

The confusion matrix in Table 1 presents the correct classification ratio ($\mathcal{CCR}$) of subject-dependent (Subject A, B, and C) and subject-independent (All) classification where the features of all the subjects are simply merged and normalized. We used the leave-one-out cross-validation method where a single observation taken from the samples is used as the test data while the remaining observations are used for training the classifier. This is repeated such that each observation in the samples is used once as the test data.

Table 1.: Recognition results in rates (*error 0.00* = $\mathcal{CCR}$ *100%*). # of samples: 120 for each subject and 360 for All

Subject A ($\mathcal{CCR}$ % = *81%*)

	EQ1	EQ2	EQ3	EQ4	*total**	*error*
EQ1	***22***	4	1	3	*30*	*0.27*
EQ2	3	***26***	1	0	*30*	*0.13*
EQ3	1	2	***23***	4	*30*	*0.23*
EQ4	3	0	1	***26***	*30*	*0.13*

Subject B ($\mathcal{CCR}$ % = *91%*)

	EQ1	EQ2	EQ3	EQ4	*total**	*error*
EQ1	***27***	3	0	0	*30*	*0.10*
EQ2	3	***25***	1	1	*30*	*0.17*
EQ3	0	2	***28***	0	*30*	*0.07*
EQ4	0	1	0	***29***	*30*	*0.03*

Subject C ($\mathcal{CCR}$ % = *89%*)

	EQ1	EQ2	EQ3	EQ4	*total**	*error*
EQ1	***28***	0	2	0	*30*	*0.07*
EQ2	0	***30***	0	0	*30*	*0.00*
EQ3	0	0	***24***	6	*30*	*0.20*
EQ4	0	0	5	***25***	*30*	*0.17*

All: Subject-independent ($\mathcal{CCR}$ % = *65%*)

	EQ1	EQ2	EQ3	EQ4	*total**	*error*
EQ1	***62***	9	8	11	*90*	*0.31*
EQ2	15	***57***	13	5	*90*	*0.37*
EQ3	9	6	***58***	17	*90*	*0.36*
EQ4	8	5	21	***56***	*90*	*0.38*

*: Actual total # of samples

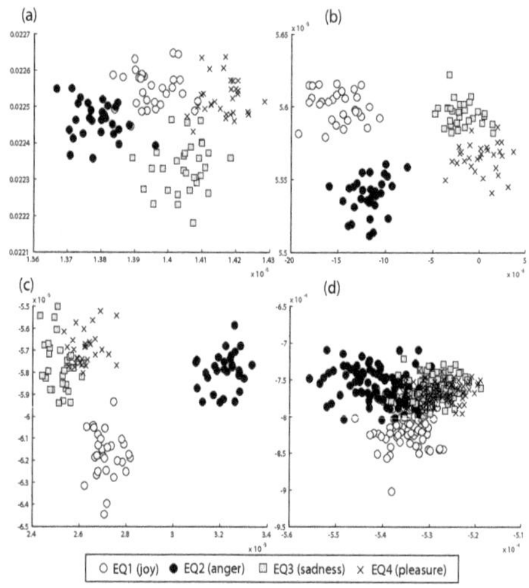

Fig. 10.: Comparison of feature distributions of subject-dependent and subject-independent case. (a) Subject A, (b) Subject B, (c) Subject C, (d) Subject-independent

The table shows that the $\mathcal{CCR}$ varies from subject to subject. For example, the best accuracy was 91% for subject B and the lowest was 81% for subject A. Not only does the overall accuracy differ from one subject to the next, but the $\mathcal{CCR}$ of the single emotions varies as well. For example, EQ2 was perfectly recognized for subject C while it caused the highest error rate for subject B. It was three times mixed up with EQ1 which is characterized by opposite valence. As the confusion matrix shows, the difficulty in valence differentiation can be observed for all subjects. Most classification errors for Subject A and B lie in false classification between EQ1 and EQ2 while an extreme uncertainty can be observed in the differentiation between EQ3 and EQ4 for Subject C. On the other hand, it is very meaningful that relatively robust recognition accuracy is achieved for the classification of emotions that are reciprocal in the diagonal quadrants of the 2D emotion model, i.e., EQ1 vs. EQ3 and EQ2 vs. EQ4. Moreover, the accuracy is much better than that of arousal classification. The $\mathcal{CCR}$ of subject-independent classification was not comparable to that obtained for subject-dependent classification. As shown in Figure 10, merging the features of all subjects does not refine the discriminating information related to the emotions, but rather leads to scattered class boundaries.

We also tried to differentiate the emotions based on the two axes, arousal and valence, in the 2D emotion model. The samples of four emotions were divided into groups of negative valence (EQ2+EQ3) and positive valence (EQ1+EQ4) and into groups of high arousal (EQ1+EQ2) and low arousal (EQ3+EQ4). By using the same methods, we then performed a two-class classification of the divided samples for arousal and valence separately. It turned out that emotion-relevant ANS specificity can be observed more conspicuously in the arousal axis regardless of subject-dependent or independent cases. Classification of arousal achieved an acceptable $\mathcal{CCR}$ of 97-99% for the subject-dependent recognition and 89% for the subject-independent recognition, while the results for valence were 88-94% and 77%, respectively.

5.2. *Finding the best emotion-relevant ANS features*

In most literature dealing with emotion-relevant ANS specificity, a tendency analysis of physiological changes has been used to correlate ANS activity with certain emotional states, e.g. EQ1 with increased heart rate or anxiety with increased skin conductivity. Even for multiclass classification problems, however, such a direction analysis of physiological changes is not sufficient to capture accompanying multimodal ANS reactions that are cross-correlated with each other when using multichannel biosensors. Therefore, we tried to first identify the significant features for each classification problem and thereby to investigate class-relevant feature domain and interrelation between the features for a certain emotion.

In Table 2, the best emotion-relevant features, which we determined by ranking the features selected for all subjects (including Subject All) in each classification problem, are listed in detail by specifying their values and domains. One interesting result is that each classification problem respectively links together with a certain

Table 2.: Best emotion-relevant features extracted from four channel physiological signals. Arousal classes: EQ1+EQ2 vs. EQ3+EQ4, Valence classes: EQ1+EQ4 vs. EQ2+EQ3, Four classes: EQ1/EQ2/EQ3/EQ4.

Classes	**Best emotion-relevant Features** (Ch_value_domain, C: ECG, R: RSP, S: SC, M: EMG)
Arousal	C_std(diff)_HRVtime, C_**sd2_PoincareHRV**, C_powerLow_HRVspec, R_**meanEnergy_SubSpectra**, R_**sd2_PoincareBRV** R_**mean_MSE**, S_mean_RawLowpassed, S_std_RawLowpassed, M_#occurrenceRatio_RawLowpassed M_mean_RawNormed
Valence	C_**sd2_PoincareHRV**, C_meanEnergy_SubSpectra, C_ratioLH_HRVspec, C_mean_MSE, C_mean(diff)_MSE R_**meanEnergy_SubSpectra**, R_mean(diff)_SubSpectra, R_sd1_PoincareBRV, R_**sd2_PoincareBRV**, R_**mean_MSE** S_mean(diff)_RawNormed, M_mean(diff)_RawNormed
Four Emotions	C_mean_HRVtime, C_std_HRVtime, C_std(diff)_HRVtime, C_mean(diff)_MSE, C_mean_MSE, C_mean_SSE C_**sd2_PoincareHRV**, C_mean_SubSpectra, R_**meanEnergy_SubSpectra**, R_mean_SSE, R_mean_BRVtime R_sd1_PoincareBRV, R_**sd2_PoincareBRV**, R_**mean_MSE**, R_power_BRVspec, S_std_RawLowpassed S_mean(diff)_RawNormed, S_mean(diff(diff))_RawLowpassed, S_mean_RawNormed, S_#occurrence_RawLowpassed M_mean(diff)_RawNormed

: overall selected features are printed in bold

feature domain. The features obtained from the time/frequency analysis of HRV time series are decisive for the classification of arousal and for the classificaton of the four emotions, while the features from the MSE domain of ECG signals are a predominant factor for correct valence differentiation. More particularly, mutually sympathizing correlate between HRV and BRV (firstly proposed in this paper) has been clearly observed in all the classification problems by the features from their time/frequency analysis and Poincaré domain, *_PoincareHRV* and *_PoincareBRV*. This reveals a manifest cross-correlation between respiration and cardiac activity with respect to emotional state. This is one of the most important findings for future work. In fact, in biomedicine, it is commonly accepted that the respiratory mechanism mediates high frequency components of HRV, but its specific role in affective ANS reactions has so far not been satisfactorily explained. When inhaling, the vagus nerve is impeded and the heart rate begins to increase, whereas this pattern is reversed when exhaling, i.e., the activation of the vagus nerve typically leads to a reduction in heart rate, blood pressure, or both[c]. Apart from its influence on the heart rate, the vagus nerve is also responsible for sweating, several muscle movements in the mouth, and even for speech. It means that most physiological channels we used are innately correlated with each other and respond together as a chain reaction to emotional stimulation. For example, when the parasympathetic nerves overcompensates a strong response from the sympathetic nervous system innervating the sinoatrial node, which occurs in cases of extreme stress or fear, the reduction in heart rate and blood pressure becomes proportionally faster to the intensity of the emotion.

Our feature analysis proves that the correlation between heart rate and respiration is obviously captured by the features from the HRV power spectrum (*_HRVspec*), the fast/long-term HRV/BRV analysis using the Poincaré method, and the multiscale variance analysis of HRV/BRV (*_MSE*). It also demonstrates that the peaks of high frequency range in the HR subband spectrum (*_SubSpectra*) provide information about how the sinoatrial node responds to vagal activity at certain respiration frequencies.

5.3. *Emotion-specific multilevel dichotomous classification*

Most common classifiers are best-suited to handle two-class problems. The pLDA we used is no exception to this and assumes that the covariance matrices of each class are the same or at least close to each other for multiclass ($c > 2$) classification. Consequently, the performance of pLDA in multiclass classification could be suboptimal depending on the difference between the covariance matrices of each class. In our work, we actually used the averaged covariance to directly solve the multiclass

[c]The influence of breathing on the flow of the sympathetic and vagus impulses to the sinoatrial node causes the so called respiratory sinus arrhythmia (RSA). The degree of fluctuation in heart rate is also significantly controlled by regular impulses from the baroreceptors in the aorta and carotid arteries.

problem using a single pLDA classifier. One straightforward way to handle a multiclass problem by using binary classifiers is to decompose the multiple categories into a set of complementary two-class problems.

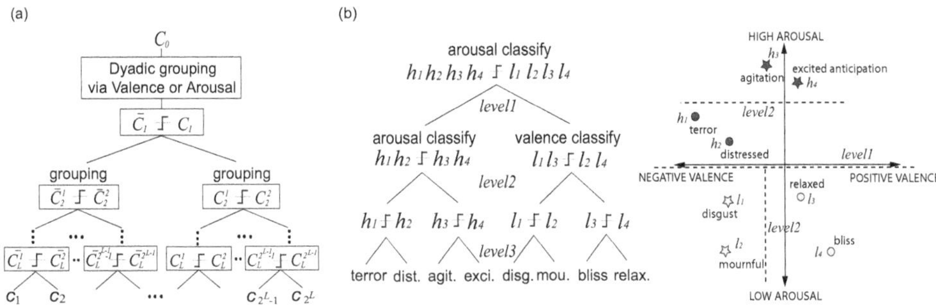

Fig. 11.: Framework of emotion-specific multilevel dichotomous classification (EMDC). (a) Diagram of decomposition process, (b) Decomposition example for an eight-class problem

By taking advantage of supervised classification (where we know in advance which emotion types have to be recognized) we developed an emotion-specific multilevel dichotomous classification (EMDC) scheme. This scheme exploits the property of the dichotomous categorization in the 2D emotion model and the fact that arousal classification yields higher $\mathcal{CCR}$ than valence classification or direct multiclass classification. This proves true in almost all previous works and according to our results as well. Figure 11 illustrates the EMDC scheme and provides an example of the dyadic decomposition for the eight-class problem.

First, the entire training patterns are grouped into two opposing *"superclasses"* (on the basis of valence or arousal), $\bar{C}$ consisting of all patterns in some subset of the class categories and C as all remaining patterns, i.e., $\bar{C} \cap C = \{\}$. This dyadic decomposition using one of the two axes is serially performed until one subset contains only two classes. The grouping axis can be different from each dichotomous level. Then multiple binary classifiers for each level are trained from the corresponding dyadic patterns. Therefore, the EMDC scheme is obviously emotion-specific and effective for a 2D emotion model. Note that the performance of the EMDC scheme is limited by a maximum $\mathcal{CCR}$ of first level classification and makes sense only if the $\mathcal{CCR}$ for one of the two superclasses is higher than that for direct multiclass classification (theoretically this always holds true for our case). Because we used four emotion classes in our experiment, we needed a two-level classification based on arousal and valence grouping for both superclasses in parallel.

Table 3 shows the dichotomous contingency table of recognition results by using the novel EMDC scheme. The best feature sets shown in Table 2 are used for the binary classification at each level. As expected, the $\mathcal{CCR}s$ significantly improve for all class problems. For the classification of four emotions, we obtained an average $\mathcal{CCR}$ of 95% for subject-dependent and 70% for subject-independent classification.

Table 3.: Results using EMDC scheme with the best features

Subject A ($\mathcal{CCR}$ % = 94%, 113/120)

	EQ1 & EQ2			EQ3 & EQ4		
EQ1 & EQ2	**58**			2		
		EQ1	EQ2			
	EQ1	**27**	1			
	EQ2	1	**29**			
EQ3 & EQ4	2			**58**		
					EQ3	EQ4
				EQ3	**29**	0
				EQ4	1	**28**

Subject C ($\mathcal{CCR}$ % = 94%, 113/120)

	EQ1 & EQ2			EQ3 & EQ4		
EQ1 & EQ2	**60**			0		
		EQ1	EQ2			
	EQ1	**30**	0			
	EQ2	0	**30**			
EQ3 & EQ4	1			**59**		
					EQ3	EQ4
				EQ3	**27**	2
				EQ4	4	**26**

Subject B ($\mathcal{CCR}$ % = 98%, 117/120)

	EQ1 & EQ2			EQ3 & EQ4		
EQ1 & EQ2	**60**			0		
		EQ1	EQ2			
	EQ1	**30**	0			
	EQ2	0	**30**			
EQ3 & EQ4	1			**59**		
					EQ3	EQ4
				EQ3	**29**	1
				EQ4	1	**28**

Subject All ($\mathcal{CCR}$ % = 70%, 251/360)

	EQ1 & EQ2			EQ3 & EQ4		
EQ1 & EQ2	**155**			25		
		EQ1	EQ2			
	EQ1	**62**	13			
	EQ2	15	**65**			
EQ3 & EQ4	16			**164**		
					EQ3	EQ4
				EQ3	**64**	19
				EQ4	21	**60**

$\mathcal{CCR}$ Comparison between pLDA and EMDC

	Subj. A	Subj. B	Subj. C	Subj. All	*Average (ABC)*
pLDA	81%	91%	89%	65%	*87%*
EMDC	**94%**	**98%**	**94%**	**70%**	***95%***

Compared with the results obtained for pLDA, the EMDC scheme achieved an overall $\mathcal{CCR}$ improvement of about 5%-13% in each class problem.

6. Discussion

We achieved an overall $\mathcal{CCR}$ of 95%, which is more than three times higher than chance probability, for four emotional states from three subjects. This should be sufficient to support the view that emotions, either produced or perceived while listening to music, exist and are accompanied by physiological differences in both the arousal and valence dimensions such that they can eventually be recognized by the machine. At the same time, however, some issues remain in relation to the processing stages of our recognition system.

Recording physiological changes using biosensors is still invasive since the subjects, for example, have to be in physical contact with adhesive electrodes. Furthermore, most biosensors using such electrodes are very susceptible to motion artifacts which we could observe in almost all signals of our dataset. For practical HCI applications, it is therefore necessary to develop non-invasive biosensors, preferably with built-in denoising filters in wirelessly miniaturized form. We expect that today's nano-technology will help design such hardware soon. This would then improve not only the signal quality and the usability of the technology, but also reduce computational costs in the preprocessing stage.

Our analysis results based on the best emotion-relevant features are incontrovertibly useful findings; for example, the consistent tendency of the feature contents to valence and arousal differentiation separately and the proven efficiency of new feature domains that are firstly considered in this paper. We should, however, note that the effectiveness of the best features might not be universally guaranteed for other datasets or classifiers. First, only three subjects might not be sufficient to generalize the features. Second, the SBS as well as most algorithms for feature selection use a criterion based on a specific classifier and are therefore effective only if the classifier used is known in advance. In addition, such sequential algorithms may lead to suboptimal subsets due to their unidirectional property, i.e. once a feature is added or removed, this action can never be reversed.

By dividing given patterns using the arousal and valence axis in the 2D emotion model, we proposed the EMDC scheme which contributed to a significant improvement of the recognition results. The scheme may, however, still be adjusted in several ways. For instance, since it needs multiple classifiers to be trained for each level, the combination of different classifiers seems to be feasible. By taking advantage of the fact that EMDC enables us to view the classification results of each level in multiresolution aspect (see Table 3), the scheme could be more sophisticatedly designed thanks to the parametric refining of each binary classifier depending on the level.

The reason for the great disparity of $\mathcal{CCR}$ between subject-dependent and independent classification can be explained in many different ways indeed. We mention that one of the main factors in the difficulty of subject-independent classification is the intricate variety of non-emotional individual contexts among the subjects, rather than an individual ANS specificity in emotion. A naive idea for improving the performance of the user-independent system for practical applications would be to first identify the user, prior to starting the recognition process, and then to classify a user's emotion in a user-dependent way. Of course, this is feasible only if the number of users is finite and the users are known to the system, or if the system can cumulatively collect the data of each user in a learning phase.

7. Future Work

One of the most challenging issues in the near future will be to explore multimodal analysis for emotion recognition. We humans use several modalities jointly to interpret emotional states, since emotion affects almost all modes- audiovisual (facial expression, voice, gesture, posture, etc.), physiological (respiration, skin temperature etc.), and contextual (goal, preference, environment, social situation, etc.) states in human communication. In the recent literature, findings concerning emotion recognition by combining multiple modalities have been reported; mostly by fusing features extracted from audiovisual modalities such as facial expression and speech. However, we note that combining multiple modalities by equally weighting them does not always guarantee improved accuracy. The more crucial issue is how to *complementarily* combine the additional modalities. An essential step towards a human-like analysis and finer resolution of recognizable emotion classes would therefore be to find the innate priority among the modalities to be preferred for each emotional state. Then, an ambitious undertaking might be to decompose an emotion recognition problem into several refining processes using additional modalities, for example: arousal recognition through physiological channels, valence recognition by using audiovisual channels, and then resolving of subtle uncertainties between adjacent emotion classes.

References

[1] R. Cowie, E. Douglas-Cowie, N. Tsapatsoulis, G. Votsis, S. Kollias, W. Fellenz, and J. G. Taylor, Emotion recognition in human-computer interaction, *IEEE Signal Processing Mag.* **18**, 32–80, (2001).

[2] R. Picard, E. Vyzas, and J. Healy, Toward machine emotional intelligence: Analysis of affective physiological state, *IEEE Trans. Pattern Anal. and Machine Intell.* **23** (10), 1175–1191, (2001).

[3] F. Nasoz, K. Alvarez, C. Lisetti, and N. Finkelstein, Emotion recognition from physiological signals for presence technologies, *International Journal of Cognition, Technology, and Work - Special Issue on Presence.* **6(1)**, (2003).

[4] J. J. Gross and R. W. Levenson, Emotion elicitation using films, *Cognition and Emotion.* **9**, 87–108, (1995).

[5] A. Haag, S. Goronzy, P. Schaich, and J. Williams. Emotion recognition using biosensors: First steps towards an automatic system. In *ADA 2004*, pp. 36–48, (2004).

[6] Center for the Study of Emotion and Attention [CSEA-NIMH]. *The International Affective Picture System: Digitized Photographs.* Gainesville, FL: Center for Research in Psychophysiology, University of Florida, (1995).

[7] K. H. Kim, S. W. Bang, and S. R. Kim, Emotion recognition system using short-term monitoring of physiological signals, *Medical & Biological Engineering & Computing.* **42**, 419–427, (2004).

[8] A. Jain, R. Duin, and J. Mao, Statistical pattern recognition: A review, *IEEE Trans. Pattern Anal. and Machine Intell.* **22**(1), 4–37, (2000).

[9] M. B. Mcllory and M. D. Cheithin, *Clinical Cardiology.* (VLANGE Medical Book, 1990), 5th edn edition.

[10] J. Pan and W. Tompkins, A real-time qrs detection algorithm, *IEEE Trans. Biomed. Eng.* **32**(3), 230–323, (1985).

[11] M. Costa, A. L. Goldberger, and C.-K. Peng, Multiscale entropy analysis of biological signals, *Phys. Rev.* **E 71**(021906), (2005).

[12] A. Jain and D. Zongker, Feature selection: Evaluation, application, and small sample performance, *IEEE Trans. Pattern Anal. and Machine Intell.* **19**, 153–163 (Feb., 1997).

[13] D. L. Swets and J. Weng, Using discriminant eigenfeatures for image retrieval, *IEEE Trans. Pattern Anal. and Machine Intell.* **18**(8), 831–836 (Aug, 1996).

CHAPTER 3.7

AUTOMATED DETECTION AND CLASSIFICATION OF SOLAR IMAGE FEATURES

Frank Y. Shih

College of Computing Sciences, New Jersey Institute of Technology, University Heights, Newark, NJ 07102, USA
E-mail: shih@njit.edu

The techniques of automated detection and classification of solar image features based on image enhancement, segmentation, pattern recognition, and mathematical morphology methods are presented in this chapter. The solar features include flares and filament. These methods have been applied to full-disk Hα (Hydrogen-Alpha) mages successfully. The solar H-α images were obtained from the Big Bear Solar Observatory in California, USA. The methods presented can lead to real-time solar features detection and classification using advanced image processing and pattern recognition techniques.

1. Introduction

The sun is the source of space weather – the disturbances whose radiative, field, and particle energy directly impact earth. Solar activity changes the radiative and particle output of the sun, producing corresponding changes in the near-earth space environment as well as at the earth surface. The most dramatic events on the sun, insofar as space weather effects are concerned, are solar flares, filament, and coronal mass ejections (CMEs).

A flare is defined as a sudden, rapid, and intense variation in brightness. A solar flare occurs when magnetic energy that has built up in the solar atmosphere is suddenly released. Radiation is emitted across virtually the entire electromagnetic spectrum, from radio waves at the long wavelength end, through optical emission to X-rays and gamma rays at the short wavelength end [1]. The amount of energy released is equivalent to millions of 100-megaton hydrogen bombs exploding at the same time. Figure 1 shows a full disk Hα (Hydrogen-Alpha) image with a flare detected on the low-left corner. Three principal techniques, multi-layer perceptron (MLP), radial basis function (RBF), and support vector machines (SVM) classifiers were developed for solar-flare detection [2]. This work was further extended to automatically detect flares and applying image segmentation techniques to compute their properties [3].

A chromospheric filament usually marks a boundary between two opposite magnetic regions. Filaments may last for multiple days, changing their manifestation, but eventually they vanish. Their disappearance may end up with large CME associated with geomagnetic storms that affect the quality of terrestrial communication, reliability of power systems and safety of space missions [4]; therefore, the filament tracking is an important task of solar observation. In solar images, filaments appear as dark elongated objects of different shape and size, and may be broken into separate parts, as shown in Figure 2. Qu *et al.* [5] developed an automatic solar filament detection algorithm based on image enhancement, segmentation, pattern recognition, and mathematical morphology [6]. Shih and Kowalski [7] developed a novel method

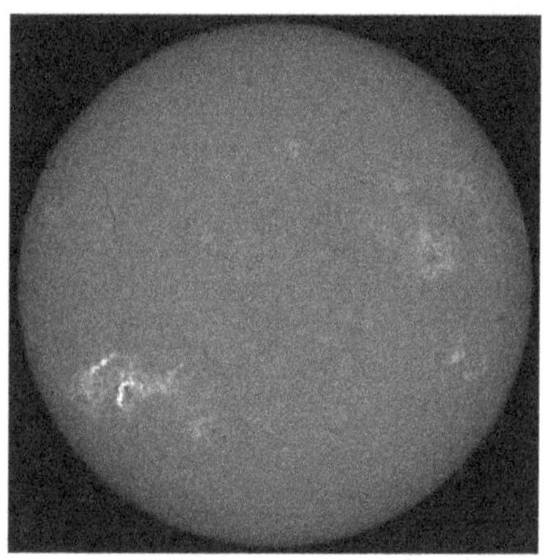

Figure 1. Solar flares on the left side of the image.

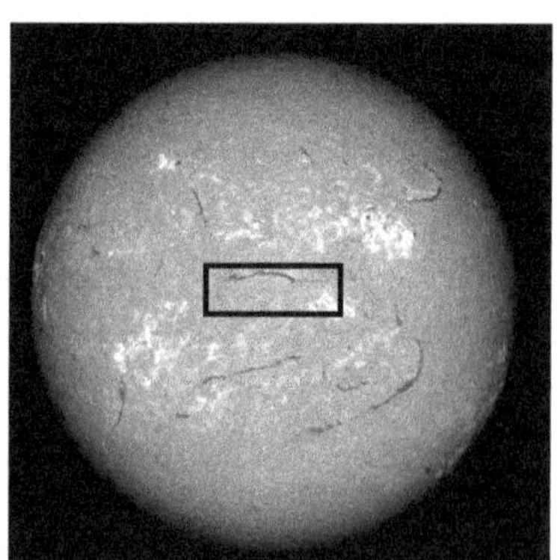

Figure 2. A fragment with a solar filament is marked by a rectangle.

which allows for the automatic extraction and tracking of the evolution of filaments in solar images.

A CME is due to a large scale rearrangement of the solar magnetic field which can induce a geo-effective solar wind structure. Each CME may carry away a mass of up to 1013 kg and release up to 1025 J of energy from coronal magnetic fields [8]. Following their discovery, CMEs were soon found to be correlated with the occurrence of geomagnetic storms. CMEs are easily observed at the solar limb, where they are seen against a dark background. However, the earth-directed disk CMEs are much harder to detect – although they are the ones that have the most important geomagnetic effects. Therefore, detecting the early manifestation of CMEs on the solar disk is essential to understanding earth-directed CMEs themselves. Establishing the correlation among the occurrence of CMEs and some other solar surface phenomena, such as filament eruptions and flares, is a long-standing problem in solar physics. After firm relationships are established, the correlated solar surface phenomena can provide an early warning of the occurrence of earth-directed CMEs and geo-effectiveness of CMEs. CMEs have been observed from Large Angle and Spectrometric Coronagraph (LASCO) images by Solar and Heliospheric Observatory (SOHO). LASCO images include three different coronagraphs: C1, C2, and C3, whose ranges are from 1.1 to 3, 1.5 to 6, and 3.7 to 30 solar radii, respectively. Coronagraph C1 has been inoperable since 1999. A detailed description of LASCO can be found in Brueckner *et al.* [9]. In Figure 3, examples of strong and weak CMEs in LASCO C2 images are

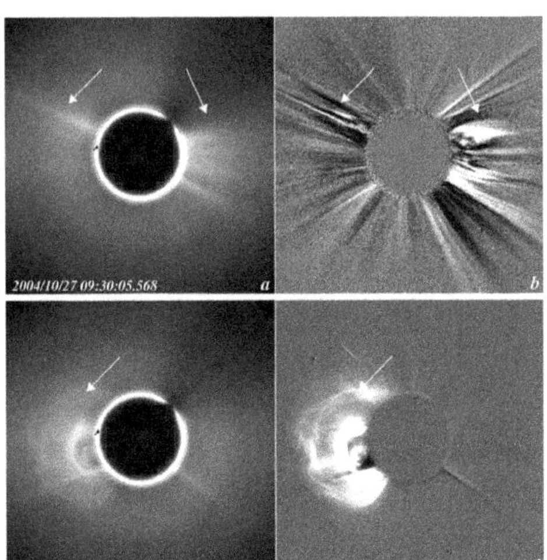

Figure 3. (a) Weak CMEs in a LASCO C2 image on October 27, 2004. (b) Same weak CMEs in the running difference image. (c) A strong CME in a LASCO C2 image on September 12, 2004. (d) Same strong CME in the running difference image. CMEs are pointed by *arrows*.

pointed by arrows. Due to page limitation, the classification of CMEs is not included in this chapter. Readers can refer to Qu *et al.* [10].

2. Solar Flare Detection

There are many ways to detect solar flares. When a solar flare first appears on the Sun's surface, there is often a radiation covering the entire range of electromagnetic waves. For example, light curves of radio waves and X-rays could be used to indicate a set of flares. However, location and properties of flares cannot be demonstrated in light curves. In the mean time, solar flares also emit high velocity charged particles that take one or two days to reach Earth. Our objective is to detect and characterize solar flares automatically in real time. Finally, most solar flares can also be detected using Hα. We start with solar images observed in Hα using an 8-inch telescope in the Big Bear Solar Observatory (BBSO) in California, operated by New Jersey Institute of Technology.

Automatic solar flare detection is the key to space weather monitoring. It is very challenging since the features of solar flares are complicated. We need a system that can handle the solar flares' complexity and scalability. Veronig *et al.* [11] proposed a method for automatic flare detection by applying a combination of region-based and edge-based segmentation methods. Simple region-based methods are applied for a tentative assignment of flare activity, making use of one of the decisive flare characteristics: the high intensities. Borda *et al.* [12] presented a method for automatic detection of solar flares using the neural network technique. The network used is multi-layer perceptron (MLP) with back-propagation training rule. They used a supervised learning technique and required a lot of iterations to train the network.

Qu *et al.* [2] compared three principal techniques: multi-layer perceptron (MLP), radial basis function (RBF), and support vector machines (SVM) classifiers, for solar-flare detection. The preprocessing step is to obtain nine principal features of solar flares for the classifiers. They show that by using SVM, the best classification rate of solar flares can be obtained. Qu *et al.* [3] further extended their work to automatically detect flares and applying image segmentation techniques to compute their properties. A solution is obtained for automatically tracking the apparent separation motion of two-ribbon flares and measuring their moving direction and speed in the magnetic fields. From these measurements with certain assumptions, the reconnection of the electric field as a measure of the rate of the magnetic reconnection in the corona can be derived.

2.1. *Feature Analysis and Preprocessing*

Borda *et al.* [12] used seven features for solar flare detection. After extensive investigations, we modify them into nine features as described below.

Feature 1: Mean brightness of the frame. Usually, when flares happen, the image becomes brighter. Let x_j denote the gray level of a pixel and N denote the number of pixels. The mean brightness $\bar{x}$ is represented as

$$\bar{x} = \sum_{j=0}^{N-1} x_j \tag{1}$$

Feature 2: Standard deviation of brightness. Usually, when flares happen, the standard deviation of the brightness becomes large. In Equation (2), *std* denotes the standard deviation, and V denotes the variance.

$$std = \sqrt{V} \tag{2}$$

$$V = \sum_{j=0}^{N-1} (x_j - \bar{x})^2 \tag{3}$$

Feature 3: Variation of mean brightness between consecutive images. This is the difference of mean values between the current image and the previous image.

Feature 4: Absolute brightness of a key pixel. By comparing consecutive images, we can find the key pixel that has the maximum gray level difference. Usually, when flares happen, the absolute brightness of the key pixel becomes large. The position of the key pixel k is represented as

$$k = where(\max(\mathrm{Im}_{pre} - \mathrm{Im}_{current})) \tag{4}$$

where Im_{pre} denotes the vector of the previous image, and $\mathrm{Im}_{current}$ denotes the vector of the current image.

Feature 5: Radial position of the key pixel.

Features 1-5 focus on the global view of the consecutive images. But these features are insufficient to obtain the information for small solar flares. Therefore, it is necessary to include the local view of solar images.

Feature 6: Contrast between the key pixel and the minimum value of its neighbors in a 7×7 window. When flares occur, this contrast becomes large.

Feature 7: Mean brightness of a 50×50 window, where the key pixel is at the center. Since we assume that the key pixel is one of the pixels on the flare, the 50×50 window will include most of the flare. This region is most effective for the solar flare detection.

Feature 8: Standard deviation of the pixels in the aforementioned 50 × 50 window.

Feature 9: Difference of the mean brightness of the 50 × 50 window between the current and the previous images.

See samples of the input nine features and the target value in Table 1.

Table 1. Samples of the nine features and the target value.

1	2	3	4	5	6	7	8	9	Target value
3994.25	2628.22	45	12474.00	431.23	3635.00	2011.43	1978.60	427.35	-1
3807.42	2458.64	187	16383.00	437.13	6484.00	9460.98	2430.92	7449.55	1

2.2. *Experimental Results*

The solar Hα images of 2032 × 2032 pixels were obtained from BBSO. We selected the images of flare events starting from January 1, 2002 to December 31, 2002. The first step taken was to carefully divide the images into two classes: flare state and non-flare state. We tried to cover all the flare events that can be recognized by human eyes. We labeled corresponding images as "flare state" and included the same number of "non-flare state" images. Because some micro-flares are difficult to be recognized even by human eyes, we regarded them as "non-flare state". Secondly, we computed the nine features for a given solar image and put it into either the training or the testing data set. We used the training data set to train the networks and used the testing data set to obtain the classification rate. Finally, we obtained flare properties using image segmentation techniques, since we are also interested in obtaining the properties of flares such as size, lifetime, and motion.

We developed the programs in Interactive Data Language (IDL) by Research Systems, Inc. The program runs on a DELL Dimension L733r with CPU time 733 Mhz and memory of 256 Mbytes under Windows 2000. There are three steps in our program:

a) Preprocessing to obtain the nine features of solar flares.
b) MLP, RBF and SVM training and testing programs used for solar flare detection.
c) Region growing and edge detection methods for obtaining the flare properties.

For the experiments of solar flare detection, we capture one image per minute and process the current image in comparison with the previous image to obtain the nine features. Firstly, we filter the noise of full-disk images using a Gaussian function, and center these two images using the IDL FIT_LIMB function. Secondly, we compute the mean brightness and standard deviation for the two images. Thirdly, we divide them to obtain the pixel with the maximum gray level difference. It is regarded as the key pixel to compute other features. After preprocessing steps, we use the nine features of the key pixel as the input to the neural network and SVM. At last, we use region growing combined with Canny edge detector to analyze the region of flares.

We tested the MLP, RBF and SVM methods based on 240 events including pre-flare, main phase and post-flare, half of them being used as the training set and the rest half as the testing set. For MLP, after about 1000 iterations for the training set of 120 events, we tested this network with 120 events. The fraction of misclassified events is about 5%. MLP uses supervised training and needs a lot of iterations to obtain the satisfied classification rate. For RBF, we obtain a better classification rate and a faster training speed. For SVM, we obtain the best performance among the three methods, such that the misclassification rate is 3.3%. The results are shown in Table 2.

Table 2. Classification report based on 120 training events and 120 testing events.

Methods based on events	Classification rate	Training time	Testing time
MLP with 1000 iterations	94.2%	60.20 seconds	0.01 second
RBF	95%	0.27 second	0.01 second
SVM	96.7%	0.16 second	0.03 second

The MLP architecture is to perform a full non-linear optimization of the network. Therefore, a lot of iterations to train the system to get the best result are needed. In Table 3, it is demonstrated that the result performs better with more iterations. The number of hidden nodes determines the complexities of the neural networks. We can adjust the number of hidden nodes to obtain the best performance. In Table 3, we show the comparison of MLP with 11 hidden nodes and MLP with other number of hidden nodes. Test correctness of MLP with 11 hidden nodes is comparable to MLP with more hidden nodes. MLP with 11 hidden nodes has better performance than MLP with more or less hidden nodes.

Table 3. Comparison of different MLP iterations and hidden nodes on 120 training events and 120 testing events.

Methods based on events	Classification rate	Training time
MLP with 1000 iterations and 11 hidden nodes	94.2%	60.2 seconds
MLP with 100 iterations and 11 hidden nodes	69.2%	6.2 seconds
MLP with 3000 iterations and 11 hidden nodes	94.2%	179.3 seconds
MLP with 1000 iterations and 11 hidden nodes	94.2%	60.2 seconds
MLP with 1000 iterations and 6 hidden nodes	92.5%	46.7 seconds
MLP with 1000 iterations and 20 hidden nodes	94.2%	115.6 seconds

RBF neural network, as compared with the MLP, is the possibility of choosing suitable parameters for each hidden unit without having to perform a full non-linear optimization of the network. Usually, the number of hidden nodes equals the number of classes. For our case, we want to use the RBF classifier to separate two classes. We try RBF with different number of hidden nodes and find that RBF with two hidden nodes is good for our pattern classification. The parameters of RBF are center vector, the weight and the radius. The training architecture can be quickly constructed by K-mean algorithm.

SVM is based on the idea of "minimization of the structural risk". To obtain the best classification rate, we consider the confidence interval and empirical risk. The complex system has high confidence interval and low empirical risk. We test our flare events using a linear SVM and the non-linear SVMs respectively with polynomial kernel classifier and Gaussian RBF kernel classifier. Experimental results are shown in Table 4.

Table 4. Comparison of different SVM training strategies on 120 training events and 120 testing events.

Methods based on events	Classification rate	Training time	Testing time
Linear SVM	96.7%	0.16 second	0.03 second
SVM with Polynomial kernel	95%	0.34 second	0.03 second
SVM with RBF kernel	90.83%	0.44 second	0.03 second

Table 4 shows that the linear SVM is better than the non-linear kernel SVM for our events. The non-linear SVM can reduce the empirical risk using complex systems, but it has a higher confidence interval. Using the linear SVM with the nine features of the solar flare, we have both low empirical risk and confidence interval to achieve the best total risk. Therefore, we prefer the linear SVM to classify the flare patterns.

3. Solar Filament Detection

The various steps of filament detection can be summarized in the flow chart depicted in Figure 4. The dark background of the Sun is first removed. Then the SIDE is used to suppress the Hα fibrils in the quiet Sun regions and enhance solar filaments. After filaments are segmented using our adaptive thresholding method, sunspots are distinguished from filaments using the SVM classifier with nine features. The filament morphology is obtained by several mathematical morphology methods. Finally, the filament disappearance is reported.

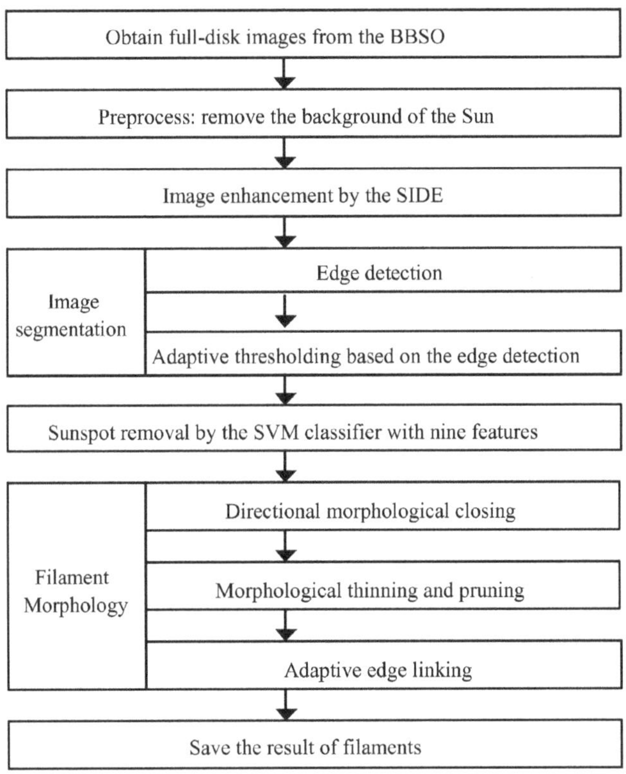

Figure 4. Flow chart of the filament detection.

3.1. *Stabilized Inverse Diffusion Equation (SIDE)*

Pollak [13] provides a comprehensive introduction to nonlinear multiscale filtering and its application to image segmentation and restoration. We use the SIDE [14] to transform a highly detailed Hα full-disk image (fibrils, filaments, sunspots, plage regions, etc.) to a coarser image with less detail that allows an easier detection of filaments. Plage regions are exceptionally bright in Hα and cannot be confused with dark filaments. We extract filaments from the quiet Sun background, which is dominated by dark Hα fibrils. These fibrils have typical sizes of about 10″. Their orientations often trace the local field of geometry; i.e., they are not randomly oriented. However, their distributions are very uniform. Therefore, Hα fibrils share many characteristics with noise that is present in images. This motivates the use of the SIDE, which has been successful in restoring images in the presence of pervasive, large-amplitude noise while preserving or even sharpening edges. The SIDE is a fast and robust image segmentation and restoration algorithm. It is robust against image blur, which affects the full-disk images due to atmospheric turbulence.

Our first goal is to suppress the Hα fibrils in the quiet-Sun regions. This can be achieved by linear shift-invariant filters such as a Gaussian filter to suppress high spatial frequencies and thus smooth sharp edges. What we need is a stabilized inverse diffusion near edges and a heat-equation-like behavior in homogeneous regions without edges.

The detailed mathematical treatment of the SIDE is given by Pollak, Willsky, and Krim [14] and we only briefly review the basic concept. An image of $M \times N$ pixels is denoted as u_{ij} with $i \in [1, \ldots, M]$ and $j \in [1, \ldots, N]$. The pixel subscripts are conveniently dropped when we refer to the entire image. If u^0 is an original image, then $u^1, u^2, \ldots, u^k$ are the successively coarser images after a series of iterative filters $A^1, A^2, \ldots, A^k$ have been applied to u^0. In this iterative filtering, the iteration number k is referred to as the scale and the succession of filtered images is part of a k-dimensional scale-space. Our algorithm is based on the semi-discrete SIDE, which is continuous with respect to the temporal evolution in the k-dimensional scale-space but discrete with respect to spatial scales.

The PDE has a simple mechanical analog, where particles with mass m_{ij} are coupled to their four neighbors by a force $F(u)$. We begin the implementation of the SIDE with the 2D PDE as

$$\dot{u}_{ij} = \frac{1}{m_{ij}}(F(u_{i+1,j} - u_{ij}) - F(u_{i,j} - u_{i-1,j}) + F(u_{i,j+1} - u_{ij}) - F(u_{i,j} - u_{i,j-1})), \quad (5)$$

where $\dot{u} = (u^{k+1} - u^k)/\Delta t$ is the temporal derivative representing the evolution in the temporal scale space. The boundary conditions for peripheral particles are satisfied if they remain stationary. The case of a linear force, e.g., Hooke's law, results in a heat-equation-like behavior.

In this implementation, the force equation is

$$F(u) = \operatorname{sgn}(u) - \frac{u}{L}, \tag{6}$$

where $\operatorname{sgn}(u) = +1$ if $u \leq 0$, and -1 otherwise. In order to satisfy the conditions of a force function, L has to be larger than the dynamic range of an image. Initially, each pixel is considered as a unique region. We apply Equation (5) with a force function given by Equation (6). Once some neighboring pixels become identical, they will remain identical throughout the iteration. Neighboring pixels are merged together, and progressively larger areas with identical values generate coarser approximations of the original image. This recursive region merging leads to a modification of Equation (5) to account for the growing blobs of pixels with identical values

$$\dot{u}_{n_i} = \frac{1}{m_{n_i}} \sum_{n_j \in A_{n_i}} F(u_{n_j} - u_{n_i}) p_{ij}, \tag{7}$$

where m_{n_i} is the number of pixels in region n_i, A_{n_i} is a set of indices of all neighbors of n_i, and p_{ij} denote the forces connecting regions n_i and n_j. Once the region merging is concluded, the next evolution starts again with Equation 7. The previous steps are repeated until the quiet-Sun background is sufficiently suppressed.

3.2. *Edge Detection Based on Adaptive Thresholding*

We propose a method for adaptive threshold selection from the results of edge detection. The edges of filaments are detected by Sobel operator which performs a 2D spatial gradient measurement on an image and emphasizes the regions of high spatial gradient that correspond to edges. It is used to find the approximate absolute gradient magnitude at each pixel [15].

Our adaptive thresholding method contains two steps. In the first step, the global threshold is computed for the entire Hα image to select filaments that have visible high contrasts relative to background. In the second step, the local thresholds are calculated for sub-images of 100 × 100 pixels to select filaments that have locally high contrasts relative to background. The threshold selection method is described as follows.

A sequence of self-adaptive thresholds $t_1 < t_2 < \cdots < t_n$ ranging from 0 to the median of pixel intensities of the image are chosen for segmenting dark regions. Suppose r_i is an individual segmented region by adopting the threshold t_i, and

r_i' is the new expanded region where $r_i' = r_i - r_{i-1}'$. Initially, we have $r_i' = r_1$. A set of gradients g_i is calculated by

$$g_i = \frac{\text{Sobel}(r_i')}{n_{r_i'}}, \tag{8}$$

where $n_{r_i'}$ is the number of pixels in the region r_i'. The best global threshold T is computed by

$$T = t_{i\max}, \tag{9}$$

where $i_{\max}$ is the index of the maximum of g_i. For example, we utilize thresholds $t_1 < t_2 < \cdots < t_n$ to obtain regions $r_1 < r_2 < \cdots < r_n$, respectively. New expanded regions are $r_1', r_2', \cdots, r_n'$. The average spatial gradient g_i for each pixel on r_i' has been calculated by Sobel operator. When a new expanded region r_i' meets the edges of filaments, the gradient g_i reaches the maximum. By finding the maximum gradient for each new expanded regions g_i, the optimal global threshold T for segmenting solar filaments is obtained.

In the second step, the local threshold is selected using the same threshold selection method with the following two criteria: the local threshold $t_{loc} = T \pm 30$; the size of the region obtained by t_{loc} is less than three times of the region obtained by T. After segmentation, the small regions whose areas are less than 10 pixels are considered as noises and are removed.

3.3. *Support Vector Machine*

An efficient feature-based classifier, called the support vector machine (SVM), is used to distinguish sunspots from filaments. We represent a sunspot by nine features and use the SVM for sunspot classification. Experimental results show that this method significantly improved the classification rate in comparison with other methods (e.g., Shih and Kowalski [7]).

The SVM is a learning system based on advances in statistical learning theory. The basic idea of the SVM is to separate the given fixed input pattern vectors into two classes using a hyperplane with the maximal margin which is the distance between the decision plane and the closest sample points [16]. Points on the decision plane are called support vectors. The support vectors can be calculated by the classical method of Lagrange multipliers. The SVM is based on the structural risk minimization (SRM) principle, which is intended to minimize the risk functional with respect to the empirical risks and the confidence interval [17].

The SVM applied in the solar flare detection is superior to the neural network classifier. In our experiment, we adopt the SVM classifier with the linear kernel. The comparisons between the linear kernel and other kernels can be found in Qu *et al.* [2]. For further information regarding the SVM, readers can refer to Guyon and Stork [18] and Vapnik [16].

According to the physical nature of sunspots, we represent them by nine features to be the input of the SVM. These nine features can be computed from the segmented results and the original Hα image. We denote the segmented regions for the whole image as A, a window of 100 × 100 pixels which is centered on a region S as W1, and a window which just encloses a region S as W2. The nine features of S are described below.

1. The number of pixels in S, say n_S.
2. The radial position of the center of S. When the radial position is far away from the center of Sun, the size and darkness of sunspots could be quite different.
3. The distance from the center of S to the equator of Sun. It is calculated as

$$d = y - y_c, \tag{10}$$

   where y and y_c denote the respective Y-coordinates of S and the equator of Sun.
4. The ratio of mean brightness between S and A is defined by

$$\text{ratio}_1 = \frac{\bar{x}_S}{\bar{x}_A}, \tag{11}$$

   where $\bar{x}$ denotes the mean brightness.
5. The ratio of standard deviation between S and A is defined by

$$\text{ratio}_2 = \frac{\text{std}_S}{\text{std}_A}, \tag{12}$$

   where std denotes standard deviation.
6. The mean brightness of the window W1.
7. The standard deviation of W1.
8. The shape feature h_1 is defined by

$$h_1 = n_S / n_{W2}, \tag{13}$$

   where n_{W2} denotes the number of pixels of W2.

9. The shape feature h_2 is defined by

$$h_2 = L_{W2} / H_{W2}, \tag{14}$$

where L_{W2} and H_{W2} denote the length and height of W2, respectively.

3.4. *Filament Morphology*

Because some adjacent regions are considered to be one component, it is necessary to check the spatial relationship between regions to determine if they should be integrated. This is achieved by morphological closing, thinning, pruning and adaptive edge linking methods.

3.4.1. *Directional Morphological Closing* - The morphological closing of A by B, denoted by $A \bullet B$, is defined as

$$A \bullet B = (A \oplus B) \ominus B, \tag{15}$$

Where $\oplus$ denotes morphological dilation and $\ominus$ denotes morphological erosion. It is used to eliminate small gaps. Shih and Kowalski [7] introduced eight directional linear 11×11 structuring elements with 0, 22.5, 45, 67.5, 90, 112.5, 135, and 157.5° slopes.

3.4.2. *Morphological Thinning and Pruning* - Morphological thinning is defined as

$$A \otimes B = \mathrm{A} - (A \circledast B), \tag{16}$$

Where $\circledast$ denotes morphological hit-and-miss transformation [19]. For thinning to be symmetrical, we use a sequence of structuring elements which are described in Figure 5a: $B^1, B^2, B^3, \cdots, B^8$, where B^i is a rotated version of B^{i-1}. The entire process is repeated until no further changes occur.

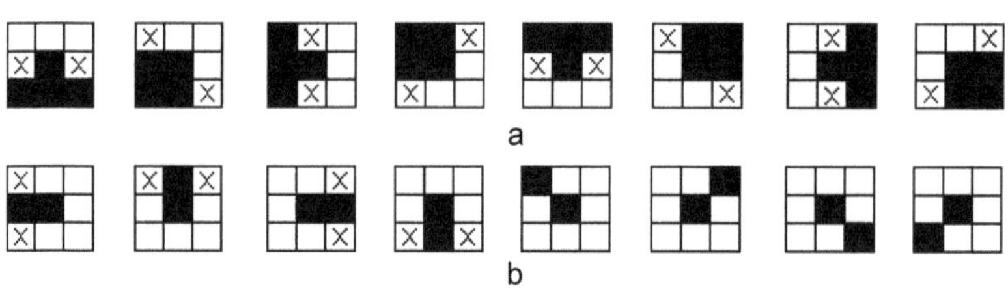

Figure 5. (a) Structuring elements for thinning; (b) structuring elements for pruning. The *black pixel* denotes "1", the *white pixel* denotes "0", and the "×" symbol denotes "don't care" [19].

We use four steps of the morphological pruning to yield X_1, X_2, X_3, and X_4 by Equations 17-20 [19]:

$$X_1 = A \otimes \{B^*\}, \tag{17}$$

where A is an input set and B^* is the structuring element used for pruning as described in Figure 3b,

$$X_2 = \cup(X_1 \circledast B^*), \tag{18}$$

$$X_3 = (X_2 \oplus C) \cap A, \tag{19}$$

where C is a 3 × 3 structuring element of 1's,

$$X_4 = X_1 \cup X_3. \tag{20}$$

The original pruning method is used to remove the fixed length of a parasitic component. For filament pruning, we propose an adaptive pruning method by monitoring X_3. Step 1 to step 3 are repeated to remove pixels until X_3 has two points left. Then the fourth step is invoked to compute the final results. The pruned results are lines which represent the spines of filaments. The two end-points in X_3 are called the footpoints of a filament.

3.4.3. *Adaptive Edge Linking* - Big gaps of broken filaments could not be filled in by directional morphological closing. Therefore, we adopt the adaptive edge linking method developed by Shih and Cheng [20]. The basic idea is to connect edges based on the orientation of spines which is obtained by the edge slope direction of the footpoints. Adaptive edge linking is performed in an iterative manner, so that the spines of broken filaments can be linked up gradually and smoothly. The final results of spines and footpoints of filaments are used for the detection of filament disappearances.

3.5. *Detection of Filament Disappearance*

Finally, we detect the filament disappearance by comparing the results of two successive days. A computer program has been developed to detect the filament disappearance automatically and send the report to our researchers as described in Figure 6. The Hα images of two consecutive days are loaded. They are matched after differential solar rotation correction (SSW; Freeland and Handy, 1998). Our enhancement, segmentation, and morphology methods are applied to obtain the spines and footpoints of filaments. To detect the filament disappearances on the two images, we propose the following method:

1. The spines of filaments are dilated with a structuring element of 20×20 pixels.
2. When two filaments are overlapped, we match them according to the size of spines and the intensity of filaments. The intensity of a filament is calculated by the mean brightness of the filament. If the size of a spine on the current day is greater than 40% of the previous day, we call their sizes as matched. If the intensity of a filament on the current day is less than 1.5 times of the previous day, we call their intensities as matched.
3. Filaments are matched if both sizes and intensities are matched. Unmatched filaments on the previous day are reported as disappearance.

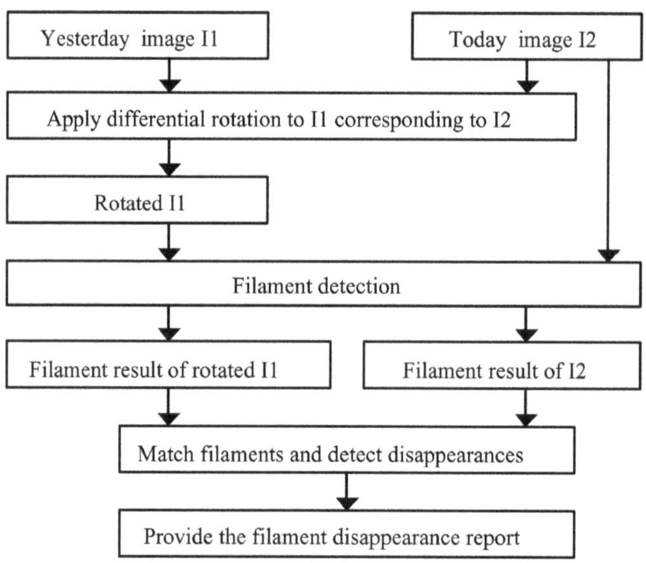

Figure 6. Flow chart of the detection of filament disappearance.

4. Conclusions

We have presented comparisons of flare detection using three advanced techniques which are multi-layer perceptron (MLP), radial basis function (RBF) and support vector machine (SVM). Through experiments, we have found that SVM is the best for the solar-flare detection because it offers the best classification result and the training and testing speed are relatively fast. The second choice is RBF. Because its training data are presented only once, its training time is the shortest. MLP is not a well-controlled learning machine. The sigmoid function has a scaling factor that affects the quality of the approximation, and the convergence of the gradient-based method is rather slow. The result of MLP is not as good as the RBF and SVM. After the solar-flare detection, the properties of flares are analyzed using region growing and Canny edge detector. Because this work is well done by Veronig *et al.* [11] we do not give an explanation here. For future study, we will apply SVM, RBF and MLP to more solar phenomena such as filaments, sunspots, and plages. We believe that we may create the best model to detect the solar phenomena by combining SVM and RBF classifiers.

A new automatic filament detection algorithm based on image enhancement, segmentation, pattern recognition, and mathematical morphology methods has been successfully applied to Hα full-disk filtergrams. For the first time, the SIDE has been adopted to enhancing filaments from the quiet-Sun background. Filaments are extracted using an adaptive threshold selection method. The SVM is an essential tool to distinguish between filaments and sunspots using nine features. Filament properties such as footpoint, spine, position, and disappearance are subsequently retrieved by directional morphological closing, thinning, pruning, directional edge linking, and component matching methods. These properties will be stored in a database for context-based data query and BBSO's space weather forecast and prediction.

Another solar feature, called prominence, is clouds of relatively cool and dense gas in the solar atmosphere. Fu, Shih, and Wang [21] presented a method to detect and characterize the prominence eruptions. The input is a sequence of consecutive solar images, and the output is a list of prominence eruption events detected. The limb events are extracted and their associated properties are measured by applying image processing techniques. First, image normalization and noise removal are performed. Then, the limb objects are isolated and the prominence features are identified. Finally, pattern recognition techniques are applied to classify the eruptive prominences. The characteristics of prominence eruptions, such as brightness, angular width, radial height and velocity are measured.

References

1. H. Zirin, *Astrophysics of the Sun*, Cambridge University Press, Cambridge, 343 (1988).
2. M. Qu, F. Y. Shih, J. Jing, and H. Wang, *Solar Phys.*, **217**, 157 (2003).
3. M. Qu, F. Y. Shih, J. Jing, and H. Wang, *Solar Phys.*, **222**, 137 (2004).

4. P. C. Martens and C. Zwaan, *The Astrophysical Journal*, **558**, 872 (2001).
5. M. Qu, F. Y. Shih, J. Jing, and H. Wang, *Solar Phys.*, **228**, 119 (2005).
6. F. Y. Shih and O. R. Mitchell, *IEEE Trans. Pattern Analysis and Machine Intelligence*, **11**, 31 (1989).
7. F. Y. Shih and A. Kowalski, *Solar Phys.*, **218**, 99 (2003).
8. R. A. Harrison, *A&A*, **304**, 585 (1995).
9. G. E. Brueckner, R. A. Howard, M. J. Koomen, C. M. Korendyke, D. J. Michels, J. D. Moses, D. G. Socker, K. P. Dere, P. L. Lamy, A. Llebaria, M. V. Bout, R. Schwenn, G. M. Simnett, D. K. Bedford, and C. J. Eyles, *Solar Physics*, **162**, 357 (1995).
10. M. Qu, F. Y. Shih, J. Jing, and H. Wang, *Solar Phys.*, **237**, 419 (2006).
11. A. Veronig, M., Steinegger, W., Otruba, A., Hanslmeier, M., Messerotti, M., Temmer, G., Brunner, and S. Gonzi, *ESA*. **463**, 455 (2000).
12. R. A. Borda, P. D. Mininni, C. H. Mandrini, D. O. Gómez, O. H. Bauer, and M. G. Rovira, *Solar Phys.* **206**, 347-357 (2001).
13. I. Pollak, *IEEE Signal Proc. Mag.* **19**, 26 (2002).
14. I. Pollak, A. S. Willsky, and H. Krim, *IEEE Trans. Image Proc.* **9**, 2 (2000).
15. B. Jahne, *Digital Image Processing*, Springer, New York (1997).
16. N. V. Vapnik, in S. Haykin (ed.), **Statistical Learning Theory**, John Wiley & Sons, Inc., NY, USA, 401 (1998).
17. N. V. Vapnik and A. J. Chervonenkis, *Theory of Pattern Recognition*, Nauka, Moscow (1974).
18. I. Guyon and D. G. Stork, *Linear discriminant and support vector machine*, The MIT Press, Cambridge, 147 (2000).
19. R. C. Gonzalez and R. E. Woods, *Digital Image Processing*, Prentice Hall, Upper Saddle River, New Jersey (2002).
20. F. Y. Shih and S. Cheng, *Inform. Sci.* **167**, 9 (2004).
21. G. Fu, F. Y. Shih, and H. Wang, *IEEE Trans. on Circuits and Systems for Video Technology*, **17**, 79 (2007).

Part 3

Recognition Applications

CHAPTER 4.1

BIOMETRIC TEMPLATE PROTECTION: TOWARDS A SECURE BIOMETRIC SYSTEM

Yicheng Feng and Pong C. Yuen

Department of Computer Science, Hong Kong Baptist University
HONG KONG
E-mail: {ycfeng, pcyuen}@comp.hkbu.edu.hk

Biometric template security is one of the most important issues in developing a practical biometric system. In this chapter, we discuss the threats to biometric authentication systems and give a review on existing approaches for protecting biometric templates. The schemes are classified into three categories, namely biometric cryptosystem approach, transform-based approach and the hybrid approach. Pros and cons of each approach are also discussed.

1. Introduction

Biometric recognition is a reliable, robust and convenient way for person authentication [3] [11] [13] [14] and is growing rather fast. Several large biometric systems such as US-VISIT program have been successfully deployed. Additionally, biometric systems are being developed for other applications [10] such as banking (for ATM machine), credit card industry and physical access control. With the growing use of biometrics, there is a rising concern about the security and privacy of the biometric data itself. Since each person is claimed to have a unique biometric (e.g. fingerprint, face and iris), if this biometric data is compromised, it is impossible to have a replacement. Therefore, biometric data (template) security [11]–[13] is one of the most important issues in developing a practical biometric system (biometric template refers to the extracted biometric features stored in a central database or a smartcard). Moreover, in e-passport [10], face and fingerprint biometrics are compulsory while iris and other biometrics are optional. Recent studies [19] [20] have shown that the raw biometric data can be retrieved from the biometric template stored in the database. As a result, protection of biometric template in biometric systems application is indispensable. To overcome this problem, the general idea is that the original biometric template should never be stored in the system, but a transformed one generated by a template protection algorithm. And, since there is no "perfect" protection scheme, the transformed secure templates should be designed to be "cancelabe". That is, if a secure template is compromised by attackers, a new secure template can be generated to replace the compromised one such that the attacker can not use the old data to access the system. This property is called "cancelability" of the protection scheme.

The straightforward solution to protect the biometric templates is to apply password-like encryption algorithm on biometric template and the matching process is performed in the encrypted domain. However, due to the intra-class biometric data variations, a small change of biometric data input will result in a large change of the encrypted data. As a result, the system's accuracy will be degraded. Therefore, directly applying password-like encryption method is not feasible. On the other hand, if the encrypted secure template is decrypted for matching, it is susceptible to interception by an impostor.

In order to solve the intra-class template variation problem while maintaining the template security, many algorithms have been developed in the last few years and can be categorized into three approaches, namely biometric cryptosystem approach, transform-based approach and hybrid approach. The basic idea of all approaches is that instead of storing the original template, it stores the transformed/encrypted template which intends to be more secure. In case that the transformed/encrypted template is stolen or lost, it is computationally hard to reconstruct the original template and the original raw biometric data from the transformed/encrypted template.

In the biometric cryptosystem approach, the *error correcting coding techniques* are employed to handle intra-class variations. Two popular techniques, namely fuzzy commitment scheme [24] and fuzzy vault scheme [25] have been proposed. The advantage of this approach is that, since the output is an encrypted template, its security level is high. However, the error correcting ability of these schemes may not be strong enough to handle large intra-class variations such as variations in face images captured under different illuminations and poses. Also, this approach is not designed to be revocable. Finally, the error correcting coding techniques require input with certain formats (e.g. binary strings [24] or integer vectors within a limited range [25]), which some biometrics like face are hard to satisfy.

In the transform-based approach, a transformed template is generated using a "one-way" transform and the matching is performed in the transformed domain. The transform-based approach has a good cancelability (revocability), but the drawback of this approach is that a trade-off between performance and security is normally required.

Combining the advantages of these two approaches, hybrid approach [1] [5] [6] [36] [37] [40] has also been proposed.

The rest of this chapter is organized as follows: Section 2 describes the threats to biometric recognition systems and explains why the biometric templates need to be protected. Section 3 gives a brief introduction of how the existing algorithms protect biometric templates. Finally, discussion and conclusion of this chapter is reported in Section 4.

2. Vulnerabilities of Biometric Recognition Systems

2.1. *Attacks on Biometric Systems*

Jain et al. [15] classified attacks on biometric systems in three categories: administration attack, nonsecure infrastructure and biometric overtness. Administration attack considers the attack performed by insiders, such as system administrator. For example, the attacker may be enrolled as a legitimate user, or the administrator tries to illegally modify the system data. Also, attackers may force the legitimate users or administrators to cooperate with them to attack the system. The nonsecure infrastructure, on the other side, means that the attackers aim at the vulnerable components of the system facilities. Attackers may try to override the devices such that they may generate data and decisions that they desire, retrieve the information stored/transferring in the system, or modify the data stored in database for system access etc. Ratha. et al. [11] have performed an analysis and reported seven types of the nonsecure infrastructure attack (type 2-8) as shown in Figure 1.

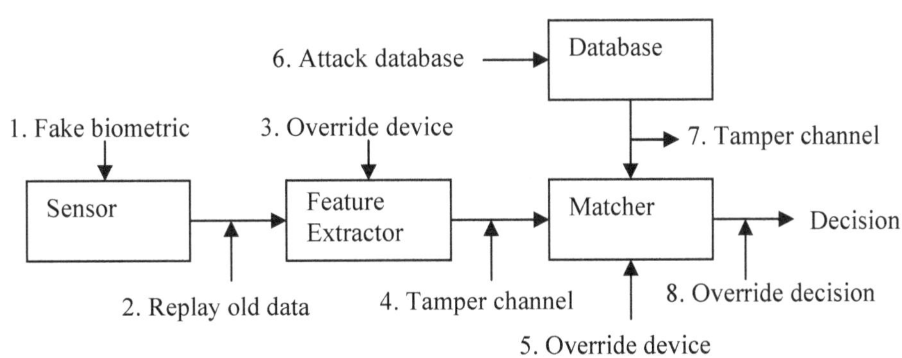

Fig. 1. Attacks aim at the biometric system (Ratha et al. [11]).

The nonsecure infrastructure attack shown in Figure 1 basically has three ways [15]: overriding the device, transferring channels tampering, and database attack. To override the device, the attacker may apply a trojan-horse attack to rewrite the programs in the devices. The attacker may also try to tamper the data transferring channel, either overwrite the data or steal the useful information that the attacker needs. In a database attack, the stored templates in the database may be modified or stolen. The attacker may use the stolen biometric templates to generate fake biometrics, or just replay the stolen biometric templates to get access to the system. Or the attacker may overwrite the original templates with his/her own template, pretending to be an enrolled legitimate user.

In biometric overtness attack, a fake biometric is used as shown in Figure 1 (attack 1). Artificial biometrics are fabricated to fool and get access to the system. The fabrication is based on some genuine user's biometric data which is secretly collected by attackers. For example, attackers may collect the fingerprints of the legal users from the objects that the user touches, or covertly take face pictures of the users.

The administration attack does not depend on the security strength of the biometric system, but other external factors such as the professional ethics of the system administrator/user. How to design and construct secure devices and transferring channels is another issue. Here we mainly focus on the attacks aim at biometric templates. That is, the database attack and biometric overtness attack.

2.2. *Template Attack*

Masquerade attack [17] is one kind of template attacks. In a masquerade attack, a reference template is stolen and the characteristic of this reference template is analyzed. A synthetic template is generated based on the analysis. Directly capturing the raw biometric data is another way to fabricate a forged template without tampering the database. Then an artificial biometric can be constructed from the captured data (biometric overtness attack). Unfortunately, raw biometric data is often not a secret to attackers. A biometric data may be captured without the owner's knowledge. A fingerprint can be captured from objects that the target person touches. Capturing a face image can be even more convenient by taking pictures stealthily.

Since a biometric overtness attack has no knowledge of the reference templates, it does not know which characteristics of the covertly captured biometric data should be used. Therefore it can only construct the artificial biometric as exquisite as possible. On the other side, a masquerade attack attempt to extract the reference templates from the biometric system and use the information of the reference templates. The help of the reference templates not only decrease the computational complexity but also enhance the probability of success. So the masquerade attack is a more serious threat to biometric systems.

A practical template attack namely hill-climbing attack [16] has been proposed by Soutar, which is a significant threat to biometric template security. This attack is powerful because it does not require the reference template stored in database. Instead, it proceeds similarly to the brute-force attack which tries different templates repetitively. Different from the brute-force attack, the templates are not generated randomly, but adjusted slightly with the matching scores of the previous attempt. In [16], Soutar presented the main idea of the hill-climbing attack. After that, Alder [18] proposed a practical hill-climbing attack algorithm and applied it on face biometric system. Alder's algorithm does not modify images pixel by pixel, but adds a small constant c times an eigenface image to the original image. The constant c varies in order to find the one with maximum matching score. The updated eigenfaces, which are used to represent the modified images, are trained within the attacker's local face database. The final generated image not only could be recognized as the target via the recognition system, but also visually similar to the target because the trained eigenfaces maintain some appearance features of the individuals. Experimental results show that after about 4000 iterations, the generated image will achieve a rather high (about 99.9%) confidence of matching scores to the target.

Uludag and Jain [21] proposed a hill-climbing attack on a fingerprint recognition system. Since the fingerprints are represented by a set of minutiae, their scheme is much different from the alder's approach. The initial testing fingerprint template is randomly generated. For each iteration, one of the following three operations is performed, (i) updating existing minutia in the testing fingerprint template, (ii) replace one minutia with a new one and (iii) randomly delete/increase one minutia. The experimental results show that within 1000 iteration, this attack can break all the accounts in the MSU-VERIDICOM database (with the fingerprint authentication algorithm proposed by Jain et al. [4]).

Through the above discussion, it can be seen that hill-climbing attack is a crucial problem to biometric template security. Especially it does not need the information of the reference templates, thus the reference template protection schemes would provide no effect on it. Along this line, some schemes [11] [16] have been proposed to confront the hill-climbing attack via changing the decision or data input process of the biometric systems.

2.3. *Template Protection: An Important Part of Biometric System Security*

While biometric systems are facing many threats, template security is an extremely important issue in biometric system security. A biometric template directly relates to human identity. If someone has his biometric template compromised, his identity may be utilized for illegal actions. If someone has managed to compromise one template stored in database, all the other templates are in danger since they may be compromised with the same way, which could lead to severe damage. Especially, the biometric data is unique to each individual. If someone's biometric data is compromised, he/she cannot offer another one to replace it unless another kind of biometric is used.

With the template attack, a cross-match attack can be performed. The fake biometric reconstructed from one biometric system may be used to access another system. Since there are more and more products of biometric systems deployed in commercial and government applications, this attack highly aggravates the threats of template attack.

Also, biometric data may contain some private information. A privacy invasion may occur if the biometric data is revealed. Take fingerprint for example, if someone (say Alice) gets a fingerprint sample of the user Bob from a biometric system database, Alice may be able to track Bob via testing whether Bob's fingerprint has been left in certain places. Biometric data may contain even more personal information, like the age and health condition of the owner. Hence, biometric template protection not only avoids identity theft, but also prevents from the leakage of private information of the users.

3. Security Enhancing Schemes for Biometric Recognition Systems

This section will discuss existing schemes for template protection. The existing schemes are divided into three approaches, namely biometric cryptosystem approach, transformation-based approach and hybrid approach. Before going into the details of each approach, we would like to discuss the requirements for a template protection algorithm.

3.1. *Requirements for Template Protection Schemes*

A template protection scheme should satisfy the following requirements:

1) **Security:** the original feature templates should never be exposed in the authentication process. It requires the protecting scheme to transform the original feature templates into another form such that the original feature templates cannot be retrieved from the transformed ones.

2) **Discriminability:** since the original feature templates are transformed into another form, the discriminability may be affected. An ideal protecting scheme should at least preserve or even enhance the discriminability of the original feature templates after transform.

3) **Cancelability:** Since there is still some likelihood that the secured templates are compromised, the secured templates should be able to be cancelled. That is, new secured templates can be generated to replace the compromised ones without changing the whole algorithm (revocability). Since a user may be enrolled in different applications, the protection scheme should be able to generate different secured templates for different applications, so as to prevent from the double-cross attack.

As we know, different acquisitions of the biometric data from the same individual will not be exactly the same due to the intra-class variations. It makes the traditional cryptographic techniques (e.g. RSA or AES encryption) not feasible for biometric template protection since these schemes are sensitive to variations. Even small changes in the biometric templates will cause large differences in the encrypted data and the matching in the encrypted domain will cause large errors. If the matching is performed with decrypted templates, the security problem still remains since the original templates are exposed in authentication. A successful template protection scheme should never expose the original feature templates in the system. In view of this, one who wants to protect biometric templates should find ways to deal with the intra-class variations, which is one of the most significant challenges in this subject. In the following sections, we will see it is not an easy task to find a template protection scheme that solves the intra-class variation problem and provides a high security level.

3.2. *Biometric Cryptosystem Approach*

The biometric cryptosystem approach combines cryptography with biometric systems. This approach encrypts the original biometric templates for protection. As we discussed, the intra-class variations remain a significant problem in the encryption process. Error-correcting codes are applied to overcome such a problem. The whole architecture of the biometric cryptosystem approach is illustrated in Figure 2.

During enrollment, after the reference template v is extracted via the sensor and feature extractor, it is encoded via an error-correcting process, and a helper data K is released with a cryptographic key S. The helper data K contains information about how to deal with the intra-class variations. The cryptographic key S is hashed ($H(S)$) for further protection, and then stored in database with K.

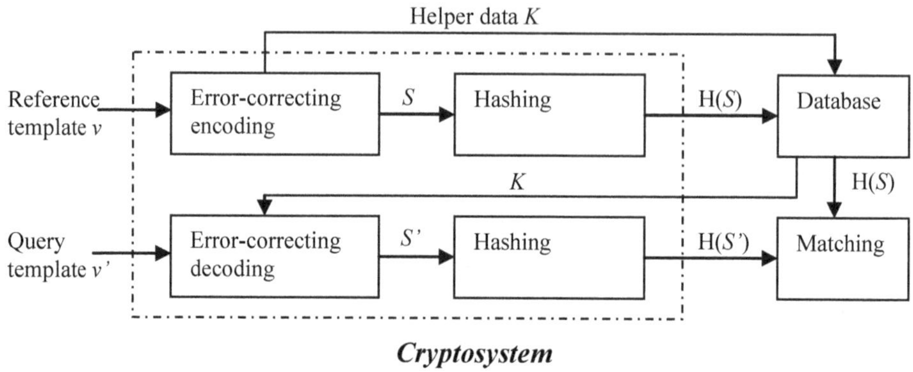

Fig. 2. Structure of the biometric cryptosystem.

In authentication, after the query template *v'* is presented, the corresponding helper data *K* is released from the database and an error-correcting decoding is proceeded with *v'* and *K* to generate a cryptographic key *S'*. The decrypted key *S'* is supposed to be as same as *S* if the query template *v'* is close enough to the reference template *v*. It is then hashed and matched with *H*(*S*).

Depending on the methods in generating the error-correcting key and the cryptographic key, the methods in biometric-cryptosystem approach can be further categorized into two classes [15]: key-binding and key-generation. In a key-binding scheme, the cryptographic key is unrelated to the original feature template. The error-correcting encoding process combines the key with the original feature template to generate an encoded template and a helper data. In a key-generation scheme, the cryptographic key is directly generated from the original feature template.

3.2.1. *Key-binding Biometric Cryptosystem*

The typical key-binding structure is illustrated in Figure 3(a). In enrollment, a cryptographic key *S* is linked with the input reference biometric template and a binding function is used to generate a new identity *K*. The key is then hashed for protection and stored to database with the identity. In authentication, the query template is used to retrieve the cryptographic key from the identity. Then the retrieved key is hashed and compared with the stored one to make a decision. The identity sometimes is called the helper data. It should not reveal the contents of either the reference template or the cryptographic key.

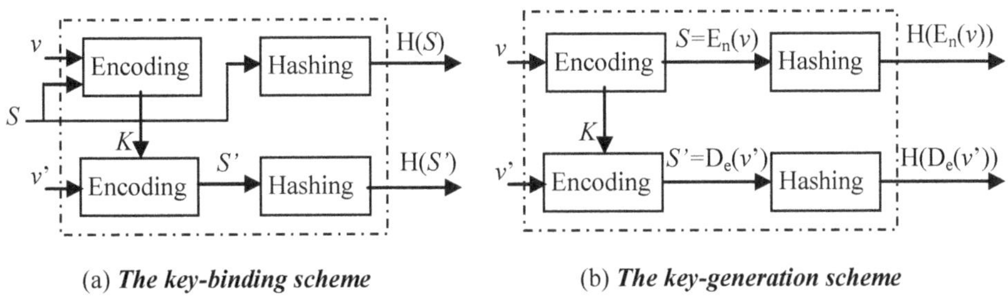

Fig. 3. The key-binding scheme (a) and key-generation scheme (b) in biometric cryptosystem approach.

The first example of the key-binding biometric cryptosystem approach is the off-line biometric identification scheme proposed by Davida et al. [23]. In the off-line scheme, an error correcting code is employed to generate a check data *K* and hash the original template *v* to *H*(*v*). In authentication, an error correcting decryption process using *K* is employed to correct the query biometric template from *v'* to *v''*. *v''* is then hashed and compared with *H*(*v*) and a decision is obtained. Since the error tolerance is not very strong, a multiple scan was required to minimize the intra-class variations of the original biometric templates. In addition, the check vector may leak some messages because it is linked to the template. Although there are some limitations in their algorithm, it provides a good foundation in protecting biometric template.

Following the off-line biometric identification scheme, Juels and Watternberg [24] proposed a key-binding algorithm namely the fuzzy commitment scheme as illustrated in Figure 4. It treats the biometric template itself as a corrupted codeword *C* of a binary error-correcting code (e.g. BCH code). The security of this method is linked to the number of codewords in the error-correcting code. The scheme encrypts the original template *v* (binary) to a pair (*H*(*C*), *v*-*C*) where the binary codeword *C* is randomly chosen from the error-correcting code. In authentication, with query template *v'*, the method computes *C'* = *v'* - (*v*-*C*) and corrects it to the closest codeword *C''*. If v is close to *v'*, noting that *v'*- (*v*-*C*) = *C* + (*v'*-*v*), C'' will be corrected to *C* (due to the property of error-correcting codes). That is, *C''* =*C*. *H*(*C''*) and *H*(*C*) are compared to obtain the decision. This algorithm can tolerate a relatively large error rate with strongly enhanced security level. However, the paper did not present any experimental results. Also, both the off-line scheme and the fuzzy commitment scheme require a binary input.

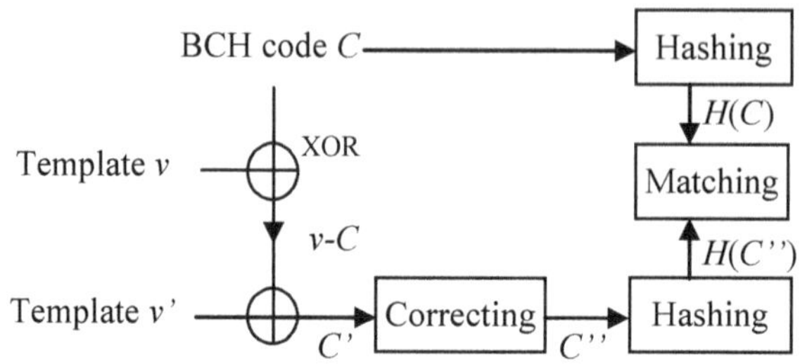

Fig. 4. The flowchart of the fuzzy commitment scheme.

Juels and Sudan [25] proposed a fuzzy vault scheme. The fuzzy vault scheme applies a polynomial f (the cryptographic key) to encrypt a set of integer values (with limited range) $S_A=\{x_1, x_2 \ldots x_n\}$ into a set of points $\{(x_1, f(x_1)), (x_2, f(x_2)) \ldots (x_n, f(x_n))\}$ which is called a vault. Chaff points are inserted to the vault in order to confuse attackers. If someone wants to retrieve the secret f, he should present a set of integer values which is close to the reference one such that the useful points can be chosen from the vault. Reed-Solomon decoding algorithm is applied to retrieve the secret f from the chosen points. When applying to biometrics, the biometric template should be represented by an integer set with limited range. This scheme has been further modified and successfully applied to fingerprint biometric [26] [28]. However, there are some limitations in this method. First, the vault may leak some messages because all important information is stored in the vault without transform/encryption. Second, the inserting of chaff points will significantly increase the data size. Finally, the Reed-Solomon coding scheme assumes that the input data set S_A only contains integers within a limited range. Experimental results [26] [28] show that applying fuzzy vault scheme on fingerprint leads to a relatively large false rejection rate (20%-30%). Improvement versions of the fuzzy vault scheme have been proposed such as the fingerprint securing scheme by Uludag and Jain [7]. Helper data was employed for the alignment of the fingerprint biometric data such that the accuracy of authentication can be improved. Also, Nagar et al. [8] proposed a fusion scheme combining the fuzzy vault scheme and the fuzzy commitment scheme. The fuzzy commitment scheme was used to encrypt the ordinate values in the fuzzy vault with a minutiae descriptor data. Even if the attacker can find the validate points in the fuzzy vault, he/she can not extract the secret key.

Tuyls et al. [32] [33] proposed an algorithm to protect real number biometric templates. This scheme is an extension of the fuzzy commitment scheme. It transforms the real number original feature vector v into Y with a δ-contracting quantization function G. For any binary codeword C and the original reference feature template v, a helper-data K can be constructed satisfying $G(v\text{-}K)=C$. In authentication, for any query template v' close to v ($||v'-v||\leq\delta$), $G(v'\text{-}K)=C$. Note that here the helper-data K has the similar effect as the term $v\text{-}C$ in the fuzzy commitment scheme. The function $Y=G(X)$ is constructed as follows:

$$y_i = \begin{cases} 1 & \textbf{if} \quad 2nq \leq x_i < (2n+1)q \\ 0 & \textbf{if} \quad (2n-1)q \leq x_i < 2nq \end{cases}$$

where y_i and x_i are corresponding elements in Y and X respectively, n is any integer and q is the quantization interval. The advantage of Tulys's scheme is that it can protect real number value feature templates. An analysis of data leakage is also discussed.

Soutar et al. [22] proposed a bioscrypt scheme which transforms the original template v to a new representation $f(v)$ with a filter function and embeds a secret to $f(v)$ for protection. Hao et al. [29] evaluated different types of errors which are introduced in iris recognition and then proposed a two-layer error correction coding method. They combined the Reed-Solomon codes and the Hadamard codes for handling burst error and random error respectively. Draper et al. [30] applied the distributed source coding to protect fingerprint and Sutcu et al. [31] analyzed and implemented the sketch, which is an error tolerant cryptographic technique, for face biometric.

3.2.2. *Key-generation Biometric Cryptosystem*

The key-generation scheme is illustrated in Figure 3(b). In a key-generation scheme, the cryptographic key is directly generated from the biometric. The advantage of this approach is that, it does not need an extra cryptographic key, thus no extra concern of the security of that key. A stable cryptographic key is rather important in the key-generation approach since it should be hashed for protection. A small variation in the cryptographic key may cause a large error after hashing. So, how to generate a stable cryptographic key from a biometric template with intra-class variation is the key point.

The early key-generation schemes apply quantization process to remove the intra-class variation during the key extraction process. Vielhauer et al. [9] proposed a signature biohash scheme. It extracts 24 feature parameters from the original signature data. For each parameter, it trains an upper bound and a lower bound to determine a quantization process. After quantization, the parameters are expected to be the same within the same class. Dodis et al. [27] proposed the idea of secure sketch and fuzzy extractor. In a secure sketch, a helper data will be used to reconstruct the reference template from the query template. In a fuzzy extractor, a new data will be extracted from the original template, while templates with small variations can still generate an identical new data. However, only the theory of the scheme was presented without any experiments/ applications. Some researchers [34] [35] followed this approach and proposed their own secure sketch and fuzzy extractor schemes.

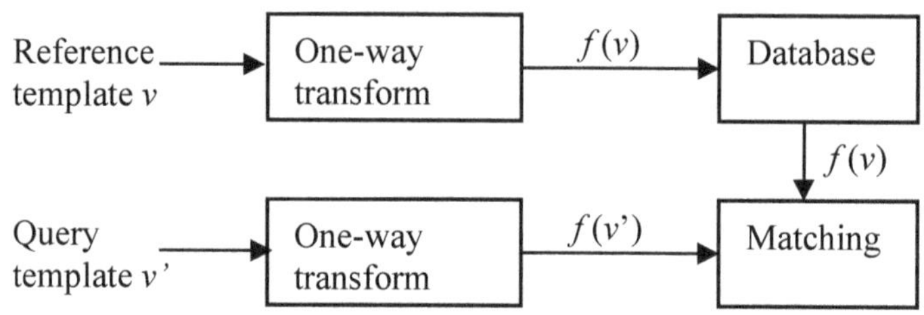

Fig. 5. The structure of the transform-based approach.

3.3. *Transform-based Approach*

On the other hand, the transform-based approach does not encrypt but transform the original feature templates into a new domain. A "one-way" function is applied such that the original feature template cannot be retrieved from the transformed one. Since there are parameters in the transform, the scheme easily acquires cancelability. If a transformed template is compromised, a new one can be generated by changing the transform parameters. Figure 5 illustrates the basic structure of the transform-based approach.

3.3.1. *Non-invertible Transform*

In a non-invertible transform, a one-way function f is applied to transform the original template v into $f(v)$ and it is computationally hard to retrieve v from $f(v)$. Thus the original feature template is protected. This approach was first proposed by Ratha et al. [11], which introduced the concept of cancelable transform. Cancelable transform is a kind of non-invertible transform that applies one-way function with random parameters. If some transformed templates are compromised, new transformed templates will be issued with changed parameters to replace the compromised ones. Generally speaking, when applying non-invertible transform approach to protect biometric templates, researchers will always try to design a cancelable transform since cancelability is an important criterion for system security. Ratha et al. [11] further classified the non-invertible transform approach into two categories: the signal-domain transform and feature-domain transform. In a signal-domain transform, the one-way transform is directly applied to the raw biometric data without feature extraction. This is always a distortion or permutation process. In a feature-domain transform, some one-way function will be applied to the extracted features from the raw data.

Two distortion transforms in signal and feature domains were discussed in the paper of Ratha et al. [11] to construct the cancelable template. The transform in the signal domain includes the grid morphing and block permutation. In feature domain, high-order polynomials are used. With these transforms, the authors claimed that the original template cannot be reconstructed from the cancelable template. Ratha et al. [11] mainly provided a basic idea of the cancelable biometrics but did not perform analysis on the intra-class variation problem. Along the line, Ratha et al. [47], [48] proposed cancelable biometrics on fingerprint. Three different kinds of distortion transforms (domain-remained) are proposed, namely Cartesian transformation, radial transformation, and functional transformation. Their algorithms provide high security but the experimental results show that the recognition performance using the cancelable template will be degraded.

Tulyakov et al. [45] followed the cancelable biometrics idea and proposed a symmetric hash function (polynomials) algorithm on fingerprint as a distortion transform. To handle the intra-class variation, sets of local features were extracted and a global matching of fingerprint was achieved by comparing with all local feature sets. Ang et al. [46] proposed a key-dependent algorithm to transform the original fingerprint template into new domain. They divided the original 2-D feature space into two halves based on the key and reflect one half to another half. Since it is hard to know which minutiae are reflected, the transform is hard to be inverted. However, these two algorithms do not perform well in handling the intra-class variation problem, resulting in relatively large performance degeneration.

Sutcu et al. [43] proposed a functional distortion of the original template. With the extracted feature vector, they estimated the distribution of each element in the training data, and approximated by the Gaussian functions. Fake Gaussian functions are combined together to confuse attackers so as to increase the non-invertibility. However, the detailed performance analysis cannot be found.

3.3.2. *Salting*

Salting means that user-specific passwords are applied to increase the between-class separation of the original templates thereby enhancing the discriminability.

A one way function is applied to transform the original feature templates into a new domain. It takes the original biometric template and the user-specific key as inputs. Since different users obtain different keys, the value retrieved by the function will be differentiated by individuals, which in turn results in an increase in between-class separation. Moreover the transformed templates can be cancelled and re-issued via changing the key.

The biohashing algorithm [38] [39] [42] can be classified in this approach. Different versions/variations of Biohashing algorithms have been proposed in the last few years. Teoh et al. proposed a two-factor authentication algorithm [39] for fingerprint and a random multispace quantization (RMQ) algorithm [42] for face biometric. In both methods, the original feature vectors are transformed into another domain using a random projection. The transformed template is further converted into binary strings by a thresholding technique. In RMQ algorithm, with the help of user data, user specific random projection matrix is adopted so as to reduce the false acceptance error. The authors claimed that a zero false acceptance rate can be achieved by the multi-subspaces projection using user data. However, there may have a security issue if the user data is stolen. Without the user data, the performance of this scheme will be degraded.

3.4. *Hybrid Approach*

Since a single approach is hard to fulfill all three requirements stated in Section 3.1, it is necessary and natural to combine the advantages of each approach and the hybrid approach is proposed. Figure 6 shows the block diagram of one popular combination of transform-based approach and cryptosystem approach. The basic idea is that transform-based method will provide the cancelability while cryptosystem method will provide security.

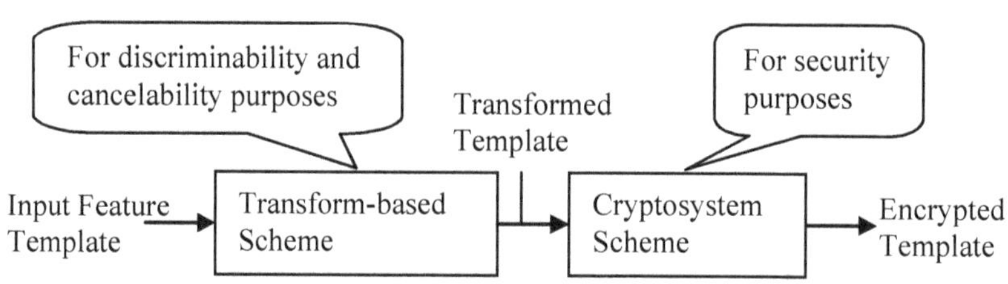

Fig. 6. The structure of the two-step framework.

Monrose et al. [36] [37] first proposed the idea of the two step approach on voice and keystroke data. The original biometric templates (keystroke or voice) are irregular features which are hard to be encrypted with error-correcting codes. So they applied a binary transform to the biometric templates. And the transformed binary strings are easily encrypted for protection. In another scheme proposed by Nandakumar et al. [1], the transform is based on a user-specific password to enhance the security of the fuzzy vault scheme. However, no matter what purpose it is designed for, the transform-based step always benefits from a common advantage – enhancing the cancelability. If there is a random parameter in the transform, it can easily generate a new transformed template via changing the random parameters, thus the transformed templates are cancelable. Transform schemes without random parameters does not have cancelability. But in most cases, it is not difficult to add random parameters to a designed transform.

In [36] [37], a cryptographic key generation scheme from biometrics was proposed. The error-correcting codes are applied after a binary transformation on the feature vectors. In the first stage, the biometric data is transformed into a binary string "feature descriptor" via thresholding. In the second stage, a cryptographic key is generated from the feature descriptor with a cryptographic algorithm. This scheme provides an interesting approach to protect templates which have large intra-class variations and transforms the original feature data into binary strings such that error-correcting schemes can be applied. However, the security level is insufficient (at most 60 bits) and the false rejection rate is relatively high (almost 20%). Another two-stage scheme with binarization was proposed by Goh and Ngo [40]. The biohashing algorithm is used in the first step. The original template is transformed using a random projection and then thresholded to a binary representation (string). An error correcting encryption namely cryptographic key interpolation was employed to encrypt the binary string. This scheme has been further modified and reported in [41] [44]. However, the discriminability of the binarized templates was not discussed. Feng and Yuen [2] [5] proposed a binary template generation scheme for face biometrics. After binarization, the binary templates are protected by cryptosystem approaches such as the fuzzy commitment scheme. A number of distinguishing points are chosen for binarization. The distances between the original face templates and the distinguishing points are computed. The set of distances are threshold to form a binary string, which is used as the binary template.

In all of the above-mentioned two-step framework schemes, a binarization process is implemented in the first step. This process is used in order to transform the real-valued original biometric feature templates to a binary form so that it can be easily encrypted in the second step. There are also some hybrid algorithms that design the first step with different purposes. Nandakumar et al. [1] proposed a two-step framework scheme which incorporates the fuzzy vault scheme with

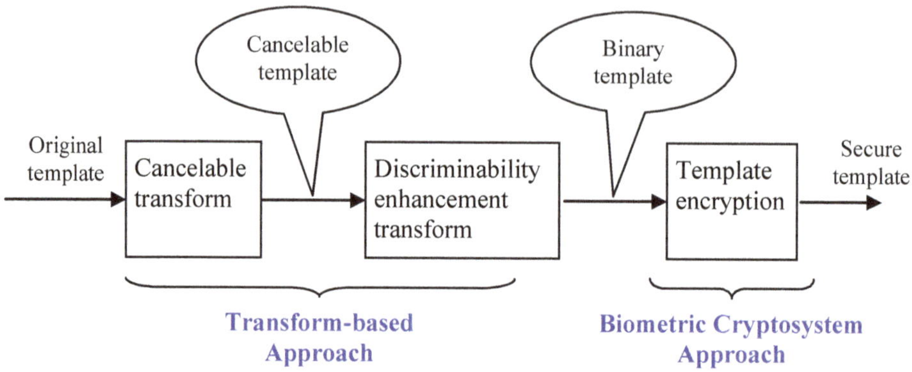

Fig. 7. The three-step hybrid framework (Feng et al. [6]).

password harden algorithm. First, an addition operation is implemented with the user-specific password and the original fingerprint template. After transform, the fuzzy vault scheme is employed. The fuzzy vaults are encrypted with the password for further security. The password not only provides cancelability to the templates, but also provides additional variability between templates in different classes. Furthermore, it provides a template distribution more close to uniform distribution to prevent from attacks at fuzzy vault. It will enhance the discriminability and security of the templates. Boult et al. [49] proposed a robust biotoken scheme. It first transforms (the first step) the original fingerprint templates into an integer part and a fraction part. The integer part is acquired by quantization while the fraction part is acquired with a folding technique such that data which are close to each other will be transformed into close fraction part. Only transformed templates that have the same integer part and close fraction part can have a successful match. After transform, the integer part is encrypted with AES and CRC algorithms. And chaff data are inserted to the encrypted data for further protection.

There is always a trade-off between security and discriminability in the two-step methods which employ a binarization process. The key problem is that most of the existing algorithms performed an *ad hoc* binarization process without much attention on the discriminability lost. To overcome this problem, Feng et al. [6] proposed a three-step framework as shown in Figure 7. Different from the traditional 2-step approach, a new discriminability enhancement transform is proposed in order to (i) enhance the discriminability of the input cancelable templates and (ii) transform the cancelable template into a binary string. In this way, the system not only will give a high security of the biometric template, but also preserve the system performance. Along this line, a hybrid 3-step algorithm

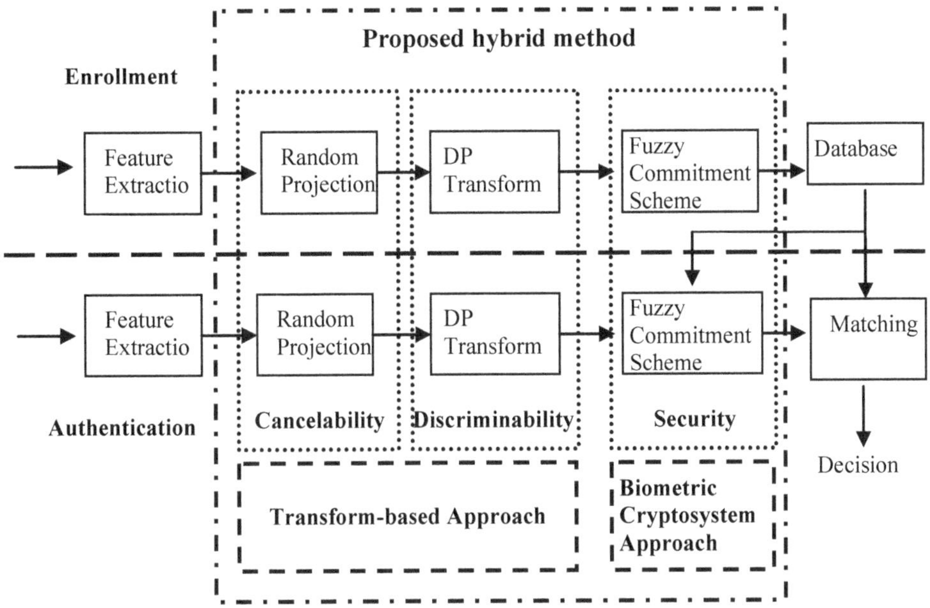

Fig. 8. The example for the three-step hybrid framework (Feng et al. [6]).

[6] was proposed as shown in Figure 8. The first step applies a random projection for cancelability while the second step makes use of distinguishing points binarization scheme (DP transform) [2] [5] to binarize the cancelable template. The binary template is finally protected using fuzzy committee scheme in the third step. In the DP transform, an optimization is performed to find the locations of the distinguishing points such that the binarized templates can mostly preserve the discriminative power of the original templates.

4. Discussions and Conclusion

In this section, we will discuss and compare the four different approaches for protecting biometric templates as reported in Section 3, based on three criteria, namely discriminability, security and cancelability. The summary is shown in Table 1. After that, we will draw a conclusion.

4.1. *Discriminability*

Among the four approaches discussed in Section 3, salting is the only one that not only can preserve the discriminability of the original feature templates, but even enhance it. A user-specific password ***p*** is used for the original feature template $\boldsymbol{v}$. This would further increase the between-class separation. Therefore the discriminability of the secured templates is high. The existing non-invertible transform, on the other hand, suffers from a trade-off between discriminability and non-invertibility. It is rather difficult to design algorithms possessing with high discriminability and non-invertibility simultaneously. Most non-invertible transform schemes will loss original template discriminability in order to ensure a high security level, and therefore, the discriminability of the secured templates is low.

The biometric cryptosystem approach, including the key-binding approach and the key-generation approach, can handle the intra-class variations to a certain extent using the error-correcting coding theory. So it is possible to preserve the discriminability of the original feature templates. And the discriminability of the secured templates is medium.

4.2. *Security*

The security of existing salting approach is relatively low because most of the information is not protected (e.g. the transformed binary strings in the biohashing algorithm is stored without encryption). Attacker can make use of the information to compromise the template. Another thing is, the introduction of user-specific key may cause extra security issue. So the salting approach suffers from a security risk. On the other hand, non-invertible transform approach provides a higher security while it suffers from template discriminability lost.

The security of the biometric cryptosystem approach depends on two factors: (i) the security of the cryptographic key and (ii) the data leakage from the helper data. A long cryptographic key would be harder for attackers to guess. But since the cryptographic key should be linked with the biometric template to construct a helper data, the length of the cryptographic key may be data-dependant. The helper data, which is constructed from the original template and the cryptographic key, does contain information of the original template. It may release some information to the attackers. Even so, the existing biometric cryptosystem schemes can provide a high security level.

4.3. *Cancelability*

Despite of the good performance on the first two criteria, biometric cryptosystem approach generally lack of cancelability because originally, the cryptosystem schemes were not designed for cancelability.

On the other hand, cancelability is one of the basic requirements for developing a transform-based approach. The transform can easily acquire cancelability by setting some random parameters in it. Therefore, both non-invertible transform and salting in transform-based approach provide good cancelability.

Table 1. Advantages and disadvantages of different template protection approaches.

	DISCRIMINABILITY	SECURITY	CANCELABILITY
Key-binding	Medium	High	Low
Key-generation	Medium	High	Low
Non-invertible Transform	Low	High	High
Salting	High	Low	High

4.4. *Conclusion*

In conclusion, this chapter has discussed the vulnerabilities of biometric systems and the security concern, in particular on the biometric template security. A brief survey of existing approaches to protect biometric templates, including the biometric cryptosystem approach and transform-based approach, and also the advantages and disadvantages of these approaches have been discussed. While many algorithms have been proposed in the last few years, the biometric system security problems have not been fully solved. For instant, hybrid approach seems to be a promising approach for template security. But still, more analysis and studies are required. Moreover, another less explored area is the security strength evaluation and measurement of an algorithm. Since the key problem is security, standard evaluation methodology and security strength measurement are required.

Acknowledgements

This project was supported by Hong Kong RGC General Research Fund 210908 and the Science Faculty Research Grant of the Hong Kong Baptist University.

References

[1] K Nandakumar, A Nagar, and A K Jain, "Hardening fingerprint fuzzy vault using password," Proceedings of 2nd International Conference on Biometrics, pp. 927-937, 2007.

[2] Y C Feng and P C Yuen, "Class-Distribution Preserving Transform for Face Biometric Data Security," Proceedings of IEEE International Conference on Acoustics, Speech, and Signal Processing (ICASSP), pp. 141-144, 2007.

[3] L O'Gorman, "Comparing Passwords, Tokens, and Biometrics for User Authentication," Proceedings of the IEEE, Vol. 91, No. 12, pp. 2019-2040, 2003.

[4] A K Jain, L Hong, S Pankanti, and R Bolle, "An identity authentication system using fingerprints," Proceedings of the IEEE, Vol. 85, No. 9, pp. 1365-1388, 1997.

[5] Y C Feng and P C Yuen, "Selection of Distinguish Points for Class Distribution Preserving Transform for Biometric Template Protection," Proceedings of IEEE International Conference on Biometrics (ICB), pp. 636-645, 2007.

[6] Y C Feng, P C Yuen and A K Jain, "A Hybrid Approach for Face Template Protection", Proceedings of International Society for Optical Engineering (SPIE), Vol. 6944, 2008.

[7] U Uludag, and A K Jain, "Securing fingerprint template: fuzzy vault with helper data", in Proceedings of Computer Vision and Pattern Recognition Workshop, pp. 163-163, 2006.

[8] A Nagar, K Nandakumar, and A K Jain, "Securing Fingerprint Template: Fuzzy Vault with Minutiae Descriptors," in Proceedings of the International Conference on Pattern Recognition, 2008.

[9] C Vielhauer, R Steinmetz, and A Mayerhöfer, "Biometric Hash based on Statistical Features of Online Signatures", in Proceedings of the 16th International Conference on Pattern Recognition, Vol. 1, 2002.

[10] A K Jain and S Pankanti, "A Touch of Money", IEEE Spectrum, pp. 22-27, July 2006.

[11] N Ratha, J Connell and R Bolle, "Enhancing security and privacy in biometric-based authentication systems," IBM Systems Journal, Vol. 40. No. 3, pp. 614 - 634, 2001.

[12] S Prabhakar, S Pankanti, and A K Jain, "Biometric Recognition: Security and Privacy Concerns," IEEE Security and Privacy Magazine, Vol. 1, No. 2, pp. 33-42, March-April 2003.

[13] U Uludag, S Pankanti, S Prabhakar, and A K Jain, "Biometric cryptosystems: issues and challenges," Proceedings of the IEEE, vol. 92, no. 6, pp. 948-960, 2004.

[14] A K Jain, A Ross and S Pankanti, "Biometrics: A Tool for Information Security", IEEE Transactions on Information Forensics and Security, Vol. 1, No. 2, pp. 125-143, 2006.

[15] A K Jain, K Nandakumar, and A Nagar, "Biometric Template Security," EURASIP Journal on Advances in Signal Processing, 2008.

[16] C Soutar, "Biometric System Security", http://www.bioscrypt.com/assets/security soutar.pdf.

[17] C Hill, "Risk of Masquerade Arising from the Storage of Biometrics," B.S. Thesis, Australian National University, 2001. http://chris.fornax.net/biometrics.html.

[18] A Alder, "Sample images can be independently restored from face recognition templates," Electrical and Computer Engineering, vol. 2, pp. 1163-1166, 2003.

[19] A Adler, "Images can be regenerated from quantized biometric match score data", Proceedings of Canadian conference of Electrical and Computer Engineering, pp. 469-472, 2004

[20] A Alder, "Vulnerabilities in Biometric Encryption Systems," Proceedings of the IEEE International Conference on Audio- and Video-Based Biometric Person Authentication, vol. 3546, pp. 1100-1109, 2005.

[21] U Uludag and A K Jain, "Attacks on biometric systems: a case study in fingerprints," Proc. SPIE, vol. 5306, pp. 622-633, 2004.

[22] C Soutar, D Roberge, A Stoinav, G Gilroy, V Kumar, "Biometric Encryption Using Image Processing," Proceedings of SPIE, vol. 3314, pp. 174-188, 1998.

[23] G Davida, Y Frankel and B Matt, "On enabling secure applications through off-line biometric identification," IEEE Symposium on Privacy and Security, pp. 148-157, 1998.

[24] A Juels, M Wattenberg, "A fuzzy commitment scheme", Proceedings of the Sixth ACM Conf. on Comp. and Comm. Security, pp. 28-36, 1999.

[25] A Juels and M Sudan. "A Fuzzy Vault Scheme", IEEE International Symposium on Information Theory, 2002.

[26] T C Clancy, N Kiyavash, and D J Lin, "Secure smartcard-based fingerprint authentication", Proceedings of ACMSIGMM Multimedia, Biometrics Methods and Applications Workshop, pp. 45-52, 2003.
[27] Y Dodis, L Reyzin, and A Smith, "Fuzzy extractors: How to generate strong keys from biometrics and other noisy data," Proceedings of Advances in Cryptology—Eurocrypt, 2004.
[28] U Uludag, S Pankanti, and A K Jain, "Fuzzy Vault for Fingerprints," Proceedings of the Sixth Int'l Conf. Audio and Video-Based Biometric Person Authentication, pp. 310-319, 2005.
[29] F Hao, R Anderson and J Daugman, "Combining cryptography with biometric effectively," Technical Report, University of Cambridge, UCAM-CL-TR-640, ISSN 1476-2986, 2005.
[30] S Draper, A Khisti and E Martinian, A Vetro and J Yedidia, "Using distributed source coding to secure fingerprint biometric," Proceedings of IEEE International Conference on Acoustics, Speech and Signal Processing, 2007.
[31] Y Sutcu, Q Li and N Memon, "Protecting biometric template with sketch: theory and practice," IEEE Transactions on Information Forensics and Security, Vol. 2, No. 3 Part 2, pp. 503-512 , 2007.
[32] P Tuyls and J Goseling, "Capacity and Examples of Template-Protecting Biometric Authentication Systems," ECCV Workshop BioAW, pp. 158-170, 2004.
[33] J Linnartz and P Tuyls, "New Shielding Functions to Enhance Privacy and Prevent Misuse of Biometric Templates," Proc. Fourth Int'l Conf. Audio and Video-Based Biometric Person Authentication, pp. 393-402, 2003.
[34] Y Sutcu, Q Li, and N Memon, "Protecting biometric templates with sketch: theory and practice," IEEE Transactions on Information Forensics and Security, vol. 2, no. 3, pp. 503C512, 2007.
[35] A Arakala, J Jeffers, and K J Horadam, "Fuzzy extractors for minutiae-based fingerprint authentication," in Proceedings of the 2nd International Conference on Biometrics, pp. 760C769, August 2007.
[36] F Monrose, M K Reiter and S Wetzel, "Password Hardening Based on Key Stroke Dynamics," Proceedings of ACM Conference on Computer and Communication Security, pp. 73-82, 1999.
[37] F Monrose, M Reiter, Q Li and S Wetzel, "Cryptographic Key Generation from Voice," Proceedings of IEEE Symp. Security and Privacy, pp. 202-213, 2001.
[38] A Teoh and C Ngo, "Cancelable Biometrics Featuring with Tokenised Random Number," Pattern Recognition Letters, vol. 26, no. 10, pp. 1454-1460, 2005.
[39] A Teoh, D Ngo and A Goh, "Biohashing: Two Factor Authentication Featuring Fingerprint Data and Tokenised Random Number," Pattern Recognition, vol. 37, no. 11, pp. 2245-2255, 2004.
[40] A Goh and D C L Ngo, "Computation of cryptographic keys from face biometrics", Proceedings of the 7th IFIP TC6/TC11 Conf. Commun. Multimedia Security, vol. 22, pp. 1-13, 2003.
[41] D Ngo, A Teoh and A Goh, "Biometric Hash: High-Confidence Face Recognition", IEEE transactions on circuits and systems for video technology, vol. 16, no. 6, 2006.
[42] A Teoh, A Goh and D Ngo, "Random Multispace Quantization as an Analytic Mechanism for BioHashing of Biometric and Random Identity Inputs," IEEE Transactions on Pattern Analysis and Machine Intelligence, vol. 28, no. 12, pp. 1892-1901, 2006.
[43] Y Sutcu, H Sencar, N Nemon, "A Secure Biometric Authentication Scheme Based on Roubst Hashing, Proceedings of the Seventh Workshop Multimedia and Security, pp. 111-116, 2005.
[44] O T Song, A B J Teoh, and D C L Ngo, "Application-Specific Key Release Scheme from Biometrics," International Journal of Network Security, Vol. 6, No. 2, pp. 127-133, 2008.

[45] S Tulyakov, V Chavan, and V Govindaraju, "Symmetric Hash Functions for Fingerprint Minutiae," Proceedings of the Int'l Workshop Pattern Recognition for Crime Prevention, Security, and Surveillance, pp. 30-38, 2005.
[46] R Ang, R Safavi-Naini, L McAven, "Cancelable Key-Based Fingerprint Templates," ACISP 2005, pp. 242-252.
[47] N Ratha, J Connell, R Bolle, S Chikkerur, "Cancelable biometrics: A case study in Fingerprints", Proceedings of International Conference on Pattern Recognition, 2006.
[48] N Ratha, S Chikkerur, J Connell, R Bolle, "Generating Cancelable Fingerprint Templates", IEEE Transactions on Pattern Analysis and Machine Intelligence, Vol. 29, pp. 561-752, 2007.
[49] T E Boult, W J Scheirer, and R Woodworth, "Revocable Fingerprint Biotokens: Accuracy and Security Analysis," IEEE Conference on Computer Vision and Pattern Recognition, pp. 1-8, 2007.

CHAPTER 4.2

FUSION OF MANUAL AND NON-MANUAL INFORMATION IN AMERICAN SIGN LANGUAGE RECOGNITION

Sudeep Sarkar[1], Barbara Loeding[2], and Ayush S. Parashar[1]

[1] *Computer Science and Engineering, University of South Florida, Tampa, Florida 33647*
sarkar@cse.usf.edu

[2] *Special Education, University of South Florida, Lakeland, Florida 33803*
bloeding@poly.usf.edu

We present a bottom-up approach to *continuous* American sign language recognition without wearable aids, but with simple low-level processes operating on images and building realistic representations that are fed into intermediate level processes, to form sign *hypotheses*. At the intermediate level, we construct representations for both manual and non-manual aspects, such as hand movements, facial expressions and head nods. The manual aspects are represented using Relational Distributions that capture the statistical distribution of the relationships among the low-level primitives from the body parts. These relational distributions, which can be constructed without the need for part level tracking, are efficiently represented as points in the Space of Probability Functions (SoPF). Manual dynamics are thus represented as tracks in this space. The dynamics of facial expressions along with a sign are also represented as tracks, but in the expression subspace, constructed using principal component analysis (PCA). Head motions are represented as 2D image tracks. The integration of manual with non-manual information is sequential, with non-manual information refining the manual information based hypotheses set. We show that with just image-based manual information, the correct detection rate is around 88%. However, with the use of facial information, accuracy increases to 92%. Thus face contributes valuable information towards ASL recognition. 'Negation' in sentences is correctly detected in 90% of the cases using just 2D head motion information.

1. Introduction

While speech recognition has made rapid advances, sign language recognition is lagging behind. With the gradual shift to speech-based I/O devices, there is a great danger that persons who rely solely on sign languages for communication will be deprived access to state-of-the-art technology unless there are significant advances in automated recognition of sign languages.

Reviews of prior work in sign language recognition appear in[1] and.[2] From these reviews, we can see that work in sign language recognition initially focused on the recognition of static gestures, e.g.[3–5] and isolated signs, e.g.[6] Starner and Pentland[7] were the first to seriously consider *continuous* sign recognition*. Using HMM based representations, they achieved near perfect recognition with sentences of fixed structure, i.e. containing personal pronoun, verb, noun, adjective, personal pronoun in that order. Vogler and Metaxas[8–10] were instrumental in significantly pushing the state-of-the-art in automated ASL recognition using HMMs. In terms of the basic HMM formalism, they have explored many variations, such as context dependent HMMs, HMMs coupled with partially segmented sign streams, and parallel HMMs. The wide use of HMM is also seen in foreign sign language recognizers.[1] While HMM based methods perform very well with limited vocabulary, they are haunted by scalability issues and requirements for large training data.

Many work in continuous sign language recognition has avoided the very basic problem of segmentation and tracking of hands by using wearable devices, such as colored gloves, data gloves, or magnetic markers, to directly get the location features. For example Vogler and Metaxas[8–10] have used 3D magnetic tracking system, Starner and Pentland[7] have used colored gloves while Ma *et al.*[11,12] have used Cybergloves. However, since this is unnatural for signers, in our research we restrict ourselves to plain color images, without the use of any augmenting wearable devices.

Non-manual information, which refers to information from facial expressions, head motion, or torso movement, convey linguistic information in ASL.[10,13] Many work in sign language recognition has concentrated on just using hand motion, i.e. manual information. Although, some work in automated understanding of sign language facial expressions is under way.[14–17] Non-manual information can provide vital cues. For example, head motion can be used to detect whether the ASL sentence has any 'Negation'. For instance, the sentence *'I don't understand'* is manually signed exactly same as *'I understand'*, except that there is distinct 'head shake' indicating 'Negation' in the sentence *'I don't understand'*†. There has been some work on detecting 'head shakes' and 'nods',[15,18,19] but there is no result in continuous sign language recognition. In this paper, we use non-manual information to decrease insertion and deletion errors, and to find whether there is 'Negation' in the sentence using the motion trajectories of the head.

---

*Note that Sign Language is different from Signed English, the later is an artificial construct that employs signs but using English language grammatical structure.

†We use following ASL conventions in the paper. Text in italics indicate sentence in English. For example *'I can lipread'*. Text in capitalized letters indicate ASL gloss. For example 'LIPREAD CAN I'. Or, the ASL gloss for sign *'lipread'* is 'LIPREAD'. Negation in a sentence signed using non-manual markers is indicated by $\widehat{\text{NOT}}$ or 'Negation'. Multiword gloss for a single sign in ASL is indicated by a hyphen. For example 'DONT-KNOW' is a multiword gloss for a single sign in ASL.

We concentrate on the problem of recognition of continuous sign language, i.e. signs in sentences and not isolated signs or finger-spelled signs. We adopt a bottom-up approach with simple low-level processes operating on images to build realistic representations that are fed into intermediate level processes integrating manual and non-manual information. Drawing on the emerging wisdom in computer vision that simple methods are usually found to be the most robust ones when tested on large data sets, we use simple components. The low-processes are fairly simple ones involving detection of skin, motion pixels, and face detection by correlation with an eye template. This would work for signs again simple backgrounds, which is the most commonly considered scenario. However, for more complex backgrounds, alternative strategies, such as described in,[20–22] could be considered. The intermediate level consists of modeling the hand motion using relational distributions, which are efficiently represented as points in the Space of Probability Functions (SoPF). This captures the placement of the hands with respect to the body, but does not capture hand shape accurately. Many signs can be recognized based on just this global information. For signs that are very hand-shape dependent, alternative methods, such as described in,[23] can be used. The expression subspace, derived using PCA, is used to represent the dynamics of facial expression. This level also deals with integrating non-manual with manual information to reduce the deletion and insertion errors. The third or topmost level, which we do not explore in this paper, would consist of using context and grammatical information from ASL (American Sign Language) phonology, to constrain and to prune the hypotheses set generated by the intermediate level processes.

The primary contribution of this work is the demonstration that, even with fairly simple 2D representations, the use of non-manual information can improve ASL recognition. The integration of facial expression with manual information is confounded by the fact that expression events may not coincide exactly with the manual events. This work also constrains itself in relying on pure image-based inputs and does not require external wearable aids, such as gloves, to enhance the low-level primitives.

2. Data Set

One of the issues of ASL recognition research is the data set used in the study. We constrained the domain to sentences that would be used while communicating with deaf people at airports. An ASL interpreter signed for the collection and helped with creating the ground truth. Two digital video cameras were used for collecting data; one captured images of the upper body, and the other synchronously captured face images. Fig. 1 shows one sample view.

Some statistics of the data set are as follows. The dataset includes 5 instances of 25 distinct sentences. In total there are 125 sentence samples spanning 325 instances of ASL signs. There are 39 distinct ASL signs. Each sentence has 1 to 5 signs. On

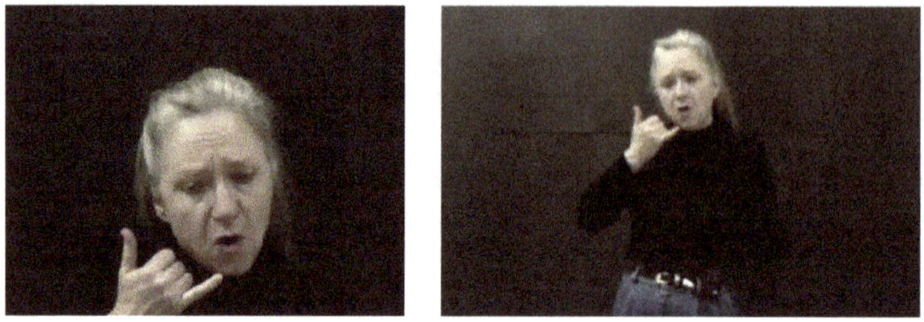

(a) (b)

Fig. 1. Sample images in the dataset. (a) shows an image of face taken for the sign 'PHONE', as captured by camera 'A' and (b) shows a synchronous image of the upper body as captured by camera 'B'.

an average 2.7 signs is present per sentence. The number of frames in a sentence varies from 51 to 163. The longest sentence is, 'AIRPLANE POSTPONE AGAIN, MAD I', comprised of 163 frames. The smallest sentence is, 'YES', made up of 51 frames. Average number of frames in a sentence is about 90. Sign length varied from 4 frames for the sign, 'CANNOT', to 71 frames for sign 'LUGGAGE-HEAVY'. On an average a sign has 18 frames spanning about 0.6 second.

There are significant variations among the 5 instances of some of the sentences. For example the sentence *'If the plane is delayed, I'll be mad'* was signed as 'AIRPLANE POSTPONE AGAIN, MAD I' as well as 'AIRPLANE AGAIN POSTPONE, MAD I'. Also in one of the instance of the sentence *'I packed my suitcase'*, the ASL sign 'I' was not present. This was also true for some other sentences. The reason, as given by the signer, was that signs like 'I' are implicit while conversing in ASL and hence can be excluded. For some sentences, 'Negation' is conveyed only through 'head shakes'. For example, for the sentences *'I understand'* and *'I don't understand'*, the ASL glosses is the same ('I UNDERSTAND'). The only difference is that in sentence *'I don't understand'*, there is presence of a 'head shake' i.e. a non-manual expressions to convey the presence of 'Negation' in the sentence.

3. Low Level Processing

In many previous work in *continuous* ASL, detection and tracking of hands have been simplified using colored gloves[7] or magnetic markers.[8] Even foreign sign language recognizers have used colored gloves or data gloves. There has been recent effort to extract information and to track directly from color images, without the use of special devices,[6,20–22] but with added computational complexity. Our intermediate level representation, as we shall see later, does not require the *tracking* of hands. We just need to segment the hands and the face in each frame. Since segmentation is not the focus of this work, we have used fairly simple ideas based on skin color to detect the hands and face. We cluster the skin-pixels by using the Expectation-Maximization (EM) algorithm based on a Gaussian model for the

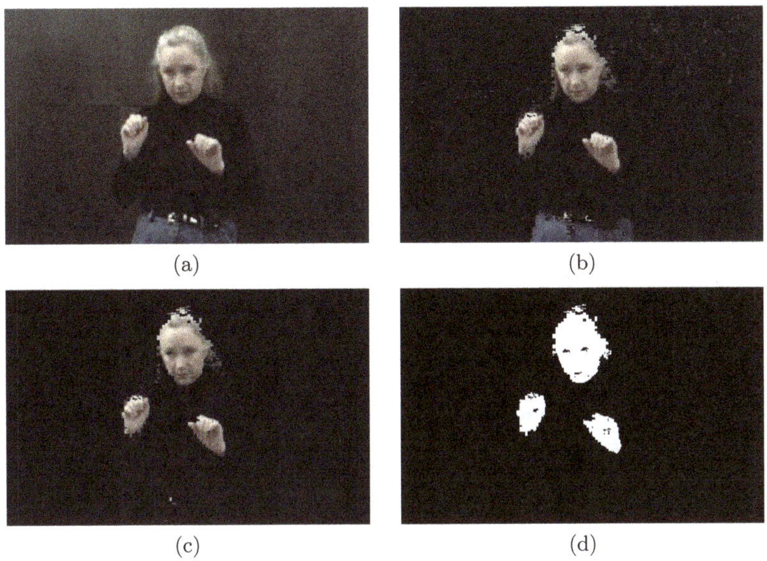

Fig. 2. Segmentation of skin pixels using EM: (a) shows original color frame, (b) shows the pixels obtained after first application of EM, (c) shows the skin pixels obtained after second application of EM, and (d) final blobs corresponding to hand and face.

clusters in the "Lab" color space. We use the 2-class version of the EM algorithm twice: first to separate the background and second time to separate the clothing of the signer from the skin pixels. Fig. 2(b) shows the segmentation after EM clustering of (a), where the skin and clothing pixels are separated from the background. Fig. 2 (c) shows the output after the second application of the EM to separate skin color from the clothing. Blobs of size greater than 200 pixels are kept. This helps to remove some pixels that are closer to skin color but do not form blobs big enough to be a part of hand or face. Fig. 2(d) shows an example of the final blobs. These blobs, along with the color values and the edge pixels detected within them, comprise the low-level primitives.

4. Intermediate Level Representations

We have separate representation for manual movement (or hold), facial expression, and head motion. We would like these representations, in particular, the manual motion, to be somewhat robust with respect to low-level errors. The manual motion (including no movement) representation *does not* require the need for tracking hands or fingers and emphasizes the 2D spatial relationships of hands and face. The face expression representation is a 2D view-based one. And, head motion model just takes into account projected 2D motion. Even without the use of 3D information, we demonstrate robust ASL recognition possibilities.

4.1. *Manual Movement*

Grounded on the observation that the *organization* or *structure* or *relationships* among low-level primitives are more important than the primitives themselves, we focus on the statistical distribution of the relational attributes observed in the image, which we refer to as *relational distributions.* Such statistical representation also alleviates the need for primitive level correspondence or tracking or registration across frames. Such representations have been successfully used for modeling periodic motion in the context of identification of a person from gait.[24] Here, we use it to model aperiodic motion in ASL signs. Primitive level statistical distributions, such as orientation histograms, have been used for gesture recognition.[25] However, the only uses of relational histograms that we are aware of are by Huet and Hancock,[26] who used it to model line distributions in the context of image database indexing. The novelty of *relational distributions* lies in that it offers a strategy for incorporating dynamic aspects.

We refer the reader to[24] for details of the representation. Here we just sketch the essentials. Let $F = \{f_1, \cdots, f_N\}$ represent the set of N primitives in an image. For us these are the Canny edge pixels inside the low-level skin-blobs described earlier. Let F_k represent a random $k-$tuple of primitives, and the relationship among these k-tuple primitives be denoted by R_k. Let the relationships R_k be characterized by a set of M attributes $\mathbf{A_k} = \{A_{k1}, \cdots, A_{kM}\}$. For ASL, we use the distance of the two edge pixels in the vertical and horizontal directions *(dx,dy)* as the attributes. We normalize the distance between the pixels by a distance D, which is inversely related to the distance from the camera. The shape of the pattern can be represented by joint probability functions: $P(\mathbf{A_k} = \mathbf{a_k})$, also denoted by $P(a_{k1}, \cdots, a_{kM})$ or $P(\mathbf{a_k})$, where a_{ki} is the (discretized, in practice) value taken by the relational attribute A_{ki}. We term these probabilities as the *Relational Distributions.*

One interpretation of these distributions is: given an image, if you randomly pick k-tuples of primitives, what is the probability that it will exhibit the relational attributes $\mathbf{a_k}$? What is $P(\mathbf{A_k} = \mathbf{a_k})$? Given that these relational distributions exhibit complicated shapes that do not readily afford modeling using a combination of simple shaped distributions, we adopt non-parametric histogram based representation. However, to reduce the size that is associated with a histogram based representation, we use the Space of Probability Functions (SoPF).

As the hands of the signer move, the relational distributions will change. Motion of hands will introduce non-stationarity in the relational distributions. Figure 3 shows some more examples of the 2-ary relational distributions for the sign 'CAN'. Notice the change in the distributions as the hands come down. The change in the vertical direction in relational distributions can be seen clearly as the hands are coming down, while there is comparatively less change in the relational distributions in other direction.

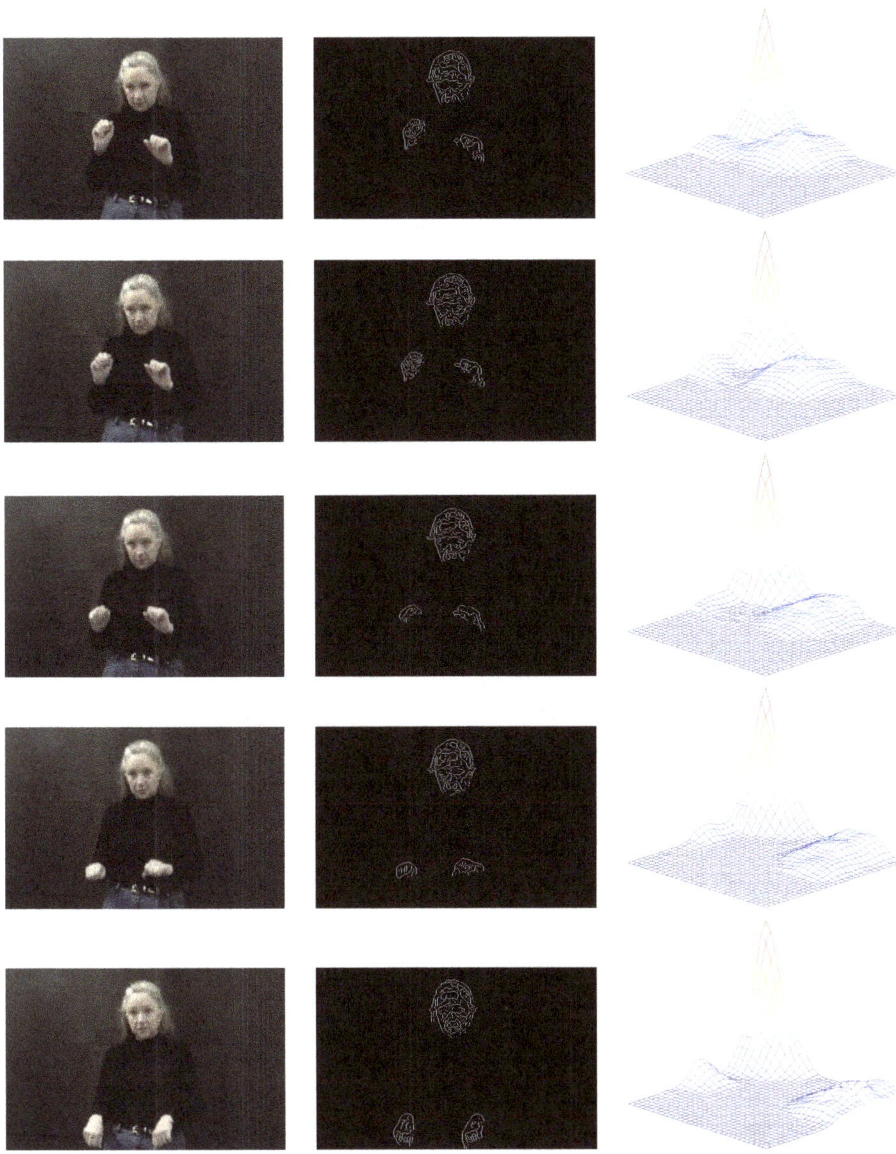

Fig. 3. Variations in relational distributions with motion. The left column shows the image frames in sign 'CAN'. The middle column shows in the edge pixels in the skin-blobs. The right column shows the relational distributions.

Let $P(\mathbf{a_k}, t)$ represent the relational distribution at time t. Let

$$\sqrt{P(\mathbf{a_k}, t)} = \sum_{i=1}^{n} c_i(t)\Phi_i(\mathbf{a_k}) + \mu(\mathbf{a_k}) + \eta(\mathbf{a_k}) \tag{1}$$

describe the *square root* of each relational distribution as a linear combination of orthogonal basis functions, where $\Phi_i(\mathbf{a_k})$'s are orthonormal functions, the function $\mu(\mathbf{a_k})$ is a mean function defined over the attribute space, and $\eta(\mathbf{a_k})$ is a function capturing small random noise variations with zero mean and small variance. We refer to this space as the Space of Probability Functions (SoPF).

We use the square root function so that we arrive at a space where the distances are not arbitrary ones but are related to the Bhattacharya distance between the relational distributions, which is an appropriate distance measure for probability distributions. More details about the derivation of this property can be found in.[24]

Given a set of relational distributions, $\{P(\mathbf{a_k}, t_i)|i = 1, \cdots, T\}$, the Space of Probability Functions SoPF can be arrived at by principal component analysis (PCA). In practice, we can consider the subspace spanned by a few ($N << n$) dominant vectors associated with the large eigenvalues. Thus, a relational distribution can be represented using these N coordinates ($c_i(t)$s), which is more compact representation than a normalized histogram based representation. The ASL sentences form traces in this Space of Probability Functions (SoPF).

The eigenvectors of the SoPF associated with the largest eigenvalues are shown in Figure 4. The space was trained with 39 distinct signs. The size of each relational distribution is 30×30. The vertical axes of the images plot the distance attribute dy, and the distance attribute dx is along the horizontal axes. Brightness is proportional to the component magnitude. The first eigenvector shows 3 modes in it. The bright spot in the second eigenvector emphasizes the differences in the attribute dx between the two features. The third eigenvector is radially symmetric, emphasizing the differences in both the attributes. Most of the energy of the variation is captured by the 15 largest eigenvalues.

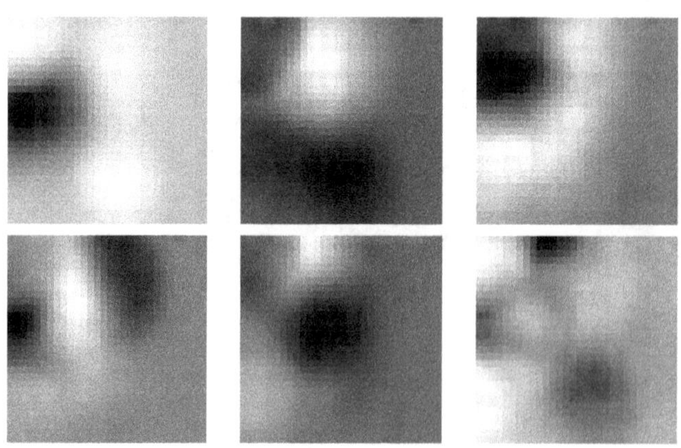

Fig. 4. Dominant dimensions of the learned SoPF, modeling the manual motion.

4.2. *Non-manual: Facial Expression*

The first step is the localization of the face in each frame. There are various sophisticated approaches to detecting faces.[27–29] Here we adopt a very simple approach that relies on eye localization using eye template matching. The eye template, which is a rectangular region enclosing the two eyes, is the average image from 4 persons, different from the ASL signer, but similar imaging geometry as used in the ASL signs. The correlation is calculated on the whole image, only for the first image frame of the sentence. For the subsequent images, the center of the rectangular box bounding the eye is found by correlating around the neighborhood of center found in previous image. A window of 10 pixels in width and height is considered for the neighborhood search. After the detection of eyes, we demarcate the face with an elliptical structure. We use the golden ratio for face,[30] to mask the face with two elliptical structures, one for the top part and the other for the bottom. Fig. 5 shows example outputs of the eye detection and facial demarcation steps.

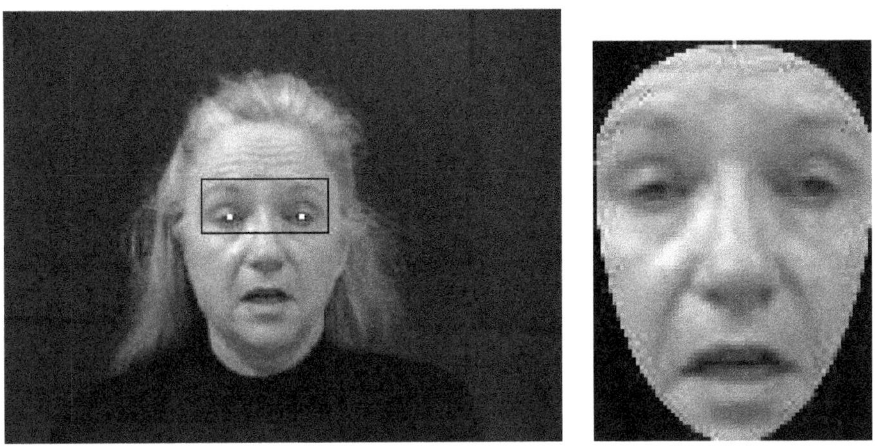

(a) (b)

Fig. 5. (a) Output of eye detection. (b) Extracted elliptical facial region.

We adopt a view-based representation of expression, modeled as traces in expression sub-space, which is computed using Principal Component Analysis (PCA) of 4 expression examples for each of the 39 signs. We have found that 20 largest eigenvalues capture most of the energy in the expression variations, at least in this dataset. It is interesting to note that various aspects of facial expression that are important to ASL, such as motion of eye brows, cheek puffing, lip movement, and nose wrinkles,[13] are captured by the dominant eigenvectors, some of which are show in Figure 6. It is observed that lip movements are emphasized by most of the eigenvectors because the interpreter was also mouthing the English equivalent of the signs. This might not be true for native signers.

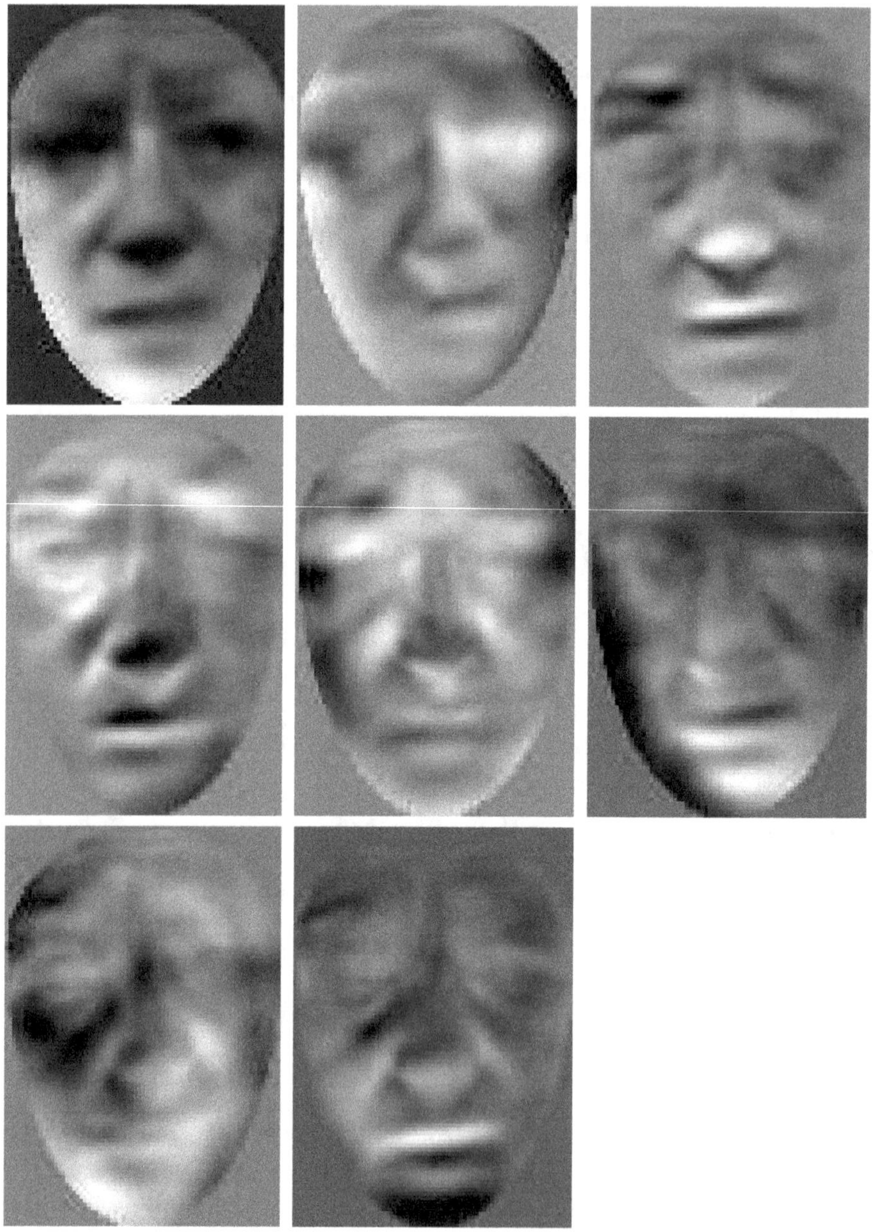

Fig. 6. Dominant dimensions of the learned facial expressions over the 39 signs.

4.3. *Non-manual: Head Motion*

Head motion is represented by the sequence of the average of the two eye locations, which are detected during face localization. The 2D trajectories are defined with respect to the location in the starting frame. Also, Fig. 7 shows some example head trajectories. Figs. 7 (a), (b) and (c) clearly shows the presence of 'Negation' in the sentences, Fig. 7(d) shows the vertical motion of face, indicating a 'head nod' while

(a) 'DONT-KNOW I'

(b) 'I NOT HAVE KEY'

(c) 'NO'

(d) 'YES'

(e) 'YOU UNDERSTAND ME'

(f) 'SUITCASE I PACK FINISH'

Fig. 7. Head motion trajectories. (a),(b) and (c) show the motion trajectories for sentences with negation. (d) shows the motion trajectory for the sign 'YES'. (e) and (f) show motion trajectories for sentences that do not convey negative meaning.

Figs. 7 (e) and (f) shows the motion trajectories for the sentences in which there is no positive or negative meaning conveyed through them.

5. Combination of Manual and Non-Manual

We have used the facial expression information to reduce the deletion and insertion errors while head motion information is used to find whether the sentence contains 'Negation' or not. The combination of the non-manual with manual information is not trivial because (i) the non-manuals are not time-synchronized with manuals. This is not a video synchronization issue. The facial event might lag or lead the manual event. Also, (ii) presence of presence of a strong non-manual indicating 'Assertion' or 'Negation' in the sentence makes it hard to extract facial information for some frames.

Fig. 8 shows the information flow architecture. We process manual information, facial expressions, and head motion as independent channels, which are then combined as follows

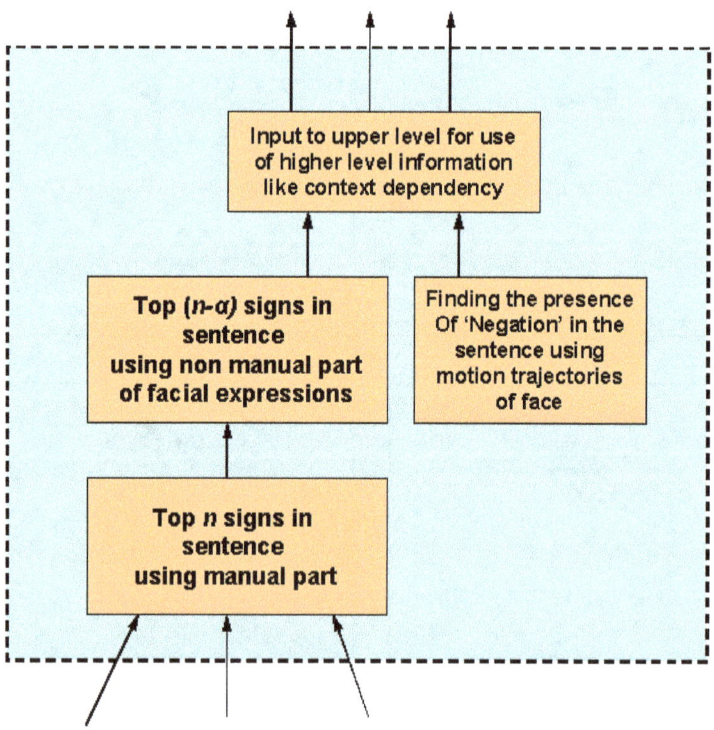

Fig. 8. A bottom-up architecture for fusing information from facial expressions and head motion with manual information to prune the set of possible ASL sign hypotheses.

(1) Find the n signs with least distances in the sentence using manual information.
(2) Find the distances for same n signs found in Step 1, using non-manual information.
(3) Sort these signs in ascending order of distances obtained from non-manual information.
(4) Discard α signs having maximum distances from sorted list obtained from Step 3.
(5) Keep the remaining $n - \alpha$ signs from Step 1.

And, finally head motion is used to detect whether the sentence has 'Negation' in it. The selection of n and α is a function of the number of signs in a sentence and the computational costs of high level processes.

Distances between SoPF traces quantify motion involved in the manual information, while the distances in the face space quantify changes in facial expression. In this work, we adopt a simple distance measure between two traces to find a sign in the sentence using manual and non-manual information. For the manual information, we cross correlate the SoPF trace of the training sign with the trace of given sentence and pick the value and shift that results in the minimum distance (Euclidean). For non-manual information, we correlate the trace of trained sign *near the time neighborhood* where the smallest distance for manual information has been found.

Let us look at some example of correlation of signs with sentences. Fig. 6(a) plots the correlations of manual information of 'LIPREAD' with the sentence 'LIPREAD CAN I'. Lower values of distance indicate matches, which in this case occurs around frame 12. Similarly, Fig. 6(b) & Fig. 6(c) shows the correlation of the manual information of the signs 'CAN' & 'I' with the same sentence. The actual position of the signs can be seen in Fig. 6(d). The *epenthesis movement*[31] indicated by E, can also be clearly seen in between the signs.

6. Experiments

In this section, we show results to demonstrate the efficacy of the proposed approach using the data described earlier. Given that we have 5 instances of 25 distinct sentences, we use 5-fold cross validation to evaluate the effectiveness of (a) manual motion modeling, (b) the integration of facial expression with manual information, and (c) the detection of negation in sentences. Four instances of each sentence are used for training and one is used for testing. Note that some signs occur multiple times in the training data in different arrangements with other signs. There are 65 sign instances, making up the 25 test sentences, to be recognized. Before we present results, a few words about the performance measures are in order.

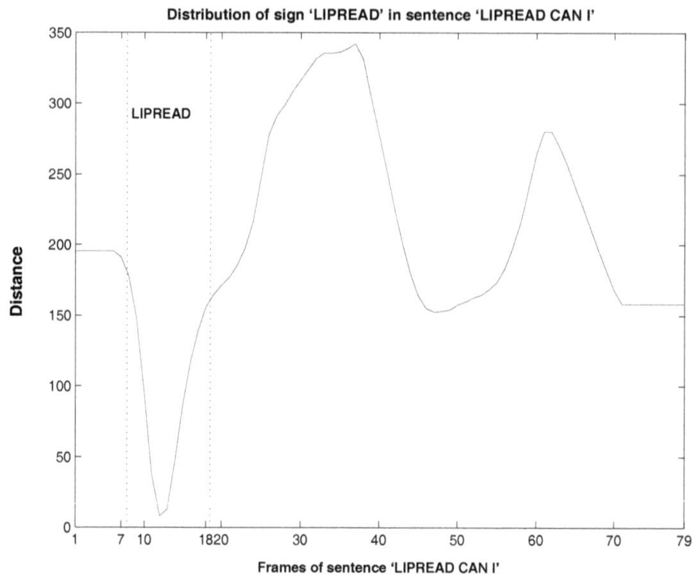

(a)

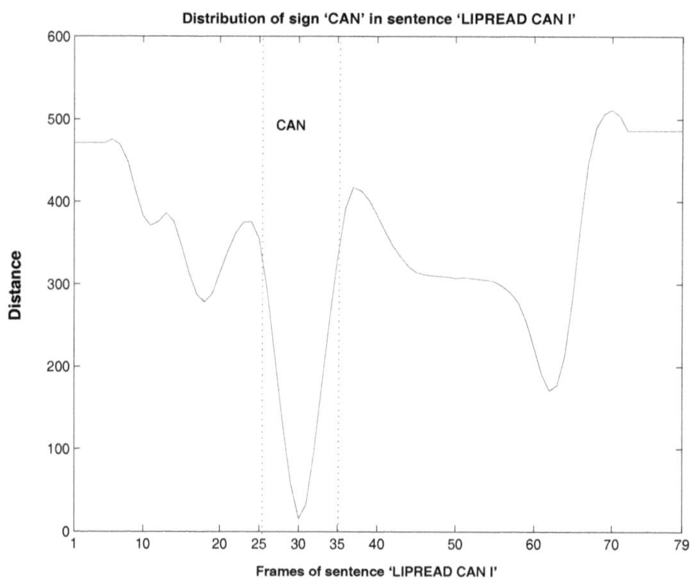

(b)

Fig. 9. Cross-correlation of sign with sentences. (a),(b) shows the correlation of signs 'LIPREAD' and 'CAN' with the sentence 'LIPREAD CAN I'. Lower values indicates closer matches.

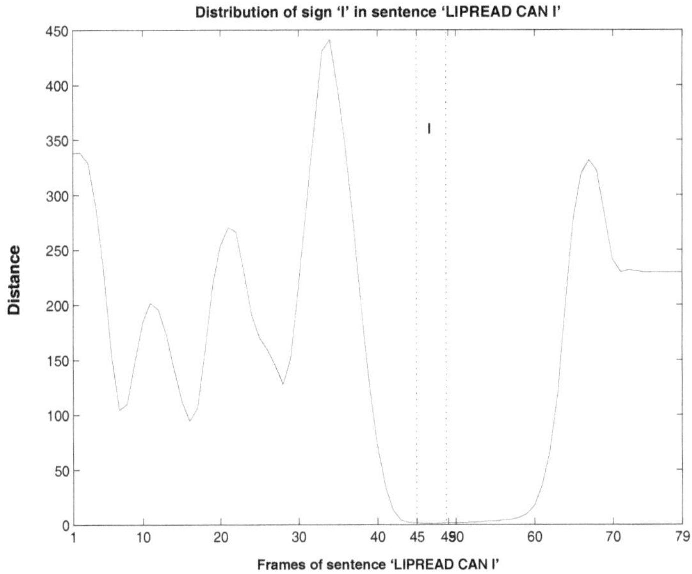

(c)

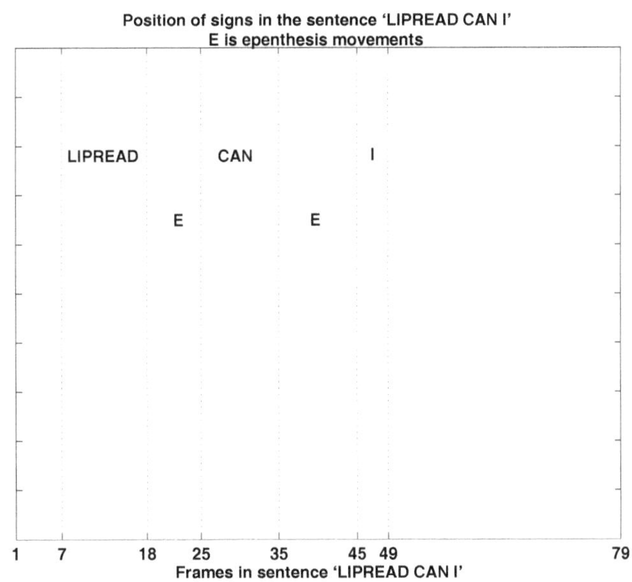

(d)

Fig. 10. Cross-correlation of sign with sentences (contd.). (c) shows the correlation of signs 'I' with the sentence 'LIPREAD CAN I'. Lower values indicates closer matches. (d) shows the ground-truth position of the signs in the sentence. E indicates the *epenthesis movements* present between two signs.

6.1. *Quantifying Performance*

Since we are considering the output of an intermediate level process that would typically be further refined using grammar constraints, the performance measures should reflect the tentative nature of the output. We sort the signs on the basis of minimum distance of a sign to the sentence. Then we choose n signs with the least distances. If a sign is part of a sentence, but not present in these n signs, then a *deletion error* has occurred. The number of deletion errors depends on n; as n increases, errors go down but the cost of high level processing increases since it has to consider more possibilities. Since the maximum number of words in a sentence is 5, we report results with $n = 6$, which is also 10% of the number of possible sign instances in the test set. The correct detection rate or accuracy is 100 minus the deletion rate. A sign that is not a part of sentence but has distance less than *the last correctly detected sign* is declared to be wrongly inserted in the sentence. We have an *insertion error.* Insertion errors can be reduced using the context knowledge or by grouping the signs that are very similar. Here we have reduced the insertion errors using facial expression information in the sentence. It is harder to recover from deletion errors.

6.2. *Use of Non-Manual Information*

To study the effect of the use of manual information, we start with the top 8 signs, as determined by manual information, and then prune out 2 signs ($\alpha = 2$) based on facial information, so as to finally arrive at 6 hypothesized signs per sentence. We compare the final insertion and deletion error rates with hypothesizing 6 possible signs per sentence based on just manual information.

The deletion error rate for the top $n = 6$ matches based on just manual information as captured by the SoPF traces, with 5-fold validation, ranges from 9% to 15% with an average of about 12%. Thus, the average correct detection rate from just manual information is 88%. Table 1 shows the improvement in rates to about 92% correct detection when face information is added. We also see a corresponding reduction in the insertion error rates.

The percentage of sentences that are perfectly recognized, i.e. all the correct signs are among the topmost ranks (zero deletion and insertion errors), is around 46%. This is a sentence level performance measure, i.e. the percentage of sentences that are perfectly recognized. This is a very strict measure. A sentence can be

Table 1. Deletion and insertion errors for n=6, with and without facial information. The range of error rates along with the average error rate, based on 5-fold validation, are reported in each case.

Error	Manual	Manual + Facial
Deletion	[9% to 15%], 12%	[4% to 11%], 8%
Insertion	[12% to 19%], 15%	[7% to 15%], 11%

misunderstood even if there is only one sign not correctly recognized. We contend that it is important to also report this number, even if it is low. Note that with the use of grammatical constraints, this performance can be further improved.

6.3. *Head Motion to Find 'Negation'*

'Negation' in a sentence is indicated by the 'head shake'. We use aspect ratio of the aggregate 2D track of the *whole sentence*, i.e the ratio of width to height of the entire trajectory, as the features to recognize the presence of 'Negation' in the sentence. We consider sentences whose motion trajectories have aspect ratio greater than 1.25, width greater than 40 pixels, and height less than 50 pixels, to have 'Negation' in them. Using this logic, out of 30 sentences in the database that have 'Negation' in them 27 were correctly recognized while there were 18 false alarms out of the remaining 95 sentences. The false alarms were mainly because of sentences like 'GATE WHERE', which also have motion of face in horizontal direction. Also, the missed detections are in sentences that specifically have sign the 'ME' in them (example: *'You don't understand me'*). This is because in sign 'ME', there is a natural downward movement of face, and hence the height of motion trajectory increases to a good extent, causing aspect ratio to decrease. In future, this performance can be increased by looking for portions in the sentences where the negation occurs and not using the entire trajectory over the whole sentence.

7. Conclusion

We presented a framework for continuous sign language recognition that combined non-manual information from faces with manual information to decrease both the deletion and insertion errors. Unlike most previous approaches that are top-down HMM based, ours is a bottom-up approach that relies on simple low-level processes feeding into intermediate level processes that hypothesize signs that are present in a sentence. The approach also does not bypass the segmentation problem, but relies on simple, yet robust, low-level representations. The manual dynamics were modeled by capturing the statistics of the relationships among the low-level features via the concepts of relational distributions embedded in Space of Probability Functions. Facial dynamics were captured using an expression sub-space, computed using PCA.

Even with fairly simple vision processes, embedded in a bottom-up approach, we were able to achieve good performance from pure image based inputs. Using 5-fold cross-validation on a data set of 125 sentences over 325 sign instances, we showed that the accuracy of individual sign recognition was about 88% with just manual information. The use of non-manual information increased the accuracy from 88% to 92%. We were also able to correctly detect negations in sentences 90% of the time.

8. Acknowledgment

This work was supported in part by the National Science Foundation under grant IIS 0312993. Any opinions, findings and conclusions or recommendations expressed in this material are those of the author(s) and do not necessarily reflect those of the National Science Foundation.

References

1. B. Loeding, S. Sarkar, A. Parashar, and A. Karshmer. Progress in automated computer recognition of sign language. In *Lecture Notes in Computer Science*, vol. 3118, pp. 1079–1087, (2004).
2. C. Sylvie and S. Ranganath, Automatic sign language analysis: A survey and the future beyond lexical meaning, *IEEE Transactions on Pattern Analysis and Machine Intelligence*. **27**(6), 873–891 (Jun, 2005).
3. Y. Cui and J. Weng, Appearance-based hand sign recognition from intensity image sequences, *Computer Vision and Image Understanding*. **78**(2), 157–176 (May, 2000).
4. M. Zhao and F. K. H. Quek, RIEVL: Recursive induction learning in hand gesture recognition, *IEEE Transactions on Pattern Analysis and Machine Intelligence*. **20** (11), 1174–1185, (1998).
5. J. Triesch and C. von der Malsburg. Robust classification of hand postures against complex backgrounds. In *International Conference on Automatic Face and Gesture Recognition*, pp. 170–175, (1996).
6. M. H. Yang, N. Ahuja, and M. Tabb, Extraction of 2d motion trajectories and its application to hand gesture recognition, *IEEE Transactions on Pattern Analysis and Machine Intelligence*. **24**, 168–185 (Aug., 2002).
7. T. Starner and A. Pentland. Real-time American Sign Language recognition from video using hidden Markov models. In *Symposium on Computer Vision*, pp. 265–270, (1995).
8. C. Vogler and D. Metaxas. ASL recognition based on a coupling between HMMs and 3d motion analysis. In *International Conference on Computer Vision*, pp. 363–369, (1998).
9. C. Vogler and D. Metaxas. Parallel hidden Markov models for American Sign Language recognition. In *International Conference on Computer Vision*, pp. 116–122, (1999).
10. C. Vogler and D. Metaxas, A framework of recognizing the simultaneous aspects of American Sign Language, *Computer Vision and Image Understanding*. **81**, 358–384, (2001).
11. J. Ma, W. Gao, C. Wang, and J. Wu. A continuous Chinese sign language recognition system. In *International Conference on Automatic Face and Gesture Recognition*, pp. 428–433, (2000).
12. C. Wang, W. Gao, and S. Shan. An approach based on phonemes to large vocabulary Chinese sign language recognition. In *International Conference on Automatic Face and Gesture Recognition*, pp. 393–398, (2002).
13. B. Bahan and C. Neidle. Non-manual realization of agreement in American Sign Language. Master's thesis, Boston University, (1996).
14. R. Wilbur and A. Martinez, Physical correlates of prosodic structure in American Sign Language, *Meeting of the Chicago Linguistics Society, April.* pp. 25–27, (2002).

15. U. M. Erdem and S. Sclaroff. Automatic detection of relevant head gestures in American Sign Language communication. In *International Conference on Pattern Recognition*, pp. 460–463, (2002).
16. C. Vogler and S. Goldenstein, Facial movement analysis in ASL, *Universal Access in the Information Society*. **6**(4), 363–374, (2008).
17. U. Canzler and T. Dziurzyk, Extraction of Non Manual Features for Videobased Sign Language Recognition, *IAPR Workshop on Machine Vision Application (MVA2002)*. pp. 11–13, (2002).
18. M. L. Cascia, S. Sclaroff, and V. Athitsos, Fast, reliable head tracking under varying illumination, *IEEE Transactions on Pattern Analysis and Machine Intelligence*. **21** (6) (June, 1999).
19. A. Kapoor and R. W. Picard. A real-time head nod and shake detector. In *Workshop on Perspective User Interfaces* (Nov., 2001).
20. R. Yang, S. Sarkar, and B. Loeding. Enhanced level building algorithm for the movement epenthesis problem in sign language recognition. In *Computer Vision and Pattern Recognition*, (2007).
21. R. Yang and S. Sarkar, Coupled grouping and matching for sign and gesture recognition, *Computer Vision and Image Understanding*. (2008).
22. J. Alon, V. Athitsos, Q. Yuan, and S. Sclaroff. Simultaneous localization and recognition of dynamic hand gestures. In *IEEE Workshop on Motion and Video Computing*, vol. 2, pp. 254–260, (2005).
23. L. Ding and A. Martinez, Modelling and recognition of the linguistic components in American Sign Language, *Image and Vision Computing*. (2009).
24. I. Robledo and S. Sarkar, Representation of the evolution of feature relationship statistics: Human gait-based recognition, *IEEE Transactions on Pattern Analysis and Machine Intelligence* (Feb. 2003).
25. W. Freeman and M. Roth. Orientation histograms for hand and gesture recognition. In *International Workshop on Face and Gesture Recognition*, pp. 296–301, (1995).
26. A. Huet and E. Hancock, Line pattern retrieval using relational histograms, *IEEE Transactions on Pattern Analysis and Machine Intelligence*. **12**(13), 1363–1370, (1999).
27. H. A. Rowley, S. Baluja, and T. Kanade, Neural network-based face detection, *IEEE Transactions on Pattern Analysis and Machine Intelligence*. **20**, 23–38, (1998).
28. A. Colemnarez and T. Huang, Face detection with information based maximum discrimination, *Computer Vision and Pattern Recognition*. pp. 782–787, (1997).
29. K. K. Sung and T. Poggio, Example-based learning for view-based human face detection, *IEEE Transactions on Pattern Analysis and Machine Intelligence*. **20**, 39–51, (1998).
30. L. G. Frakas and I. R. Munro, *Anthropometric Facial Proportions in Medicine*. (Charles C. Thomas, Springfield, IL, 1987).
31. S. K. Liddell and R. E. Johnson, American Sign Language: The phonological base, *Sign Language Studies*. **64**, 195–277, (1989).

CHAPTER 4.3

A DEFORMABLE TEMPLATE MODEL WITH FEATURE TRACKING FOR AUTOMATED IVUS SEGMENTATION

Prakash Manandhar[1], Chi Hau Chen[2], Ahmet Umit Coskun[3]

[1]*University of Massachusetts Dartmouth*
E-mail: pmanandhar@umassd.edu

[2]*University of Massachusetts Dartmouth*
E-mail: cchen@umassd.edu

[3]*Northeastern University, Mechanical & Industrial Engineering Department*
E-mail: acoskun@coe.neu.edu

Intravascular Ultrasound (IVUS) has been established as a useful tool for diagnosis of coronary heart disease (CHD). Recent developments have opened the possibility of using IVUS to create a 3D map from which preventative prediction of CHD can be attempted. Segmentation of IVUS images is an important step in this process. However reliable automated segmentation has been elusive, in part because of the variety of image features that are invariably present in the image that distract from the main segmentation objectives. Active contour models (ACM)s have been used successfully for automated segmentation of IVUS images. However, the accuracy of the segmentation is still not adequate for clinical use. Here we describe a new approach of a constrained deformable template model (DTM) that improves on the standard ACM algorithm by (1) detecting other distracting image features (2) using tracking algorithms to get a better estimate of the positions of these features (3) including the knowledge of these positions to eliminate distortions in the ACM due to these features. In addition, semantic constraints are inbuilt into the DTM so that computational time is not wasted in improbable segmentation results. Our results show that this is a promising approach to achieving fully automated segmentation with accuracy comparable to manual segmentation.

1. Introduction

Starting with its introduction in the late 1980s [1], Intravascular Ultrasound (IVUS) has been used successfully as a diagnostic tool for various coronary heart diseases. More recently, the IVUS technique has been used for "vascular profiling" where a 3D representation of the coronary artery is formed by combining a series of IVUS images with location data from x-ray angiography This process is accurate and reproducible [2], but labor-intensive since a single profile can consist of thousands of IVUS frames. Automated segmentation algorithms have been improving. However, a fully automated algorithm that achieves accuracy at the level of manual segmentation has been elusive. Numerous automated IVUS segmentation algorithms have been reported ranging from knowledge-based graph search [3], texture-based [4], active-contour models [5] or a combination of these techniques [6]. These techniques do not directly address the presence of artefacts in the typical IVUS image, of which two of the major ones are due to the guidewire and the catheter (Fig. 2) [7].

Observation of the manual segmentation by human experts hinted that the experts are well aware of the presence of these artefacts and actively allowed for their effects to create a good segmentation. Additionally, the positions of the artefacts are strongly correlated between adjacent frames which provide further hints for the segmentation process. Here we present evidence that detecting artefacts and tracking them to later compensate is a good strategy for increased accuracy in automated segmentation.

Blood vessels are composed of several layers of tissues. The lumen through which blood flows is surrounded by the intima, which is surrounded by the media and then the adventitia. The media and adventitia are separated by the external elastic membrane (EEM). In addition there can be deposits of plaque of different types within the intimal layer. Multilayered tissue creates an ultrasound image that is non-trivial to interpret. Adding to this complexity is the scatter from flowing blood in the lumen.

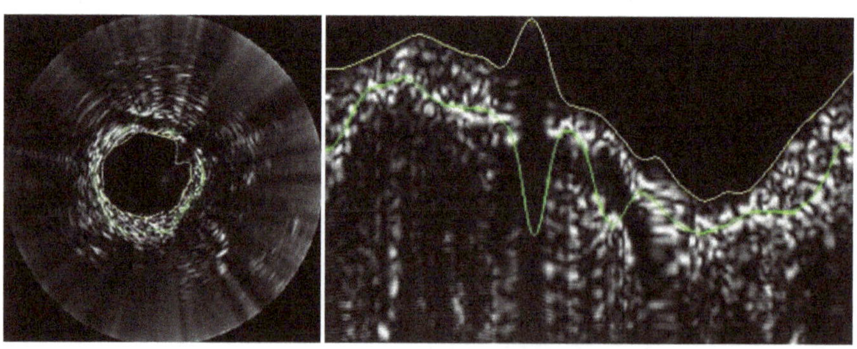

Fig. 1. Distortion in contour initialization due to guidewire and its shadow (The left side shows a cross-sectional slice in normal view while the right side is 'polar' view). The contour initialization was done using the algorithm described in [5].

2. Guidewire detection and tracking

A guidewire (usually 0.038 inch) is inserted and manipulated inside the blood-vessel to the point of interest. The IVUS catheter is threaded over the guidewire, either co-axially or in a monorail fashion. The guidewire in the monorail system creates an acoustic shadow (along with a bright reverberation from the metal wire itself). When an interventionalist is in an interactive diagnosis session, the IVUS catheter can be rotated to reveal the hidden area, however in a pullback sequence, this can not be done. [8]

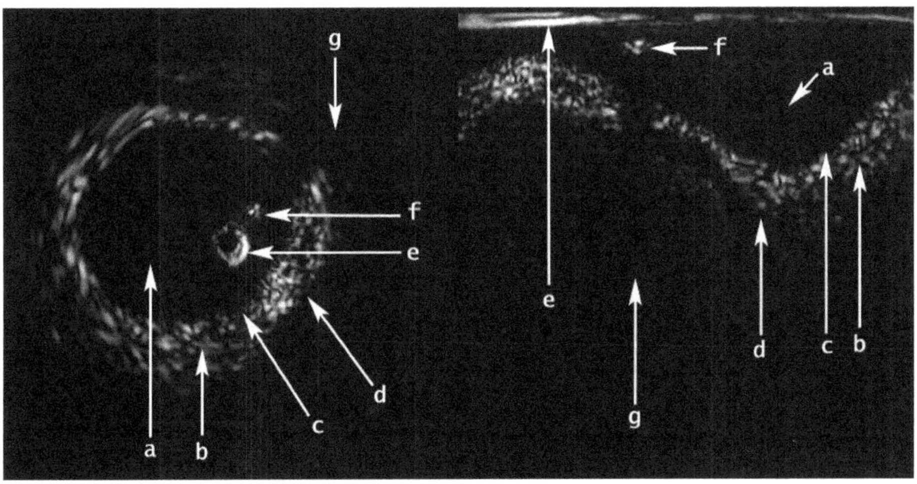

Fig. 2. Typical IVUS image (left: rectangular view, right: polar view) showing (a) lumen (b) media (c) intima (d) EEM (e) catheter artefact (f) guidewire artefact (g) guidewire shadow.

The guidewire feature was investigated from all frames of 2 different pull-back sequences consisting of about 10,000 frames in total. A typical frame of data is presented in Fig. 2. It can be seen that near to the catheter (low r-index, refer to Fig. 10) the guide wire produces a distinct echo-dense region with reverberations followed by a long narrow shadow region. As it is possible to have shadow regions due to other features such as calcified deposits and branches in the vessels, the salient features of the guidewire artefact were identified. Based on this a template matching algorithm was implemented where a series of templates consisting of Gaussian blobs at various r-index positions were matched with slices of the image. However, this only resulted in 80% accuracy for guidewire position. A simpler method of ordering "energy" levels in a certain radial range, combined with tracking produced better results.

Since the guidewire is a continuous wire, it can be expected from the nature of the physical process that the change in guidewire position in subsequent frames is minimal. Often, to reduce the effects of the moving heart an ECG set is gathered simultaneously with the IVUS data. The IVUS frames are then 'gated' so that only the frames from the same part of the cardiac cycle are considered. This means that a majority of the data is discarded. We found that using the full data set is helpful in tracking the guidewire position since changes in position are minimal between adjacent frames.

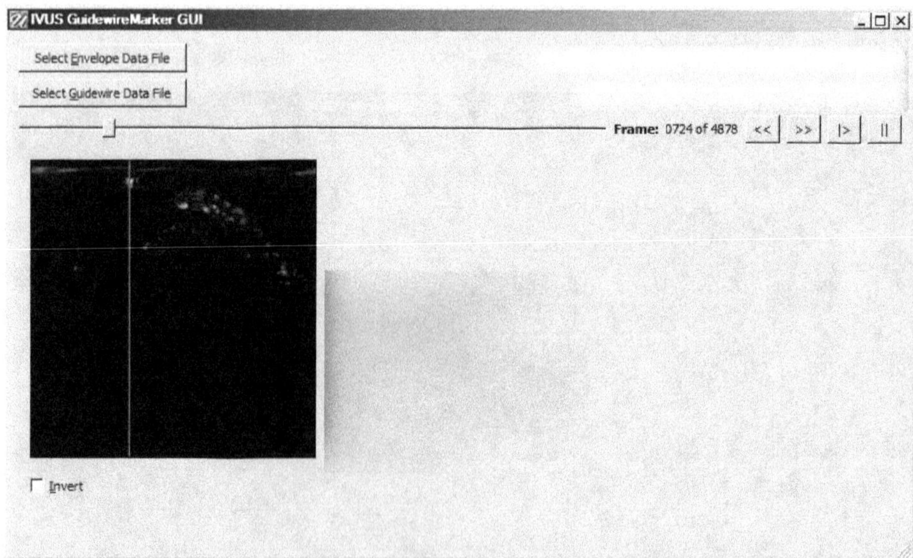

Fig. 3. Screenshot of custom software developed to quickly mark the guidewire position in an IVUS pullback sequence. After the user loads the data file and the guidewire marker file which contains, for each frame a flag indicating whether the guidewire has been marked or not and if yes, the position of the guidewire. The when the mouse pointer is inside the image area, a line follows the pointer and when the user clicks the position is marked the frame moves forward; existing marks are shown in a different color. This allows the user to quickly mark a large number of frames – at the rate of about 5000 frames in several hours.

To study the character of the data a reference set of data where the guidewire position was manually marked was needed. Since this data is not typically recorded during routine manual segmentation, custom software was designed to quickly mark the guidewire position in a pullback sequence manually (Fig. 3). Four typical pullback sets consisting of about 5000 frames were marked using this software. Various statistics of the vectors along the guidewire area were compiled from these data to determine an appropriate algorithm for accurate guidewire detection and tracking algorithm. Fig. 5 shows the manually marked guidewire position for a particular pullback sequence. Analysis of the data showed that the guidewire position can be considered to be an autoregressive process where the next position is the previous position plus white Gaussian noise. This suggests the use of Kalman filtering for tracking the guidewire, however the presence of clutter (shadowy regions other than those due to the guidewire) resulted in reduced accuracy. Using the ConDensation algorithm gave much better results (Table 1).

Comparison between the vectors with guidewire and without guidewire in the manually marked data showed that a simple process to compare the intensity values after radius index from $r_{min} = 50$ and $r_{max} = 200$ might be just as effective as using a template matching algorithm. Indeed the simpler algorithm was able to track the guidewire with improved accuracy (Table 1). N hypotheses (where N = 5) were generated using the following ordering method:

$$x_{n,t} = \min_{\theta,n} \sum_{r=r_{min}}^{r_{max}} f(r,\theta);\ n = 1,2,\ldots N$$

where

$f(r,\theta)$	is	the image pixel value at radius r and angle θ
t	is	the frame index
$x_{n,t}$	is	the nth hypothesis that the guidewire is located at $\theta = x_{n,t}$

Multiple observation hypotheses generated by this process are fed into a particle filtering tracking algorithm [9] based on a variation of the conditional density propagation (ConDensation) tracking algorithm [10]. This algorithm performs better than the Kalman filtering algorithm and can handle multiple hypotheses simultaneously due to extensive clutter (Fig. 4). The accuracy of the results was evaluated by considering the difference in θ between the tracked guidewire position and the manually marked position.

We tried several different methods for guidewire tracking and a variation of the ConDensation algorithm that we have named particle filtering with weighted tunneling has given us the best results. In this algorithm the Bayesian reasoning box in Fig. 4 gives higher weight to the more probable observations and lower weights to the less likely ones. Additionally a "tunneling probability" is also included which keep the possibility that all the observations are false taking into account the probability of non-detection. This probability allows the guidewire estimates from the previous frames to carry on even when none of the new observations match. Table 1 compares results between choosing the best hypothesis without tracking and using various tracking algorithms and shows that particle filtering with weighted tunneling is clearly a more robust method.

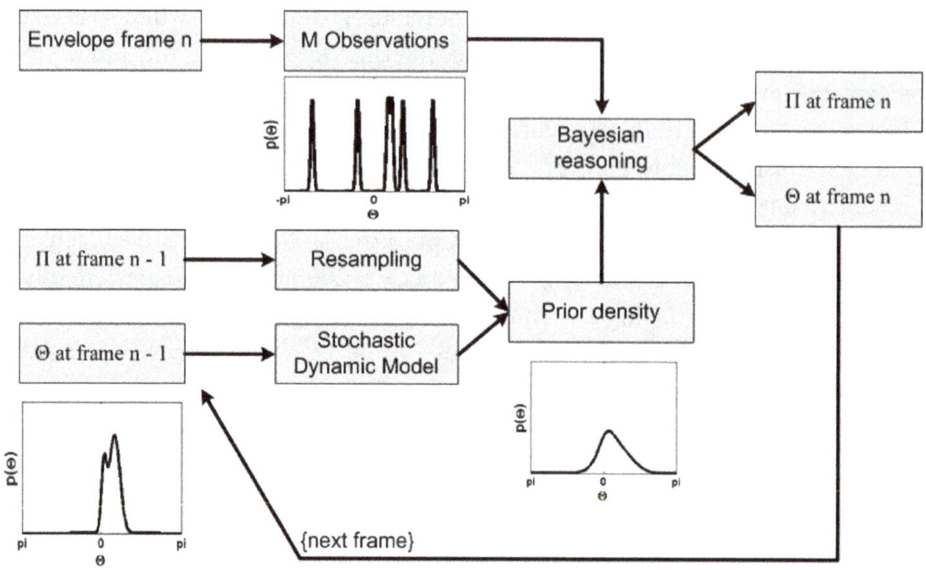

Fig. 4. Conditional density tracking (a.k.a. sequential Monte Carlo particle filtering [11]) used for accurate tracking of guidewire in cluttered observation. Θ and Π are the sets of samples (particles) and their probability weights. The dynamic model that we have used is an AR-1 model where the estimate for the next iteration is assumed to be the estimate for the present iteration with some noise added. Re-sampling and applying the stochastic model diffuses the PDF of the prior and also prevents degeneracy. Note the presence of many peaks in the observation density due to 'clutter'. The observation and prior are combined to get the new densities from which the new estimate of guidewire position can be derived in a maximum-apriori (MAP) sense or by taking the expectation of the distribution.

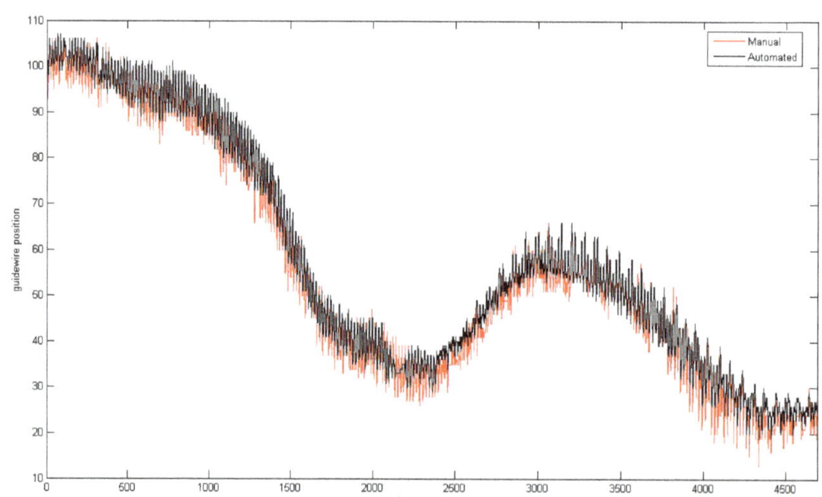

Fig. 5. Match between ConDensation tracking estimate and manual marked guidewire position.

Table 1. Comparison of error rates among various algorithms that we developed for guidewire tracking. BL – baseline algorithm just chooses the best estimate from the observation of the current frame. KP – Kalman filtering with position model is a Kalman filter implementation where the next guidewire position is estimated to be the previous position with Gaussian noise. KV – similar to KP, but includes a velocity model. PF – simple particle filtering without tunneling. PT – particle filtering with tunneling. PW – particle filtering with tunneling and weighting of observations. Details of the algorithm implementation have been left out as they are out of the scope of this chapter.

		Error >±10						Error >±5					
Dataset	Frames	BL	KP	KV	PF	PT	PW	BL	KP	KV	PF	PT	PW
1	5211	18	3	11	14	28	**16**	406	367	490	704	805	494
		0.35%	0.06%	0.21%	0.27%	0.54%	**0.31%**	7.79%	7.04%	9.4%	13.51%	15.44%	9.48%
2	4691	88	99	100	136	10	**2**	311	286	389	596	535	275
		1.88%	2.11%	2.13%	2.90%	0.21%	**0.04%**	6.63%	6.10%	8.29%	12.71%	11.40%	5.86%

3. Catheter position and vessel diameter estimate using the Circular Hough Transform (CHT)

"Ring-down" artefacts observed as bright halos of variable thickness surround the catheter. These are produced by acoustic oscillations in the transducer and obscure the area immediately adjacent to the transducer [7].

The image is easier to segment when the catheter lies near the centre of the lumen so that it is not touching the vessel intima. A human observer can quite easily determine the catheter position and identify regions within the image where the ring-down artefacts can reduce the certainty of conclusions about the position of the intima and the other vessel wall layers.

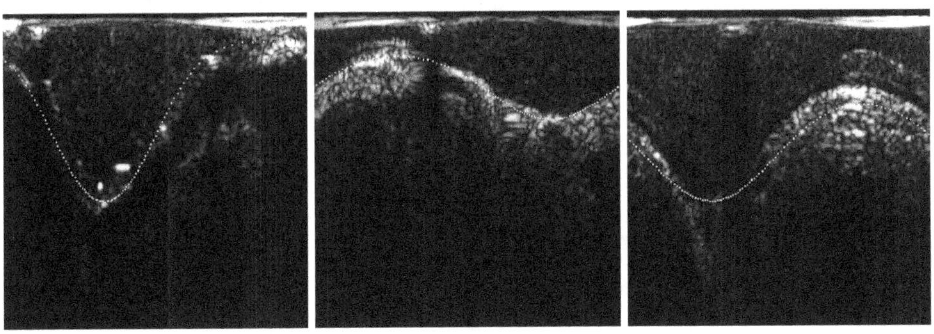

Fig. 6. Circular Hough Transform results for three different catheter positions. These results also show how the ring-down artefacts can be a hindrance in subsequent active contour model processing. For example in the second the third figure the catheter is sufficiently close to the lumen wall that the active contour model is highly likely to get confused distorting the final result.

Although it is clear that the lumen is most often not a perfect circle, it was found that in most frames the circle approximation does provide an approximate location of the catheter with respect to the vessel wall. Rosales [12] also use a circular model for tracking the catheter position within the vessel, while Gil [13] use an eclipse template. A circle only has three parameters (centre coordinates and radius), an eclipse has 5 parameters which dramatically expands the search space increasing the time and memory resources needed for the algorithm. Since we are not considering the result of this step to be the final segmentation and still need to apply the deformable model, a circular model does suffice in our case. Huang [14] have also found tracking of CHT parameters directly in the Hough domain across video frames using a ConDensation like probabilistic data association filter, to be feasible for real time tracking of circular objects in clutter for robotic vision. Elsewhere, French [15] have found tracking in the Hough space to be suitable for vehicle tracking using parallel lines.

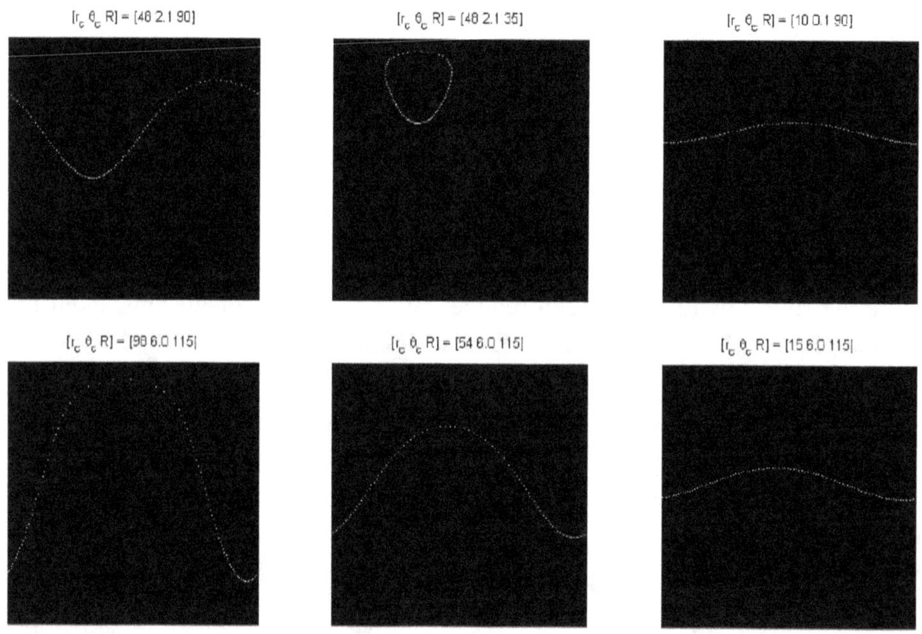

Fig. 7. Simulated catheter position and vessel radius variations. Note that for the catheter to be within the vessel, the radius of the vessel has to be greater than the radial position of the center of the catheter; the top center image shows an example of a violation of this principle. When the catheter is close to the center of the vessel we get an almost straight line for the wall (right column), while an a highly curved structure means that the catheter is positioned close to the vessel wall.

Fig. 7 shows the trace of the shape of the curve for circles in polar co-ordinates. These traces are very similar to the vessel wall outlines we see in the polar IVUS images hinting that CHT in the polar domain might be a good process to identify them.

Since the guidewire position has already been estimated, the image regions distorted by the guidewire can be ignored while performing the Hough transform. This reduces the resources required by the algorithm and increases the accuracy of the method. Fig. 8 depicts a case where the CHT fails to find the correct wall when there is too much distraction from the guidewire artifact. Inclusion of the guidewire position knowledge corrected the mistake in the CHT output.

Fig. 6 shows some results of the application of CHT to IVUS images. Observation of the catheter position shows that it is also highly correlated

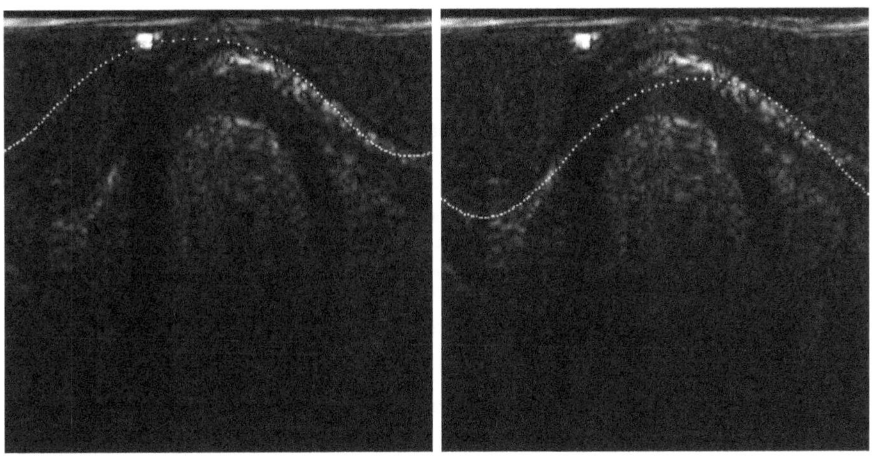

Fig. 8. Left – error in registration by CHT due to presence of guidewire artefact. Right – correction of this error by incorporation of unreliable data avoidance with guidewire tracking.

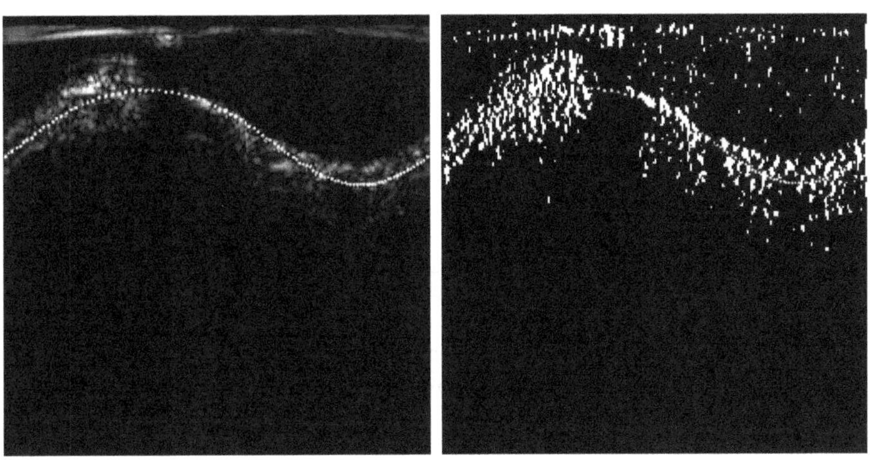

Fig. 9. Left: CHT result superimposed over original frame image. Right: the Sobel edge detected image with adaptive threshold based on which the CHT is performed.

between adjacent frames. The ConDensation algorithm is general enough to track more than one parameter at once. Incorporation of tracking along with avoidance of artefacts by using results from guidewire tracking shows a marked improvement in registration of vessel wall without error (Fig. 8).

Curve fitting algorithms are usually applied on binary images and their statistical performance has also been evaluated based on the assumption of a binary image [16]. It is possible to directly work on the grayscale image, however to reduce the effect of noise and focus on the changes in intensity rather than on constant graylevel areas. We have used an adaptive Sobel method to create the binary image from the envelope data (`Algorithm 1`).

Algorithm 1 Sobel edge filter with adaptive thresholding

$$f_i \ : \ \text{input grayscale image}$$
$$f_b \ : \ \text{output binary image}$$
$$H = \begin{bmatrix} 1 & 2 & 1 \\ 0 & 0 & 0 \\ -1 & 2 & -1 \end{bmatrix} \text{(vertical Sobel edge filter)}$$
$$t = 1.05 \text{ (threshold ratio)}$$
$$f = H * f_i$$
$$m = \text{median}(f)$$
$$f_b = f > (t \times m)$$

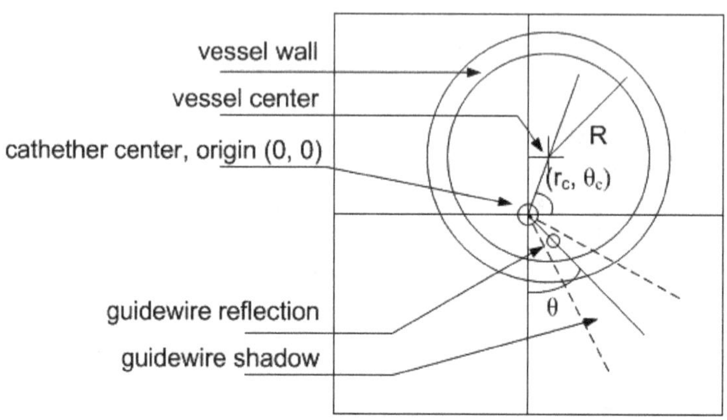

Fig. 10. Geometry of tracked parameters. The angle at which guidewire artefact appears (θ) is tracked using particle filtering (or ConDensation) algorithm with observations generated by ordering of energy values along data vectors as described in the text (section 2). The parameters (R, r_c, θ_c) corresponding to the position and radius of the vessel wall modeled as a circle are tracked separately using a similar particle filtering method with observations generated by ordering bins in a Circular Hough Transform space. All parameters are indexed into bins, for example θ is indexed into 256 bins of 360/256 = 1.4 degrees.

4. Deformable template model for IVUS

Since its introduction in the late 1980s in the computer vision literature, active contour models or deformable models in general have gained widespread popularity in the medical image processing community [17].

Here we devise a deformable template model with a goal to refine the contour estimated by the CHT in the previous step so that a more accurate contour for the lumen and adventitia are produced. There are some additional constraints that need to be satisfied at all times, so that the artefacts do not disturb the contour and some natural domain constraints are satisfied:

1. extra edges and gradients in the gap due to the guidewire and other gaps due to branches should be avoided
2. extra edges due to the catheter should be avoided
3. the catheter should always be located inside the lumen
4. the lumen should always be located inside the EEM

```
Data Structure 1 IVUS template
Constants:
  • N [Integer, power of 2]: number of pixels in IVUS image in the
    r and theta directions, so that we have a NxN image [N == 256].
  • M [Integer, power of 2]: number of node points per contour. [M
    == 8].
  • delM [Integer, power of 2]: = N/M
  • rMin, rMax [Integer, rMin < rMax < N]: rMin [== 30] defines the
    region close to the catheter where catheter artefacts are
    prominent. rMax [== 225] defines the region far away from
    catheter where there is no usable data.
  • minWall [Integer]: minimum radial distance between lumen and
    adventitial giving a minimum thickness of vessel wall.
  • gradSigma [Real]: parameter for Gaussian smoothing to be
    applied before calculating the gradient.
  • gwPos [Integer between 1 and N]: location of guidewire. This
    parameter is initialized once for each frame.
  • circleParams [Real x Real x Real]: triplet giving r_c (center
    of catheter radial co-ordinate), theta_c (center of catheter
    theta co-ordinate) and R (radius of vessel). This parameter is
    initialized once for each frame, and then is used to initialize
    luNodes and avNodes.
Variables:
  • luNodes, avNodes [array of M integers]: r indices of lumen and
    adventitia respectively. Together these form the 2M parameters
    to be optimized in a constrained space.
```

The problem of adding hard constraints has been dealt with variously in the literature. For example, Fua [18] have devised an optimization method where the gradient is projected onto the constraint to speed up convergence in the direction of the constraint. They have argued that the simpler method of just including a penalty in the cost function results in an ill behaved optimization problem with poor convergence and also requires careful consideration for commensurate choice of weights. Jain [19] show that a global model based approach is suitable for ensuring consistency due to constraints and bridging between gaps in the model. The deformable template model that we have devised has served as a good example of these properties. In addition, compared to a simple active contour model, we have achieved good convergence while minimizing the number of parameters to be optimized.

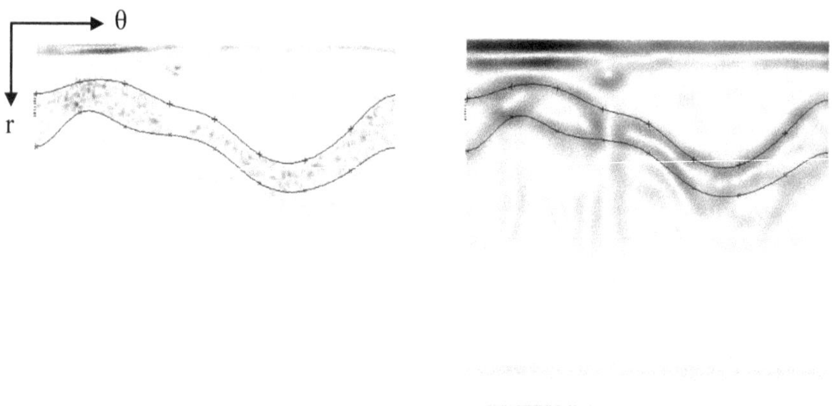

Fig. 11. Result of fitting deformable template model with two contours – lumen (top) and EEM (bottom). The cost function is actually calculated on the Derivative of Gaussian image (right). The left image is the actual frame data.

We have used a template model based on closed 3rd degree B-splines [20]. The method has been extended by encoding into the model parameters for two separate contours: lumen and EEM in the same template and then optimizing simultaneously. The constraints defined above are naturally incorporated in this approach. The deformable template model data structure is listed in box: Data Structure 1. It is convenient to choose `M` and `N` to be powers of 2 for computational convenience and simplicity in algorithms. The node points are uniformly divided along the theta direction, with the distance between them as `delM` (this is simplified if `N` and `M` are both powers of 2). A larger `M` allows the contours to follow the features more closely, but this also means the smoothness of the resulting curve is compromised. A small `M` results in a contour that is too smooth. By choosing an appropriate `M`, we are in effect implementing an internal energy function in the traditional snake model [20]. A multi-resolution approach can also be implemented by varying `M`, `minWall` and `gradSigma`.

```
Algorithm 2 Deformable template optimization
T: optimized template = optimize (f: image frame, To: old template)
fg = DoG (f, gradSigma) : take the derivative of Gaussian
initializeNodes : see Algorithm 4
del = 1 : search stepsize
Iterate until no change in current cost
       1. For each node (n[i]) in luNodes and avNodes check
          (sequencially) minimum cost of n[i] + del, n[i] - del, and
          n[i], avoiding the rMin, rMax and minWall constraints and
          assign the lowest cost to luNodes or avNodes respectively
       2. For each node pair (l[i], a[i]) in (luNodes, avNodes) check
          (sequencially) minimum combined cost of (l[i] + del, a[i] +
          del), (l[i] - del, a[i] - del) and (l[i], a[i]), subject to
          the constraints. Assign lowest cost as new contours
```

```
Algorithm 3 Generate exclusion mask
Ze = exclusionMask (T: template)
Ze = zero matrix of size T.N x T.N
Ze = 1 where r <= T.rMin and r >= T.rMax
Ze = 1 where theta is between plus or minus thWidth (= 10) of T.gwPos
```

```
Algorithm 4 Initialize Nodes
delR = 10 is the spacing between the circle and initial luNodes and
avNodes initialization

c  = r_c*r_c - R*R
th = array of M theta values equally spaced from -pi to pi
b  = -2*r_c*cos(th - theta_c)
r  = round((b + sqrt(b*b - 4*c))/2)
luNodes = r - delR; avNodes = r + delR
if luNodes[i] < rMin, luNodes[i] = rMin
if avNodes[i] > rMax, avNodes[i] = rMax
```

```
Algorithm 5 Cost function for template model optimization
c : Real = cost (nodes: luNodes or avNodes, fg: NxN gradient image)
r_array = interpolate between nodes and round to get from list of nodes
to radius index of the contour for all theta.
c = negative of sum over theta of fg[r_array[theta], theta]] except at
exclusion mask positions generated by Algorithm 3.
```

Although B-splines are more naturally implemented as control points, we use nodes (nodes pass through the B-spline curve, while control points do not) to simplify the process of initializing and updating the contours. An efficient algorithm described in [21] is used to convert between the representations and also to interpolate between the nodes. These give rise to the cost function (`Algorithm 5`) to be optimized along with the frame data. The frame data is processed through a Derivative of Gaussian filter [22] to eliminate speckle noise effects to get a gradient map; costs are calculated as line integrals on this map

(Fig. 11). Note that we use the negative of the integral so that we have a minimization problem.

Once we have the cost function, we perform a gradient descent search on the parameters until the cost can no longer be reduced. This simple greedy approach [23] produces good results since the approximation of the vessel wall from which we start is already a good approximation. (Refer to box: for more details.) Note that in each iteration we are tweaking each of the 2M parameters in turn towards decreasing cost. In effect we are following a space alternating gradient descent method similar to the space alternating expectation maximization algorithm described in [24]. In our case, we are also simultaneously enforcing geometric constraints to further constrain the search space. It is also possible to do a true gradient descent or conjugate gradient search. Brigger [20] outline a method to achieve this when we have B-spline nodes as parameters, by first converting them to control points and taking the necessary partial derivatives with respect to these parameters. However, due to the large parameter space and the presence of noise in the gradient, we find that this method is less effective than a space alternating search.

Improvements in the search strategy can be made by incorporating non-greedy steps in the search process, or by incorporating more flexible greedy steps onto the process. An example of the former could be the incorporation of large jumps in non-optimal directions every few iterations which can be backtracked if no optimal solution is found in that direction. An example of the later could be to incorporate search at `2*del` and `10*del` positions away from the current positions into each iteration. We have left these improvements for further work.

5. Conclusion

Comparing the traditional snake based method with our deformable template method, it can be seen that the contours are not distracted (Fig. 9, Fig. 11) due to the presence of the guidewire or catheter artifact (compare with Fig. 1). Additionally, these contours converged from the CHT based initialization in only 10 iterations, compared to several hundred iterations required for the inferior results in Fig. 1. As the template model is severely constrained according to domain knowledge, it is not surprising that better results can be obtained in fewer iterations.

We have outlined a framework for fully automated IVUS image segmentation from a pullback sequence which can potentially overcome the limitations of current methods. However, a very high accuracy level in all the frames is required if the process is to be fully automated. For example even if only 1% of the frames have mistakes in them, in a whole set of 5,000 frames it would be difficult to sort through manually and find the 50 frames and fix them. The time it would take to do this would be almost as large as that required manual segmentation and prohibitive for routine diagnostic use. Further verification and improvement needs to me made in the current algorithm to make it fit for clinical use. Some results for various cases are shown in Fig. 13 and Fig. 14. While the fully automated segmentation results in these figures are near to what we would expect from manual segmentation, there is room for improvement. Verification using manually segmented data is required for full evaluation.

Fig. 12 summarizes the whole dataflow pipeline combining the various methods outlined in this chapter. It is conceivable that further improvements in the results can be achieved if errors due to convergence to local minima in the final step of fitting the deformable contour model can be removed. One possible way to achieve this is to use a deformable surface model and fit across frames. Godbout [25] describe one such approach that can be adopted to fit the method described in this chapter.

In summary, deformable template models where the model embodies the constraints due to the domain can be a feasible way to modify the simple active contour model. Tracking by particle filtering of multiple features can be a feasible preprocessing step for improving accuracy of segmentation by deformable models by simply changing the cost function to reflect data about the

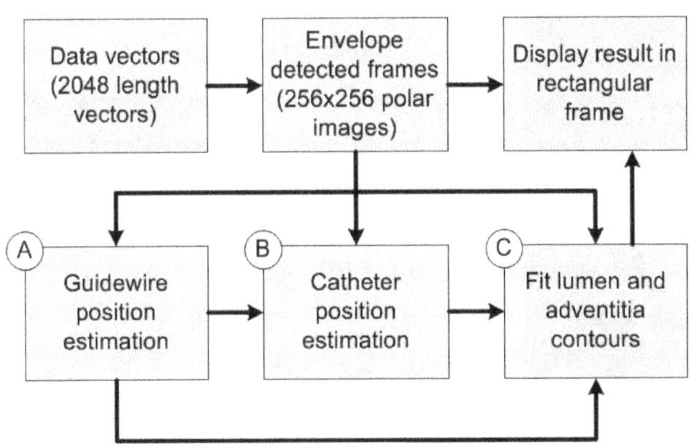

Fig. 12. Dataflow in our automated IVUS segmentation algorithm showing its three major components: (A) Guidewire position estimation (B) Catheter position estimation and (C) lumen and adventitia contour fitting using a deformable template model. All three processes get their data from the envelope detected frame. Steps (B) and (C) use the data from (A) to create a data unreliability region where data in the envelope frame is not used (to avoid distortion due to artefacts). The contours (deformable template parameters) in (C) are initialized with data from (B), the deformable template then optimizes its precise position with data from the envelope frame.

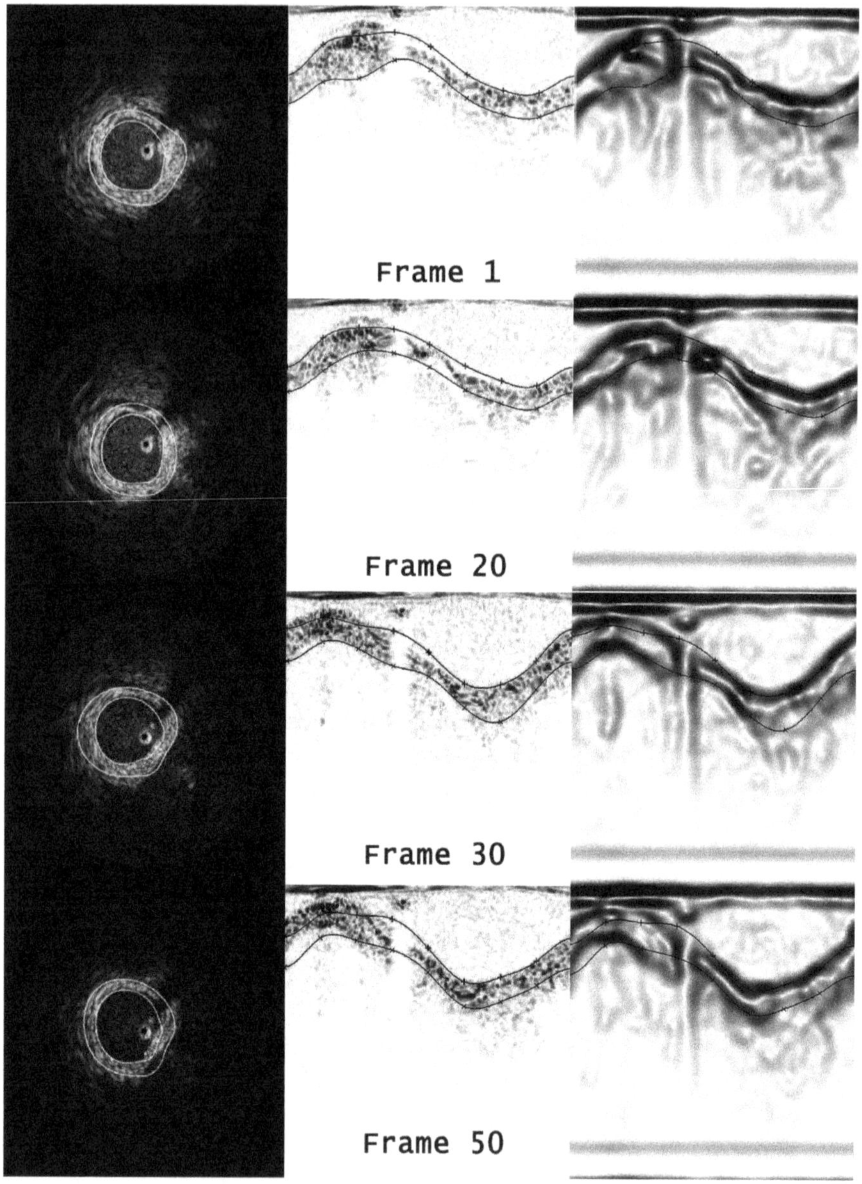

Fig. 13. Several representative frames of fully automated segmentation. In frame 50, there is a non-uniform rotational distortion and the algorithm was successful in smoothing over this distortion. Left column: segmentation result plotted on rectangular image, middle column: segmentation result plotted on polar image (image inverted for clarity), right column: segmentation results plotted on gradient image.

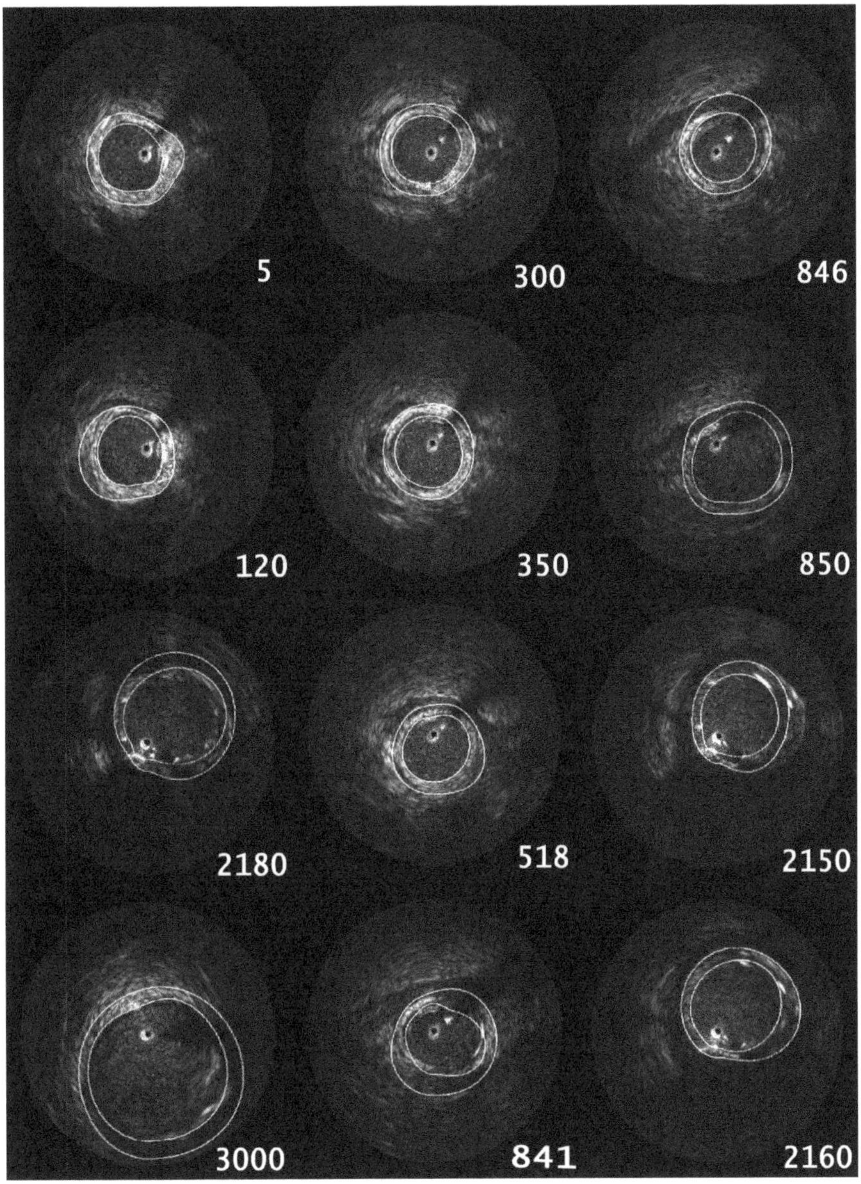

Fig. 14. A selection of segmentation results from a pull-back sequence. The segmentation is mostly correct but can miss parts of the intima or EEM pointing to room for improvement.

position of the tracked features. This can be especially helpful when the features to be tracked are persistent across frames and their positions correlated between frames. Fig. 10 depicts the parameters that are tracked in the current implementation in relation to the geometry of the problem. Further improvements could be made by tracking more parameters. For example the radial location of the guidewire location can be tracked along with the angle at which the guidewire artefact occurs. The tracking algorithm can also be further refined since we know that the motion of the catheter within the vessel and the guidewire with respect to the catheter is cyclic in nature resulting from the beating heart.

Acknowledgements

We thank the help of Gail MacCallum, Michelle Lucier and Nicholas Cefalo of the Vascular Profiling Laboratory at the Brigham & Women's Hospital, Boston, MA for their help with the IVUS image data.

References

[1] P. Yock, D. Linker, O. Saether, H. Thapliyal, J. Arenson, N. White and others, "Intravascular Two-Dimensional Catheter Ultrasound: Initial Clinical Studies," *Circulation,* vol. 78, pp. 11-21, 1988.

[2] A. U. Coskun, Y. Yeghiazarians, S. Kinlay, M. E. Clark, O. J. Ilegbusi, A. Wahle, M. Sonka, J. J. Popma, R. E. Kuntz, C. L. Feldman and P. H. Stone, "Reproducibility of coronary lumen, plaque, and vessel wall reconstruction and of endothelial shear stress measurements in vivo in humans," *Catheterization and Cardiovascular Interventions,* vol. 60, pp. 67-78, 2003.

[3] M. Sonka, X. Zhang, M. Siebes, M. Bissing, S. Dejong, S. Collins and C. McKay, "Segmentation of intravascular ultrasound images: a knowledge-basedapproach," *Medical Imaging, IEEE Transactions on,* vol. 14, pp. 719-732, 1995.

[4] M. Papadogiorgaki, V. Mezaris, Y. Chatzizisis, I. Kompatsiaris and G. Giannoglou, "A fully automated texture-based approach for the segmentation of sequential IVUS images," in *13th International Conference on Systems, Signals \& Image Processing (IWSSIP 2006), Budapest, Hungary,* 2006, pp. 461-464.

[5] G. D. Giannoglou, Y. S. Chatzizisis, V. Koutkias, I. Kompatsiaris, M. Papadogiorgaki, V. Mezaris, E. Parissi, P. Diamantopoulos, M. G. Strintzis and N. Maglaveras, "A novel active contour model for fully automated segmentation of intravascular ultrasound images: In vivo validation in human coronary arteries," *Comput. Biol. Med.,* vol. 37, pp. 1292-1302, 2007.

[6] D. Gil, A. Hernandez, O. Rodriguez, J. Mauri and P. Radeva, "Statistical Strategy for Anisotropic Adventitia Modelling in IVUS," *IEEE Trans. Med. Imaging,* vol. 25, pp. 768, 2006.

[7] G. Mintz, S. Nissen, W. Anderson, S. Bailey, R. Erbel, P. Fitzgerald, F. Pinto, K. Rosenfield, R. Siegel, E. Tuzcu and others, "ACC Clinical Expert Consensus Document on Standards for Acquisition, Measurement and Reporting of Intravascular Ultrasound Studies (IVUS). A report of the American College of Cardiology Task Force on Clinical Expert Consensus Documents," *J. Am. Coll. Cardiol.,* vol. 37, pp. 1478-1492, 2001.

[8] P. Neglén, "Venous stenting using intravascular ultrasound," in *Noninvasive Vascular Diagnosis*, 2nd ed.A. F. AbuRahma and J. J. Bergan, Eds. London: Springer, 2007, pp. 406-413.

[9] S. Das, D. Lawless, B. Ng and A. Pfeffer, "Factored particle filtering for data fusion and situation assessment in urban environments," in *Information Fusion, 2005 7th International Conference on,* 2005.

[10] M. Isard and A. Blake, "CONDENSATION—Conditional Density Propagation for Visual Tracking," *International Journal of Computer Vision,* vol. 29, pp. 5-28, 1998.
[11] M. Arulampalam, S. Maskell, N. Gordon, T. Clapp, D. Sci, T. Organ and S. Adelaide, "A tutorial on particle filters for online nonlinear/non-GaussianBayesian tracking," *Signal Processing, IEEE Transactions on [See also Acoustics, Speech, and Signal Processing, IEEE Transactions on],* vol. 50, pp. 174-188, 2002.
[12] M. Rosales, P. Radeva, O. Rodriguez-Leor and D. Gil, "Modelling of image-catheter motion for 3-D IVUS," *Med. Image Anal.,* 2008.
[13] D. Gil, P. Radeva, J. Saludes and J. Mauri, "Automatic segmentation of artery wall in coronary IVUS images: aprobabilistic approach," *Computers in Cardiology 2000,* pp. 687-690, 2000.
[14] C. M. Huang, C. W. Lai and L. C. Fu, "Visual tracking with probabilistic data association filter based on the circular hough transform," in *Robotics and Automation, 2006. ICRA 2006. Proceedings 2006 IEEE International Conference on,* 2006, pp. 4094-4099.
[15] A. French, S. Mills and T. Pridmore, "Condensation tracking through a hough space," in *Pattern Recognition, 2004. ICPR 2004. Proceedings of the 17th International Conference on,* 2004,
[16] K. Kanatani, "Statistical Optimization for Geometric Fitting: Theoretical Accuracy Bound and High Order Error Analysis," *International Journal of Computer Vision,* vol. 80, pp. 167-188, 2008.
[17] T. McInerney and D. Terzopoulos, "Deformable models in medical image analysis: a survey," *Med. Image Anal.,* vol. 1, pp. 91-108, 1996.
[18] P. Fua and U. Brechbuehler, "Imposing hard constraints on soft snakes," *LECTURE NOTES IN COMPUTER SCIENCE,* pp. 495-506, 1996.
[19] A. K. Jain, Y. Zhong and M. P. Dubuisson-Jolly, "Deformable template models: A review," *Signal Process,* vol. 71, pp. 109-129, 1998.
[20] P. Brigger, J. Hoeg and M. Unser, "B-spline snakes: a flexible tool for parametric contour detection," *Image Processing, IEEE Transactions on,* vol. 9, pp. 1484-1496, 2000.
[21] M. Unser, A. Aldroubi and M. Eden, "B-spline signal processing. II. Efficient design and applications," *Signal Processing, IEEE Transactions on [See also Acoustics, Speech, and Signal Processing, IEEE Transactions on],* vol. 41, pp. 834-848, 1993.
[22] M. Sonka, V. Hlavac and R. Boyle, *Image Processing, Analysis, and Machine Vision.* ,2nd ed.PWS Pub., 1999, pp. 770.
[23] T. H. Cormen, C. E. Leiserson, R. L. Rivest, C. Stein and T. H. Cormen, *Introduction to Algorithms.*, 2nd ed.MIT Press, 2001, pp. 1180.
[24] J. Fessler and A. Hero, "Space-alternating generalized expectation-maximization algorithm," *Signal Processing, IEEE Transactions on [See also Acoustics, Speech, and Signal Processing, IEEE Transactions on],* vol. 42, pp. 2664-2677, 1994.
[25] B. Godbout, J. A. de Guise, G. Soulez and G. Cloutier, "3D elastic registration of vessel structures from IVUS data on biplane angiography1," *Acad. Radiol.,* vol. 12, pp. 10-16, 2005.

Chapter 4.3

active contour model (ACM), 497, 507
angiography, 498
automated segmentation, 497
autoregressive model, 500
B-spline, 509
catheter, 499
Circular Hough Transform (CHT), 503
Conditional Density Propagation (ConDensation), 501, 504
coronary heart disease (CHD), 497
curve fitting, 506
deformable template model (DTM), 497, 507
Derivative of Gaussian (DoG), 508, 509
gradient descent search, 510
guidewire, 498
Intravascular Ultrasound (IVUS), 497
Kalman filtering, 501, 503
particle filtering, 503
registration, 506
Sobel edge detection, 505
space alternating search, 510
tissue ultrasound, 498
tunneling probability, 501
vascular profiling, 498

CHAPTER 4.4

USER-ASSISTED SEGMENTATION FOR MEDICAL IMAGES

Vasco Gervásio and Joaquim A. Jorge
INESC-ID/IST, Technical University of Lisbon,
R. Alves Redol, 9, 1000-029 Lisboa, Portugal,
vmrg@vimmi.inesc-id.pt, jaj@inesc-id.pt

Medical images are an important area of application for image segmentation methods. However, despite considerable research into automatic segmentation methods, many difficult medical images are still processed using manual or manually-assisted methods. This is because automatic methods still do not yield reliable results in many cases and often require tweaking parameters to achieve satisfactory performance. This chapter describes a new technique combining manual and automatic techniques to approach this problem. In our method, users start the segmentation by drawing a rough sketch around the area of interest. Then this sketch is automatically adapted to the desired contour. Our technique allows users to refine the desired solution iteratively using simple strokes in a localized manner. In this way it is possible to correct imperfections and overcome algorithmic limitations while taking advantage of the expeditious convergence afforded by conventional image processing methods. Our experiments show that users are able to produce good quality results in an expedite fashion using few, simple and easy interactions. By doing away with the need for parameter tweaking users can segment images which would otherwise be difficult to process via "purely automatic" methods in a simple and controlled manner.

1. Introduction

Image segmentation separates an image into different parts. To accomplish this task there are many techniques ranging from manual approaches, performed by trained operators, for example in medical images, to a plethora of automatic algorithms.

Segmenting medical images is a difficult problem to solve given the quality of many of these images and the requirement to separate different areas to a high degree of precision. For these reasons, even today, it is very common in medical imaging for operators to manually segment images in critical applications by defining a contour around areas of interest, point by point, in a slow and tedious process.

In this chapter we propose a system that accelerates this process, by using a more advanced segmentation algorithm, while maintaining the same level of precision that can be achieved manually through user interactions. From the many different image

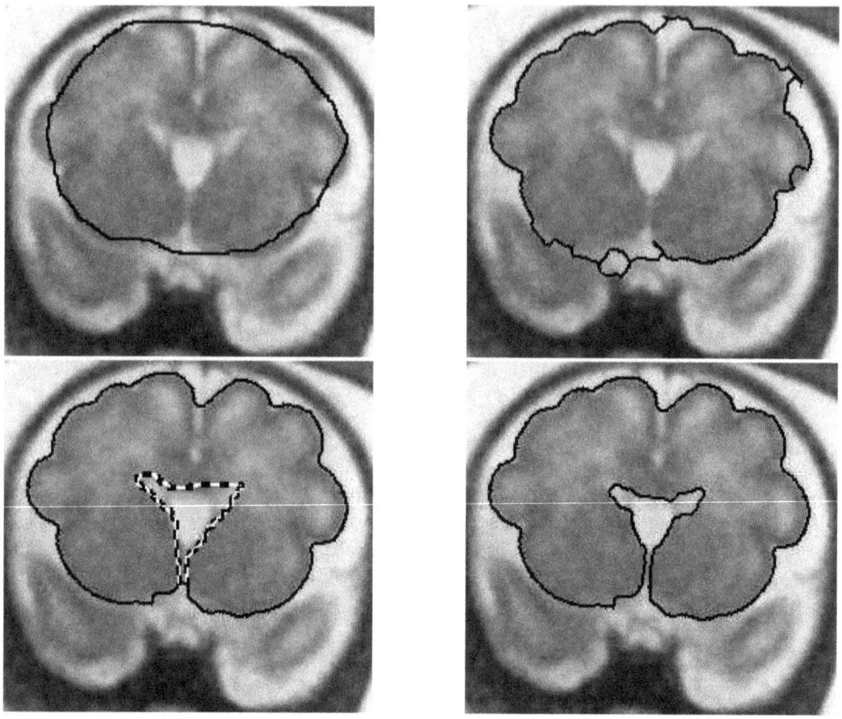

Fig. 1. From top to bottom, left to right: Initial sketch; Result of applying GVF snakes to the sketch; The user sketched an inner extension (dashed line) to the contour after issuing three corrections to improve the outer curve; Final result.

algorithms we have chosen active contour models, or snakes,[1] because they provide a simple and efficient solution, that requires manual initialization of a contour in order to segment a determined region of an image. This contrasts to methods that segment all the image automatically[2],[3] but whose precision is difficult to control, since they are mostly based in thresholds. However, as we will show in the following sections, active contour models, also exhibit limitations in the quality of the segmentation obtained due to convergence problems and sensitivity to noise. To overcome these we present a new technique, which we call *oversketching snakes*.[4]

Our method comprises five different steps. First the user sketches an initial approximation to the desired contour, using a quick stroke. Then we apply a snake algorithm to adjust the contour. The user corrects the resulting contour by sketching over selected areas, as necessary. Each of these local corrections is adjusted by applying a local algorithm that implicitly adapts its parameters, according to the type of correction specified by users in a transparent way, i.e. users need not be aware of this adaptive behavior. Users continue refining the contour until they are satisfied with the result (see **Figure 1**). This method allows us to obtain a solution much quickly than an exclusively manual approach would afford, and closer to the optimal result than what was possible using only snakes. Furthermore the method is much simpler and easier to understand than purely "automatic" techniques, which require tweaking with explicit parameters to achieve reasonable results.

The remainder of this chapter is organized as follows: In section 2 we present related work and interactive techniques to improve segmentation. In section 3 we describe the two most common snake algorithms, regular snakes[1] and Gradient Vector Field (GVF) snakes.[5] In section 4 we explain our algorithm in greater detail and the way it can improve on current results. In section 5 we analyze experimental segmentation results. Finally, in section 6 we present conclusions and future work.

2. Related Work

The many different algorithms used in medical image segmentation can be divided in four classes. **Threshold based methods** are very simple techniques where arbitrary values are chosen to divide the image in several parts, according to the different intensity values of the pixels in the image. These simple methods are often used as auxiliary to others,[6] although more complex variants are used on their own right.[7] The big limitation of this method is the difficulty in choosing the appropriate threshold values.

Region growing[8] methods use a small seed that is then expanded to similar neighboring points until it fills the region of interest.[9] The placement of the seeds is usually done manually. These methods exhibit problems with contour leaks.

Probabilistic models span a large group of algorithms divided in three main categories: First, algorithms that model specific structures such as the brain,[10] whether simple or composite. Second category includes techniques that use feature extraction,[11] where the value of features obtained in the training set are used to identify the regions in new images. Finally, evolutionary algorithms.[12] The major disadvantages of these algorithms are that they require previously segmented images, that are obtained manually, and have problems adapting to cases outside the training examples.

Finally **active contour models** are the most known variation of deformable models[13] and will be the object of this chapter. One variation of these algorithms uses Level Sets[14] and is called **geometric contour models**. These have the advantage to adapt to topology changes, namely structures with holes, and do not require manual initialization. However, because the segmentation results are defined by an implicit function, these are not so easy to address as the regular implementation, where the result is described by control points. If we add to this difficulty the increased complexity and inefficiency of such algorithms, we realize that they are not very amenable to interactive correction as described in this chapter. Paragios[15] describes an interactive technique for level sets which uses fixed points to suitably constrain the implicit function. However, this method is still cumbersome, whereas our technique is more flexible, familiar to users and straightforward to apply.

While much work has been devoted to image segmentation methods, few publications discuss interaction with active contour models and similar algorithms, which in spite (or in virtue) of the necessary user initialization, are particularly suited to address difficult domains such as medical images. Two such references are[16] and.[17]

Foo's survey[16] presents several semi-automatic segmentation methods, mostly applied to three–dimensional images, that are divided in three classes. The first is **user steering**, by which the user can issue constant changes and corrections during the segmentation process. Secondly, **user intervention**, requires that changes are only done when incorrect data is detected, at which point the segmentation process can be restarted or simply allowed to proceed. Finally, in **correction or manipulation of final results**, users are allowed to make corrections after the segmentation process is over. The interactive improvements, as presented in this chapter, show characteristics of this third class. In a similar vein, Hug et al[17] describe interactive techniques to improve segmentation, which like ours are specific to snakes.

Many of the techniques presented in[16] are only accessible to expert users, because they require intimate knowledge of the way specific algorithms work, so that users can change their parameters. In contrast, other algorithms described in the same survey, as in[17] allow detailed manipulation of contours, but on a point–by–point basis. As a consequence, these techniques require lengthy interactions, are unnatural, and can be tricky in the choice of the point to move, which makes them almost as effective as purely manual methods.

Straying away from medical imaging to digital photography where often it is desirable to extract objects from the background of color pictures, there are greater concerns with interaction. Li et al[18] and Rother et al[19] perform segmentation using variations of the Graph Cut algorithm[20] and introduce different ways to improve the resulting segmentation result. In the first technique[18] an user can draw a band and the contour is perfected only within that band. The second method,[19] uses Graph Cut, allowing users to draw strokes to either add or remove parts from the object of interest.

However, because these algorithms extract objects from the background based on color differences, they are not very suited to all images. In particular, for many medical images, often there is not a significant color difference between the object in to be segmented and the background. This makes techniques such as Graph Cut infeasible in practice. Thus, approaches such as those based on snakes become more attractive.

3. Snakes

In the next two subsections we introduce the two snake-based segmentation algorithms (both regular and GVF) that are at the core of our approach and discuss some of their features and limitations as applied to medical images.

3.1. *Regular Snakes*

A normal snake is a curve $x(s) = [\mathrm{x}(s), \mathrm{y}(s)]$ defined as to minimize the total energy

$$E = \int_0^1 \left[\frac{1}{2}(\alpha|x'(s)| + \beta|x''(s)|) + E_{ext}(x(s)) \right] ds \quad (1)$$

where α and β are weights that control, respectively, the tension and rigidity of the curve. The external energy is obtained from the image $I(x, y)$ and should assume the lowest values at the points in the boundary of the region of interest. Because we are interested in approximating contours, a good choice will be $E_{ext}(x, y) = -|\nabla [G_\sigma(x, y) * I(x, y)]|^2$, where $G_\sigma(x, y)$is the Gauss function with standard deviation σ and ∇ is the gradient operator, because the contour points are those which have the highest gradient value and therefore the lowest external energy.

A snake that minimizes (1) should satisfy the Euler equation

$$\alpha \cdot x'' - \beta \cdot x''' - \nabla E_{ext} = 0 \quad (2)$$

that can be interpreted as a balance of forces, where the internal force (first two terms) is responsible for shrinking and smoothing the snake while the external force pushes the curve to the contour points.

By solving Equation 2 we obtain a numerical formula that is applied iteratively until the result stabilizes (for more details the reader should refer to[1]). The resulting formula has three constants that can be easily changed, the α and β mentioned above and a new one, γ, that controls how external forces influence the curve.

Figure 2 shows some limitations of snakes. On one hand regular snakes need to

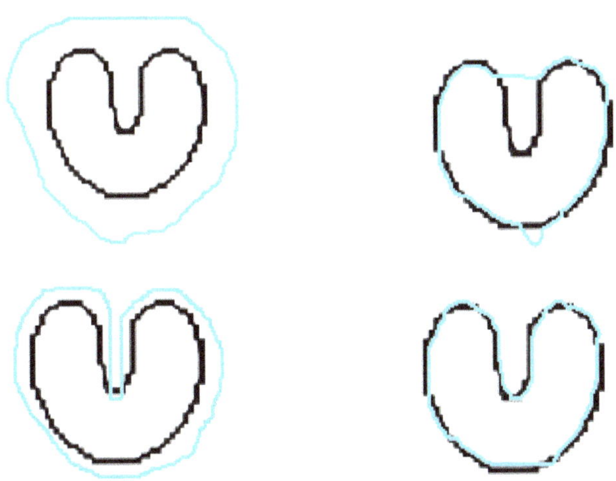

Fig. 2. The left images show the initial contour while the right images display the contour obtained after the application of the regular snakes algorithm.

be initialized very close to the desired contour, due to the low range of external forces (see **Figure 3**). On the other hand they have difficulties converging into concave sections of the desired contour, because of the direction of the external forces in those areas, which make snake curves converge to the top contours, but not inside the concavity (see **Figure 3**). Finally, snakes experience difficulties tracking sharp corners due to using C2-splines to approximate contours, although this tends not to be the major limitation when processing medical images.

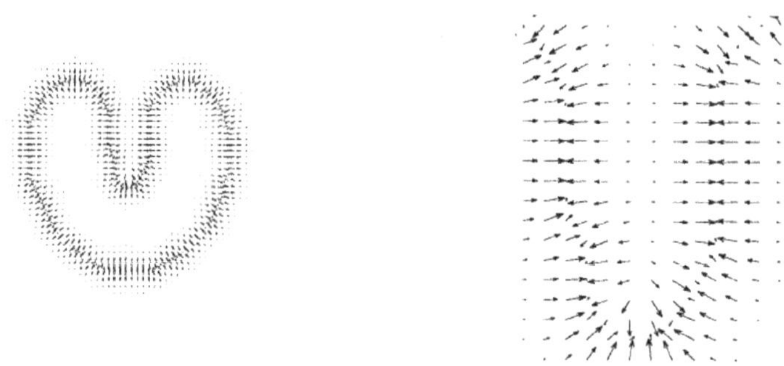

Fig. 3. The left image shows the normal snakes external forces and the right one a close-up of these forces within the boundary concavity.

3.2. *Gradient Vector Field (GVF) Snakes*

GVF snakes attempt to solve the limitations of regular snakes, which we presented in the previous section. To this end they use a new formulation for the external forces, while keeping the same approach to internal forces.

In a GVF snake the component $-\nabla E_{ext}$ is replaced by the GVF field $w(x,y) = [u(x,y), v(x,y)]$ which is obtained by minimizing the function

$$\varepsilon = \iint \mu \left(u_x^2 + u_y^2 + v_x^2 + v_y^2\right) + |\nabla f|^2 |w - \nabla f|^2 dxdy \tag{3}$$

where f is a contour map derived on the image $I(x,y)$, which satisfies the property that its value has to be the highest along the desired contour points. Among the many functions possible the authors chose $f = |\nabla \left[G_\sigma(x,y) * I(x,y)\right]|^2$.

The solution to minimize (3) is obtained by solving the following equations

$$\mu\nabla^2 u - (u - f_x)\left(f_x^2 + f_y^2\right) = 0$$
$$\mu\nabla^2 v - (v - f_y)\left(f_x^2 + f_y^2\right) = 0$$

where ∇^2 is the Laplacian operator. Xu and Prince[5] show how to find numerical solutions to obtain the values of the components u and v for each pixel in the image, through an iterative formula.

By virtue of this modification we can observe that it becomes possible to achieve better results than those possible with normal snakes (see **Figure 4**), because the

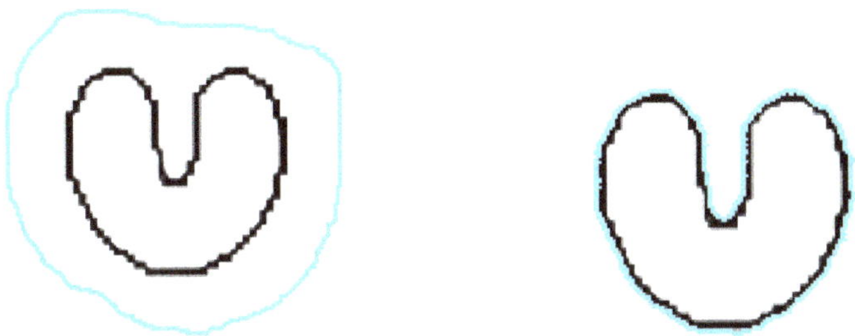

Fig. 4. Application of GVF snakes.

external forces increase both in magnitude and reach and they now point to the insides of concavities as shown in **Figure 5**.

Fig. 5. The left image shows external forces in GVF snakes and the right one shows a close-up of these forces along the boundary of the concavity.

However, these results become worse when either the contours are less clear or in images for which there are many contours next to each other. This is because some parts of the snake can become attracted to one contour and other parts to a different contour (see **Figure 6**).

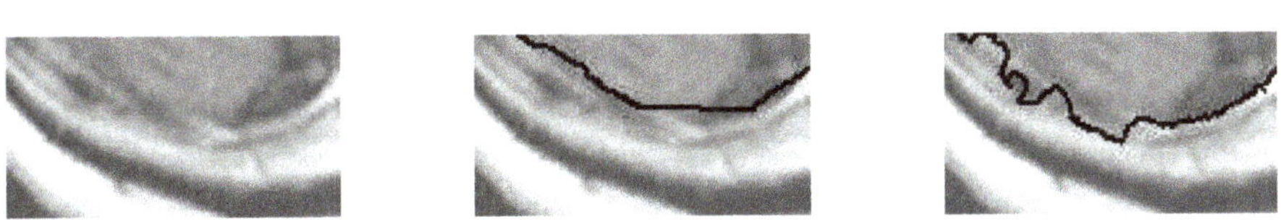

Fig. 6. The first two images show the original image and a user-specified initial contour. The image to the right shows the result of applying GVF snake techniques without further human interaction.

4. Our Method

As we have seen from the previous sections snake-based approaches provide flexible techniques to segmenting images. However, convergence and stability problems make it difficult to apply these techniques in a completely autonomous manner, especially in images where there is a low signal-to-noise ratio, the image gradients are not well-defined or where there are many different areas with similar levels of gray lying adjacent to one another and where there are very stringent requirements on precision. Indeed in critical applications such as brain-scan images, human intervention is required to precisely identify areas of interest such as tumors. Whereas the conventional approach would be to resort to purely manual segmentation, we introduce a hybrid technique, which we call *oversketching snakes*. Our method combines snake-based automatic segmentation guided by human-specified loose sketches with interactive local corrections. In this way human operators are able to correct problem areas by *sketching over contours*, using a tablet PC with an electronic stylus or mouse. These simple interactions allow combining human skill with the excellent performance of snake methods to provide an expedite approach to manually segmenting otherwise problematic images.

4.1. *Motivation*

The two main problems of regular snakes are their limited range in capturing contours and difficulty in converging to concave spots. While these are mostly addressed by GVF snakes, we found by experimentation that even the regular snake-based algorithms can be steered to overcome these problems if users redraw the initial contour a bit more carefully (see **Figure 2**). However, this requires that the segmentation process is repeated over again, and translates into an increased difficulty in drawing the initial contour, so that it could be very similar to the desired curve. Which is worse, this process provides no guaranteed convergence. This is because even though the resulting contour can improve in some places it could actually get worse in other areas of the image due to numerical instability.

We developed a technique which only requires users to redraw the incorrect parts of a contour instead of re-specifying the initial contour many times over. This entailed developing a new approach that allows users to sketch over previously drawn contours to correct them, without the need to erase and redraw. The algorithm would then identify which parts of the contour were modified and apply a modified version of the snake refinement method *to those portions only* while keeping the remaining contour unmodified. We call this technique *oversketching snakes* (see **Figure 7**). The next section describes our method in detail.

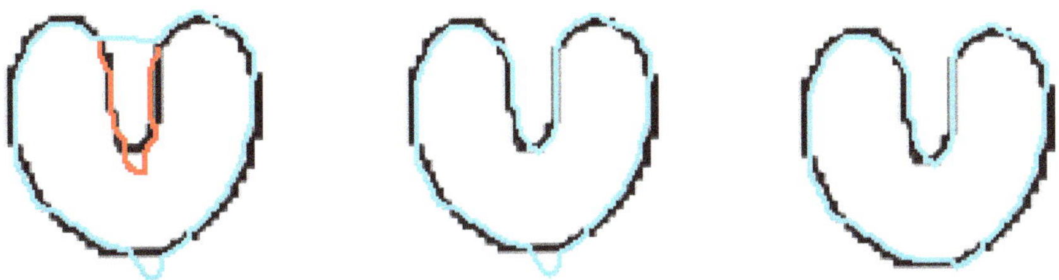

Fig. 7. The left image is the oversketching of one of the result images in Figure 2 and the middle one its result. The right image shows the final result after another correction.

4.2. *Oversketching snakes*

Our procedure for image segmentation works as follows: after the image is loaded we obtain for each pixel its gradient value, using Sobel operators. After, we apply a high–pass filter and ignore the pixels whose gradient falls below a given threshold which we determined by trial and error. Indeed, these pixels amount to noise in low–variation areas. Next, the edges of the image are enhanced by increasing the gradient value of the remaining pixels. We chose these filtering operations, because they require comparatively few resources to compute in contrast to other traditional approaches to noise reduction.[21] Furthermore, their effect in the images is easily predicted, which makes it easier to prevent involuntary elimination of small details, that can be crucial to correctly identifying regions of interest in medical images.

In a second phase, we compute the u and v components required by the GVF snake algorithm. We chose to apply these operations to all image pixels as a pre-processing step, instead of locally, each time the user enters a correction, because this way, even though we spend a little more time processing pixels immediately after loading the image, the interactive part of the process will run faster, which users identified as a most desirable feature.

After the image is loaded users specify a rough approximation to the contour desired by drawing a quick sketch. This contour is then adjusted by applying a conventional GVF snake algorithm (see **Figure 8**).

This step provides users with a closer estimation of the desired contour that serves as a starting point to possible corrections. In this initial phase, it is most important to have at least parts of the initial sketch attracted by the desired contours. Accordingly, we place emphasis on the external force component of the GVF snake algorithm by increasing the value of γ to double its default setting. Furthermore, it is important to try to capture as many salient features in an image as possible, namely corners. To allow for that we reduce the value of β to 0 (see the left image of **Figure 8**) in this second step. Setting $\beta = 0$ allows snakes to be less smooth and therefore to better adjust to sharp variations in the contour. After applying the GVF the resulting contour can be inspected for quality. If the user is satisfied with the obtained curve, the process ends.

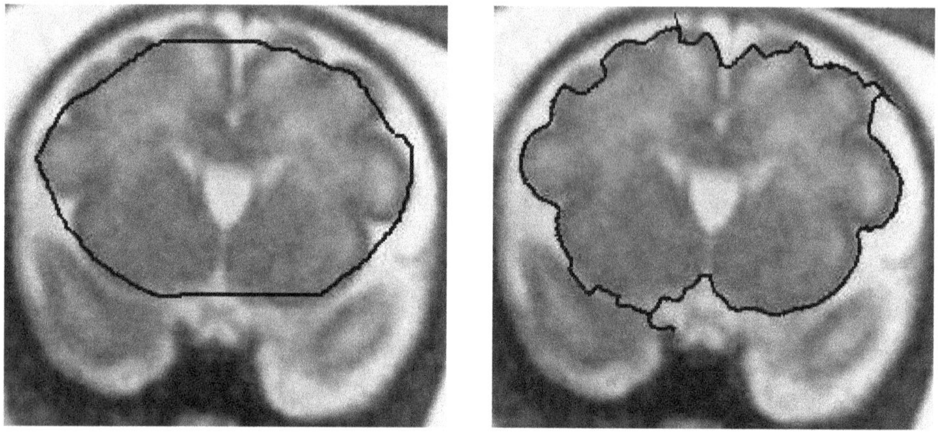

Fig. 8. The left image shows the initial sketch and the right the result after the application of the GVF snake.

Otherwise, he or she sketches corrections to the contour which are closer to the desired result, thus replacing part of the previous snake, without the need to erase the portions sketched over. This is an important feature, since it makes corrections easier to specify and makes the dialogues more fluid. This works as follows: the algorithm obtains the points of the previous contour that are closest to the starting and end points of user-specified sketch and replaces the smallest number of points between them with the points of the new contour, to obtain a closed curve (see **Figure 9**) such that the remaining points of the initial contour stay unchanged.

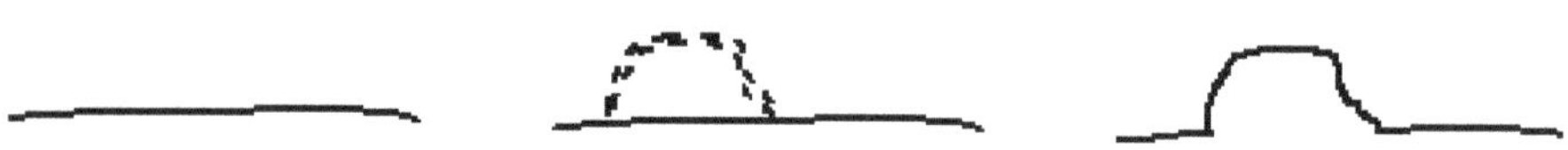

Fig. 9. Example of the oversketch mechanism. Left shows original contour. Middle shows correction in a dashed line. Right shows corrected contour.

Next we apply a *modified* GVF snake to *the points in the user sketch*, in order to adjust the correction to the contour (see **Figure 10**). To this end, we have performed four main modifications to the GVF snake algorithm which we describe below.

First, we modify the GVF snake to work with open contours, so that we do not recompute the contour which is unchanged by user-specified corrections. Since the GVF computations require neighbor information for each point, we consider the two points where the sketched correction joins the previous contour as the missing neighbors. These new points are fixed and therefore the algorithm is only applied to the points between those two and not to the remaining contour.

Second, we reduce the reach of the GVF snake, by bounding the number of iterations in computing u and v components to ten rather than the usual *eighty*.

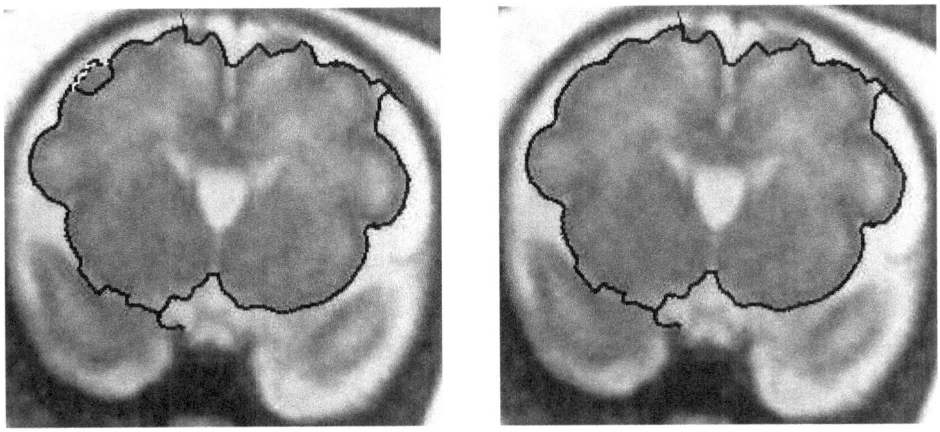

Fig. 10. The left image shows a correction (dashed line) and the right the result after it is adapted.

This is because the user is correcting a (hopefully smaller) part of the result and therefore it is more likely that the new user-sketched segment is closer to the desired contour. Thus the influence of other edges farther from the correction should be prevented (see **Figure 11**).

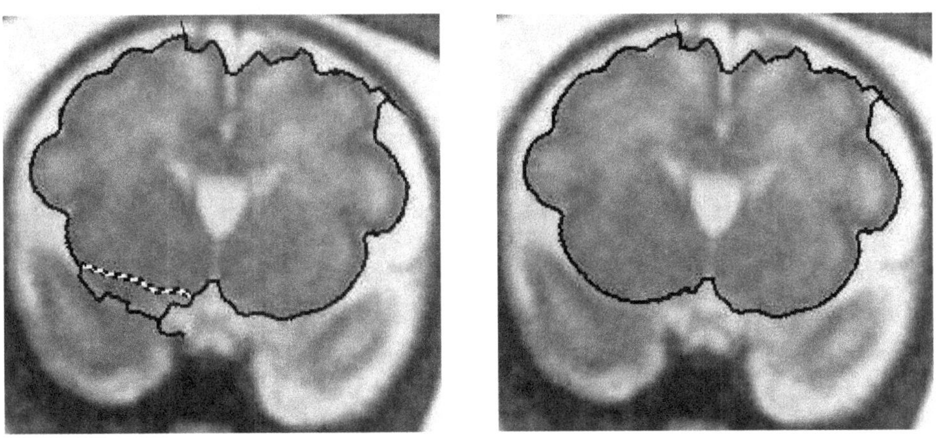

Fig. 11. The left image shows a correction (dashed line) and the right the result after it is adapted.

Third, in order to obtain a smoother and better-looking contour the value of β is increased (see **Figure 12**).

Finally, if the length of the new sketch is very small as compared to the previous contour, it is likely that the new segment is very close to the desired contour and the user is entering final minor details. Since, the GVF algorithm is not very accurate with very few points, the points of the new sketch are simply approximated by a spline. **Figure 13** illustrates this.

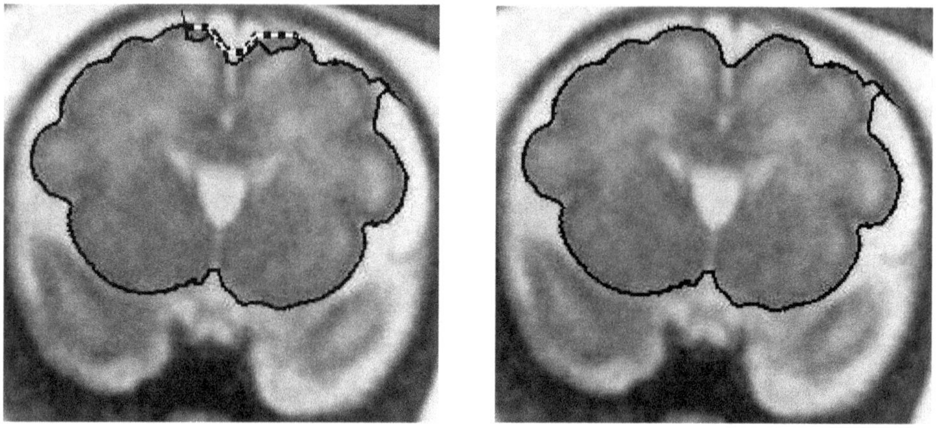

Fig. 12. The left image shows a correction (dashed line) and the right the result after it is adapted.

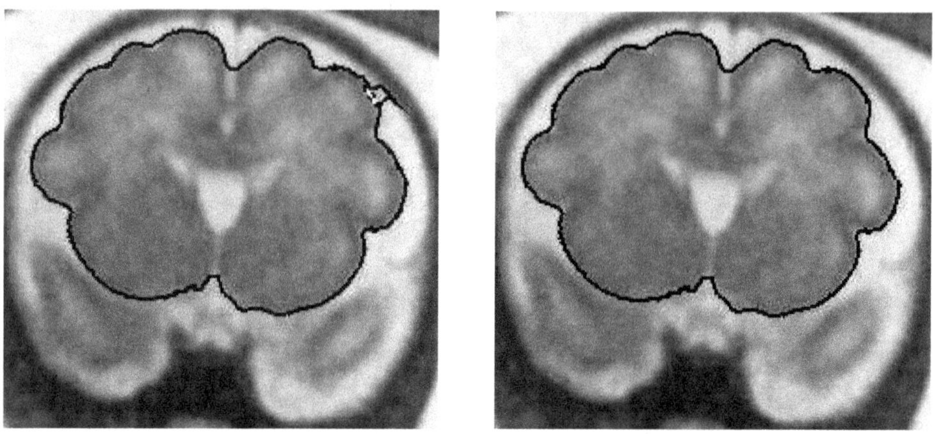

Fig. 13. The left image shows a correction (dashed line) and the right the result after it is adapted.

Furthermore, our method also allows users to refine the contour, rather than just correcting algorithmic problems. Therefore, instead of making a complex initial contour the user can specify a simpler one (e.g. convex) through a rough sketch, correct it as desired and then add more details (e.g. the concave part), to improve the final result as shown in **Figure 14**.

The process ends when the user is satisfied with the result. In practice we have found out that half-dozen iterations can suffice for an experienced user to obtain a satisfying contour for images with average complexity.

4.3. *Analysis*

The algorithm described in the previous section allows us to overcome not only flaws of the regular snakes but also those typical of GVF snakes:

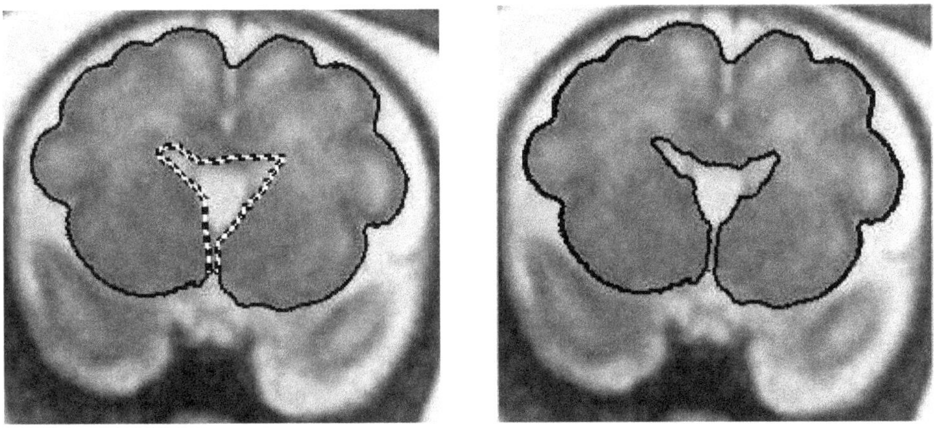

Fig. 14. The left image shows a user-sketched refinement of the contour (dashed line). The right image shows the final result.

- Convergence — Our method allows users to iteratively change the boundary curve after each segmentation, by performing local corrections, that only affect the desired parts of the contour. Because of the imprecise nature of sketches, users can make bad user corrections. Thus the result of each new change can become worse than the previous segmentation, and we cannot claim absolute convergence. However, a bad correction can be canceled by pressing the Undo button. We can plausibly argue that this feature makes convergence feasible;
- Efficiency — Our corrections are expeditious due to three main reasons. First, the components necessary for the GVF snake (gradient, u and v vectors) are calculated in a preprocessing step while the image was loaded. Second, all corrections are local, which means that only the points that were added to the contour, with the new correction, are used in subsequent computations, whilst all other points in the previous contour remain unchanged. Finally, the number of points in each correction is usually much smaller than all the points in the contour and with each subsequent correction this number decreases, because more of the contour becomes correct;
- Ease of use — The user interacts in a simple and natural way by drawing few strokes, in contrary to other implementations of snake algorithms where the segmentation is initialized with predefined shapes (circles) and the contour can only be manipulated by manually moving each of the points that define the contour - a tedious process at best. Furthermore, the user does not need to know the algorithm in detail, because all parameter changes are done automatically, without human intervention;
- Robustness — One of the limitations of algorithms based on active contour models lies in their dependence on the initialization, whereas different initial

sketches yield different results. Even though our method shares this problem with other techniques in our case this is solved by allowing the users to perform extra corrections. This makes our approach less dependent on a good initial sketch.

While our technique is also sensitive to the initial gradient estimates as GVF snakes, where this influence depends equally on the value of the gradient (higher values signify more defined contours) and the distance of a contour to the initial sketch, in subsequent user-defined strokes we increase the importance of distance to the contour by reducing the scope of the GVF snake, as mentioned in the previous subsection. In this way, poorly-defined contours are more easily captured when users sketch corrective strokes close to them. Thus we have developed a suitable technique to overcome limitations of snakes by means of user interaction.

5. Results

In this section we analyze the segmentation results from several images using the algorithm described in the previous section. For each of the images we show the initial sketch, the contour after applying the GVF snake, with the following corrections highlighted in dashed lines, and the resulting final contour.

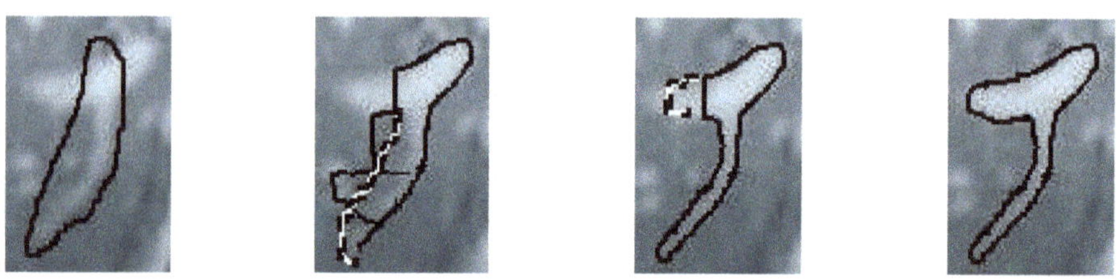

Fig. 15. From left to right: initial sketch, result of the GVF snake (black line) with the first correction (dashed line), the next contour with another correction and the final result.

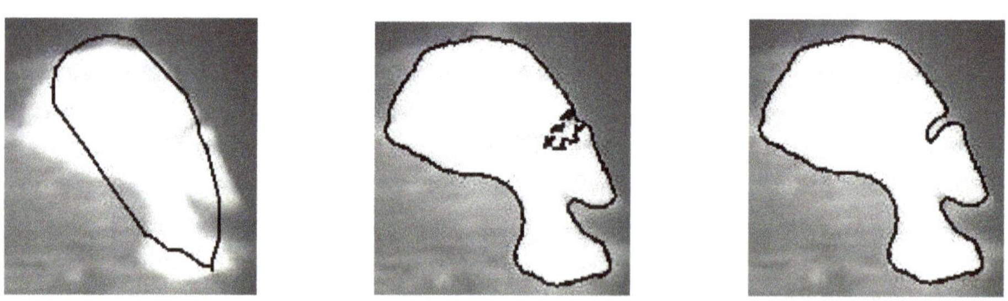

Fig. 16. From left to right: initial sketch, the result of GVF snake (black line) with the correction (dashed line) and the final result.

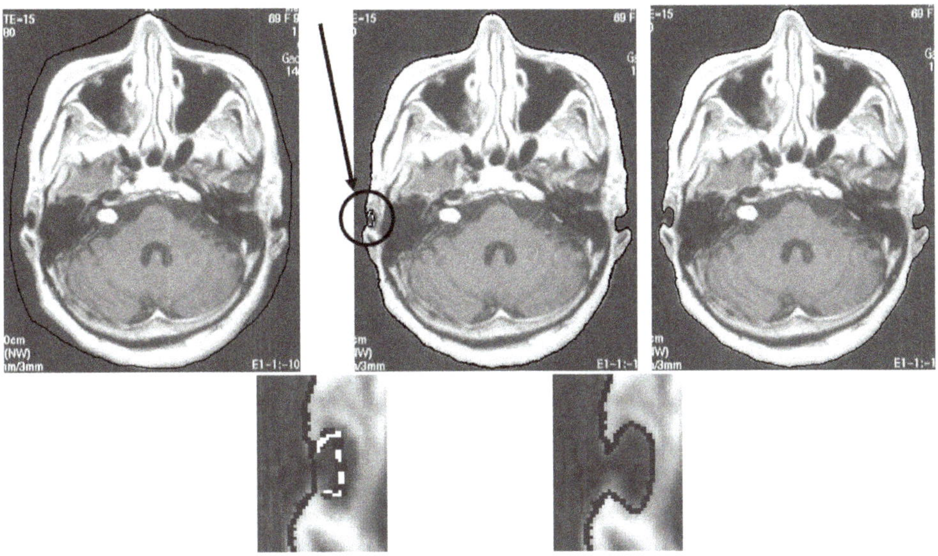

Fig. 17. The two left images on top show the initial sketch and the result of applying the GVF snake (black line) with a correction (dashed line) highlighted by the circle. The two images below show an augmented view at the previously selected correction site and its result. The image on top to the right shows the final result.

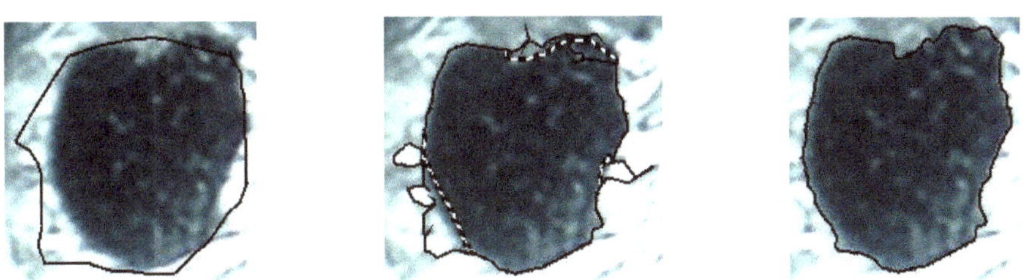

Fig. 18. From left to right: initial sketch, result of GVF snake (black line) with three corrections (dashed lines) and the final result.

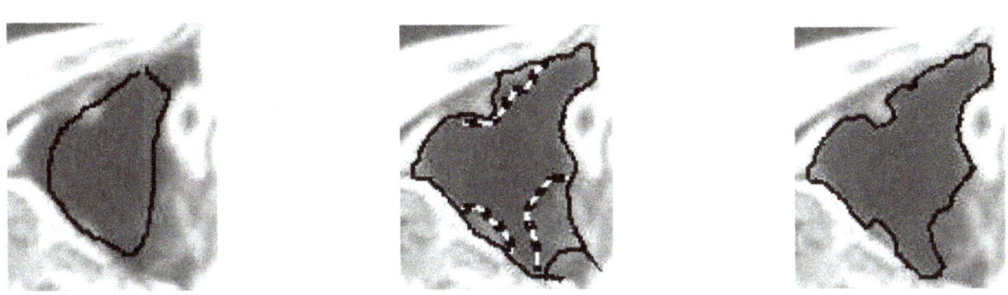

Fig. 19. From left to right: initial sketch, result of GVF snake (black line) with three corrections (dashed lines) and the final result.

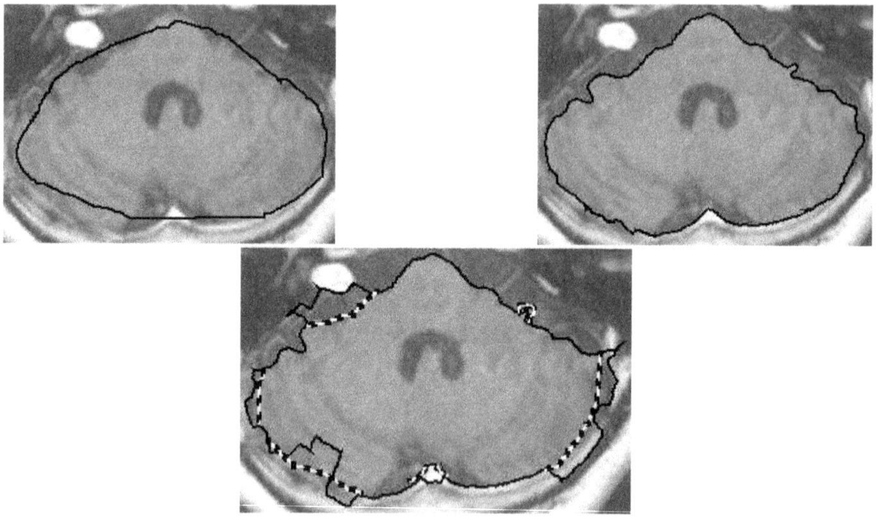

Fig. 20. Top row: initial sketch and final result. Bottom row: intermediate result of GVF snake (black line) with six corrections (dashed lines).

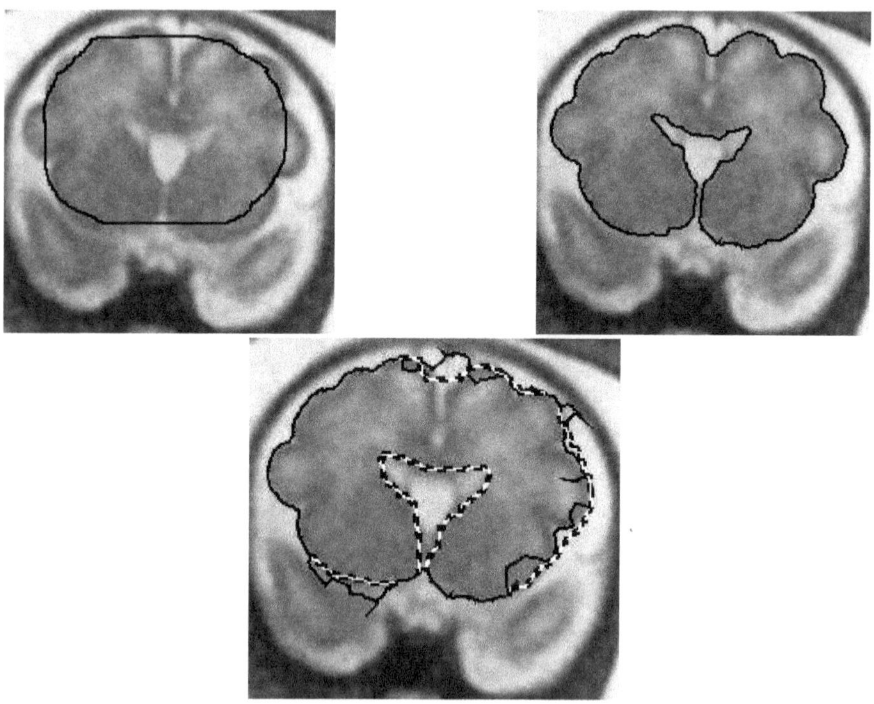

Fig. 21. Top row: initial sketch and final result. Bottom row: intermediate result of GVF snake (black line) with four corrections (dashed lines).

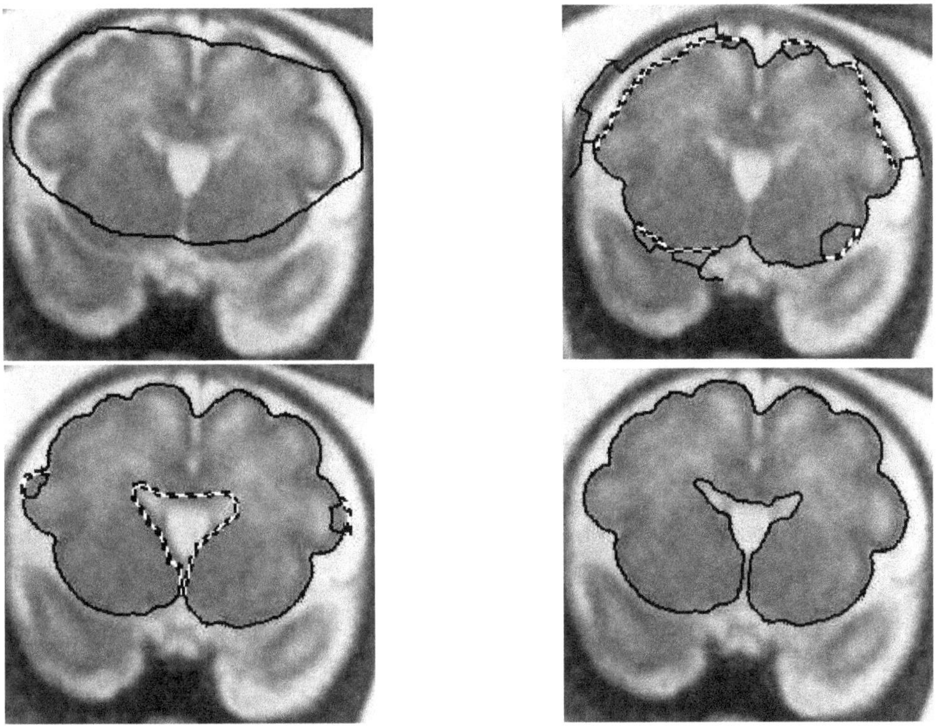

Fig. 22. Top row: initial sketch and result of GVF snake (black line) with five corrections (dashed lines). Bottom row: result after the previous corrections (black line) with three more corrections (dashed lines) and the final result.

Table 1. Different measures for the various segmentation results.

	Number of points of the initial contour	Time for GVF snake (in sec.)	Number of corrections	Total time for corrections (in sec.)
Figure 15	355	0.351	2	4.486
Figure 16	356	0.170	1	2.624
Figure 17	1917	3.515	1	2.293
Figure 18	577	0.210	3	7.876
Figure 19	310	0.260	3	9.216
Figure 20	903	1.352	6	19.090
Figure 21	887	1.303	4	15.436
Figure 22	754	0.751	5+3	28.571

Table 1 presents, for each of the images, the number of points in the initial sketch, filtered such there is no less more than two pixels between consecutive points, the time required by the initial GVF snake run, the number of corrections performed to obtain the final result and the time it took to make all the corrections.

The results show that, even though the GVF snake algorithm is efficient, in several occasions it needs to be partially corrected, particularly in situations with undefined contours (**Figures 15**, **19** and **20**), or with different contours close to each other (**Figures 18**, **20** and **22**) and to correct local details (**Figures 16** and **17**). Depending on the initialization (**Figures 21** and **22**) the number of necessary corrections increases, but even though it takes a few more interactions the technique still converges to the desired result.

Each of these corrections takes, on average, between two and four seconds, depending on its complexity. These runtimes are acceptable, because there are many cognitive steps required to make a correction (the user has to identify the need, apply the oversketching stroke and wait for the contour to adjust) and only a few corrections are necessary to achieve good results with quality comparable to manual segmentations.

6. Conclusions and Future Work

We have shown an hybrid technique combining modified snake algorithms with sketched input to overcome limitations of automatic approaches in segmenting medical images. This technique allows improving segmentation results, obtained by other algorithms, both in quality and complexity, through user interaction. Although this intervention may slow the process down, the extra time required is largely compensated, especially in domains such as radiology which yield low-contrast noisy images because this "semi-automatic approach" combines the best of two worlds: on one hand considerably more expeditious than a purely manual solution and on the other it improves on the precision of the final segmentation result, by allowing an expert to improve the solution through simple and local actions.

Future areas of exploration may include user tests to further evaluate this work, by comparing the results with manual segmentations, applying oversketching in conjunction with more advanced algorithms,[22] to reduce the number of corrections, while maintaining the same final precision, and assess the possibility of using different algorithms for obtaining the initial contour approximation and to compute new contours after each correction. We feel that there is a fertile ground for the application of these techniques in many useful and important areas of medical imaging.

Acknowledgment

Vasco Gervásio was supported by the Portuguese Foundation for Science and Technology, grant SFRH/BD/16014/2004.

References

1. M. Kass, A. Witkin, and D. Terzopoulos, Snakes: Active contour models, *International Journal of Computer Vision.* **1**(4), 321–331 (Jan., 1988).
2. R. H. Vu, R. M. Rangayyan, and G. S. Boag. Multi-seed segmentation of the primary tumor mass in neuroblastoma using opening-by-reconstruction. In *BioMed'06: Pro-*

ceedings of the 24th IASTED international conference on Biomedical engineering, pp. 242–249, Anaheim, CA, USA, (2006). ACTA Press. ISBN 0-88986-578-7.
3. M. Antonelli, B. Lazzerini, and F. Marcelloni. Segmentation and reconstruction of the lung volume in CT images. In *SAC '05: Proceedings of the 2005 ACM symposium on Applied computing*, pp. 255–259, New York, NY, USA, (2005). ACM Press. ISBN 1-58113-964-0. doi: http://doi.acm.org/10.1145/1066677.1066738.
4. V. Gervásio and J. A. Jorge, Oversketching snakes for medical image segmentation, *International Journal of Tomography & Statistics*. (to appear).
5. C. Xu and J. L. Prince, Snakes, shapes, and gradient vector flow., *IEEE Transactions on Image Processing*. **7**(3), 359–369 (Mar., 1998).
6. S. Hu, E. A. Hoffman, and J. M. Reinhardt, Automatic lung segmentation for accurate quantitation of volumetric X-ray CT images, *IEEE Trans. Med. Imaging*. **20**(6), 490–498, (2001).
7. B. Vasilic and F. W. Wehrli, A novel local thresholding algorithm for trabecular bone volume fraction mapping in the limited spatial resolution regime of in vivo mri, *IEEE Trans. Med. Imaging*. **24**(12), 1574–1585, (2005).
8. R. Adams and L. Bischof, Seeded region growing, *IEEE Trans. Pattern Anal. Mach. Intell.* **16**(6), 641–647, (1994).
9. R. Martin, N. Bordes, T. Hugh, and B. Pailthorpe. Semi-automatic feature delineation in medical images. In *APVis '04: Proceedings of the 2004 Australasian symposium on Information Visualisation*, pp. 127–131, Darlinghurst, Australia, Australia, (2004). Australian Computer Society, Inc. ISBN 1-920682-17-1.
10. P.-Y. Bondiau, G. Malandain, S. Chanalet, P.-Y. Marcy, J.-L. Habrand, F. Fauchon, P. Paquis, A. Courdi, O. Commowick, I. Rutten, and N. Ayache, Atlas-based automatic segmentation of MR images: validation study on the brainstem in radiotherapy context, *International Journal of Radiation Oncology*Biology*Physics*. **61**(1), 289–98 (January, 2005). doi: doi:10.1016/j.ijrobp.2004.08.055.
11. M. Loog and B. van Ginneken, Segmentation of the posterior ribs in chest radiographs using iterated contextual pixel classification, *IEEE Trans. Med. Imaging*. **25**(5), 602–611, (2006).
12. R. Li, M. T. Emmerich, J. Eggermont, and E. G. Bovenkamp. Mixed-integer optimization of coronary vessel image analysis using evolution strategies. In *GECCO '06: Proceedings of the 8th annual conference on Genetic and evolutionary computation*, pp. 1645–1652, New York, NY, USA, (2006). ACM. ISBN 1-59593-186-4. doi: http://doi.acm.org/10.1145/1143997.1144268.
13. T. McInerney and D. Terzopoulos, Deformable models in medical image analysis: a survey., *Medical Image Analysis*. **1**(2), 91–108, (1996).
14. S. Osher and J. A. Sethian, Fronts propagating with curvature-dependent speed: algorithms based on Hamilton-Jacobi formulations, *J. Comput. Phys.* **79**(1), 12–49, (1988). ISSN 0021-9991. doi: http://dx.doi.org/10.1016/0021-9991(88)90002-2.
15. N. Paragios. User-aided boundary delineation through the propagation of implicit representations. In *MICCAI 2003: Proceedings of the 6th International Conference on Medical Image Computing and Computer-Assisted Intervention*, pp. 678–686, (2003).
16. J. L. Foo. A survey of user interaction and automation in medical image segmentation methods. Technical report, Iowa State University - Human Computer Interaction (Jan., 2006).
17. J. Hug, C. Brechbuhler, and G. Szekely. Tamed snake: A particle system for robust semi-automatic segmentation. In *MICCAI*, pp. 106–115, (1999).
18. Y. Li, J. Sun, C.-K. Tang, and H.-Y. Shum, Lazy snapping, *ACM Trans. Graph.* **23**(3), 303–308, (2004). ISSN 0730-0301. doi: http://doi.acm.org/10.1145/1015706.1015719.

19. C. Rother, V. Kolmogorov, and A. Blake. "Grabcut": interactive foreground extraction using iterated graph cuts. In *SIGGRAPH '04: ACM SIGGRAPH 2004 Papers*, pp. 309–314, New York, NY, USA, (2004). ACM Press. doi: http://doi.acm.org/10.1145/1186562.1015720.
20. Y. Y. Boykov and M.-P. Jolly, Interactive graph cuts for optimal boundary & region segmentation of objects in N-D images, *Eighth IEEE International Conference on Computer Vision, 2001. ICCV 2001. Proceedings.* **1**, 105–112, (2001). doi: http://doi.ieeecomputersociety.org/10.1109/ICCV.2001.10011.
21. J. Weickert. A review of nonlinear diffusion filtering. In *SCALE-SPACE '97: Proceedings of the First International Conference on Scale-Space Theory in Computer Vision*, pp. 3–28, London, UK, (1997). Springer-Verlag. ISBN 3-540-63167-4.
22. J. Cheng and S. W. Foo, Dynamic directional gradient vector flow for snakes., *IEEE Transactions on Image Processing.* **15**(6), 1563–1571 (Jun., 2006).

CHAPTER 4.5

AUTOMATED QUANTIFICATION OF CELLULAR PROCESSES WITH APPLICATIONS TO HIGH-CONTENT SCREENING

Vannary Meas-Yedid[1], Alexandre Dufour[1], Bo Zhang[3], Alexandre Grassart[2], Nathalie Sauvonnet[2] and Jean-Christophe Olivo-Marin[1]

1. Institut Pasteur, Quantitative Image Analysis Unit, CNRS URA 2582, 25-28 rue du Dr. Roux, 75724 Paris Cedex 15, France
vmeasyed@pasteur.fr

2. Institut Pasteur, Biology of Cell Interactions Unit, CNRS URA 2582, 25-28 rue du Dr. Roux, 75724 Paris Cedex 15, France

3. Medisys Research Lab, Philips Healthcare, 33 rue de Verdun, BP 313, 92156 Suresnes Cedex, France

Recent advances in fluorescence microscopy provide us with an unprecedented powerful way to observe many biological objects (pathogens, vesicles, single molecules, etc.) and events (intracellular trafficking, adhesion, migration, pathogens entry) in the living cell under multiple experimental conditions. To understand the underlying biological phenomena, modern microscopy imaging confront researchers to the challenges of extracting and analyzing biological-relevant information from huge amounts of data. In this chapter, we focus on recent advances on automated image analysis frameworks to quantify specific cellular processes. Our aim is to demonstrate that systematic application of such frameworks on thousands of images produce statistically relevant measures of several interesting biological phenomena at the cell population level. Methodologically, we shall emphasize on a recently published robust cell segmentation approach using coupled shape-constrained active contours, and a statistical spot extraction method based on a multiscale variance stabilizing transform. We finally present an example of such approaches to the quantification of cellular endocytosis that demonstrates its usefulness to screening applications in biology.

1. Introduction

Recent advances in fluorescence-based bio-imaging techniques and laboratory automation devices have led high-content screening campaigns to produce exponentially increasing data amounts, boosting the need for automated and robust image analysis techniques.

Among the numerous bioimaging challenges to solve, we focus on two recurrent problems that pose a real problem in terms of image analysis: cell segmentation and spot detection. These tasks are of crucial importance in many applications involving cell counting or spot detection and localization within cells. In the latter case, both cell segmentation and spot extraction are conjointly performed in order to perform quantification on a cell-to-cell basis. Cells and spots visually appear at different scales in a given image. Consequently, their extraction requires different sets of methods, each adapted to the appropriate scale. Following this scale-dependent process, we shall start by giving in section 2 an overview of the techniques tailored for automated cell segmentation. We shall then review in section 3 a number of methods developed for efficient spot detection and extraction in challenging conditions. In section 4, a concrete application is presented where we propose to adapt cell segmentation and spot detection approaches for the study of cellular endocytosis, as recently published in [16]. We conclude the chapter in section 5 by discussing other types of applications relying on high-content screening and automated image processing to quantify biological phenomena at the cell level.

2. Cell segmentation

2.1. *Goals and challenges*

We give below a review of existing solutions available for cell segmentation. The general purpose of image segmentation is to partition the image into groups of pixels belonging to the same entity (i.e. background or cells). As we shall see along this section, this task can be performed in very diverse fashions. For easier reading, the state-of-art methods shall be presented in order of increasing computational complexity.

2.2. *Standard approaches*

2.2.1. *Intensity thresholding*

The simplest approach to cell segmentation is to classify image pixels with respect to their intensity according to one or more threshold values. Intensity thresholding requires that the entire cells exhibit everywhere a different intensity than the background. This requirement imposes the use of combined nucleic and cytoplasmic markers rather than membrane markers. Threshold values can be determined automatically using various methods, generally based on a statistical analysis of the intensity histogram [14, 34, 53].

Naive thresholding algorithms relying solely on pixel intensities are generally very fast, and are therefore widely used in numerous time-critical applications including high-throughput imaging setups. Yet, global thresholding is strongly subject to image noise and background variations, therefore correct segmentation can only be achieved on images with a flat background and a high signal-to-noise ratio. While the former condition is rarely met, the latter can be fulfilled at the cost of increasing exposure time and cell photo-damage. To overcome this problem, one may turn to adaptive thresholding methods, aided by various pre-processing filters and post-processing operators such as mathematical morphology. The other and most important limitation of such methods is that they are restricted to the segmentation of non-touching cells. In practice, this condition is rarely met, as cells may obviously touch, separate or even divide over time. Clusters of cells exhibiting a similar intensity distribution cannot be set apart automatically, and more complex algorithms must be involved to determine the separation between touching cells.

2.2.2. *The watershed transform*

A more advanced set of segmentation techniques rely on the watershed transform [21]. Here the image is considered as a topographic map defined by the pixel intensities. The principle is to iteratively *flood* this map from its local minima, so as to divide the image into regions (*the catchment basins*) and their delimiting contours (*the watersheds*). In the ideal case, this method can separate touching cells, provided each cell has a local minimum located approximately in its center. The major drawback of watershed transforms is their tendency to yield over-segmentation (i.e. producing more regions than there actually are in the image), mostly due to noise in the image. As in the case of thresholding methods, pre-filtering algorithms can be applied to smooth the image before flooding, and post-processing operators can be used to merge regions that are considered part of the same entity, according to empirical parameters such as region size or mean intensity. Other enhancements of the watershed transform for segmentation involve modifying the flooding speed locally, i.e. according to local image gradient. Applications of the watershed technique for cell segmentation can be found in [30, 36, 48, 52].

2.2.3. *Template matching*

Another technique that has received significant attention for cell segmentation is based on template matching. The idea is to define a template of the cell to segment (e.g. based on its intensity profile), and find the best match or matches of this template in the observed image. This technique has been employed to track various types of cells such as amoebas [3], algal cells [53], leukocytes [1, 25], and more recently keratocytes [49] and blood stem cells [23]. The major drawback of template matching methods is their lack of flexibility regarding cell deformation. Indeed, deforming cells do not conform with the template anymore, and a full database of templates should then be created in order to recognize each cell during its generally continuous deformation. Since single template matching can be a computationally expensive process, dealing with multiple templates is clearly undesirable for practical applications.

2.3. *Active contours*

2.3.1. *Background*

The last decade has witnessed a growing trend for deformable models and their applications to segmentation and tracking tasks, thanks to their flexibility, robustness to noise and quantitative interpretability [13, 17, 29, 32, 38, 43, 58]. The principle is to deform an initial contour towards the boundary of the cell by minimizing a pre-defined energy functional. Deformable models have received extensive attention from the computer vision community thanks to the numerous advantages they offer regarding object displacements and deformations monitoring. Indeed, various shape descriptors can be directly computed from a curve or surface, and these descriptors can be straightforwardly compared from frame to frame, and in our case from cell to cell. Another advantage resides in the implicitness of the association step mentioned earlier. The final position of a contour on a given frame can be directly used as the initial position of the model on the next sequence frame, thus providing a good initial guess of the cell's future position [17, 27, 32, 39, 56]. This strategy may further improve the contour deformation or its monitoring, e.g. via shape constraints [38]. This procedure however requires that cells do not move significantly between consecutive frames, such that their successive positions have at least a small overlap region. If this requirement is not met, one can turn to more sophisticated methods allowing a contour to reach an object lying far from its initial position. e.g. extended gradient vector flows [39] or incorporating known or estimated cell dynamics [13, 43]. Finally, in case cells move into contact, one may rely on the prior assumption that cells do not merge and thus introduce a coupling term that prevents contours from intersecting [17, 19, 28, 55, 57].

The problem can be formalized mathematically as a sum of several energy terms E_{image}, E_{regul}, E_{coupling}, E_{prior}, (see description of each below) that needs to be minimized for the contour to be stabilized at a position corresponding to the borders of the objects. To solve the energy-minimization process, one may employ a number of mathematically sound frameworks, among which variational and statistical approaches are the most popular [2, 44]. The key issue actually boils down to designing adequate energy terms so as to deform the contour(s) towards the cell boundaries in the most robust fashion. In some cases, this can turn out to be strongly application-dependent. We give below a more detailed description of each set of terms.

- Data attachment terms E_{image} describe the relation between the contour and the image data in order to compute a driving force that moves the contour towards the cell boundary. One may classify these terms in two categories: 1) *edge-based* terms attract the contour towards local intensity gradients [24]. These terms are suited for phase-contrast microscopy, where boundaries are characterized by steep intensity variations. The major drawback of local gradient terms is that distant boundaries cannot be reached, imposing the initial contour to be set close from the solution. A simple walkaround to this issue consists in inducing a permanent inflation (or deflation) to the contour, in order to 'grab' the cell boundary lying far away [12]. Clearly though, such solutions fail when the initial contour sits across the actual cell boundary. More sophisticated methods rather use gradient vector flows to extend the influence of local gradients over the entire image domain [40, 50]; 2) in fluorescence microscopy, edge information is generally weak due to low signal-to-noise ratios. One must then turn to *region-based* terms, attracting the contour towards the position that best separates homogeneous image regions (e.g., in terms of average intensity, texture, etc.). In fluorescence microscopy, region-based terms perform best with fully-stained cells, which requires either to employ combined nuclear and cytoplasmic fluorescent markers (e.g. in confocal imaging), or to observe cell auto-fluorescence when available (e.g. in two-photon imaging).

- Image segmentation in general is an *ill-posed* problem in the sense that small variations in the observed data quickly induce large variations in the segmentation result. To control this instability, classical solutions to inverse problem regularization consist in minimizing the first derivative(s) of the solution, which are incorporated into the general functional [46]. Minimizing the terms E_{regul} regularizes the convergence process, increasing the overall robustness of the method by reducing risks of falling into local energy minima for a small trade-off in segmentation quality. As an example, typical regularization terms used in active contour frameworks minimize the global length or curvature of the deforming contour, so as to avoid the contour to segment spurious noise artifacts in the image.
- Coupling terms E_{coupling} aim at preventing a contour or set of contours from considering touching cells as a unique object, inducing a wrong segmentation and consequently an invalid tracking, since the identity of all touching objects is lost upon contact. Coupling may be achieved using various strategies, depending on the mathematical formulation of the contour, as we shall see below.
- The final set of terms E_{prior} regroups all the other type of information that can be exploited to enhance either segmentation or tracking. For instance, prior information can be used to enhance boundary localization of touching cells. Although the coupling terms ensure that contours do not overlap, indeed determines a separation, there is no guarantee that this separation reflects the reality. A simple solution is to propagate previous information such as cell volume, exploiting prior knowledge that cell volume does not change significantly over time [17]. This assumption is especially true in 3D imaging, while in 2D volume may visually vary for crawling cells such as amoebas. A recent extension of this idea involves estimating the

best frontier between the touching cells by directly exploiting local image data [19].

The combination of these terms offers a great flexibility to the approach, as shown by its diverse applications [58]. In the following we shall give a particular emphasis on our work on multi-cell segmentation and tracking using deformable models, and describe a general energy functional combining both edge-based and region-based information, together with implementation details. Note that others have also focused on combining such information to improve segmentation [10].

Finally, a novel approach combining active contours and the graph-cut theory was recently applied to segment cells in RNAi screening campaigns [11]. The general principle of graph-cut segmentation consists in representing the image as a weighted undirected graph linking a source (e.g. a pixel within of an object) to a sink (e.g. a pixel in the background) through the entire pixel grid, and find the best partition of this graph according to a min-cut / max-flow strategy. Here also an energy functional must be defined to set the graph weights between neighboring pixels. This theory has the advantage of converging towards the global energy minimum while reducing the overall computation time. With this idea in mind, the theory can be easily transcribed to the active contour formalism (see [51] for an example). Yet, not all energies are suitable for minimization using graph-cuts, making the framework less flexible than the general formulation presented above.

2.3.2. *Application to the segmentation of clustered cells*

In numerous applications, cells in culture often appear clustered on the images, yielding several challenges such as determining the number of cells automatically and locating the frontier between touching cells in a precise manner. In an active contour context, the former issue requires particular attention, as the number of cells directly determines the number of contours to initialize in the image. The latter is due to the fact that cells globally exhibit similar intensity profiles, therefore intensity alone is not sufficient to find the correct frontier between touching cells. To solve this issue, one must generally incorporate additional prior knowledge to enhance boundary localization. We present below a recently published framework for fully automatic cell segmentation using coupled active contours that solves the above-mentioned issues by introducing a robust initialization scheme and a new shape constraint term in the energy functional [16].

Active contours intialization Using multiple active contours requires a specific initialization for each contour in the image, often provided manually for clustered objects. This step can be automatized using prior knowledge on the biological protocol. In our protocol (cf. section 4), cells are stained such that the nucleus appears brighter than the cytoplasm. Ideally, clustered cells would be distinguished by a simple threshold with the nuclei intensity. However, due to imperfect imaging conditions, intensity varies across the image, causing a global threshold to fail. Instead, we start by quantizing the image into N intensity classes using the K-Means algorithm, and produce a binary image for each threshold. Then, for each threshold in ascending order, objects smaller than a given value (e.g. the maximum nucleus size) are extracted from the binary image, polygonized into a contour, and erased from the remaining binary images to avoid duplicate extractions. This method is fast and robust since the maximum nucleus size is known *a priori*, and initialization results converge for increasing values of N ($N > 6$ in our experiments).

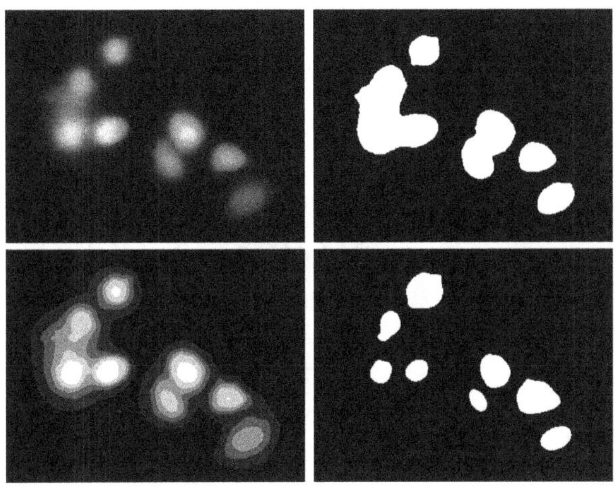

Fig. 1. Initialization scheme. Top-left: original image. Top-right: thresholded image using 2-class K-Means intensity clustering. Bottom-left: 6-class image quantization using K-Means intensity clustering. Bottom-right: final thresholded image obtained from the bottom-left image (as described in text).

Shape constrained deformation As described earlier, the contours deformation is obtained by minimizing an energy functional of the form:

$$E(C_{1..n}, I) = \lambda E_{\text{image}} + \mu E_{\text{regul}} + \gamma E_{\text{coupling}} + \delta E_{\text{prior}}, \tag{1}$$

where $C_{1..n}$ are the contours evolving on image I and $\lambda, \mu, \gamma, \delta$ are non-negative empirical weights. E_{image} is a data attachment term combining region information through the Mumford-Shah functional and edge information through image intensity gradients. E_{regul} is a regularizer (in the Tikhonov sense) that smooths the contours geometry during its deformation to avoid local energy minima. E_{coupling} is a coupling term that penalizes overlaps between contours. These three first terms have been successfully employed in biological imaging ([56] [57]), but are not sufficient to correctly determine the contact location between cells (cf. figure 2). To improve contact localization, we have introduced in the prior term of (1) a shape constraint that will monitor the contour shape during its deformation. This term expresses our prior biological knowledge that Hep2β cells are generally elongated, and thus constrains each contour deformation along its major axis. For each contour, this term reads

$$E_{\text{prior}}(C) = e(C) \int_0^1 g\left(\left|\vec{N}(p) \cdot \vec{a}_1\right|\right) dp, \tag{2}$$

where g is a positive definite non-increasing function, $\vec{N}$ is the outer normal to the contour at point p and e is the contour eccentricity ratio computed from the major axis $\vec{a}_1$ and minor axis $\vec{a}_2$ by $e \triangleq \sqrt{||\vec{a}_1||^2 - ||\vec{a}_2||^2}/2$. Minimizing this term tends to constrain contour points to remain close to the main contour axis. The eccentricity ratio regularizes the shape constraint by cancelling its effect for circular contours (i.e. non-elongated cells).

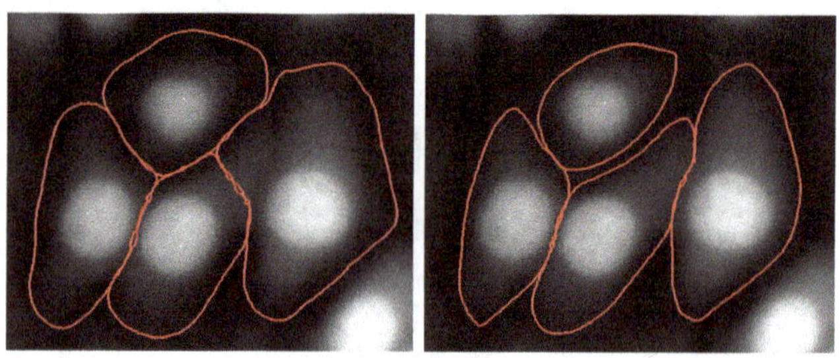

Fig. 2. Segmentation results on touching cells. Left: without shape constraint. Right: with shape constraint.

Contour evolution is then performed using a classical Euler-Lagrange gradient descent scheme, yielding an iterative deformation process where each contour point is progressively displaced by a force combining the four terms of Eq. (1). Convergence is detected individually for each contour when the overall deformation falls under a small value ϵ close to 0.

Figure 3 presents segmentation results on the HCS labeled image (figure 5-top-left). Since the fluorescence marking is inhomogeneous and difficult to see, we superimpose the result on a maximum gradient map of the image, where real cell contours are easier to distinguish visually. One can see that the contours fit the cell boundaries correctly, while touching cells are well distinguished. This is due to the incorporation of the new shape constraint, which improves the detection of elliptic shaped objects, as shown in figure 2. Without shape constraint (figure 2-left), contours fail to detect cell edges and contact locations correctly. This behavior is successfully corrected using the shape-constraint described in Eq. (2) (figure 2-right).

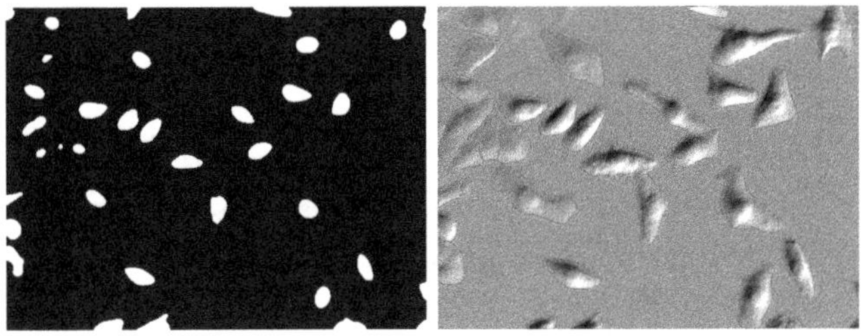

Fig. 3. Left: initialization on the HCS labeled image. Right: final contours superimposed on a gradient map of the image.

3. Spot detection

3.1. *Goals and challenges*

The joint development of bright fluorescence probes, optical microscopes and sensitive cameras in the context of live-cell imaging provides us with a set of standard tools to study subcellular processes. However, the methods for analyzing these images and extracting accurate quantitative data are much less developed.

In this section we shall only consider the problem of extracting small biological particles. These particles can be either pathogens, vesicles, single molecules or subcellular structures. In this context, the fluorescent-labeled objects generally appear as small specific spots superimposed on an unspecific background. Such spots have no clear border and their intensity profile is both diffuse and higher than that of the immediate neighborhood. Microscopy images are subject to numerous disturbances such as photon and camera readout noises. The requirement of short exposure times to avoid photo-damage of the sample results in images with typical SNRs of 15 or less. Consequently, the reliable detection of fluorescence spots is a challenging task. Extracting useful information about the spots such as their number, localization, intensity and spatial distribution of the spots is of major importance for biologists to understand the biological information encoded by the image.

We give below a review of the existing methods for spot detection in biological fluorescence images, with a particular focus on computation speed and algorithm genericity.

3.2. *State of the art*

3.2.1. *Automated thresholding*

As for cell segmentation, a classical and fast method to detect objects is by global intensity thresholding [37]. This method however requires a high-SNR image with a bimodal histogram, which is generally not the case of biological images where a structured background is often present.

A more sophisticated approach has been proposed in [15], based on fuzzy set theory. In the context of a colocalization analysis, image sequences of the labeled-proteins are modeled as a realization of a fuzzy temporal random set. The proposed procedure allows the selection of thresholds in an unsupervised manner. It is based on the definition of Vorob'ev fuzzy set average [4]. The main idea is to define the average of a given fuzzy set by using different definitions of the mean of a random compact set. Using fuzzy set averages leads to an automatic, systematic segmentation which does not require human intervention. However, as this method still uses a global threshold, it induces over-segmention on images with inhomogeneous background.

As soon as the background is uneven, spurious detections quickly arise, disturbing the whole quantification process. In such cases, it is clear that the processing should rather consider local information instead of the global image intensity profile.

To overcome this issue, a number of methods have been proposed in order to enhance the image signal by flattening the background while enhancing the signal of interest to increase the detection performance.

3.2.2. *Filtering and template matching*

Another intuitive method to detect spots is to associate them to local intensity maxima. In a noise-free image, each spot would produce a single local intensity maximum. In practice though, images are noisy due to fluorochrome concentration, photobleaching, non-equivalent electronic channel settings, etc.. Consequently, an image with a low signal-to-noise ratio contains a number of local intensity maxima that does not reflect the actual number of spots in the image. Therefore, before the detection, noise should be reduced by applying a pre-filtering procedure. Typical strategies rely on conventional image filtering techniques followed by some type of maximum search strategy in which the position of a spot is associated with a peak in the filter response field.

Manders applied uniform filters of different sizes [31]. The sizes are chosen smaller than the minimum distance between neighboring spots, which is pre-estimated by visual inspection. Then, an individual threshold level is chosen for each local maximum to determine the spot domain. The spot domain is computed using the Largest Contour Segmentation (LCS) procedure, which is an iterative region-growing algorithm. For each local maximum the LCS procedure finds the largest contour surface that best delineates the corresponding spot. The drawbacks of this method are that several parameter values have to be preset interactively and the time-consuming due to region growing.

The top-hat filter [42] and its variation, the top-hat box filter [9], implement a local analysis through the use of feature masks. After having properly set some ad hoc parameters, the image is scanned and spots are located at positions where the response of the filters is above a pre-defined threshold. In [33] the authors perform a threshold on a top-hat transform to extract isolated spots, and apply a non-linear Laplace filter followed by a thresholding to detect overlapping spots. These techniques are effective for detecting small round spots that respond strongly to circle- and box-shaped filters, but become less effective in the presence of objects with different shapes or size.

In [7] images are convolved with a median filter to remove photon noise and the mean intensity is substracted to remove background.

In [47], the authors perform an anisotropic diffusion filtering that removes noise without disturbing edge information.

Despite an adequate filtering to overcome the noise sensitivity, such basic approaches are not adapted to biological images with fluorescent background structures.

When the observed particles have sizes smaller than the microscope resolution, one may resort to more sophisticated approaches based on fitting a model of the microscope's point spread function (PSF), since the spots can be deemed as point sources. There are several ways to define the adequate PSF model, the most popular being empirical and analytical strategies.

Empirical PSFs In the empirical case, the PSFs of the investigated fluorochromes are measured by recording images of fluorescent microspheres also called 'beads'.

Bornfleth *et al.* [8] perform a two step image segmentation through a region-growing procedure that includes prior knowledge from the measured PSF: 1) local maxima in the image are determined by means of a top-hat filter designed in a shape that approaches filter isotropy. 2) all spots are assigned an initial threshold of 82% of their maximum intensity value, corresponding to the initial assumption of point-like spots. The spots are allowed to grow in shells around the spot center according to the model.

As this method relies on iterative region-growing, it is time-consuming and hence not suitable for time-constrained applications.

Theoretical PSFs In the approaches proposed by [45], [18] the authors apply analytical point-spread function-model fitting in the detection and the localization procedure.

Thomann *et al.* [45] pre-detect the spots using a Gaussian curvature score in widefield microscopy. Then, each spot profile h is described by the convolution of the underlying spot intensity distribution with the microscope PSF. The Gaussian approximation simplifies the model fitting procedure, which enables super-resolution localization. However the computation time is prohibitive when a large number of spots are present.

A similar method is used in [18] where the authors propose to use a Gaussian approximation adapted to a line-scanning confocal microscope. The authors then apply a global threshold [37] on the mean curvature score computed from the Hessian matrix. This strategy reduces signicantly computation time at the expense of sub-resolution capabilities.

3.3. *Multiscale approach*

In [35] a multi-resolution algorithm is proposed for the detection of bright spots based on an undecimated wavelet representation of the image and on the selective filtering of wavelet coefficients.

The undecimated wavelet transform is implemented by the "à trous" algorithm:

$$\begin{cases} A_j = \bar{h}^{\uparrow j-1} \star A_{j-1} \\ d_j = A_{j-1} - A_j, \end{cases} \tag{3}$$

where h is the 3rd order B-Spline low-pass filter, A_j and d_j are respectively the approximation and the wavelet coefficients at scale j ($\leq J$), $h^{\uparrow k}[l] = h[l]$ if $l/2^k \in \mathbb{Z}$ and 0 otherwise, $\bar{h}[n] = h[-n]$ and "$\star$" denotes convolution.

This transform is redundant and translation-invariant, which is fast and ideal for representing isotropic objects. In the wavelet domain, the spots are characterized by a small number of coefficients, that are both large and correlated across the scales. On the contrary, the noise has smaller coefficients that do not propagate across levels.

From this observation, the correlation images P_n are defined as the multiscale spatial product between the detail images. On this correlation map, spots exhibit high peaks which can be easily segmented out through a threshold.

This method is robust to the variability of biological images and to high level of noise present in typical microscopy images. Furthermore, it is also accurate, fast and easy to implement, making it a very attractive spot detection method and as shown by its numerous biological applications [41], [22], [26].

Nevertheless, image acquisition is a stochatic process due to the inherent Poisson noise of light emission. Therefore it is more relevant to model this stochastic process by a Mixed-Poisson-Gaussian (MPG) model [54], and the detection of the significant wavelet coefficients is achieved using a statistical test.

3.3.1. *VST for a filtered MPG process*

A MPG process $\mathbf{x} := (X_i)_{i \in \mathbb{Z}^d}$ is defined as:

$$X_i = \alpha U_i + V_i, \quad U_i \sim \mathcal{P}(\lambda_i), \quad V_i \sim \mathcal{N}(\mu, \sigma^2) \tag{4}$$

where $\alpha > 0$ is the overall gain of the detector, U_i is a Poisson variable modeling the photon counting, V_i is a normal variable representing the readout noise, and all $(U_i)_i$ and $(V_i)_i$ are assumed mutually independent. A commonly used technique is to apply a variance stabilizing transform (VST), in order to yield a near Gaussian process so that each sample is near-normally distributed with an asymptotically constant variance. Hence, the VST allows to apply standard denoising and deconvolution methods on the transformed data. Then, the final estimate is obtained by inverting the VST on the processed data.

3.3.2. *Detection of significant coefficients by FDR*

We detect the significance of coefficients by testing the binary hypothesis: $\forall d,\ H_0:\ d_j = 0$ vs. $H_1:\ d_j \neq 0$. Since a wavelet has a zero mean, if d_j comes from a signal having a constant intensity within the wavelet support, then d_j would be zero, if no noise was present. Thus $d_j \in H_0$ in this case. In order to control a global statistic error rate, multiple hypothesis tests should be used. Therefore, the false discovery rate (FDR) [5] [6] is chosen to control the significance of the wavelet coefficients. The FDR is the average fraction of false detections over the total number of detections. The FDR control has two advantages: 1) it usually has a greater detection power; 2) it can easily handle correlated data. The latter point is important since the isotropic undecimated wavelet transform is over-complete.

Finally, we compute the correlation images which corresponds to the multiscale spatial product between the detail images. Once the correlated image is thresholded and binarized, the connected components are considered as putative bright spots (cf. figure 4).

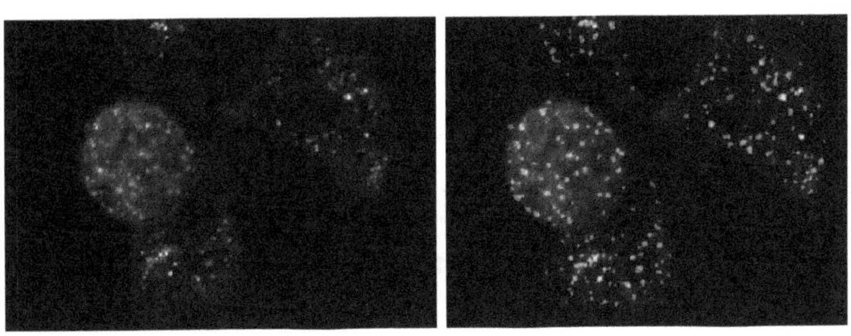

Fig. 4. Spot detection results. Left: Original image of IL-2Rβ endosomes (stained using a red fluorescent marker). Right: extracted spots (green dots) superimposed on a grayscale version of the original image.

4. Practical case study: cellular endocytosis

Cellular endocytosis is an important research topic in cell biology. Receptor-mediated endocytosis allows cells to communicate with their environment via membrane receptors which transport macromolecules from the extracellular medium to intracellular compartement named endosomes. This phenomenon is now easily observable using fluorescence microscopy, however due to the lack of adapted algorithms and softwares, image analysis is usually performed by visual examination, which is a tedious task, time consuming, user-biased and provides poor quantitative measurements. Although new tools have been developped to quantitatively measure multiple features on individual cells (e.g. cycle, size, intensity, localization of sub-cellar compartments, etc.), none of them currently focuses on celullar endocytosis in an efficient, intuitive and automated manner.

Here we present a concrete application of the techniques presented above for the quantification of cellular endocytosis of the IL-2 receptor. The goal of this work is to compare the internalization of this receptor between a wild-type Hep2β cell context and Rac1-mutant cell context.

4.1. *Experimental protocol*

The biological protocol is conducted as follows. A dominant negative mutant of Rac1 (myc-Rac1T17N) was transfected in Hep2β cells as described in [20]. Endocytosis assays were performed 24 hours after the transfection. IL-2R endocytosis was monitored for 15 min at 37°C, by the addition a monoclonal antibody (561) directed against it and conjugated to Cy3 fluorochrome (561-Cy3) (Amersham). Then, cells were fixed, permeabilized and to detect myc-Rac1T17N, cells were reacted with an anti-myc and an anti-mouse antibodies coupled to an AlexaFluor488 (Molecular Probe). To label the cell nucleus and cytoplasm, Hep2β cells were stained with HCS CellMask Alexa Fluor 350 (Invitrogen). Images were obtained on a motorized Zeiss Axiovert 200 epifluorescence microscope, equipped with a 25X objective (0,8 NA) and a Roper Scientific Coolsnap HQ camera. A total of 3×100 images were automatically acquired under the same acquisition settings, yielding around 1200 cells. Figure 5 shows a sample image acquired via this protocol. The cells appear

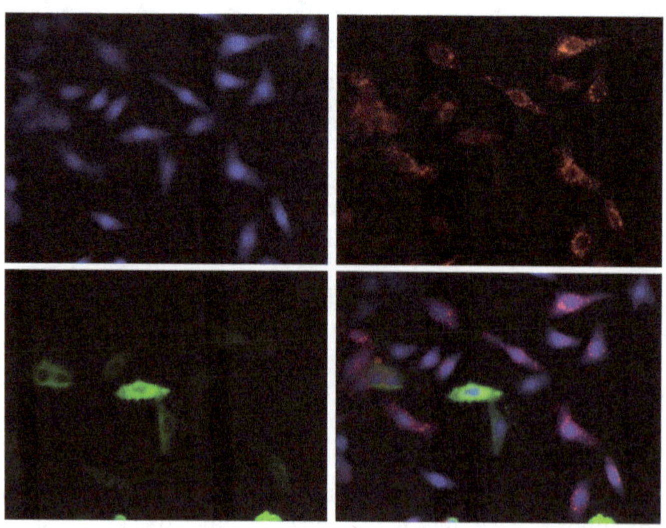

Fig. 5. Sample image of IL-2 receptor endocytosis in Hep2β cells. Top-left: cells. Top-right: endosomes. Bottom-left: myc-Rac1 mutant. Bottom-right: merged view.

in the top-left hand image, the endosomes in the top-right hand image, the bottom-left hand image shows transfected cells expressing the myc-Rac1 mutant, and the bottom-right hand image presents a merged view of all color channels.

4.2. *Results*

Fig.6-top-right shows spot detection results, which are robust to the variability of biological images and to the high level of noise. We would like to notice that we have only to specify the scale and the statistical value of FDR. To validate the method, we quantified the endocytosis of IL-2R for two different cell populations, and compare the results with those published in [20]. The first population correspond to transfected cells expressing a myc-Rac1 mutant (known to inhibit endocytosis). The second population corresponds to control cells which do not express the mutant. To automatically sort the two populations, we use the K-means algorithm to cluster the cells mean intensity into 3 classes, so as to capture weakly transfected cells (an example of cell transfection is shown in fig.6-bottom-left). For the three experiments, 437 cells were found to be transfected and 788 were not, and the expression of myc-Rac1 mutant inhibited 45.8%±10.5 (mean ± standard deviation) of the total endosome entry as compared with the non-transfected cells. These results are similar to the previous published results [20] based on manual analysis.

Processing time ranged from 5 to 10 seconds per image using a 2.4 GHz dual-

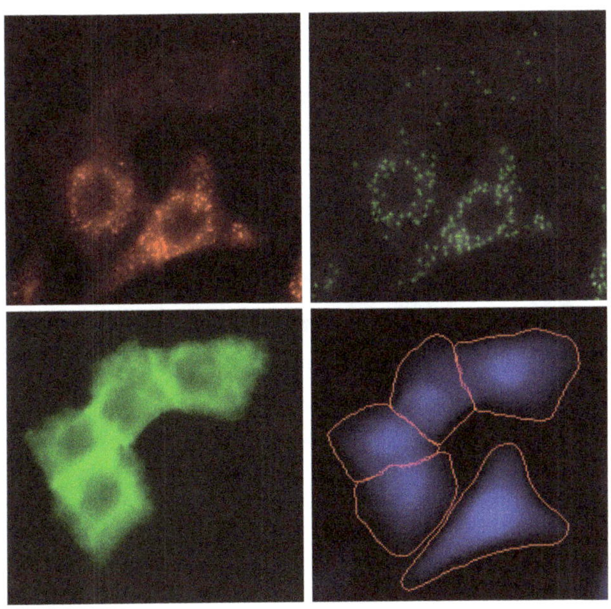

Fig. 6. Results on clustered cells. Top-left: IL-2R endosomes. Top-right: spot detection results. Bottom-left: myc-Rac1 mutant transfection. Bottom-right: cell segmentation results.

core cpu, including initialization, cells segmentation and endosomes extraction, by comparison with manual analysis which takes from 5 to 10 minutes. The time mostly depends on the number of cells per image. 10 seconds is the worst case time needed for an image full of cells (around 30 cells).

Results show that the method is sufficiently robust to detect cell boundaries and extract spots within each cell independently in a fast and automated manner, even in case of touching cells. Compared to manual analysis, the described method is more accurate and provides more specific information about the endosomes such as their number and intensity, which is a great interest for biologists. Note that this data can be further exploited to quantify more information such as the spatial distribution of the endosomes.

5. Conclusion

Many biological studies have been greatly facilitated over the past decades thanks to numerous advances in automated imaging systems, able to acquire thousands of images of a given phenomenon in multiple colors and dimensions in a robust and reproducible manner. Their systematic analysis requires automated algorithms that are robust against the inherent variablity of biological processes and against images with low signal-to-noise ratio. Moreover, these algorithms should have very few tunable parameters for simplicity of use and reproducibility purposes. In this chapter we have browsed a number of methods in the literature suitable for two important quantification tasks in biology, namely cell segmentation and spot detection. Through a concrete application, we showed that robust and flexible frameworks generally offer the best comprise between genericity and adaptability to a given problem, yielding fast and robust tools applicable for day-to-day analysis of high-content screening data.

References

[1] Acton, S. T., Wethmar, K. and Ley, K. (2002). Automatic tracking of rolling leukocytes in vivo, *Microvascular Research* **63**, pp. 139–148.

[2] Aubert, G. and Kornprobst, P. (2001). *Mathematical problems in image processing* (Springer).

[3] Awasthi, V., Doolitle, W., Parulkar, G. and McNally, J. (1994). Cell tracking using a distributed algorithm for 3d image segmentation, *Bioimaging* **2**, 2, pp. 98–112.

[4] Ayala, G., León, T. and Zapater, V. (2005). Different averages of a fuzzy set with an application to vessel segmentation, *IEEE T. Fuzzy Systems* **13**, 3, pp. 384–393.

[5] Benjamini, Y. and Hochberg, Y. (1995). Controlling the false discovery rate: a practical and powerful approach to multiple testing, *J. Roy. Statist. Soc. ser. B* **57**, 1, pp. 289–300.

[6] Benjamini, Y. and Yekutieli, D. (2001). The control of the false discovery rate in multiple testing under dependency, *Ann. Statist.* **29**, 4, pp. 1165–1188.

[7] Bornfleth, H., Edelmann, P., Zink, D., Cremer, T. and Cremer, C. (1999). Quantitative motion analysis of subchromosomal foci in living cells using four-dimensional microscopy, *Biophysical Journal* **77**, pp. 2871–2886.

[8] Bornfleth, H., Satzler, K., Eils, R. and Cremer, C. (1998). High-precision distance measurements and volume-conserving segmentation of object near and below the resolution limit in three-dimensional confocal fluorescence microscopy, *Journal of Microscopy* **189**, pp. 118–136.

[9] Breen, X., Joss, G. and Williams, K. (1991). Locating objects of interest within biological objects: the top-hat box filter, *Comp. Ass. Microsc.* **2**, 3, pp. 97–102.

[10] Chakraborty, A., Staib, L. and Duncan, J. (1996). Deformable boundary finding in medical images by integrating gradient and region information, *IEEE Transactions on Medical Imaging* **15**, 6, pp. 859–870, doi:10.1109/42.544503.

[11] Chen, C., Li, H., Zhou, X. and Wong, S. (2008). Constraint factor graph cut-based active contour method for automated cellular image segmentation in rnai screening, *Journal of Microscopy* **230**, 2, pp. 177–191.

[12] Cohen, L. D. (1991). On active contour models and balloons, *CVGIP: Image Underst.* **53**, 2, pp. 211–218.

[13] Debeir, O., Camby, I., Kiss, R., Ham, P. V. and Decaestecker, C. (2004). A model-based approach for automated in vitro cell tracking and chemotaxis analyses, *Cytometry Part A* **60A**, p. 2940, doi:10.1002/cyto.a.20040.

[14] Demou, Z. and McIntire, L. (2002). Fully automated three-dimensional tracking of cancer cells in collagen gels, *Cancer Research* **62**, 18, pp. 5301–5307.

[15] Diaz, M., Ayala, G., Leon, T., Zoncu, R. and Toomre, D. (2008). Analyzing protein-protein spatial-temporal dependencies from image sequences using fuzzy temporal random sets, *Journal of Computational Biolology: a journal of computational molecular cell biology* **15**, 9, pp. 1221–1236.

[16] Dufour, A., Meas-Yedid, V. and Olivo-Marin, J. (2008). Automated quantification of cell endocytosis using active contours and wavelets, in *IAPR International Conference on Pattern Recognition* (Tampa, FL, USA), to appear.

[17] Dufour, A., Shinin, V., Tajbaksh, S., Guillen, N., Olivo-Marin, J. and Zimmer, C. (2005). Segmenting and tracking fluorescent cells in dynamic 3d microscopy with coupled active surfaces, *IEEE Transactions on Image Processing* **14**, 9, pp. 1396–1410, doi:10.1109/TIP.2005.852790.

[18] Dusch, E., Dorval, T., Vincent, N., Wachsmuth, M. and Genovesio, A. (2007). Three-dimensional point spread function model for line-scanning confocal microscope with high-aperture objective, *Journal of Microscopy* **228**, pp. 132–138.

[19] Dzyubachyk, O., Niessen, W. and Meijering, E. (2008). Advanced level-set based multiple-cell segmentation and tracking in time-lapse fluorescence microscopy images, in A. L. J.-C. Olivo-Marin, I. Bloch (ed.), *Proceedings of the IEEE International Symposium on Biomedical Imaging: From Nano to Macro (ISBI 2008)* (IEEE).

[20] Grassart, A., Lazarow, A. D. P., Dautry-Varsat, A. and Sauvonnet, N. (2008). Clathrin-independent endocytosis used by the IL-2 receptor is regulated by Rac1, Pak1 and Pak2, *EMBO Reports* **9**, pp. 356–362, doi:10.1038/embor.2008.28.

[21] Grau, V., Mewes, A., Alcaniz, M., Kikinis, R. and Warfield, S. (2004). Improved watershed transform for medical image segmentation using prior information, *IEEE Transactions on Medical Imaging* **23**, 4, pp. 447–458, doi:10.1109/TMI.2004.824224.

[22] Jaqaman, K., Loerke, D., Mettlen, M., Kuwata, H., Grinstein, S., Schmid, S. L. L. and Danuser, G. (2008). Robust single-particle tracking in live-cell time-lapse sequences. *Nature methods*, pp. 695-7702, doi:10.1038/nmeth.1237.

[23] Kachouie, N., Fieguth, P., Ramunas, J. and Jervis, E. (2006). Probabilistic model-based cell tracking, *International Journal of Biomedical Imaging* **1**, pp. 1–10.

[24] Kass, M., Witkin, A. and Terzopoulos, D. (1988). Snakes : active contour models, *International Journal of Computer Vision* **1**, pp. 321–331.

[25] Krooshoop, D., Torensma, R., van den Bosch, G., Nelissen, J., Figdor, C., Raymakers, R. and Boezeman, J. (2003). An automated multi well cell track system to study leukocyte migration, *Journal of Immunological Methods* **280**, 1-2, pp. 89–102.

[26] Law, Y. N., Ogg, S., Common, J., Tan, D., Lane, E. B., Yip, A. M. and Lee, H. K. (2008). Automated protein distribution detection in high-throughput image-based sirna library screens, *Journal of Signal Processing Systems* **55**, pp. 1–13.

[27] Leymarie, F. and Levine, M. (1993). Tracking deformable objects in the plane using an active contour model, *IEEE Transactions on Pattern Analysis and Machine Intelligence* **15**, 6, pp. 617–634, doi:10.1109/34.216733.

[28] Li, B. and Acton, S. T. (2008). Automatic active model initialization via poisson inverse gradient, **17**, 8, pp. 1406–1420, doi:10.1109/TIP.2008.925375.

[29] Li, K., Miller, E. D., Chen, M., Kanade, T., Weiss, L. E. and Campbell, P. G. (2008). Cell population tracking and lineage construction with spatiotemporal context. *Med Image Anal* **12**, 5, pp. 546–566, doi:10.1016/j.media.2008.06.001, URL `http://dx.doi.org/10.1016/j.media.2008.06.001`.

[30] Lin, G., Adiga, U., Olson, K., Guzowski, J., Barnes, C. and Roysam, B. (2003). A hybrid 3D watershed algorithm incorporating gradient cues and object models for automatic segmentation of nuclei in confocal image stacks, *Cytometry Part A* **56**, pp. 23–36.

[31] Manders, E., Hoebe, R., Strackee, J., Vossepoel, A. and Aten, J. (1996). Largest Contour Segmentation: A Tool for the Localization of Spots in Confocal Images, *Cytometry* **23**, pp. 15–21.

[32] Mukherjee, D., Ray, N. and Acton, S. (2004). Level set analysis for leucocyte detection and tracking, *IEEE Transactions on Image Processing* **13**, 4, pp. 562–572, doi:10.1109/TIP.2003.819858.

[33] Nelson, C. and Plosser, C. (1997). Fish and chips: Automation of fluorescent dot counting in interphase cell nuclei, *Cytometry* **28**, pp. 1–10.

[34] Olivo, J.-C. (1994). Automatic threshold selection using the wavelet transform, *CVGIP: Graphical Models and Image Processing* **56**, 3, pp. 205–218.

[35] Olivo-Marin, J.-C. (2002). Extraction of spots in biological images using multiscale products, *Pattern Recognition* **35**, 9, pp. 1989–1996.

[36] Ortiz, C., Garcia-Rodriguez, E., Jones, A., Pinkel, D., Gray, J., Sugar, D. and Lockett, S. (1999). Segmentation of confocal microscope images of cell nuclei in thick tissue sections, *Journal of Microscopy* **193**, 3, pp. 212–226, doi:10.1046/j.1365-2818.1999.00463.x.

[37] Otsu, N. (1979). A threshold selection method from gray level histograms, *IEEE Transactions on Systems, Man and Cybernetics* **9**, pp. 62–66, minimize inter class variance.

[38] Ray, N. and Acton, S. (2002). Active contours for cell tracking, in *Proceedings of the Fifth IEEE Southwest Symposium on Image Analysis and Interpretation.*

[39] Ray, N. and Acton, S. (2005). Data acceptance for automated leukocyte tracking through segmentation of spatiotemporal images, *IEEE Transactions on Biomedical Engineering* **52**, 10, pp. 1702–1712, doi:10.1109/TBME.2005.855718.

[40] Ray, N., Acton, S. and Ley, K. (2002). Tracking leukocytes in vivo with shape and size constrained active contours, *IEEE Transactions on Medical Imaging* **21**, 10, pp. 1222–1235, doi:10.1109/TMI.2002.806291.

[41] Rus, C., Vanninen, M. and Ruotsalainen, U. (2007). Automatic approach for analyzing the endosome movement in fluorescence video microscopy, in *Proceedings of Eurapsip*, pp. 1122–1126.

[42] Serra, P. (1982). *Image Analysis and Mathematical Morphology* (Academic Press, New York).

[43] Shen, H., Nelson, G., Nelson, D., Kennedy, S., Spiller, D., Griffiths, T., Paton, N., Oliver, S., White, M. and Kell, D. (2006). Automated tracking of gene expression in individual cells and cell compartments, *J. R. Soc. Interface* **3**, p. 787.

[44] Staib, L. and Duncan, J. (1992). Boundary finding with parametrically deformable models, *IEEE Transactions on Pattern Analysis and Machine Intelligence* **14**, 11, pp. 1061–1075, doi:10.1109/34.166621.

[45] Thomann, D., Rines, D., Sorger, P. and Danuser, G. (2002). Automatic fluorescent tag detection in 3D with super-resolution: application to the analysis of chromosome movement, *Journal of Microscopy* **208**, 1, pp. 49–64.

[46] Tikhonov, A. (1963). Regularization of incorrectly posed problem, *Soviet Math.* **4**, pp. 1624–1627.

[47] Tvarusko, W., Bentele, M., Misteli, T., Rudolf, R., Kaether, C., Spector, D. L., Gerdes, H. H. and Eils, R. (1999). Time-resolved analysis and visualization of dynamic processes in living cells, *Proc. Natl. Acad. Sci. USA* **96**, 14, pp. 7950–7955, 0027-8424 Journal Article.

[48] Wahlby, C., Sintor, I.-M., Erlandsson, F., Borgefors, G. and Bengtsson, E. (2004). Combining intensity, edge and shape information for 2D and 3D segmentation of cell nuclei in tissue sections, *Journal of Microscopy* **215**, 1, pp. 67–76, doi:10.1111/j.0022-2720.2004.01338.x.

[49] Wilson, C. and Theriot, J. (2006). A correlation-based approach to calculate rotation and translation of moving cells, *IEEE Transactions on Image Processing* **15**, 7, pp. 1939–1951, doi:10.1109/TIP.2006.873434.

[50] Xu, C. and Prince, J. (1998). Generalized gradient vector flow external forces for active contours, *Signal Processing* **71**, 2, pp. 131–139, doi:10.1016/S0165-1684(98)00140-6.

[51] Xu, N., Bansal, R. and Ahuja, N. (2003). Object segmentation using graph cuts based active contours, in *IEEE Conference on Computer Vision and Pattern Recognition*, Vol. 2, pp. 46–53.

[52] Yang, X., Li, H. and Zhou, X. (2006). Nuclei segmentation using marker-controlled watershed, tracking using mean-shift, and kalman filter in time-lapse microscopy, *IEEE Transactions on Circuits and Systems* **53**, 11, pp. 2405–2414, doi:10.1109/TCSI.2006.884469.

[53] Young, S., Tzeng, W., Kuo, Y., Hsiao, M. and Chiang, S. (1996). Real-time tracing of spermatozoa, *IEEE Engineering in Biology and Medecine Magazine* **15**, 6, pp. 117–120, doi:10.1109/51.544519.

[54] Zhang, B., Fadili, J., Starck, J.-L. and Olivo-Marin, J.-C. (2007). Multiscale variance-stabilizing transform for mixed-poisson-gaussian processes and its applications in bioimaging, in *Proc. IEEE International Conference on Image Processing ICIP 2007*, Vol. 6, pp. VI–233–VI–236, doi:10.1109/ICIP.2007.4379564.

[55] Zhang, B., Zimmer, C. and Olivo-Marin, J.-C. (2004). Tracking fluorescent cells with coupled geometric active contours, in *International Symposium on Biomedical Imaging* (Arlington), pp. 476–479, doi:10.1109/ISBI.2004.1398578.

[56] Zimmer, C., Labruyere, E., Meas-Yedid, V., Guillen, N. and Olivo-Marin, J.-C. (2002). Segmentation and tracking of migrating cells in videomicroscopy with para-

metric active contours: a tool for cell-based drug testing, *IEEE Transactions on Medical Imaging* **21**, 10, pp. 1212–1221, doi:10.1109/TMI.2002.806292.

[57] Zimmer, C. and Olivo-Marin, J.-C. (2005). Coupled parametric active contours, *IEEE Transactions on Pattern Analysis and Machine Intelligence* **27**, 11, pp. 1838–1842, doi:10.1109/TPAMI.2005.214.

[58] Zimmer, C., Zhang, B., Dufour, A., Thebaud, A., Berlemont, S., Meas-Yedid, V. and Olivo-Marin, J.-C. (2006). On the digital trail of mobile cells, *Signal Processing Magazine* **23**, 3, pp. 54–62.

CHAPTER 4.6

ADVANCED EVOLUTIONARY ALGORITHMS FOR INTELLIGENT MICROARRAY IMAGE ANALYSIS

Eleni Zacharia and Dimitris Maroulis

Dept. of Informatics and Telecommunications,
University of Athens GR 15784, Ilisia, Greece
E-mail: eezacharia@gmail.com, dmaroulis@di.uoa.gr

cDNA microarrays, one of the most fundamental and powerful biotechnology tools, is being utilized in a variety of biomedical applications as it enables scientists to simultaneously analyze the expression levels of thousands of genes over different samples. One of the most essential processes of cDNA microarray experiments is the image analysis one, which is divided into three phases, namely, gridding, spot-segmentation and intensity extraction. Although its two former phases appear relatively straightforward, they are in fact rather challenging procedures due to the nature of microarray images. For their implementation, the currently available software programs require human intervention, which significantly affects the biological conclusions reached during microarray experiments. In this chapter, the basic process of analyzing a microarray image is described and advanced evolutionary algorithms implementing the automatic gridding and segmentation processes are presented. In reality, both of these algorithms are based on optimization problems which are solved by using evolutionary genetic algorithms. Contrary to existing software systems, the proposed methods are fully automatic as they do not require any human intervention; they are also noise resistant and yield excellent results even under adverse conditions. Last but not least, they outperform other software programs as well as established techniques.

1. Introduction

In the last decade, microarray technology is being increasingly applied in numerous fields of biomedical research such as cancer research, pharmacology research, toxicology research, infectious disease diagnosis and treatment, and agricultural development.[1] The reason for its broad use and success can be attributed to its main revolutionary feature: the ability to simultaneously analyze the expression levels of thousands of genes over different samples.

The process of a microarray experiment[2] starts with the selection of a set of DNA probes that are of particular interest. A robot places the selected DNA probes on a glass slide, creating an invisible array of DNA dots. Two distinct populations of mRNAs (messenger RNAs) are then isolated from a control sample (i.e a cell developed under normal conditions) and a test sample (i.e. a cell developed under a specific treatment). The mRNA populations are reversely transcribed into cDNA (complementary DNA) populations which in turn are colored with separate fluorescent dyes of different wavelengths (i.e. Cy3 and Cy5). The dyed cDNA populations are mixed with water and the solution is placed on the glass slide in order for the cDNA populations to be hybridized with the slide's DNA dots. Finally, the hybridized glass slide is fluorescently scanned twice; one scan for each dye's wavelength. Hence, two digital images are produced, one for each population of mRNA. Each digital image contains a number of spots (corresponding to the DNA-cDNA dots) of various fluorescence intensities. Given that the intensity of each spot is proportional to the hybridization level of the cDNAs and the DNA dots, the gene expression information is obtained by analyzing the digital images.

Ideally, the existing spots inside the microarray image have a circular 2D shape with fixed diameters, while their intensity peaks at their central region and declines at regions further from their centre. Moreover, they are aligned in 2D array layouts, namely blocks. Likewise, all blocks inside the microarray image are arranged in a 2D array layout too and each contains an equal number of spots. As a result, adjacent blocks and adjacent spots are clearly separated inside the microarray image. However, the edge-to-edge distance between two adjacent blocks is larger than the distance between two adjacent spots within the same block (Fig. 1).

As stated by Yang et al.,[3] the process of analyzing a microarray image can be divided into three main phases namely: "Gridding", "Spot-Segmentation"

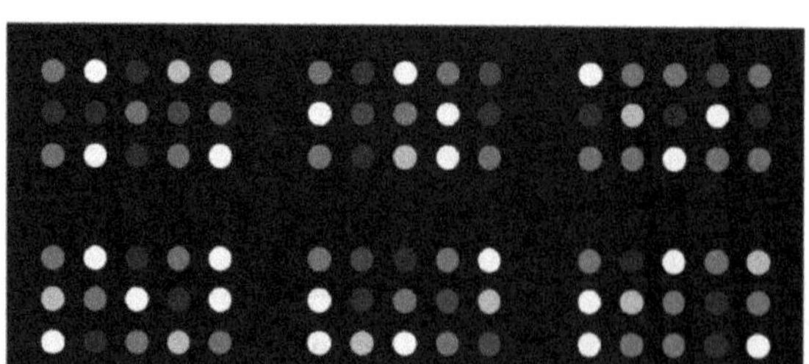

Fig. 1. An illustration of a synthetic microarray image: This image is composed of 6 blocks in a 3x2 layout. Each block contains 15 spots which are located in a 5x3 layout.

and "Spot-Intensity extraction". During the 1st phase, the microarray image is segmented into numerous compartments, each containing one individual spot and background. During the 2nd phase each compartment is individually segmented into a spot area and a background area, while during the 3rd phase the brightness of each spot is calculated. The expression-levels of the genes in these spots result from their individual brightness.

There is a considerable number of software systems and techniques that have been developed and proposed to date in order to analyze the microarray images. However, gridding microarray images and segmenting microarray spots remain two rather challenging and complicated procedures, and therefore they still require human intervention. This is attributed to the nature of microarray images. More precisely, the quality of images is often degraded due to the existence of noise and/or artifacts as well as due to uneven background.[4] Furthermore, microarray images may contain low-intensity spots that are so poorly contrasted that are not clearly visible.[5] Furthermore, there may be rotations, misalignments and local deformations of the ideal rectangular grid[6] as spots are not ideally aligned in a 2D array layout. In addition, many spots are rather different to the ideal ones as they vary in size, shape and position due to imperfect sample-preparation and hybridization processes.

Examples of software systems which are frequently used for analyzing microarray images include the ScanAlyze,[7] Dapple,[8] ImageGene,[9] and Spot.[10] Most of these software programs perform the gridding procedure by enabling manual template matching. In particular, the user defines the 'optimal' template – by specifying a number of parameters – with the prospect of maximizing the inclusion of spot pixels inside their corresponding compartments of the template. Moreover, most of them perform the segmentation procedure by using either the circle segmentation algorithm (fixed or adaptive), or the adaptive shape-segmentation algorithm. The former algorithm assumes that microarray spots have a circular shape while the latter algorithm can segment regions of irregular shapes by implementing a watershed algorithm.

Apart from these software systems, a considerable number of techniques have been proposed to date in order to analyze microarray images. Some well-known approaches to gridding microarray images are based on axis projections,[11] or on morphological filtering.[12,13,14] For example, Liew et al.[12] has estimated the grid from some guide-spots, which are detected by using adaptive thresholding and morphological processing steps. Other gridding methods use graph models[15] in which spots are represented as ε-graphs, with up, down, left, and right edges. There are also many gridding methods trying to solve rotation and misalignment problems,[16-20] which improve the accuracy results of the previous ones. Several methods have been also developed to segment microarrays spots. For instance, Chen et al.[21] suggested a threshold method based on the statistical Mann-Whitney test which relies on the appropriate choice of background samples. Clustering algorithms, such as K-means, hybrid K-means and, fuzzy C-means (FCM) have been additionally applied in order to determine which pixels belong to the spot area and which ones to the background area.[22-25] Another segmentation method, based on the clustering of pixels' values, is the model-based segmentation algorithm, proposed by Li et al.[26] This method uses a removal technique of the components that are spatially connected in order to exclude small disconnected clusters which are assumed to be artifacts.

All aforementioned techniques require human intervention in order to define input parameters or to correct the gridding or segmentation results. This apparent lack of automation can be disadvantageous during microarray image analysis. Indeed, human intervention may inevitably modify the actual results of the microarray experiment and lead to erroneous biological conclusions. As a result, the automation of the gridding procedure as well as the automation of the spot-segmentation procedure remains an important issue to resolve.

In this chapter, accurate algorithms implementing the automatic gridding and segmentation processes are presented. Both these algorithms are based on the optimization technique of evolutionary algorithms and particularly of genetic algorithms. Consequently, prior to describing the two methods, a brief introduction of the evolutionary genetic algorithms' technique is apposed. The remainder of this chapter is organized in five sections: In section 2 a brief description of the evolutionary genetic algorithms' optimization technique is provided. In Sections 3 and 4, advanced evolutionary genetic algorithms are presented for the gridding and segmentation phase respectively. Section 5 illustrates experiments, while in section 6 conclusions are apposed.

2. Evolutionary Genetic Algorithms' Optimization Technique

Genetic algorithm is a powerful optimization search methodology based on the principles of natural selection and evolution.[27] Compared to conventional heuristic methods, it is a promising alternative as it exhibits significant advantages: it can optimize a large number of variables of extremely complex cost surfaces and solve problems for which there is little or no *a priori* knowledge of the underlying processes.

A conventional genetic algorithm[28] begins its search by constructing a finite number of potential solutions encoded as alpha-numerical sequences called chromosomes. These chromosomes, which constitute an initial population Pop_1, are evaluated using a fitness function. Subsequently, the population Pop_1 evolves into a new population Pop_2 using the following three genetic operators: reproduction, crossover, and mutation. This evolutionary cycle of a current population Pop_n to its next Pop_{n+1} (where n stands for the consecutive number of populations), continues until a specific termination criterion is satisfied. In conclusion, the following four basic elements are deemed necessary for the determination of a particular problem: chromosome representation, chromosome evaluation, evolutionary cycle, and termination criteria.

In the following sections, the evolutionary genetic algorithms used in gridding and spot-segmenting phases enclose the same evolutionary cycle and termination criteria. More precisely, a new population Pop_{n+1} is created from the current Pop_n by applying the following stages: (i) Reproduction stage: P_r% of the best chromosomes of the current population Pop_n are carried over to the new population Pop_{n+1}, and (ii) Crossover-Mutation stage: The chromosomes needed to complete the new population Pop_{n+1} are produced through iterations; Four chromosomes of the population Pop_n are selected using the tournament selection method.[29] These chromosomes are subsequently subjected in turn to the joint application of the BLX-a,[30] and the dynamic heuristic crossover operator,[30] according to a P_c% probability and then to the wavelet mutation operator,[31] according to a P_m% probability. The best two of the four resulting chromosomes proceed to the new population Pop_{n+1}. Moreover, the genetic algorithms are executed up to a maximum number of populations G_{Max}, or up to a maximum number of populations G_{Fit} for which the best fitness value has remained unchanged.

3. Gridding Microarray Images

Normally, gridding a microarray image is a two-stage procedure: Firstly, the microarray image is cut into distinct, quadrilateral in shape, blocks. Subsequently, each of these blocks is cut into numerous, quadrilateral in shape, single-spot compartments (Fig. 2). All of the above procedures are accomplished by the determination of the line-segments constituting the borders of blocks (1st stage) or spots (2nd stage).

Based on the aforementioned remark, our gridding method utilizes two similar stages; its 1st stage determines the borders of each block, while its 2nd one determines the borders of each spot inside the block. Let G be a quadrilateral which represents a microarray image or block. Each of the aforementioned stages include the determination of the line-segments whose members are defined by the two horizontal sides of G (Fig. 3a, "BC" and "AD" sides), and the

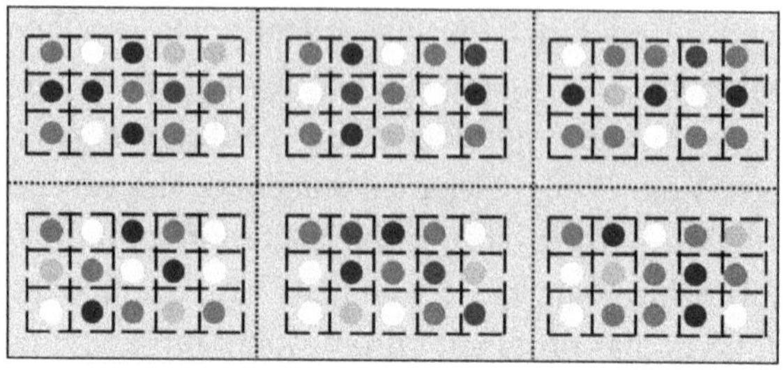

Fig. 2. An illustration of a gridding result of a synthetic microarray image.

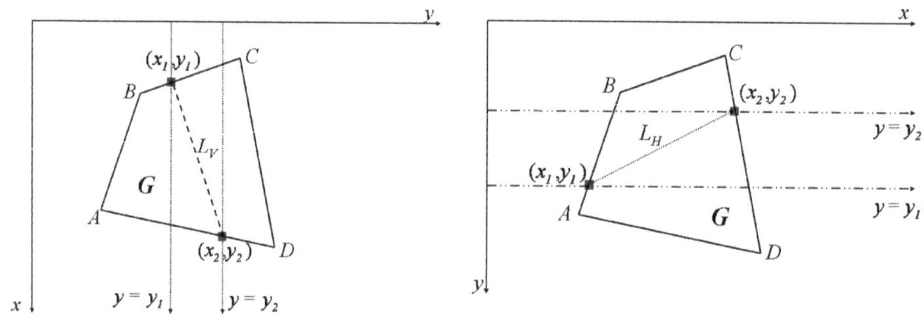

(a) (b)

Fig. 3. Illustrations of the quadrilateral in shape G: Cartesian systems for the representation of the line-segments belonging to the (a) SV set and (b) S_H set.

determination of the line-segments whose members are defined by the two vertical sides of G (Fig. 3b, "*BA*" and "*CD*" sides). Let S_V be the former set of line-segments, while S_H be the latter set of line-segments. The determination of each of the S_V or S_H sets of line-segments is expressed as an optimization problem which is tackled by using an evolutionary genetic algorithm.

3.1. *Attributes of the line-segments contained to the S_V or S_H sets*

Any line-segment on a 2D Cartesian plane can be defined by its end-points using the following equation:

$$y = y_1 + \frac{y_2 - y_1}{x_2 - x_1}(x - x_1), \tag{1}$$

where (x_1, y_1), (x_2, y_2) denote the coordinates of the two end-points of the line-segment, and $x \in [x_1, x_2]$ is the independent variable of the equation.

In the case of $x_1 = x_2$, the aforementioned equation (Eq. 1) is not valid. Consequently, in order for any line-segment L_i belonging to the S_V or S_H set to be able to be algebraically described by Eq. 1, the 2D Cartesian plane of Fig. 3a or Fig. 3b respectively is utilized.

Moreover, the line-segments L_i belonging to the S_V or S_H sets intersect with the two sides of the quadrilateral in shape *G*. Thereby, the x-coordinates (x_1 and x_2) of their end-points can be computed from the y-coordinates (y_1 and y_2). For instance, the x-coordinates of the line-segment "L_V" belonging to the S_V set (Fig. 3a) can be computed given that: (i) the intersection point of the line $y = y_1$ and the line-segment ("*BC*") is the point (x_1, y_1), and (ii) the intersection of the line $y = y_2$ and the line-segment ("*AD*") is the point (x_2, y_2). Likewise, the x-coordinates of the line-segment "L_H" (Fig. 3b) belonging to the S_H set can be also computed in the same manner.

3.2. *Genetic Algorithm for Gridding Microarray Images*

3.2.1. Chromosome Representation

The chromosome *m* represents all line-segments L_i, i=1,..., $N(m)$ belonging either to the S_V set or the S_H set. Due to the nature of the alignment of blocks inside the microarray image and the arrangement of spots inside the blocks, the chromosome encodes a line-segment (belonging to the S_V set or the S_H set) and the distance *d* between adjacent line-segments of the corresponding set, instead of encoding each line-segment. More precisely, the chromosome *m* has been encoded as a string of three real values (Fig. 4); the two y-coordinates (y_1 and y_2) of the end-points of a line-segment and the distance *d* between two adjacent line-segments. The y-coordinates (y_1 and y_2) are located on the horizontal sides ("*BC*" and "*AD*") or the vertical sides ("*BA*" and "*CD*") of the quadrilateral in shape *G*.

Fig. 5 depicts a quadrilateral in shape *G* as well as the line-segments constituting the borders between adjacent blocks or spots. In the case when the genetic algorithm searches for the exact values of the variables of the optimal line-segments belonging to the S_V set, its chromosome will encode the y-coordinates of the end-points of L_1 and the distance d_V. Respectively, for the S_H set, its chromosome will encode the y-coordinates of the end-points of L_4 and the distance d_H.

Variables of a line-segment (y-coordinates)		Distance between adjacent line-segments
y_1	y_2	d

Fig. 4. A schematic illustration of chromosome's encoding.

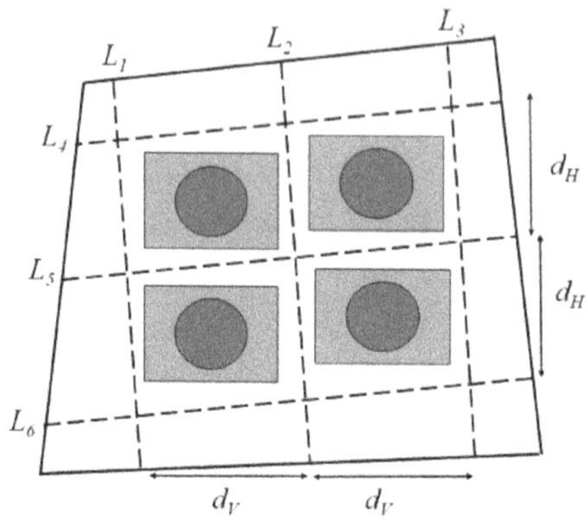

Fig. 5. Illustration of the line-segments constituting the grid structure of a quadrilateral in shape G.

3.2.2. Chromosome Evaluation

It is self-explanatory that a line-segment which is part of the grid is located in the background area existing between adjacent blocks or spots. The pixels of this area have intensities which are generally lower than the intensities of the pixels constituting spots. As a result, the probability $P(L_i)$ of a line-segment L_i to be part of the grid is defined by the following equation:

$$P(L_i) = f_B^{Ri}(L_i) - f_S^{Ri}(L_i), \tag{2}$$

where R_i denotes the region of G which contains those pixels whose distance from the line-segment L_i is less than a margin w (Fig. 6). The real-valued function $f_B^{Ri}(L_i)$ expresses the percentage of pixels of the region R_i whose intensity is lower than a value I_B (background pixels), while the real-valued function $f_S^{Ri}(L_i)$ expresses the percentage of pixels of the region R_i whose

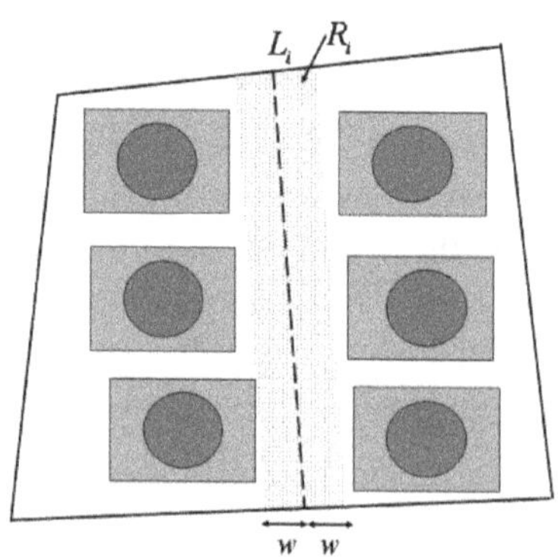

Fig. 6. Illustration of the region R_i (dotted gray area) near the dashed line-segment L_i.

intensity is higher than the value I_B (spot pixels). I_B is an intensity value which is defined as the value which is present in most pixels of G. Any pixel with intensity value lower than I_B is considered to belong to the background.

The Fitness Function $F_G(m)$ of a chromosome m that encodes a possible solution to the particular optimization problem is defined by the following equation:

$$F_G(m) = \begin{cases} S_p(m) \cdot N(m), \ if\ m \equiv efficient \\ S_p(m), otherwise \end{cases} \tag{3}$$

where the real-valued function $S_p(m)$ denotes a total sum of the probabilities $P(L_i)$ of the line-segments L_i, i=1,…,$N(m)$, that are represented by the chromosome m, and have a higher than a threshold P_{MAX} probability $P(L_i)$ to be part of the grid. P_{MAX} is a threshold which expresses the minimum acceptable probability of a line-segment to be part of the grid. Therefore, it controls which of the line-segments L_i participate in the sum $S_p(m)$. More precisely, $S_p(m)$ is defined as:

$$S_p(m) = \sum_{i=1}^{N(m)} P(L_i) \cdot q_i\ , \tag{4}$$

where

$$q_i = \begin{cases} 1, \ \ if\ P(L_i) > P_{MAX} \\ 0, \ \ otherwise \end{cases} \tag{5}$$

A chromosome m is efficient if most of the line-segments that it represents are well-defined, by being located in background areas. Therefore, we define the percentage $f_{LS}(m)$ of the line-segments L_i, i=1,…,$N(m)$, that are represented by the chromosome m, and have a lower than or equal to a threshold P_{Low} probability $P(L_i)$ to be part of the grid:

$$f_{LS}(m) = \frac{\sum_{i=1}^{N(m)} k_i}{N(m)}, \tag{6}$$

where

$$k_i = \begin{cases} 1, \ \ if\ P(L_i) \le P_{Low} \\ 0, \ \ otherwise \end{cases} \tag{7}$$

If the $f_{LS}(m)$ percentage of the chromosome m is lower or equal to a threshold f_{Max}, it means that the line-segments, represented by the chromosome m, are well-defined because they are located in background areas, some of which may be contaminated with noise. Therefore the chromosome m is efficient. Otherwise, the chromosome m is inefficient.

Using the Fitness Function $F_G(m)$, the genetic algorithm can assign to an efficient chromosome m a higher fitness value than to an inefficient one. Moreover, the multiplication $S_p(m)\cdot N(m)$ in Eq. 3 prevents the genetic algorithm from converging to a local solution that would be not an optimal one (this would lead to termination without determining all the line-segments belonging to the S_H or the S_V sets). Indeed, the higher the fitness value of the chromosome is, the greater number $N(m)$ of the line-segments L_i which have a high probability $P(L_i)$ to be part of the grid it represents.

3.3. *The Refinement Procedure*

The genetic algorithm is based on the observation that the line-segments - having the same direction and constituting the borders of blocks (or spots) - are ideally equidistant. Although this observation is true for an ideal microarray image, in reality due to rotations, misalignments and local deformations of the ideal rectangular grid, it may not be valid. As a result, the determined line-segments may slightly vary from the optimal ones.

In order to tackle this problem, each line-segment L_i belonging to the S_V or S_H sets is replaced by a new one, L_i', so long as the following are valid:

- the line-segment L_i' is located inside the region R_i of G ($L_i' \in R_i$)
- the probability $P(L_i')$, of the line-segment L_i', to be part of the grid, is higher than the equivalent probability $P(L_i)$ of L_i, by more than a threshold T_p ($P(L_i') - P(L_i) > T_p$).

An example of the refinement procedure is depicted in Fig. 7.

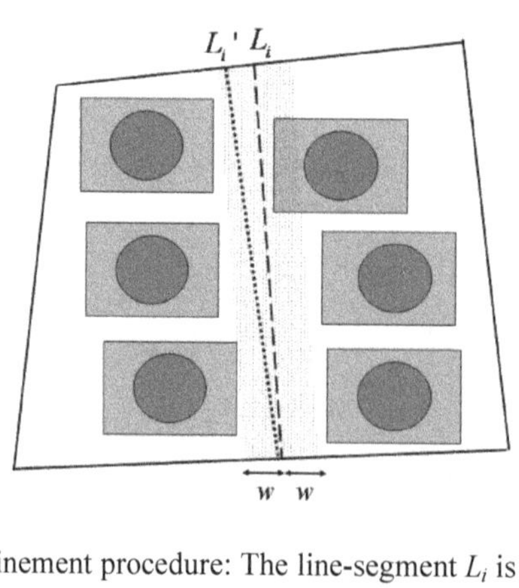

Fig. 7. Illustration of the refinement procedure: The line-segment L_i is replaced by the line-segment L_i'.

4. Spot-Segmenting Microarray Images

Based on empirical observations, microarray spots have some common characteristics. One of the most significant is that their shape is approximately elliptical and it can be discriminated into the following four categories[32]: peak, plateau, volcano and doughnut (Fig. 8).

Spot-segmentation is the second essential phase of microarray image analysis, which follows the gridding one. Due to the aforementioned common spots' characteristic, it may be more efficient to deal with the segmentation of a microarray spot thanks to its 3D modeling, using intensity as the third dimension. This is due to the fact that the contour of a microarray spot can be depicted in the image plane by drawing the contour of its spot-model. In the proposed method, the determination of the 3D spot-model of a specific real-spot is expressed as an optimization problem which is solved by using an evolutionary genetic algorithm.

4.1. *3D mathematical model for a microarray compartment*

4.1.1. Spot-model and its components

Let I_{REAL} be a compartment of a real microarray image containing one individual spot S_{REAL} and its background. The real-spot S_{REAL} can be represented by a spot-model S_{MODEL} which is expressed as a 3D-curve constituting of two components: The first component S_{MB} represents the 3D-curve of the main-body of the spot-model, while the second one S_{ID} represents the 3D-curve of the inner-dip of the spot-model.

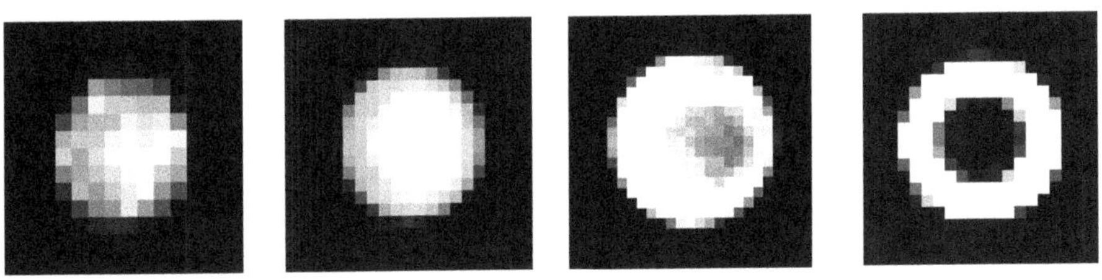

(a) (b) (c) (d)

Fig. 8. Examples of real microarray spots: (a) a peaked-shaped spot, (b) a plateau-shaped spot, (c) a volcano-shaped spot, and (d) a doughnut-shaped spot.

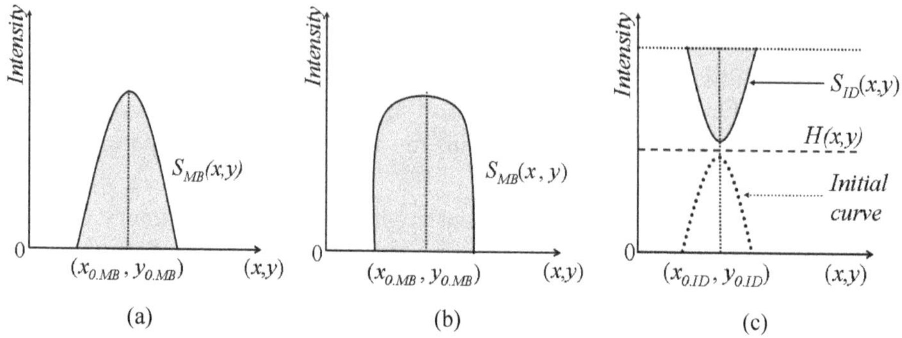

Fig. 9. Cross-sections of the 3D curves of the components of the spot-model: (a) a Gaussian SMB(x,y) 3D curve, (b) a plateau SMB(x,y) 3D curve and (c) a $S_{ID}(x,y)$ 3D curve.

The main-body $S_{MB}(x,y)$ of the spot-model can be expressed as the diffusion function $C(x,y)$ proposed by Bettens et al.[33] This 3D-curve resembles the 3D Gaussian or plateau curve (Fig. 9a,b). Likewise, the inner-dip of the spot-model $S_{ID}(x,y)$ can be defined as a symmetrical 3D-curve to an 'initial $C(x,y)$ 3D-curve', in respect to an horizontal surface $H(x,y)$ (Fig. 9c). The spot-model $S_{Model}(x,y)$ is constructed by combining the $S_{MB}(x,y)$ and $S_{ID}(x,y)$ 3D-curves as the following equation indicates:

$$S_{MODEL}(x,y) = Min\left[S_{MB}(x,y), S_{ID}(x,y)\right] \tag{8}$$

A graphical explanation of Eq. 8 is depicted in Fig. 10. The resulting total-models depend on the 3D curves of their corresponding S_{MB} and S_{ID} components and especially on the relative position of their centers and heights. More precisely, in the case of the distance between the S_{MB} and S_{ID} centers being large, the resulting total-model resembles a peak-shaped spot (Fig. 10a and Fig. 10b). In the case of the distance between the S_{MB} and S_{ID} centers being small, the resulting total-model resembles a volcano-shaped spot (Fig. 10c) or a doughnut-shaped spot (Fig. 10d), according to the height of the S_{ID} 3D curve.

4.1.2. *Compartment-model and its components*

Likewise to the spot-model S_{MODEL}, let I_{MODEL} be a compartment-model which represents, in a 3D space, the real one I_{REAL} by modeling the latter's intensities' values. The I_{MODEL} can be expressed as a 3D-curve constituting of two components: i) a surface representing the average background intensity B_{AV} of the compartment-model and, ii) a 3D-curve representing the spot-model S_{MODEL}.

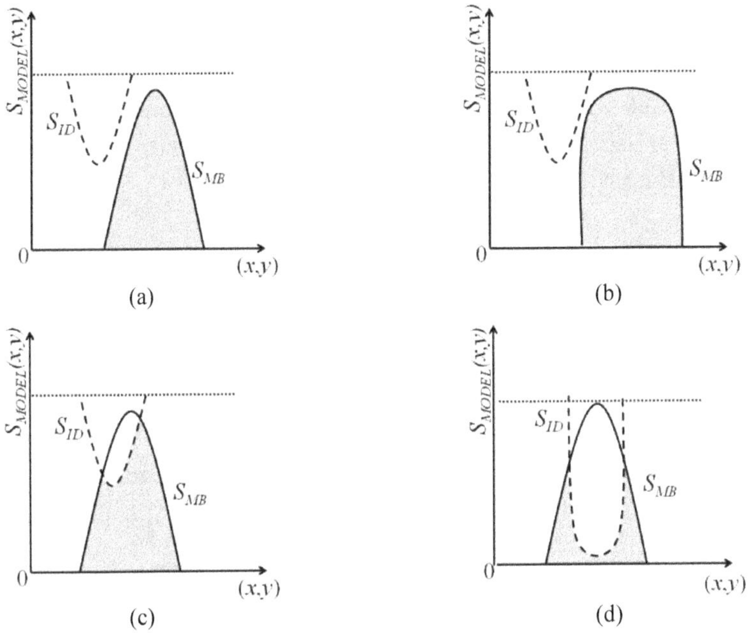

Fig. 10. 2D illustrations of four spot-models by portraying the cross-sections of their $S_{MB}(x,y)$ and $S_{ID}(x,y)$ 3D-curves. The resulting spot-models $S_{MODEL}(x,y)$ which are the surfaces coloured in grey are the following: (a) a peak-shaped spot, (b) a plateau-shaped spot, (c) a volcano-shaped spot, (d) a doughnut-shaped spot.

The compartment-model $I_{MODEL}(x,y)$ corresponds to a 3D-curve whose equation is defined as:

$$I_{MODEL}(x,y) = Max\left[B_{AV}, S_{MODEL}(x,y)\right], \tag{9}$$

where B_{AV} denotes the average background intensity of the compartment-model and it corresponds to a threshold of the lowest values of the $S_{MODEL}(x,y)$. Pixels whose values are lower than B_{AV} belong to the background and their values are set equal to B_{AV}. A graphical explanation of Eq. 9 is depicted in Fig. 11.

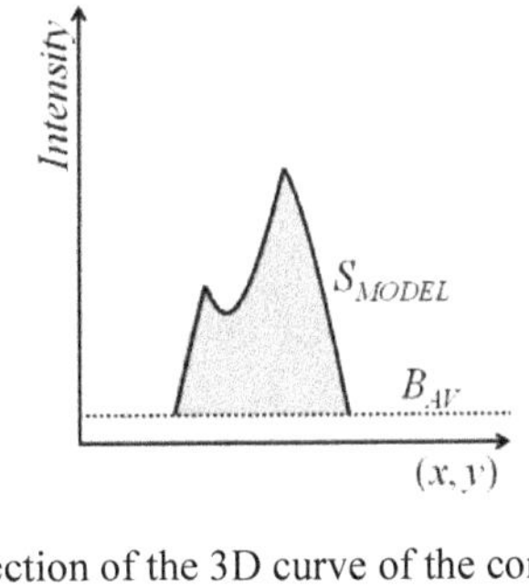

Fig. 11. Cross-section of the 3D curve of the compartment model.

4.2. *Evolutionary genetic algorithm for segmenting microarray images*

Given a specific compartment I_{REAL}, the genetic algorithm determines the compartment-model which represents optimally the real-one I_{REAL}. Thus, the contour of the real-spot is depicted by drawing the contour of the spot-model inside its compartment.

4.2.1. Chromosome Representation and Characterization

A chromosome *m* represents in a 3D space a specific compartment-model I^{m}_{MODEL}. It has been encoded as a numerical sequence consisting of three segments (Fig. 12). The first segment encodes the value of the average background intensity B^{m}_{AV} of the compartment-model. The second segment encodes the values of the variables of the main-body S^{m}_{MB}, while the third segment encodes the values of the variables of the inner-dip S^{m}_{ID} of the spot-model S^{m}_{MODEL}.

B^{m}_{AV}	*variables of* S^{m}_{MB}	*variables of* S^{m}_{ID}

Fig. 12. Illustration of a chromosome *m*.

Each chromosome *m* is characterized as '*inefficient*' or '*efficient*'. The former denotes a chromosome which represents a compartment-model I^{m}_{MODEL} that hardly resembles the real-one I_{REAL}. In this case, the S^{m}_{MB} component of its spot-model S^{m}_{MODEL} approximates deficiently the real-spot S_{REAL}. The latter denotes a chromosome which represents a compartment-model I^{m}_{MODEL} that resembles – to a degree – the real-one I_{REAL}. In this case, the S^{m}_{MB} component of the spot-model S^{m}_{MODEL} approximates – to a degree – the real-spot S_{REAL}. Fig. 13 illustrates examples of '*inefficient*' and '*efficient*' chromosomes.

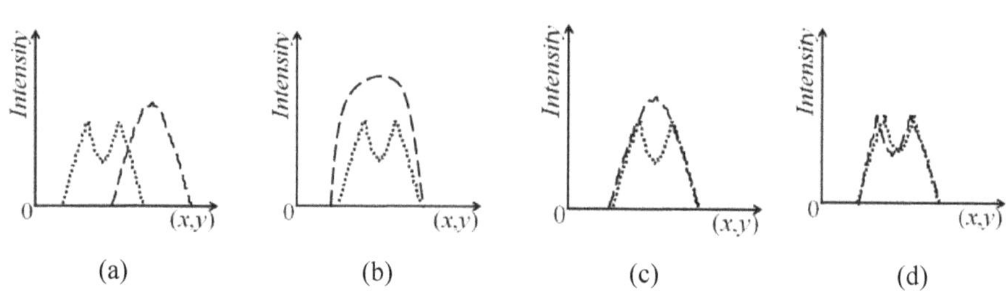

Fig. 13. 2D illustrations of four chromosomes: (a,b) two inefficient chromosomes, (c,d) two efficient chromosomes.

4.2.2. *Chromosome Evaluation*

The key concept of chromosome evaluation is that the higher the resemblance of a compartment-model I^m_{MODEL} to the real-one I_{REAL} is, the higher the value of the fitness function of the chromosome m becomes. Therefore, the relative intensity error $E(x,y)$ between the intensity of the model-compartment's pixel (x,y) and the corresponding one of the real-compartment's pixel is calculated as follows:

$$E(x,y) = \frac{|I^m_{MODEL}(x,y) - I_{REAL}(x,y)|}{I_{REAL}(x,y)} \tag{10}$$

The fitness function $F_S(m)$ of a chromosome m is defined by the following equation:

$$F_S(m) = -\iint_{x,y \in Compartment} E(x,y)\ dxdy \tag{11}$$

Using the Fitness Function $F_S(m)$, the genetic algorithm can progressively assign – from left to right – a higher fitness value to the chromosomes representing the compartment-models in Fig. 13.

5. Results

Several experiments on various real and synthetic cDNA microarray images were conducted so as to evaluate the performance of our methods in gridding and spot-segmentation.

5.1. *Gridding Results*

The microarray images used for the evaluation were obtained from the Stanford Microarray Database (SMD),[31] which is publicly available and broadly used. These microarray images have been produced by comprehensively analyzing the gene expression profiles in 54 specimens of acute lymphoblastic leukemia, 37 positive and 17 negative to BCR-ABL.[35] BCR-ABL is a fusion gene product resulting from translocation between the 9th and the 22nd chromosomes. In all the conducted experiments for gridding evaluation, the parameters have been experimentally adjusted once, and thus they remained constant during all experiments.[36] Thus, the whole experimental procedure on real cDNA microarray images took place without any human intervention.

The efficiency of the proposed method was evaluated by means of the statistical analysis described by Blekas et al.[36] More precisely, each microarray spot was classified in one of the following three categories: '*perfectly*', '*marginally*' and '*incorrectly*' gridded. A spot was '*perfectly*' gridded if the entire spot area was contained inside the equivalent compartment of the grid. A spot was '*marginally*' or '*incorrectly*' gridded - respectively - if more or less than 80% of the entire spot area was contained inside the equivalent compartment of the grid. It should be noted that the ground truth is given in the SMD.

Using the gridding method, 95.1% of spots were perfectly placed inside the compartment, 4.3% were very nearly gridded, while only 0.6% were gridded incorrectly. It should be pointed out that the gridding method outperforms established techniques,[36] such as the one proposed by Blekas et al.,[36] as well as popular software programs such as ScanAlyze and SpotFinder. Fig. 14 depicts the gridding result of a noisy and rotated microarray sub-image containing several spots of various intensities and sizes. This example indicates that the effectiveness of the proposed method is not influenced by spot intensities and sizes neither by rotations and misalignments of the ideal rectangular grid.

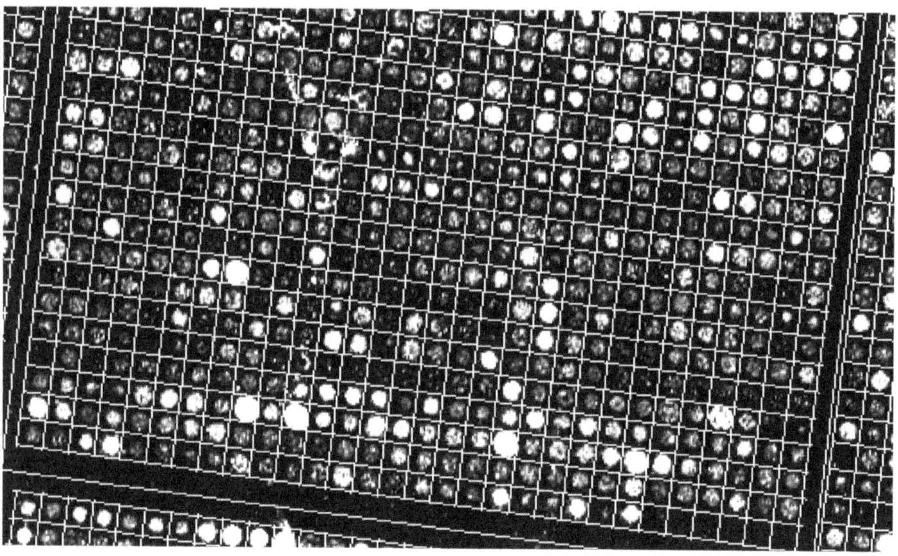

Fig. 14. Example of a gridding result.

5.2. *Spot-Segmentation Results*

Since ground truth for the segmentation's results on real microarray images does not exist – even for the SMD database –, synthetic microarray images were used for the evaluation of the proposed segmentation approach. More precisely, the microarray images used for the evaluation were obtained on the Internet.[38] They have been produced by the microarray simulator of Nykter which generates synthetic microarray images with realistic characteristics[39] and varied quality. Half of the images (good quality images) have low variability in spot sizes and shapes, and their noise level is reasonably low. The rest of them (low quality images) contain spots whose shape and size vary substantially. In addition, the noise level is sufficiently higher in the low quality images. Finally, each image is digitized at 330 x 750 pixels and contains 1000 spots.

Using the presented spot-segmentation approach, 91.5% of spots were very efficiently segmented, and no spurious spot has been detected. Fig. 15 illustrates the segmentation result of a microarray block taken from a good-quality synthetic image, while Fig. 16 illustrates the segmentation result of a microarray block taken from a low-quality synthetic image. On these segmentation results, one can observe that the proposed method has optimally segmented all the microarray spots of Fig. 15, and nearly all the microarray spots of Fig. 16.

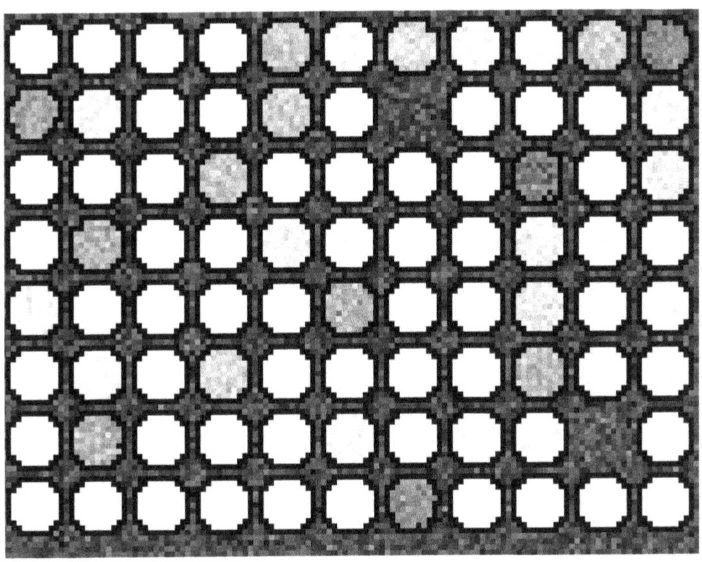

Fig. 15. Examples of spot-segmentation results on a good quality sub-image.

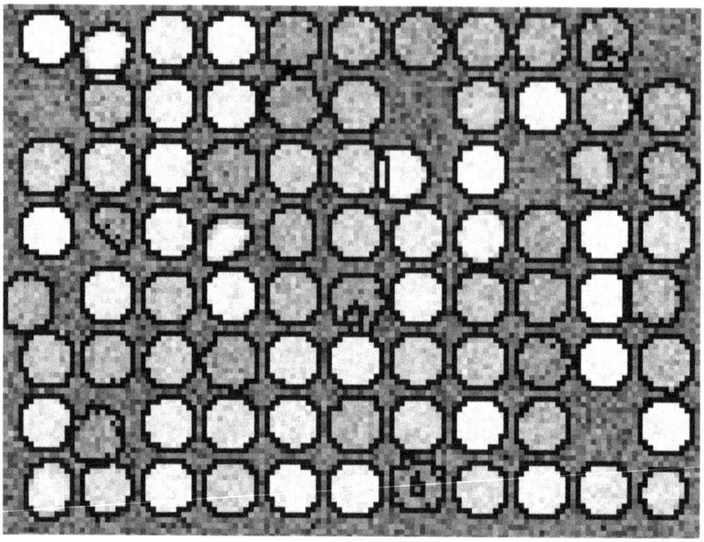

Fig. 16. Examples of spot-segmentation results on a low quality sub-image.

6. Conclusions

In the process of microarray image analysis, gridding and spot-segmenting remain undoubtedly the most challenging stages. Although much progress has been made, they still require human intervention which can affect significantly the biological conclusions reached during microarray experiments.

In this chapter a method for gridding and spot-segmenting microarray images is presented which is based on the concept of evolution. The proposed method is fully automatic since all needed parameters have been adjusted once and they have been kept fixed during all the experiments. Consequently, there is no requirement of any input parameter or human intervention in order to determine properly the gridding and the contours of microarray spots. The experimental results over synthetic and real images firmly confirm the validity of our method, as well as its robustness and effectiveness.

References

1. E. K. Lobenhofer, P. R. Bushel, C. A. Afshari and H. K. Hamadeh, "Progress in the Application of DNA Microarrays," *Enviromental Health Perspectives*, vol. 109, no. 9, pp. 881–891, Sept. 2001.
2. A. M. Campbell and L. J. Heyer, *Discovering Genomics, Proteomics & Bioinformatics*, 2nd ed., Pearson Benjamin Cummings, 2007, pp. 233-238.
3. Y. H. Yang, M. J. Buckley, S. Dudoit, and T. Speed, "Comparison of methods for image analysis on cDNA microarray data," *Journal of Computational & Graphical Statistics*, vol. 11, no. 1, pp. 108–136, Mar. 2002
4. X. H. Wang, and R. S. H. Istepanian, "Microarray image enhancement by denoising using stationary wavelet transform," *IEEE Trans. of Nanobioscience*, vol. 2, no. 4, pp 184-189, Dec. 2003.
5. W. B. Chen, C. Zhang, and W. L. Liu, "An Automated Gridding and Segmentation Method for cDNA Microarray Image Analysis," in *Proc. 19th IEEE Symp. Computer-Based Medical Systems*, Salt Lake City, 2006, pp. 893-898.
6. Y. Tu, G. Stolovitzky, and U. Klein, "Quantitative noise analysis for gene expression microarray experiments," in *Proc. National Academy of sciences*, USA, 2002, pp.14031-14036.
7. M. B. Eisen. (1999). ScanAlyze. [Online]. Available: http://rana.lbl.gov/EisenSoftware.htm
8. J. Buhler, T. Ideker, and D. Haynor, "Dapple: improved techniques for finding spots on DNA microarrays," UW CSE Technical Report UWTR 2000-08-05, pp. 1-12, Aug. 2000.
9. Biodiscovery Inc. (2005). ImaGene. [Online]. Available: http://www.biodiscovery.com/imagene.as
10. Buckley MJ (2000) The Spot User's Guide, CSIRO mathematical and information sciences. http://www.cmis.csiro.au/IAP/Spot/ spotmanual.htm
11. N. Deng and H. Duan, "The Automatic Gridding Algorithm based on projection for Microarray Image", in *Proc. Int. Conf. Intelligent Mechatronics and Automation*, Chendu, 2004, pp. 254-257.
12. A. W. C. Liew, H. Yan, and M. Yang, "Robust adaptive spot segmentation of DNA microarray images," *Pattern Recognition*, vol. 36, no. 5, pp. 1251–1254, May 2003.
13. J. Angulo and J. Serra, "Automatic analysis of DNA microarray images using mathematical morphology," *Bioinformatics*, vol. 19, no. 5, pp. 553–562, 2003.
14. R.Hirata, J. Barrera, R. F. Hashimoto, and D. O. Dantas, "Microarray gridding by mathematical morphology," in *Proceedings of 14th Brazilian Symposium on Computer Graphics and Image Processing (SIBGRAPI '01)*, pp. 112–119, Florianopolis, Brazil, October 2001.
15. H.-J. Jin, B.-K. Chun, and H.G. Cho, "Extended epsilon regular sequence for automated analysis of microarray images," in The IEEE Computer Society Conference on Computer Vision and Pattern Recognition (CVPR '05), the Workshop on Computer Vision Methods for Bioinformatics (CVMB), San Diego, Calif, USA, June 2005.
16. P. Bajcsy, "Gridline: automatic grid alignment in DNA microarray scans," IEEE Trans. Image Processing, vol. 13, no. 1, pp. 15–25, Jan. 2004.
17. M. Steinfath, W. Wruck, H. Seidel, H. Lehrach, U. Radelof, and J. O'Brien, "Automated image analysis for array hybridization experiments," *Bioinformatics*, vol. 17, no.7, pp. 634–641, Jul. 2001.

18. N. Brandle, H. Bischof, and H. Lapp, "Robust DNA Microarray image analysis," *Machine Vision and Applications*, vol. 15, no.1, pp. 11–28, Oct. 2003.
19. J. Ho, W. L. Hwang, H. H. S. Lu, D. T. Lee, "Gridding Spot Centers of smoothly distorted microarray images," *IEEE Trans. Image Processing*, vol. 15, no. 2, pp. 342-353, Feb. 2006.
20. G. Antoniol and M. Ceccarelli, "Microarray image gridding with stochastic search based approaches," *Image and Vison Computing*, vol. 25, no. 2, pp. 155-163, Feb.2007.
21. Y. Chen, E. R. Dougherty, and M. L. Bittner, "Ratio-based decisions and the quantitative analysis of cDNA microarray images," *Journal Biomedical Optics*, vol. 2 no. 4, pp. 364-374, 1997.
22. D. Bozinov, and J. Rahnenfuhrer, "Unsupervised technique for robust target separation and analysis of DNA microarray spots through adaptive pixel clustering," *Bioinformatics*, vol. 18, no. 5, pp. 747–756, 2002.
23. J. Rahnenfuhrer, and D. Bozinov, "Hybrid clustering for microarray image analysis combining intensity and shape features," *BMC Bioinformatics*, vol. 5, no. 5, p. 47-58, Apr. 2004.
24. E. Ergut, Y. Yardimci, E. Mumcuoglu, and O. Konu, "Analysis of microarray images using FCM and K-Means clustering algorithms," *in Proc. Int. Conf. Signal Processing*, 2003, pp. 116–121.
25. R. Nagarajan, "Intensity-based segmentation of microarray images," I*EEE Trans. Medical Imaging*, vol. 22, no. 7, pp. 882–889, Jul. 2003.
26. Q. Li, C. Fraley, R. E. Bumgarner, K. Y. Yeung, and A. E. Raftery, "Donuts, scratches and blanks: robust model-based segmentation of microarray images," *Bioinformatics*, vol. 21, no. 12, pp. 2875-2882, Apr. 2005.
27. M. Mitchell, An Introduction to Genetic Algorithms, MIT Press, Cambridge, 1999.
28. D. E. Goldberg, *Genetic Algorithms in Search, Optimization & Machine Learning*, Boston: Addison-Wesley, Reading, 1989, ch. 1.
29. B. L. Miller, and D. E. Goldberg, "Genetic Algorithms, Tournament selection, and the Effects of Noise," Complex Systems, vol. 9, no. 3, pp. 193-212, 1995.
30. F. Herrera, M. Lozano, and A. M. Sanchez, "Hybrid crossover operators for real-coded genetic algorithms: An experimental study," *Soft Computing*, vol. 9, no. 4, pp. 280-298, Apr. 2005.
31. S. H. Ling, and F. H. F. Leung, "An improved genetic algorithm with average-bound crossover and wavelet mutation operations," *Soft Computing*, vol. 11, no. 1, pp. 7-31, 2007.
32. H. Y. Kim *et al.*, "Characterization and simulation of cDNA microarray spots using a novel mathematical model," *BMC Bioinformatics*, vol. 8, pp. 485-496, March 2007.
33. E. Bettens, P. Scheunders, D. van Dyck, L. Moens, and P. van Osta, "Computer analysis of two-dimensional electrophoresis gels: a new segmentation and modeling algorithm," *Electrophoresis*, vol. 18, pp. 792-798 1997.
34. Standford Microarray Database. [Online]. Available: http://genome-www5.stanford.edu/
35. D. Juric *et al.*, "Differential Gene Expression Patterns and Interaction Networks in BCR-ABL-Positive and -Negative Adult Acute Lymphoblastic Leukemias," *Journal of Clinical Oncology*, vol. 25, no. 11, pp. 1341-1349, April 2007.
36. E. Zacharia, D. Maroulis, "Microarray Image Analysis based on an Evolutionary Approach", *in Proc of the 19th International Conference on Pattern recognition* (*ICPR*), Tampa, Florida, USA, 2008.

37. K. Blekas, N. P. Galatsanos, A. Likas, and I. E. Lagaris, "Mixture Model Analysis of DNA Microarrray Images," *IEEE Trans. Medical Imaging*, vol. 24, no. 7, pp. 901-909, July 2005.
38. Microarray Database. [Online]. http://www.cs.tut.fi/sgn/csb/spotseg/
39. M. Nykter, *et al.*, "Simulation of microarray data with realistic characteristics," *BMC Bioinformatics*, vol. 7, pp. 349-366, Jul. 2006.

CHAPTER 4.7

COMPUTATIONAL INTELLIGENCE APPROACHES FOR PROTEIN FOLD RECOGNITION

Yuehui Chen[1], Yaou Zhao[2] and Jack Y. Yang[3]

[1]*Computational Intelligence Lab., School of Information Science and Engineering University of Jinan, Jinan 250022, P.R.China*
E-mail: yhchen@ujn.edu.cn

[2]*Computational Intelligence Lab., School of Information Science and Engineering University of Jinan, Jinan 250022, P.R.China*
E-mail: yaou_zhao@yahoo.com.cn

[3]*Harvard UniversityCambridge, Massachusetts USA*
E-mail: dr.jack.yang@gmail.com

New approaches based on the implementation of support vector machine (SVM) and flexible neural trees (FNT) with the error correcting output codes (ECOC) are presented for recognition of multi-class protein folds. ECOC is used for reducing multi-class classification problem to multiple binary classification problems. The SVM and FNT trained by particle swarm optimization algorithm are then employed as the base classifier. The experimental results show that the proposed method can improve prediction accuracy by 4%-15% on two datasets containing 27 SCOP folds.

1. Introduction

Protein fold recognition represents an important process in understanding the associations between sequence and structure as well as possible functional and evolutionary relationships. Over the past three decades several methods have been proposed for predicting protein structural classes[1]. These methods can be divided into two methodological approaches: (a) the informatics based methods that involve the sequence-based methods[2,3,4,5,6,7,8,9,10,11] and the structure based methods[12,13,14] and (b) the bio-physics based methods[15,16]. Sequence based methods use protein sequence or predicted secondary structure information to perform sequence comparison and detect whether two proteins share a fold or not. Structure based or threading methods create an energy function describing how well a probe sequence matches a target fold. In fold recognition by threading, we must take the amino acid sequence of a protein and evaluate how well it fits into one of the known three-dimensional (3D) protein structures. Besides purely sequence based or structure based methods, a combination of them is also possible[17]. On the other hand, methods based on biophysics perform ab initio structure prediction. They detect a native conformation or ensemble of conformations of the protein that are at or near the global free-energy minimum[18].

Sequence-based methods are very popular in fold recognition. Computational intelligence or Machine learning techniques, i.e., genetic algorithms[5], neural networks[6], support vector machines[7,33,34], hidden Markov models[9,10], segmentation conditional random fields[11], information retrieval[19], neural network ensemble[21] and linear discriminate analysis (LDA)[20] have been adopted to exploit protein sequence or secondary structure information.

Although significant improvement has been made in the field of fold recognition, the accuracy of the existing methods remains limited and there is a need to develop new methods.

In this paper, new approaches were proposed for multi-class protein fold classification problem. The approach includes four parts: feature extraction, ECOC, SVM or FNT trained by particle swarm optimization[31] and compare system. The obtained results are very encouraging; our results improved the average predictive accuracy obtained in previous published work.

The paper is organized as follows. In Section 2, the error correcting output cods for multi-class classification problems is introduced. A support vector machine and flexible neural tree trained by particle swam optimization algorithm is given in Section 3 and Section 4. The proposed method and experimental results are presented in Section 5 and Section 6. Finally, the conclusion and future work are presented.

2. Error Correcting Output Codes

ECOC represents a general framework based on a coding and decoding (ensemble strategy) technique to handle multi-class problems. One of the most well-known properties of the ECOC is that it improves the generalization performance of the base classifiers. Moreover, the ECOC technique has demonstrated to be able to decrease the error caused by the bias and the variance of the base learning algorithm[22].

In the ECOC technique, the multi-class to binary division is handled by a coding matrix. Each row of the coding matrix represents a codeword assigned to each class. On the other hand, each column of the matrix (each bit of the codeword) shows a partition of the classes in two sets. The ECOC strategy is divided into two parts: the coding part, where the binary problems to be solved have to be designed, and the decoding technique, that given a test sample looks for the most similar codewords.

The basis of the ECOC framework is to create a codeword for each of the N_c classes. Arranging the codewords as a rows of a matrix we define the "coding matrices" *M*, where $M \in \{-1,1\}^{N_c \times n}$, and *n* is the code length. From view point of learning, matrix *M* represents *n* binary learning problems (dichotomies), and each corresponding to a matrix column. Joining classes in sets, each dichotomy defines a partition of classes (coded by {+1,-1} according to their class membership). Applying the *n* trained binary classifiers, a code is obtained for each data defined in the matrix *M*, and the data point is assigned to the class with the "closest" codeword. To design an ECOC system, we need a coding and decoding strategy. When the ECOC technique was first developed it was believed that the ECOC code matrices should be designed to have certain properties to generalize well. A good error correcting output code for a *k-class* problem should satisfy that rows, columns are well separated from the rest in terms of Hamming distance. Usually there is no general method to construct ECOC for any number of classes. Following the method proposed[23], for *27-class* protein fold prediction problem, a ECOC is created with the length of code 24 and the minimum Hamming distance 12 (see Table 1).

Table 1. The Matrix M used in 27-class protein fold classification problem

-1	-1	-1	-1	-1	-1	-1	-1	-1	-1	-1	-1	-1	-1	-1	-1	-1	-1	-1	-1	-1	-1	-1	-1
1	1	1	1	1	1	1	1	1	1	1	1	-1	-1	-1	-1	-1	-1	-1	-1	-1	-1	-1	-1
1	1	1	1	1	1	-1	-1	-1	-1	-1	-1	1	1	1	1	1	1	-1	-1	-1	-1	-1	-1
-1	-1	-1	-1	-1	-1	1	1	1	1	1	1	1	1	1	1	1	1	-1	-1	-1	-1	-1	-1
1	1	1	-1	-1	-1	1	1	1	-1	-1	-1	1	1	1	-1	-1	-1	1	1	1	-1	-1	-1
-1	-1	-1	1	1	1	-1	-1	-1	1	1	1	1	1	1	-1	-1	-1	1	1	1	-1	-1	-1
-1	-1	-1	1	1	1	1	1	1	-1	-1	-1	-1	-1	-1	1	1	1	1	1	1	-1	-1	-1
1	1	1	-1	-1	-1	-1	-1	-1	1	1	1	-1	-1	-1	1	1	1	1	1	1	-1	-1	-1
1	-1	-1	1	1	-1	1	1	-1	1	-1	-1	1	1	-1	1	-1	-1	1	-1	-1	1	1	-1
-1	1	1	-1	-1	1	-1	-1	1	-1	1	1	1	1	-1	1	-1	-1	1	-1	-1	1	1	-1
-1	1	1	-1	-1	1	1	1	-1	1	-1	-1	-1	-1	1	-1	1	1	1	-1	-1	1	1	-1
1	-1	-1	1	1	-1	-1	-1	1	-1	1	1	-1	-1	1	-1	1	1	1	-1	-1	1	1	-1
-1	1	1	1	1	-1	-1	-1	1	1	-1	-1	-1	-1	1	1	-1	-1	-1	1	1	1	1	-1
1	-1	-1	-1	-1	1	1	1	-1	-1	1	1	-1	-1	1	1	-1	-1	-1	1	1	1	1	-1
1	-1	-1	-1	-1	1	-1	-1	1	1	-1	-1	1	1	-1	-1	1	1	-1	1	1	1	1	-1

Table 1. Continued

-1	1	1	1	1	-1	1	1	-1	-1	1	1	1	1	-1	-1	1	1	-1	1	1	1	1	-1
-1	1	-1	1	-1	1	1	-1	1	-1	1	-1	1	-1	1	-1	1	-1	-1	1	-1	1	-1	1
1	-1	1	-1	1	-1	-1	1	-1	1	-1	1	1	-1	1	-1	1	-1	-1	1	-1	1	-1	1
1	-1	1	-1	1	-1	1	-1	1	-1	1	-1	-1	1	-1	1	-1	1	-1	1	-1	1	-1	1
-1	1	-1	1	-1	1	-1	1	-1	1	-1	1	-1	1	-1	1	-1	1	-1	1	-1	1	-1	1
1	-1	1	1	-1	1	-1	1	-1	-1	1	-1	-1	1	-1	-1	1	-1	1	-1	1	1	-1	1
-1	1	-1	-1	1	-1	1	-1	1	1	-1	1	-1	1	-1	-1	1	-1	1	-1	1	1	-1	1
-1	1	-1	-1	1	-1	-1	1	-1	-1	1	-1	1	-1	1	1	-1	1	1	-1	1	1	-1	1
1	-1	1	1	-1	1	1	-1	1	1	-1	1	1	-1	1	1	-1	1	1	-1	1	1	-1	1
1	1	-1	-1	1	1	-1	1	1	1	1	-1	-1	1	1	1	1	-1	1	1	-1	-1	1	1
-1	-1	1	1	-1	-1	1	-1	-1	-1	-1	1	-1	1	1	1	1	-1	1	1	-1	-1	1	1
-1	-1	1	1	-1	-1	-1	1	1	1	1	-1	1	-1	-1	-1	-1	1	1	1	-1	-1	1	1

3. Support Vector Machine Trained with PSO

3.1. *SVM*

The SVM proposed by Vapnik[24] has been studied extensively for classification, regression and identification. SVM maps the input patterns into a higher dimensional feature space through some nonlinear mapping chosen a priori. A linear decision surface is then constructed in this high dimensional feature space. Thus, SVM is a linear classifier in the parameter space, but it becomes a nonlinear classifier as a result of the nonlinear mapping of the space of the input patterns into the high dimensional feature space. For a two-class problem, assuming the optimal hyperplane in the feature space is generated, the classification decision of an unknown pattern x will be made based on

$$f(x) = \operatorname{sgn}\left(\sum_{i=1}^{N} a_i y_i K(x_i, x) + b\right)$$

where $\alpha_i \geq 0, i = 1, 2, \ldots, N$ are nonnegative Lagrange multipliers that satisfy

$$\sum_{i=1}^{N} \alpha_i y_i = 0, \{y_i \mid y_i \in -1, +1\}, i = 1, 2, \ldots, N$$

where y_i are class labels of training patterns $\{x_i \mid x_i \in R^N\}$, and $K(x_i, y)$ for $i = 1, 2, \ldots, N$ represents a symmetric positive definite kernel function that defines an inner product in the feature space.

Training the SVM is a quadratic optimization problem. The construction of a hyperplane $W^T X + b = 0$ (W is the vector of hyperplane coefficients, b is a bias term) so that the margin between the hyperplane and the nearest point is maximized and can be posed as the quadratic optimization problem. SVM has been shown to provide high generalization ability. A proper kernel function for a certain problem is dependent on the specific data and till now there is no good method on how to choose a kernel function[24]. In this study, the choice of the kernel functions was studied empirically and optimal results were achieved using radial basis function (RBF) kernel function.

$$K(x, x_j) = \exp\left(-\frac{|x - x_j|^2}{2\lambda^2}\right)$$

where λ is also a constant specified by users.

3.2. *PSO*

The PSO conducts searches using a population of particles that correspond to individuals in an Evolutionary Algorithm (EA)[25,26]. To get the parameters of SVM, particle swarm optimization (PSO) algorithm is employed. All free parameters in the SVM constitute a particle. Initially, a population of particles is randomly generated. Each particle represents a potential solution and has a position represented by a position vector x_i. A swarm of particles moves through the problem space with the moving velocity of each particle represented by a velocity vector v_i. At each time step, a function f_i (representing a quality measure) is calculated by using x_i as input. Each particle keeps track of its own best position, which is associated with the best fitness it has achieved so far in a vector p_i. Furthermore, the best position among all the particles obtained so far in the population is kept track of as p_g. In addition to this global version, another version of PSO keeps track of the best position among all the topological neighbors of a particle. At each time step t, by using the individual best position $p_i(t)$, and the global best position, $p_g(t)$, a new velocity for particle i is updated by

$$v_i(t+1) = v_i(t) + c_1\phi_1(p_i(t) - x_i(t)) + c_2\phi_2(p_g(t) - x_i(t))$$

where c_1 and c_2 are positive constants and ϕ_1 and ϕ_2 are uniformly distributed random numbers in [0,1]. The term c_i is limited to the range of V_{max} (if the velocity violates this limit, it is set to its proper limit). Changing velocity this way enables the particle i to search around both its individual best position, p_i, and global best position, p_g. Based on the updated velocities, each particle changes its position according to

$$x_i(t+1) = x_i(t) + v_i(t+1)$$

4. Flexible Neural Tree Classifier

4.1. *Flexible Neuron Instructor and FNT Model*

The function set F and terminal instruction set T used for generating a FNT model are described as follows:

$$S = F \cup T = \{+_2, +_3, \ldots, +_N\} \cup \{x_1, \ldots, x_n\}$$

where $+_i (i = 2, 3, \ldots, N)$ denote non-leaf nodes' instructions and taking i arguments. $x_1, x_2, \ldots, x_n$ are leaf nodes' instructions and taking no other arguments. The output of a non-leaf node is calculated as a flexible neuron model (see Fig. 1). From this point of view, the instruction $+_i$ is also called a flexible neuron operator with i inputs. In the creation process of neural tree, if a non-terminal instruction, i.e., $+_i (i = 2, 3, 4 \ldots, N)$ is selected, i real values are randomly generated and used for representing the connection strength between the node $+_i$ and its children. In addition, two adjustable parameters a_i and b_i are randomly created as flexible activation function parameters. Some examples of flexible activation functions are shown in Table 2.

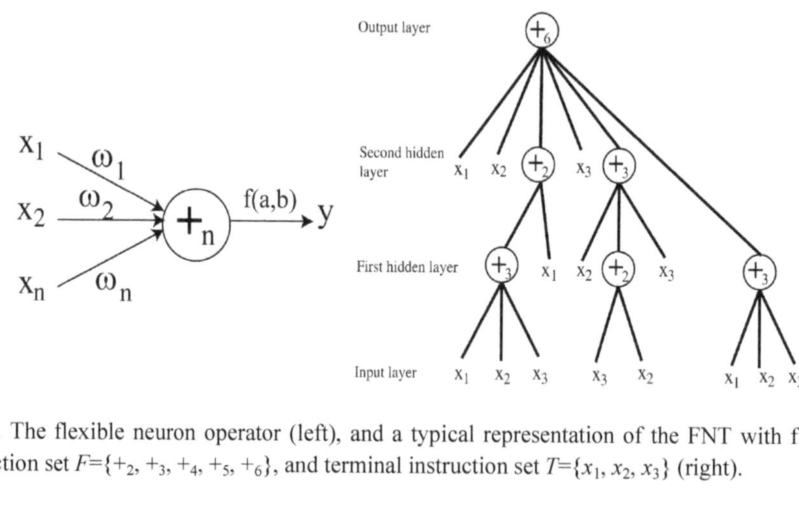

Fig. 1. The flexible neuron operator (left), and a typical representation of the FNT with function instruction set $F=\{+_2, +_3, +_4, +_5, +_6\}$, and terminal instruction set $T=\{x_1, x_2, x_3\}$ (right).

Table 2. The flexible activation functions

Gaussian Function	$f(x,a,b)=e^{-\frac{(x-a)^2}{b^2}}$
Unipolar sigmoid function	$f(x,a)=\frac{2\|a\|}{1+e^{-2\|a\|x}}$
Bipolar sigmoid function	$f(x,a)=\frac{1-e^{-2ax}}{a(1+e^{-2ax})}$
Nonlocal radial coordinates	$f(x,a,b)=(b^2+(x-a)^2)^{-\alpha}(\alpha>0)$
General multi-quadratics	$f(x,a,b)=(b^2+(x-a)^2)^{\beta}(0<\beta<1)$
Thi-plate s-spline function	$f(x,a,b)=(b\\|x-a\\|)^2\ln(b\\|x-a\\|)$

For developing the FNT classifier, the following flexible activation function is used.

$$f(a_i,b_i,x)=e^{-(\frac{x-a_i}{b_i})^2}$$

The output of a flexible neuron $+_n$ can be calculated as follows. The total excitation of $+_n$ is

$$out_n=f(a_n,b_n,net_n)=e^{-(\frac{net_n-a_n}{b_n})^2}$$

A typical flexible neuron operator and a neural tree model are illustrated in Fig. 1. The overall output of flexible neural tree can be computed form left to right by depth-first method, recursively.

4.2. *The Optimization of FNT Model*

The optimization of FNT includes the tree-structure and parameter optimization. The parameters are optimized with PSO algorithm. The process is the same as SVM parameter optimization.

Finding an optimal or near-optimal neural tree-structure is formulated as a product of evolution. In our previous work, the probabilistic incremental program evolution (PIPE) and ant programming (AP) algorithm have been employed to find a near-optimal neural tree[35,36]. In this study, the crossover and selection operators used are same as those of standard GP. A number of neural tree mutation operators are developed as follows:

1. Changing one terminal node: randomly select one terminal node in the neural tree and replace it with another terminal node.
2. Changing all the terminal nodes: select each and every terminal node in the neural tree and replace it with another terminal node.
3. Growing: select a random leaf in hidden layer of the neural tree and replace it with a newly generated subtree.
4. Pruning: randomly select a function node in the neural tree and replace it with a terminal node.

4.3. *Feature / Input Selection with FNT*

It is often a difficult task to select variables (features) for the classification problem, especially when the feature space is large. A fully connected NN classifier usually cannot do this. In the perspective of FNT framework, the nature of model construction procedure allows the FNT to identify important input features in protein fold that is computationally efficient and effective.

The mechanisms of input selection in the FNT constructing procedure are as follows. (1) Initially the input variables are selected to formulate the FNT model with same probabilities; (2) The variables which have more contribution to the objective function will be enhanced and have high opportunity to survive at next generation by an evolutionary procedure; (3) The evolutionary operators i.e., crossover and mutation, provide a input selection method by which the FNT should select appropriate variables automatically.

4.4. *Procedure of the General Learning Algorithm*

The general learning procedure for constructing the FNT model can be described as follows:

1. Create an initial population randomly (FNT trees and its corresponding parameters);
2. Structure optimization is achieved by the neural tree variation operators as described in Section 4.2
3. If a better structure is found, the go to step 4, otherwise go to step 2;
4. Parameter optimization is achieved by the PSO algorithm. In this stage, the architecture of FNT model is fixed, and it is the best tree developed during the end of run of the structure search. The parameters (weights and flexible

activation function parameters) encoded in the best tree formulate a particle.

5. If the maximum number of local search is reached, or no better parameter vector is found for a significantly long time then go to step 6; otherwise go to step 4;
6. If satisfactory solution is found, then the algorithm is stopped; otherwise go to step 2.

5. The Proposed Algorithm

The proposed algorithm in this paper is composed of four parts: feature extraction, ECOC, classifiers and compare system. This framework is shown in Fig. 2.

The algorithm can be described as follows.

1. Feature extraction. For protein fold recognition, many features, i.e., amino acids composition, predicted secondary structure, hydrophobicity, normalized van der waals volume, polarity, polarizability or their different combinations can be employed to construct the classification model. In this paper, all above features are used for constructing the classification system;
2. Error Correcting Output Codes. A search method is used to find the matrix M to represent the N_c classes. So every column in matrix M can be the target output of the training-data according to their class;
3. Classifier. Multi-class recognition problem is divided into many binary class recognition problems using the ECOC. And then SVM or FNT is used to find the better plane to divide every two classes;

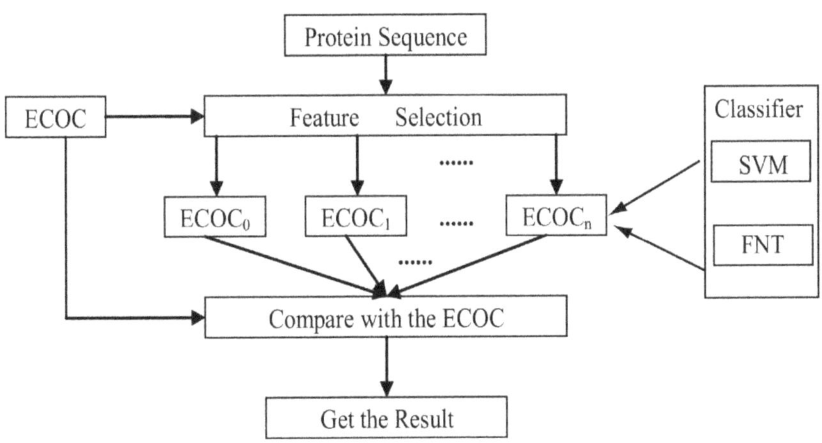

Fig. 2. The proposed protein fold prediction system.

4. Compare system. An output vector $X = \{x_0, x_1, \ldots, x_n\}$ is compared to the every rows of the matrix M by counting the Hamming distance, and then get a result according the nearest distance, where x_i is the ith SVM's (FNT's) output.

6. Experimental Results

Two datasets derived from the Structural Classification of Proteins (SCOP) database were used[32]. These datasets were available on line (http://crd.lbl.gov/~cding/protein/) and their detailed description could be found in [6]. Each dataset contained the 27 most populated folds represented by seven or more proteins and corresponding to four major structural classes. The first set consisted of 313 protein folds having (for each two proteins) no more than 35% of the sequence identity for aligned subsequences longer than 80 residues. The second dataset of 385 proteins, also known as independent test dataset, was composed of protein sequences of less than 40% identity with each other and less than 35% identity with the proteins of the first dataset. In fact, 90% of the proteins of the second dataset had less than 25% sequence identity with the proteins of the first dataset.

For constructing fold recognition system, six features, amino acid composition, polarity, polarizability, normalized van der Waals volume, hydrophobicity and predicted secondary structure were employed in this paper (see Table 3). All (except the first) features had dimensionality 21, whereas the composition had dimensionality 20. Thus, in total, a feature vector combining six features had 125 dimensions (or components). Please note that the features were normalized to [0, 1]. For comparison purpose, two experiments had been accomplished: (1) for the first dataset, the system was trained on the first dataset, and then tested on the second set to measure the average predictive accuracy; (2)

Table 3. The Description of features used for protein fold recognition

Symbol	Feature Name	Dimension
C	Amino Acid Composition	20
S	Predicted Secondary Structure	21
H	Hydrophobicity	21
P	Polarity	21
V	Normalized Van Der Waals Volume	21
Z	Polarizability	21

for the second dataset, we estimated the average predictive accuracy by 10-fold cross-validation. Note that for the first dataset, we ran the SVM and FNT with ECOC 10 times (without cross validation) to measure the average predictive accuracy. In a cross-validation run, 10 folds of approximately the same size were created. Then, 9 folds out of 10 were used as the training set and the remainder as the testing set. At the end of training, the folds were shifted by one position.

The performance measure adopted in this paper is the Q percentage accuracy[6]. Suppose we have $N = n_1 + n_2 + \ldots + n_t$ test proteins (n_l are observed to belong to class 1 of etc. and t is the number of classes). Suppose that out of n_l proteins, c_1 are correctly recognized as belonging to 1, etc., so that total $C_n = c_{1p} + c_{2p} + \ldots + c_{tp}$ proteins are correctly recognized. The total accuracy is

$$Q = \frac{C_p}{N}.$$

The result of the proposed method for protein fold recognition is showed in Table 4. For comparison purpose, the total accuracy with different classification methods and same datasets are also presented in Table 4. From simulation results, the SVM+ECOC method achieves the best prediction accuracy of 64.8% and

Table 4. Comparison, using the average predictive accuracy, among different methods

HKNN [27]	57.3	47.1
FLC [21]	52.4	44.8
RS1 HKNN K25 [21]	60.3	48.7
RS1 HKNN K125 [21]	60.0	49.1
BS2 FLC K125 [21]	59.2	48.0
DIMLP-B [28]	61.2	45.0
DIMLP-A [28]	59.1	46.7
MLP [29]	48.8	NA
GRNN [29]	44.2	NA
RBFN [29]	49.4	NA
SVM [29]	51.4	NA
RBFN [30]	51.2	NA
SVM [6]	56.0	45.4
SVM+ECOC (This paper)	64.8	55.4
FNT+ECOC (This paper)	70.1	62.3

55.4%, and FNT+ECOC method achieves the best prediction accuracy of 70.1% and 62.3%. All the results shows that our proposed methods compares well with other systems in the literature that accomplish the task of fold recognition.

7. Conclusions

Multi-class protein folds recognition becomes more and more important in recent years, especially with the rapid development of bioinformatics. In this work, an error-correcting output coding and ensemble classifiers (SVM or FNT) are proposed to recognize the multi-class protein fold. Our method compares well with other systems in the literature that accomplish the task of fold recognition.

However, several improvements could be considered in order to increase its efficiency. In our future work, the following topics should be addressed: (1) It is not determined that which features are important for the protein fold recognition, and then feature extraction should play a key role in the protein fold recognition; (2) A number of classification algorithms have been applied to the fold recognition, but it is still needed to find the new classification algorithms for improving the classification accuracy and generalization ability.

Acknowledgment

This research was supported by the NSFC (60573065), the the Natural Science Foundation of Shandong Province (Y2007G33), and the Key Subject Research Foundation of Shandong Province.

References

1. Chen Y., Ding F., Nie H., Serohijos, A.W., Sharma S., Wilcox K.C., Yin, S., Dokholyan N.V., (2007) Protein folding: Then and now, Archives of Biochemistry and Biophysics, (In press).
2. Fischer D, Eisenberg D. (1996). Protein fold recognition using sequence- derived predictions. Protein Sci. 5, 947-955.
3. Di Francesco V., Geetha V., Garnier J., Munson P.J. (1997). Fold recognition using predicted secondary structure sequences and hidden Markov models of proteins folds. Proteins: Struct Funct Genet. 1, 123-128.
4. Hargbo J., Elofsson A. (1999). Hidden Markov models that use predicted secondary structures for fold recognition. Proteins 36, 68-87.
5. Dandekar T., Argos P. (1996). Identifying the tertiary fold of small proteins with different topologies from sequence and secondary structure using the genetic algorithm and extended criteria specific for strand regions. J. Mol. Biol. 256, 645-660.
6. Ding C., Dubchak I. (2001). Multi-class protein fold recognition using support vector machines and neural networks. Bioinformatics. 17, 349-358.

7. Shi J.Y., Pan Q., Zhang S.W., et al. (2006). Protein fold recognition with support vector machines fusion network. Prog Biochem Biophys. 33(2), 155-162.
8. Karplus K., Karchin R., Shackelford G., Hughey R. (2005). Calibrating E-values for hidden Markov models using reverse-sequence null models. Bioinformatics. 21, 4107-4115.
9. Lindahl E., Elofsson A. (2000). Identification of related proteins on family, superfamily and fold level. J Mol Biol. 295, 613-625.
10. Karchin R., Cline M., Mandel-Gutfreund Y., Karplus K. (2003). Hidden Markov models that use predicted local structure for fold recognition: alphabets of backbone geometry. Proteins. 51, 504-514.
11. Liu Y., Carbonell J.,Weigele P. (2006). Protein fold recognition using segmentation conditional random fields (SCRFs). J Comput Biol. 13(2), 394-406.
12. Flockner H., Domingues F., Sippl M.J. (1997). Proteins folds from pair interactions: a blind test in fold recognition. Proteins: Struct Funct Genet. 1, 129-133.
13. Xu J. (2005). Fold recognition by predicted alignment accuracy. IEEE/ACM Trans Comput Biol Bioinform. 2(2), 157-165.
14. Sander O., Sommer I., Lengauer T. (2006) Local protein structure prediction using discriminative models. BMC Bioinform. 7(14).
15. Murzin A.G. (1999). Structure classification based assessment of CASP3 predictions for the fold recognition targets. Proteins: Struct Funct Genet. 37, 88-103.
16. Ortiz A.R., Kolinski A., Rotkiewicz P., Ilkowski B., Skolnick J.. (1999). Ab initio folding of proteins using restraints derived from evolutionary information. Proteins: Struct Funct Genet. 3, 177-185.
17. Elofsson A., Fischer D., Rice D.W., LeGrand S.M., Eisenberg D. (1996). A study of combined structure-sequence profiles. Folding & Design. 1, 451-461.
18. Bonneau R., Baker D. (2001). Ab initio protein structure prediction: progress and prospects. Annu Rev Bioph Biom. 30, 173-189.
19. Cheng, J. and Baldi, P. (2006). A Machine Learning Information Retrieval Approach to Protein Fold Recognition. Bioinformatics. 22(12), 1456-1463.
20. Gromiha M.M., Thangakani A.M. and Selvaraj S. (2006) FOLD-RATE: prediction of protein folding rates from amino acid sequence. Nucleic Acid Res.. 34, 70-74.
21. Nanni L. (2006). Ensemble of classifiers for protein fold recognition. Neurocomputing. 69, 850-853.
22. Ludmila I.K. (2005). Using diversity measures for generating error-correcting output codes in classifier ensemble. Pattern ecognition Letters. 26, 83-90.
23. Jiang Y.H., Zhao Q.L., Yang X.J. (2004). A General Coding Method for Error- Correcting Output Codes. Lecture Notes in Computer Science. 3056, 648-652.
24. Vapnik V. (1995). The Nature of Statistical Learning Theory, Springer-Verlag, New York.
25. Kennedy J. (1995). Particle swarm optimization. In: Proc IEEE Int Conf on Neural Networks. Vol IV, 1942-1948.
26. Yoshida H., Kawata K., Fukuyama Y., Takayama S., Nakanishi Y. (2000). A particle swarm optimization for reactive power and voltage control considering voltage security assessment. IEEE Trans Power Syst. 15, 1232-1239.
27. Okun O., (2004). Proteins fold recognition with klocal hyperplane distance nearest neighbor algorithm. in: Proceedings of the Second European Workshop on Data Mining and Text Mining in Bioinformatics. September, Pisa, Italy, 51-57.
28. Bologna G., Appel R.D. (2002). A comparison study on protein fold recognition. in: Proceedings of the Ninth International Conference on Neural Information Processing, November, Singapore, 2492-2496.
29. Chung I.-F., Huang C.-D., Shen Y.-H., Lin C.-T. (2003). Recognition of structure classification of protein folding by NN and SVM hierarchical learning architecture. in:

Artificial Neural Networks and Neural Information Processing ICANN / ICONIP. June, Istanbul, Turkey, 1159-1167.

30. Pal N.R., Chakraborty D. (2003). Some new features for protein fold recognition. in: Artificial Neural Networks and Neural Information Processing ICANN / ICONIP. June, Istanbul, Turkey, 1176-1183.
31. Sun C., Gong D. (2006). Support Vector Machines with PSO Algorithm for Short- Term Load Forecasting. IEEE International Conference on Networking, Sensing and Control. April, 676-680.
32. Murzin A.G., Brenner S.E., Hubbard T., Chothia C. (1995). SCOP: a structural 560 classification of proteins database for the investigation of sequences 561 and structures. J Mol Biol. 247, 536-540.
33. Zhao X.M., Cheung Y.M., Huang D.S. (2005). A novel approach to extracting features from motif content and protein composition for protein sequence classification, Neural Networks, vol.18, no.8, 1019-1028.
34. Huang D.S., Zhao X.M., Huang G.B., Cheung Y.M., (2006). Classifying protein sequences using hydropathy blocks, Pattern Recognition, vol.39, no.12, 2293–2300.
35. Chen Y., Yang B., Dong J., (2004). Nonlinear system modeling via optimal design of neural trees, Int. J. Neural Syst., vol.14, no. 2, 125-137.
36. Chen Y., Yang B., Dong J., (2004). Evolving flexible neural networks using ant programming and PSO algorithm, International Symposium on Neural Networks (ISNN'04), Lecture Notes on Computer Science, vol. 3173, 211-216.

bioinformatics, 590
error correcting output codes, 580
flexible neural tree classifier, 584
particle swarm optimization (PSO), 583
protein fold recognition, 579
  with bio-physics based method, 579
  with sequence-based method, 579
  with structure based method, 579
support vector machine (SVM), 582

Part 5

System and Technology

CHAPTER 5.1

A METHOD FOR SYNTHESIS OF A 3D FACE USING A SINGLE 2D IMAGE

Martin D. Levine[1] and Yingfeng Yu

Dept. of Elec. & Comp. Eng. & Center for Intelligent Machines
McGill University, Montreal, Quebec, Canada

3D facial reconstruction systems create 3D facial computer models of individuals from their 2D photographic images or video sequences. Currently, published face recognition systems are mostly based on large training sets of 2D facial images although there has been an increase in interest in using 3D data input instead. An intermediate approach is to synthesize a 3D face making use of a single 2D image. In this paper, we present such a method that does not require complicated optimization steps or user-defined parameters, which distinguishes it from existing 3D face reconstruction methods. The method has been used in 2D face recognition experiments to generate a 3D Morphological Model (3DMM) from a single facial training image.

Given a single 2D facial image, a small set of morphological feature points are selected. Their corresponding three-dimensional indices in the morphological model are obtained using the so-called Fast Marching Method. Shape alignment between the 2D input image and the 3D morphological model is achieved using Newton's Method to solve a single nonlinear equation to obtain a scalar parameter. This yields all of the shape parameters of the 3DMM. Texture recovery for the model involves bridging the 2D image representation of the 3DMM using the so-called UV space as an intermediary. Thus the advantage of the proposed method over others in the literature is that a computationally complex geometric problem is simplified and transformed to a 2D-to-3D warping problem. This enhances both the efficiency and accuracy of texture recovery.

We also introduce a quantitative shape measure based on the Bending Invariant Canonical Form to determine the quality of the 3D reconstruction. This is shown to produce a reconstruction error of 5% over the database of 100 stored 3D faces. In addition, the reconstructed 3D faces at various head poses have been

---

[1] Address for correspondence: Centre for Intelligent Machines & Dept. of Elec. & Computer Eng., McGill University, Montreal, Quebec, Canada H3A 2A7. Tel. 514-398-7115, fax: (514) 398-7348, email: levine@cim.mcgill.ca.

used to create 2D images of faces that form the training set for face recognition experiments reported elsewhere.

Keywords: 3D faces, single 2D image, 3D face reconstruction, shape measure, shape correction, Radial Basis Function, Fast Marching Method, 3D Morphable Model, texture, Bending Invariant Canonical Form

Acknowledgements: The authors would like to acknowledge the financial support of the Natural Sciences and Engineering Research Council of Canada (NSERC).

1. Introduction

This paper describes a 3D facial reconstruction system with the ability to recover a 3D human face from a *single* 2D frontal so-called natural or standard image. The latter is intended to be used as a training image in a face recognition system. Consequently, it was assumed that the initial enrolment process involved just a single frontal image, as is currently conventionally being done for passport photos. In developing this approach, we have followed the "3D supported 2D model" framework, which consists of two stages: shape reconstruction and texture recovery.[2] These are computed separately. It is not necessary that every image pixel be involved in the 3D shape reconstruction process and we use only a small number of specific morphologic feature points in this computation. The feature points are defined by a Feature Point Definition Table (FPDT), which contains 57 user-defined points from the 2D human facial image. This table is sufficient to represent the most important morphological information on a human face and is obtained offline by a pre-processing stage discussed in the appendix.

The offline pre-processing stage requires the determination of a set of 57 landmarks. Although there seems to be a pervasive supposition by the computer vision research community that the problem of *automatically* detecting facial landmarks is a solved problem, this is actually not the case. Readers are referred to the recent paper by Gu and Kanade [2] for a short historical overview on this topic. They report an overall misalignment error for so-called "clean" images of 6.6%, 4.9% and 2.9% using the Active Appearance Model (AAM) [3], Bayesian Tangent Shape Model [4], and their model, respectively. Riopka and Boult [5] have demonstrated experimentally that eye localization has a major effect on the ultimate recognition results. Clearly, then, detecting even a relatively small set of landmarks within a face remains a challenging but important issue. That is why, for example, a current paper [6] on adaptive AAM also employs manual specification for the required landmarks. Another recent paper by Asteriadis et al.

[2] See [1].

[7] indicates a mean average error for just four facial landmarks of 5.4% of the inter-ocular distance. Thus, in order to pursue the research discussed in this paper, visual selection of the landmarks was adopted. Both in-plane and scale invariance are implicit in the use of these landmarks.

The main task of the pre-processing stage is to determine the correspondence between the three-dimensional morphological model (3DMM) and the Feature Point Definition Table (FPDT). This correspondence plays an essential role in both shape reconstruction, discussed in Section 2, and texture recovery, discussed in Section 3. Section 4 presents synthesized image results acquired using our reconstruction system. Section 5 introduces a method for 3D shape comparison of actual scanned 3D faces and their reconstructions, which yields a 5% error for a 100 face database. Finally, we conclude by comparing our system with other "3D supported 2D model" approaches [8].

The basic concept behind our 3D facial reconstruction system is shown in Figure 1. An arbitrary frontal 2D facial image is input to the 3D facial reconstruction system and first processed for background removal and visual selection of a small set of 2D facial landmarks. Secondly, a person's 3D shape and texture information are recovered in the shape reconstruction stage and texture information retrieval stage, respectively. Finally, the shape and texture are combined to produce a synthesized 3D face as the output.

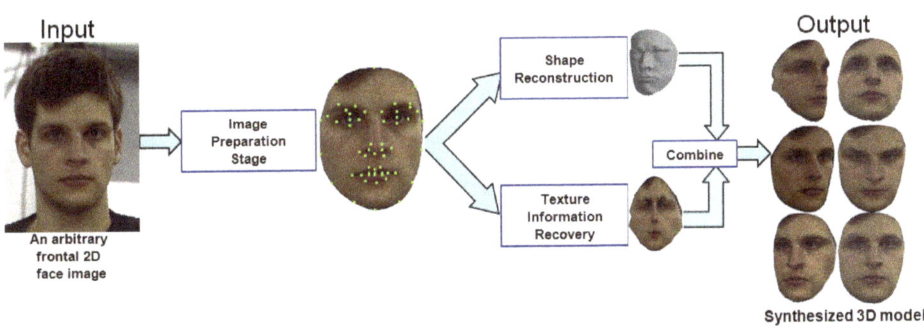

Figure 1. 3D face reconstruction: an arbitrary frontal 2D face image is the input; the output of the system is the synthesized 3D model containing shape and texture information.

There are two approaches to 3D facial reconstruction: (1) "3D supported 2D model" (2) "analysis-by-synthesis" [8]. Both use a 3D Morphable Model; however, the former reconstructs shape and texture *separately*, while the latter achieves this *simultaneously*. Because of this, the latter requires less than 10 feature points to initialize the reconstruction, while the former needs 50 to 90. This trade-off necessitates that "analysis-by-synthesis" simultaneously computes all of the pixels in the 2D image that are involved in the reconstruction process and thus takes a significant amount of time to compute the optimal reconstruction. On the other hand, the "3D supported 2D model" approach makes use of only the initial feature points and roughly, yet efficiently, estimates the 3D face using shape alignment, followed by shape correction.

2. Shape Reconstruction

A 2D image is a projection of a 3D face on a 2D plane, and therefore using the 2D image to retrieve the original 3D face is impossible unless some additional information is available. The purpose of offline pre-processing is to determine a correspondence between a set of specific user-defined landmarks on the 2D image and predefined morphological vertices on a 3D so-called Generic Mean Face (GMF), as described below.

The GMF was obtained by Blanz and Vetter using 100 laser-scanned human heads as a set of exemplars [9; 10]. Each head was saved in an exactly 75,972 vertex, 150,958 triangle-facet format[3]. The authors in [9; 10] assume that *any* 3D face can be realistically modeled by a convex combination of shape and texture vectors of a set of exemplar faces, which in their case were defined as follows:

$$\begin{aligned} S_i &= (x_{i,1}, y_{i,1}, z_{i,1}, \ldots, x_{i,75972}, y_{i,75972}, z_{i,75972})^T \in \Re^{227,916\times 1} \\ T_i &= (R_{i,1}, G_{i,1}, B_{i,1}, \ldots, R_{i,75972}, G_{i,75972}, B_{i,75972})^T \in \Re^{227,916\times 1} \end{aligned} \tag{1}$$

where S_i and T_i have dimension 227,916(=3*75,972). These data are then reduced in number by performing a Principle Component Analysis. Based on the observation that *all heads have the same vertices and facets*, we define

$$\begin{aligned} &S_0 = \frac{1}{100}\sum_{i=1}^{100} S_i \,; T_0 = \frac{1}{100}\sum_{i=1}^{100} T_i \\ &\mathbf{S} = [S_1, S_2, \ldots, S_{100}] \in \Re^{(3\times 75972)\times 100}; \mathbf{T} = [T_1, T_2, \ldots, T_{100}] \in \Re^{(3\times 75972)\times 100} \end{aligned} \tag{2}$$

where (S_0, T_0) represents the mean face of the 100 exemplar faces and is referred to as the GMF. The appendix explains how to obtain the Feature Point Definition Table (FPDT) using the GMF and the corresponding index for each vertex on the GMF.

---

[3] Generally, we would expect that each 3D laser scanned face would be saved using a different number of vertices and type of triangular connectivity, based on the characteristic features of an individual face. However, to facilitate their method [10], Blanz and Vetter chose to use optic flow to force every face to have exactly 75,972 vertices and share identical 150,958 facets.

We briefly review the concept of the 3DMM. Assume that any arbitrary 3D face can be expressed in terms of the 3DMM as follows:

$$S_j = S_0 + \sum_{i=1}^{100} \frac{\alpha_i}{\lambda_{s,i}} \mathbf{e}_i^s \tag{3}$$

$$T_j = T_0 + \sum_{i=1}^{100} \frac{\beta_i}{\lambda_{t,i}} \mathbf{e}_i^t \tag{4}$$

where $\mathbf{e}_i^s$ and $\mathbf{e}_i^t$ correspond to the i-th eigenface and $\lambda_{s,i}$ and $\lambda_{t,i}$ stand for the corresponding eigenvalues of the shape and texture, respectively, of the 3DMM. For convenience, we write all eigenvalues in vector form as $\lambda_s = \left[\lambda_{s,1},\ldots,\lambda_{s,100}\right]$ and $\lambda_T = \left[\lambda_{t,1},\ldots,\lambda_{s,100}\right]$. Let $\overline{\alpha} = \left[\alpha_1,\ldots,\alpha_{100}\right]$ and $\overline{\beta} = \left[\beta_1,\ldots,\beta_{100}\right]$ denote the shape and texture parameters, respectively, that characterize a specific 3D face. As opposed to the "analysis-by-synthesis" approach, the "3D supported 2D models" framework does not require a $\overline{\beta}$ estimation step to recover texture information (precise details are found in Section 3). Moreover, only the input 2D feature points (indicated by the primed symbols below) and their associated morphological feature vertices in the 3DMM are required for shape alignment.

Given a single 2D input image, the 57 morphological feature points, which are defined in the Feature Point Definition Table (FPDT), are found manually. The 2D coordinates of each of these 57 feature points are recorded and their corresponding 3DMM indices are found as discussed in the appendix. The objective of shape alignment is to align these in 2-space and recover the associated shape parameters from the 2D input image using only these feature points. Clearly, the shape alignment algorithm is based on the assumption that we can reconstruct an arbitrary human face using just 100 eigenfaces and the GMF S_0, as provided by the 3DMM. The arbitrary human face has an associated shape parameter vector $\overline{\alpha}$.

An arbitrary face S_i in the 3DMM is given by $S_i = S_0 + P\alpha$, where

$$\mathbf{P} = [\lambda_{s,1}^{-1}\mathbf{e}_1^s,\ldots,\lambda_{s,i}^{-1}\mathbf{e}_i^s,\ldots,\lambda_{s,100}^{-1}\mathbf{e}_{100}^s] = [\mathbf{e}_1^s,\ldots,\mathbf{e}_i^s,\ldots,\mathbf{e}_{100}^s]\underbrace{\begin{bmatrix} \lambda_{s,1}^{-1} & 0 & 0 \\ 0 & \ddots & 0 \\ 0 & 0 & \lambda_{s,100}^{-1} \end{bmatrix}}_{\Lambda} \in \Re^{(3\times 75972)\times 100}$$

The eigenface matrix, $\boldsymbol{\lambda}_S = [\lambda_{s,1},\ldots,\lambda_{s,100}]$, contains the eigenvalues which are associated with their corresponding eigenvectors (eigenfaces) $\mathbf{e}_1^s, \mathbf{e}_2^s, \ldots, \mathbf{e}_{100}^s$.

Before continuing, consider $\mathbf{e}_1^s, \mathbf{e}_2^s, \ldots, \mathbf{e}_{100}^s$, $\mathbf{P}$ and S_0. All of these have the same format in three dimensions. Thus

$$e_i^s = \begin{bmatrix} e_{ix}^s \\ e_{iy}^s \\ e_{iz}^s \end{bmatrix} \in \Re^{3\times 75972\times 1}$$

where, for example,

$$e_{ix}^s = \begin{bmatrix} e_{ix}^{s,1} \\ \vdots \\ e_{ix}^{s,j} \\ \vdots \\ e_{ix}^{s,75972} \end{bmatrix} \in \Re^{75972)\times 1}$$

and $\mathbf{e}_{iy}^s, \mathbf{e}_{iz}^s$ share the same format as $\mathbf{e}_{ix}^s$. Moreover, S_0 also shares the same format as $\mathbf{e}_i^s$. Thus it may be written as

$$S_0 = \begin{bmatrix} S_{0x} \\ S_{0y} \\ S_{0z} \end{bmatrix} \in \Re^{(3\times 75972)\times 1}$$

where S_{0x}, S_{0y} and $S_{0z} \in \Re^{75972\times 1}$. For convenience, we use S_{0x}, S_{0y} to indicate *subsets* of the matrix S_0 and the superscript "*xy*" to represent the *subset* of matrices and vectors. For instance,

$$S_0^{xy} = \begin{bmatrix} S_{0x} \\ S_{0y} \end{bmatrix} \in \Re^{(2\times 75972)\times 1}.$$

The same notation is applied to $\mathbf{P}^{xy}$.

As indicated earlier, the shape alignment algorithm may be classified as being within the "*3D supported to 2D models*" framework. Thus, during shape parameter estimation, *only the N feature points defined in the FPDT are involved.* This implies that a different mean shape and eigenface matrix are required that correspond only to these specific N feature points. These are referred to as $\mathbf{S}_f^{xy}$ and $\mathbf{P}_f^{xy}$. We use the subscript "*f*" to denote the matrices and vectors, which relate to these N feature points.

Having dealt with the 3DMM, we consider the 2D input image. In this case, three quantities, defined as $S'_{f,j}$, $\overline{S}'_f$ and $\mathbf{S}'_f$. $S'_{f,j}$ represent any feature point in the 2D input image, while the average of all $S'_{f,j}$ is denoted by $\overline{S}'_f$. The normalized matrix form of all feature points is defined as $\mathbf{S}'_f$. The relationship among $S'_{f,j}$, $\overline{S}'_f$ and $\mathbf{S}'_f$ can then be written as:

$$\mathbf{S}'_f = \begin{bmatrix} S'_{fx,1} - \overline{S}'_{fx} \\ \vdots \\ S'_{fx,N} - \overline{S}'_{fx} \\ S'_{fy,1} - \overline{S}'_{fy} \\ \vdots \\ S'_{fy,N} - \overline{S}'_{fy} \end{bmatrix}$$

Finally, the objective is to find the specific $\overline{\alpha}$ which is best able to fit $\mathbf{S}^{xy}_f + \mathbf{P}^{xy}_f \vec{\boldsymbol{\alpha}}$ to $\mathbf{S}'_f$. Let us assume a scale value *c*. Also we observe that there is no rotation matrix involved since our assumptions are based on using a 2D frontal facial image.[4] Hence, we can define the following constrained optimization problem:

$$\min_{c,\vec{\boldsymbol{\alpha}}} \| S'_f - c(S^{xy}_f + \mathbf{P}^{xy}_f \vec{\boldsymbol{\alpha}}) \|^2$$
$$subject\ to: \vec{\boldsymbol{\alpha}}^T \vec{\boldsymbol{\alpha}} = 100$$

This can be restated as:

$$\min_{\vec{n}} \| M\overline{\chi} - S^{xy}_f \|^2$$
$$subject\ to:\ \overline{\chi}^T \Pi \overline{\chi} = 100$$

where $$\overline{\chi} = \begin{bmatrix} r \\ \overline{\alpha} \end{bmatrix}; \quad r = 1/c$$

$$r(S'_f - S^{xy}_f) - P^{xy}_f \overline{\alpha} = \underbrace{[(S'_f - S^{xy}_f), -P^{xy}_f]}_{M} \begin{bmatrix} r \\ \overline{\alpha} \end{bmatrix}$$

$$\chi^T \Pi \overline{\chi} = \begin{bmatrix} r \\ \overline{\alpha} \end{bmatrix}^T \underbrace{\begin{bmatrix} 0 & 0 \\ 0 & I_{100'100} \end{bmatrix}}_{\Pi} \begin{bmatrix} r \\ \overline{\alpha} \end{bmatrix}$$

[4] This is not a necessary restriction.

To perform the optimization subject to the constraint, we define a Lagrangian Function

$$\psi(\bar{\chi},x)=\left\|M\bar{\chi}-S_f^{xy}\right\|^2+\xi(\bar{\chi}^T\Pi\bar{\chi}-100)$$

where the gradient update required is given by

$$\frac{\partial\psi(\bar{\chi},\xi)}{\partial\bar{\chi}}=0\Rightarrow\bar{\chi}(\xi)=(\mathbf{M}^T\mathbf{M}-\zeta\mathbf{\Pi})^{+}\mathbf{M}^T\mathbf{S}_f^{xy} \qquad (5)$$

The next step is to find ξ, which is a constant in the Lagrangian equation above. Using a Singular Value Decomposition (SVD) of M, i.e., $\mathbf{M}=\underset{2N\times 2N}{\mathbf{U}}\ \underset{2N\times 101}{\mathbf{D}}\ \underset{101\times 101}{\mathbf{V}}^T$, the rank of $\mathbf{M}$ satisfies the condition of being less than or equal to 101. Here, we assume the rank of $\mathbf{M}$ equals 101. Hence, the matrix $\mathbf{D}$ has a format

$$\mathbf{D}=\begin{bmatrix}\underset{101\times 101}{\mathbf{D}_2}\\ \underset{(2N-101)\times 101}{\mathbf{0}}\end{bmatrix} \text{ and } \mathbf{D}_2=\begin{bmatrix}\sigma_1 & & 0\\ & \ddots & \\ 0 & & \sigma_{101}\end{bmatrix}$$

where the σ_i, $i=1,\cdots,101$ are the singular values of $\mathbf{M}$. It can be shown that Equation (5) can be simplified significantly to [11]:

$$\sum_{i=2}^{101}\frac{g_i^2\sigma_i^2}{(\sigma_i^2-\xi)^2}-\frac{g_1^2}{\sigma_1^2}=100 \qquad (6)$$

where the g_i are the components of the vector $\bar{\mathbf{g}}=\mathbf{U}^T\mathbf{S}_f^{xy}$. This nonlinear equation can be solved numerically using Newton's method for the only unknown ξ, thereby determining $\bar{\boldsymbol{\alpha}}$.

All of the quantities defined above are listed in Table 1 while the algorithm is detailed in Figure 2. The only input to the shape alignment algorithm is the matrix $\mathbf{M}$ that is constructed from $\mathbf{S}_f'$, $\mathbf{S}_f^{xy}$ and $\mathbf{P}_f^{xy}$ according to the relationship:

$$\mathbf{M}=[\mathbf{S}_f'-\mathbf{S}_f^{xy},-\mathbf{P}_f^{xy}]$$

Once Newton's Method [14] has yielded the value of ξ, it is subsequently used as an intermediate step (highlighted in yellow in Figure 2) to obtain the final value of the shape parameter vector $\bar{\alpha}$ of the 3DMM. The computational time

Table 1. Information regarding each symbol used in the shape alignment algorithm given in Figure 2.

Notation		Meaning	Mathematical Expression (or Value)	Dimension
3DMM	$S_{f,j}$, $\forall j = 1,..,N$	Feature vertex in GMF	$S_{f,j} = \left[S_{f,jx} \quad S_{f,jy} \quad S_{f,jz} \right]^T$	$\Re^{3\times 1}$
	S_f	Matrix form of all feature vertices in GMF	$\mathbf{S}_f = \begin{bmatrix} S_{f,1x} \\ \vdots \\ S_{f,Nx} \\ S_{f,1y} \\ \vdots \\ S_{f,Ny} \\ S_{f,1z} \\ \vdots \\ S_{f,Nz} \end{bmatrix}$	$\Re^{3N\times 1}$
	$\mathbf{e}^s_{f,i}$	Subset of e^s_i	–	$\Re^{3N\times 1}$
	$\mathbf{E}_{f,S}$	Matrix form of all $\mathbf{e}^s_{f,i}$ $\forall i = 1,..,100$	$\mathbf{E}_{f,S} = [\mathbf{e}^s_{f,1}, \mathbf{e}^s_{f,2}, \ldots, \mathbf{e}^s_{f,100}]$	$\Re^{3N\times 100}$
	$\mathbf{P}_f$	Eigenface shape matrix (See text for details)	$\left\langle \mathbf{E}_{f,S}, \boldsymbol{\lambda}_s^{-1} \right\rangle =$ $[\lambda_{s,1}^{-1}\mathbf{e}^s_{f,1}, \ldots, \lambda_{s,i}^{-1}\mathbf{e}^s_{f,i}, \ldots, \lambda_{s,100}^{-1}\mathbf{e}^s_{f,100}]$	$\Re^{3N\times 100}$
	$\mathbf{P}^{xy}_f$	Subset of $\mathbf{P}_f$, only containing data about the *x,y* direction of $\mathbf{P}_f$	$\mathbf{P}_f = [\mathbf{P}_{f,x} \quad \mathbf{P}_{f,y} \quad \mathbf{P}_{f,z}]^T = [\mathbf{P}^{xy}_f \quad \mathbf{P}_{f,z}]^T$	$\Re^{2N\times 100}$
	$\mathbf{S}^{xy}_f$	Subset of S_f which only contains data about the x, y direction of S_f	$\mathbf{S}_f = [\mathbf{S}_{f,x} \quad \mathbf{S}_{f,y} \quad \mathbf{S}_{f,z}]^T = [\mathbf{S}^{xy}_f \quad \mathbf{S}_{f,z}]^T$	$\Re^{2N\times 1}$
2D Input Image	$S'_{f,j}$, $\forall j = 1,..,N$	Feature points in the 2D image	$S'_{f,j} = \left[S'_{fx,j} \quad S'_{fy,j} \right]^T$	$\Re^{2\times 1}$
	$\overline{S'_f}$	Mean feature points in the 2D image	$\frac{1}{N}\sum_{j=1}^{N} S'_{f,j} = \left[\overline{S'_{fx}} \quad \overline{S'_{fy}} \right]^T$	$\Re^{2\times 1}$
	$\mathbf{S}'_f$	Normalized matrix form of all feature points in the 2D image	$\mathbf{S}'_f = \begin{bmatrix} S'_{fx,1} - \overline{S'_{fx}} \\ \vdots \\ S'_{fx,N} - \overline{S'_{fx}} \\ S'_{fy,1} - \overline{S'_{fy}} \\ \vdots \\ S'_{fy,N} - \overline{S'_{fy}} \end{bmatrix}$	$\Re^{2N\times 1}$
M		Input to the shape alignment algorithm proposed in Figure 9	$[\mathbf{S}'_f - \mathbf{S}^{xy}_f, -\mathbf{P}^{xy}_f]$	$\Re^{2N\times 101}$

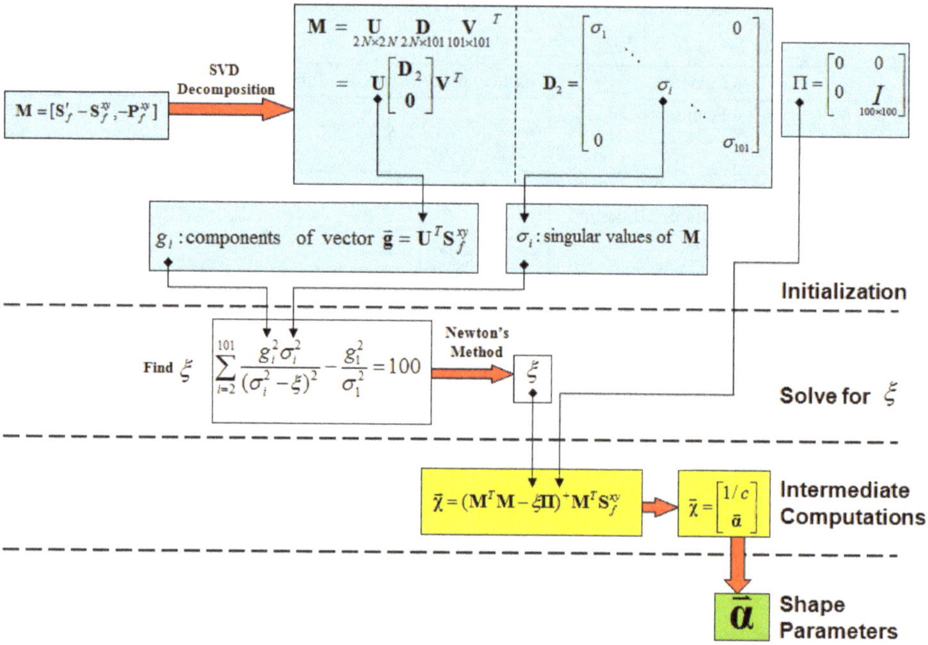

Figure 2. Shape alignment algorithm: The matrix M, defined in Table 2, is the input to algorithm. Shape parameter $\vec{\alpha}$ is the output.

required for the whole algorithm is largely dependent on solving the nonlinear Equation (6), although Newton's Method is very efficient. A comparison of this algorithm with the other four existing methods can be found in [1]. Readers interested in further details regarding the shape alignment algorithm should consult [11].

An important point regarding the shape alignment algorithm shown in Figure 2 is that it requires only one equation (shown in the white box) to be solved.

The shape alignment process discussed above produces a *rough* shape estimate requiring the result to be refined further. This is argued by Jiang et al. [15] who showed that there is a significant linear displacement in the x, y plane of feature points between the original 2D image and the 3D reconstructed surface[5]. Moreover, we note that both the texture information and non-feature points do not contribute to the shape reconstruction (the shape alignment step) in

[5] This may be attributed to the fact that accuracy of $\vec{\alpha}$ generated by the shape alignment algorithm is heavily dependent on the rank of the input matrix $\mathbf{M}$. The rank of $\mathbf{M}$ in turn is limited by either the number of feature points or the number of faces in the database. In our case, since only 100 faces exist and an arbitrary number of feature points can be chosen, we can assume that the shape space is limited by the size of the 3D face database.

the "3D supported 2D model" framework for estimating the shape parameter vector $\bar{\alpha}$. Thus, the accuracy of the shape reconstruction is inferior to that obtained by the "analysis-by-synthesis" approach. In order to compensate for this limitation, the authors in [15] have suggested that interpolation be used to post-process the non-feature vertices on the 3D reconstructed surface. This is achieved by aligning the *x*, *y* coordinates of the feature vertices on the 3D reconstructed surface with the feature points on the 2D image. We exploit radial basis functions (RBF) [16] to perform this interpolation by computing the displacement of the non-feature points on the 3D reconstructed surface using just the displacements of the feature vertices.

3. 3D Texture Information Recovery

Texture in the context of this paper refers to the RGB value at a 2D pixel in an image or on a 3D surface. The "painting process" of transferring 2D information to the surface requires a so-called UV image. The latter is not a normal realistic human facial image but a convenient intermediate image for retrieving 3D texture information from a lower dimensional embedded space, i.e., the UV map. The concept of a UV image is reviewed in [8]. It can be construed as a 2D mesh, where each 2D pixel coordinate, (u_k, v_k), maps directly to a specific 3D index k on the 3DMM. Due to this correspondence between the 2D pixels (u_k, v_k) on the UV map and the 3D vertex indices on the 3DMM, complete retrieval of the UV image of an individual's face is equivalent to recovering the pixel data for every 3D vertex.

In addition, we assume that a 2D image resides in 2D image space (x, y), where each pixel coordinate (x_i, y_i) has a value that represents the texture information at that specific location. Since we also maintain a Feature Point Definition Table for every landmark feature point in the 2D image (x_p, y_p), where $p=1...N$ (the total number of landmark feature points), we know the corresponding landmark feature point index k on the 3DMM. As a result, the "painting" task of placing the 2D texture information associated with these feature points on the 3DMM is trivial. However, we do not know the corresponding indices on the 3DMM for the other *feature points on the 2D face that are not landmarks.* Thus, the task of mapping all of the texture data of the 2D image to the 3DMM is complex and can realistically only be achieved by means of a UV map. This is so since *every* 2D pixel in the UV map (u_k, v_k) directly corresponds to a unique index on the 3DMM. Therefore, the process of applying 2D texture data from a complete 2D image onto the 3DMM can be simplified by first interjecting a 2D-to-2D mapping step involving the UV map.

In summary, the objective of texture data recovery translates into bridging the 2D image and the 3D space of the 3DMM by using the UV space as an intermediary. Figure 3 summarizes this idea.

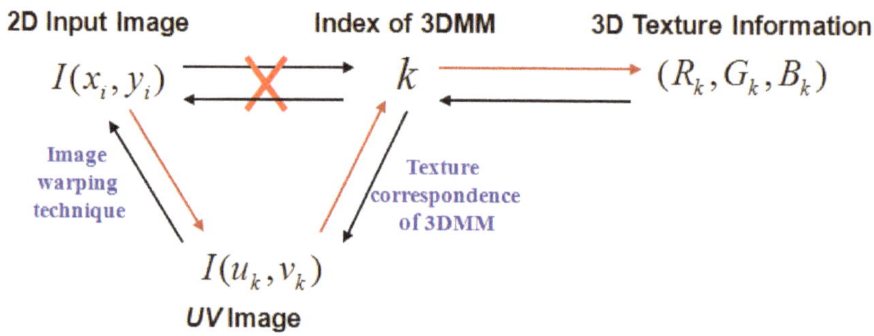

Figure 3. The concept of texture data transfer from any 2D image to a 3DMM. There is no direct way to know the correspondence between the 2D image pixel $I(x_i, y_i)$ and the vertex (u_k, v_k) of the 3DMM. An alternative is to warp the 2D image to UV space (e.g., using the RBF method) and then employ texture correspondence to obtain (u_k, v_k) once the (x_i, y_i)'s corresponding UV coordinate (u_k, v_k) has been obtained. The three red arrows represent the computational flow for recovering the 3D texture data from a 2D input image.

Let us consider the image warping method which bridges $I(x_i, y_i)$ and $I(u_k, v_k)$. To avoid confusion, we use the notation I_{source} to represent the input 2D image and I_{target}, the corresponding UV image.

To begin, given an image I_{source}, we obtain the coordinates (x, y) of all the feature points specified in the FPDT in Table 5. Using the shape correspondence concept discussed in Section 2, the corresponding 3D shape index of each feature point in the feature definition table is obtained. Following this, we use shape and texture correspondence to obtain the accurate UV coordinates (u,v in I_{target}) of each feature point in the feature definition table I_{target}.

To perform the above procedure, we again use radial basis functions [16] to warp the images from (x, y) to (u, v) space. The two groups of feature points (x, y coordinates in I_{source} and u, v coordinates in I_{target}) are used as control points to train an RBF network. Subsequently, the non-feature points in I_{source} are warped to I_{target} according to the RBF network functions.

Once the UV image is obtained (i.e., every pixel in the 2D image has been warped into UV space), the 3D texture information at each pixel can be directly "painted" onto the 3DMM since the latter's indices corresponding to every pixel (u_k, v_k) on the UV image are known.

Figure 4 displays the real faces as rectangular images that should be compared with the synthesized images to their right. In Figure 5, images of six individuals, representing different genders, races and age groups, are shown and contrasted with their synthesized images. The red boxes indicate the 2D input images used by our 3D facial reconstruction system. By closely examining the real and synthesized images, we observe that some distortions do occur when the human head is rotated to a large yaw or pitch angle.

It is interesting to compare our approach for 3D texture information recovery with that of Jiang et al. [15]. In their paper, there is no clear explanation of the process for obtaining the texture information. The only sentence pertaining to this subject is "the 2D image is projected orthogonally to the 3D geometry to generate the texture". To the best of the authors' knowledge, an input image cannot directly be projected orthogonally[6] to a 3D surface as stated in the above quotation. The problem of mapping 2D pixel data to 3D space is well known in

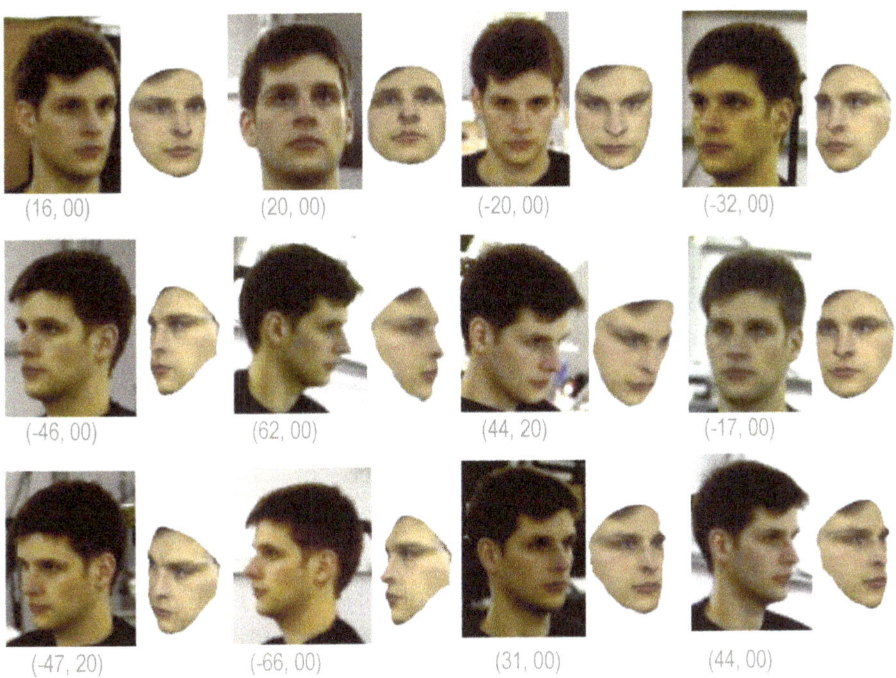

Figure 4. Synthesized images with different head poses compared to real images. The number in the bracket below each real image indicates the approximate camera view (yaw and pitch angle).

[6] The problem of direct orthogonal projection is discussed in Eckstein et al. [17] (see their Figure 1).

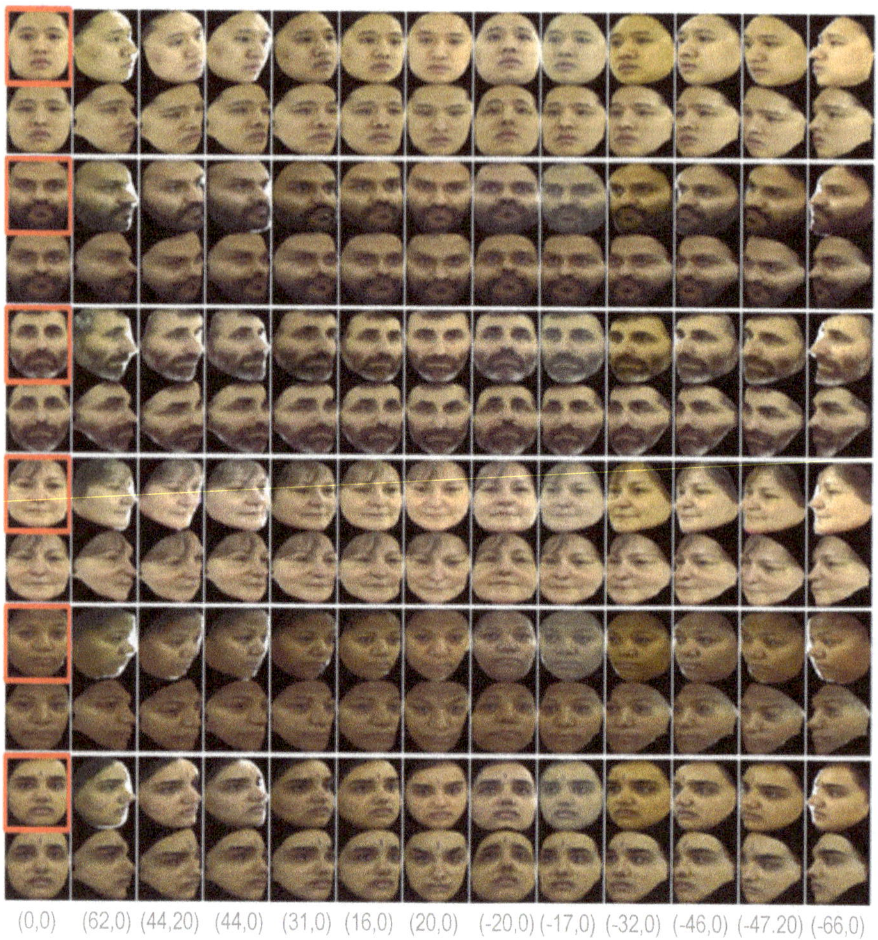

Figure 5. Synthesized images for six individuals. Images of each individual are presented in two rows. The upper row shows the individual's real images and the lower row shows the corresponding synthesized images. The images highlighted in red are the ones input to our 3D facial reconstruction system.

the field of computational geometry. Several papers [17; 18] have attempted to provide solutions to this problem and all emphasize the fact that direct recovery from a 2D image is difficult. Moreover, the implementation of these techniques requires complex computations and complicated data structures since they use a geometry-dependent flattening process (unlike 3DMM which is geometry-free[7]).

[7] The coordinates of the UV map do not depend on the geometric information of any input 2D image.

The advantage of the method proposed in this paper is that it simplifies a computationally complex geometric problem to a 2D-to-2D image-warping problem, thus enhancing the efficiency and accuracy of texture recovery. It is not possible to recover texture information simply by just using UV mapping without incorporating the previously discussed correspondence established a priori between the three-dimensional morphological model and the UV plane. The success of our method is rooted in the fact that we take advantage of this correspondence.

Finally, we emphasize the advantage of the method discussed here compared to "analysis-to-synthesis" methods, which attempt to estimate a single texture parameter vector directly in order to represent an individual's complete facial texture pattern. Since these parameters are also only an approximation, this often results in poor image renditions of 3D facial appearance. On the other hand, the method proposed in this paper directly applies real texture information from the 2D image onto the 3DMM via an image warping technique.

4. 3D Shape Similarity Based on Intrinsic 3D Geometric Structure

It is insufficient to visually compare real and synthesized versions of faces to evaluate the performance of a reconstruction process. We recall that in this paper the input is a single 2D image, which is then used to reconstruct a 3D model. Since the ultimate intention is to perform 2D face recognition based on projections from the 3D model, it is desirable that the quality of the latter be measureable. Thus this section focuses on *quantitatively* evaluating the performance of the 3D reconstruction system by considering a similarity measurement between real and synthesized models in 3D. This is achieved by extracting *intrinsic* shape signatures using geometric transformations that capture the inherent shape properties for various poses of the shape. The geometric transformation used for this is the Bending-Invariant Canonical Form (BICF) method introduced in [19]. A dissimilarity score of two BICFs is then computed using the Iterative Closest Point (ICP) method [20] for surface matching.

Three-dimensional face recognition algorithms usually treat the human face as rigid objects; in other words, they tend to ignore facial expressions, which clearly negate this assumption. On the other hand, the BICF method [19] captures the *intrinsic* 3D geometric structure using geodesic computing. The most important aspect in the determination of the BICF is the computation of geodesic distances between vertices on the facial surface and this is accomplished by using the Fast Marching Method in the triangulated domain (FMMTD) [10]. The fast marching method produces a distance map

$\mathbf{T} = [T(v_1), T(v_2), \ldots, T(v_N)] \in \Re^N$, which intersects every vertex in a 3D shape. The physical meaning of the distance map **T** is that it records how much time it takes for the geodesic propagation wave to travel from the initial vertex v_0 to a vertex v on the mesh surface. Geodesic extraction retrieves a geodesic path ***P***, which is an ordered list, from the distance map. The geodesic distance $gd(v_0, v)$ between two vertices v_0 and v can be expressed as

$$gd(v_0, v) \cong |v_0 v_1^p| + \sum_{i=1}^{n-1} |v_i^p v_{i+1}^p| + |v_n^p v|$$

if we assume that the geodesic path ***P*** is constructed by a set of vertices $\{v_0, v_1^p, v_2^p, \ldots, v_{n-1}^p, v_n^p, v\}$. The computational process for obtaining the geodesic path is found in [21]. Figure 6 illustrates the result of the geodesic computation for a 3D head model (2D mesh surface) where we observe that FMMTD propagates like a wave front on the 3D mesh surface. The different colors represent the level sets of the distance function. A sample geodesic path from the starting to the end point is highlighted in red.

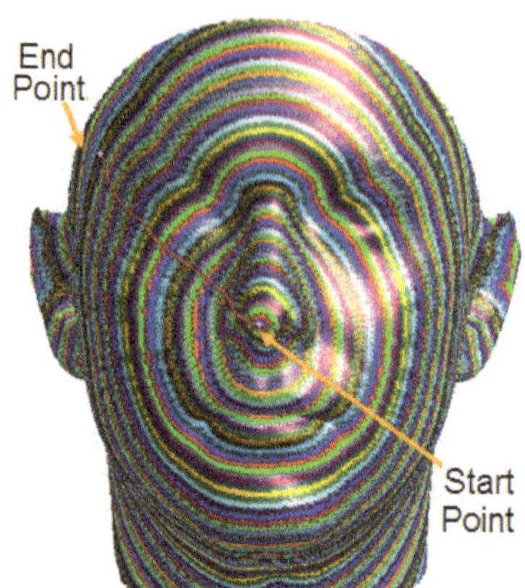

Figure 6. Computing geodesic paths on a triangulated manifold using FMMTD.

The human face is assumed to be a deformable object, with facial expressions producing the deformations. The authors in [19; 22] showed that facial expressions could be modeled as near-isometric transformations[8] and

---

[8] If a surface S_1 is ***isometric*** to a surface S_2, then S_1 deforms to S_2 through all length-preserving transformations or deformations (e.g., geodesic distance computing), including all possible bending (e.g., changes facial expressions). Such deformations must also satisfy the conditions of no tearing or stretching.

used the so-called Bending Invariant Canonical Form (BICF) to represent all equivalent surfaces (e.g., facial expression surfaces) obtained through isometric transformations. We employ three steps to construct the BICF of the human head:

- First, FMMTD produces the distance map **T**. (We omit the final step given in the original formulation which determines the geodesic distance between every pair of points on the surface.)
- Second, an accelerated algorithm, Geodesic Uniform Sampling (GUS) [23], is employed to construct the BICF. This step was not originally included in Bronstein et al. [1].
- Finally, the non-linear dimensionality reduction technique, Local Linear Embedding (LLE) [24] or multi-dimensional scaling (MDS) [25]) is applied to map the intrinsic geodesic structure of the surface obtained in the previous steps to a lower Euclidean space.

Using the FMMTD to exhaustively compute the BICF by extracting the pairwise geodesic distances between every two points on the surface requires a considerable amount of computation. For instance, the generic face model contains approximately 40,000 vertices and would need approximately 1.2 million computations to determine all of the geodesic distances using the FMMTD. To avoid this, we used the approach of Peyre [26], which proposed a sampling method that determines the most important features on a 3D mesh. This algorithm returns an adequate set of vertices uniformly distributed on the surface according to the geodesic distance. Figure 7 illustrates an example of using GUS, which illustrates how it automatically captures uniformly distributed vertices on a mesh surface.

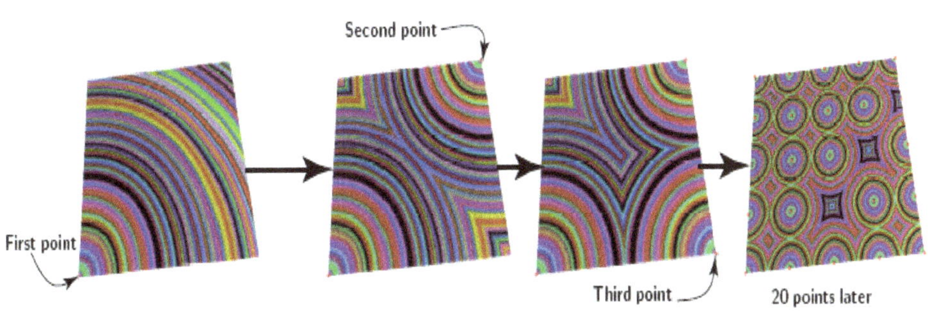

Figure 7. Example of geodesic uniform sampling (GUS). The algorithm starts at one point, propagating using FMMTD. The point associated with the largest distance map value is chosen as the second sampled propagating source point. This continues until, after 20 iterations, 20 points are uniformly distributed on the surface according to the geodesic distance. Colors represent the vertices sharing the same distance values from the source point.

The algorithm is given as follows:

- **Initialization**: Once an arbitrary vertex v is chosen to initiate FMMTD, it propagates and calculates the distance map values for all vertices. The vertex associated with the largest distance map value is selected as the second sampled vertex (new vertex).
- **Update**: The Fast Marching Method is applied to the new vertex and the geodesic distance vector $\mathbf{T}_x = [T_x(v_1), T_x(v_2), \ldots, T_x(v_N)] \in \Re^N$ is returned as the output of FMMTD. The current distance vectors $\mathbf{T}$ and $\mathbf{T}_x$ are compared, and each component of $\mathbf{T}$ is updated in the following manner:

$$T(v_i) \leftarrow \min\{T(v_i), T_x(v_i)\} \tag{7}$$

  The vertex with the largest distance value according to the newest $\mathbf{T}$ is chosen as the new sampled vertex. The update step is then repeated.
- **Stop**: Once GUS has collected enough vertices, updating is terminated.

Figure 8 illustrates the output of the algorithm for a 3D head.

Non-linear dimensionality reduction, the third stage in the construction of the BICF, helps transform a surface in a non-Euclidean metric space to a low-dimensional Euclidean space. Bronstein et al. [19; 22] exploited classical multidimensional scaling (MDS) to accomplish this task. However, local linear embedding (LLE) is proposed as an alternative in this work. Both have been well studied in the literature [24; 26; 28].

This section has concentrated on using the BICF [19] as a measure of 3D face similarity. BICF is a geometric transformation technique used to extract implicit spatial information from non-rigid 3D surface data. Section 5 focuses on experiments using this transformation that were performed to measure 3D facial similarity.

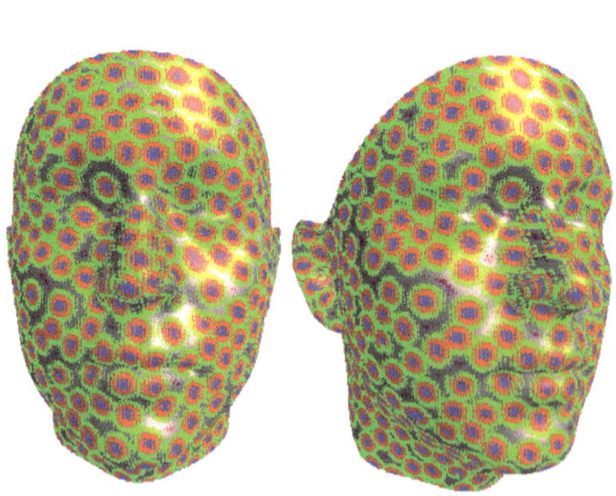

Figure 8. Geodesic uniform sampling (GUS) results after 3,000 iterations.

5. Experiments with 3D Shape Similarity Measures

The objective of the experiments discussed in this section is to *quantitatively* evaluate the dissimilarity between real 3D facial surfaces obtained using a 3D scanner and the synthesized 3D facial surfaces obtained using the reconstruction system described in this paper. We use three different similarity measurements, namely, the traditional surface-matching algorithm Iterative Closest Point method (ICP) and two novel Bending Invariant Canonical Form (BICF) techniques: BICF-LLE+ICP and BICF-MDS+ICP. To avoid confusion, three shortened forms are proposed in Table 2 to represent these three similarity measurements.

Table 2. The shortened forms for the three similarity measurement methods.

Notation	**Shortened Form**	**Nonlinear Dimensionality Reduction of Bending Invariant Canonical Form**	**Surface-Matching Algorithm**
M_1	ICP-only	No	Iterative closed point (ICP) method
M_2	BICF-MDS+ICP	Multidimensional Scaling (MDS)	Iterative closed point (ICP) method
M_3	BICF-LLE+ICP	Local Linear Embedding (LLE)	Iterative closed point (ICP) method

Before beginning the discussion of these experiments, we address two concerns that later will be seen to dominate their design. The first arises due to limitations in performance of the similarity measurements. The second relates to the consistency of different similarity measures.

First Concern: Limitations in performance of the similarity measurements

We use the notation $\Delta(S_i, S_j)_{M_k}$ to represent the similarity score between two 3D surfaces S_i and S_j obtained using a specified similarity measure M_k. We propose two scenarios to explain this issue:

Scenario A: Assume a similarity measure M_k is used to measure the similarity between surface S_1 and *itself.* Theoretically, the similarity score $\Delta(S_1, S_1)_{M_k}$ should be 1 (we assume 1 is the highest and 0 is the lowest similarity score). Furthermore, every $\Delta(S_1, S_i)_{M_k}$, where S_i represents any arbitrary surface but not S_1, cannot be higher than 1 (i.e., no other surface S_i will be more similar to S_1 than itself). Unfortunately, in practice, we may find a surface S_i such that $\Delta(S_1, S_i)_{M_k} \geq \Delta(S_1, S_1)_{M_k} = 1$. Such an error is due to the limitations of the similarity measure M_k.

Scenario B: Assume a measure M_k is used to compute the similarity between a real surface S_1 and its corresponding synthesized surface S_1', and produces a similarity score $\Delta(S_1, S_1')_{M_k} = c_1$, which is low. The lower the similarity score, the more dissimilar S_1' is compared to S_1, indicating a poorly designed facial reconstruction system. However, the similarity score may not accurately represent the quality of the reconstructed surface S_1'. Let us assume we find a similarity score of a real surface S_1 *with itself* using a similarity measure M_k and the similarity score equals $\Delta(S_1, S_1)_{M_k} = c_2$. Ideally, c_2 should always be greater than c_1 but this will not always happen in practical situations. Thus, c_2 plays the role of an upper bound on the similarity score c_1 using similarity measure M_k. This leads to the conclusion that we should focus our attention on the disparity $d = c_2 - c_1$ instead of c_1 itself.

Second Concern: Consistency of different similarity measures

Another interesting question would be to assess if different similarity measures $M_1, \ldots, M_k, \ldots, M_n$ produce similar scores $\Delta(S_i, S_i')_{M_k}$ when comparing a real 3D surface S_i and its corresponding synthesized surface S_i'.

Our objective is to investigate the limitations of the three aforementioned similarity measures, M_1, M_2 and M_3, using Confusion Matrix Analysis (CMA) [29] and the Equal Error Rate (EER) [29] to determine the upper bound[9] of performance for each measurement M_k.

Confusion Matrix Analysis

Suppose each row and column of the confusion matrix represents one 3D facial surface in the Freiburg 3D face database. The *(i, j)*-th element of the matrix contains the dissimilarity scores between the *(i, j)*-th person's synthesized and real 3D face obtained using a specific similarity measure. Intuitively, the matrix diagonal should contain elements with the lowest dissimilarity scores whereas the non-diagonal elements should contain higher dissimilarity scores, as they are comparing two different person's real 3D facial surfaces. Also, based on visual inspection, the more the diagonal components contrast with the off-diagonal components in the confusion matrix, the more that the particular similarity measure used can distinguish between the two sets of 3D facial surfaces.

---

[9] Practically speaking, the three similarity measurements M_1, M_2, M_3 measure *dissimilarity* as opposed to similarity. Therefore, we seek to determine the lower bound of dissimilarity as opposed to the upper bound of similarity.

Table 3 shows the confusion matrices for the three similarity measures. We observe that the confusion matrices of both cases using the BICF measure show highly contrasting diagonal components, thus leading to the conclusion that BICF enhances the ability of ICP to measure the recognition capability. Quantitative comparisons are discussed next.

We group the confusion matrix into two sets, diagonal G1 and off-diagonal components G2. The diagonal components are a consequence of comparing two 3D facial surfaces of the same person while the off-diagonal components are a consequence of comparing two 3D facial surfaces belonging to different people. We plot a separate histogram for G1 and G2 and fit a Gaussian to each. We compute the sample mean $\tilde{\mu}$, standard deviation $\tilde{\sigma}$, and the 95% confidence intervals (CI) corresponding to G1 and G2. The area of intersection between the two Gaussian-fitted curves is also computed since it can serve as an excellent criterion for evaluating the degree of separation between the two sets.

Table 4 presents a statistical summary of the three confusion matrices. We observe that BICF-LLE+ICP, with the lowest area of intersection, appears to perform the best among the three measures, closely followed by BICF-MDS+ICP and finally Only-ICP.

As an additional test of the performance, the Equal Error Rate (EER) was used in a simulated face verification experiment. Each 3D face in the database was verified against every other face surface, including itself. The comparison between each facial surface and all others was accomplished using dissimilarity scores obtained using the three similarity measures and the Equal Error Rate (EER) was chosen as the optimal threshold. The EER values for the Only-ICP, BICF-LLE+MDS, and BICF-LLE+ICP were 12%, 7%, and 5%, respectively. Again, the BICF-LLE+ICP method produced the superior result.

In this section, we have compared two BICF methods (BICF-MDS+ICP and BICF-LLE+ICP) and the traditional 3D shape similarity measure ICP [30]. From our results, we observe that the BICF method is superior with regard to distinguishing two sets of surface data due to its ability to extract intrinsic geometric information from the surfaces. Between the two BICF methods, BICF-LLE+ICP performed slightly better than BICF-MDS+ICP.

Another objective was to assess the performance of our 3D facial reconstruction system by *quantitatively* measuring the disparity between *real* 3D facial surfaces and their corresponding *synthesized counterparts*. The results, which were based on the two different figures of merit discussed in this section, proved that our 3D reconstruction system was capable of synthesizing highly realistic 3D facial surfaces.

Table 3. Confusion matrices for the three similarity measures; A: Iterative Closet Point (ICP); B: Bending Invariant Canonical Form (BICF)-Multidimensional Scaling(MDS)+Iterative Closet Point(ICP); C: Bending Invariant Canonical Form (BICF)+Locally Linear Embedding(LLE)+ Iterative Closet Point(ICP).

Methods	**Real versus Real**

M1:ICP

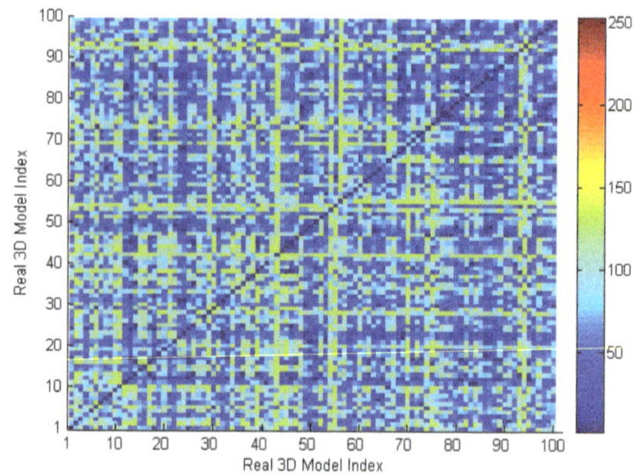

A

M2:BICF-MDS+ICP

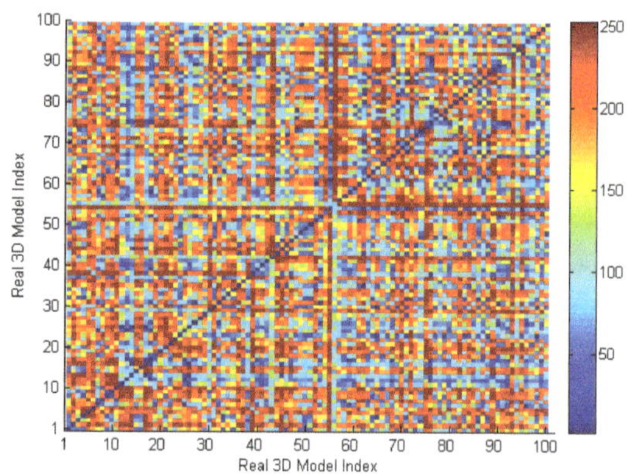

B

M3:BICF-LLE+ICP

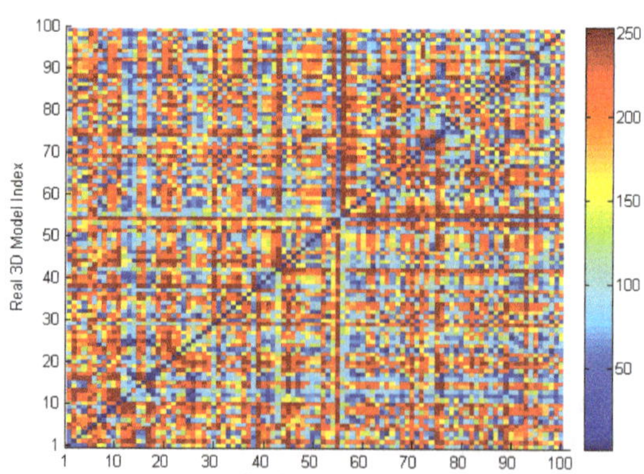

C

Table 4. Statistical summary of the three confusion matrices: The mean $\tilde{\mu}$, standard deviation $\tilde{\sigma}$, and 95% confidence intervals ($\tilde{\mu}$ CI) plus the intersection data for the two Gaussian-fitted G1 and G2.

Real data vs. Real data		Only-ICP	BICF-MDS+ICP	BICF-LLE+ICP
G1	$\tilde{\mu}$	0.5077	0.1409	0.1355
	$\tilde{\sigma}$	1.0444	1.0091	0.9652
	$\tilde{\mu}$ CI	[0.3005, 0.7150]	[-0.0791, 0.3610]	[-0.0561,0.3270]
G2	$\tilde{\mu}$	2.8407	2.9671	2.96702
	$\tilde{\sigma}$	0.5372	0.7575	0.7575
	$\tilde{\mu}$ CI	[2.8258, 2.8557]	[2.9459, 2.9882]	[2.9459, 2.9881]
Intersection	Point of Intersection	1.8821	1.7041	1.6596
	Area	0.1313	0.1084	0.0993

6. Discussion

In this chapter, we have proposed a facial reconstruction system using a single frontal 2D image as the input. We can compare our system with the other four existing methods for facial reconstruction:

1) Stochastic Newton Optimization method (SNO) [10; 11; 31].
2) Inverse Compositional Image Alignment (ICIA) applied to 3D Morphable Models (3DMM) [32] initially introduced by Baker and Matthews [33]. This algorithm was later adopted by Romdhani and Vetter with regard to 3DMM [9].
3) Linear Shape and Texture Fitting (LiST) [34], introduced by Romdhani, Blanz and Vetter.
4) Shape Alignment and Interpolation Method Correction (SAIMC) [15] introduced by D. Jiang, Y. Hu, S. Yan, L. Zhang, H. Zhang, and W. Gao to retrieve a 3D facial model from a single frontal 2D image.

The first three can be classified as falling within the "analysis-by-synthesis" design framework and the last one within the "3D supported 2D model" framework. Readers may consult [8] for a discussion of the differences between the two frameworks. The method presented in this paper follows the approach of the "3D supported 2D model". There are several similarities between our method and that of Jiang et al., [15] which also falls within this category. For instance, both employ a shape alignment procedure to estimate shape parameters and then incorporate a correction to deal with the inaccuracy of the resulting reconstructed shape. However, the SAIMC method of [15] raises some significant issues:

1) No explicit objective function is provided and no detailed explanation of how the authors perform the shape alignment is shown in their flowchart.

2) Two ambiguous parameters Λ and λ are shown in the flowchart and different suspected meanings for them were investigated. We assumed that Λ was a vector of eigenvalues of the 3D morphable model and λ, of user-defined parameters. However, on further investigation, by assuming Λ as the eigenvalues of the 3DMM, it was seen that several different persons' synthesized results converged to the mean shape of the 3DMM (GMF). We also varied the λ values over a large range to determine whether acceptable surfaces could be obtained. Unfortunately, none of these attempts worked.
3) No explanation for texture retrieval is expounded in [15]. The paper only states that "the 2D image is projected orthogonally to the 3D geometry to generate the texture".

The reconstruction process described in this paper, although similar in many respects to SAIMC, was designed to overcome the above problems. The advantages of our methodology compared to SAIMC can be summarized as follows:

1) Shape alignment is formulated as a formal optimization problem solution.
2) The shape alignment algorithm of SAIMC requires one ill-specified user-defined parameter λ [35]. The shape alignment algorithm in this paper is fully automatic and no user-defined parameters are required.
3) We present a simple texture recovery procedure using 2D-to-2D image warping techniques. SAIMC lacks any details regarding texture recovery.

Finally, with respect to computational efficiency, we have indicated elsewhere that the method reported in this paper is the fastest (roughly speaking) of the five reported in the literature [8].

Appendix

The FPDT is a table containing landmarks on a 2D image of a face, which together represent the most significant morphological information on the face. These need to be manually indicated by the user. The choice of which landmarks to use is not unique, however, once the table has been created, the order of the landmarks and their definitions cannot be changed. We have defined 57 morphological landmarks divided into two groups; namely, contour and internal landmarks. Twelve features, referred to as contour landmarks, are located on the face contour. The latter are obtained using a Graphical User Interface (GUI), which prompts the user to separately select 5 and 7 landmarks, respectively for the upper and lower contours. For the lower contour, the user is prompted to select two anthropomorphic landmarks at the bottom of the ears and the tip of the chin, as indicated on a cartoon example on the GUI. The remaining 4 landmarks were chosen to symmetrically fit the chin area. With respect to the upper contour, 2 landmarks are chosen at the top of the ears and 1 at the top of the forehead. The remaining 2 landmarks are selected as a symmetric pair. Since the resulting curves through these landmarks were constrained by third order splines, the user is forced to provide the best representation of the head subject to all of the stated geometrical limitations. The remaining 45, describing the eyebrows, eyes, pupils, nose and lips are referred to as internal landmarks.

Figure 9 shows a 2D face with the 57 landmarks used in this paper and Table 5 provides the details.

Using visual inspection, it is obviously trivial to determine the corresponding indices of the 45 two-dimensional internal landmarks on the 3D GMF model. However, manual determination of the landmarks on a contour is more difficult. This arises for two reasons: Firstly, the contour of the face may vary depending on the angle of view and the interpretation of what constitutes the actual contour. Secondly, an arbitrary point visually selected as the 3D correspondence for a 2D landmark could be one of several vertices, since the 3DMM is very densely sampled. Thus, many options for selecting a unique 3D vertex exist which correspond to a particular 2D landmark. This is obviously undesirable, as the

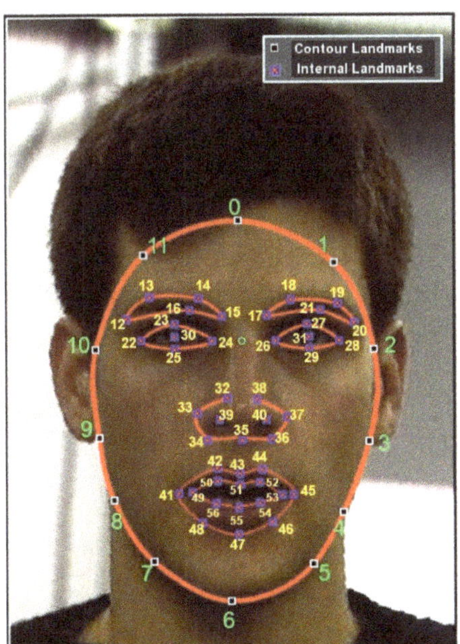

Figure 9. The 57 user-defined landmarks specified on a 2D facial image. The green labels are contour and the yellow ones internal landmarks.

Table 5. Landmark descriptions. Note that each of the landmarks shown in 2D in Figure 5 is *linked to the index of a specific 3D landmark* on the 3D Generic Mean Face. See [8] for an explanation of the significance of this arrangement.

	Description	Landmark indices
Contour Landmarks	Forehead	0, 1, 2, 11, 10
	Chin	3, 4, 5, 6, 7,8,9
Internal Landmarks	Left Brow	12,13,14,15,16
	Right Brow	17,18,19,20,21
	Left Eye	22,23,24,25
	Right Eye	26,27,28,29
	Left Pupil	30
	Right Pupil	31
	Nose	32,33,34,35,36,37,38
	Nostril	39,40
	Outer Lip	41,42,43,44,45,46,47,48
	Inner Lip	49,50,51,52,53,54,55,56

morphable model requires a unique one-to-one correspondence between a 2D feature point and its corresponding 3D vertex. Therefore, we propose a geodesic contour-fitting algorithm to address this issue.

Geodesic contours can be computed by the Fast Marching Method (See [12] for details), which not only helps crop non-crucial facial regions on the GMF, but also provides an automatic repeatable method for detecting the 3D contour. Figure 10 illustrates how the output of the geodesic contour method crops the GMF. Briefly, the FMM can be interpreted as a method for propagating a geodesic wavefront on a mesh surface with a set of fixed initial and final vertices. The initial points are chosen as the tip of the nose[10] and the nose apex (the nasion, the topmost point of the nose bone)[11]. Algorithmic determination of these two facial points is straightforward by examining unique features such as curvature for the nose apex and height for the tip of the nose. A vertex near the chin of the GMF, index #38,985, is set as the FMM wavefront stopping point. Any vertex residing within the region not covered by the propagating FMM wavefront is then disregarded. By using this algorithm, we can determine the 3D contour of the GMF while retaining only the most important regions of the 3D face.

[10] Vertex index 37,814

[11] Vertex index 38,100

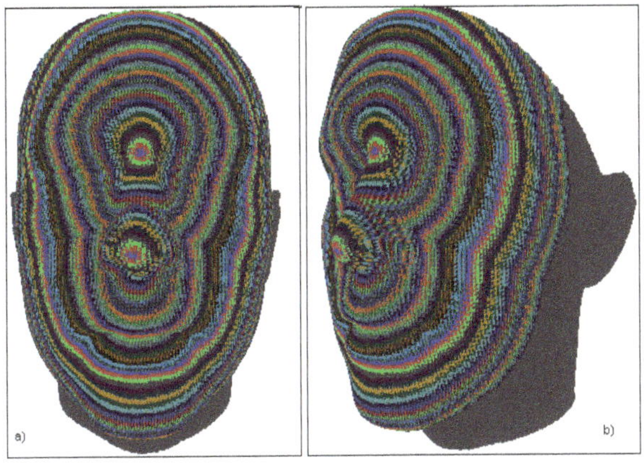

Figure 10. Geodesic contours obtained using the Fast Marching Method: The grey area contains the discarded vertices.

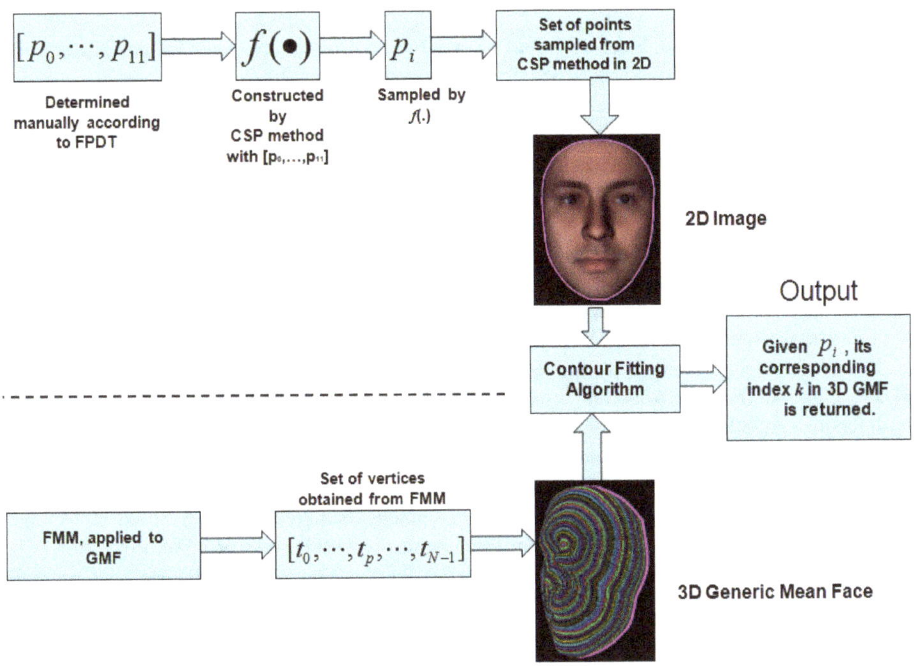

Figure 11. Flow chart showing the procedures used for obtaining corresponding pairs of landmarks in the 2D image and the 3D GMF, which are the two inputs to the system. The index k in the 3D GMF corresponding to the 2D feature point p_i is returned as the output. See text for details.

After automatically cropping the face, it is necessary to obtain the corresponding index for each vertex on the GMF, which is also accomplished employing the FMM. Figure 11 illustrates the procedure used to obtain associated points in the 2D image and the 3D GMF, which are the two inputs. Two contours in the figure are highlighted in violet: the contour in the 2D image and the one on the GMF. The former is represented by the 2D contour function $f(\bullet)$ and is constructed using the Cubic Spline Polynomial (CPS) method [36] on the twelve 2D feature points $P = [p_1, \ldots, p_{12}]$ defined in the FPDT as being on the contour of the face[12]. Thus any point p_i on the contour can be determined by sampling the curve given by the contour function $f(\bullet)$. On the other hand, the associated contour, highlighted in violet, on the 3D GMF is found by the FMM, which determines all vertices of interest on the 3D contours. The vertices are saved as an ordered sequential list $L = [t_0, \ldots, t_p, \ldots, t_{N-1}]$. It is then necessary to find the correspondence between these two contours and this is accomplished by a simple contour-fitting algorithm as described in [13].

References

[1] M.D. Levine and Y. Yu, State-of-the-art of facial reconstruction **methods** for face recognition based on a single 2D input image. Pattern Recognition Letters, 2009, in press.

[2] L. Gu, and T. Kanade, A Generative Shape Regularization Model for Robust Face Alignment, Proceedings of the 10th European Conference on Computer Vision: Part I, Springer-Verlag, Marseille, France, 2008.

[3] T.F. Cootes, G.J. Edwards, and C.J. Taylor, Active Appearance Models, Proceedings of the 5th European Conference on Computer Vision-Volume II - Volume II, Springer-Verlag, 1998.

[4] Y. Zhou, L. Gu, and H. Zhang, Bayesian Tangent Shape Model: Estimating Shape and Pose Parameters Via Bayesian Inference, CVPR03, 2003, pp. 109-116.

[5] T.P. Riopka, and T. Boult, The Eyes Have It, Proceedings of ACM SIGMM Multimedia Biometrics Methods and Applications Workshop, Berkeley, 2003, pp. 33-40.

[6] J. Sung, and D. Kim, Adaptive active appearance model with incremental learning. Pattern Recogn. Lett. 30 (2009) 359-367.

[7] S. Asteriadis, N. Nikolaidis, and I. Pitas, Facial feature detection using distance vector fields. Pattern Recogn. 42 (2009) 1388-1398.

[8] Y. Yu, A Method For Synthesis Of A Three Dimensional Head With A Single 2D Image, Department of Electrical and Computer Engineering, McGill Univerisity, Montreal, 2007.

[9] V. Blanz, and T. Vetter, A Morphable Model for the Synthesis of 3D Faces. Computer Graphics Proc. SIGGRAPH (1999) 187-194.

[10] V. Blanz, and T. Vetter, Face Recognition Based on Fitting a 3D Morphable Model. IEEE Transactions on Pattern Analysis and Machine Intelligence 25 (2003) 1063-1074.

[11] Ibid. [8], p. 129.

[12] Ibid. [8], p. 133.

[13] Ibid. [8], p. 127.

[14] T.K. Moon, and W.C. Stirling, Mathematical Methods and Algorithms for Signal Processing, Prentice Hall, New York, 2000.

---

[12] See Table 1.

[15] D. Jiang, Y. Hu, S. Yan, L. Zhang, H. Zhang, and W. Gao, Efficient 3D Reconstruction for Face Recognition. Pattern Recognition 38 (2005) 787-798.
[16] S. Haykin, Neural Networks: A Comprehensive Foundation, Prentice Hall, 1998.
[17] I. Eckstein, V. Surazhsky, and C. Gotsman, Texture Mapping with Hard Constraints. proceedings of Eurographics 20 (2001) 95-104.
[18] B. Levy, and J.-L.M. Mallet, Non-Distorted Texture Mapping for Sheared Triangulated Meshes. Conference Proceedings SIGGRAPH 98 (1998) 343-352.
[19] A. Bronstein, M. Bronstein, and K. Ron, Expression-Invariant 3D Face Recognition. Proceedings of Audio & Video-based Biometric Person Authentication (AVBPA) 2688 (2003) 62-70.
[20] K.W. Bowyer, K. Chang, and P. Flynn, A Survey of Approaches and Challenges in 3D and Multi-Modal 3D + 2D Face Recognition Computer Vision and Image Understanding 101 (2004) 1-15.
[21] A. Quarteroni, R. Sacco, and F. Saleri, Numerical Mathematics Springer, 2000.
[22] A. Bronstein, M. Bronstein, and R. Kimmel, Expression-Invariant Representations of Faces. IEEE Trans. Image Processing 16 (2007).
[23] G. Peyré, and L.D. Cohen, Geodesic Remeshing Using Front Propagation. International Journal of Computer Vision 69 (2006) 145-156.
[24] L.K. Saul, and S.T. Roweis, Think Globally,Fit Locally: Unsupervised Learning of Low Dimensional Manifolds. Jornal of Machine Learning Research 4 (2003) 119-155.
[25] E.L. Schwartz, A. Shaw, and E. Wolfson, A Numerical Solution to the Generalized Mapmaker's Problem: Flattening Nonconvex Polyhedral Surfaces. IEEE Trans. PAMI 11 (1989) 1005-1008.
[26] G. Peyre, and L. Cohen, Geodesic Computation for Adaptive Remeshing. IEEE Computer Society Conference on Computer Vision and Pattern Recognition 2 (2005) 1193-1197.
[27] S. Roweis, and L. Saul, Nonlinear Dimensionality Reduction by Locally Linear Embedding. Science 290 (2000) 2323-2326.
[28] B. Ingwer, and G. Patrick, Modern Multidimensional Scaling: Theory and Applications Springer-Verlag, New York, 1997.
[29] R.O. Duda, P.E. Hart, and D.G. Stork, Pattern Classification, Wiley-Interscience, 2000.
[30] S. Rusinkiewicz, and M. Levoy, Efficient Variants of the ICP Algorithm. Proceedings of the 3rd International Conference on 3-D Digital Imaging and Modeling (2001) 145-152.
[31] V. Blanz, Automatische Rekonstruction der dreidimensionalen Form von Gesichtern aus einem Einzelbild, Universitat Tubingen, 2001.
[32] S. Romdhani, and T. Vetter, Efficient, Robust and Accurate Fitting of a 3D Morphable Model. IEEE International conference on Computer Vision 2003 2 (2003) 59-66.
[33] S. Baker, and I. Matthews, Equivalence and Efficiency of Image Alignment Algorithms. CVIP 2001. Proceedings of the 2001 IEEE Computer Society Conference 1 (2001) 1090-1097.
[34] S. Romdhani, V. Blanz, and T. Vetter, Face Identification by Fitting a 3D Morphable Model Using Linear Shape and Texture Error Functions. Computer Vision -- ECCV'02 4 (2002) 3-19.
[35] D. Jiang, Y. Hu, S. Yan, L. Zhang, H. Zhang, and W. Gao, Efficient 3D Reconstruction for Face Recognition. Pattern Recognition 38 (2005) 787-798.
[36] M. Gandhi, A Method for Automatic Synthesis of Aged Human Facial Images, Computer and Electrical Engineering, McGill, Montreal, QC, Canada, 2004, pp. 117.

CHAPTER 5.2

AUTOMATIC ANNOTATION OF BIOLOGICAL IMAGES

Jie Zhou[1], Hanchuan Peng[2]

[1]*Department of Computer Science, Northern Illinois University*
DeKalb, IL 60115, USA
E-mail: jzhou@cs.niu.edu

[2]*Janelia Farm Research Campus, Howard Hughes Medical Institute*
Ashburn, VA 20147, USA
E-mail: peng@janelia.hhmi.org

Annotation of biological images is to label biological images or parts of the images, often using ontological or anatomical vocabularies. The task has become increasingly important along with the availability of large data sets of multidimensional and multiscale biological images in recent years. Examples include the annotation of the anatomical regions of gene expression patterns of Drosophila melanogaster (fruitfly), or the annotation of cellular or subcellular structures for other model organisms at different developmental stages. To automate the process in order to increase efficiency and consistency, the annotation problem can be formulated as a pattern recognition task with ontological or anatomical terms viewed as the targets to be recognized in the image. The task presents various objectives: we may annotate the entire image or many Regions-Of-Interest (ROI) in an image; the problem may be mutual-exclusive (when the image or region corresponds to one target) or multi-objective (when several annotation targets co-exist in an image). The task also has unique challenges such as skewed data distribution, morphological variety and big image size. In this chapter, we discuss algorithms and applications for automated annotation of biological images. We detail the algorithms for *in situ* fruitfly gene expression patterns during embryogenesis which is a multi-objective annotation problem. We also cover mutual-exclusive and ROI annotation applications. We demonstrate that extracting and selecting a concise set of good image features are essential for such applications and show effectiveness of our proposed algorithms.

1. Introduction

Annotation of biological images is to label biological images or parts of the images, often using various ontological or anatomical vocabularies in an automatic or semi-automatic way. In this chapter, we discuss the automatic way of annotation by formulating it as a pattern recognition task with ontological or anatomical terms viewed as the pattern targets to be recognized in the image.

Annotation of images, as a means to assign semantic information to the raw images, is a common task in many applications. However, concurrent with the recent availability of large amount of complicated molecular and cellular microscopic images, annotating biological images is of high demand and presents unique challenges and opportunities. Advancement in imaging methods such as the confocal or two-photon/multi-photon laser scanning microscopy (LSM)[1,2], scanning or transmission electron microscopy (EM)[3,4] has made it possible to produce images at different resolutions from millimeters to nanometers. Meanwhile, high-throughput biological imaging projects are making a large number of biological images available in an unprecedented scale. Examples of these image datasets/databases include Berkeley Drosophila Genome Project (BDGP) which contains *in situ* embryogenesis gene expression patterns of 6000 Drosophila melanogaster (fruitfly) genes[5] and Allen Brain Atlas which integrates the genome wide RNA *in situ* hybridization gene expression information of 20,000 mouse genes[6]. Other projects such as Cell Centered DataBase (CCDB)[7] and Gene Expression Nervous System ATlas (GENSAT)[8] have also generated/ collected multi-dimensional gene or protein expression images of mouse or other model organisms at different resolutions and different developmental stages[9].

Annotation plays an essential and sometimes critical role in effective use of these biological data. Without annotations that transform the images to meaningful information, the valuable knowledge embedded in raw images cannot be readily stored, searched or reused[10]. Traditionally, manual annotation has been performed on biological datasets. For example, manual annotation using a controlled vocabulary is done on 70,000 images for fruitfly image gene expression patterns in BDGP[5]. The manual annotation was done by single person during early phase of the project, and later changed to two annotators. Annotation terms were grouped into the six developmental stage ranges. For each developmental stage-range, the curators are presented with batches of images and annotations (terms such as *cellular blastoderm*) for each gene in a list. Images were manually selected, annotated and reviewed. As another example, CCDB uses an ontology-based annotation to annotate the celluar/subcelluar electron tomographic images of the nervous system[11]. Curators perform a manual segmentation of each image, then label each segmented object an ontology class (such as *Axon_0000*) that represents a subcellular anatomy structure.

Such manual annotation is very time consuming and can be a formidable task for some of these new large scale data sets. Internal inconsistency is also a problem hard to avoid with manual annotation due to inconsistency among different annotators as well as with the same annotator (caused by learning curve or other human factors). Automatic annotation of biological images, on the other hand, has high potential in both dramatically increasing efficiency as well as in improving annotation consistency. With the availability of a large number of images in biological databases, an automatic and systematic annotation approach becomes highly desirable.

To automate the annotation process, the annotation problem can be formulated as a pattern recognition task with ontological or anatomical terms as the targets to be recognized in the image. Despite the potential importance of automatic annotation of biological images, there are several challenges for automating the task due to unique properties of these recent biological images. For example, the distribution of images of various annotation targets can be very biased: Some ontology terms are commonly used to annotate a large percentage of images while others are associated with a relatively small number of images; Many annotation targets can coexist in the same image; Morphological variety of the same annotation target among different images is common, which can be caused by the biological nature (such as structure morphology) or the imaging technique (such as gene expression staining and image orientation); Different biological image datasets are of various quality and some make preprocessing and registration a challenge; Inconsistency or missing annotations during manual annotation can introduce obstacles for generating an objective evaluation dataset; The sizes of biological image data sets can be large, as well as the dimension of a single biological image. As a result, devising an efficient and effective automatic annotation system is a nontrivial task.

Nevertheless, some studies have begun to tackle the above problems of automatic annotation of biological images and gained promising results[12,13,14]. In this chapter, we introduce the approaches of formulating the automatic annotation problem into a pattern recognition and machine learning task. Depending on the objective of the application, we may annotate the entire image by assigning a term or a set of terms, or annotate Regions of Interest (ROI) in a image. Annotation tasks can be mutual-exclusive or multi-objective. From pattern recognition viewpoint, a mutual-exclusive annotation means that each image (or a ROI in an image) corresponds to one and only one annotation target, while a multi-objective case means that several annotation targets may co-exist for an image. We discuss the algorithms and applications for these categories of annotation problems, with the focus on multi-objective annotations that are assigned to an entire image.

In Sec. II, we introduce our methodology of automatically annotating gene expression patterns during embryogenesis of fruitfly by formulating it as a multi-objective problem with a parallel series of binary predictors. We detail the Embryo-wavelet feature we use to extract features from pattern images, and explain the min-Redundancy Max-Relevance (mRMR) feature selection[15] used to select annotation-specific features. We also conduct a comparison against other features using two synthetic data sets. Sec. II also briefly discusses methods for mutual-exclusive and Region-of-Interest (ROI) annotation in biological images. In Sec. III, concrete annotation results are presented. Section IV discusses other related issues such as preprocessing and future work. Section V is the conclusion.

2. Methodology

Problem formulation for the annotation task varies based on the objective of annotation. As mentioned in Introduction, an automatic annotation problem can be multi-objective or mutual-exclusive when formulated as a pattern recognition problem. In this section, we will focus on the problem of multi-objective annotation. In particular, we will introduce our methods used for automatic annotation of fruitfly *in situ* gene expression during embryogenesis in Sections 2.1 – 2.5. We will then briefly discuss other types of annotation problems in Section 2.6.

2.1. *Background for the Multi-objective Annotation Problem*

As a case study for multi-objective annotation problem, we discuss algorithms for annotating fruitfly's gene expression pattern images during embryogenesis obtained by *in situ* mRNA hybridization. (See Figure 1 and Table 2 for some examples of these images.) These images contain body structures that emerge during certain developmental stages and provide spatial-temporal information that is important for understanding the complicated relationship of genes and their functions.

For an arbitrary input image of gene expression pattern, the embryonic tissues in which a target gene can be expressed are often unknown beforehand, and it may have many different ontological annotations associated. From the viewpoint of pattern recognition, this means that each input sample may correspond to multiple class labels. In other words, given the set of K class labels $\Phi = \{c_1, \ldots, c_K\}$, automatic annotation of a fly embryo gene expression image is multi-objective: an input images x_i corresponds to a target *set* $a_i \subset \Phi$.

For the multi-objective problem of embryonic image annotation, features corresponding to an unknown number of tissue structures co-exist in the same image. Meanwhile, when each image pixel is often regarded as a dimension, the total dimensionality of the image is large. So how to effectively extract features is critical for identifying the discriminating features correlated with different structures. In addition, the classifier used for annotation needs to be aware of the fact that the sample distribution is heavily skewed for different annotations.

In Sec. 2.2 – 2.5, we will introduce our methods for the problem. It is worth noting that although the algorithms are demonstrated on fruitfly expression annotations, the methods presented in this section can also be applicable to annotating other multi-objective biological images.

2.2. *Extracting Wavelet-embryo Features*

With a large number of features coexisting in a single embryonic gene expression image, the task of feature extraction is to produce a representation that can best characterize this embryo image.

For automatic annotation of embryo images, we propose using the multi-resolution wavelet representation. Wavelet representation lies between the spatial and Fourier domains in the sense that wavelets are localized in both space and frequency whereas the standard Fourier transform is only localized in frequency[16]. Multi-resolution representations based on wavelet decomposition are effective for identifying and analyzing local and multi-scale features from signals or images[17]. With wavelet-decomposition, a fly embryo image is projected to a feature space where information is decomposed into various frequency bands and different levels of resolutions so that features of different structures may be effectively separated.

We use two-dimensional discrete wavelet transform (DWT) to extract multi-resolution image features. 2D DWT decomposes an image using orthonormal wavelet basis functions. Let the set of wavelet basis functions be $\{\Psi_{k,n}(x_1,x_2), k, n \in Z\}$, where n is the translation factor and k is the dilation factor which denotes a particular resolution level. Let $I(x_1, x_2)$ represent the intensity values of embryo image pixels indexed by the location vector (x_1, x_2). We can obtain the wavelet transform coefficient $d_{k,n}$ as below[16,18] :

$$d_{k,n} = \iint \overline{\Psi_{k,n}(x_1,x_2)} I(x_1,x_2)dx_1dx_2 \tag{1}$$

To analyze 2D image $I(x_1,x_2)$, we can extend the 1D analysis by defining the 2D space as the tensor product of two 1D spaces. In 1D counterpart of multiresolution analysis theory, the wavelet function $\psi(x)$ has a companion, the scaling function $\varphi(x)$, which approximates the signal/image at a specific level of resolution. The sets of scaling and wavelet functions are built by translation and dilation. 2D orthonormal wavelet basis function $\Psi(x_1,x_2)$ can factorized using the scaling function $\varphi(x)$ and mother wavelet function $\psi(x)$. Schematically, the factorization represents a two-step transform process: 1) perform a 1D wavelet transform on each row (indexed by x_1) of the image; and then 2) perform the transform on each column (indexed by x_2) of the result of step one. In our experiments, Daubechies-1 (HAAR) wavelet and scaling functions are used.

Mallat (1999)[16] linked orthogonal wavelets and scaling functions to quadrature mirror filters in signal processing theory with which we can also view each subspace as a two-dimensional filter that transforms the original image into one of the four components: the low-low (LL), low-high (LH), high-low (HL) and high-high (HH) parts of the transform. The LL quadrant is a smoothing of the original image. The other three parts are detail images. The set of coefficients $\{d_{k,n}\}$ of all sub-images at all resolution levels are used as features to characterize an gene expression pattern. In our experiment, $k = 0, 1$ (2 levels of resolutions); $n = 1, \ldots |x_1|*|x_2|$, where $|x_1|$ and $|x_2|$ are width and height of the sub-image at a specific resolution level. As the result of 2D DWT, the total number of coefficients (features) is about the same as the number of pixels of original image (there could be a small discrepancy if the number of pixels is not exactly the power of 2). In the rest of the Sec. II, we will use wavelet-embryo features to refer to features generated from embryonic images by wavelet decomposition.

2.3. *mRMR-based Feature Selection*

The dimensionality of the wavelet-embryo features is still big. For a multi-objective problem, this feature set contains the information of all objectives. Therefore, the next step is to select a most characterizing subset of features that can best discriminate the patterns corresponding to a specific annotation target in the embryo images. We consider generating a compact feature set with strong discriminating strength for specific annotations using the min-redundancy max-relevance (mRMR) feature selection method[15]. The mRMR feature selection algorithm is theoretically formulated as that features should be selected to maximize the statistical dependency between the annotation-distribution of image samples and the joint distribution of the selected features. Based on information theory, this can be factorized as finding features that are mutually far away from each other (minimum redundancy) but also individually most similar to the distribution of annotations (maximum relevance).

We use mutual information to measure the level of similarity between features. Let S denote the features subset that we are seeking and Ω the pool of all candidate features $\{f_i\}$ (they equal $\{d_{k,n}\}$ in the case of wavelet-embryo features). The minimum redundancy condition is

$$\min_{S\subset\Omega}\frac{1}{|S|^2}\sum_{i,j\in S}I(f_i,f_j)$$

where $|S|$ is the number of features in S and $I(f_i, f_j)$ is mutual information between f_i and f_j,

$$I(f_i,f_j)=\iint p(f_i,f_j)\log\frac{p(f_i,f_j)}{p(f_i)p(f_j)}df_i df_j\,.$$

We again use mutual information $I(c, f_i)$ between the target class $c \in \{c_1,\ldots, c_K\}$ and the feature f_i to quantify the relevance of f_i for the classification task. The maximum relevance condition is to maximize the total relevance of all genes in S:

$$\max_{s\subset\Omega}\frac{1}{|S|}\sum_{i\in S}I(c,f_i)$$

To obtain mRMR features, we optimize these two conditions simultaneously, in quotient form by

$$\max_{S\in\Omega}\left\{\sum_{i}I(c,f_i)\Big/[\frac{1}{|S|}\sum_{i,j\in S}I(f_i,f_j)]\right\} \tag{2}$$

The solution of Eq. (2) can be computed efficiently in O($|S|\cdot N$) time, with N being the total number of features in Ω.

2.4. *Classification and Rankscore Assignment*

Since the problem of annotating images is multi-objective, we solve it by decomposing it into multiple binary subproblems, for each of which we train a classifier to predict a particular annotation term based on the *one-vs-others* method. In another word, for a particular annotation term, the training samples are associated with labels 1 or 0 depending on whether this image sample has this annotation or not in the training data. In the testing phase, this trained classifier is responsible for predicting whether or not an input image should be associated with this image ontology annotation term. In this way, a series of parallel classifiers will determine if an image has any of the annotations in the training set.

The trained classifiers also calculate a probabilistic confidence score in order to give users a quantitative measure of predictions. Since classifiers for different annotations are trained independently, we are able to add annotation-specific classifiers based on user requirement and/or system complexity constraints, without altering previously trained classifiers.

One challenge of our problem is that an annotation term usually corresponds to only a small portion of images. For example, in the evaluation set of genes we collect, even the most common annotation terms are associated with only about 20% image samples. As a result, for the one-vs-others binary classifier, the sample numbers for class 1 and class 0 are very unbalanced, as most samples are labeled as class 0. Therefore, we need to choose a classifier that is able to predict more accurately for the small-sample-number class instead of being biased toward the large class. We use Linear Discriminant Analysis (LDA) to implement the binary annotation classifier. It is reported in literature that LDA may give better performance than the Quadratic Discriminant Analysis (QDA) when true covariance matrices are unknown and the sample sizes are small[19,20]. Our empirical comparison with the Support Vector Machine[21,22] shows LDA has a relatively balanced performance on both large and small classes. These observations help decide our choice of classifier.

Denote the posterior probability of an image $\boldsymbol{x}$ having an annotation as $p(y=1|\boldsymbol{x})$, it can be shown that[19], based on the assumption of multivariate normal density of LDA, we have

$$p(y=1\,|\,\mathbf{x})+p(y=0\,|\,\mathbf{x})=1$$

and

$$\log p(y=1\,|\,\mathbf{x})-\log p(y=0\,|\,\mathbf{x})=g_1(\mathbf{x})-g_0(\mathbf{x}),$$

where

$$g_1(x)=\log p(y=1)-\tfrac{1}{2}\mu_1^T\Sigma^{-1}\mu_1+x^T\Sigma^{-1}\mu_1$$

with μ_1 is the sample mean of class 1, Σ is the pooled within-class covariance matrix. $g_0(\boldsymbol{x})$ for class 0 can be calculated similarly. We can then compute $p(y=1|\boldsymbol{x})$ and $p(y=0|\boldsymbol{x})$.

For embryonic image annotation, the body part structure with gene expression is considered to be present if 1) $p(y=1|x) > p(y=0|x)$; and 2) $p(y=1|x) > TH$, where TH is a threshold set to 0.6 in our experiment. $p(y=1|x)$ is the confidence score of the annotation.

Because a gene may be associated to multiple annotations, we can aggregate all individual annotation confidence scores as an overall rank score,

$$rankscore = \frac{\sum_i^M p_i \times f(p_i - TH)}{\alpha + \sum_i^M f(p_i - TH)} \tag{3}$$

where M is the number of total possible annotations, p_i is the confidence score for a particular annotation term, $f(\cdot)$ is the hard-limiting function with $f(u)=1$ if $u > 0$ and otherwise $f(u)=0$; the parameter α on the denominator is mainly for avoiding division by zero when no annotation is found to be associated to the image. When α is 0, the *rankscore* is simply the average of the individual probabilities for various annotations. With the *rankscore,* users will have a global quantitative control of the confidence level of each annonation and the predicted annotation results can be ranked/sorted based on these scores.

The algorithms presented in Sec. 2.2 – 2.4 show that we use the wavelet decomposition scheme to project the original pixel-based embryonic images to a new feature domain in order to reveal features at different resolutions and frequency bands. We apply the mRMR feature selection algorithm to identify the optimal distinguishing features for a specific annotation. We then construct a series of parallel bi-class predictors to solve the multi-objective annotation problem, with each predictor working with one class label. The annotation system produces three pieces of information for each testing image: 1) The decision whether the specific annotation is considered "present" in this image; 2) The estimation confidence score of each annotation; 3) The *rankscore* of all annotations given to this image. These algorithms are summarized in Figure 1.

2.5. *Evaluating the Feature Extraction and Selection Methods*

Other options to extract features from digital images exist, such as using raw pixels or the Eigen-embryo approach that uses principal component analysis to extract image features. With the high dimensionality of embryonic images, transformed features are preferred over raw pixels have been considered due to its smaller computational load and information redundancy[23,24,25]. While Eigen-embryo features in conjunction with graph partition methods have led to very interesting results in clustering gene expression patterns in an unsupervised way, the method seems to be less appropriate for the purpose of automatic annotation where features with different levels of prominences need to be revealed to characterize various structures coexisting in the same image.

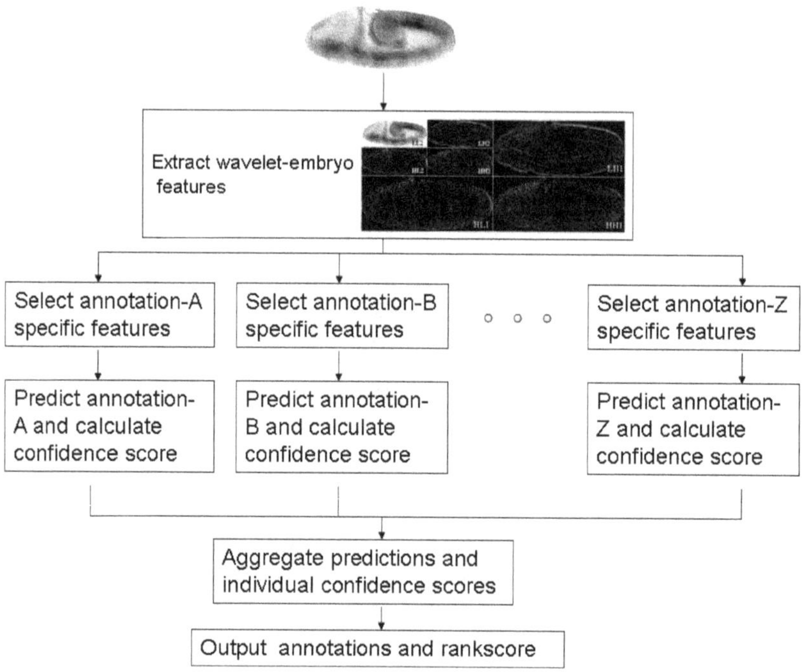

Figure 1. Algorithm flow for annotating *in situ* fruitfly gene expression during embryogenesis. The wavelet-embryo features shown in the flow are obtained by applying 2-Level 2D wavelet decomposition.

For comparison purpose (e.g. with Eigen-embryo features, which in conjunction with graph partition methods have led to interesting results in clustering gene expression patterns in an unsupervised way[23]), we built 10 testing sets using real images drawn from five stage ranges. For each stage-range, we built two data sets: in one set, each image has only one of the two possible annotations; these annotations will never correspond to one image sample at the same time. In another word, these two annotations are mutually exclusive. In the second data set, some images may correspond to two annotations at the same time, i.e., annotations overlap and the problem is multi-objective. We thus call the first set "M.E. set" and the second one "M.O. set". In addition, as we need a scheme working well with unbalanced sample sizes, the synthetic sets were built in the way that one annotation class has a lot more images than the other: the larger class usually doubles or triples the number of images of the smaller class.

Table 1 shows the result of comparing wavelet-embryo features against eigen-embryo features on the 10 sets. Input images have the size 50*100 pixels. For eigen-embryo features, we used the top 20 principal components based on the

Table 1. Recognition Rates (%) of Feature Extraction/Selection Methods on Synthetic Data Sets with *Mutually Exclusive (M.E.)* Annotations and *-Objective (M.O.)* Annotations. Annotation Short Terms: CB - cellular blastoderm; TMA - trunk mesoderm anlage; PEA - posterior endoderm anlage; TMP - trunk mesoderm primordium; PEP - procephalic ectoderm primordium; HPP - hindgut proper primordium; PMP - posterior midgut primordium; EBNS - embryonic central nervous system; VNC - ventral nerve cord.)

Stage Range	Set Size	M.E. Set Details	Score	Eigen-Embryo	Wavelet-Embryo	Wavelet Embryo + mRMR	M. O. Set Details	Set Size	Eigen-Embryo	Wavelet-Embryo	Wavelet Embryo + mRMR
4-6	98	78CB+ 20Subset	R_S	70	70	100	78CB+20Subset+ 54Both	152	49	49	76
			R_A	82	82	98			64	64	82
7-8	69	49TMA+ 20PEA	R_S	60	60	100	49TMA+20PEA+ 20Both	89	45	45	83
			R_A	67	67	100			40	40	85
9-10	69	49TMP+ 20PEP	R_S	65	65	100	49TMP+20PEP+ 18Both	87	42	39	79
			R_A	71	71	100			49	47	80
11-12	61	41HPP+ 20PMP	R_S	45	55	95	41HPP+20PMP+ 14Both	75	38	38	88
			R_A	59	64	98			33	37	89
13-16	32	20EBNS+ 12VNC	R_S	83	83	100	20EBNS+12VNC+ 65Both	97	57	57	87
			R_A	84	84	100			55	55	86

best performance tested in the range of 10-50 principal components. Without feature selection, the dimension of raw wavelet features is 5050. We also selected 20 top wavelet-embryo features using mRMR. Table 1 reports recognition rates using Leave-One-Out Cross-Validation (LOOCV) and the LDA classifier. Since the data sets are unbalanced, we report both the recognition rate on the small annotation class only (R_S in Table 1) and the overall recognition rate (R_A in Table 1). We observe from Table 1 that: 1) On both M.E. and M.O. sets, the combination of wavelet-embryo features and mRMR feature selection delivers a superior performance; 2) A multi-objective recognition problem is harder than the mutually exclusive situation; 3) R_A and R_S recognition rates are often very close to each other. This indicates that with the good features, the LDA classifier works effectively on unbalanced data sets.

In a summary, we find that the wavelet-embryo feature extraction and selection method for fly embryonic image features is suitable for the multi-objective problem case when features with different levels of prominences need to be revealed, in order to characterize various structures coexisting in the same image so that these structures can be effectively decomposed and recognized. Results of the system will be presented in Section III.

2.6. *Methods for Mutual Exclusive and ROI Annotation*

If an image or a region in an image corresponds to one and only one label, then the classification problem is mutual-exclusive. In such cases, the problem is a mutual-exclusive multi-class pattern recognition problem. We can use one multi-class predictor to do the automatic annotation. In this section, we discuss the algorithms for ROI annotation as a multi-exclusive pattern recognition problem.

Regions-Of-Interest (ROIs) in a biological image can be important landmarks. At different resolutions, they can be interesting tissue structures, particular types of cells (such as neurons, muscle or epithelia), or various subcellular patterns that exist at certain regions of an image. In such cases, the annotation can be done at the level of ROI instead of the entire image.

The automatic annotation of ROI can also be formulated as a pattern recognition problem. In ROI annotation, it is typical that one region would be associated with one and only one label, thus it is a mutual-exclusion annotation problem with trainable models that can predict a set of annotation targets. A trainable model needs to learn via training examples which are image regions associated with known labels. Typically these labels of training samples were manually annotated by an expert using a tagging tool. The goal of ROI annotation is that, with a reasonable number of manually annotated region images, the automatic annotation system can generalize the knowledge and annotate a large number of unseen images by finding out ROIs in those images and assigning labels to them.

The process starts with tagging. A tagging tool is typically provided as part of the annotation pipeline to assist the experts to label the regions in the training images. After the set of tagged training images are obtained, the annotation system can be applied which includes the training process and the annotation process. The training process extracts features of each tagged region of certain size (e.g. 100*100). Wavelet-based features described in Section 2.2 can be used. Shape based features can also be useful for some applications. A mutual-exclusive classifier (such as multi-class SVM or LDA) can then be trained based on these features.

The trained and automated annotation system scans through the images to be annotated. It moves a window of the size of ROI region on the image, extract features from the current window, and send the features to the trained classifier to get a decision on annotation target for current ROI region. The shifting increment of the window determines the number of decisions made during annotation.

Assume the size of the image to be annotated is $|x_1|$ and $|x_2|$ for width and height. The ROI is a rectangle with width w and height h. The window is shifted i pixels each time. The annotation system scans from top to down, then from left to right. Then the first region to be annotated in the image has (0, 0) as the left-top corner and (w, h) as the right-bottom corner; while the next region to be annotated is from (0, i) to (w, $h+i$). Total number of regions to be annotated is $((|x_1| - w)/i + 1)*((|x_2| - h)/i + 1)$. If i is set to 1, then the window shifts one pixel each time, and each pixel of the image is associated with an annotation tag.

While problem formulation is straightforward for a mutual-exclusive annotation problem, it is worth noting that some special algorithms may be necessary due to image size or quality etc.. For example, three or higher dimensional images are typically big. Many 3D images can be hundreds of megabytes and there can be hundreds of these images. For such multi-dimensional images, extracting a concise set of features is important if annotating system is to be run on a computer with typical memory configuration. For example, if wavelet transform is used as feature extractor for 3D images, one natural option is to conduct a full version 3D-DWT. However, considering the computational expense, an alternative is to treat the 3D image as a stack of 2D images and conduct 2D DWT on each image in the stack. The final set of features are obtained from a weighted sum among the stack. This approach is specially suitable for images obtained from 3D imaging techniques that have relatively low resolution on the z-direction. It generates the same number of coefficients as 2D DWT. Once the features are obtained from multi-dimensional images, consequent steps can be similar as 2D cases.

The annotation conclusions are typically subject to further post-processing. They can also be rendered on the original image for visualization and manual editing. Section III will give an example of ROI annotation of mosaic protein labeling image of rat brain.

3. Experiments and Applications

This section discusses results of automatic annotation in two biological applications.

3.1. *Annotation of Drosophila in situ Gene Expression during Embryogenesis*

We applied the multi-objective annotation system described in Sections 2.1 – 2.5 to a data set of mRNA gene expression pattern data set of 463 genes[24]. With a few exceptions, each gene has a representative expression pattern from each stage range over the course of fly embryogenesis (stages 1-3, 4-6, 7-8, 9-10, 11-12, and 13-16). So for any stage range, there are about 463 gene expression pattern images and our system automatically assigns annotations terms to these image. Top 20 mRMR-selected wavelet-embryo features were used in training each LDA classifier. As a quantitative assessment of the results, LOOCV is used to produce the prediction accuracy.

Table 2 presents the annotation results on stage range 7-8. Genes are sorted so that embryo images with the most confident annotations predictions are shown at the top. Entries with a probabilistic confidence score lower than 0.6 are marked with a dash "-". Due to space limit, Table 2 shows the top 10 ranked genes with their probabilities of having any of the most popular 11 annotations. Authors' websites have a complete list of annotation results. Table 2 shows that the majority of our top ranked automatic annotations are consistent with the expert's annotations stored in the BDGP database. Our system can successfully recognize patterns and annotations for some pretty poor images in term of the blur and deformation, such as the patterns of gene CG3123.

3.2. *ROI annotation of protein labeling map of rat brain*

This section describes a ROI annotation application that is formulated as a mutual-exclusive pattern recognition problem.

Table 2. Predicted annotations for multi-objective problem (Images are from BDGP of fruitfly developmental stages 7-8).

Gene Name	Pattern Image	Rank Score	Annotation Terms											Our Automatic Annotations	Manual Annotations from BDGP Experts
			VEA	TMA	POEA	PREA	HMA	HA	AEA	DEA	VNA	MA	FA		
CG4798		0.96	0.92	1	0.98	-	0.98	-	1	-	-	-	-	VEA TMA POEA HMA AEA	VEA TMA POEA HMA AEA
CG8960		0.96	-	0.98	0.97	-	-	0.98	0.98	-	-	-	1	TMA POEA HA AEA FA	TMA POEA HMA HA AEA FA
CG11988	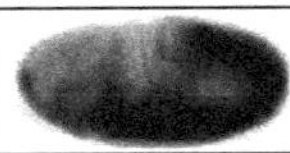	0.96	0.91	0.99	0.95	0.93	0.99	-	0.99	-	1	1	1	VEA TMA POEA PERA HMA AEA VNA MA FA	VEA TMA POEA PERA HMA AEA VNA MA FA
CG7413	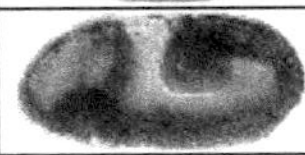	0.95	0.99	0.99	0.99	0.87	0.99	-	-	-	-	-	-	VEA TMA POEA PERA HMA	VEA TMA POEA PREA HMA AEA
CG5661	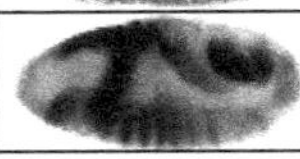	0.95	0.99	-	-	0.96	-	-	-	1	-	-	-	VEA PREA DEA	VEA DEA
CG10281		0.95	0.99	0.87	0.92	0.96	-	0.99	0.94	0.99	-	1	1	VEA TMA POEA PREA HA AEA DEA MA FA	VEA TMA POEA PERA HMA HA AEA DEA VNA MA FA
CG7354	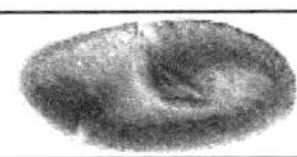	0.95	0.98	0.97	0.99	-	0.95	-	0.93	-	-	-	-	VEA TMA POEA HMA AEA	VEA TMA POEA HMA AEA
CG12094	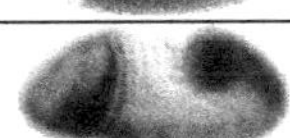	0.95	-	-	1	-	-	-	-	-	1	-	-	POEA VNA	POEA VNA
CG2941		0.94	0.99	0.93	0.97	0.75	1	-	0.97	-	1	1	-	VEA TMA POEA PREA HMA AEA	VEA TMA POEA PREA HMA AEA
CG3132		0.94	-	-	-	0.97	-	-	-	-	-	1	-	PREA MA	PREA MA

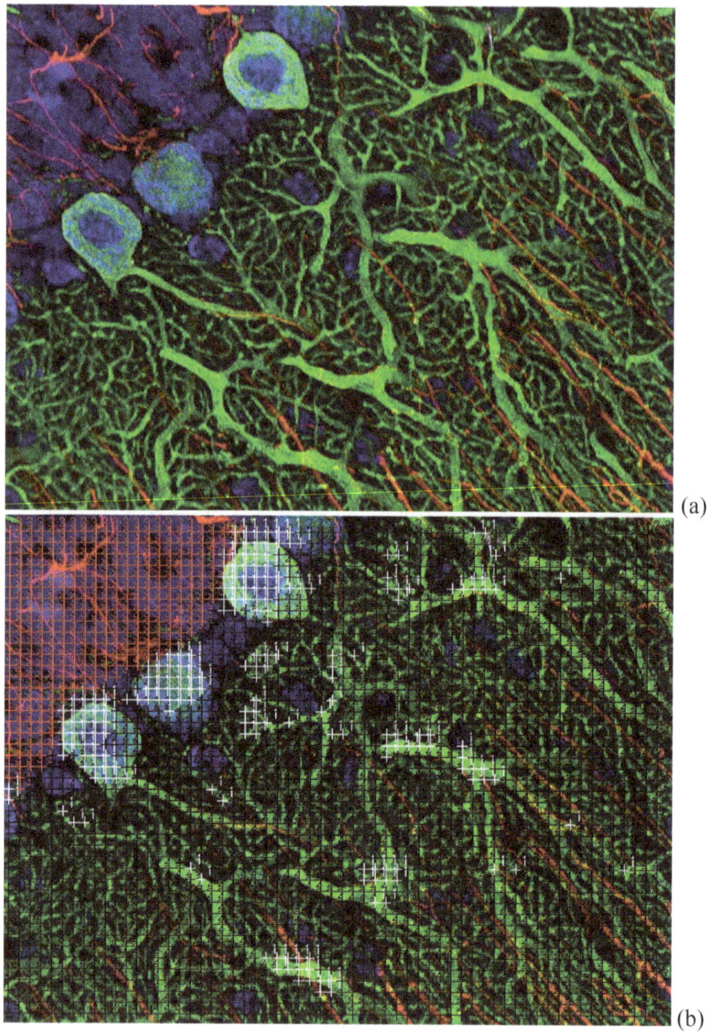

Figure 2. Protein labeling map of selected regions of a rat brain. (a) original image; (b) ROI annotation results overlaid on the image. Crosses of three types denote three annotation targets. The raw image is from CCDB[7].

Figure 2(a) is a protein labeling map in the cerebellum region of rat brain obtained using multiphoton microscope. The imagse is multi-channel with green for IP3 receptor expression, blue for Hoescht stain, and red from the expression of protein GFAP. We are interested in three annotating targets: neuron cell bodies, dendrites, and background. We use the green channel for our purpose.

The size of the image to be annotated in Fig. 2 is 703*535. and the size of ROI is 100*100. 37 ROI training images were generated from similar but different images with 23, 7, and 7 images for each of the three targets respectively. Wavelet features are extracted from ROI. Support vector classifier with linear kernel and default parameter is used.

Fig. 2(b) shows the result of ROI annotation overlaid on the original image. The shifting increment is 10 pixels. The annotate results are rendered using crosses of three colors, with white crosses represent class 1 (cell body), black crosses represent class 2 (dendrite) and red crosses represent class 3 (background). The crosses are the center locations of the ROI windows. A total of 3294 decisions were made to generate the rendering showed on Figure 2(b).

We can see from Fig. 2 that the proposed AOI annotation algorithm can reasonably detect and annotate the regions of cell bodies on the images. It can also accurately detect the background region. There are some false positives on cell body annotation, which we believe can be improved when a balanced set of more training images are available.

4. Discussions

4.1. *Preprocessing and Other Candidate Modules in an Annotation System*

The experiments demonstrated the promising results of automatic annotation of biological images and the effectiveness of our approach for annotation. However, it is worth noting that the success of annotation of biological images is related to other modules of the pipeline.

Before the annotation is conducted, we need to perform preprocessing. Image alignment, rotation and size normalization are typical examples of such preprocessing steps. When the annotation is based on an entire image, alignment may be an essential step. Preprocessing such as rotation, size normalization are important for feature extraction algorithms with low tolerance on variance. Embryo-wavelet features, for example, are based on such preprocessing steps.

Image segmentation, although not necessarily considered as just a preprocessing step, is another related issue for annotation. As we know, segmentation is the process of separating objects of interest on an image from each other and the background. It is a widely studied topic and many methods exist[26]. However, it is also known that the segmentation task can often be nontrivial depending on the application and image quality. For example, in the case of *in situ* gene expression patterns of fuitfly embryogenesis from BDGP, the quality of the image makes segmentation difficult and in many cases impossible since different body parts (such as different parts of guts or neural system) are connected by nature. This is an important reason why we formulated a multi-objective annotation task for the problem. For applications that a reliable segmentation is possible, then segmentation may be able to change a multi-objective image annotation to a mutual-exclusive ROI annotation which may lead to better annotation performance. On the other hand, the ROI annotation described in Sec. 2.6 can be used for more accurate segmentation: Many segmentation algorithms[26,27] need manual interaction to provide seed information before applying the algorithm. Automatically locating the numbers/positions of objects of interests can provide such information for the segmentation algorithm.

4.2. *Other Issues and Future Directions*

Automatic annotation of biological images is still relative young and efforts to solve the unique challenges are underway but far from mature. For example, the issue of annotating large number of multi-dimensional images and how to visualize these annotation results is still a largely unsolved problem. Conducting more experiments on larger datasets and multi-dimensional data is an important direction of our future effort.

Beside the technical difficulties, we also realize that the success of biological image annotation is a multi-disciplinary effort and many issues are being addressed by both computer scientist and biologists. While pattern recognition researchers are overcoming the cross-disciplinary learning curve to understand the biological problems accurately, biologists are also working on issues that can help the development of the field. As an example, it is a recent consensus by biologists that consistent definition of biological ontology, which is not yet always the case depending on the biological subfield, is essential for sharing the biological images, for collaboration among different laboratories, and for providing reliable validation set of automated annotation[10].

5. Conclusions

When the task of automatic annotation is formulated as a pattern recognition problem, many unique issues need to be addressed for annotating the newly available biological images obtained by various imaging techniques. We discussed algorithms and applications for several cases of automatic annotation of biological images with the focus on annotating fruitfly gene expression pattern images during embryogenesis. Applications are discussed for multi-objective vs. mutual-exclusive annotation, and ROI vs. entire image annotation. We showed promising results on these applications, demonstrated the effectiveness of our proposed algorithms, and pointed out future directions for improvement.

References

1. Pawley, JB (editor). (2006) Handbook of Biological Confocal Microscopy, 3rd ed., Berlin: Springer.
2. Konig, K. 1, (2000) Multiphoton microscopy in life sciences, Journal of Microscopy. 200(2):83-104.
3. Bozzola, J., and Russell, L.D., (1999) Electron Microscopy, 2nd ed., Jones & Bartlett Publishers.
4. Suzuki, E. (2002). High-resolution scanning electron microscopy of immunogold-labelled cells by the use of thin plasma coating of osmium, Journal of Microscopy, 208 (3): 153–157.
5. Tomancak, P., Berman, B. P. , Beaton, A, Weiszmann, R., Kwan E., Hartenstein V., Celniker, S. E. and Rubin, G. M., (2007) Global analysis of patterns of gene expression during Drosophlia embryogenesis, Genome Biology 2007, 8:R145 doi:10.1186/gb-2007-8-7-r145.
6. Lein, E., Hawrylycz, M., et al. (2007) Genome-wide atlas of gene expression in the adult mouse brain, Nature 445, 168-176.
7. Martone, M.E., Tran, J, Wong, W. W., Sargis, J., Fong, L., Larson, S., Lamont, S.P., Gupta, A., Ellisman, M.H. (2007) The Cell Centered Database project: An update on building community resources for managing and sharing 3D imaging data. J Struct Biol. 2007 Oct 16.
8. The Gene Expression Nervous System Atlas (GENSAT) Project, NINDS Contracts N01NS02331 & HHSN271200723701C to The Rockefeller University (New York, NY).
9. Peng, H. (2008) Bioimage Informatics: A new area of engineering biology. Bioinformatics. 24(17):1827-1836; doi:10.1093/bioinformatics/btn346
10. Larson, S., Fong, L., Gupta, A., Condit, C., Bug, W. and Martone, M. (2007) A formal ontology of subcellular neuroanatomy. Front. Neuroinform. 1:3. doi:10.3389/neuro.11/003.2007
11. Fong, L.L., Larson, S.D., Gupta, A., Condit, C., Bug, W.J., Chen, L., West, R., Lamont, S., Terada, M., Martone, M.E. (2007) "An Ontology-Driven Knowledge Environment for Subcellular Neuroanatomy", OWL: Experiences and Directions, Innsbruck, Austria, CEUR Workshop Proceedings, ISSN 1613-0073, http://CEUR-WS.org/Vol-258/, June 6-7, 2007
12. Zhou, J., and Peng, H., (2007) Automatic recognition and annotation of gene expression patterns of fly embryos. Bioinformatics, 23(5), 589-596.
13. Peng,H., Long,F., Zhou,J., Leung,G., Eisen,M., and Myers, E.W. (2007) Automatic image analysis for gene expression patterns of fly embryos. BMC Cell Biology, 8(Supp.1), S7.
14. Ji, S, Sun, L, Jin, R., Kumar, S., and Ye, J. (2008) Automated annotation of Drosophila gene expression patterns using a controlled vocabulary, Bioinformatics, 24(7): 1881-1888.
15. Peng, H., Long, F., and Ding, C., (2005) Feature selection based on mutual information: criteria of max-dependency, max-relevance, and min-redundancy. IEEE Trans. Pattern Analysis Machine Intelligence, 27(8), 1226-1238.
16. Mallat, S. G.(1999) A wavelet tour of signal processing. Academic Press.
17. Daubechies I. (1992) Ten Lectures on Wavelets, Science for Industrial and Applied Mathematics. Society for Industrial and Applied Mathematics
18. Resnikoff, H.L. and Wells,R.O. (1998) Wavelet Analysis, the Scalable Structure of Information, Springer New York.
19. Webb, A (2002) Statistical Pattern Recognition, 2nd Edition, John Wiley & Sons.
20. O'Neill, T.J. (1992) Error rates of non-Bayes classification rules and the robustness of Fisher's linear discriminant function. Biometrika, 79(1), 177-184

21. Vapkin, V (1995) The nature of statistical Learning theory. Berlin: Springer-Verlag.
22. Chang, C. and Lin C., (2001) LIBSVM: a library for support vector machines, Software available at http://www.csie.ntu.edu.tw/~cjlin/libsvm
23. Peng, H. and Myers,E.W. (2004) Comparing in situ mRNA expression patterns of Drosophila embryos. Proc. RECOMB 2004, pp.157-166.
24. Peng, H., Long,F., Eisen,M. and Myers,E. (2006) Clustering gene expression patterns of fly embryos. Proc. ISBI 2006, pp.1144-1147.
25. Pan, JY, Balan, A, Xing, E.P., Traina, A., and Faloutsos, C., (2006) Automatic Mining of Fruit Fly Embryo Images, Proc. ACM SIGKDD.
26. Linda G. Shapiro and George C. Stockman (2001). Computer Vision, pp 279-325, New Jersey, Prentice-Hall, ISBN 0-13-030796-3.
27. Mortensen, E. N.; Barrett, W. A. (2005) Intelligent scissors for image composition, Proceedings of the 22nd annual conference on Computer graphics and interactive techniques. New York, NY, USA: ACM Press, 1995. p. 191–198. ISBN 0-89791-701-4.

CHAPTER 5.3

DUAL ENERGY COMPUTED TOMOGRAPHY FOR CHECKED BAGGAGE SCREENING

Zhengrong Ying and Ram Naidu

Analogic Corporation
8 Centennial Drive, Peabody, MA, USA
E-mail: {zying, rnaidu}@analogic.com

Over the past decades, Computed Tomography (CT) theories, technologies, and applications have undergone rapid development and deployment. Although CT was developed as an imaging technique primarily for medical diagnosis, its applications in areas outside the medical field have gained increasing attention. In particular, CT combined with pattern recognition techniques has been used in Explosives Detection Systems (EDS) for Automated Threat Detection (ATD) in checked baggage screening. In this chapter, we describe dual energy x-ray CT technologies and associated ATD techniques used for checked baggage screening at airports.

Keywords: Computed tomography, dual energy, image reconstruction, automated threat detection, explosives detection systems, checked baggage screening.

1. Introduction

Checked baggage screening is one of many aviation security programs instituted to prevent terrorists from accessing civil aviation facilities [1]. In the aftermath of September 11th, 2001, the US government mandated the deployment of certified EDS to examine checked baggage for aircraft at US airports by the end of 2002 [2]. As a result, there were approximately 1600 CT-based EDS machines deployed in the US by the end of 2006 [3].

Figure 1 shows a block-diagram illustrating a checked baggage screening process at airports. When a passenger checks in a piece of baggage over an airline ticket counter, the baggage goes through an x-ray CT scanner. The CT scanner illuminates the baggage with an x-ray beam and generates projection data. The image reconstructor reconstructs the projection data into CT images of the scanned baggage and sends the CT images to the ATD for threat analysis. If the ATD clears the baggage, the baggage is routed to its designated airplane; otherwise, the CT images of the baggage are reviewed by a Transportation Security Officer (TSO) to decide whether the potential threats alarmed (detected) by the ATD are false alarms. If the TSO can't resolve any of the alarmed objects of the baggage, the baggage is sent to a search room for further examination. The further examination usually consists of either performing a hand search or conducting an Explosive Trace Detection (ETD) on the baggage. Extreme actions such as calling a bomb squad and evacuating a terminal may be needed if the hand search or the ETD result does not clear the potential threats alarmed by the ATD.

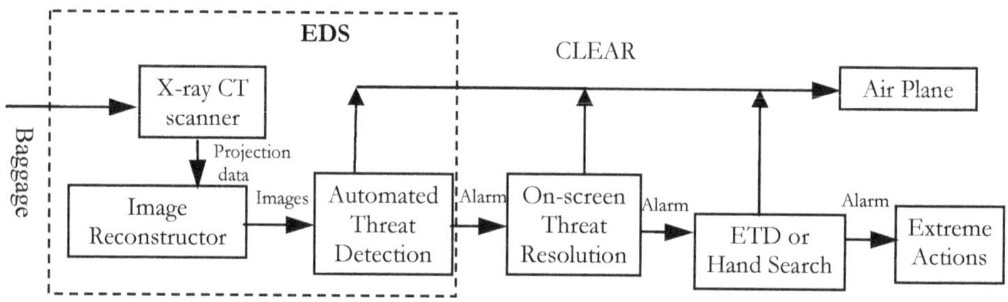

Figure 1: A block diagram illustrating checked baggage screening process.

As shown in Figure 1, an EDS machine is in the first stage of the screening process, and its performance is critical to the entire screening process. Therefore, an EDS machine must be certified to meet certain performance requirements before it can be deployed. The key performance requirements of an EDS machine are the probability of detection of all categories of explosives, false alarm rates, and throughput rates (number of pieces of baggage scanned per hour) [4]. So far CT is the only technology that has passed the EDS certification tests.

An EDS machine is comprised of an x-ray CT scanner, an image reconstructor, an ATD system, and a workstation for on-screen threat resolution. The EDS CT scanners can be divided into two types based on the scanning scheme: one is step-and-shoot scanning, which is called axial scanning in medical applications, and the other is helical scanning. The throughput of a step-and-shoot EDS machine is lower compared to a helical EDS machine. Examples of commercially deployed EDS machines include CTX 9000 by General Electric, Examiner 3DX 6000 by L-3 communications, and CT-80 by Reveal Imaging Technologies.

False alarm rates of EDS machines have an significant impact on operational costs in checked baggage screening. All alarms must be resolved and the resolution requires some sort of manual inspection, adding to labor and other system costs [5]. Dual energy CT has been demonstrated as an effective improvement over existing single energy CT in terms of the false alarm reduction [6, 7]. A single energy CT only provides the density measurements of scanned objects, while a dual energy CT provides the atomic number measurements in addition to the density measurements. The atomic number measurement is usually called "effective atomic number" [8], since it describes the atomic number of a hypothetical single element that gives the same x-ray attenuation as a compound or mixture being measured. Note that the two terms "atomic number" and "effective atomic number" are used interchangeably in this chapter.

The atomic number measurements provide an additional dimension to the density measurements for characterizing the physical properties of scanned materials. Figure 2 shows a scatter plot of density and atomic number of explosives and non-explosives scanned from a dual energy CT based EDS machine. The atomic number information from the dual energy CT scanner provides an additional dimension for discriminating explosives from non-explosives, i.e. additional capability of separating two objects with same densities but different atomic numbers. This enables dual energy CT based EDS machines with capabilities to reduce false alarm rates while maintaining or improving detection performance. Using the increased detection capability from

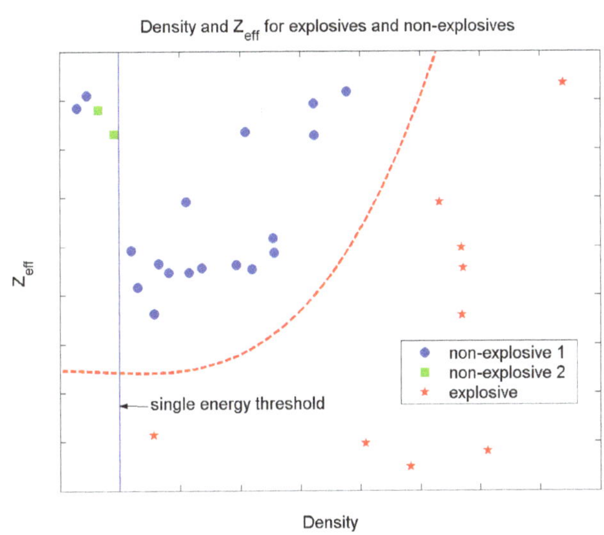

Figure 2: A plot of density and atomic number distribution of explosives and non-explosives. The blue solid line indicates a detection boundary based on density measurements using a single energy CT scanner; the dashed red curve indicates a detection boundary based on both density and atomic number measurements using a dual energy CT scanner.

dual energy CT scanners also allows EDS machines to address new and evolving threats such as home made explosives and liquid explosives [5].

2. Dual Energy CT

Interactions between x-rays with photon energy ranging from 30 keV to 200 keV and materials are dominated by Compton scatter and the photoelectric effect [9]. Compton scatter and the photoelectric effect are both dependent on material and x-ray photon energy; and each of them can be modeled as a product of a material dependent coefficient and an energy dependent term, as follows:

$$\mu(\vec{r},E) = \underbrace{a_c(\vec{r})f_{\mathrm{KN}}(E)}_{\text{Compton scatter}} + \underbrace{a_p(\vec{r})f_p(E)}_{\text{photoelectric effect}} \tag{1}$$

where $\vec{r}$ represents the position of the material under consideration, $\mu(\vec{r},E)$ is the total x-ray attenuation caused by the material at location $\vec{r}$, and E is the incident x-ray energy. The term $f_p(E) = E^{-3}$ approximates the energy dependence of the photoelectric interaction, and the term $f_{\mathrm{KN}}(E)$ is the Klein-Nishina cross section for Compton scatter given by:

$$f_{\mathrm{KN}}(E) = \frac{1+a}{a^2}\left[\frac{2(1+a)}{1+2a} - \frac{1}{a}\ln(1+2a)\right] + \frac{1}{2a}\ln(1+2a) - \frac{1+3a}{(1+2a)^2} \tag{2}$$

where $\alpha = E/510.975$ keV. Coefficients $a_c(\vec{r})$ and $a_p(\vec{r})$ are the material dependent Compton coefficient and the material dependent photoelectric coefficient, respectively. Both $a_c(\vec{r})$ and $a_p(\vec{r})$ are functions of the physical properties of scanned materials, the mass density ρ, and effective atomic number Z, as follows,

$$a_c \approx k_1 \frac{\rho}{A} Z, \qquad a_p \approx k_2 \frac{\rho}{A} Z^n \tag{3}$$

where k_1 is Avogadro's number, and k_2 is a constant including Avogadro's number, A is atomic weight, and $n \approx 4$.

The nonlinear interaction as shown in Eqs. (1) and (3) is a function of two unknowns, ρ and Z, thus their solution requires at least two measurements at known energies to solve. A dual energy CT scanner is designed to use two different incident x-ray spectra, a low energy spectrum $S_L(E)$ and a high energy spectrum $S_H(E)$, to acquire two sets of logarithmic projection data, low energy projection data P_L and high energy projection data P_H, as follows,

$$P_L = -\ln\left\{\int S_L(E)\exp\left[-A_p f_p(E) - A_c f_{\mathrm{KN}}(E)\right]dE\right\} + \ln\int S_L(E)dE \tag{4}$$

$$P_H = -\ln\left\{\int S_H(E)\exp\left[-A_p f_p(E) - A_c f_{\text{KN}}(E)\right]dE\right\} + \ln\int S_H(E)dE \qquad (5)$$

where $A_c = \int a_c(\vec{r})dl(\vec{r})$ and $A_p = \int a_p(\vec{r})dl(\vec{r})$ are line integrals of the Compton coefficient, $a_c(\vec{r})$, and the photoelectric coefficient, $a_p(\vec{r})$, with respect to an x-ray line path $l(\vec{r})$, respectively. The second term of the right hand side of Eqs. (4) and (5) represents the x-ray detector readings without presence of any object in the x-ray beam; while the first term represents the x-ray detector readings with presence of scanning objects; thus the difference of these two readings represents the x-ray attenuation absorbed by the scanning objects.

The dual energy projection data, P_L and P_H, can be obtained by three categories of data acquisition mechanisms: 1) generating two x-ray spectra by one or more x-ray sources; 2) filtering x-ray transmission through different materials or thickness of materials at x-ray detectors (sandwich detectors) [10]; and 3) using energy resolving detectors [11]. The generation of two x-ray spectra by x-ray sources includes using a single x-ray tube with high voltage power supply alternating between a low voltage, e.g., 90 kV, and a high voltage, e.g., 180 kV at two consecutive projection angles (also known as kVp switching) [12]; or using dual x-ray sources with each one dedicated to a fixed voltage, e.g. Dual Source CT (DSCT) [13]. The kVp switching mechanism is generally a better method than the sandwich detector mechanism for dual energy CT applications, particularly from SNR perspective [14]. Figure 3 shows an example of dual

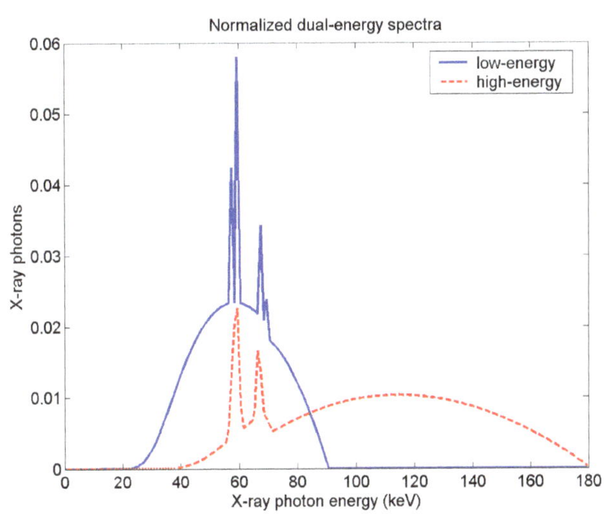

Figure 3: An example of dual energy x-ray spectra of an EDS scanner. Note that the spectra are normalized, i.e. $\sum_E S(E) = 1$.

energy x-ray spectra of an EDS scanner using kVp switching mechanism; the low energy is at 90 kV, and the high energy is at 180 kV.

Given the dual energy projection data, P_L and P_H, generated from a dual energy CT scanner, there are two classes of methods that can be used to decompose the projection data and reconstruct dual energy CT images: post-reconstruction decomposition and pre-reconstruction decomposition.

2.1. *Post-reconstruction Decomposition*

In this decomposition method, the low energy and high energy projection data, P_L and P_H, are first reconstructed into low energy CT images and high energy CT images, I_L and I_H; then dual energy decomposition is performed on a voxel by voxel basis as shown in Figure 4 to generate dual energy output images $\vec{I}_{out}$. The dual energy output images $\vec{I}_{out}$ usually include the low energy CT images I_L, high energy CT images I_H, and any linear combinations of the low and high energy CT images $wI_L + (1\text{-}w)I_H$, $0 \le w \le 1$. In order to increase the sensitivity of the dual energy data for diagnosis [15], the Dual Energy Index (DEI) image, which is the difference between the low energy and high energy CT images, is generated as one of the dual energy output images below,

$$DEI = \frac{I_L - I_H}{I_L + I_H + 2000HU} \tag{6}$$

The post-reconstruction decomposition method is easy to implement without the need to access the image reconstructor of a CT scanner, which is usually proprietary to the manufacturer of the CT scanner. The decomposition of the dual energy data in this method is somewhat heuristic (such as the generation of DEI using Eq. (6)) and does not conform to the physical model as described in Eqs. (4) and (5). Therefore, the post-reconstruction decomposition method does not fully extract and represent the information contained by a dual energy scanning, which limits its applicability for medical diagnosis and automated threat detection. The output images using this decomposition method also suffer from the same artifacts as the single energy CT images such as the beam hardening effect.

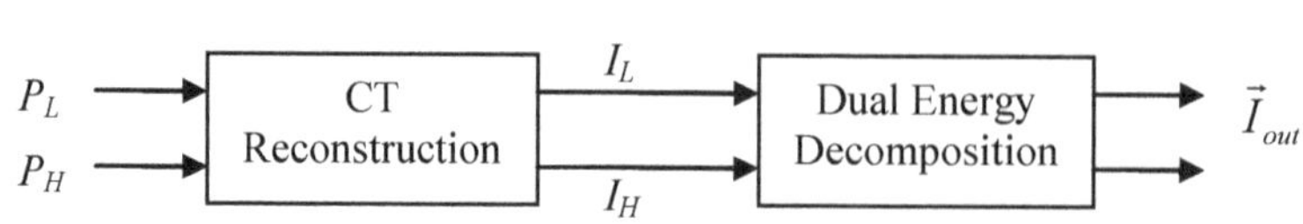

Figure 4: A block diagram illustrating the dual energy CT post-reconstruction decomposition method.

2.2. *Pre-reconstruction Decomposition*

In the pre-reconstruction decomposition method, the decomposition is first performed in the projection data as shown in Figure 5. The low energy and high energy projection data, P_L and P_H, are decomposed into Compton projection data A_c and photoelectric projection data A_p, or any linear combinations of A_c and A_p, determined by using different basis functions [16]. The decomposed dual energy projection data, A_c and A_p, are then reconstructed into Compton image I_{ac} and photoelectric image I_{ap}. In the last step, different types of output images such as density image, atomic number image, bone image, tissue image, etc. can be generated based on particular requirements of specific applications.

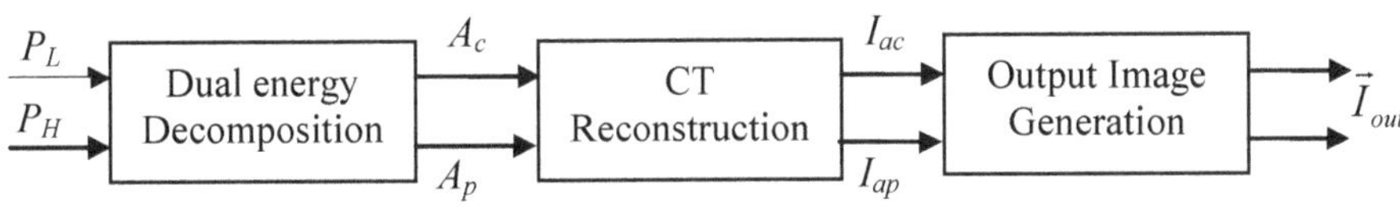

Figure 5: A block diagram illustrating the dual energy CT pre-reconstruction decomposition method.

The pre-reconstruction decomposition method is difficult to implement, because it requires the access to the image reconstructor of a CT scanner. However the decomposition of the dual energy data in this method conforms to the physical model as described in Eqs. (4) and (5). Therefore, the pre-reconstruction decomposition method offers improved accuracy than the post-reconstruction decomposition method. The dual energy reconstruction discussed below only refers to this pre-reconstruction decomposition method.

2.3. *Unique Issues in Dual Energy CT Reconstruction*

The additional information that can be obtained from a dual energy CT scanning is the atomic number of a scanned item. As indicated by Eq. (3), the atomic number information is more prominent in the photoelectric coefficient. However, the photoelectric effect is a small portion of the total x-ray attenuation of scanned materials at typical x-ray energies for CT. For example, the photoelectric effect only accounts for about 2% of the total x-ray attenuation for water at 80 keV. Therefore, the dual energy CT reconstruction, to some extent, can be stated as simply to reconstruct a weak signal (photoelectric coefficient) from large signals (total x-ray attenuation) that are corrupted by many sources of noise, errors, and inaccuracies; including,

- Inaccuracy in dual energy spectrum estimation;

- Spatial inconsistency of dual energy projection acquisition;
- Decomposition inaccuracy;
- Scatter;
- Image artifacts; and,
- X-ray spectral drifts.

The effects of the above listed issues on the performance of dual energy CT reconstruction are discussed below.

In order to solve for the Compton and photoelectric projection data, A_c and A_p, from Eqs. (4) and (5), the low energy and high energy x-ray spectra, $S_L(E)$ and $S_H(E)$, have to be estimated. Different computational methods based on empirical or semi-empirical models and sophisticated Monte Carlo calculations have been proposed in the literature, and all these estimation models produce differences between the estimated spectra and measured spectra that range from 2% to 20% [17].

The dual energy projection data, P_L and P_H, have to be obtained for the same line integral path (spatial consistency) to generate a valid decomposition. This is not a problem for a scanner that uses sandwich detector mechanism for dual energy data acquisition, because the low energy and high energy projection data are simultaneously obtained for each line integral path. Scanners using kVp switching mechanism require that the acquired dual energy projection data be interpolated to obtain the spatial consistency before decomposition. In DSCT, the two x-ray sources may follow different spiral trajectories, and interpolating the low energy and high energy projection data may not resolve this spatial inconsistency. A two-pass algorithm involving backprojecting the dual energy projection data into dual energy CT images and forward-projecting the dual energy CT images to obtain spatially consistent dual energy projection data for decomposition has been proposed [18].

Polynomial approximations of Eqs. (4) and (5) are widely used in obtaining the solution of A_c and A_p given P_L, P_H, $S_L(E)$, and $S_H(E)$ [9,18,19]. It has been shown in [6] that the polynomial approximation methods can yield over 200% approximation error in estimating the photoelectric effect. Special cares have to be taken to obtain non-negative solution of A_c and A_p according to their physical definition. Simplistic truncation of the negative values of A_c and A_p causes image artifacts.

Scatter is already a known issue for single energy CT scanners, and most, if not all, of single energy CT scanners use anti-scatter plates, each of which is vertically placed between two neighboring detectors, to block scattered x-ray photons. Dual energy CT is even more sensitive to scatter, because the nonlinear dual energy decomposition amplifies scatter effect: a small amount of scatter may cause large artifacts, such as a cupping effect in the photoelectric image.

The streak artifact is another prominent artifact in dual energy CT reconstruction. The streak artifact is caused by inconsistencies in the projection data. The inconsistencies could be the result of noise, scatter, and a large dynamic range of the photoelectric coefficient. For example, the photoelectric coefficient is about 3000 keV^3/cm for plastics and is 1,600,000 keV^3/cm for iron. Therefore, objects with a large photoelectric coefficient generate streak artifacts and affect the measurements of objects with a small photoelectric coefficient when located nearby during scanning. This is similar to the streaks generated by metals in single energy CT images [20].

X-ray spectral drift causes errors in the decomposition, and the x-ray spectra can drift over time. The drift can result from variations in thermally sensitive components of the high voltage power supply (HVPS), and the drift usually includes a short term drift on the order of hours and a long term drift on the order of days [6]. Such drifts can cause errors in the density and atomic number measurements. Additionally x-ray spectra can also vary from one scanner to another scanner because of variations in different beamline components. Therefore, a calibration and a corresponding correction are needed to compensate for the spectral variations across scanners so that each scanner produces density and atomic number measurements within designed tolerances.

2.4. *A Dual Energy Reconstruction System*

Figure 6 shows a block diagram of a dual energy reconstruction system of a dual energy CT-based EDS for checked baggage screening [6, 7]. The dual energy CT scanner is a third generation cone beam scanner with an x-ray source capable of producing a cone beam at two distinct energy levels. The data acquisition system (DAS) gathers signals generated by detector array at discrete angular positions of the rotating gantry, and passes the signals to the preprocessing unit. The preprocessing unit corrects the data from the DAS for detector temperature, gain, offset, intensity of the primary beam, and other deterministic error factors. The preprocessing unit extracts the data corresponding to high-energy views and the data corresponding to low-energy views.

An adaptive scatter correction algorithm is applied to the dual energy projection data to reduce scatter effect. The algorithm estimates the amount of scatter from projection data based on a low-pass filter with an adaptive spatial support, which accounts for the shape and density of objects that produce scatter.

The dual energy decomposition is performed by using a constrained decomposition method to generate photoelectric projection data. In this constrained decomposition method, the decomposition is posed as a constrained nonlinear optimization problem as follows,

$$(A_c, A_p) = \arg\min_{(A_c, A_p)} \underbrace{[P_L(A_c, A_p) - P_L]^2 + [P_H(A_c, A_p) - P_H]^2}_{Q(A_c, A_p)} \tag{7}$$

subject to the constraints: $A_c \geq 0$ and $A_p \geq 0$. In Eq. (7), P_L and P_H are measured low energy and high energy projection data, $P_L(A_c, A_p)$ and $P_H(A_c, A_p)$ are the modeled low energy and high energy projection data as described in Eqs. (4) and (5) respectively, and $Q(A_c, A_p)$ is the cost function for optimization. Note that (A_c, A_p) denotes a pair of Compton projection, A_c, and photoelectric projection, A_p, to be optimized. The optimization is solved using the Newton-Raphson method numerically. It has been experimentally shown that the optimal solution exists uniquely; however, the proof of the uniqueness remains an open problem.

A nonlinear destreaking algorithm is applied to the photoelectric projection data to remove outlier projection data. The destreaking algorithm consists of outlier detection and nonlinear filtering, which forces the detected outlier to become consistent. The nutating slice reconstruction (NSR) [21] is used to reconstruct CT images from high energy projection data and photoelectric images from photoelectric projection data. The reconstructed CT images and photoelectric images are post-processed on a voxel-by-voxel basis to generate effective atomic number images (Z images).

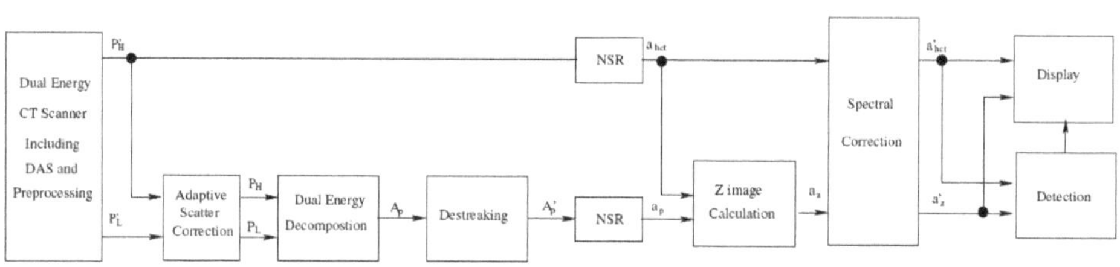

Figure 6: A block diagram of a dual energy reconstruction system of a dual energy CT based EDS.

A correction algorithm is applied to both CT images and Z images to compensate for the x-ray spectral drift along time and variations across scanners. In order to measure and track the x-ray spectral drift, a filter (mask) made of copper is mounted on a corner of the x-ray detector array. The ratios of the detector readings covered by the copper filter to the detector readings not covered by the copper filter are used to monitor the spectral drift. A linear model is used to relate the density and atomic number measurements to the copper ratios. Correction coefficients are obtained by performing an x-ray spectral calibration using materials of known densities and atomic numbers, and the linear model with these obtained coefficients is used with real-time implementation to compensate for the x-ray spectral drift.

Automated threat detection, which will be described in the next section, is then performed using both CT images and Z images to analyze images of scanned baggage. Together with the detection results, the CT images and the Z images are displayed to operators for on-screen threat resolution.

3. Automated Threat Detection

3.1. *CT Image Quality*

The certification performance criteria of an EDS machine require that an EDS machine meet a minimum probability of detection P_D and a maximum probability of false alarm P_{FA} [4]. Although the detection performance of an EDS machine (P_D and P_{FA}) is related to the design of an ATD algorithm and its associated parameters, the underlying detection capability of an EDS machine is intrinsically determined by the image quality of a CT scanner.

The image quality factors related to the threat detection performance include image resolution, noise, and image artifacts. The image resolution includes axial in-plane resolution, which is measured by MTF (Modulation Transfer Function), and inter-slice resolution, which is measured by SSP (Slice Sensitivity Profile). The image resolution is determined by the design parameters of a CT scanner, and the image reconstruction voxel sizes are usually chosen to be comparable with the image resolution of a CT scanner for the purpose of threat detection.

Image artifacts such as metal artifacts, scatter, streaking, photon starvation, and partial volume effect significantly affect the threat detection performance. In some sense, it is not exaggerated that the design of an ATD algorithm and the tuning of its parameters are essentially to compensate for variations of the measured material properties caused by different kinds of image artifacts of a CT scanner.

Most of the EDS CT scanners have image quality inferior to medical CT scanners. The reasons are two-folded. The first is technical: the contents in checked baggage have a much larger variability than human bodies, from clothing, kitchenware, computers, electronics, and anything that can be imagined. However, the design and the reconstruction algorithms of the EDS CT scanners are "borrowed" from the medical CT scanners, and limited efforts have been made to improve the image quality for the purpose of checked baggage screening. The second fold is non-technical: the certification performance criteria of an EDS machine were set not specifically targeting the CT-based EDS, but for all possible technological modalities in early 90's; though the CT is the only modality which has passed the certification tests so far. The manufacturers of the CT based EDS have demonstrated that the existing CT image quality is sufficient to pass the certification tests, and there is no monetary reward for a certified EDS machine with better detection performance and lower false alarm rate, resulting in limited research and development activities directed forward for improving the CT image quality.

3.2. *Overview of an ATD Algorithm*

Figure 7 shows a block diagram of an ATD algorithm of a dual energy CT-based EDS machine. The reconstructed CT image (density measurement) and Z image (atomic number measurement) of a piece of scanned baggage are sent to the ATD system. Because of the image artifacts, such as partial volume effect and streaks, two or more physical objects in close proximity may grow together into a single object during the segmentation. Morphological operations such as erosions are used to increase the separation of these close-by objects [22]. The CFAR (Constant False Alarm Rate) method is used to obtain a CFAR image with each voxel representing a CFAR score, which indicates statistical likelihood of a voxel being a thin object [23]. The CCL (Connected Component Labeling) method [24] is used to segment both the eroded CT image and the CFAR image into potential objects of interest [22, 23]. Features such as density, mass, and atomic number of each object are extracted, and then discrimination based on the extracted features is applied to each object to determine whether the object is a threat or not. When multiple physical objects are segmented together into a compound object, a compound object splitting algorithm [25] using a modified mean shift method [26] is employed to separate the compound object into individual objects.

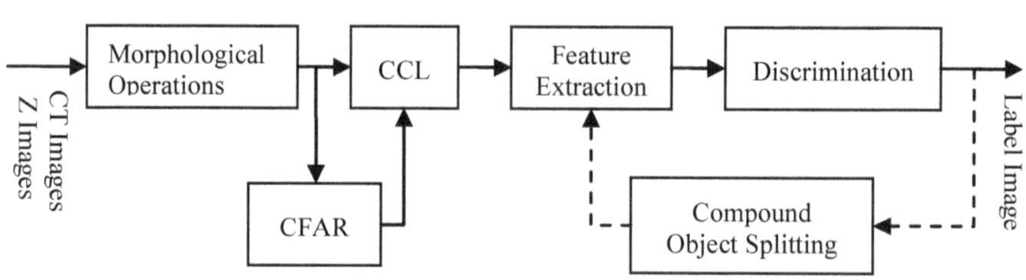

Figure 7: A block diagram illustrating an ATD algorithm of a dual energy CT-based EDS machine.

3.3. *Erosion*

The purpose of erosion operation is to remove or to break weak connections between two or more physical objects in close proximity due to the CT partial volume effect or because objects are physically touching each other. To solve this problem, erosion is employed to effectively remove surface layers of voxels from objects to increase the separation among objects before segmentation.

The erosion for processing CT images has been extended and generalized from the standard erosion operation, which is typically applied to a binary image with a mask. Let $C(i,j,k)$ denote a 3D CT image at voxel (i,j,k), where $i=0,\cdots,I-1$; $j=0,\cdots,J-1$; $k=0,\cdots,K-1$; let $C_e(i,j,k)$ denote the eroded 3D CT image, which is obtained with an erosion operation performed on the CT image $C(i,j,k)$ as follows,

$$C_e(i,j,k)=\begin{cases} C(i,j,k), & \sum_{i',j',k'=-1}^{1} R(C(i+i',j+j',k+k'),e_1,e_2)\geq E \\ 0, & \sum_{i',j',k'=-1}^{1} R(C(i+i',j+j',k+k'),e_1,e_2)< E \end{cases} \tag{8}$$

where $R(x, r_1, r_2)$ is a rectangular box function as follows,

$$R(x,r_1,r_2)=\begin{cases} 1, & r_1\leq x\leq r_2 \\ 0, & \text{otherwise} \end{cases} \tag{9}$$

Thresholds e_1 and e_2 specify the CT number range of interest for particular types of threats for detection, and E is a threshold, which is compared with the number of neighbors of the voxel (i,j,k) including itself that is within the CT number range of interest specified by $[e_1, e_2]$. Since Eq. (8) uses a three-by-three neighborhood, the maximum threshold E is 27. The larger E is, the more voxels are eroded. The erosion operation can be applied as many times as desired, the more times the erosion operation is applied, the more layers of surface voxels of an object are eroded.

3.4. *CFAR*

Because of the limited resolution of a CT scanner and its corresponding partial volume effect, thin objects are difficult to segment if processed in the same manner as thick objects. The CFAR method [23] is designed to obtain a statistical score, referred to as CFAR score, for each voxel indicating the likelihood of being a voxel of a thin object. Only voxels with CFAR scores of interest are then used in the CCL step to obtain segmentations of thin objects.

The CFAR method was originally devised to detect anomalous pixel regions by computing a local sample mean and standard deviation around each pixel [27]. An anomalous pixel is identified as statistical outlier with respect to the estimated mean and standard deviation of its neighboring pixels. Using a Gaussian distribution model, the threshold for determining an outlier pixel is adapted by the mean and standard deviation to maintain a constant probability of false alarm.

The CFAR score of voxel (i,j,k) of a CT image, denoted by $S(i,j,k)$, is computed as follows,

$$S(i,j,k)=\frac{C(i,j,k)-\mu(i,j,k)}{\sigma(i,j,k)+\varepsilon} \tag{10}$$

where $C(i,j,k)$ is the CT number of voxel (i,j,k), $\mu(i,j,k)$ and $\sigma(i,j,k)$ are the mean and standard deviation of a local region set of voxels, denoted by $\Theta(i,j,k)$, surrounding the voxel (i,j,k), and ε is a small regularization constant to prevent the division from blowing up when $\sigma(i,j,k)$ is small. A typical choice of the local region set $\Theta(i,j,k)$ is a set of surface voxels of a cube with a radius (half-side length) of τ centered by the voxel (i,j,k). The parameter radius τ is determined by the maximum thickness of a thin object of interest for detection as illustrated in Figure 8 and Figure 9.

Figure 8 shows that the CFAR method produces a higher CFAR score for voxel (i,j,k) when the voxel belongs to a thin object. In this case, the local region set $\Theta(i,j,k)$ has a mean CT number, which is different from the CT number of the voxel (i,j,k); the more uniform (the smaller standard deviation) the local region set is, the higher CFAR score is generated for the voxel (i,j,k). Figure 9

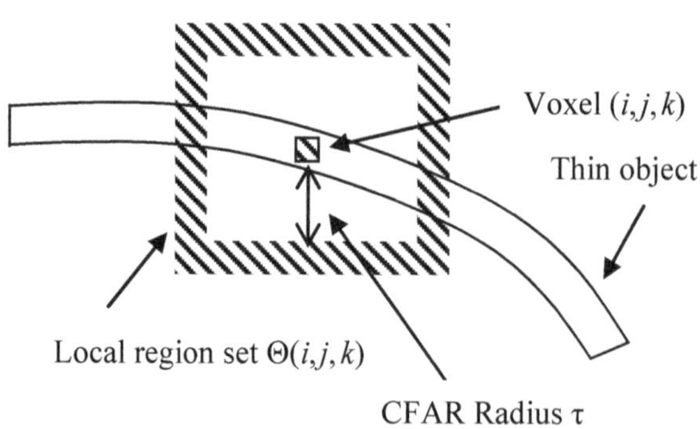

Figure 8: CFAR method in the presence of a thin object yields a high CFAR score for voxel (i,j,k).

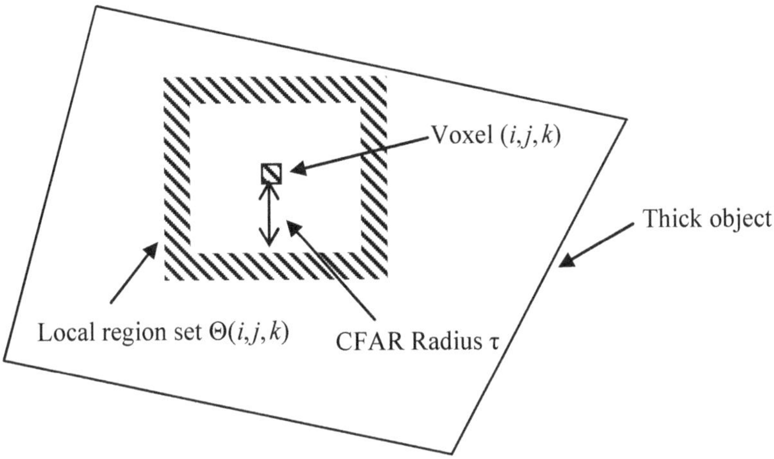

Figure 9: CFAR method in the presence of a thick object yields a low CFAR score for voxel (i,j,k).

shows that the CFAR method produces a lower CFAR score for voxel (i,j,k) when the voxel belongs to a thick object (thicker than 2τ). In this case, both the voxels in the local region set $\Theta(i,j,k)$ and the voxel (i,j,k) belong to the same thick object, thus the mean of $\Theta(i,j,k)$ is very close to the CT number of the voxel (i,j,k), resulting in a small CFAR score.

From Eq. (10), it is expected that isolated voxels and voxels from string-like objects generate highest CFAR scores; that voxels from thin objects, or surface of thick objects generate medium CFAR scores; and that voxels from thick objects have the lowest CFAR scores. Therefore for the detection of voxels of thin objects, a lower threshold is used to exclude voxels from thick objects, and an upper threshold is used to exclude isolated voxels or voxels from string-type of objects. The CT numbers of the voxels that have CFAR scores lower than the lower threshold or higher than the upper threshold are set to zero so that only the voxels from thin objects are grouped together and labeled as objects using the CCL algorithm described below.

3.5. *CCL*

Connected component labeling (CCL) was originally designed to label different components in a binary image [28]. A connected component of a binary image is a maximal set of pixels that are connected to each other. Two pixels are considered to be connected when a path between these two pixels can be found. A path between two pixels is a sequence of pixels starting with one of the two pixels and ending with the other of the two pixels, in which any pixel has a value of one and any consecutive two pixels are neighbors. The CCL algorithm has been generalized to obtain labeled components for different types of image representations [29].

When applying the CCL algorithm to the 3D CT images for the purpose of segmentation, connectivity, denoted by $\Gamma(\vec{r}_1,\vec{r}_2)$, between two neighboring voxels $(\vec{r}_1)$ and $(\vec{r}_2)$ with their corresponding CT numbers $C(\vec{r}_1)$ and $C(\vec{r}_2)$ is defined as follows,

$$\Gamma(\vec{r}_1,\vec{r}_2) = U(\delta - |C(\vec{r}_1) - C(\vec{r}_2)|) \tag{11}$$

where function $U(x)$ is a step function,

$$U(x) = \begin{cases} 1, & x \ge 0 \\ 0, & x < 0 \end{cases} \tag{12}$$

and δ is a pre-determined parameter based on the CT image properties of different types of threats of interest for detection. When $\Gamma(\vec{r}_1,\vec{r}_2)$ is one, the two neighboring voxels $(\vec{r}_1)$ and $(\vec{r}_2)$ are connected; when $\Gamma(\vec{r}_1,\vec{r}_2)$ is zero, the two neighboring voxels $(\vec{r}_1)$ and $(\vec{r}_2)$ are not connected. The bigger δ is, the more voxels are connected as a single label.

A typical implementation of the CCL algorithm requires two passes over a 3D CT image: the first pass is to assign temporary labels and record equivalences of these temporary labels; and the second pass is to replace each temporary label by the label of its equivalent label [30]. On the first pass:

1. Raster scan through each voxel of the CT image data by column, by row, then by slice;
2. If the voxel is not within the CT number of interest for detection, label the voxel with background label (e.g. 0); otherwise:
   a. Get the labeled neighboring voxels that are connected to the current voxel;
   b. If there are no labeled neighbors that are connected to the current voxel or simply no neighbors that are connected to the current voxel, uniquely label the current voxel (usually by incrementing the label number) and continue;
   c. Otherwise, find the connected neighbor with the smallest label and assign that smallest label to the current voxel;
   d. Store the equivalence among all the connected neighboring labels

On the second pass:

1. Raster scan through each voxel of the CT image data by column, by row, then by slice;

2. If the voxel is not the background label, relabel the voxel with the smallest equivalent label.

The output of the CCL algorithm is called a label image, in which all the voxels of a segmented object (a labeled component) are assigned a same unique positive integer number. For example, if there are three segmented objects in a CT image, the corresponding label image will have labels from one to three indicating the first, second, and the third segmented object; the voxels of the first object are all assigned with a label value of one in the label image, and so on; the voxels that do not belong to any segmented object are assigned a label value of zero (background).

3.6. *Feature Extraction and Discrimination*

Features of each segmented object are computed for the purpose of discrimination. Typical features include density, mass, mean atomic number, second or higher order statistics, and dimensions of a segmented object. To compute the dimensions of a segmented object, principal axes of an object are obtained first. The principal axes of an object are computed using Principal Component Analysis (PCA) [31] on all the voxels of a segmented object, thus the principal axes of an object align with longest, medium, and shortest directions of the object.

Developing discriminatory boundaries to detect threats from non-threats usually involves: i) obtaining threat data set and non-threat data set; ii) identifying labels of threat objects in the threat data set; iii) defining discriminatory boundaries using the labeled threat objects and non-threat objects. Non-threat data can be obtained by scanning checked baggage at airports, and threat data can be obtained by scanning live explosives at the Transportation Security Laboratory (Atlantic City, NJ) [32]. Discriminatory boundaries can be obtained by automatic training methods such as support vector machine, neural network, genetic algorithms, or manually.

3.7. *Compound Object Splitting*

A segmented object is usually a compound object if it is significantly large in dimensions. A compound object typically consists of several physical objects that are grown together due to the CT artifacts or simply because they are physically touching together. Features of a compound object are less specific than the features of individual objects, and reduce the discriminatory capability of an ATD algorithm. Although partially occluded object recognition method [33] can be used to improve the performance of an ATD algorithm, it is advantageous to split a compound object into different parts, each of which corresponds to a physical object, for discrimination.

In the compound object splitting algorithm, a 2D density and atomic number histogram of a compound object is generated. The 2D histogram is then segmented by using a variant of the mean shift method [26]. The mean shift algorithm is essentially a clustering algorithm, which convolves a Gaussian kernel with a discrete histogram and takes advantage of the continuous properties of the convolved histogram to search for the modes (maxima) of the 2D histogram. The segmented 2D histogram is then mapped back to the voxels of the compound object. A post-processing step is used to fine-tune the split results using the proximity relationship of the voxels so that isolated or disconnected voxels are removed or reconciled within split parts. The features of the split compound objects are then extracted and used to discriminate each individual component as described in Section 3.6.

3.8. *Examples of ATD Results*

Among all the deployed EDSs in the U.S., about a half of them acquire volumetric data and the other half acquire data in a step-and-shoot manner. In volumetric data acquisition, a whole bag is acquired and can be analyzed and visualized; while in the step-and-shoot data acquisition, only the portion of a bag that is suspected to contain threats is acquired. Therefore, volumetric data acquisition allows a volume rendering of a scanned bag, and provides a volume rendered image in addition of CT slice images to operators for examination. Figure 10 shows examples of the rendered CT images of two suitcases with the detection results of thin threat objects, which are colored in red.

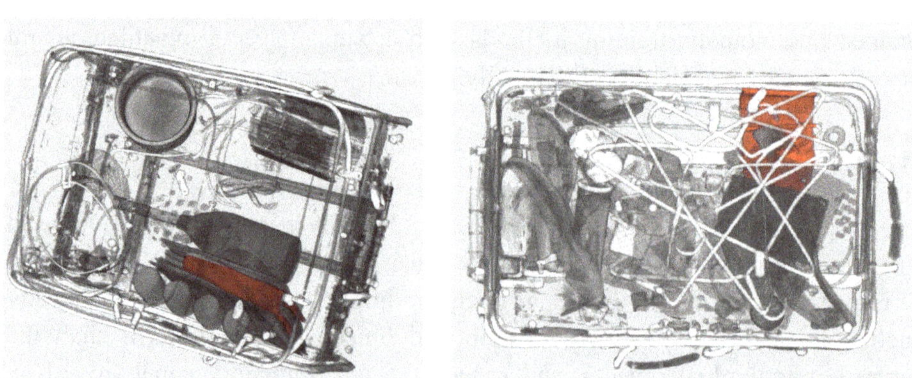

Figure 10: Examples of ATD results of thin threat objects.

4. Discussion and Conclusions

In this chapter, dual energy CT technology and associated ATD techniques have been discussed in the context of checked baggage screening. Dual energy CT technology will continue to advance and show advantages in the checked baggage screening and other security applications to detect new and evolving threats, such as home made explosives and liquid explosives.

Because of the large variety of contents in checked baggage screening applications, the CT design and the reconstruction technology will have to evolve along a new path that deviates from medical CT in order to improve the detection performance for next generations of CT-based EDS. Carbon nanotube based x-ray tube technology [34], which has the potential for miniaturization of x-ray tubes, might enable such new CT designs.

Among different mechanisms for implementing dual energy CT data acquisition, it seems clearer now that the kVp switching mechanism offers better SNR with reasonable spatial consistency of the dual energy projection data. However, the choice of x-ray spectra optimal to a particular application is still a topic in ongoing research.

Solid state detectors such as CdZnTe offer a potential realization of multi-energy CT technology [35]; however, it remains to be determined how much performance the multi-energy CT technology can be gained over the dual energy CT technology.

The real-time implementation requirement of the ATD algorithms limits the implementation of more complex ATD algorithms, particularly for segmentation, which is the key to the performance of an ATD algorithm; however, with fast development of the Graphics Processor Unit (GPU) technology for general purpose computation, it is foreseeable that next generations of EDS will take advantage of this readily available massive computing power to improve the ATD performance.

References

1. Singh, S. and Singh, M. (2003). Explosives detection systems (EDS) for aviation security, Signal Processing, 83, pp. 31–55.
2. US PUBLIC LAW 107-71, (2001). Aviation and Transportation Security Act.
3. U.S. Government Accountability Office (GAO), (2007). Aviation Security: DHS Has Made Progress in Securing the Commercial Aviation System, but Key Challenges Remain, GAO-08-139T.
4. Committee on Commercial Aviation Security, Commission on Engineering and Technical systems, and National Research Council, (1993). Detection of Explosives for Commercial Aviation Security. (National Academies Press).

5. Zarur, G. L., (2007). Next generation EDS CT systems, 23rd Annual National Test & Evaluation Conference, (Hilton Head Island, SC).
6. Ying, Z., Naidu, R., and Crawford, C. R., (2006). Dual energy computed tomography for explosive detection, Journal of X-Ray Science and Technology, 14, 4, pp. 235-256.
7. Ying, Z., et al., (2007). Dual energy volumetric x-ray tomographic sensor for luggage screening, IEEE Sensors Applications Symposium, (San Diego, California).
8. Weber, J. and van den Berge, D. J., (1969). The effective atomic number and the calculation of the composition of phantom materials, The British Journal of Radiology 42, 497, pp. 378–383.
9. Alvarez, R. E. and Macovski, A., (1976). Energy selective reconstructions in x-ray computed tomography, Phys Med Biol, 21, 5, pp. 733–744.
10. Ritchings, R. T. and Pullan, B. R., (1979). A technique for simultaneous dual energy scanning, J Comput Assist Tomogr, 3, 6, pp. 842–846.
11. Alvarez, R. E., Seibert, J. A., and Thompson, S. K., (2004). Comparison of dual energy detector system performance, Med Phys, 31, 3, pp. 556–565.
12. Gordon, B. M., et al., (1997). Dual Energy Power Supply, U.S. Patent 5,661,774.
13. Flohr, T. G., et al., (2006). First performance evaluation of a dual-source CT (DSCT) system, European Radiology, 16, 2, pp. 256-268.
14. Heismann, B. J., and Wirth, S., (2007). SNR performance comparison of dual-layer detector and dual-kVp spectral CT, IEEE Nuclear Science Symposium, 5, pp. 3820-3822.
15. Barreto, M., et al., (2008). Potential of dual-energy computed tomography to characterize atherosclerotic plaque: ex vivo assessment of human coronary arteries in comparison to histology, Journal of Cardiovascular Computed Tomography, 2, 4, pp. 234-242.
16. Lehmann, L. A., et al., (1981). Generalized image combinations in dual KVP digital radiography, Med Phys, 8, 5, pp. 859-867.
17. Ay, M. R., et al., (2005). Assessment of different computational models for generation of x-ray spectra in diagnostic radiology and mammography, Med Phys, 32, 6, pp. 1660- 1675.
18. Knaup, M., Stenner, P., and Kachelrieß, M., (2007). Rawdata-based dual energy CT (DECT) from inconsistent scans, IEEE Nuclear Science Symposium, 6, pp. 4457-4459.
19. Chuang, K. S., and Huang, H. K., (1987). A fast dual-energy computational method using isotransmission lines and table lookup, Med Phys, 14, 2, pp. 186-192.
20. De Man, B., et al., (1999). Metal streak artifacts in x-ray computed tomography: A simulation study, IEEE Trans Nucl Sci, 46, 3, pp. 691–696.
21. Larson, G. L., Ruth, C., and Crawford, C. R., (1998). Nutating slice CT image reconstruction apparatus and method, U.S. Patent, 5,802,134.
22. Simanovsky, S., et al., (2000). Apparatus and method for detecting sheet objects in computed tomography data, U.S. Patent 6,026,143.
23. Hiraoglu, M., et al., (2000). Apparatus and method for detecting sheet objects in computed tomography data, U.S. Patent 6,111,974.
24. Eberhard, J. W., and Hsiao, M. L., (1999). X-ray computed tomography (CT) system for detecting thin objects, U.S. Patent 5,905,806.
25. Simanovsky, S., Ying, Z., and Crawford, C. R., (2007). Method of and system for splitting compound objects in multi-energy computed tomography images, U.S. Patent 7,539,337.
26. Comaniciu, D., and Meer, P., (2002). Mean shift: A robust approach toward feature space analysis, IEEE trans. on Pattern Analysis and Machine Intelligence, 24, 5, pp. 603-619.
27. Frosgate, C. H., et al., (1997). Multiscale segmentation and anomaly enhancement of SAR imagery, IEEE Trans. on Imag. Proc., 6, 1, pp. 7-20.
28. Horn, B. K. P., (1986). Robot Vision. (MIT Press).
29. Dillencourt, M. B., Samet, H., and Tamminen, M., (1992). A general approach to connected-component labeling for arbitrary image representations, J. ACM, 39, 2, pp. 253-280.

30. Shapiro, L., and Stockman, G. (2001). Computer Vision. (Prentice Hall).
31. Crawford, C. R., et al., (2002). Apparatus and method for processing object data in computed tomography data using object projections, U.S. Patent 6,345,113.
32. Hallowell, S., (2008). Transportation security laboratory: overview, 2008 Homeland Security S&T Stakeholders Conference-West, (Los Angeles, CA).
33. Ying, Z. and Castañon, D., (2002). Partially occluded object recognition using statistical models, International Journal of Computer Vision, 49, 1, pp. 57-78.
34. Zhang, J., et al., (2006). A multi-beam x-ray imaging system based on carbon nanotube field emitters, Proc. Of SPIE, Medical Imaging, 6142.
35. Claesson, T., et al., (2001). A CT demonstrator based on a CZT solid state detector, IEEE Nuclear Science Symposium, 3, pp. 1644-1646.

CHAPTER 5.4

ASSESSMENT OF FOOD INTAKE AND PHYSICAL ACTIVITY: A COMPUTATIONAL APPROACH

Mingui Sun, John D. Fernstrom, Wenyan Jia, Ning Yao, Steven A. Hackworth, Xiaoyu Liu, Chengliu, Qiang Liu,Yuecheng Li, Madelyn H. Fernstrom, Robert J. Sclabassi

Departments of Neurosurgery, Psychiatry, Electrical & Computer Engineering, and Bioengineering, University of Pittsburgh
3520 Forbes Ave., Suite 202, Pittsburgh, PA 15213, USA
E-mail: drsun@pitt.edu

Obesity is one of the most rapidly evolving public health problems in many parts of the world. Currently, self-reporting is the main method for diet and physical activity assessment. The lack of objective measurement technology has significantly hampered obesity research and clinical treatment of patients. We present an image-based computational approach to monitor both dietary and physical activity continuously for extended periods. Our goal is to produce quantitative measurements of energy balance for overweight and obese individuals. A new wearable device has been designed to be worn below the neck. This device contains a miniature camera which records video from a viewpoint similar to that of the wearer and sensors to estimate food portion size and other variables. The image data, along with other sensor measurements, are compressed and stored in a multi-gigabyte flash memory. The recorded data are uploaded to a computer where advanced multimedia data processing algorithms, including food portion size and physical activity estimations, are utilized to compute calories and nutrients in foods and energy expenditure during physical activity. This multimedia approach helps people to manage their nutritional and caloric intake and maintain sufficient amounts of physical activity in their particular working and living environments, providing an effective tool for aiding in curbing the progression of the obesity epidemic.

1. Introduction

Obesity has become a serious public health problem in many parts of the world. According to the World Health Organization [1], over one billion people in the world are overweight. Among them, at least 300 million are obese. Obesity affects human health in many ways, causing high blood pressure, high cholesterol, birth defects, type 2 diabetes, etc. [2]. Obesity tends to run in families. Some individuals tend to gain weight more easily than others. Research in genetics has already identified genetic markers associated with metabolic disorders and obesity, including the adrenergic-receptor gene β_3 and a group of receptor genes involved in appetite regulation [3]. It has also been found that environmental factors, such as energy intake and expenditure, culture, and lifestyle play critical roles in obesity development [4]. While the genetic study of obesity can be performed in laboratory settings based on field-acquired blood or tissue samples, environmental studies must use "free-living" data acquired from individuals. The acquisition of such data poses a daunting task due to the insufficient methodologies suitable for assessment of the general population. Currently, self-reporting using questionnaires is the main method for evaluating energy intake and expenditure. However, numerous studies have demonstrated that the data obtained using questionnaires are biased and inaccurate [5]. Many people do not actively control the amounts of food consumed and attend the nutritional contents. Similarly, they tend to under-perform physical exercises that they have originally planned. Without accurate information from objective measurement of energy balance, it is difficult to develop effective strategies to help people choose healthy foods that best fit their dietary preferences and maintain sufficient amounts of physical activity in their particular working and living environments.

The recent advances in microelectronics, pattern recognition, and multimedia systems can provide powerful solutions to the objective dietary physical activity assessment problems. Through the support of the Genes, Environment, and Health Initiative (GEI), funded by National Institutes of Health (NIH) of the United States, our research team is developing a new wearable device that digitally acquires free-living data for days or weeks to constantly monitor energy balance of overweight and obese individuals. Specifically, our device records food intake and physical activity videos and other sensor data, stores the data within a massive flash memory on the device, and then uploads the data to a main computer for processing. Multimedia data processing algorithms are utilized to extract relevant information. This information is then computationally compiled into an electronic chronicle, and coded by dietitians and other health care professionals. The ultimate goal of our work is to develop devices to support various studies in exposure biology and human behavior.

In this chapter, we will first describe the structural and electronic designs of our wearable system. Next, we will address an important issue in food dimension measurements using special sensors and image processing models and algorithms. Then, we will present a method to convert the food volumetric measurement into gram weight using a fuzzy logic approach. Finally, we will present a single, but effective computational method to characterize walking and jogging, the most common forms of physical activities. Since this study involves several hardware and software components, we describe these components in different sections and present our experimental data after each method's description. We summarize our study at the end of this chapter.

2. System Design

The conceptual design of the wearable device is shown in Fig. 1. A pendant-like front unit is connected by two thin cables using a miniature connector (on the right side). The front unit consists of a miniature camera (top), a range sensor (bottom), and optionally other sensors that require exposure to the air or light. Most electronic circuits, including those for camera and sensor interfaces, data compression, and flash memory, are also placed within the front unit. The viewing angle of the camera is adjusted such that, at normal head positions, the central visual fields of the human eyes and the camera lens overlap. To stabilize the device during body motion, an adhesive or fabric-hooking pad in the back of the main unit (not shown) is used to affix the unit to the clothes. The back unit (the white block in the upper part of the figure) consists of a large-capacity rechargeable battery providing power for the front unit through a miniature connector. Since the back unit can firmly contact the lower neck due to the

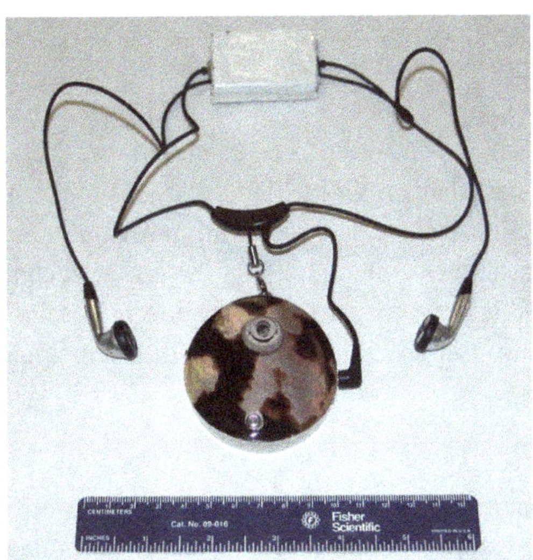

Figure 1. Wearable system design.

weight of the front unit, sensors that require skin contact can be placed within the back unit. Our design also includes two or more additional devices extending from the back unit. These devices can reach other locations of the body, such as the ears (to provide voice-based user interface using earphones), the scalp (to acquire the electroencephalogram (EEG) or electrooculogram (EOG) using electrodes), and the upper body (to monitor respiratory or cardiac function using accelerometers or electrocardiographic (ECG) electrodes). To obtain geolocational measurements, a GPS sensor may be placed in either the front or back units. Although not all the sensors included in the design are useful for dietary and physical activity assessment, we envision a platform design capable of simultaneous collection of multiple variables. This will greatly benefit future studies involving interactions of these variables, such as diet, physical activity, physiological conditions, and environmental exposure.

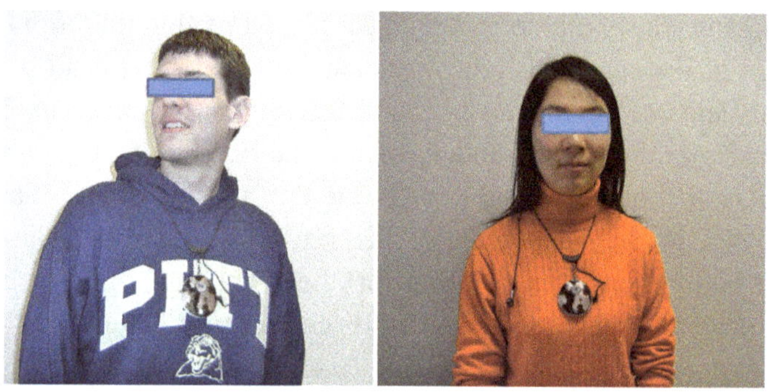

Figure 2. Our system aims to provide users the maximum convenience with a stylish wearable design.

An important issue is end-user acceptance of the device. We meet this challenge with a design that provides both style and maximum convenience to users. It also has a high-tech style resembling a wearable music player that is acceptable to both men and women (see Fig. 2). As the technology advances we will substantially reduce the size of the device and use different colors and designs to enhance end-user acceptability within different age groups.

3. Electronic Design

The primary components of our wearable sensor system are image acquisition, data compression, and storage modules. In the current version (see Fig. 3 (a)), the image data acquired by a CCD camera are digitized and stored within a field-programmable gate array (FPGA). The data are read by a DSP chip where image compression is performed, and then fed through a serial peripheral interface (SPI) to a controller-area network (CAN) controller which allows data storage in a 4GB NAND flash memory. Within the system, the image compression module requires the most computation. We implement this module using the TMS320C6416 DSP chip (Texas Instruments) which is one of the highest-performing fixed-point DSPs in the TMS320C6000 family. Since the frame rate in our system varies significantly, we utilized the JPEG [6], a frame-based image compression standard, for image compression.

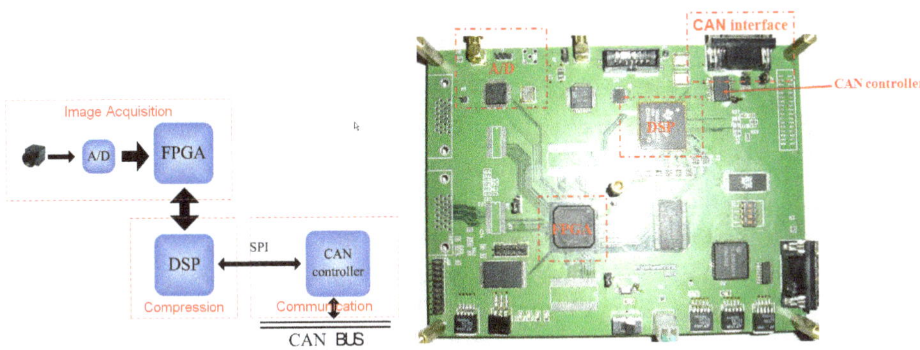

(a) (b)

Figure 3. Block diagram (a) and actual implementation (b) of image acquisition, compression and storage memory interface modules.

The first-version system modules (Fig. 3 (b)) have been successfully tested to verify their functionality. In the next design phase, we will consolidate system functions and use more integrated camera module and interface chips. Future designs will reduce the size and power requirements of the device.

4. Food Portion Size Measurement

The wearable camera can be used to obtain food images; however, the volume of the food in the image cannot be determined without a known reference. This is a critical issue in dietary assessment because, without knowing the food size, neither the nutrient nor the caloric content in the food can be estimated. We have investigated three methods to solve this problem: 1) including a reference or fiducial object (e.g., a checkerboard) in the picture; 2) emitting three laser beams from the front of the device to the food while taking images; and 3) installing a light emitting diode (LED) on the front surface of the device producing a spotlight pattern on the food image. The study of different methods is necessary because a fiducial object is often missing from pictures, and the construction of sensors is limited by hardware design. Currently, our measurements are not entirely deterministic and three-dimensional. However, with the knowledge of food shape (e.g., an ellipsoid, a cylinder, or a mound) in a container (e.g., a cup or a plate), food volume can be estimated in most practical cases. We describe these methods and compare their results as follows.

4.1. *Food Size Measurement Using Reference Object*

In this method, we use a common pin-hole camera model, shown in Fig. 4, to study the geometrical relationship between a camera and the objects in space [7]. A point P in space has a 3D coordinate of (x_w, y_w, z_w) in the world coordinate system. The same point has a coordinate of (x_c, y_c, z_c) in the camera coordinate system with the origin O_c fixed at the camera's optical center. The image plane is perpendicular to the z_c axis, which is the optical axis of the camera. Point P_I, which lies at the intersection between the image plane and line PO_c, provides the image of P. For object measurement, a camera calibration process is required which maps a point in the world coordinate system to the pixel index on the image plane. This procedure determines two sets of parameters related to the camera, the intrinsic and extrinsic parameters [8][9]. The intrinsic parameters include focal length of the camera, pitch size of the CCD chip within the camera, image center index, radial distortion parameters, and image skewness. The extrinsic parameters include rotation matrix and a translation vector of the image plane with respect to the world coordinate frame. For each camera location, the extrinsic parameters must be recalculated from the acquired image. The intrinsic parameters, on the other hand, are constant which means that they can be determined offline. With the pin-hole model and the sets of parameters, the mapping of a point in the world coordinate to its pixel index can be determined by a matrix expression. The details of the derivation are described in [10][12].

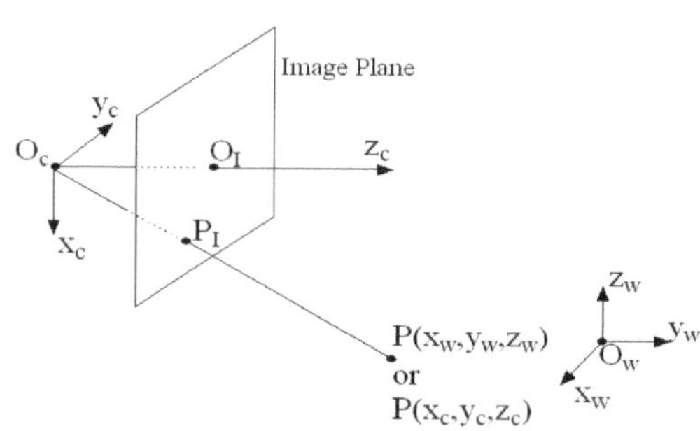

Figure 4. Camera geometry with perspective projection in the pin-hole camera model.

The estimation of intrinsic parameters requires multiple images. Fortunately, these parameters only need to be estimated once for each of the constructed devices during the calibration process. There is, however, one exception that focal length can vary if the camera has an auto-focus function, which is highly desirable in our application. In order to calculate the focal length, we take advantage of the observed patterns of a checkerboard and utilize the concept of vanishing-point as shown in Fig. 5. We calculate the converging point of the perspective projections of two parallel lines according to the detected corners on the checkerboard. The major assumption of this method is that a pair of vanishing points can be estimated from the food image. With this pair of points, the camera focal length can be calculated by a triangulation formula [10].

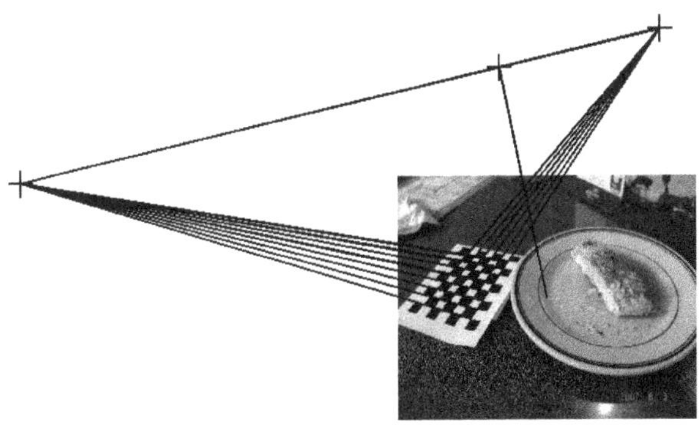

Figure 5. Two sets of lines were determined by fitting the detected corners of the checkerboard. The focal length of the camera is determined by triangulation.

We performed experiments involving three computational components: semi-automatic image segmentation, camera calibration (with corner extraction), and portion size estimation.

In order to segment food from the image, the outline of the food was roughly sketched, which initialized the segmentation process. The initial region was then refined to produce a precise segmentation automatically. We employed an active contour approach based on the level-set theory [11], which worked reasonably well (see Fig. 6). The calibration card was segmented with the same process. The corners of the checkerboard patterns were then extracted automatically using the Harris corner detector [12]. We then calculated the intrinsic parameters using five images taken from different perspectives [9]. As mentioned previously, the calculations of these intrinsic parameters are required only once. The distortion parameters were set to zero since the images were acquired at a reasonable distance from the objects.

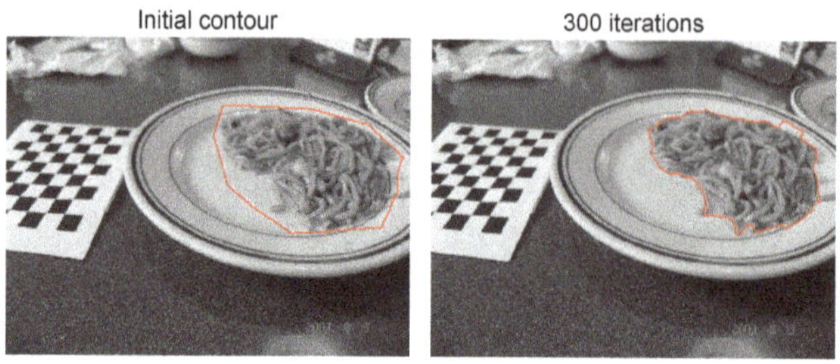

Figure 6. Left: Initial segmentation of food; Right: final segmentation result.

With the intrinsic parameters determined, the rotation matrix and translation distance were calculated for each test image [12] and the focal length was estimated using the vanishing point method described previously.

We estimated the volumes of a variety of real food items using randomly selected camera positions and orientations. After the images were obtained, we sketched an outline of each food. Next, automatic segmentation was utilized as described previously. The food volume was then calculated using the pin-hole camera model and the fiducial marker (checkerboard card in Fig. 6). More detailed descriptions of these processes are provided in [10].

The actual portion sizes, measured using a mechanical caliper tool, were used as the ground truth. We examined the diameters of plates, thicknesses of food portions, heights of drinks, and surface areas of food items. The mean error of the measured point-to-point lengths was 5.27% with a standard deviation of 3.44%. The mean error of the height measurement was 7.96% with a standard deviation of 3.14%. The mean error of the area estimation was 5.64% with a standard deviation of 4.13%. Comparing to the results obtained using the self-reporting 24 hour recall method which often uses descriptive terms (e.g., large, medium, and small), our objective results are more accurate.

4.2. *Food Size Measurement Using Laser Beams*

The previous method is highly accurate. However, it requires research participants to place a checkerboard card on the table. This procedure is not only inconvenient, but also undesirable from the research point of view because both the participants and his/her companions are reminded about the research study, which could modify the habitual behavior. We investigated an automatic measurement method using three laser beams (best to be invisible to human vision but visible to a camera) emitted from the wearable device. These beams, when appropriately directed, produce reference points in the recorded image.

Since the pattern and distances of the intersections between the emitted beams and a perpendicular target are both known (e.g., an equilateral triangle of a certain size), the distortion of this triangular pattern on the image provides information about the distance and orientation of the surface that contains the triangle. This concept is shown in Fig. 7 where the geometric model of the measurement method is provided. It can be observed that triangle ABC at the device surface is mapped to another triangle A'B'C' at the physical surface. Based on this model, we established mathematical relationships that uniquely determine the distance between any arbitrary pair of points in the recorded image and calculate their real-world distance (e.g., distance between D' and E' in Fig. 7).

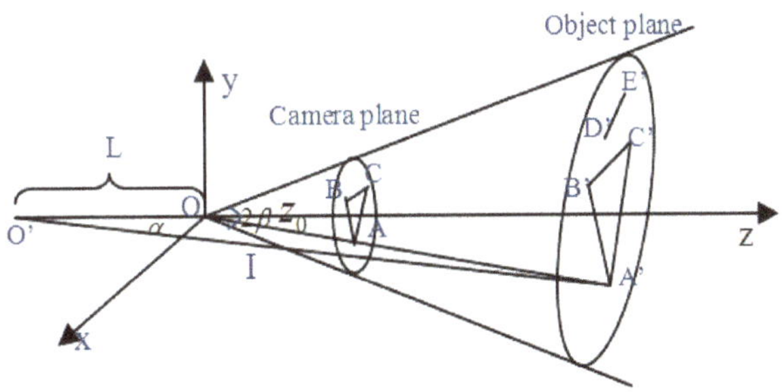

Figure 7. Measurement system model using three laser beams.

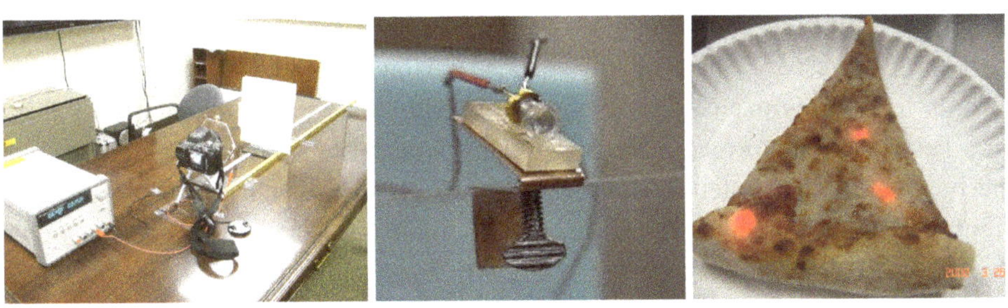

(a) (b) (c)

Figure 8. (a) Measurement system setup with a Plexiglass plate mounted with three laser beam emitters, a camera behind the Plexiglass, a target, and a rail to facilitate distance and rotation measurements; (b) Close-up view of one of the laser emitters; (c) Three laser fiducial markers on the plate and food object forming a measurement plane.

In our experiments, we estimated the depth of both artificial objects (planner patterns) and food objects on arbitrarily oriented flat plates. Fig. 8 shows our system setup, a self-made laser beam emitter, and an example of a food image with laser-generated fiducial markers. The performances of our system for four

Table 1. Measurement Results Using Laser Beams

	Bread	Cheese slice	Round steak	Noodle Mass
Number of trials	13	15	12	14
Measured length (mm)	107	76	80	130
Average estimated length (mm) :	109.3	81.4	81.4	135.2
Mean and S.D. of relative error (%)	2.1±4.7	7.1±5.4	1.8±1.0	4.0±5.0
Mean error (mm)	2.3	5.4	1.4	5.2

types of food objects are provided in Table 1. Our experimental results indicate an overall average error of less than 5%.

4.3. *Food Size Measurement Using an LED Spotlight*

The laser beam method has a relatively simple mathematical model, and its measurement results are satisfactory. However, its practical implementation is a technical challenge. Since the thickness of our wearable system is small, it is difficult to find commercial laser diodes and lenses that have sufficiently small depths. In addition, calibration of the laser beams and the maintenance of this calibration by research participants are also difficult. In an alternative approach, we investigated the use of a spotlight produced by an LED mounted on the front unit. This spotlight formed a single fiducial pattern on the object being imaged. The beam angle and intensity of the LED were both selected carefully so that a circular pattern of appropriate size and intensity can be observed on a perpendicularly oriented target at a known distance to the spotlight (Fig. 9).

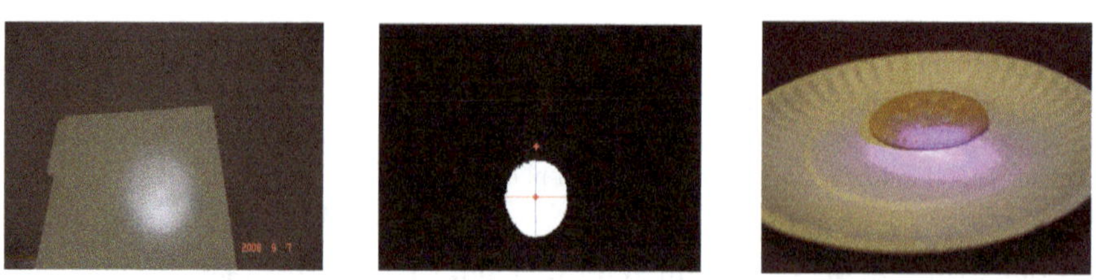

(a) (b) (c)

Figure 9. (a) Spotlight pattern on a planar target; (b) Segmented spotlight pattern; (c) Spotlight pattern on food object.

When the distance and the orientation of the object vary, the shape and size of the spotlight also vary. As in the previous case, these variations can be mathematically modeled and, based on this model, the orientation of the object and its physical dimensions can be measured.

The geometric model of the system is shown in Fig. 10 where O_c and O_l are the focuses of the camera lens and the location of the LED, respectively. The LED is permanently positioned besides the camera. Its optical axis is parallel to the optical axis of the camera at a fixed distance and the two parallel optical axes are in the same plane. The spotlight, which forms a conic region, is projected from O_l onto the object plane. The small and large ellipses, respectively, represent the light pattern and its image.

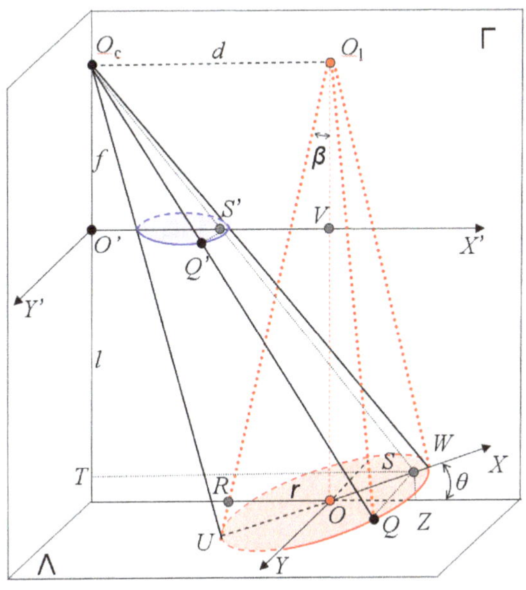

Figure 10. Measurement system model using a single spotlight.

Two experiments were performed to evaluate the LED based measurement system using a planar target. The first experiment was to estimate the distance while keeping the object plane perpendicular to the camera optical axis. In the second experiment, we estimated both the oblique angle and distance for the tilted target. The intrinsic parameters of the camera were calibrated before the experiments. In each image, the center position and deformation of the light spot were extracted after binarization using a threshold, as in Fig. 9(b). The experimental results are listed in Table 2 and Table 3. It can be observed that, as the plane is tiled from the orthogonal position, the error of estimation increases significantly. It is also technically challenging to segment and analyze the spotlight pattern distorted by the geometric shape of the food object (see Fig. 9(c)), which is under our investigation.

Table 2. Estimated Distance from Perpendicular Object Plane

Distance l (cm)	20	25	30	35	40	45
Estimated l (cm)	20.65	25.21	29.97	34.71	40.14	45.22

Table 3. Estimated Distance and Angle from Tilted Object Plane

True / Estimated	l=30cm			l=40cm		
	θ=30°	θ=40°	θ=50°	θ=30°	θ=40°	θ=50°
L (cm)	29.10	28.87	28.95	38.33	38.46	38.33
θ (°)	24.3	36.2	48.3	23.4	41.7	51.9

5. Fuzzy Logic Study on Food Density and Nutrient Analysis

Although digital food images provide objective data for accurate dietary assessment, without additional information, these images can only be used to identify foods and measure their volumes. A food database, such as the United States Department of Agriculture (USDA) Food and Nutrient Database for Dietary Studies (FNDDS) [13], is required to convert food volumes to values of calories and nutrients. However, the currently available databases were designed for different applications using digital food images. For many foods, the gram weight is the measurement unit based on which the nutritional information is listed in the database. Without knowing the food density value, the volumetric measurement obtained from digital images cannot be used to obtain calories and nutrients. Re-supplying the density information in the database implies re-performing food preparation and measurement procedures, which are both expensive and time consuming. In addition, many foods, especially those consumed by minority ethnic groups in the U.S. and by people in other parts of the world, are often missing in currently available databases.

5.1. *Fuzzy Inference of Food Density*

Since an average adult spends approximately 1.5 hours per day eating meals [14], people who cook by themselves have rich knowledge about the foods that they frequently consume, including their components, water contents, physical variation during cooking, etc. However, this type of knowledge is highly descriptive in nature, and is mostly in the form of linguistic terms rather than mathematical expressions. These descriptive terms are not exact, as opposed to numerical descriptors which can be accurate up to certain decimal places; they are also loosely defined (i.e., fuzzy), intuitive and summative, as opposed to the highly restrictive definitions of partial differential equations (PDEs) that engineers frequently use to describe a physical system. In fact, in nutrition and food science, fuzzy descriptors are already widely used, such as the terms "miniature", "small", "medium", "large", "a loaf", or "a roll" to describe a portion size. It has also been observed that an experienced restaurant chef can well predict the size of cooked food from components and ingredients that he or she uses. These observations indicate that, by developing an inexact knowledge system consisting of a suitable set of fuzzy descriptors and employing fuzzy logic theory [15], food density is predictable by using human experience with foods in addition to the existing empirical formulas and expressions in the food engineering field. Although in general this estimation could be rough and less accurate when compared to the traditional analytical and numerical methods, the fuzzy logic approach has distinct advantages of simplicity and robustness, and the accuracy of this approach is expected to be acceptable for our particular application.

Specifically, our fuzzy logic method involves the following steps: 1) developing sets of fuzzy variables, membership functions, and rules for certain food components; 2) developing fuzzy inference systems that map the fuzzy rules, membership functions, and logical operations into volumetric and density variables; 3) verifying the accuracy of fuzzy inference system outputs by using the known ground truth, such as experimental results.

5.2. *Experimental Results*

In the following, we demonstrate a specific example of using a fuzzy inference system to estimate the density of French fries. This density, which includes the effect of air spaces among fry pieces, provides the desired conversion between the gram weight and the volumetric measure from French fry images. Our system contains membership functions, fuzzy rules, and an output function of the density value. In our design, we assume the use of pre-processed, partially cooked frozen fries of uniform thickness. For this industry-supplied material, we found that the dominant variables that influence density are the frying time and temperature, although other variables, such as the time of oil draining after frying, also affect the output density. Fig. 11 shows the input membership functions utilized by our fuzzy system (panels (a) and (b)), and the output membership functions of estimated French fry density (panel (c)) in variation with the frying time and temperature. Note that, in each panel, multiple membership functions are plotted to save space. Both types of membership functions and the fuzzy rules were constructed empirically as a reflection of human experience. The centroid-based defuzzification method [15] was used to produce the crisp output values.

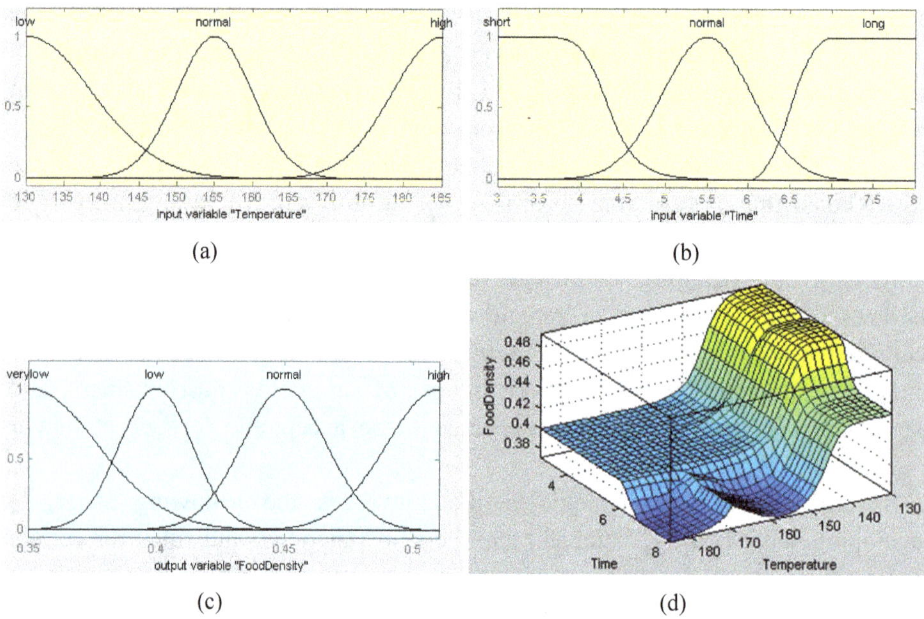

Figure 11. (a) Input membership functions of temperature; (b) Input membership functions of frying time; (c) Output membership function of French fry density; (d) Output function (crisp) of density vs. temperature and frying time obtained by our fuzzy inference system.

In order to evaluate the accuracy of the fuzzy system, we compared the computed and experimentally obtained densities of French fries as shown in Table 4. Our results of different temperatures (six variations) and frying times (two minutes (left side) and seven minutes (right side)) indicate an overall average error of $6.95\% \pm 3.75\,\%$. It can be observed that the average error of 4-minute frying time ($4.97\% \pm 2.29\%$) is much smaller than the error of seven-minute frying time ($8.93\% \pm 4.03\%$), and the errors in both cases are mostly single-sided. These indicate that, by improving system design and incorporating more variables into the system, the errors could be reduced further.

Table 4: Comparisons of Computed and Experimentally Obtained French Fry Density Values

Celsius Degree	Estimate Density (g/cm*3)	Real Density (g/cm*3)	Error (%)	Celsius Degree	Estimate Density (g/cm*3)	Real Density (g/cm*3)	Error (%)
130	0.494	0.506	2.4	130	0.451	0.443	1.8
140	0.487	0.472	3.1	140	0.449	0.395	12.0
150	0.411	0.394	4.1	150	0.388	0.351	9.5
160	0.401	0.374	6.7	160	0.375	0.347	7.5
170	0.401	0.381	5.0	170	0.395	0.343	13.2
180	0.401	0.367	8.5	180	0.374	0.338	9.6

5.3. *Future Perspectives*

Unlike conventional mathematical models which are analytical in nature, the fuzzy system is very intuitive, closely relying on human experience. Using a similar approach described previously which translates the human experience into desired numerical outputs, fuzzy models may be built individually for elementary food components. Then, higher levels of fuzzy inference systems may be constructed which allow combinations of food components according to recipes that utilize these components. In this way, we expect to estimate the densities and nutrients for a subclass of foods at acceptable accuracies.

6. Physical Activity Evaluation

It is well-known that, besides genetic factors, overweight and obesity are mainly caused by overeating and lack of physical activity. Like the dietary case, current assessment of physical activity is mainly based on detailed records of activity type and duration, such as aerobic exercise, strength exercise, and sports participation. Considerable errors exist due to self-reporting [16][17]. However, in contrast to the dietary case where few measurement tools are available, there exist a number of wearable sensors which measure body motion and dynamics [18]. Commercial systems are also available. For example, the SenseWear system (Bodymedia, Inc., Pittsburgh, PA) is convenient to use but has limited measurement functions [19]. Another system IDEEA (MiniSun LLC, Fresno, CA) consists of five sensors attached on the chest, two on the front thighs, and two on the feet [20]. The IDEEA system has better measurement functions, but it requires wiring harnesses at different body locations connecting to a central unit, discouraging its real-life application because of its inconvenience to users.

6.1. *Video-Based Approach to Physical Activity Evaluation*

In our approach, a camera is affixed on the body of a person. Since the camera is always looking outward, the person never appears in the recorded video. However, when there is a change in body position, a corresponding scene motion is recorded in the video. By analyzing the motion pattern and the scene itself, physical activities can be identified by direct observation and/or characterized by computation. When compared with the traditional approaches, our method does not require affixing multiple accelerometers at different body locations, yet it can still identify a variety of physical activities.

We utilize the wearable camera to evaluate walking and jogging which are among the most common forms of physical activity. Since the scene motion is our main interest rather than the details of the objects in the video, a high resolution is not required. However, a relatively high frame rate is desired in order to characterize fast motion. Therefore, we used a screen size of 320×240 pixels and a frame rate of 30 frames per second.

6.2. *Computational Methods*

Computational algorithms are utilized to process data after compressed video is uploaded to the computer. In order to reduce the amount of data to be processed, we project each video frame $f(x,y)$ to the x- and y-axes, i.e., by computing two vectors $f_x(y)$ and $f_y(x)$:

$$f_x(y) = \sum_x f(x,y) \text{ and } f_y(x) = \sum_y f(x,y)$$

For the walking and jogging case, the motions observed in the video are mostly in the x-direction. Therefore, we need only to analyze $f_y(x)$. A piled view of 10 seconds (300 frames) of jogging data is shown in Fig. 12 (a) where the features of steps can be observed. In order to extract these features computationally, we compute the cross-correlation between adjacent projection vectors:

$$R(s) = \frac{1}{N} \sum_{x=1}^{N} f_{y,t}(x) f_{y,t-1}(x-s)$$

where N is the dimension of the image in the x-direction, and we use an additional subscript t to denote the projection vector computed at time index t. It can be shown [21], under the assumption of white and uncorrelated noise, that the maximum likelihood estimate of the shift in adjacent projection vectors is given by

$$\hat{s} = \arg\max_{s}(R(s)).$$

Our computational result of $\hat{s}$ for all 300 frames is show in Fig. 12 (b) from which motion features can be extracted. For example, the gait speed v can be estimated easily by the maximum spectral value calculated from the discrete Fourier transform (DFT)

$$v^* = \arg\max_{v}(|DFT(\hat{\mathbf{s}})|^2)$$

where $\hat{\mathbf{s}}$ is a vector containing all shifts.

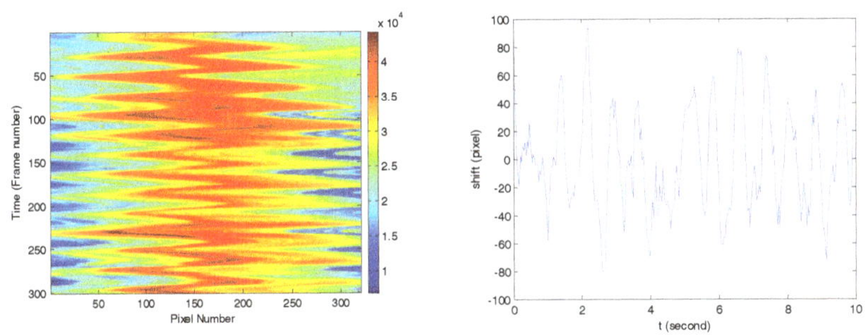

Figure 12. (a) Piled plot of 300 x-axis projections corresponding to 10 seconds of jogging. Color represents intensity values; (b) extracted motion features from projection data.

6.3. *Experimental Results*

We performed experiments on three types of physical activity videos, including outdoor jogging, outdoor walking on a hilly road, and indoor walking on a treadmill. Our algorithms described previously were utilized to obtain motion time series. Then, the spectra of the series were computed from which the gait speeds were estimated. Our results are shown in Fig. 13. The estimated speed values were approximately 2.9, 2.5, and 2.2 steps/second for the outdoor jogging, outdoor walking, and indoor walking cases, respectively. These results were close to the true speed values counted by human (3.0, 2.2 and 2.2 steps/sec, respectively). Once the speed values are estimated, the energy expenditure reflecting these physical activities can be calculated along with other inputs, such as the step size and the body weight.

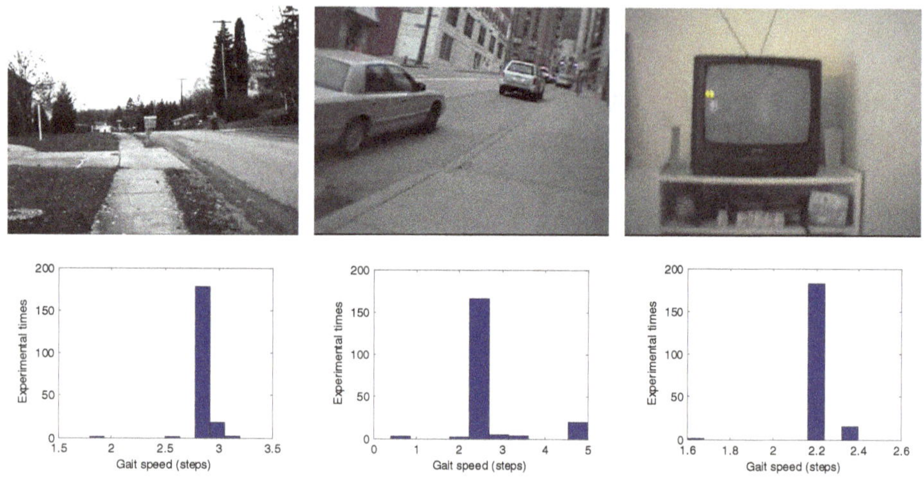

Figure 13. Top row: Snapshots of outdoor jogging, outdoor walking on a hilly road, and indoor walking on a treadmill. Bottom row: Results of gait speed estimation by DFT after extracting motion features.

6.4. *Application to Other Physical Activities*

Walking and jogging are relatively easy to analyze because of their essentially one-dimensional motion in the video and high level of periodicity. The projection method is applicable to many other physical activities; however, projections in both x- and y-directions may be necessary if the motion in the video involves the other dimension. In certain cases, projections in optimal directions, which are not necessarily horizontal or vertical, may be considered.

Complex physical activities, such as gardening and many types of sports, may require understanding of the context in the background video. In these cases, automatic analysis is challenging since the current state-of-the-art in computer vision and pattern recognition is limited. However, these activities are usually well recognizable by a human observer assisted by computer-selected key video frames. Therefore, at least for the present time, the construction of a human-centric system for data analysis with computer assistance seems to be the most practical solution.

7. Discussion

The explosive technological advances in microelectronics, information technology, and electrical and biomedical engineering over the past decades have provided unique opportunities to develop new electronic devices and methods for objective dietary and physical activity assessment. Our group has been translating these advances to assist the research community in nutrition and food science to improve methods of dietary assessment. Although many of these developments are still in initial stages, and most of them have not yet been complete, the results of these studies have been very encouraging. We believe that, as the technologies advance further, it is possible to conduct objective diet assessment and totally eliminate or greatly reduce the use of self-reporting questionnaires. We also believe that the electronic technology has much broader applications to the studies of exposure biology and human behavior, such as psychosocial stress and intake of addictive substances.

This paper has covered only several sub-domains of our efforts in new system developments. Several other important sub-domains are also being studied in our group, such as data scanning and user interface methods which enable quick search for relevant data, privacy protection algorithms for automatic filtration of raw data to eliminate human faces and sensitive scenes, and automatic food and physical activity identification using pattern recognition and computer vision techniques. These new developments will be integrated with the ones described herein to form a more sophisticated system.

Acknowledgement

This work was supported by National Institutes of Health grant U01 HL91736. The authors are thankful to Amy Subar, Catherine Loria, and Phyllis Stumbo for their valuable suggestions and proofreading of the manuscript.

References

1. World Health Organization, Obesity and Overweight, (2006), available at http://www.who.int/dietphysicalactivity/publications/facts/obesity/en/
2. A. Must, J. Spadano, E. H. Coakley, A. E. Field, G. Colditz, and W. H. Dietz, JAMA, 1523 (1999).
3. P. G. Kopelman, Nature, 635 (2000).
4. G. S. Barsh, I. S. Farooqi, and S. O'Rahilly, Nature, 644 (2000).
5. A. H. Goris, M. S. Westerterp-Plantenga, and K. R. Westerterp, Am. J. Cl. Nutr., 130 (2000).
6. JPEG Standard, JPEG ISO/IEC 10918-1 ITU-T Recommendation T.81 (1992).
7. R. Hartley and A. Zisserman, Multiple view geometry in computer vision (Cambridge University Press, 2nd ed., 2003).
8. J. Heikkilä and O. Silven, Proc. CVPR, 1106 (1997).
9. Z. Zhang, Proc. ICCV, 666 (1999).

10. M. Sun, Q. Liu, K. Schmidt, J. Yang, N. Yao, J.D. Fernstrom, M. H. Fernstrom, J. P. DeLany, and R. J. Sclabassi, IEEE Proc. EMBC, 871(2008).
11. C. Li, C. Xu, C. Gui, and M. D. Fox, Proc. CVPR, 430 (2005).
12. California Institute of Technology, Camera Calibration Toolbox for Matlab, available at http://www.vision.caltech.edu/bouguetj/calib_doc/.
13. USDA Food and Nutrient Database for Dietary Studies, Beltsville Human Nutrition Research Center, Agricultural Research Service, USDA (2004).
14. Bureau of Labor Statistics (BLS), U.S. Department of Labor, American time use survey -2005, USDL 06-1276.
15. E. Cox, The fuzzy systems handbook: a practitioner's guide to building, using, maintaining fuzzy systems (AP Professional, Boston, 1994).
16. K. F. Janz, Br.J. Sports Med., 191 (2006).
17. N. J. Wareham and K. I. Rennie, Int. J. Obes, S30 (1998).
18. Y. Schutz, R. L. Weinsier, and G. R. Hunter, Obesity Research, 368 (2001).
19. Bodymedia Inc., http://www.bodymedia.com/ .
20. K. Zhang, P. Werner, M. Sun, F. X. Pi-Sunyer, and C. N. Boozer, Obesity Research, 33 (2003).
21. N. Yao, R. J. Sclabassi, Q. Liu, J. D. Fernstrom, M. H. Fernstrom, J. Yang, and M. Sun, Proc. IEEE ISCAS, 2582 (2008).

CHAPTER 5.5

INTELLIGENT WATERMARKING SYSTEMS: A SURVEY

Eduardo Vellasques, Eric Granger and Robert Sabourin

Laboratoire d'Imagerie, de Vision et d'Intelligence Artificielle
École de Technologie Supérieure,
1100 rue Notre-Dame Ouest, Montreal, Canada H3C 1K3,
evellasques@livia.etsmtl.ca, {eric.granger, robert.sabourin}@etsmtl.ca

Digital watermarking techniques allow for the insertion of an imperceptible multi-bit payload into digitized multimedia data, such as video, image and audio. In contrast to other information hiding techniques, the payload carries information about the cover work, and apply to enforcement of authenticity and of integrity, media enhancement, etc. One of the key challenges to the practical application of digital watermarking, regards adapting these techniques to variations across different types of images. Early research in the area focused on exploring properties of each specific watermarking technique to improve performance. However, recent trends in research indicate that computational intelligence can be very beneficial for designing adaptive or intelligent watermarking techniques. Some evolutionary computation techniques have been developed to automatically set user-defined parameters of watermarking systems. In addition, machine learning techniques have recently been used to learn processes of these systems. These techniques affect a trade-off between robustness (against attacks), image quality, and system resources. In this chapter these techniques are reviewed and a case study is presented as a demonstration of their effectiveness.

1. Introduction

Managing digital versions of documents like bank cheques, invoices and printed forms has a significant role in modern economy. BancTec Inc[a] claims that its customers process 50 million documents a day, in 50 countries across the globe. The most common process involves transforming a continuous physical document into digitized image using an acquisition equipment (like a scanner), so they can be latter processed accordingly. These images are known as document images. Each specific application poses different requirements on the quality of these images. Some applications require high-definition, color images (e.g. over 16 million colors). In others, when storage and computational resources are limited, grey-level (e.g. 256 tones of grey) images are adequate. In many applications, black-and-white (bi-tonal) images

[a]http://www.banctec.com/

are adequate, which allow saving even more storing space and computational effort. This type of image is known as bi-tonal image. Bi-tonal images account for a significant share in the document management industry. According to BancTec, 95% of the 50 million document images processed by its customers daily are bi-tonal.

Enforcing the (1) integrity (has tampering occurred) and (2) authenticity (who is the author) of document images is considered a strategic issue by the financial industry, policy-makers and high-tech industry. A technique named digital watermarking allows enforcing these aspects, it comprises the covert embedding of information in an image through modifications on its pixel values.

The applicability of digital watermarking for the enforcement of aspects (1) and (2) has been shown in the literature.[1–3] The most common approach is to use a watermark embedder to add side information in a subtle way so it can be latter read with the use of a detector. The integrity is usually achieved through robust watermarking – a watermark that can be still detected after the image has been modified (assuming that the modification has not affected the commercial value of the image) – while the authenticity is achieved through fragile watermarking – a watermark that is destroyed in the case of tampering.

In a digital watermarking system, substantial efforts are required to adjust system parameters to obtain an optimum trade-off between the robustness against attacks and the noise introduced by the watermarking process. Usually, an increase in robustness leads to an increase in the noise rate. Optimizing such parameters is not a trivial task. The most common strategy is to perform this optimization on each image with the use of evolutionary techniques such as Genetic Algorithms (GA)[4] and Particle Swarm Optimization (PSO).[5]

Another issue with digital watermarking refers to making the processes involved in the embedding and detection of a watermark more adaptable to variations across different images. This leads to another strategy which involves the application of suppervised learning in order to model these processes through regression.

Although many of the general concepts apply to different media like audio, video or images, there are many issues specific to each type of media. In this chapter, the state of the art in the use of computational intelligence in the watermarking of images is reviewed.

A method to optimize a system for bi-tonal image watermarking with the use of evolutionary computing is proposed as a case study. An adaptive watermarking system based on PSO is proposed for tuning the parameters used for watermarking of bi-tonal images. This baseline system embeds two watermarks – a robust one to enforce the integrity and a fragile one to enforce the authenticity. Bi-tonal images have some particularities. Since its pixels can only assume two values – black or white – the embedding must be carefully performed in order to preserve the imperceptibility of the watermark. This poses some constraints to the embedding capacity, since it is usually based on a trade-off between perceptibility and robustness against noise.

This chapter is divided into five sections. In Section 2 the main techniques and challenges of digital watermarking are presented. Section 3 covers the main aspects regarding the use of computational intelligence to devise adaptive digital watermarking systems. In Section 4, these concepts are illustrated in the optimization of a bi-tonal watermarking system[6] with the use of PSO.[5] Finally, Section 5 concludes this chapter.

2. Digital watermarking

A watermark is an imperceptible (or minimally perceptible) mark, embedded into an image through modifications on pixel intensity values. There are numerous applications for digital watermarking such as broadcast monitoring, owner identification, proof of ownership, transaction tracking, authentication, copy control and device control.[1] It has two main objectives. The first is to ensure authenticity, and for this reason it must be robust to attempts of reproducing, removing or replacing. The second is to ensure integrity – any change to watermarked image should create modifications also in the watermark so tampering could be latter detected.

There are two alternatives for adding a watermark to a digital image. The first is through the use of visible (but translucid) marks. Since visible watermarks affect the commercial value of an image, this option will not be considered in the scope of this research. The alternative, consists of adding side information, in an imperceptible manner, usually with some a perceptual model. The imperceptible mark can be added either with or without the partition of the host image into blocks, to allow the embedding of more than one bit. Regarding the domain, the encoding can be performed either by directly changing pixel values (spatial domain) or by mapping the image to a different domain (e.g. wavelet) and then changing the coefficients of this domain.

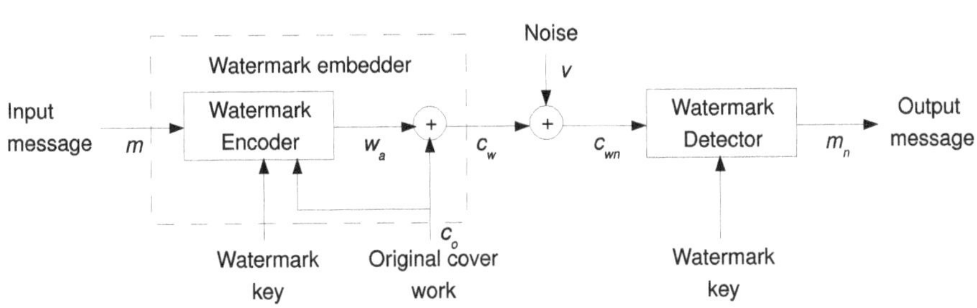

Fig. 1. Communication model of digital watermarking.[1]

The typical structure of a digital watermarking system can be seen in Figure 1. The main components of such system are the embedder and the detector. Since in digital watermarking, a message is embedded into a media (image) and then recovered from that same image with the use of a detector, the most common approach is to model watermarking as a form of communication system[1] (as depicted in Figure 1). In this figure, a message (m) is encoded into an appropriate signal, which is the watermark (w_a). The watermark is then embedded into a host or cover image (c_o), resulting in a marked image (c_w). The marked image is then compressed and/or processed and/or attacked. Then, a detector extracts the watermark from the watermarked/attacked image c_{wn} (here m_n, which might have been influenced by the compression/processing/attack) and uses it accordingly, for copyright control, tampering detection, etc. Data can be detected in two possible ways – with or without the use of cover image. The first is called **informed detection** and requires a sample of the cover media. The second is called **blind detection** and does not require.

The fundamental problem of digital watermarking is to embed a certain amount of data into a cover image in accordance with two conflicting objectives – watermark robustness and image quality. That is, it is possible to make the embedded watermark more robust against certain types of attacks by increasing the power of the watermark signal. But this usually requires introducing more visual artifacts to the watermarked work.

Through the rest of this section, a survey of watermarking techniques will presented. Since watermarking is limited by perceptual and robustness constraints, a review of the metrics and techniques employed in the evaluation the perceptual impact as well as the robustness of the watermarking process will also be presented. Finally, this section will be closed with a discussion about the main challenges concerning digital watermarking of images.

2.1. *Survey of Watermarking Techniques*

Although digital watermarking research is still new, many efforts have been devoted to structure its fundamentals. Some of the concepts involved came from other areas of research as communications theory. Cox *et al*[1] describe the main properties of a watermarking system.

A common approach is to model watermaking as a communication problem. Here, the watermark is treated as a message and the cover media is treated as communication channel. Through this approach it is possible to add layers to cover aspects like:[6,7]

- Security;
- How to embed and detect one bit;
- How to embed more than one bit using multiplexing/modulation;
- How to deal with parts of the host data that cannot embed data;
- How to detect which part of the data should be changed in order to ensure imperceptibility;
- What data to embed;
- What processing/embedding domain to use;

2.1.1. *Embedding effectiveness*

The embedding effectiveness of a watermarking system is related with the capacity of successfully adding a watermark into a cover image. That is, it is the probability that the output of the embedder will be watermarked.[1]

2.1.2. *Fidelity*

The fidelity of a watermarking systems is related with the similarity between the watermarked and the original image. Usually it comes at a price. A trade-off between fidelity and another property like embedding effectiveness or embedding rate must be considered when designing a watermarking system.

2.1.3. *Embedding rate*

Different images can present different embedding capacity (or payload). Some images contain smooth areas which make the embedding of data more difficult. With this in mind, there are two possible options in defining the payload of the watermarking system. One is to fix the embedding rate as low as possible, to deal with the cases where the image contains huge smooth areas. This approach is called Fixed Embedding Rate (FER). The other approach is to change the embedding rate accordingly and is called Variable Embedding Rate (VER). The problem with this approach is that control (side) information must be included, and it reduces the capacity of encoding watermark data.

2.1.4. *Blind or informed embedding*

During the embedding process, information about the cover image can be used, in order to improve system performance (imperceptibility, make watermark stronger to noise, etc). This approach is named informed embedding. For some other applications, there is no such huge demand on performance and for this reason, the embedding can be done without the use of cover image information. This type of embedding is called blind embedding.

2.1.5. *Informed coding*

During message coding, a source message, which is usually related with an specific watermarking application, is mapped into a message mark. This message mark is later embedded into the cover work through an addition or multiplication operation. Since it has been demonstrated in the literature that the embedding performance for a given cover work may vary for different messages, a very useful strategy is to use a mesage coding which uses information about the cover work and performs a one-to-many message-to-watermark mapping in order to improve the trade-off between the imperceptibility and robustness.

2.1.6. *Reliability*

The reliability of a watermarking system relates with the capacity of detecting an embedded watermark. A very useful tool to assess it is the Receiving Operating Characteristics (ROC) curve analysis. A ROC curve presents the False Positive versus False Negative results for a sequence of experiments. The analysis of such curves allows understanding the effect of a given parameter (e.g. capacity) in detection performance.

2.1.7. *Robustness*

Robustness refers to the ability to detect the watermark after common signal processing operations.[1] It is assessed empirically, by evaluating the watermark detection probability after the application of distortion. The use of benchmarking tools for evaluating robustness is widely accepted by the digital watermarking community. There are many benchmarking tools available such as Stirmark[b], Checkmark[c], Optimark[d] and Certimark[e].

2.1.8. *Bi-tonal images*

The watermarking of bi-tonal images is a particular class of watermarking problem. The main issue concerning such type of watermarking regards the range of values a pixel can assume. In a grey-scale image, a pixel can usually assume an integer value between 0 and 255. In a bi-tonal image instead, a pixel can assume only two values: 0 or 1. For this reason, modifications in pixel values in a bi-tonal image are likely to be more perceptible for a human viewer than modifications in pixel values in grey-scale or colour images. Numerous works have been devoted to this particular type of watermarking.[8–15] Chen *et al*[3] provide a survey of such type of technique. Most of these methods are either limited to a certain class of application like printed text or to a certain class of watermarks (robust or fragile). Wu and Liu[16] proposed a general block-based method which allows embedding more than one watermark in the same image at the same time, with different levels of robustness. This approach allows, for example, adding at the same time a robust watermark to enforce the authenticity of an image and a fragile watermark to enforce the integrity.

Despite the specific issues regarding the watermarking of bi-tonal images, it is also possible to convert the image to a grey-scale representation and perform the embedding with the use of more general techniques, followed by a post-binarization.[17] Furthermore, most digital watermarking systems share a common modular framework, both in terms of embedding and detection. Despite the particularities of bi-tonal watermarking, it is possible to consider a general framework for water-

[b] http://www.watermarkingworld.org
[c] http://watermarking.unige.ch/Checkmark
[d] http://poseidon.csd.auth.gr/optimark
[e] http://www.certimark.org

marking. In such framework, each individual module (or group of modules) can be replaced accordingly in order to improve the watermarking system or adapt it to new applications (e.g. watermarking of color images).

The rest of this subsection presents a general framework for watermarking and state-of-the art techniques for each of its modules.

2.1.9. *Embedder*

Although each application has its own specificity, the general structure of an embedder is depicted in Figure 2.

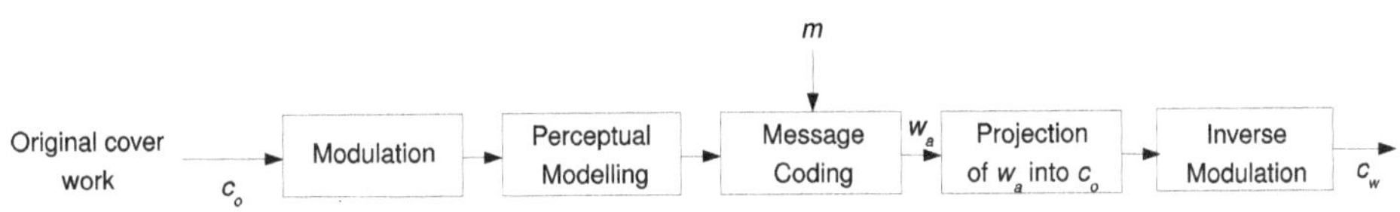

Fig. 2. Structure of a watermark embedder.

Modulation: Modulation relates to choosing an appropriate representation for the image so information can be embedded into it. There are two main families of techniques – those that rely in the pixel representation of the image, namely spatial domain modulation and those that rely on a frequency representation of the image, namely frequency (or transformed) domain techniques. In the spatial domain techniques the pixel values are changed in order to embed one[1] or many[6,16] bits.

In the transformed domain techniques, the image is converted from its spatial representation to a frequency representation, using techniques such as Discrete Cosine Transform (DCT),[18] Discrete Wavelet Transform (DWT)[19] and Discrete Fourier Transform (DFT).[20] These techniques apply better to grey-scale[18,21] and color[11] images, since in the case of bi-tonal images, the post-binarization of the watermarked image can lead to loss of the embedded information. However, through appropriate choice of frequency spectrum and binarization algorithm, this technique can be successfully applied to the watermarking of bi-tonal images.[17]

It is a common practice in bi-tonal watermarking to shuffle image pixels (with the use of a shuffling key) to distribute the flippable pixels through the image.[6,16] Figure 3 from Wu and Liu[16] gives an example of the effect of shuffling pixel positions in the distribution of flippable pixels.

Perceptual modeling: Since the visual impact caused by the embedding process is one of the main constraints in most digital watermarking systems, an appropriate choice of frequency band[18] or flippable pixels[6,13,15,16] is crucial.

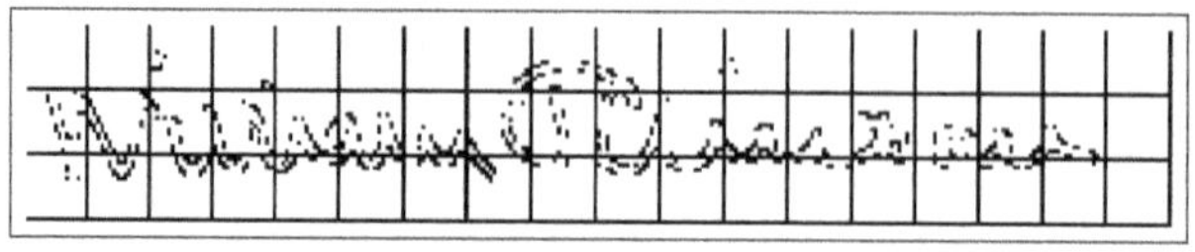

(a)

(b)

Fig. 3. Effect of shuffling in the distribution of highly flippable pixels.[16] (a) Highly flippable pixels before shuffling. (b) Highly flippable pixels after shuffling.

The Structural Neighbourhood Distortion Measure (SNDM) flippability metric proposed by Muharemagic[6] illustrates the principle of perceptual modeling. This method uses a reciprocal distance matrix D_b in order to compute the flippability of a bi-tonal pixel, based on its $b \times b$ neighbourhood. A D_3 reciprocal distance matrix can be seen in Table 1.

Table 1. A 3×3 reciprocal distance matrix as seen in.[6] Each element corresponds to the distance from the central element.

0.7071	1.0	0.7071
1.0	0	1.0
0.7071	1.0	0.7071

The SNDM of a candidate pixel (cp) is computed as follows:

$$SNDM_{cp} = \frac{(cp \oplus N_b) \bullet D_b}{\mid D_b \mid} \tag{1}$$

where N_b represents the $b \times b$ neighbourhood of cp, D_b is $b \times b$ reciprocal distance matrix and $|D_b|$ is its number of pixels.

Message coding: Message coding is the process in which a message (m), which can be either a bit or a sequence of bits is transformed into a watermark that can be then, appropriately inserted into the modulated image. There are two main families of message coding techniques – direct message coding and multi-symbol message coding.[1] In direct message coding, a single bit is transformed into a message mark that is later embedded into the image through a sum or multiplication operation. Usually, a pre-defined reference mark with the same size of the cover image is required in both, embedding and detection.

In multi-symbol message coding, more than one bit must be encoded. There are three different approaches for transforming a multi-bit sequence into a message mark. The first is to break the multi-bit problem in many one-bit problems and apply direct coding to each bit. This approach is known as Time/Space Division Multiplexing and is practical only for small problems, since each one of the possible representation of the bit sequence requires a separate reference mark, that is, 2^N reference marks are required to encode a sequence of N bits. The second approach is frequency division multiplexing. In this approach the frequency domain is partitioned into several disjoint bands and a reference mark for each bit is encoded and then embedded on each band. Spread spectrum techniques rely on this type of encoding.[18,22] The third approach is named code division. In this approach, several uncorrelated reference marks are embedded in the same work.

Projection: Projection is the effective modification of image pixels (or frequency coefficients) required to insert the coded watermark into the cover image. Although the intrinsic mechanisms by which the pixels or the frequency coefficients are modified is related with each coding technique, there are two main strategies to do these modifications.[22] In the first (Type-I), the watermark signal is injected directly into the host signal, by either a sum or multiplication of the host signal (which can be a grey-level pixel value, a DCT coefficient) with the watermark signal. In the second (Type-II), the watermark is embedded by manipulating a given relationship within the host signal (e.g. ratio of black/white pixels).

Inverse modulation: In this step, the modulation process applied in the beginning must be reversed. In some situations, like in the shuffling case, it might be desirable to keep the image transformed (shuffled), and then in the detection side, reverse the shuffling upon a successful watermarking detection, in order to enforce the confidentiality of the image.

2.1.10. *Detector*

The detector basically extracts the watermark by applying the reverse process used on embedding. Generally, a detector has the structure shown in Figure 4.

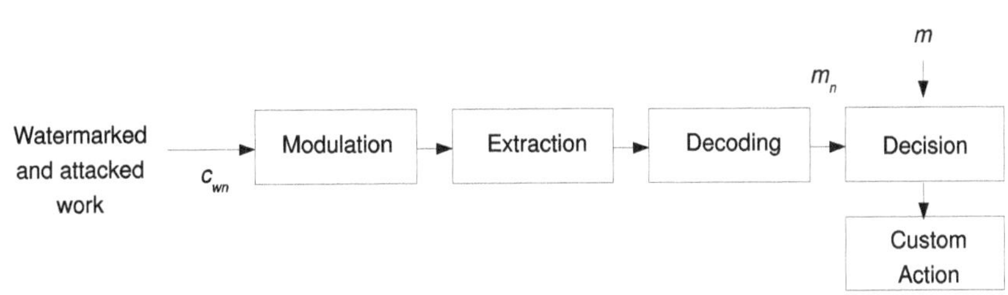

Fig. 4. Structure of a watermark detector.

Modulation: The modulation process is the same that was applied on embedding. If because of optimization either more than one modulation technique and/or parameters (e.g. shuffling key) are employed on embedding, the chosen technique/parameter must be known on detection.

Extraction: The extraction process is in general, the inverse operation of projection. There are two main approaches – informed detection and blind detection. Informed detection requires a copy of the cover image. The difference between those images either in the spatial or frequency[18] domain is employed in order to extract the watermark signal. In the blind detection otherwise, the original image is not required.

Decoding: Decoding can be seen as the inverse process of encoding. Here the extracted watermark signal is transformed into one or more bits.

Decision: On this step, the extracted mark is compared against a reference mark and then a decision is made. There are two possible outcomes of this decision – watermark is valid or watermark is invalid.

Custom action: During custom action, the extracted watermark and the decision are used to perform an application related task like preventing and image of being copied, reporting that the image is tampered or does not come from a certified sender.

2.2. *Evaluation of visual impact*

The visual impact of a watermark can be evaluated by two different approaches – fidelity and quality. Fidelity is a measure of similarity between two signals (from a digital watermarking standpoint, the cover and watermarked signals, or more specifically, images). However, due to some particularities of the human visual system (HVS), the fidelity of a given image does not necessarily relates with the perceived quality by a human viewer. For example, it is a known issue that when the watermarking is performed in the frequency domain of an image (like the Discrete Cosine Transform), the modifications in lower frequencies are less perceptible by human viewers. This makes possible producing images with same fidelity but different quality.

Fidelity is computed using distortion metrics. Bellow, the most common distortion metrics are presented, where C_w is the watermarked image, C_o is the original image, $C_o[i]$ and $C_w[i]$ are the ith pixels of C_o and C_w, respectively and $|C_w|$ is the number of pixels in C_w

(1) Mean Squared Error (MSE):

$$MSE(C_w, C_o) = \frac{1}{|C_w|}\sum_{i=1}^{|C_w|}(C_w[i] - C_o[i])^2 \quad (2)$$

(2) Signal-to-Noise Ratio (SNR):

$$SNR(C_w, C_o) = \frac{\sum_{i=1}^{|C_w|} C_o^2[i]}{\sum_{i=1}^{|C_w|}(C_w[i] - C_o[i])^2} \quad (3)$$

(3) Peak Signal-to-Noise Ratio (PSNR):

$$PSNR(C_w, C_o) = max_{|C_w|}\left(\frac{\sum_{i=1}^{|C_w|} C_o^2[i]}{\sum_{i=1}^{|C_w|}(C_w[i] - C_o[i])^2}\right) \quad (4)$$

The quality of a watermarked image can be evaluated either by human observers (using standard test procedures, such as the *two alternatives, forced choice*), or by computational techniques that model the behaviour of the HVS.

For example, one of such techniques is the Distance Reciprocal Distortion Measure (DRDM).[6] This metric has been specifically created to evaluate the difference between two bi-tonal images in terms of quality. Modifications in a bi-tonal image may affect the structure of elements within that image, affecting drastically the quality of the image. For this reason, care must be taken in order to avoid such modifications. The DRDM is based on the assumption that modifications in pixels close to viewer's focus are more noticeable. Also, due to particularities of human visual system, modifications in diagonal neighbours of a pixel are less noticeable than modifications on its immediate vertical and horizontal neighbours (4-neighbourhood).

A normalized weight matrix W_d, with size $d \times d$ is used to compute the distortion between two bi-tonal images. Each element of this matrix represents the reciprocal distance, relative to the center pixel. The distortion between two bi-tonal images is calculated as:

$$DRDM = \frac{\sum_{k=1}^{|C_w|} DRDM_k}{K} \quad (5)$$

where K is the number of non-uniform blocks (blocks that are neither all black nor all white) and $DRDM_k$ is a local distortion, calculated for each pixel, based on its $d \times d$ neighbourhood

$$DRDM_k = \sum_{d \times d}[|a_d - b_d| \times W_d] \quad (6)$$

2.3. *Evaluation of robustness*

As mentioned before, robustness refers to the ability to detect the watermark after the watermarked image has suffered common signal processing operations. These operations can be intentional or not. The intentional use of such type of operation in a watermarked image is called an attack. There are four main families of attacks: removal, geometric, cryptographic and protocol attacks.[23] In a removal attack, the embedded watermark is partially or completely removed either by a source of noise or with the use of image processing techniques such as de-noising, lossy compression, cropping, etc. In a geometric attack by its way, the watermark is not removed but instead, the synchronization between the embedder and detector is affected with the use of affine transformations, such as rotation. In a cryptographic attack, the intention is to crack the security mechanisms employed on watermarking (such as the watermarking key). Finally, in a protocol attack, the objective is to threaten the validity of the system rather than its functionality. For example, in an protocol attack known as invertible watermark, an attacker extracts his own watermark from a watermarked image and claims he is the owner.

Intelligent watermarking usually aims at improving the robustness against removal attacks, since it is possible to increase the robustness against such attacks by adjusting embedding parameters (at the cost of adding more visible artifacts). Geometric attacks can be addressed either by detecting and inverting the distortion in the detector[1,22] or by embedding the data in a domain resistant to affine transformations such as the Discrete Fourier Transform (DFT).[20] Cryptographic attacks can be made unfeasible by using large watermark keys. Finally, protocol attacks can be minimized by embedding signal-dependent watermarks, for example, a signature of the cover work.[14]

Robustness against removal and geometric attacks is assessed empirically, by evaluating how does the watermark detector performs after the watermarked image has been attacked, that is, how similar are the embedded and detected watermarks. Fidelity metrics (such as the MSE) are employed to this end. Since an attack is only considered a concern when it does not affect the commercial value of the watermarked work, the embedded mark does not have to be resistant against attacks that affect the quality of the watermarked work. Usually, watermark-to-noise ratio (WNR), which gives the ratio between the power of the watermark signal and that of the noise introduced by attacks[24] is used in order to define the limit of the robustness

$$WNR(C_w, C_{wn}) = \frac{\sum_{i=1}^{|C_w|}(C_w[i] - C_o[i])^2}{\sum_{i=1}^{|C_w|}(C_{wn}[i] - C_w[i])^2} \tag{7}$$

where C_w is the watermarked image, C_o the original image, C_{wn} is C_w after processing/attack.

2.4. *Challenges of watermarking*

The use of digital watermarks makes possible the embedding of side information into a cover image in an imperceptible way. The embedding must be performed according to a trade-off between robustness and image quality. Watermarking can thus, be considered an optimization problem.

The main advantage of securing a document image with a digital watermark is that the protection provided is not ostensive. Depending on the perceptual model employed, document's authenticity and integrity is protected in an invisible manner. Despite these advantages, there are many known attacks to digital watermarking systems. For example, if a watermark detector is widely available, an attacker could use detection information to repeatedly make small changes to the watermarked work until the detector fails to detect the watermark.[6] Moreover, in a type of attack named ambiguity attack, someone can add a watermark to an already watermarked work in such a way that it would appear that this second watermark is the true watermark. In another type of attack named geometric attack, rotation, scale and translation transformations are applied to the watermarked image in a way that the synchronization between the embedded and detected watermark signal is lost, what could be a threat for an authenticity application.

The use of a robust watermark can mitigate the effects of most of these attacks (except for geometric attacks, which must be tackled with the use of registration marks[1,22]), at the expense of adding more visual artifacts. This makes robust watermarks very attractive for authenticity applications. A fragile watermark can be very useful in the detection of intentional or unintentional modifications in the cover image (integrity enforcement). A watermark can be added in a fragile manner, and once its detection fails, it can be assumed that the image was tampered. The side effect of using fragile watermarks is that its detection will be affected by small variations in the image due to compression, processing or channel noise (here the cover image is considered a source of noise to the watermark signal). A balance between tampering protection and noise robustness must be considered.

Given these aspects, the combined use of fragile and robust watermarks may provide a very efficient way to protect both, the authenticity and the integrity of an image. However, the two main challenges in digital watermarking are (1) coping with variations across different types of images and (2) fine tune the embedding parameters to find an optimum balance between robustness and quality. As mentioned before, computational intelligence can be used in order to mitigate these problems.

3. Intelligent watermarking

In this section the main strategies concerning the use of computational intelligence in digital watermarking systems will be reviewed. The interference of channel and external noise in the message being transmitted is a known problem in information theory. There are several alternatives to tackle this problem. The most obvious is to increase the power of the message signal. However, in most channels, the power of the message signal is subject to constraints. This is specially the case in digital watermarking, where the power of the signal is subject to fidelity constraints. Another alternative is to spread the message signal through the host signal (spread spectrum).[18] Since modifications in certain frequencies are less perceptible than in others, it is possible to increase the energy of the message signal in those frequencies without affecting the fidelity constraints. However, as demonstrated by Wu,[22] although spread spectrum minimizes the influence of secondary sources of noise (attacks), it performs very poorly in what regards channel noise. Costa[25] demonstrated that if the properties of the host signal are known, it is possible to adapt the message coding to the host signal, minimizing the interference. These two examples show us that it is possible to explore properties of the cover work (side information) during embedding in order to make the watermarking process more adaptive to different cover works and types of attacks. In the literature, there are two main strategies to improve the adaptiveness of a watermarking system. The first is to use statistical or neural network classifiers for supervised learning of either a watermarking process, e.g. detection, or the evaluation of a given property of watermarking, e.g. imperceptibility. The second is to use evolutionary optimization in order to find a set of embedding parameters that result in near-optimal performance, according to one or more objectives such as robustness and fidelity. Both approaches are problem specific and thus, must be adapted to the specific watermarking systems. But the literature provides some guidelines for each of these approaches.

3.1. *Supervised learning*

In supervised learning, labelled samples collected for a problem are employed to estimate the parameters of a neural or statistical classifier. Assuming the samples have been assigned with two or more class labels, the trained model will provide a mapping of the samples into two or more regions corresponding to classes. This mapping is defined by the use of a linear or non-linear decision function. Once the parameters have been estimated with the use of training samples, it is possible to assign a class to unlabelled query samples in a task known as classification.

Classifiers such as the Multilayer Perceptron (MLP)[26] and Support Vector Machines (SVM)[27] allow the estimation of very complex decision functions. These non-linear classifiers are very suitable to regression, as they can be considered universal function approximators. Thus, they can be applied in the task of learning a specific process of watermarking (e.g. detection) based on a set of labelled (training) data (e.g. a set containing cover works and their respective watermarked images). Moreover, they can be applied to the task of learning how to analyze a given property of a watermarked image (e.g. quality) based on labelled images.

The MLP is a very popular type of neural network classifier. It contains one input layer, one or more hidden layers and one output layer which produces either a label assignment or a function estimation. Each layer consists of one or more units, named neurons, which are connected to units in other layers by weights. Despite the simplicity of the heuristic employed, the ability of MLPs to learn any arbitrary decision function have been formally demonstrated.[28]

SVM is a large margin statistical classifier. It is based on the principle that given any two populations of labelled samples, the optimal decision function will maximize the distance between the hyperplane separating the two sets of samples and the samples that are closest to this hyperplane. In SVM, a preprocessing phase projects the data to a higher dimensionality. Through preprocessing, non-linearly separable sets of samples become linearly separable. This allows the use of such type of classifier in the regression of non-linear functions.

As illustrated in Figures 2 and 4, a digital watermarking system is modular. The implication of this modularity is that each module usually handles a very specific aspect of the watermarking process. The modular nature of a watermarking system makes possible to isolate some of these modules and train non-linear classifiers to implement their functionality.

Although digital watermarking is based on a solid theoretical framework, one of its weaknesses is that the noise sources (both, host channel and external) are always assumed to have a given form (usually Gaussian). The alternative found was to use non-linear classifiers, to make some of the watermarking processes more adaptable to the real form of the noise sources. The use of a classifier in this process is straightforward. A classifier is trained with labelled data, where the raw data is usually the same data that the real module receives as an input while the label (or target data) is the output. There are two approaches in what regards target data. The first is to use the data provided by the module to be replaced (i.e. someone could pick the message coding module described in sub-section 2.1.9 and generate a set of target data for a given range of input) or use data provided by humans (e.g. a score for the perceptual modeling module).

3.1.1. *MLP*

The watermarking system proposed by Chang and Lin[29] illustrates the first approach. In the baseline watermarking system, a watermark is embedded in the Discrete Wavelet Transform (DWT) coefficients of an image. The DWT decomposition breaks the image into a hierarchical set of coefficients, where each coefficient has four children (quad-tree). Each level of this quad-tree corresponds to a given level of resolution. A pseudo-random number sequence (based on a seed) is employed in the task of choosing the set of coefficients where the embedding will be performed. Given a coefficient s_k, the embedding is performed by adding (to embed a '1') or subtracting (to embed a '0') a constant α to each of the four children of s_k. For example, given an hypothetical coefficient $s_k = C_3$ in Figure 5, four bits are embedded by adding or subtracting this constant to D_1, D_2, D_3 and D_4. In this Figure, each letter (A, B, C and D) corresponds to a given resolution while each index (1, 2, 3 and 4) corresponds to each of the subbands for that resolution.

However, after modifying these coefficients and inverting the DWT transform, a reference image is necessary in order to compute the difference between both coefficients and detect the embedded data. Instead of this, the authors use a MLP in order to map the relationship between the coefficients. Basically, this network contains eight input neurons - the parent of s_k (B_3), the three siblings of its parent

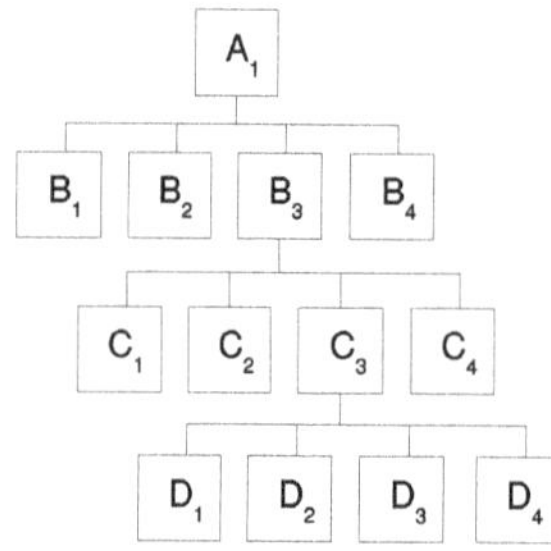

Fig. 5. Example of a quad-tree structure.

(B_1, B_2 and B_4), the three siblings of s_k (C_1, C_2 and C_4), and s_k itself (C_3). The output neurons are the four children of s_k (D_1, D_2, D_3 and D_4). The MLP learns the mapping between a given coefficient and its children so the data can be detected latter without the use of the cover image.

During detection, the trained MLP is employed in order to recover the previous coefficient values (that is, their value before embedding) and the data is extracted from these coefficients by computing the difference between the output of the MLP and the children of each coefficient.

3.1.2. *SVM*

Chang and Lin used this principle in the task of creating a SVM-based perceptual modeling module.[30] In the baseline watermarking system technique, the embedding is performed by manipulating the pixel values in blocks of 3×3 pixels. However, the extent at which each pixel can be manipulated is limited by image quality constraints. To cope with this, a SVM is used in the task of providing a score which will define the amount of modification each pixel in any given 3×3 window can suffer. In this technique, the four most significant bits of each one of the nine pixels in a given 3×3 are employed in the task of training a SVM classifier. The target data, which is a score, is manually provided by an human specialist.

Tahir *et al*[31] also employed a SVM on digital watermarking but in the decoding side. The basic principle is to use SVM in order to improve the detection performance under Gaussian attack. The baseline watermarking system embeds a message in an image by manipulating its DCT coefficients. If no attack has occurred, the embedded bits will form two distinct Gaussian distributions. However, after an attack, these two distributions will overlap. SVM can be employed in order to make these two overlapped distributions separable in a higher dimension, during detection. This was the approach employed by the authors. Basically, for each bit, 22 statistical features are computed and used as a feature vector. The bit value is used as target data. After the SVM has been trained, during detection, the same features are computed from each bit and fed into the trained classifier.

Davis and Najarian[32] employed an MLP in order model the Human Visual System (HVS). In the proposed technique, each image is subdivided in blocks of 64×64 pixels. The image is transformed to a wavelet domain (DWT). These 4096 coefficients, along with a given watermark strength are used in order to train an MLP. The target data is a score provided by a human viewer. The trained MLP can be employed in the task of analyzing the visual impact of a given watermarking task.

3.2. *Optimization of watermarking parameters*

This family of techniques relies in the use of optimization in order to finetune the parameters of embedding algorithms, aiming thereby, increasing the robustness of the embedded watermark and decreasing the visual distortion caused by the embedding process.

Optimization can be categorized in three approaches. In the first approach, theoretical properties of the watermarking system are explored, using mathematical analysis (e.g. Cox *et al*[1] uses mathematical analysis in order to adjust the embedding strength parameter so the distance between embedded "1" and "0" bits can be increased).

In the second, parameters are assumed to be independent and then, local optimization (greedy algorithm) is performed on each parameter. This was the strategy employed by Muharemagic[6] in his adaptive system.

In the third approach, Evolutionary Computing (EC) techniques such as Genetic Algorithms (GA)[4] or Particle Swarm Optimization (PSO)[5] are used in order to adjust the embedding parameters according to constraints in the robustness of the watermark and/or the fidelity/quality of the watermarked image. This is the most common approach in the literature, mainly due to the simplicity of techniques and the ease in adapting them to many different types of watermarking systems. Moreover, EC, does not assume a distribution of the noise source or parameter space, as with mathematical analysis/greedy search. Figure 6 illustrates the general structure of a system based on this approach (the watermark embedder corresponds to the embedding system depicted in Figure 2 while the detector corresponds to the detection system depicted in Figure 4).

Usually, one or more embedding parameters are encoded either as a chromosome (GA) or as a particle (PSO). The objective functions usually involve at least one fidelity/quality (e.g. PSNR, as seen in Equation 4) and one robustness (e.g. MSE, as seen in Equation 2, between embedded and detected watermarks) metrics. To evaluate robustness, one or more attacks are applied to watermarked image. Then, the detected watermark is compared with the embedded one with the use of fidelity metrics. The objective of the optimization algorithm is to minimize simultaneously, (1) the visual impact caused by the embedding procedure and (2) the difference between the embedded and detected watermarks under a given set of attacks.

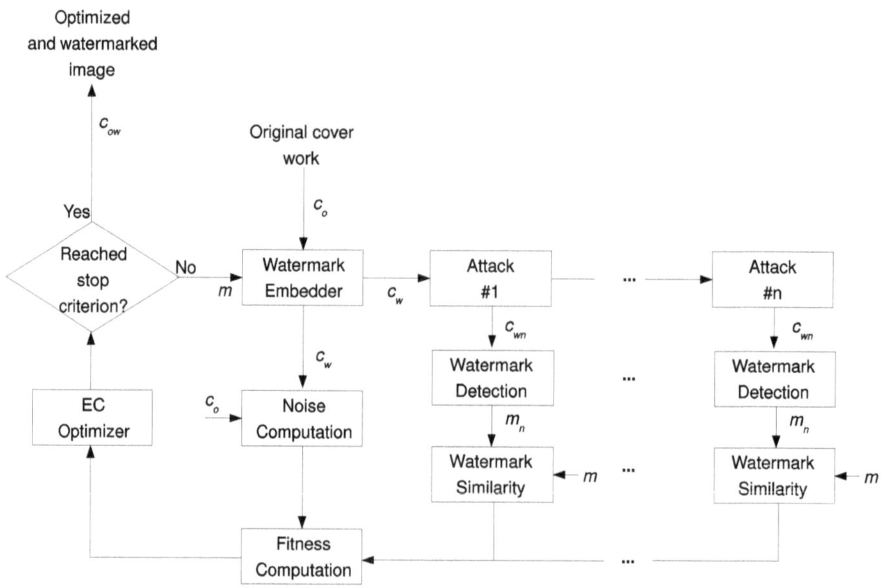

Fig. 6. General structure of a system based on EC optimization strategy (based on the model proposed in[33]).

Some authors however do not follow this multi-objective optimization approach and use only one objective function (either noise or robustness). In some methods also, although robustness is evaluated, no attack is applied to watermarked image (there is theoretical basis to assume that the cover image itself is a source of distortion to the embedded watermark[1]). The robustness evaluation consists of a simple distance computation between the detected and embedded watermarks.

Shieh *et al*[33] optimize a DCT-based watermarking system with GA. A DCT block transformation is applied to a grey-scale image. After that, the embedding is performed through the manipulation of the polarity between the watermark and the DCT coefficients. The authors employ GA to find the DCT coefficients that result in the best combination of robustness and image quality. The robustness is computed by embedding a watermark into an image, applying one of three different attacks (Low Pass Filtering, Median Filtering and JPEG compression with quality factor 80%), detecting the watermark and computing its normal correlation (NC) against the original watermark. The quality is computed with the use of PSNR.

Lee *et al*[34] employed a hybrid GA/PSO technique in the optimization of a DWT-based watermarking system. Heuristic weights are used to deal with the trade-off between robustness and fidelity in the modulation process. Authors proposed using GA and PSO in a paralel to optimize the heuristic weights. The authors applied various classes of attacks only after the optimization procedure, to evaluate the effectiveness of the proposed method (Filtering, geometrical, JPEG compression and image enhancement).

Ji *et al*[35] optimized a Least Significant Bit (LSB) substitution steganography method with the use of GA. Since the embedding procedure is based on the use of a mapping function, the authors employed GA in order to obtain a mapping function that provides robustness and fidelity, at the same time. Distortion metric is employed as fitness function.

Li and Wang[36] employed PSO in the task of optimizing a DCT-based steganographic method. This method embeds a secret image into the least significant bits of the DCT coefficients of a cover image and relies in the use of a substitution matrix during the message encoding step of the embedding process. The authors employed PSO to find an optimal substitution matrix. The objective function is based on a distortion metric (PSNR).

Wei *et al*[37] applied GA to the task of identifying the best coefficients in a spread spectrum DCT watermarking system. The combination of the similarity metrics between original and extracted mark is fed into the GA algorithm as a fitness function. Four different attacks - Low Pass Filtering, Scaling, Gaussian Noise, JPEG compression - are employed in this method.

Pan *et al*[38] applied GA to the task of optimizing a Block Pixel Statistic Manipulation (BPSM) method. In this BPSM watermarking method, the mean of the grey-level values of the 8-neighbourhood surrounding pixels of a given central pixel is computed. Then, the embedding is performed by manipulating that value. The manipulated value is stored in the central pixel. Authors used GA to search for a near optimal set of pixels, in terms of robustness and fidelity. The Bit Correct Ratio (BCR) between original and extracted watermark (JPEG compression is applied in the watermarked image) as well as the PSNR of the watermarked image are employed as fitness functions.

Sal *et al*[39] applied NSGA-II,[40] which is a Pareto-based Multi-Objective Genetic Algorithm (MOGA), in the task of optimizing a DCT-based watermarking system. The parameters being optimized are the DCT coefficients where embedding will be performed. The distortion and robustness are measured directly on the DCT coefficients. The authors do not apply attacks during optimization process.

Chen and Lin[41] employed GA in the detection of nearly optimal embedding positions on DCT blocks. The fitness evaluation is based only in the MSE between original and watermarked images. No similarity between embedded and detected watermarks is employed. Also, the proposed method uses no attack during optimization procedure.

Areef and Heniedy[42] apply GA in the optimization of a DWT-based watermarking method. Basically, the cover image is decomposed with the use of the Haar wavelet filter. Then, the watermark signal is embed into a given set of wavelet coefficients as a multiplicative watermark. GA is applied then, in order to identify a set of coefficients which maximizes the robustness against JPEG compression (BCR computation is performed on watermarked/JPEG compressed images for this purpose) and minimizes the embedding noise (measured with the use of MSE).

Shih and Wu[43] applied GA in order to create a rounding rule for DCT embedding. The basic problem is that on DCT embedding, integer pixel values are transformed into real-valued DCT coefficients. The watermark is embedded on these coefficients which are then transformed back to integer pixel values by an Inverse DCT (IDCT). During this process, information might be lost due to rounding error. The authors proposed the use of GA to tackle this problem. Basically, a gene is used for each DCT coefficient, where '1' means that the resulting value from the IDCT process must be truncated and added to 1 ($\phi_i^* = Trunc(\phi_i) + 1$) and '0' means that the resulting value must be just truncated ($\phi_i^* = Trunc(\phi_i)$), where ϕ_i is the DCT coefficient at location i. Two fitness functions - one based on the Normalized Correlation (NC) between embedded and detected watermark and another based on the PSNR between cover and watermarked images - are employed.

Kumsawat and Attakitmongcol[44] proposed the use of GA to optimize a Multilevel Wavelet Transform watermarking method. In the proposed method, GA is employed in order to identify the coefficients that improve the performance of the base method. Here, the authors make use of the Universal Quality Index (UQI) to measure the similarity between watermarked and cover images. The robustness is evaluated with the use of correlation.

Diaz and Romay[45] applied NSGA-II[40] in the optimization of a DCT-based watermarking method. Normalized correlation is applied to measure the robustness of the proposed solution against JPEG compression and smoothing. MSE is applied to measure the noise between watermarked and cover images.

Khan and Mirza[46] proposed the use of Genetic Programming to achieve an adaptive perceptual modeling algorithm for a DCT-based watermarking system. In the proposed method, genetic programming operators are employed in the task of creating a perceptual modeling function for a given embedding task. The Structure Similarity Index (a quality measure) and the Bit Correct Ratio (a robustness measure) are used as objective functions.

Wu and Shih[47] applied GA in order minimize the occurrence of statistical features that are used for steganalysis purposes. In the proposed method, the modifications to be done to a DCT block in order to embed a given message are coded as chromosomes. During optimization, a message is embedded into the DCT coefficients of a cover image. Then, Bit Error Rate (BER) is used to evaluate the difference between extracted and detected watermarks. Analysis functions based on the type of steganalysis attack the system must resist are used in order to evaluate the robustness against such attacks. These two metrics are employed as fitness functions in the GA optimizer.

3.3. *Key issues*

Among all the existing families of optimization techniques, those based on EC have been successfully employed in many different scenarios involving the optimization under uncertainty (stochastic optimization). As mentioned before, the number of parameters to be adjusted in a digital watermarking system is indeed a concern. Adjusting these parameters according to an optimum tradeoff between robustness and quality can help to make watermarking more suitable to industrial applications. But it is difficult to know the exact form of the problem beforehand. A strategy to tackle this problem is to consider watermarking as stochastic optimization problem and apply EC to such end (optimization of embedding parameters).

The main reason for the use of EC in the optimization of watermarking systems is that the objective functions are usually noisy (multi-modal). Since EC techniques are based on population of candidate solutions, it is less likely to the optimization algorithm to get stuck in a local optimum. Moreover, due to the modular nature of a watermarking system (with numerous different techniques for each module) the use of EC provides flexibility to the optimization process, since it does not require gradient information of the function under consideration.[48] There many methods based on this strategy in the literature (Table 2). Actually, the majority of the intelligent watermarking methods are based on this strategy. Regarding the number of objective functions employed, there are two main optimization strategies - one consisting of the use of a single objective function (e.g. fidelity), known as Single Objective Optimization Problem (SOOP) and another one consisting of the combination of many objective functions, known as Multi Objective Optimization Problem (MOOP). With respect to the GA or PSO algorithms employed to deal with MOOP, there are two strategies. One which consists of aggregating many objective functions into one through weighted sum - and then use classical GA and PSO - and another which consists of handling many conflicting objectives during optimization - which is the case of Multi Objective GA (MOGA) and Multi Objective PSO (MOPSO).

The optimization of a watermarking system is a multi-objective problem, since it must handle at least two conflicting objectives - fidelity/quality and robustness. However, the vast majority of research has been directed towards the use of single-objective optimization algorithms. The most common approach to handle multi-objective optimization in a single-objective optimization algorithm is to combine all fitness functions into one with the use of weighted sum. However, such approach usually favours one objective in detriment of the others. Multi-objective optimization algorithms such as the NSGA-II[40] and MOPSO[49] rely on Pareto dominance and can be employed in order to mitigate the problem of favouring one objective.

Regarding the use of supervised learning, there are two main approaches. The first is to learn a watermarking process, as in[29] where the mapping between the original and embedded DWT coefficients is performed with the use of a MLP. The main benefit of this type of approach is that it allows knowing the model of a given property of the the cover image on the detection side, which can boost detection performance, but without the burden of transmitting the cover image to the detector (informed detection). The second approach is to learn how to evaluate a given property of the watermarking process such as the visual impact or the robustness

Table 2. Summary of EC-based digital watermarking techniques.

Method	Watermarking Strategy	Optimization Method / Algorithm	Parameter	Distortion	Attack
Shieh *et al*[33]	DCT	MOOP/GA	Coefficients	PSNR	Low Pass Filtering
					Median Filtering
					JPEG Compression
Lee *et al*[34]	DWT	SOOP/ Hybrid (GA/PSO)	Coefficients	Perceptual Lossless Ratio (PLR)	Median Filtering
					Wiener Filtering
					Average Filtering
					Gaussian Filtering
					Rescaling
					Rotation
					Cropping
					Jittering
					StirMark
					JPEG Compression
					Image enhancement (6 different algorithms)
Li and Wang[36]	DCT	SOOP/PSO	Encoding (substitution matrix)	PSNR	None
Ji *et al*[35]	Least Significant Bit	SOOP/GA	Substitution matrix	PSNR	None
Wei *et al*[37]	DCT	MOOP/GA	DCT coefficients	None	Low Pass Filtering
					Scaling
					Noise
					JPEG Compression
Pan *et al*[38]	Block Pixel Statistic Manipulation (BPSM)	MOOP/GA	Embedding blocks	PSNR	JPEG Compression
Sal *et al*[39]	DCT/Hyperspectral images	MOOP/NSGA-II	DCT coefficients	Coefficient values	Low Pass Filtering
Wu and Shih[47]	DCT	MOOP/GA	Coefficient values	None	Steganalisys
Chen and Lin[41]	DCT	SOOP/GA	DCT coefficients	MSE	None
Areef and Heniedy[42]	DWT	MOOP/GA	Frequency bands	PSNR	JPEG Compression
Shih and Wu[43]	DCT	MOOP/GA	Coefficient rounding rule	PSNR	None
Kumsawat and Attakitmongcol[44]	Discrete Multiwavelet Transform	MOOP/GA	Coefficients	Universal Quality Index	JPEG Compression
					LPF
					Wiener Filtering
					Gaussian Noise
					Image Cropping
					Image Rotation
Diaz and Romay[45]	DCT	MOOP/NSGA-II	Coefficients	Coefficient Value	JPEG Compression Smoothing

Table 3. Summary of intelligent watermarking techniques based on supervised learning.

Method	Watermarking Strategy	Classifier	Feature Set
Chang and Lin[29]	DWT	MLP	DWT coefficients.
Chang and Lin[30]	Spatial domain.	SVM	Significant bits of a pixel.
Tahir *et al*[31]	DCT	SVM	Detection statistics.
Davis and Najarian[32]	DWT	MLP	DWT coefficients.

of the embedded watermark. The main benefit of this approach is that it makes possible modeling the visual impact according to evaluations provided by human viewers.

Another issue is that as mentioned before, bi-tonal images account for 95% of the use in industrial applications. Nevertheless, no single work was found in the literature, regarding the use of evolutionary computing in the optimization of bi-tonal documents. Although most of the techniques presented are based on single-channel (grey-scale) images, they can be easily adapted to multichannel images (like RGB).

In the next section, the applicability of using an evolutionary computing algorithm (PSO) to this task is demonstrated.

4. Case study - optimization of a bi-tonal watermarking system using PSO

In this section a system that optimizes the bi-tonal watermarking system is proposed based on the system of Muharemagic.[6] The adaptive technique proposed by Muharemagic[6] is based on greedy search and thus, does not consider the effect of choosing one parameter on the remaining parameters. A strategy to address this issue is to employ PSO to optimize these parameters simultaneously. Compared to the adaptive method proposed by Muharemagic, the use of PSO for this task allows a global search in the parameter space. Compared to other Evolutionary Algorithms (EA), PSO is considered to have a quick convergence.

In the proposed method, three embedding parameters (block size, shuffling seed and SNDM window size) are encoded as a particle position in the PSO algorithm. The first parameter (block size), a limited set of block sizes $B = \{b_i | i = 0, ..., (|B| - 1)\}$ is employed. The index of this set is used as one of the dimensions of the search space. For this reason, this given axis must be clipped to the $[0 - (|B| - 1)]$ range. The second parameter is the seed used to shuffle the image. The index of the set of seeds $K = \{k_j | j = 0, ..., (|K| - 1)\}$ is used as another dimension of the search space. Finally, the third parameter is the size of the SNDM window. Let us define a set of SNDM window sizes B_s. As for the other parameters, the index of the set is used as a dimension of the search space. In a canonical PSO, the search space will have though, three dimensions. An alternative is to employ a discrete PSO[50] for this task, where each parameter can be encoded as a sequence of bits. These are exactly the same parameters used by Muharemagic.

As in the baseline adaptive watermarking system, there are three objectives to be minimized - the MSE (Equation 2) between the embedded and detected fragile watermarks, the MSE (Equation 2) between the embedded and detected robust watermarks and the DRDM (Equations 5 and 6) between the cover and watermarked images. Although there are multi-objective versions of the PSO in the literature,[49] for a matter of simplicity a single objective PSO, with function aggregation, is employed in this case study. Since the objective of this work is a proof-of-concept, the Conventional Weighted Aggregation (CWA) is applied.

4.1. *Framework*

4.1.1. *Baseline watermarking system*

The bi-tonal watermarking system proposed by Muharemagic[6] will be the baseline watermarking system. In the proposed intelligent watermarking approach, two watermarks, a robust and a fragile, are embedded, along with Error Correction Code. This system embeds multi-bit messages in the spatial domain of bi-tonal images. Shuffling[16] is employed to handle uneven embedding capacity. Perceptual modeling is performed with the use of SNDM, which has been specifically developed for the bi-tonal spatial domain representation and is more flexible than its counterparts.[16]

The message coding is based on code division, which comprises partitioning the image into blocks of a same size and then, encoding one bit at each block by manipulating the number of black pixels in each block. Regarding watermarking of bi-tonal images, one of the most simple techniques is to force the number of black pixels in a block to be either even (to embed a '0') or odd (to embed a '1'). This technique is known as odd-even embedding. The problem with this type of technique is that if a single pixel is changed in a block (due to noise, compression or even an attack), the bit value changes as well. A strategy to deal with this issue is to define a fixed quantization step size Q and to force the number of black pixels in a block to be either $2kQ$ or $(2k+1)Q$ (for a given k).[51,52] This technique is known as Uniform Quantization (UQ). A larger Q will allow more pixels being randomly shifted (e.g. in the case of an attack) without changing the embedded bit value. This adds robustness to the watermark at the cost of more visual artifacts. Detection is done by checking the enforced relationship.

When compared to other bi-tonal watermarking techniques, the advantage of UQ is that it has been specifically designed for the embedding of multi-bit messages and has proven success in the watermarking of bi-tonal images.[16] Moreover, UQ is not tied to a specific application - there are numerous techniques in the literature that have been developed for specific applications like watermarking of handwritten text document images (text or character shifting), *fac-simile* and others.[3,14,53]

The projection of pixels in this baseline watermarking system is based on the manipulation of a property of the host signal (Type-II). Here, the watermark signal (w_a) computed for a block B is projected into the image by changing w_a flippable pixels on that block (black pixels are flipped to white if w_a is positive while the opposite happens if w_a is negative).

A UQ method named Scalar Costa Scheme (SCS) proposed by Eggers *et al*[52] will be employed since it is general and flexible when compared to the alternative (Chen and Wornell[51]).

In this method, in order to embed a bit m_i into a given element of the cover signal x_i (in this case, quantity of black pixels at block i), quantization of the cover signal must first be performed

$$q_i = SQ_Q\{x_i - Q(\frac{m_i}{D} + \kappa_i)\} - (x_i - Q(\frac{m_i}{D} + \kappa_i)) \tag{8}$$

where $SQ_Q\{\}$ is the scalar uniform quantization operation, Q is the quantization step size, D is the alphabet size (2 for binary encoding), and κ_i is a pseudo-random number in the $[0, 1)$ range, used for security purpose.

The watermark signal (w_a) is obtained by multiplying q by the embedding strength α

$$w_a = \alpha q \tag{9}$$

The inverse process is done during detection. Here, the number of black pixels in a given partition box is the received signal w_n, that may have been attacked, that is $w_n = x + w_a + v$ (where v is the noise signal). Then, the detected message (m_n) is extracted from a given block with the use of the uniform quantizer. Firstly, quantization is applied to the received signal

$$q_{ni} = SQ_Q\{w_{ni} - \kappa_i Q\} - (w_{ni} - \kappa_i Q) \tag{10}$$

It is necessary to use the same Q and κ_i used on embedding.

Then, the message bit is extracted from q_{ni}. In our case, since the encoding comprises the block-based use of UQ, it is necessary to know here the partitioning scheme and the quantization step size employed on embedding. Basically, if the value of q_{ni} is close to either Q or 0, it means the corresponding bit is $m_{ni} = 0$. If instead, the value of q_{ni} is close to $Q/2$, it means the corresponding bit is $m_{ni} = 1$ (see Figure 7).

This process is repeated for each partition block. In this case study, the embedded message is a logo image. Given the dimensions of this image, it is possible to reconstruct the logo with the use of the detected bitstream.

An interesting aspect of this watermarking technique is that it allows for the embedding of several watermarks, at different levels of robustness (where robustness, as mentioned before, is determined primarily by the Q parameter). This process is called multi-level embedding and it basically comprises embedding the watermarks sequentially, starting with the one with the biggest value for Q until the one with the smallest value.

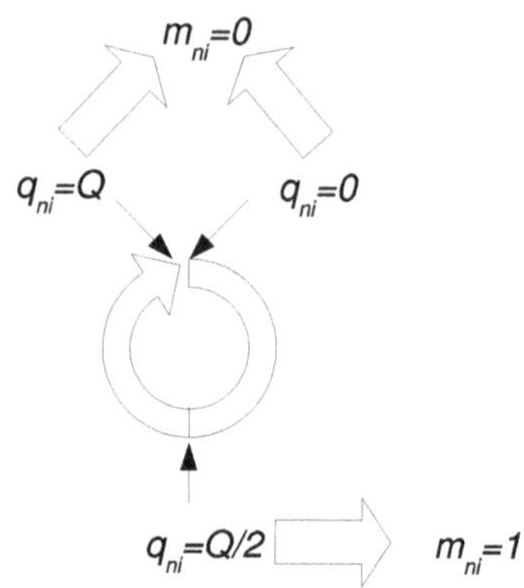

Fig. 7. Detection decision.

4.1.2. *Particle Swarm Optimization (PSO)*

The use of PSO is justified by its quick convergence towards the global minimum in comparison with other popular methods - e.g. GA.[54] The drawback is that the search performed by GA is more exhaustive. However, in the case study scenario, it is acceptable to have near-optimum solutions if they satisfy a predefined quality and robustness criteria. These two properties make PSO more suitable to the problem of fast optimization of watermarking systems.

In a seminal paper, Kennedy and Eberhart[5] proposed the Particle Swarm Optimization technique. The initial motivation behind their work was to graphically simulate the choreography of a bird flock. In PSO, a particle navigates through the search space based on two influences - the best position visited by itself (cognitive component) and the best position visited by its neighbour (social component). The neighbourhood of a particle can be restricted to a limited number of particles (L-Best topology) or the whole swarm (G-Best topology). The algorithm has passed through many different stages on its conceptual development. Parsopoulos and Vrahatis,[48] Kennedy[55] and Poli *et al*[54] provide a review of the improvements in PSO algorithm since its inception. The most popular implementation of PSO is known in the literature as the Canonical PSO. In this implementation, the velocity and position of a particle are updated at each iteration according to the best locations visited by itself and by its best neighbour. Two different factors are employed to balance the influence of both attractors - the c_1 (cognitive) and c_2 (social) acceleration constants. To provide a fine grain search in the end of the optimization process, a inertia weight ($\omega_{inertia}$) and constriction factor (χ) were added. So the velocity and position of each particle in the Canonical PSO are calculated at each iteration as:

$$V_{id} = \chi \times (\omega_{inertia} \times V_{id} + c_1 \times r_1 \times (P_{id} - X_{id}) + c_2 \times r_2 \times (P_{gd} - X_{id})) \quad (11)$$

$$X_{id} = X_{id} + V_{id} \quad (12)$$

where V_i and X_i are the velocity and position of particle i, P_i is its best visited location, P_g is the best visited location for all of its neighbours. As mentioned before, the inertia weight controls the impact of previous history of velocities on the current one and the constriction controls the magnitude of the velocities. The most common approach in the literature is to fix χ. Regarding ω, the common approach is to set a large value at the beginning and then, gradually decrease it.

The method proposed in this case study employs the same notation as in (12), but uses only the $\omega_{inertia}$ (in the proposed method case, χ will be fixed to 1.0). During optimization this parameter is initialized with a large value, and then decreased, as seen in.[48]

As mentioned, the CWA approach is used in this case study for multi-objective optimization (MOO). Digital watermarking is in essence a multi-objective optimization problem. On its inception, PSO was able to handle only single objective problems. However, through a weighted sum of the fitness values, it is possible to aggregate several fitness functions (f_i) into a global one (F). This process is known as Weighted Aggregation[48]

$$F = \sum_{i=1}^{N_{fitness}} \gamma_i f_i(x) \quad (13)$$

where γ_i is a non-negative weight, having $\sum_{i=1}^{N_{fitness}} \gamma_i = 1$ and $N_{fitness}$ is the number of fitness functions.

4.2. *Simulation results*

In this subsection the performance of the adaptive watermarking system based on the canonical PSO is evaluated for the optimization of the baseline watermarking system. This adaptive watermarking system is compared with the greedy technique proposed by Muharemagic.[6] The experiments were conducted in the CCITT database which was also used by Muharemagic.[6] This database is composed of 8 bi-tonal images (Figure 8. All images have the same dimension (2376×1728 pixels) and were scanned at 200 dpi.

The same 35×26 pixels OK and BIz binary logos (Figures 9a and b) were used as the fragile and robust watermarks, respectively. An eight-bit random number along with its CRC-4 code were appended to the logo in order to allow the search of the parameters during detection. This resulted in a payload of 922 bits for each watermark (910 bits for the logo and 12 bits of self-verifiable data). For the robust watermark, $Q = 10$ and $\alpha = 0.7$ are employed while for the fragile watermark $Q = 2$ and $\alpha = 0.95$ are employed. The values of Q were chosen based on the literature.[6]

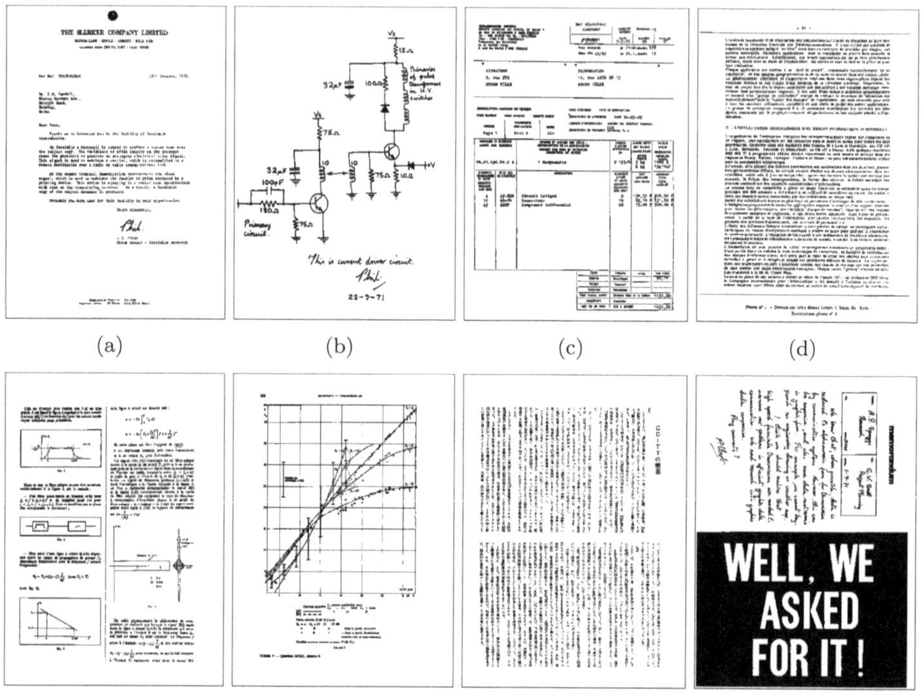

(a) (b) (c) (d)

(e) (f) (g) (h)

Fig. 8. Samples from the CCITT database (left to right, top-down, CCITT1 to CCITT8).

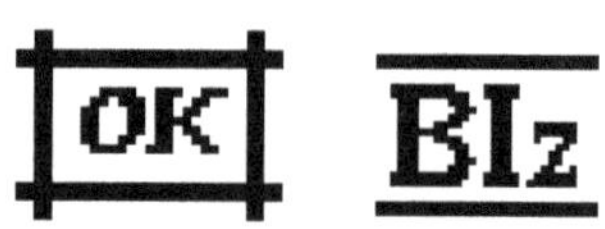

(a) (b)

Fig. 9. OK and BIz logos.[6]

For the α, a few different options were evaluated empirically, in order to find values leading to a similar DRDM as reported in[6] for the given value of Q (the DRDM values reported in Table 4 are very similar to those reported by Muharemagic).

It is interesting to observe that this baseline watermarking system provides high levels of quality and robustness for this proposed dataset and payload, even before optimization. For example, Figures 10 and 11 demonstrate the visual impact of embedding these two watermarks, using a 64×64 partition block size and a 3×3 SNDM window size into the CCITT2 image.

The robustness can be demonstrated by manipulating a given number of pixels in the watermarked image (as in[6]). In Figure 12, the watermarked CCITT1 image was manually modified by 64, 128, 192 and 256 pixels respectively. Figure 13 shows the detection of the BIz and OK logos in these scenarios.

As expected, the robust watermark was more resistant against tampering. For the four attacks (64, 128, 192 and 256 pixel modifications in the watermarked image) only 0.2%, 0.8%, 0.8% and 1.3% of the pixels in the BIz logo, respectively, were corrupted against 6.4%, 11.3%, 16.4% and 20%, respectively, for the OK logo.

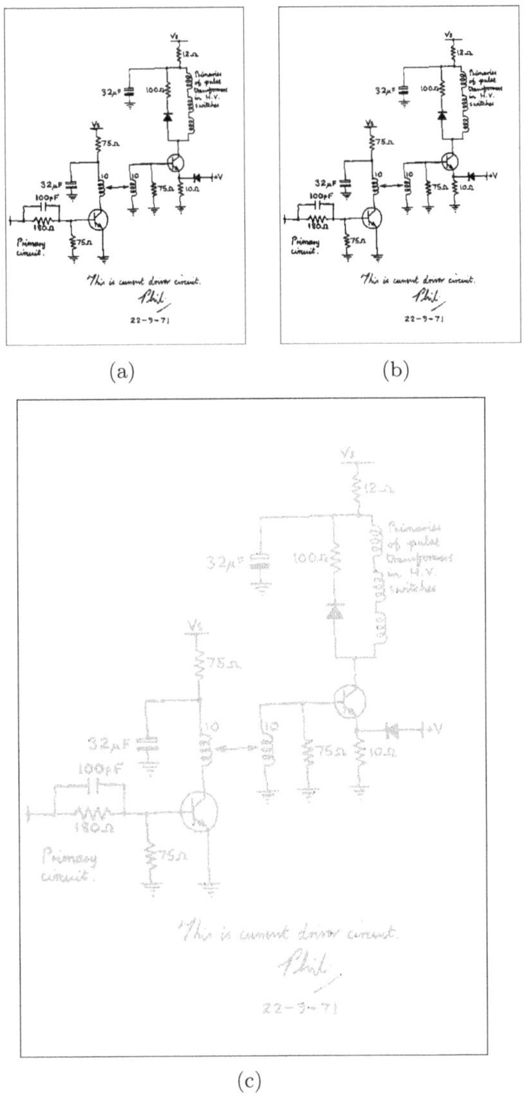

(a) (b) (c)

Fig. 10. Visual impact of multilevel embedding in the CCITT2 image. The BIz logo was embedded as a robust watermark ($Q = 10$ and $\alpha = 0.77$) while the OK logo was embedded as a fragile watermark ($Q = 2$ and $\alpha = 1$). (a) Original image. (b) Watermarked image. (c) Difference image.

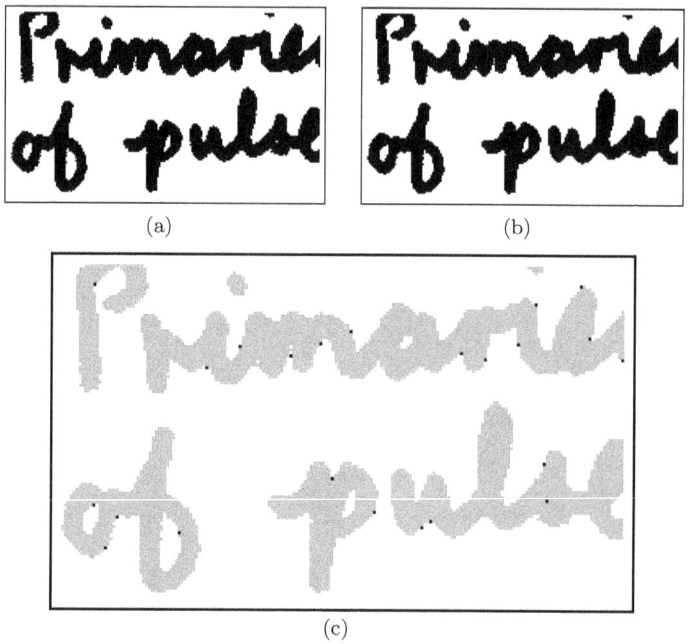

Fig. 11. Detail on visual impact of multilevel in the CCITT2 image. The BIz logo was embedded as a robust watermark ($Q = 10$ and $\alpha = 0.77$) while the OK logo was embedded as a fragile watermark ($Q = 2$ and $\alpha = 1$). (a) Original image. (b) Watermarked image. (c) Difference image.

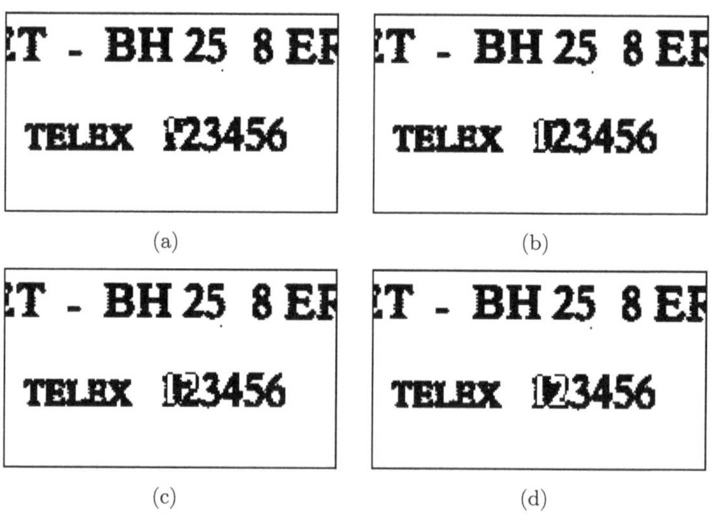

Fig. 12. Flipping pixels on CCITT1 image. The BIz logo was embedded as a robust watermark ($Q = 10$ and $\alpha = 0.77$) while the OK logo was embedded as a fragile watermark ($Q = 2$ and $\alpha = 1$). Then, four different modifications were applied to image (a) Modification of 64 pixels. (b) Modification of 128 pixels. (c) Modification of 192 pixels. (d) Modification of 256 pixels.

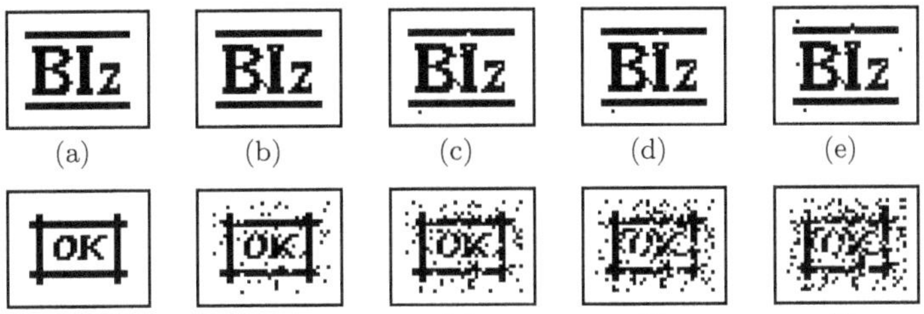

(a) (b) (c) (d) (e) (f) (g) (h) (i) (j)

Fig. 13. Detection of watermarks on watermarked/attacked CCITT1 image. A given number of pixels was modified in the watermarked image. Effect of modifying (a) No pixel, BIz watermark. (b) 64 pixels, BIz watermark. (c) 128 pixels, BIz watermark. (d) 192 pixels, BIz watermark. (e) 256 pixels, BIz watermark. (f) No pixel, OK watermark. (g) 64 pixels, OK watermark. (h)128 pixels, OK watermark. (i) 192 pixels, OK watermark. (j) 256 pixels, OK watermark.

During experiments, five different options of block size were employed, that is $B = \{8 \times 8, 16 \times 16, 32 \times 32, 64 \times 64, 128 \times 128\}$. Regarding the shuffling key, a set containing 16 different randomly generated seeds was employed. Finally, three different SNDM window sizes were considered during optimization (3×3, 5×5 and 7×7).

Experiments were performed for each image and the parameters found were reported. The optimal watermark MSE results for the fragile and robust watermarks and the SNDM were also reported.

4.2.1. *Baseline adaptive system*[6]

As mentioned, the choice of block size is based on heuristic. Therefore, since the same payload was applied to all images, the block size chosen was always the same. The results can be seen in Table 4.

Table 4. Performance of the adaptive system proposed by Muharemagic in the CCITT database.[6]

Image	Block size	Key	SNDM window	MSE fragile	MSE robust	DRDM
CCITT1	64 × 64	5	5 × 5	15	29	0.0053
CCITT2	64 × 64	15	7 × 7	13	26	0.0067
CCITT3	64 × 64	8	3 × 3	19	30	0.0017
CCITT4	64 × 64	14	3 × 3	18	23	0.0002
CCITT5	64 × 64	5	3 × 3	17	28	0.0020
CCITT6	64 × 64	12	5 × 5	17	32	0.0050
CCITT7	64 × 64	1	3 × 3	14	30	0.0006
CCITT8	64 × 64	11	5 × 5	17	29	0.0034

Table 5. Performance of the canonical PSO version of the adaptive system proposed by Muharemagic in the CCITT database.

Image	Block size	Key	SNDM window	MSE fragile	MSE robust	DRDM
CCITT1	64×64	15	5×5	16	19	0.0056
CCITT2	32×32	5	7×7	17	23	0.0034
CCITT3	64×64	2	5×5	22	13	0.0017
CCITT4	64×64	1	3×3	22	15	0.0003
CCITT5	8×8	3	7×7	20	16	0.0005
CCITT6	32×32	9	7×7	18	16	0.0028
CCITT7	16×16	9	3×3	17	13	0.0003
CCITT8	32×32	2	7×7	16	25	0.0018

4.2.2. *Adaptive system based on PSO*

The canonical PSO was employed to optimize the same three parameters (block size, shuffling key and SNDM window size). The swarm was composed of 20 particles, both cognitive and social constants were set to 2.05. The inertia weight was initialized with 1.2 and gradually decreased until 0.01. The number of iterations was set to 100. The maximum number of iterations without improvement in the global maximum was set to 10. The same aggregation weight of $\frac{1}{3}$ was employed for the three objective functions. The PSO topology chosen was the G-best (one single global best for the whole swarm). The results can be seen on Table 5.

4.2.3. *Discussion*

On average, both the robustness of the robust watermark and the quality of the watermarked image were improved when compared to the Muharemagic adaptive system. It is interesting to observe that the robustness of the fragile watermark has degraded. This happened because in the adaptive method proposed by Muharemagic there is an "hierarchy" in optimization of the robustness of the watermarks. The first objective is to find parameters that improve the robustness of the fragile watermark. For PSO in contrast, the objective is to minimize a weighted sum of the three functions equally. Considering that the robustness of the robust watermark (which naturally requires more payload than the fragile mark) was increased with an improvement in the fidelity, it was expected that some of the channel capacity employed by the fragile mark would be transferred to the robust watermark. However there is no concern regarding this decrease in the robustness of the fragile mark since this type of watermark does not have to be robust anyway.

Another factor that led to the decrease in robustness of fragile watermark is that a single-objective version of PSO was employed. As mentioned before, such type of algorithm usually favours one objective in detriment of the others. A Pareto-based multi-objective evolutionary algorithm such as the NSGA-II[56] or MOPSO[49] should probably find a more balanced trade-off between the three objectives.

It is possible to observe that the images are from different classes and this reflected in the optimal solution found by the PSO. For this reason, full optimization must be performed for each image, a costly process.

5. Conclusion

In this chapter a brief introduction to digital watermarking was provided, followed by an extensive survey on intelligent watermarking. As observed, most of the efforts in this area are concentrated in the optimization of embedding parameters with the use of evolutionary computing. One of the drawbacks of these approaches is the computational burden of optimization.

In intelligent watermarking, evolutionary computing is by far the most employed approach in what regards the optimization of embedding parameters. There are many reasons for this popularity, the simplicity of EC algorithms, their adaptability, which allows the direct application to many different types of digital watermarking techniques. However, intelligent watermarking has some drawbacks. One of them concerns the communication of embedding parameters to the detector. One of the main advantages of digital watermarking over other security techniques is the self-contained protection it offers. Notwithstanding, most watermarking systems require the knowledge of some of the embedding parameters on detection. A common approach is to use a fixed set of parameters and communicate them to all the detectors through a secure channel. But in intelligent watermarking, these parameters must be optimized according to each particular image. This can limit the application of intelligent watermarking, mainly in situations where the overhead of a secure channel to communicate these parameters is not acceptable. Another approach is to use part of the payload to either embed a training sequence, in the form of an Error Correction Code (ECC) or as a second watermark.

Another drawback of the use of EC in the optimization of digital watermarking systems is its high computational cost. Depending on the complexity of the problem, an EC algorithm such as PSO or GA can require thousands of fitness function evaluations. In intelligent watermarking this means thousands of costly embedding, detection and image processing (attack) operations. This limits intelligent watermarking to small sets of images. For this reason, decreasing the computational burden of evolutionary optimization techniques is a key issue, which can make possible the industrial use of intelligent watermarking.

As digital watermarking task comprises embedding a signal into an image in accordance with robustness and quality constraints, it can be said that it is in essence a multi-objective optimization problem. In the literature, many research works have tried to address this multi-objective problem with the use of a weighted sum of the objective functions. However, it is a known problem in evolutionary optimization that this approach usually favours one objective in detriment of the others. The alternative is instead of conducting a search for a single global solution to the problem, try to find a set of nondominated solutions, known as Pareto-optimal as in.[39,45]

Supervised learning has also been used in the context of creating adaptive watermarking systems. Most classifiers will rely in some sort of optimization (e.g. gradient descent for MLP and quadratic programming for SVM). However, differently than in EC-based intelligent watermarking, optimization is employed in the task of finding optimal parameters for the classifiers and not for the watermarking systems. These classifiers are then employed in intelligent watermarking either replacing a watermarking process or in the evaluation of a watermarking property.

As a case study, PSO was employed to the task of optimizing a bi-tonal watermarking system. It was possible to observe in this simulation that although EC can be useful in the task of finding a near-optimum trade-off between robustness and quality, the performance is bounded by theoretical limitations of the watermarking system (embedding capacity, etc). The advantage of PSO over greedy search is that it allows a paralel search for optimal parameters (that is, adjusting more than one parameter at the same time). This can make possible the optimization of more complex parameters such as the embedding strength (α) and the quantization step size (Q). Adjusting these parameters against some removal attacks can be considered a future work. Another important issues to be addressed include: decreasing the computational burden of EC and employing a Pareto-based multi-objective optimization technique.

Acknowledgements

The authors would like to thank BancTec Canada and Natural Sciences and Engineering Research Council (NSERC) Canada for the financial support and Defence Research and Development Canada (DRDC) Valcartier for the strategic collaboration.

A.1. List of Symbols

Table A.1. List of symbols.

Symbol	Description
m	Message.
w_a	Added watermark.
c_o	Cover image.
c_w	Marked image.
c_{wn}	Marked/attacked image.
m_n	Detected message.
D_b	Distance matrix.
b	SNDM window size.
cp	Candidate pixel.
N_b	$b \times b$ neighbourhood of cp.
N	Number of bits.
W_d	Normalized weight matrix.
$DRDM$	Distance reciprocal distortion metric.
a_d, b_d	Original and modified pixels, positioned at the dth location of w_r.
s_k	DWT coefficient.
α	Embedding strength.
ϕ	DCT coefficient.
B	Set of embedding block sizes.
K	Set of shuffling keys.
B_s	Set of SNDM window sizes.
x	Cover signal (number of black pixels at a given block).
q	Quantized/embedded signal.
D	Alphabet size.
$SQ_Q\{\}$	Scalar Uniform Quantizer.
Q	Quantization step size.
κ	Watermark security key.
v	Noise source.
w_n	Watermark signal (after noise/attack).
q_n	Received/quantized signal.
X	Particle position.
V	Particle velocity.
r_1, r_2	Random numbers (PSO).
P	Local best.
P_g	Neighbourhood best.
$\omega_{inertia}$	Inertia weight.
χ	Constriction factor.
c_1, c_2	Acceleration constants.
γ	Fitness function weight.
F	Weighted sum of fitness functions.
$N_{fitness}$	Number of fitness functions.

References

1. I. Cox, M. Miller, and J. Bloom, *Digital Watermarking.* (Morgan Kaufmann Publishers, 2002).
2. F. Petitcolas, R. Anderson, and M. Kuhn, Information hiding-a survey, *Proceedings of the IEEE.* **87**(7), 1062–1078, (Jul 1999).
3. M. Chen, E. K. Wong, N. Memon, and S. Adam, Recent developments in document image watermarking and data hiding, *Proc. SPIE.* **4518**, 166–176, (2001).
4. J. H. Holland, *Adaptation in natural and artificial systems.* (MIT Press, Cambridge, MA, USA, 1992). ISBN 0-262-58111-6.
5. J. Kennedy and R. Eberhart. Particle swarm optimization. In *Proc. IEEE Int'l. Conf. on Neural Networks*, Perth, Australia (November-December 1995, 1995).
6. E. Muharemagic. *Adaptive Two-Level Watermarking for Binary Document Images.* PhD thesis, Florida Atlantic University (December, 2004).
7. M. Wu and B. Liu, Data Hiding in Image and Video: Part I - Fundamental Issues and Solutions, *IEEE Transactions on Image Processing.* **12**(6), 685–695, (2003).
8. H. Pan, Y. Chen, , and Y. Tseng. A secure data hiding scheme for two-color images. In *IEEE Symposium on Computers and Communication*, pp. 750–755, (2000).
9. Y. Tseng and H. Pan, Secure and invisible data hiding in 2-color images, *Twentieth Annual Joint Conference of the IEEE Computer and Communications Societies.* **2**, 887–896, (2001).
10. I. Awan, S. A. M. Gilani, and S. A. Shah. Utilization of maximum data hiding capacity in object-based text document authentication. In *IIH-MSP '06: Proceedings of the 2006 International Conference on Intelligent Information Hiding and Multimedia*, pp. 597–600, Washington, DC, USA, (2006).
11. J. Zhao and E. Koch. Embedding robust labels into images for copyright protection. In *International Congress on Intellectual Property Rights for Specialised Information, Knowledge and New Technologies*, Vienna, Austria (21-25, 1995).
12. Q. Mei, E. K. Wong, and N. Memon. Data hiding in binary text documents. In *SPIE Proc. Security and Watermarking of Multimedia Contents III*, San Jose, USA, (Jan. 2001).
13. A. T. S. Ho, N. B. Puhan, P. Marziliano, A. Makur, and Y. L. Guan. Perception based binary image watermarking. In *2004 IEEE International Symposium on Circuits and Systems (ISCAS)*, pp. 23–26, (2004).
14. H. Yang and A. C. Kot, Binary image authentication with tampering localization by embedding cryptographic signature and block identifier, *Signal Processing Letters, IEEE.* **13**(12), 741–744, (Dec. 2006).
15. C. Zhang and Z. Qiu. Fragile watermarking with quality control for binary images. In *Proceedings of the Fourth International Conference on Machine Learning and Cybernetics*, pp. 4952–4956 (August, 2005).
16. M. Wu and B. Liu, Data Hiding in Binary Image for Authentication and Annotation, *IEEE Transactions on Multimedia.* **6**(4), 528–538, (2004).
17. H. Lu, X. Shi, Y. Q. Shi, A. C. Kot, and L. Chen. Watermark embedding in DC components of DCT for binary images. In *Multimedia Signal Processing, 2002 IEEE Workshop on*, pp. 300–303, (9-11 Dec. 2002).
18. I. J. Cox, J. Kilian, T. Leighton, and T. Shamoon. A secure, robust watermark for multimedia. In *Workshop on Information Hiding*, pp. 1–16, (1996).
19. S. Rezazadeh and M. Yazdi. An adaptive watermarking scheme based on morphological binary wavelet decomposition. In *Signal Processing, The 8th International Conference on*, (16-20 2006).

20. J. J. K. ÓRuanaidh and T. Pun, Rotation, scale and translation invariant spread spectrum digital image watermarking, *Signal Process.* **66**(3), 303–317, (1998).
21. M. Wu, H. Yu, and B. Liu, Data Hiding in Image and Video: Part II - Fundamental Issues and Solutions, *IEEE Transactions on Image Processing.* **12**(6), 696–705, (2003).
22. M. Wu. *Multimedia Data Hiding.* PhD thesis, Princeton University, (2001).
23. S. Voloshynovskiy, S. Pereira, T. Pun, J. Eggers, and J. Su, Attacks on digital watermarks: classification, estimation based attacks, and benchmarks, *Communications Magazine, IEEE.* **39**(8), 118–126, (Aug 2001).
24. M. Barni and F. Bartolini, *Watermarking Systems Engineering: Enabling Digital Assets Security and Other Applications.* (CRC Press, 2004).
25. M. Costa, Writing on dirty paper (corresp.), *Information Theory, IEEE Transactions on.* **29**(3), 439–441 (May, 1983). ISSN 0018-9448.
26. C. M. Bishop, *Neural Networks for Pattern Recognition.* (Oxford University Press, 1996).
27. V. Vapnik, *The Nature of Statistical Learning Theory.* (Springer, 1995).
28. R. O. Duda, P. E. Hart, and D. G. Stork, *Pattern Classification.* (Wiley-Interscience Publication, 2000).
29. C. Chang and I. Lin, *Intelligent Watermarking Techniques*, chapter Robust Image Watermarking Systems Using Neural Networks. World Scientific Co., (2004).
30. C. Chang and I. Lin, *Intelligent Watermarking Techniques*, chapter A Perceptually Tuned Watermarking Scheme for Digital Images Using Support Vector Machines. World Scientific Co., (2004).
31. S. F. Tahir, A. Khan, A. Majid, and A. M. Mirza, Support vector machine based intelligent watermark decoding for anticipated attack, *International Journal of applied Mathematics and Computer Sciences.* **1**(1), 7–12, (2005).
32. K. J. Davis and K. Najarian. Maximizing strength of digital watermarks using neural networks. In *Neural Networks, 2001. Proceedings. IJCNN '01. International Joint Conference on*, vol. 4, pp. 2893–2898, (2001).
33. C. Shieh, H. Huang, F. Wang, and J. Pan, Genetic watermarking based on transform-domain techniques, *Pattern Recognition.* **37**(3), 555–565, (2004).
34. Z. Lee, S. Lin, S. Su, and C. Lin, A hybrid watermarking technique applied to digital images, *Applied Soft Computing.* **8**(1), 798–808, (2007).
35. R. Ji, H. Yao, S. Liu, and L. Wang. Genetic algorithm based optimal block mapping method for LSB substitution. pp. 215–218, (Dec. 2006).
36. X. Li and J. Wang, A steganographic method based upon JPEG and particle swarm optimization algorithm, *Information Sciences.* **177**(15), 3099–3109, (2007).
37. Z. Wei, H. Li, J. Dai, and S. Wang. Image watermarking based on genetic algorithm. pp. 1117–1120, (July 2006).
38. J. Pan, H. Huang, and L. Jain, *Intelligent Watermarking Techniques*, chapter Genetic Watermarking on Spatial Domain. World Scientific Co., (2004).
39. D. Sal, M. Grana, and A. d'Anjou. A MOGA to place the watermark in an hyperespectral image. In *2006 IEEE International Geoscience and Remote Sensing Symposium*, Denver, USA (July 31, August 04, 2006).
40. K. Deb, A. Pratap, S. Agarwal, and T. Meyarivan, A fast and elitist multiobjective genetic algorithm: Nsga-ii, *IEEE Transactions on Evolutionary Computation.* **6**, 182–197, (2002).
41. C. Chen and C. Lin, A GA-based nearly optimal image authentication approach, *International Journal of Innovative Computing, Information and Control.* **3**(3), 631–640, (2007).

42. T. E. Areef, H. S. Heniedy, and O. M. O. Mansour, Optimal transform domain watermark embedding via genetic algorithms, *Information and Communications Technology, 2005. Enabling Technologies for the New Knowledge Society: ITI 3rd International Conference on.* pp. 607–617, (5-6 Dec. 2005).
43. F. Y. Shih and Y. Wu, Enhancement of image watermark retrieval based on genetic algorithms, *Journal of Visual Communication and Image Representation.* **16**(2), 115–133, (2004).
44. P. Kumsawat and K. A. amd A. Srikaew, A new approach for optimization in image watermarking by using genetic algorithms, *IEEE Transactions on Signal Processing.* **53**(12), 4707–4719, (2005).
45. D. S. Díaz and M. G. Romay. Introducing a watermarking with a multi-objective genetic algorithm. In *GECCO '05: Proceedings of the 2005 conference on Genetic and evolutionary computation*, pp. 2219–2220, New York, NY, USA, (2005). ACM.
46. A. Khan and A. M. Mirza, Genetic perceptual shaping: Utilizing cover image and conceivable attack information during watermark embedding, *Inf. Fusion.* **8**(4), 354–365, (2007). ISSN 1566-2535.
47. Y. Wu and F. Y. Shih, Genetic algorithm based methodology for breaking the steganalytic systems, *Systems, Man, and Cybernetics, Part B, IEEE Transactions on.* **36** (1), 24–31, (Feb. 2006).
48. K. E. Parsopoulos and M. N. Vrahatis, Recent approaches to global optimization problems through particle swarm optimization, *Natural Computing: an international journal.* **1**(2-3), 235–306, (2002). ISSN 1567-7818.
49. C. Coello, G. Pulido, and M. Lechuga, Handling multiple objectives with particle swarm optimization, *Evolutionary Computation, IEEE Transactions on.* **8**(3), 256–279 (June, 2004).
50. J. Kennedy and R. Eberhart, A discrete binary version of the particle swarm algorithm, *Systems, Man, and Cybernetics, 1997. 'Computational Cybernetics and Simulation'., 1997 IEEE International Conference on.* **5**, 4104–4108 (Oct, 1997).
51. B. Chen and G. Wornell, Quantization index modulation: a class of provably good methods for digital watermarking and information embedding, *IEEE Transactions on Information Theory.* **47**(4), 1423–1443, (2001).
52. J. Eggers, J. Su, and B. B. Girod, Scalar Costa Scheme for Information Hiding, *IEEE Transactions on Signal Processing.* **51**(4), (2003).
53. N. B. Puhan and A. T. S. Ho. Binary document image watermarking for secure authentication using perceptual modeling. In *2005 IEEE International Symposium on Signal Processing and Information Technology*, pp. 393–398, (2005).
54. R. Poli, J. Kennedy, and T. Blackwell, Particle swarm optimisation: an overview, *Swarm Intelligence Journal.* **1**(1), 33–57, (June 2007).
55. J. Kennedy. Some issues and practices for particle swarm. In *Proc. IEEE Swarm Intelligence Symposium*, (2007).
56. K. Deb, *Multi-Objective Optimization Using Evolutionary Algorithms.* (Wiley, 2001).

CHAPTER 5.6

A PARALLEL AND MODULAR PATTERN CLASSIFICATION FRAMEWORK FOR LARGE-SCALE PROBLEMS

Bao-liang Lu[1,2] and Xiao-lin Wang[1]

[1] *Center for Brain-Like Computing and Machine Intelligence*
Department of Computer Science and Engineering

[2] *MOE-Microsoft Key Laboratory for Intelligent Computing and Intelligent Systems*
Shanghai Jiao Tong University
800 Dong Chuan Rd., Shanghai 200240, China
E-mail: bllu@sjtu.edu.cn

The number of samples that are available on the internet to train pattern classifiers is increasing rapidly, while traditional pattern classification techniques based on a single computer system are powerless to process these large-scale data sets. This chapter presents a parallel and modular pattern classification framework for coping with large-scale pattern classification problems. The proposed framework follows a divide-and-conquer strategy that easily assigns a given large-scale problem to an available parallel and distributed computing infrastructure. The framework consists of three independent parts: decomposing training data sets, training component classifiers in a parallel way, and combining trained component classifiers. In order to evaluate the performance of the proposed framework, we perform experiments on a large-scale Japanese patent classification problem, containing about 3,500,000 patent documents. The experimental results show that our framework has the following attractive features: (a) The framework is general, and therefore any traditional pattern classification techniques such as support vector machines can be easily embedded in the framework as component classifiers. (b) The framework can incorporate explicit domain or prior knowledge into learning through the process of dividing training data sets. (c) The framework has good scalability and is easily implementable in hardware.

1. Introduction

More and more large-scale classification problems appear in the field of machine learning and pattern recognition. By large-scale we mean that the training set is extremely large, typically to such an extent that traditional classifiers cannot finish the learning task within a bearable time. For example, recently, the training of support vector machines (SVMs) on large real-world data sets is reported to take several weeks. With the emergence of these large-scale classification problems, the efficiency of classifiers, besides accuracy, has become a subject of great focus.

A large-scale classification problem has huge time and memory costs. The ultimate solution to reducing time and memory costs is parallelism, that is, splitting one large-scale classification task among several CPUs, by which the system can meet an expected efficiency requirement. Unfortunately, most traditional classifiers are sequential in spirit, so we have to expend great efforts in parallelizing them. Recently, several parallel implementations of SVMs have been released [8, 4, 14, 34, 26, 13]. There are also parallel implementations of other classifiers, such as k-NN [3].

A very useful technology for these large-scale classification problems is ensemble learning. It employs a group of classifiers for a classification task, instead of one single classifier. Ensemble learning aims to achieve high accuracy, and it shows great efficiency in large-scale problems. Some of the ensemble learning methods require the sub-classifiers to be trained sequentially, while the others allow parallel training sub-classifiers. For methods such as Adaboost, the training set of a sub-classifier relies on feedback from previous trained sub-classifiers, and thus parallel learning is infeasible. As for other parallel ensemble learning methods, they are ideal solutions for large-scale tasks.

In this chapter, we introduce a parallel and modular pattern classification framework, named min-max modular network (M^3-network) [20, 21, 24]. This framework is a kind of a parallel ensemble learning method. First we train component classifiers on the subsets of the original data set. Then we make these component classifiers produce predictions on the arriving test sample. In the end we integrate all the predictions by two module combination rules, namely the minimization principle and the maximization principle.

To solve a large-scale multi-class classification problem, the M^3-network consists of three main steps: (a) Decompose the original multi-class problem into two-class problems and further divide the two-class problems which are difficult to be learned into a series of relatively smaller and balanced two-class subproblems. (b) Train all the two-class subproblems in parallel. (c) Combine all the trained component classifiers into a hierarchical, parallel, and modular network that serves as the solution to the original problem.

Many studies prove that M^3-network is an efficient classifier, especially in solving large-scale and complex pattern classification problems [22, 10, 32, 19, 5, 6]. M^3-netowrk has the following three advantages over traditional pattern classification approaches: (a) A large-scale and complex multi-class problem can be decomposed to two-class subproblems as small as the user expects. (b) The two-class subproblems are independent to each other, therefore, they can be solved in parallel without the trouble of communication. (c) The two module combination principles are simple, which are easily implementable in software and hardware.

2. Min-Max Modular Network

Lu and Ito proposed the Min-Max Modular Network (M^3-network for short) in order to solve hard classification problems in 1997 [20, 21]. M^3-network is a sort of ensemble learning methods that employs a group of classifiers for one classification problem. The M^3-network follows the principle of divide and conquer, dividing the whole problem into small pieces and solving them one by one, and adopts a 3-step work flow: task decomposition, training component classifiers and module combination (see Figure 1).

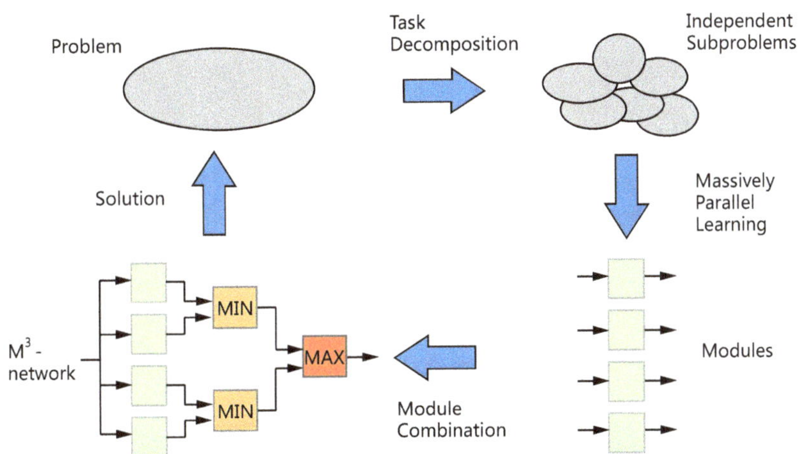

Fig. 1. Working Flow of the M^3-network

Before formally describing the M^3-network, we would like to present an illustration (Figure 2) [25]. Suppose that subfigure (a) is the two-class classification problem that needs to be solved, where small red disks represent positive samples and small blue rectangles represent negative ones. Though simple at first sight, it's actually a non-linearly separable problem. To demonstrate M^3-network learning, we divide samples of each class into two subsets, surrounded by dashed lines in subfigure (a). According to this partition of the training data set, we generate four smaller classification subproblems shown in subfigures (b) to (e). Then we train classifiers on these subproblems, where dashed lines represent the discriminant surfaces. After that we derive separately subfigures (f) and (g) from subfigures (b) through (e) with the *minimization* operator, resulting in negative zone expanding. Finally, we derive subfigure (h) from subfigures (f) and (g) with the *maximization* operator, resulting in positive zone expanding. From subfigure (h), it is obvious that we find the correct discriminant plane to the classification problem.

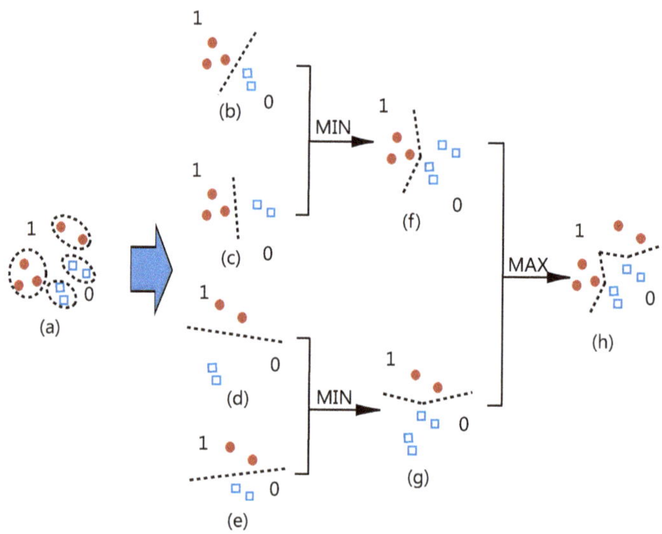

Fig. 2. Illustration of the M^3-network learning

2.1. *Task Decomposition*

Let T denote the training data set for a K-class classification problem,

$$\mathcal{T} = \{(X_i, Y_i)\}_{i=1}^{L} \tag{1}$$

where $X_i \in R^d$ is the training input, $Y_i \in R^K$ is the corresponding desired output, and L is the total number of training samples.

At the beginning of the task decomposition process, a large-scale K-class classification problem is divided into a series of relatively smaller two-class subproblems by using a one-versus-one or one-versus-rest task decomposition strategy. Suppose a K-class classification problem is decomposed by a one-versus-one strategy into the following $K(K-1)/2$ two-class subproblems:

$$\begin{gathered}\mathcal{T}_{ij} = \{(X_l^i, +1)\}_{l=1}^{L_i} \cup \{(X_l^j, -1)\}_{l=1}^{L_j} \\ \text{for } i = 1, \cdots, K \text{ and } j = i+1, \cdots, K\end{gathered} \tag{2}$$

where $X_l^i \in \mathcal{X}_i$ and $X_l^j \in \mathcal{X}_j$ are the training inputs belonging to class $\mathcal{C}_i$ and class $\mathcal{C}_j$, respectively; $\mathcal{X}_i$ is the set of training inputs belonging to class $\mathcal{C}_i$; L_i denotes the number of data in $\mathcal{X}_i$; $\cup_{i=1}^{K}\mathcal{X}_i = \mathcal{X}$; and $\sum_{i=1}^{K} L_i = L$. In this chapter, the training data in a two-class problem are called *positive* training data if their desired outputs are $+1$, and they are called *negative* training data if their desired outputs are -1.

Even though the two-class problems defined by Eq. (2) are smaller than the original K-class problem, this partition may be not adequate for parallel learning. Since the number of samples in each class varies largely, the sizes of these two-class problems can be quite different. Thus the training period can be delayed by some larger two-class problems. A large class and a small class together form an imbalanced classification problem; there are too many samples on one side, which increases the difficulty in training classifiers [2]. To speed up training and to improve classification accuracy, all the large and imbalanced two-class problems are further divided into smaller and more balanced two-class problems.

Assume that $\mathcal{X}_i$ is partitioned into N_i subsets in the form

$$\mathcal{X}_{ij} = \{X_l^{ij}\}_{l=1}^{L_i^j} \quad \text{for } j = 1, \cdots, N_i \text{ and } i = 1, \cdots, K, \tag{3}$$

where $1 \leq N_i \leq L_i$ and $\cup_{j=1}^{Ni}\mathcal{X}_{ij} = \mathcal{X}_i$.

After partitioning $\mathcal{X}_i$ into N_i subsets, every two-class problem $\mathcal{T}_{ij}$ defined by Eq. (2) is further divided into $N_i \times N_j$ smaller and more balanced two-class subproblems:

$$\begin{aligned} \mathcal{T}_{ij}^{(u,v)} &= \{(X_l^{(iu)}, +1)\}_{l=1}^{L_i^{(u)}} \cup \{(X_l^{(jv)}, -1)\}_{l=1}^{L_j^{(v)}} \\ &\quad \text{for } u = 1, \cdots, N_i,\ v = 1, \cdots, N_j, \\ &\quad i = 1, \cdots, K, \text{ and } j = i+1, \cdots, K, \end{aligned} \tag{4}$$

where $X_l^{(iu)} \in \mathcal{X}_{iu}$ and $X_l^{(jv)} \in \mathcal{X}_{jv}$ are the training inputs belonging to class $\mathcal{C}_i$ and class $\mathcal{C}_j$, respectively; $\sum_{u=1}^{N_i} L_i^{(u)} = L_i$ and $\sum_{v=1}^{N_j} L_j^{(v)} = L_j$.

2.2. *Training Component Classifiers*

After task decomposition, all of the two-class subproblems are treated as completely independent, non-communicating tasks in the learning phase. Therefore, all the two-class subproblems defined by Eq. (4) are efficiently learned in a massively parallel way.

From Eqs. (2) and (4), we see that a K-class problem is divided into

$$\sum_{i=1}^{K-1} \sum_{j=i+1}^{K} N_i \times N_j \tag{5}$$

two-class subproblems. The number of training data for each of the two-class subproblems is approximately

$$\lceil L_i/N_i \rceil + \lceil L_j/N_j \rceil, \tag{6}$$

where $\lceil z \rceil$ denotes the smallest integer than or equal to z. Since $\lceil L_i/N_i \rceil + \lceil L_j/N_j \rceil$ is independent of the number of classes, K, the size of each two-class subproblems is much smaller than the original K-class problem for the reasonable values of N_i and N_j.

Traditional machine learning algorithms and pattern classification approaches such as multilayer neural networks [21], support vector machines [23] and k-nearest neighbor algorithm [35], can be used to learn the two-class subproblems defined by

Eq. (4). In this chapter, most commonly used SVMs are selected as component classifiers and used to learn all of the two-class subproblems.

2.3. *Module Combination*

After all of the two-class subproblems defined by Eq. (4) have been learned by SVMs, all the trained SVMs are integrated into a M^3-SVM with the *Min* and *Max* units according to the minimization principle and maximization principle [23]. The function of the *Min* unit is to find a minimum value from its multiple inputs. The transfer function of the *Min* unit is given by

$$q(\mathbf{x}) = \min_{i=1}^{P} \mathrm{M}_i(\mathbf{x}) \tag{7}$$

where x denotes the input variable. The function of the *Max* unit is to find a maximum value from its multiple inputs. The transfer function of the *Max* unit is given by

$$q(\mathbf{x}) = \max_{i=1}^{P} \mathrm{M}_i(\mathbf{x}) \tag{8}$$

Minimization Principle: Suppose a two-class problem $\mathcal{B}$ is divided into P smaller two-class subproblems, $\mathcal{B}_i$ for $i = 1, \cdots, P$, and also suppose that all the two-class subproblems have the same positive training data and different negative training data. If the P two-class subproblems are correctly learned by the corresponding P individual SVMs, M_i for $i = 1, \cdots, P$, then the combination of the P trained SVMs with a *Min* unit will produce the correct output for all the training inputs in $\mathcal{B}$.

Maximization Principle: Suppose a two-class problem $\mathcal{B}$ is divided into P smaller two-class subproblems, $\mathcal{B}_i$ for $i = 1, \cdots, P$, and also suppose that all the two-class subproblems have the same negative training data and different positive training data. If the P two-class subproblems are correctly learned by the corresponding P individual SVMs, M_i for $i = 1, \cdots, P$, then the combination of the P trained SVMs with a *Max* unit will produce the correct output for all the training input in $\mathcal{B}$.

Note that the module combination strategy based on the minimization and the maximization principles is called the min-max module combination in this chapter. Figure 3 illustrates the min-max module combination for a two-class problem. Here the two-class problem is first decomposed into $N_1 \times N_2$ two-class subproblems. Then each component classifier is trained on the corresponding two-class subproblem. Finally a M^3-network is constructed by combining these component classifiers. In this case, the M^3-network consists of $N_1 \times N_2$ component classifiers, N_1 *Min* units, and one *Max* unit.

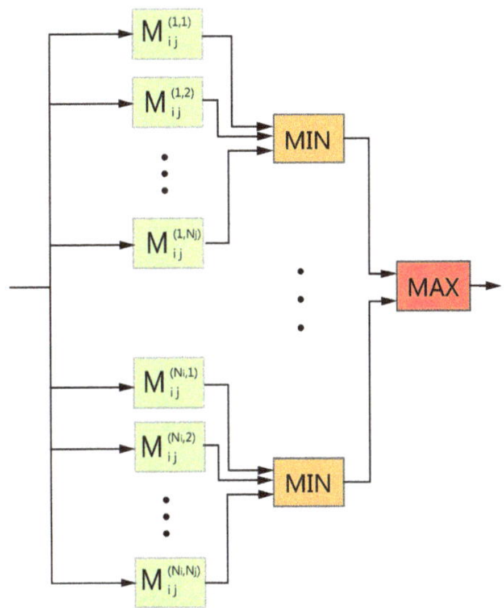

Fig. 3. Illustration of min-max module combination for a two-class problem.

3. Parallel Implementation of M^3-network

As a principle, improvement in both the feasibility and the efficiency of parallelizing in a computation process is inverse to communication among its high-level parts. The M^3-network is highly modular, that is, it has clear modules with almost no communication among them, except the final step of combining multiple results. Therefore, it is natural and efficient to run these modules simultaneously, that is, to parallelize M^3-network.

3.1. *Parallel Training of M^3-network*

The parallel implementation of M^3-network is based on its modular structure. In the training phase, we first distribute subproblems among the computer nodes, then the nodes train component classifiers independently, and finally we collect all the trained component classifiers to finish training. In the test phase, we first distribute component classifiers, namely modules, among computer nodes, then we send the test samples to nodes and receive predictions from them, and finally we get the final predictions of the test samples through both *minimization* and *maximization* operations.

Figure 4 illustrates the parallel framework for training M^3-network. From the viewpoint of parallel programming, M^3-network falls into the category of pleasant parallel problems, and therefore master-salve framework is recommended for this category [29].

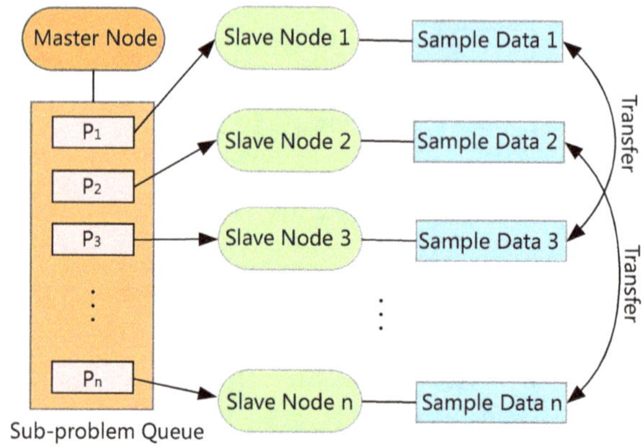

Fig. 4. The parallel framework for training M^3-network.

In the training phase, we need to provide each slave node with samples, which have long input vectors in large volumes. If we store all of the samples in the master node and then later send them into the slave nodes together with the subproblems, the master node will become a bottleneck for data transfer. As a solution, we distribute the samples into salve nodes that can communicate with one another through point-to-point communication. Thus, in assigning the subproblems, the master node only needs to send the slaves nodes the indexes of the samples, which greatly reduces the communication load.

Figure 5 illustrates the parallel framework of M^3-network in test phase. In the test phase, the operation of either *minimization* or *maximization* can be done with one call to *MPI_Reduce* procedure in MPI programming environment, which is well designed and works acceptably fast [29].

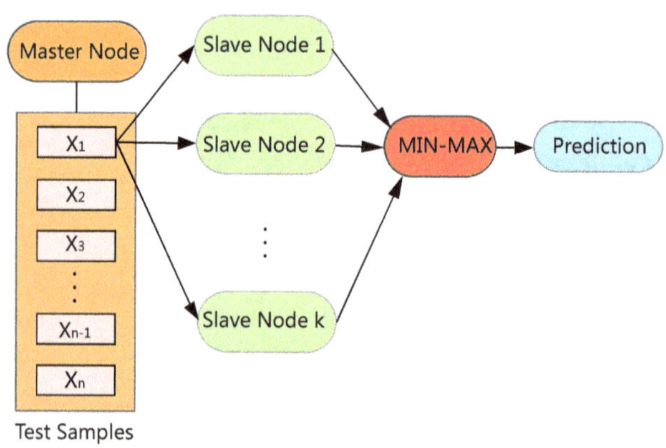

Fig. 5. The parallel framework of M^3-network in test phase.

3.2. *Asymmetric and Symmetric Module Selection Strategies*

The symmetric module selection strategy is an improved version of the min-max module combination method, namely asymmetric module selection strategy. This strategy works exponentially faster. Its main disadvantage is that it must work on concrete classifiers which output either 0 or 1, instead of continuous confidence values in the min-max module combination.

Let us review the min-max module combination method, and explain why we call it the asymmetric module selection strategy. Suppose the outputs of all of the modules on a test sample are positive 1 or negative 0 (see Figure 6(a)). Following the min-max module combination method, we first perform *minimization* operation along each row, then perform *maximization* operation among the rows' results. The final result is 1 in this example. Note that this result is caused by the fourth row, full of 1's as its elements. The translation of this process in feature space is that the test sample input is located within the subset $\mathcal{X}_{i,4}$, so that it shows positive in all the subproblems $\mathcal{T}_{ij}^{(4,k)}(k = 1, 2, \ldots, 5)$. The logical definition of the min-max module combination for M^3-network is:

- If there exists a row of all 1 in module's prediction matrix, the output is 1;
- Otherwise, the output is 0.

The min-max module combination method has a bias towards the output of 0, and that is why we call it asymmetric module selection.

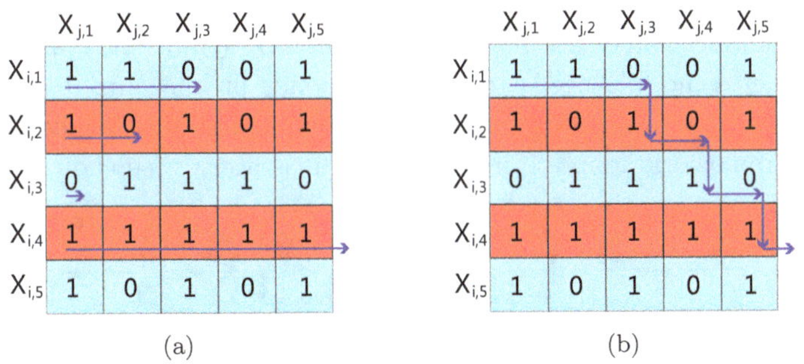

Fig. 6. Two module selection strategies: (a) asymmetric module selection and (b) symmetric module selection.

The other combination method, the symmetric module selection strategy, is illustrated in Figure 6(b) [36]. In this method, we start from the top left element. If the element is 1, we go one step right; otherwise, we go one step down. The process continues until we go beyound the boundaries of the matrix. If we pass through the right edge of the matrix, the method outputs 1; otherwise, it outputs 0. The logical definition of this method is:

- If there exists a row of all 1 in module's prediction matrix, the output is 1;

- If there exists a column of all 0, the output is 0;
- Otherwise, the output can be either 1 or 0.

The third case is rare in solving regular pattern classification problems with M^3-network, and it doesn't need to be considered [36]. From the viewpoint of feature space, the first case means that the test sample input is located within some positive subsets, and the second case for some negative subsets, so this method make sense. Moreover, this method has no bias towards either 1 or 0, that is why we call it a symmetric module selection. Indeed, this characteristic makes it likely to perform better than the asymmetric module selection method.

The symmetric module selection strategy has a great advantage on computation complexity, compared with the asymmetric one. Suppose the training data sets for two classes are divided into m and n subsets, respectively. The complexity of the asymmetric module selection is $O(mn)$, and that of symmetric module selection is $O(m+n)$; so it is clear the symmetric module selection is one order faster than the asymmetric module selection. Further, the component classifiers that are not visited do not need to be actually computed (see Figure 6), that is, the module combination method determines the complexity of whole test process of M^3-network. For these two reasons, the new symmetric module selection method can greatly accelerate the response speed of M^3-network, which is our main purpose.

3.3. *Parallel Implementation of Symmetric Module Selection*

A kind of communication exists among the modules by symmetric module selection, that is, whether a module needs to be run depends on the output of some other modules. Though we can ignore such dependency and run all the modules with the master-slave framework adopted in Section 3.1, it is not a wise choice.

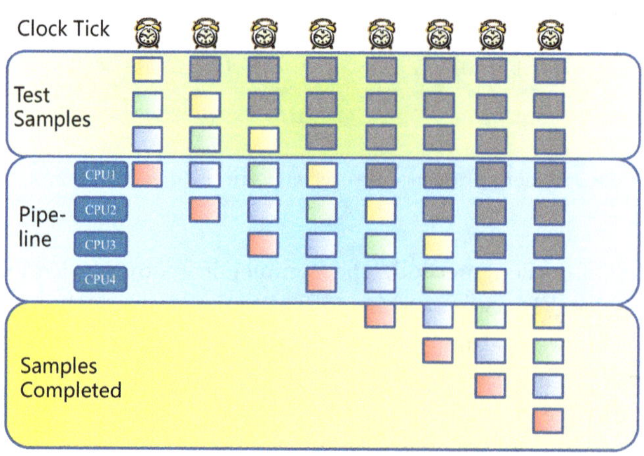

Fig. 7. Illustration of the pipeline framework

The parallel pipeline framework shown in Figure 7 fits this context well. In a pipeline, some nodes do partial processing of data and then forward the partially processed results to another processing nodes down the pipeline for further processing [29].

The key issue of applying the pipeline framework is to design the task for each stage. To implement M^3-network in the style of a pipeline, we decompose the matrix of modules along the reverse diagonals; then take the modules in the same reverse diagonal as one stage, and number them from 1 to $n+m-1$, the same as the distance to the top-left cell (see the nine dotted lines in Figure 8) [33].

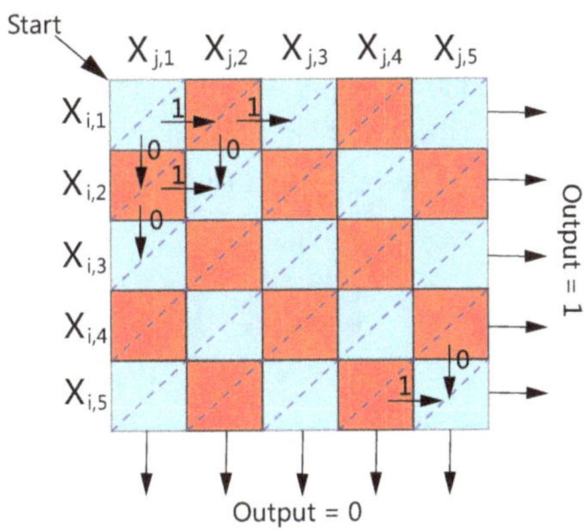

Fig. 8. Design of a pipeline for M^3-network

We build M^3-network by assigning some computer nodes to run the modules in each reverse diagonal. After that we start the system by inputting a test sample with a position status of $(1,1)$, to a computer node that is in charge of the first reverse diagonal. All the successive steps are iterative. When the test sample with the position status (r,c) arrives at the $(r+c-1)$th reverse diagonal, a computer node of that diagonal first runs the module $\mathrm{M}_{ij}^{(r,c)}$ corresponding to the two-class subproblem $\mathcal{T}_{ij}^{(r\,c)}$. Then it updates the sample's position status with $(r,c+1)$ if the module outputs 1, or with $(r+1,c)$ otherwise. At last it passes this sample to the next reverse diagonal. The advantage of such stage design is that each computer node can complete its job by running exactly one module, which simplifies workload balancing.

The number of computer nodes assigned to each reverse diagonal should agree with the number of arriving samples, for the sake of workload balance. Considering that samples leave the matrix from the right and bottom edges, from 1th to $min(n,m)$-th reverse diagonals, the number of computer nodes should be fixed; after that, the number should gradually decrease.

4. Application to Patent Documents Classification

Patents, as a social innovation system, have been an increasingly important role in science and technology improvement. As a result, automatic patent classification, as a basic data mining technique on patents, has received wide attention. A number of organizations including the European Patent Office, the Japanese Patent Office, and companies have been working on this topic [9, 12].

4.1. *Patent Documents Classification*

Patent documents classification takes the standard of the International Patent Classification (IPC) as a label system [12]. IPC is a complex hierarchical symbol system, where all the technological fields are divided into 8 Sections, 120 Classes, 630 Subclasses and approximately 69,000 Groups. We use M^3-network to classify Japanese patent documents on the Section level of the International Patent Classification taxonomy, which is a very large classification task [25].

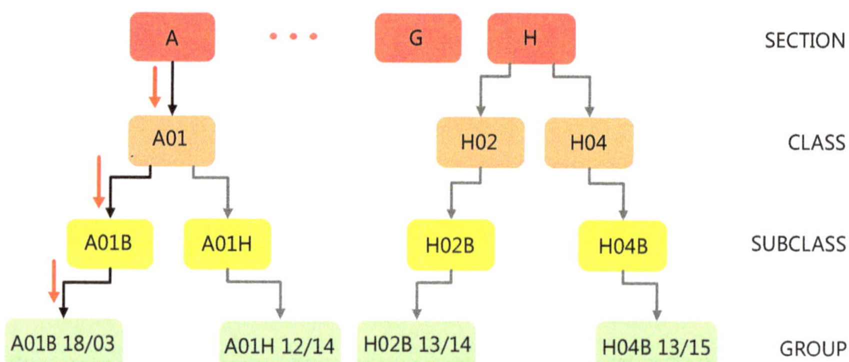

Fig. 9. Illustration of the IPC taxonomy. Here, 'A' is the Section category label, 'A01' is the Class category label, 'A01B' is the Subclass category label, and 'A01B 13/08' is the Group category label.

We work on a Japanese patent corpus called the Japanese National Information Institutes' Testing Corpus for Information Retrieval (NTCIR), which is publicly available for research purpose[1]. The corpus consists of about 3,500,000 documents of Japanese patent applications from 1993 to 2002. A patent document is a structured text with one title and three fields, Abstract, Claim and Description (see Table 1).

[1]http://research.nii.ac.jp/index-en.html

Table 1. The structure of Japanese patent documents

	PATENT-JA-UPA-1998-000001
<Bibliography>	
[publication date]	(43)【公開日】平成１０年（１９９８）１月６日
[title of invenction]	(54)【発明の名称】土壌改良方法とその作業機
<Abstract>	
[purpose]	【課題】心土破砕、特に雪上心土破砕作業の際に積雪 ...
[solution]	【解決手段】心土破砕を行うために用いるサブソイラの ...
<Claims>	
[claim1]	【請求項１】サブソイラ作業機を用いて心土破砕作業 ...
[claim2]	【請求項２】サブソイラ作業機において、そのナイフ ...
<Description>	
[technique field]	【発明の属する技術分野】本発明は、土壌改良方法とそ ...
[prior art]	【従来の技術】圃場の表面がまだ積雪に覆われている状 ...
[problem to be solved]	【発明が解決しようとする課題】心土破砕は通常春先に ...
[means of solving problems]	【課題を解決するための手段】述のような日的達成す ...
[effects of invention]	【発明の効果】以上の説明から明らかなように、本発明 ...
...	...
< Explanation of Drawing >	
[figure1]	【図１】本発明を施す圃場断面図である。
...	...

4.2. *Task Decomposition with Prior Knowledge*

Using hints greatly improves the effects of machine learning [1]. In this application we make use of the patents' hints, publish dates and hierarchical labels, to perform the task decomposition for M^3-network. The decomposition process, namely M^3-YC, consists of the following three steps (see Figure 10).

step 1 Divide the training data set of each Section by publish dates, each subset for one year;

step 2 Further divide the subsets by class, thus each subset for one Class published in one year;

step 3 Further randomly divide the remaining large subsets into smaller fix-sized subsets.

Note that with steps 1 and 2 removed, the above process becomes the standard M^3-network, namely M^3-Rand. In this research, our main concern is M^3-YC, and we take M^3-Rand and plain SVMs as the baseline methods.

In both M^3-Rand and M^3-YC, we can control the size of subproblems through the step of random decomposition, for example, setting the maximum number of samples in a subset. We must keep the subproblems in moderate size, neither too large for a single classifier to learn, nor so small that there are too many little subproblems. We finally decide on the maximum size of a subset to be 2000 samples, according to some pilot experiments.

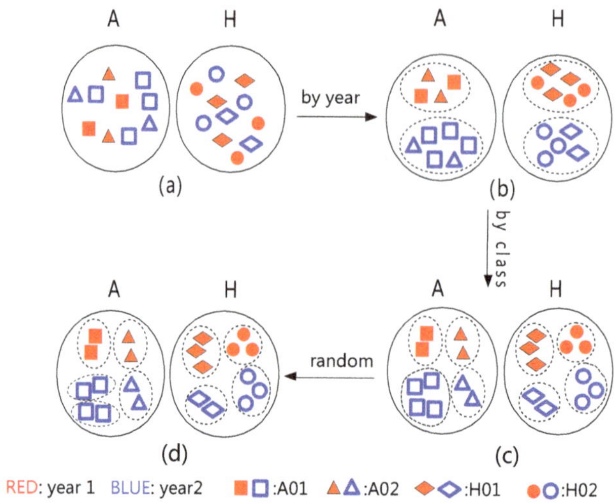

Fig. 10. Illustration of task decomposition with prior knowledge. Different attributes among the samples are shown by colors and shapes. The colors correspond to publish dates: red samples published in the same year and blue ones in another. The shapes correspond to the sub-categories: rectangles and triangles for two sub-categories of category A, and the circles and diamonds for two sub-categories of category B. The task decomposition consists of three steps, first by the prior knowledge of the publish year (b), and then by the Class category (c), and last by random for remaining large subsets (d).

4.3. *Experiment Settings*

4.3.1. *Training and Test Data Sets*

We have an extended version of NTCIR-5 Japanese patent corpus at hand, which consists of all the patent applications published by the Japanese Patent Office from 1993 to 2002. In order to truthfully evaluate the three methods, conventional SVMs, M^3-Rand and M^3-YC, we make eight pairs of training and test data sets, creating a real-world context (see Table 2).

Table 2. Description of training and test data sets

# of years for training	years		set size	
	training	test	training	test
1	2000	2001,2002	358,072	733,570
2	1999,2000	2001,2002	714,004	733,570
3	1998–2000	2001,2002	1,055,391	733,570
4	1997–2000	2001,2002	1,386,850	733,570
5	1996–2000	2001,2002	1,727,356	733,570
6	1995–2000	2001,2002	2,064,325	733,570
7	1994–2000	2001,2002	2,415,236	733,570
8	1993–2000	2001,2002	2,762,563	733,570

4.3.2. *Feature extraction and filtering*

Several steps of preprocessing need to be done before getting the vector representation of the Japanese text used by classifiers. These steps are tokenization, term filtering, and term indexing.

Tokenization generates a list of clean and informative words from raw text. We first extract the raw text from the four patent fields, Title, Abstract, Claim and Description (see Table 1), as they are most informative about the patent's content. Then we segment these texts into isolated words using the software Chasen[2]. After that, we remove the stop words (or empty words) from the results. The remaining words, namely *terms* in the research domain of text categorization, are features for the successor classification task. Table 3 shows the result we get from the example shown in Table 1.

Table 3. The terms of a Japanese patent document shown in Table 1

土壌(soil) 改良(improve) 方法(methods) 作業(working) 機(machine) % title
心土(subsoil) 破砕(crack) 雪上(snow) 心土 破砕 作業 ...% abstract
サブソイラ(subsoiler) 作業 機 心土 破砕 作業 ...%claim
発明(patent) 土壌 改良 方法 ...%description

Term Filtering removes the useless terms in the classification task. This shortens the length of the representation vectors, and thus cuts the computational cost and reduces generalization errors. We take χ^2_{avg} as filtering criterion [31]), and pick up 5000 top terms as features, according to our pilot experiments. Table 4 shows the top 10 terms sorted by χ^2_{avg}, most of which are technical terms representing patent documents.

Table 4. Top 10 terms selected by χ^2_{avg}

terms	explanation	χ^2_{avg}	terms	explanation	χ^2_{avg}
データ	data	10384.72	物	article	7528.72
情報	information	10199.42	含有	contain	7374.12
回路	circuit	9561.67	接続	connect	7324.43
信号	signal	8387.75	絶縁	insulation	7194.85
記録	record	7901.17	基板	baseplate	7076.72

Term Indexing generates the weights of feature terms for a sample with the real numerical vectors. We adopt the dominant methods of TFIDF [27].

$$tfidf(t,d) = n(t,d)\log\frac{|T_r|}{n_{T_r}(t)} \tag{9}$$

where t denotes a term, d denotes a document, T_r denotes the training corpus, $n(t,d)$ denotes the number of times t occurs in d, namely term frequency, and $n_{Tr}(t)$

[2]http://chasen.naist.jp/hiki/ChaSen/

Table 5. The vector representations of patent documents, where the format of the vectors adopted by SVM^{light} is taken, that is, $vetor := (dimension : value)+$.

No.	Vectors				
1	72 : 0.730	98 : 1.790	138 : 1.310	141 : 4.495	...
2	28 : 26.353	29 : 9.232	31 : 2.795	71 : 1.463	...
3	71 : 1.463	79 : 2.441	85 : 2.993	113 : 11.393	...
4	42 : 2.164	60 : 0.905	109 : 2.061	138 : 2.947	...
5	28 : 7.529	72 : 6.577	139 : 8.103	167 : 8.728	...

denotes the number of documents where t occurs, namely document frequency. Table 5 shows the vector representations of selected patent documents.

4.3.3. *Computational Platform*

A Lenovo cluster system consisting of three fat nodes and thirty thin nodes run all of our experiments. Each fat node has 32 GB of RAM and two 3.2-GHz quad-core CPUs, while each thin node has 8 GB of RAM and two 2.0-GHz quad-core CPUs. Experiments with the conventional SVMs were performed on the fat nodes because they need large memory, while experiments with the M^3-network were done on the thin nodes because each sub-problem was small and a lot of processors were required for parallel training.

4.3.4. *Training Component Classifier*

M^3-network is a general framework for pattern classification. The user can select suitable component classifiers according to the classification task to be solved and available computing resource. There are several alternative choices, such as multilayer neural networks [21], k-nearest-neighbor algorithm [35, 30], and SVMs [23]. Here we select conventional SVMs with linear kernel and use SVM^{light}, an implement of SVM by Joachims [15]. SVM^{light} actually plays two roles in this chapter, the baseline method and the component classifiers for M^3-network.

4.3.5. *Performance Measurement*

The conventional measurement of accuracy in machine learning domain makes less sense for most text categorization tasks; instead, we usually use the function F_1 [27]. The formular F_1 of one class is :

$$F_1 = \frac{2PR}{P+R} \tag{10}$$

$$P = \frac{TP}{TP+FP} \tag{11}$$

$$R = \frac{TP}{TP+FN} \tag{12}$$

where TP is the number of classifier's True Positive predictions, FP for False Positive, and FN for False Negative. P is named the precision, and R is named the recall. In practice, there are the two following versions of F_1, depending on the methods of integrating individual results:

$$micro - F_1 = \frac{2PR}{P+R} \tag{13}$$

$$macro - F_1 = \underset{c \in C}{avg} \frac{2P_c R_c}{P_c + R_c} \tag{14}$$

where P and R are computed with all the classes, while P_c and R_c are computed on only class c. For example, sample s actually belongs to classes c_1 and c_2, while the classifier's predictions is c_2 and c_3. Then in computing $micro - F_1$, TP, FP and FN will be increased by 1, while in computing P_{c1} and R_{c1} for $macro - F_1$, only FN will be increased by 1.

4.4. *Results and Discussions*

4.4.1. *Accuracy*

Figure 11 is the experimental results, with micro-F_1 and macro-F_1 used as accuracy measurements. The following conclusions can be drawn from our results:

(a) On the aspect of test accuracy, the two M^3-SVM methods, M^3-Rand and M^3-YC, are both superior to conventional SVMs. We can learn from the training scores that conventional SVMs with linear kernel are unable to learn the training set completely, because its micro-F_1 and macro-F_1 are only about 80%. On the contrary, as an ensemble learning algorithm, M^3-SVMs can generate a powerful classifier by combining simple classifiers. As a result, M^3-SVMs have fulfilled the learning on all the training sets with accuracies of nearly 100%.

(b) M^3-Rand and M^3-YC show superior robustness to conventional SVMs over dated samples. Along the moving backward of the starting time point of training set, more and more dated samples are added, thus the performance of conventional SVMs decreases. Contrarily, the performance of two M^3-SVMs increases at the same time. Though their performance deceases a little in the 5th year point and 7th year point. This does not affect the curves' overall trend.

(c) M^3-YC over-performs M^3-rand on each year point, which indicates that incorporating prior knowledge of the publishing date and the Class category into task decomposition will undoubtedly improve classification performance.

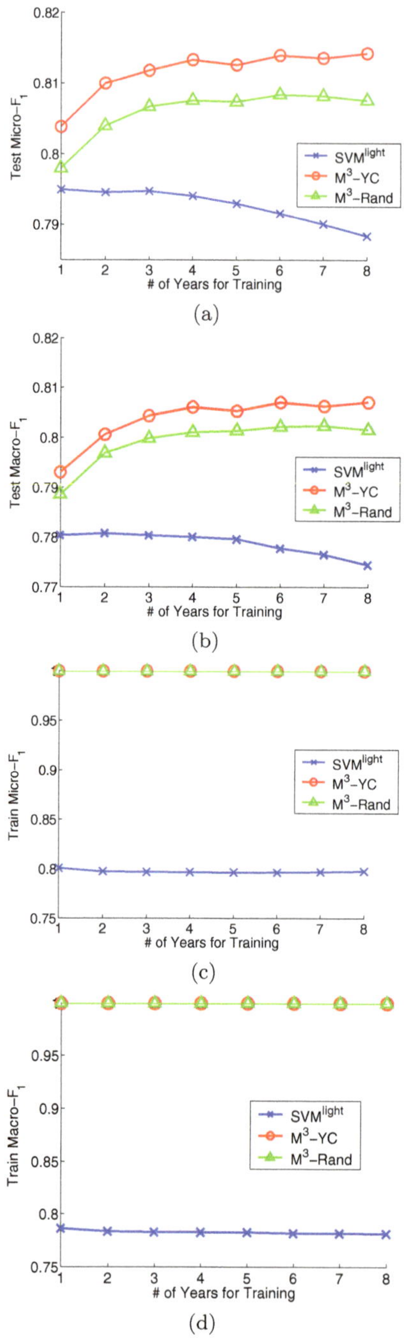

Fig. 11. Performance comparison of conventional SVMs, M^3-Rand and M^3-YC: (a) micro-F_1 on test data; (b) macro-F_1 on test data; (c) micro-F_1 on training data; and (d) macro-F_1 on training data.

4.4.2. *Time Cost*

As mentioned before, M^3-SVMs (both M^3-Rand and M^3-YC) have the merit of parallel computing, which greatly speeds up learning. However, we are sharing the Lenovo parallel computer system with other users while performing the experiments, so the accurate time cost of M^3-SVMs could not be measured. The time cost of conventional SVMs has been recorded as it didn't involve parallel running.

We evaluate the time cost of M^3-YC using the following formula:

$$t_{M^3} = \frac{n_{mod} \times t_0}{n_{cpu}} \tag{15}$$

where n_{mod} is the number of modules (or subproblems), t_0 is the average time cost of per module, and n_{cpu} is the number of CPUs. During our experiments, n_{mod} have been recorded automatically and n_{cpu} is 30 on the Lenovo cluster system. As for t_0, it must be measured by experiment, and the eventual value we get is 0.025 seconds per module.

Figure 12 is the time costs of conventional SVMs and M^3-YC in our experiments. From this figure, we can know that the M^3-YC's time cost is only about 1/10 of the SVM's, which agrees with our expectation well.

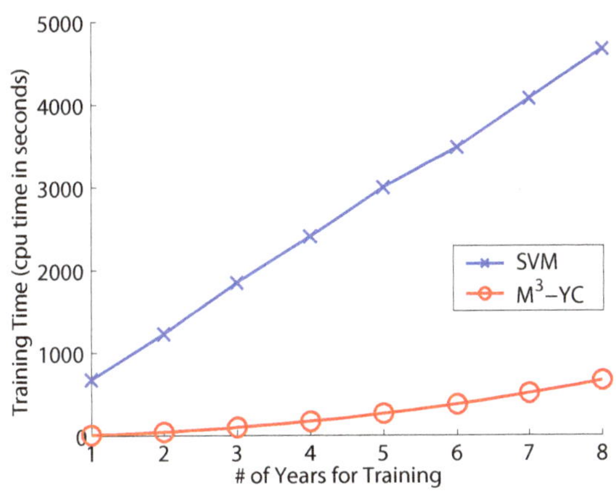

Fig. 12. Training time comparison of conventional SVMs and M^3-YC.

5. Conclusions

In this chapter, we presente a parallel and modular pattern classification framework for large-scale problems. The framework works in a modular manner, and has several advantages in solving large-scale problems. On the one hand, massively parallel and distributed training is easily implementable because of its modularity. On the other hand, it has a balanced performance on all classes because of various task decomposition strategies.

The experiments are conducted on a large-scale Japanese patent corpus. Taking into account prior/domain knowledge, we partition large training data sets into smaller subsets according to the publishing date and the subclass. The experimental results show the effectiveness of the proposed framework, and demonstrate that M^3-SVMs is superior to conventional support vector machines in solving such complex problems.

As more and more large-scale applications appear in the fields of machine learning and pattern recognition, there is greater need for parallel and distributed methods. M^3-network uses simple task decomposition methods and efficient module selection strategies and can be implemented easily in practice. We recognize that incorporating prior/domain knowledge into task decomposition is a reliable way to improve the learning efficiency and the generalization performance of M^3-network. We believe that the proposed parallel and modular framework will be very useful for solving complex classification problems with very large data sets.

Acknowledgements

This work was partially supported by the National Natural Science Foundation of China (Grant No. 60773090 and Grant No. 90820018), the National Basic Research Program of China (Grant No. 2009CB320901), and the National High-Tech Research Program of China (Grant No. 2008AA02Z315).

References

1. Abu-Mostafa, Yaser S.: Machines that learn from hints, Scientific American, vol. 272, no. 4, pp. 64–69, 1995.
2. Anand, R.; Mehrotra, K.G.; Mohan, C.K.: An improved algorithm for neural-network classification of imbalanced training sets, IEEE Transactions on Neural Networks, vol 4, no. 6, pp. 962-969, 1993.
3. Angiulli, F; Folino, G.: Distributed nearest neighbor-based condensation of very large data sets, IEEE Transactions on Knowledge and Data Engineering, vol. 19, pp. 1593-1606, 2007.
4. Cao, L. J., Keerthi, S. S., Ong, C. J., Zhang, J. Q., Periyathamby, U., Fu, X. J. and Lee, H. P.: Parallel sequential minimal optimization for the training of support vector machines, IEEE Transactions on Neural Networks, pp. 1039-1049, 2006.
5. Chen, K., Lu, B. L. and Kwok, J. T.: Efficient classification of multi-label and imbalanced data using min-max modular classifiers, Proc. of IEEE/INNS International Joint Conference on Neural Networks, pp. 1770-1775, 2006.
6. Chu, X.L., Ma, C., Li, J., Lu, B.L., Utiyama, M., and Isahara, H.: Large-scale patent classification with min-max modular support vector machines, Proc. of IEEE/INNS International Joint Conference on Neural Networks, pp. 3973-3980, 2008.
7. Collobert, Ronan; Bengio, Samy; Bengio, Yoshua.: A parallel mixture of SVMs for very large scale problems, Neural Computation, vol. 14, Issue 5, pp. 1105-1114, 2002.
8. Dong, J., Krzyzak, A. and Suen, C. Y.: Fast SVM training algorithm with decomposition on very large data sets, IEEE Transactions on Pattern Analysis and Machine Intelligence, vol. 27, no. 4, pp. 603-618, 2005.

9. Fall, C.J., T orcsvári, A., Benzineb, K., Karetka, G.: Automated categorization in the international patent classification, ACM SIGIR Forum ACM Press New York, NY, USA, vol. 37, pp. 10-25, 2003.
10. Fan, Z.G. and Lu, B.L.: Multi-view face recognition with min-max modular SVMs, Lecture Notes in Computer Science, Springer, vol. 3611, pp. 396-401, 2005.
11. Fan, R.E., Chang, K.W., Hsieh, C.J., Wang, X.R., and Lin, C.J.: LIBLINEAR: A Library for large linear classification. Journal of Machine Learning Research, vol. 9, pp. 1871-1874, 2008.
12. Fujii, A., Iwayama, M., Kando, N.: Introduction to the special issue on patent processing, Information Processing and Management, pp. 1149–1153, 2007.
13. Hazan, T., Man, A., and Shashua, A.: A parallel decomposition solver for SVM: distributed dual ascend using Fenchel duality, Proc. IEEE Conference on Computer Vision and Pattern Recognition, pp. 1-8, 2008.
14. Ferreira, L.V.; Kaszkurewicz, E.; Bhaya, A.: Parallel implementation of gradient-based neural networks for SVM training, Proc. of International Joint Conference on Neural Networks, pp. 339–246, 2006.
15. Joachims, T.: Making large-scale support vector machine learning practical. Advances in Kernel Methods: Support Vector Learning. MIT Press, Cambridge, MA, pp. 169-184, 1998.
16. Keerthi, S. S. and DeCoste, D.: A modified finite Newton method for fast solution of large scale linear SVMs, Journal of Machine Learning Research, vol. 6, pp. 341-361, 2005.
17. Kugler, M; Kuroyanagi, S., and Nugroho, A.S.: CombNET-III with nonlinear gating network and its application in large-scale classification problems, IEICE Transactions on Information and Systems, vol. E91D, no. 2, pp. 286-295, 2008.
18. Larkey, L.: Some issues in the automatic classification of US patents, Proc. of AAAI-98 Workshop on Learning for Text Categorization, Technical Report WS-98-05, pp. 87-90, 1998.
19. Liu, F. Y., Wu, K., Zhao, H. and Lu, B. L.: Fast text categorization with a min-max modular support vector machine, Proc. IEEE/INNS International Joint Conference on Neural Networks, pp. 570–575, 2005.
20. Lu, B. L. and Ito, M.: Task decomposition based on class relations: a modular neural network architecture for pattern classification, Biological and Artificial Computation: From Neuroscience to Technology, Lecture Notes in Computer Science, Springer, vol. 1240, pp. 330-339, 1997.
21. Lu, B. L. and Ito, M.: Task decomposition and module combination based on class relations: A modular neural network for pattern classification, IEEE Transactions on Neural Networks, Vol. 10, pp. 1244-1256, 1999.
22. Lu, B. L., Ma, Q., Ichikawa, M. and Isahara, H.: Efficient part-of-speech tagging with a min-max modular neural-network model, Applied Intelligence, vol. 19, no. 1, pp. 65-81, 2003.
23. Lu, B. L., Wang, K. A., Utiyama, M., and Isahara, H.: A part-versus-part method for massively parallel training of support vector machines, Proc. of IEEE/INNS International Joint Conference on Neural Networks, pp. 735-740, 2004.
24. Lu, B. L and J. Li: A Min-Max modular network with gaussian-zero-crossing function, Studies in Computational Intelligence, vol. 35, pp. 285-313, 2008.
25. Lu, B. L, Wang, X. L. and Utiyama, M.: Incorporating prior knowledge into learning by dividing training data, Journal of Frontiers of Computer Science in China, pp. 109-122, 2009.

26. Lu, Y. and Roychowdhury, V.: Parallel randomized sampling for support vector machine and support vector regression, Knowledge and Information Systems, vol. 14, pp. 233-247, 2008.
27. Sebastiani, F.: Machine learning in automated text categorization, ACM Computing Surveys, vol. 34, pp. 1-47, 2002.
28. Sun, Y. M., Kamel, M. S. and Wong, A. K. C: Cost-sensitive boosting for classification of imbalanced data, Pattern Recognition, vol. 40, pp. 3358-3378, 2007.
29. Wilkinson, B. and Allen, M.: Parallel Programming: Techniques and Applications Using Networked Workstation and Parallel Computers, Prentice Hall Inc., 1999.
30. Wu, K., Lu, B. L., Uchiyama, M., and Isahara, H.: An empirical comparison of min-max-modular k-NN with different voting methods to large-scale text categorization, Soft Computing, vol. 12, no. 7, pp. 647-655, 2008.
31. Yang, Y.M. and Pedersen, J. O.: A comparattive study on feature selection in text categorization, Proc. of International Conference on Machine Learning, pp. 187-196, 1997.
32. Yang, Y. and Lu, B. L.: Prediction of protein subcellular multi-locations with a min-max modular support vector machine, Lecture Notes in Computer Science, Springer, pp. 646-651, 2005.
33. Ye, Z. F., Lu, B. L. and Hui, C.: Patent classification using parallel Min-Max modular support vector machine, Autonomous Systems, Springer, pp. 157-167, 2008.
34. Zanni, L; Serafini, T; Zanghirati, G.: Parallel software for training large scale support vector machines on multiprocessor systems, Journal of Machine Learning Research, vol. 7, pp. 1467-1492, 2006.
35. Zhao, H. and Lu, B. L.: A modular k-nearest neighbor classification method for massively parallel text categorization, Lecture Notes in Computer Science, Springer, vol. 3314, pp. 867–872, 2004.
36. Zhao, H. and Lu, B. L.: Improvement on response performance of Min-Max modular classifier by symmetric module selection, Lecture Notes in Computer Science, Springer, vol. 3497, pp: 39-44, 2005.

CHAPTER 5.7

3-D COMPUTER VISION MODELING IN VIDEO SURVEILLANCE APPLICATIONS

Giovanni B. Garibotto

Vision Technology Consultant
ElsagDatamat spa,
Via Puccini, 2, I-16154 Genova, Italy
E-mail: giovanni.garibotto@elsagdatamat.com

Intelligent Video Analysis (also called Video Analytics, VA) represents the key function of new generation Video Surveillance Systems, to support security operators in the management of video information from a large number of cameras displaced in the scene. The chapter is addressing the problem of human detection and tracking in a variety of applications, with the objective to perform all the process of scene analysis directly in world 3D coordinates by using video-camera data mainly for hypothesis validation. This centralized supervision approach (a single brain with many eyes) is supposed to manage an arbitrary number of stationary cameras, depending on the available computational power. A fairly accurate camera calibration process is required to predict the position in time and space of the targets in the scene. A simple 3D model is proved to be sufficient to describe not only rigid objects, but also the articulated flexible shape of humans in the hypothesis of low resolution targets and wide field of view. A few examples of possible applications to video surveillance are referred including standard optics and wide-field panoramic cameras (fish-eye lens).

1. Introduction to Video Analysis Functions

Video Analytics, VA (or Intelligent Video Analysis) is the collection of processing functions aimed to automatically identify objects and events in the scene, without the need of the continuous supervision of the human operator. Space-time reasoning represents an essential tool to achieve that objective, which is possible only with an accurate 3D data representation of the involved objects and humans and their spatial relations. In the last few years it has become the most relevant and competitive feature of new generation Video Surveillance Systems. The most established and professionally effective solutions of VA are referred to perimeter violation control, license plate recognition and people counting. Other consolidated results can be found in video indexing and annotation, with time stamped markers corresponding to some basic detected event, like motion activity, face and people detection, elementary actions and movement tracking. But there is an increasing demand in the Video Surveillance and Security field, for more advanced capabilities, to reach a certain level of understanding of what is happening in the scene, and provide a better decision support to security operators, that are otherwise flooded by a continuously increasing and almost unmanageable volume of data. In spite of the great promise, professional security users are still asking about more stable and robust solution of VA to be successfully applied in real world applications. Beside cost, the main problems limiting the use and growth of VA today seem to be related to the rate of false alarm messages, and the difficulty of system set up and maintenance.

Since the goal of VA is to reduce human involvement, the objective to remove any false alarm message becomes a must, not only for the cost of any false alarm, but also for the consequence of frustration with the system. An acceptable figure may be in the range of 1 or 2 false alarms per month, which is definitely not possible for most conventional commercial solutions currently available on the market. Main limitations are related to the lack of adaptation to environmental changes and the complexity of system set up, parameters tuning and learning procedures that often require the involvement of Computer Vision specialists. Most relevant success stories are connected to human costs reduction or significant improvement of task efficiency where the automatic process may even outperform the human perception system, like on-line and mobile automatic number plate recognition, ANPR, with a real time comparison against data-bases of millions of plate strings [1]. Advanced VA technology needs adaptive solutions to manage 24 hour operation, day/night and variable environmental changes, including the possibility to drive a suitable lighting system (working both in the visible spectrum or the near infrared IR). Another challenging objective of VA technology is the need to manage and coordinate a large number of camera views, to cover a fairly wide inspection area. A main goal of new generation VA is the possibility to provide a synthetic high-level description of the imaged scene and the most important components of this abstraction are the humans that are present in the scene, their movement (trajectories) and their interactions during time (behavior), to be able to send to the video security operators only the messages related to significant events and alarms. The first step towards this very ambitious goal is the development of robust Vision techniques for human detection, tracking and trajectory analysis, by considering also the flexible nature of the articulated human motion. This chapter will be limited to address this first stage of the process, by considering a multi-cameras integration process and their calibration procedures, with the objective to perform successfully the task of people detection and trajectory tracking.

2. Human Detection and Tracking

Observation of human activity from a stationary camera is an essential component of Video Surveillance systems. There are many possible applications related to human activity, from the simple detection (people counting and crowding level estimation, mainly using overhead cameras) to basic movement classification (like standing, sitting, running, bending, etc.) up to trajectory analysis and behavior recognition with the final objective to distinguish between normal and abnormal situations. Human detection in a real case, without any control of the environment is still an open issue and is strongly limited by visibility problems (mainly due to variable lighting conditions), by occlusion (in very crowded scenes) and by change of appearance due to different viewpoints beside the great variety of possible articulations and the presence of occluding accessories like bags, briefcases, that are deeply changing the human silhouette [2] [3].

The problem of visibility and noise sensitivity is often addressed by adaptive image processing techniques. Some level of occlusion can be managed by multiple camera configurations and an accurate prediction of the expected shape and position of the target in the 3D scene. The overall problem can be partially simplified when using calibrated cameras, which means that the intrinsic parameters as well as their position and orientation with respect to some external reference system in the scene are known. In the following we will briefly recall the main problems related to camera calibration in a multi-camera system set up.

3. Camera Calibration

Camera calibration is an essential component for the implementation of video surveillance systems. The possibility to fuse together the visual information coming from different viewpoints (multi-camera) as well as the integration of different sensors (planar lens and fish eye panoramic cameras) represents a key feature towards the control of wide areas in many security scenarios. Actually surveillance camera calibration requirements are quite different from metrological robotic applications, in terms of precision. Moreover, the installation procedures must be developed in such a way to become simple and friendly enough to be used by non-expert technicians. A video surveillance system is supposed to deal with any kind of input video sequence, without a-priori knowledge of the camera system itself. The possibility to achieve high precision intrinsic camera calibration parameters, by a constrained set-up in a controlled lab environment, is very unlikely in real security applications.

The most consolidated approach to camera calibration is based on the use of fixed patterns of 3D visible points (landmark) to be properly placed, or made visible, in the scene [4]. Such reference points can be arranged in a real 3D pattern (like an orthogonal structure of points) or a series of markers that are usually placed on the ground floor (or any other planar surface) in the imaged scene. The need of such physical infrastructure and its precise measurement is not always acceptable for a practical implementation of a video surveillance system. Quite often it is necessary to deal with already installed video cameras. Moreover, focal length adjustment does often take place directly on the field. As such, the video installation process must rely only on the video recorded sequences to fully define the calibration properties of the system, and the implemented solution must be robust enough to deal with uncertainty and measurement errors. On the other hand it is highly recommended to make use of existing infrastructure components (to be used as calibration landmarks).

The development of a fully automatic self-calibration procedure is still far to be fully realized although some progress has been made along this direction [5], and the vanishing point calibration is a good approach to consider, as described in the following. The ultimate goal of any calibration procedure is the estimation of the 3D transformation (rotation & translation) between the camera reference system and a 3D reference system in world coordinates, as shown in Fig. 1, with the assumption of an ideal pin-hole camera model, where the image projection surface is supposed to be an ideal plane without any optical distortion. The main steps of a calibration process are briefly recalled in the following, starting from the estimation and correction of the optical distortion of the camera. Then a camera calibration solution based on the classical vanishing point approach will be described. It represents a good compromise in terms of precision and easy implementation to be successful in video surveillance applications. The same approach can be used also to obtain the calibration of a panoramic camera (fish-eye spherical lens) with arbitrary axis orientation [6]. The result of multi-camera calibration is a set of transformations, to relate each camera coordinates to a common map of the reference ground surface. The accuracy of this transformations is particularly important when target tracking is performed in the 3D coordinate system, since it is always possible to predict what are the cameras

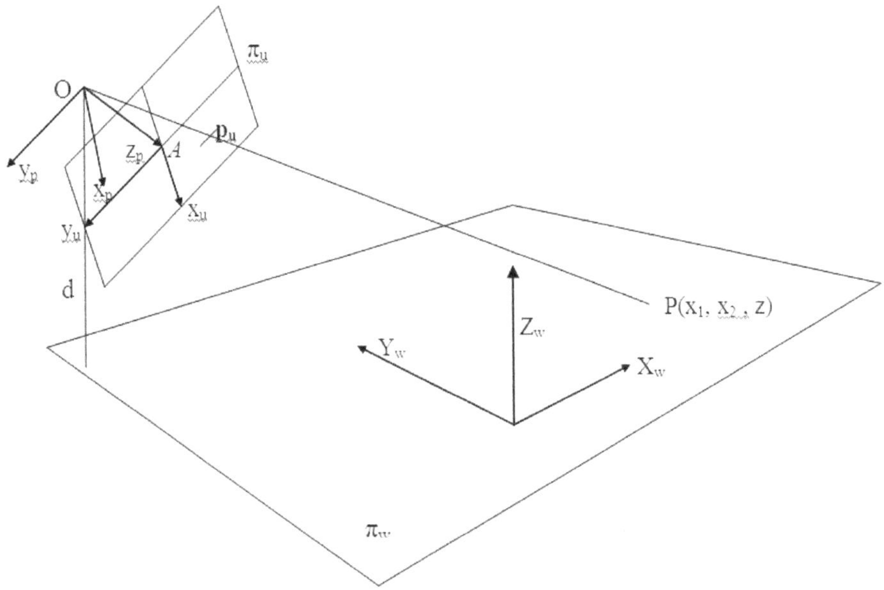

Figure 1. Direct and inverse transformation between 3D world coordinates (X_w, Y_w, Z_w) and the undistorted image plane $\pi_u(x_u, y_u)$

involved and use that visual information to confirm (or reject) the supposed position of the target in the scene.

3.1. *Optical lens distortion*

All the Computer Vision algorithms are based on the assumption that the digital image to be processed has been obtained by an ideal pin-hole camera model. Actually some optical distortion is always present in a real acquisition system, and the image data recorded by the sensor array do not always correspond to an ideal planar projection surface. Many different approaches have been proposed to obtain an effective distortion correction. One of the most popular solutions in the Computer Vision community is based on the first order approximation of the radial distortion (k_1 factor) as described in [7] where a method for the estimation of this correction factor is described. Such a correction factor is usually very small ($<10^{-4}$) with some problems of numerical instability. A new optical distortion model has been recently proposed [6], based on the same symmetrical radial distortion assumption as in [7].

In this case the image projection surface is modeled as a sector of a sphere whose radius curvature c [in pixel units] is shown in Fig. 2, where C is the centre of the sphere of radius c. This spherical image sector is supposed to be mapped

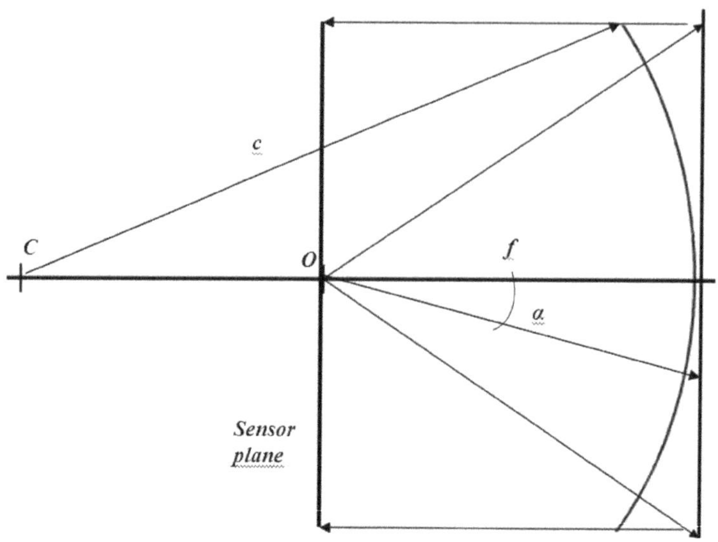

Figure 2. Spherical lens distortion model

onto the sensor array, by a parallel projection. As a general rule, when dealing with a good quality optical lens, with a narrow field of view FOV (and fairly long focal length) the distortion curvature c tends to infinity and we are approaching the ideal case of a planar projection surface (pin-hole model). On the contrary, when using wide FOV optical systems (with short focal length), the equivalent curvature c of the lens becomes closer to the value of the focal length f, as it happens when using a half-space panoramic optical system (with a fish-eye optics), where ($c \approx f$). The relations between undistorted coordinates $p_u(x_u, y_u)$ and the actual distorted coordinates $p_d(x_d, y_d)$, become:

$$p_u = p_d \cdot fd \qquad (1)$$
$$fd = (c / r_d) \tan(r_d / c)$$
$$r_d^2 = (x_d^2 + y_d^2)$$

$$p_d = p_u \cdot fi \qquad (2)$$
$$fi = (c / r_u) \operatorname{atan}(r_u / c)$$
$$r_u^2 = (x_u^2 + y_u^2)$$

where all coordinates are referred to the principal point A (i.e. the intersection of the optical axis with the ideal image plane). The estimation of both the curvature c and the principal point A is performed by minimization of the co-linearity error of a set of projected segments that are known to correspond to straight lines in the scene.

3.2. *Vanishing point Camera calibration*

The proposed calibration scheme is based on the well known approach of vanishing point calibration [8]. It is appropriate in video security applications, since it may exploit existing physical structures and parallel line geometry, when possible, or even the vertical height of walking persons, as proposed in [5]. The level of precision, when compared to the classical algebraic approach, as described in [9], has proved to be still sufficient for many practical video applications [5].

Camera calibration is obtained by the estimation of the orthogonal vanishing points from the imaged scene, in a general multi-camera configuration, as described in [10], with the assumption to deal with an almost ideal pin-hole image without optical distortion. As such, it is necessary to perform the preliminary step of distortion correction, as discussed before. The vanishing point geometry represents a unique 3D reference system in the 3D world, being related to orthogonal directions of physical structures present in the scene. As such, they provide all necessary information about the rotation parameters for each camera system. The identification of the 3D origin position in the world coordinates (and the corresponding translation vector) can be performed individually for each camera, if they are covering disjoint areas of the scene, or by using a common reference point (if all the cameras have a sufficient overlapping field of view). The position of the origin point can be selected interactively by the user, from the video images themselves. Moreover, it is necessary to identify at least one segment of known length, to get the necessary metrical scale factor for coordinate transformations. The use of the vanishing point geometry is particularly appropriate when dealing with structured environments, like buildings or man-made objects in both indoor and outdoor scenes. The vanishing point geometry at infinity is computed for each camera view, by manual selection of a set of lines (straight lines after distortion correction) converging towards the corresponding point at infinity. The identification of the 3 vanishing points provides a complete solution to the estimation of both intrinsic parameters (including focal length and principal point) and extrinsic rotation coefficients. Approximate solutions are possible [5], by including some additional constraint on the *horizon* (line at infinity) or its corresponding vertical vanishing point. Direct (and inverse) Camera to Scene transformations are easily performed by projecting the 3D point coordinate vector, representing the current image pixel, onto the 3D vanishing point geometry [6]. To recover the translation vector, it is necessary to compute the distance $\boldsymbol{d}$ of the sensor from the reference plane. This is achieved by selecting on each of the involved images, at least one segment of known length in the 3D scene.

Another widely used solution in Video Surveillance system is based on the use of panoramic cameras (fish-eye lens) as shown in Fig. 7, because of their capability to collect in a single image a very large field of view (close to half-space, 180°). Again, camera calibration is required, to recover the orientation of the 3 orthogonal axes of the world reference system, in camera coordinates, and compute both direct and inverse transformations between image coordinates and world coordinates. According to the optical model described before, the fish-eye image is a spherical mapping where the focal length f is almost equal to the curvature c. In this case the rectification process to the ideal planar image is not necessary. In fact, any direction in space (i.e. points at infinity in the pin-hole camera model) will naturally map onto the spherical image at a spatial coordinate (x_d, y_d) for any selected orientation [6].

4. People Detection and Tracking

The potential benefits of using 3D models in people detection and tracking will be briefly described in the following example. Without any claim to be the best solution to this task, the proposed approach allows to address the most relevant steps of the tracking chain. A short discussion is then referred on the most promising results as well as the open problems that are still to be solved.

The first step is the calibration of the involved cameras (multi-camera systems) which can be implemented as described before, including optical distortion correction.

The second step is the identification of the foreground features. In the following we consider a solution based on background estimation and updating with the assumption to deal with fixed stationary cameras.

Target tracking is performed using 3D object models with identification attributes. A simple state model is proposed for tracking, including the predicted target position in the 3D reference system and the first order motion vectors.

The next step is devoted to the detection of new targets in the scene, after removal of the already traced and accepted targets in the sequence.

Low-level data fusion is performed to remove local ambiguities, improve or reduce the confidence level, and minimize the number of false alarms and/or missed targets. Trajectory analysis is performed in the 3D coordinate system, to easily detect conflicting target paths, and possibly manage also some temporary occlusions.

High-level data integration and event detection represents the final step of the process being highly dependent on the specific application constraints and contextual information.

The considered application is wide-field target detection and tracking, where the target size is supposed to be quite small as compared to the image frame. In high resolution applications with a close-up view of the persons in the scene, the complexity of human body articulation requires more sophisticated and detailed models [11].

4.1. *3D model*

The proposed approach is based on the analysis in 3D space by using simple rigid models for the target objects, which are used also to roughly describe the shape of humans. The basic model is shown in Fig. 3 and consists in a series of planes corresponding to the level of the head, shoulders, body, feet. The target silhouette is obtained as the convex-hull of the projected set of points. Such model may be symmetric or not; in this second case it may keep also the information related to target heading direction. A specific model (height dependent) can be selected for each target person in the video sequence, to drive the following tracking process. Actually, this simple model is used primarily for the prediction of the search area in the different camera views, where additional corrections are implemented, to compensate for the vertical axis orientation (small bending during walking and running phases).

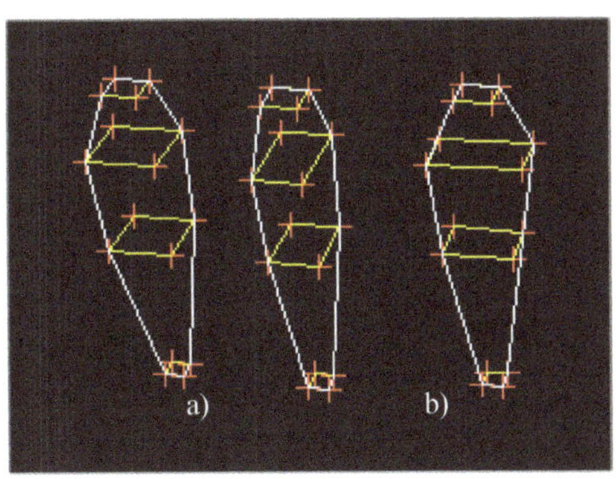

Figure 3. 3D target model; a) symmetrical; b) not-symmetrical

An example is shown in Fig. 4, where the projected shape of the target is used as a starting point for a better estimation of both head and foot position in the image view. The image gradient edge map, contained in the bounding box of the projected 3D model, provides all the necessary texture features for the correction of the body orientation.

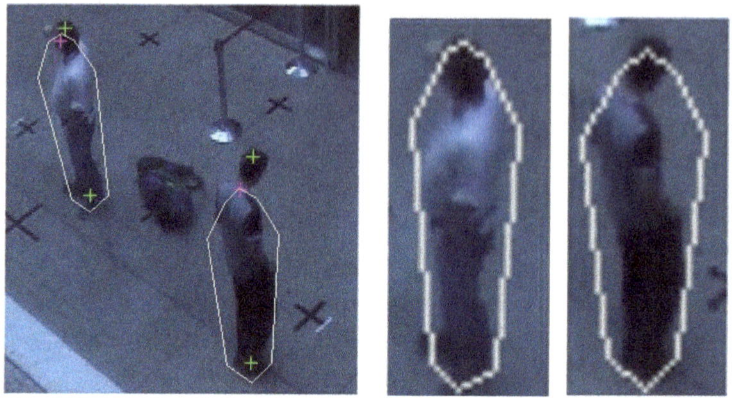

a) b)

Figure 4. Human target model projection; a) 2 targets in the scene and the estimated position of feet and head (color cross) b) normalized model map to the vertical pose

After the estimation of the head and foot position of the projected target, the corresponding color and texture features are mapped onto a normalized model of fixed size (Fig. 4b), to get a scale independent representation and allow the computation of tracking features (histogram of gradients and histogram of color texture). An alternative approach, mainly oriented to head detection and localization, is described in [12].

4.2. *Foreground detection*

The tracking process is based on the identification of a foreground map where all relevant objects (both moving and static) are supposed to be found. This is a quite common approach when dealing with stationary cameras, where a slowly changing background image can be computed and updated during time. There is an extensive literature devoted to background estimation based on a variety of statistical techniques [13]. In the following we will consider one of the most simple and fast ways of background estimation, based on time-space median filtering [14] to keep the most static (and stable) features in the video sequence. Then, a continuous update of the background image must be performed at each step of the tracking process, by leaving unchanged all image pixels that have been classified as foreground targets or shadow during the tracking process. Actually, some of the static blob shapes, in the foreground map, may come from real static targets in the scene (whose shape depends on their position in the calibrated ground map) but they can also be ghosts due to classification errors of removed objects. A suitable local analysis of the local texture of that static blob shape in both the background image and the current image of the sequence is often sufficient to solve such ambiguity. Another promising alternative is based on the use of a texture gradient representation of the dynamic foreground by subtraction of the background gradient, which is much less sensitive to light changes. The drawback of this approach is to miss the nice feature of a compact blob foreground representation, which is substituted by a fragmented foreground edge map.

As already mentioned before, there are many promising alternatives of foreground estimation in the scientific literature [15], with variable computational complexity. The main objective of this chapter is to prove the effectiveness of 3D model-based prediction and tracking, even in the case of using a simple and not-optimal background estimation process.

4.3. *3D Model based tracking*

Let's suppose that an object has been already found in the scene. The state of the target is described by the following data:

- The class of the 3D model and its attributes (3D shape parameters)
- An identification number
- Target position and motion vector prediction in 3D world coordinates
- Search area size (on the ground surface)
- confidence level of tracking

Tracking is performed by a local search of the target within a neighbor area (ROI_map) of the predicted position in the 3D reference plane. Such a region is increased in size during tracking if the confidence level is getting lower. For each predicted position, the bounding 3D volume of the target is re-projected onto the image plane of those cameras that are supposed to cover that position in space (from the calibration map). An example of such region selection is shown in Fig. 4 for a human target. Feature matching is performed on the normalized 2D template map (Fig. 4b) by computing a score measure S as the weighted sum of two components,

$$S = \lambda_c C + \lambda_t T \tag{3}$$

i.e. the Battacharrya distance of color histogram (C) and gradient histogram (T) between the current target estimation and the reference target template. This distance measure is computed within the same normalized template map (Fig. 4b). Both gradient and color histogram templates are slowly updated during tracking, with a careful control of a minimum acceptable level of texture density, to avoid the divergence of the tracking process. Of course many other alternative feature representations can be used to improve target identification [16] [17]. The "best" result of tracking is achieved as the maximum of the score function. When the score is found higher than a selected threshold, a correct tracking is confirmed and the target is removed from the foreground map to avoid confusions in multi-target tracking. This operation is repeated for all image views covering the same ground map area. Moreover, the different targets are ordered according to their distance from the camera sensor (from the closest to the farthest). In a multi-camera system, all predicted target models are re-projected onto all the camera views and the best match is used for tracking. This is quite important to manage possible occlusions of the 3D objects in the scene.

4.4. *New target detection*

While tracking is based on shape and color feature matching, new target detection is based on their shape model only. At first, the foreground map is analyzed to find all the clusters of foreground pixels having a minimum area size above a selected threshold. From such foreground connected blobs, an estimate of their base position on the ground (i.e. the foot coordinates) is computed and it is projected onto the reference map as a first hypothesis of the 3D target position.

This mapping is obtained by a simple homographic transformation between the image coordinates and the ground floor (from calibration data). A local search is performed around this predicted position of the target, to get the best match with respect to the average volumetric model of the selected object class. In this case the similarity function is limited to the shape texture term T in (3). When such score value is higher than a selected threshold, a new target is generated and included in the list of the living targets to be tracked, with a first estimate of its color histogram representation.

4.5. *Trajectory analysis*

At the end of the tracking process, the sequence of all the available targets is checked to remove possible outliers, fuse overlapping or close candidates, improve target position and update the state vector of the targets. The confidence level is increased for any successful tracking, while it is reduced when the tracking process fails. When the confidence level becomes ≤ 0, the target is removed from the list of the active tracking objects. All the tracked targets are arranged in increasing order starting from those that are closer to the camera sensor in the 3D reference system to minimize occlusion effects in the further steps of the tracking process. The analysis of all the targets in the same time-space coordinate framework has the great advantage to allow the management of possible interferences between them, and a better prediction of the future target position even in a complex situation of temporary severe occlusions.

5. Multi-camera Integration and Experimental Results

The previously described 3D model based tracking system has been tested with video-recorded sequences coming from different application scenarios. The following examples are intended to show the potential effectiveness of the proposed approach as well as to underline the main limitations that need future improvements.

5.1. *PETS data-base*

The first experimental analysis has been performed on a set of video sequences from PETS data bases [8][9]. A list of target positions has been manually selected by the operator, including the polygonal approximation of the body silhouette. This ground truth data file has been used to test the performance of direct and inverse transformations (from image to world and vice versa) obtained by the proposed camera calibration and the optical distortion correction [6]. The best selected position of the target on the ground plane map is that which receives the maximum consensus from the different image views. The test consists in the manual identification of a candidate target position from a selected camera, and the projection of the corresponding 3D model onto the other cameras to measure the amount of overlap of such a profile with the foreground image computed until that frame of the video sequence. The amount of matching is computed after template normalization as described in Fig. 4b. Actually not all the targets are visible from all the cameras and the target appearance is so variable, (both in shape and color texture) from the different viewpoints, that it almost impossible to implement an effective data fusion. The calibration test is limited to check the volumetric correspondence of the projected target shapes. In the example of Fig. 5, with a high crowding level, the detection algorithm can be successful only for a limited set of target positions and camera views, when the amount of occlusion is low.

The tracking process described before is not fully satisfactory for this cluttered video data, mainly because of a noisy foreground signal, and the targets are often lost and found again in the sequence with different attributes and different identification codes. Moreover, the large baseline distance between the cameras does not allow stereo matching, which, in other situations, has proved to be a very effective solution for people detection and counting in this kind of complex scenes [20].

Figure 5. 3D Target positions from the 4 different views of the PETS 2007 DB

5.2. *Multi-level ground map representation (railway example)*

Most of the tracking experiments that are considered in the scientific literature are based on the assumption that the human target is moving on a flat reference plane (ground floor). Actually, there are many video surveillance applications where the walking area cannot be described by a single plane. A very simple example is shown in Fig. 6, where there are at least two levels to be considered, i.e. the rail level and the platform level.

Other more complex situations may require the definition of multiple levels, and in general it may be necessary to recover the full digital elevation model of the scene. A precise definition of the vertical position of the tracking target is particularly relevant when using a 3D model, as described before, since the base position strongly affects its projected size onto the image sensory plane. The vanishing point calibration model can be successfully used in this case by exploiting the natural properties of the parallel rail structure and the knowledge of the fixed rail width.

Figure 6. 2-level ground map model of a railway scenario

The main vanishing point is found as the convergent directions of the rails and the vanishing line [5], corresponding to the *horizon* is obtained by the cross-ratio of the segments that intersects the rails (in any arbitrary direction). From the visibility of any stable vertical line in the scene it is possible to recover the vertical vanishing point Vp and complete the calibration process. In this simple case the reference plane (zero level) is assumed to be at the rail level, and the platform height (80 cm) has been introduced, as a fixed parameter. Moreover, the areas corresponding to the rail level and the platform level have been manually selected, on the camera image, by the security operator, through the GUI interface. Many different views of the same 3D person model have been tested to check the accuracy of the calibration, and the corresponding 3D model target projection. An example of the test is shown in Fig. 6 with two persons walking at both levels. In this example the 3D human model has been assumed to have the same height.

5.3. *Fish-eye tracking example*

The same 3D target projection model has been tested with a fish-eye panoramic camera. In this case the shape of the target is continuously modified by the heavy perspective distortion [6].

An example of 3D model tracking is shown in Fig. 7. The estimated positions of three human targets are displayed on the projected ground map and the outline of a simplified box-shaped 3D model is shown re-projected on the image plane.

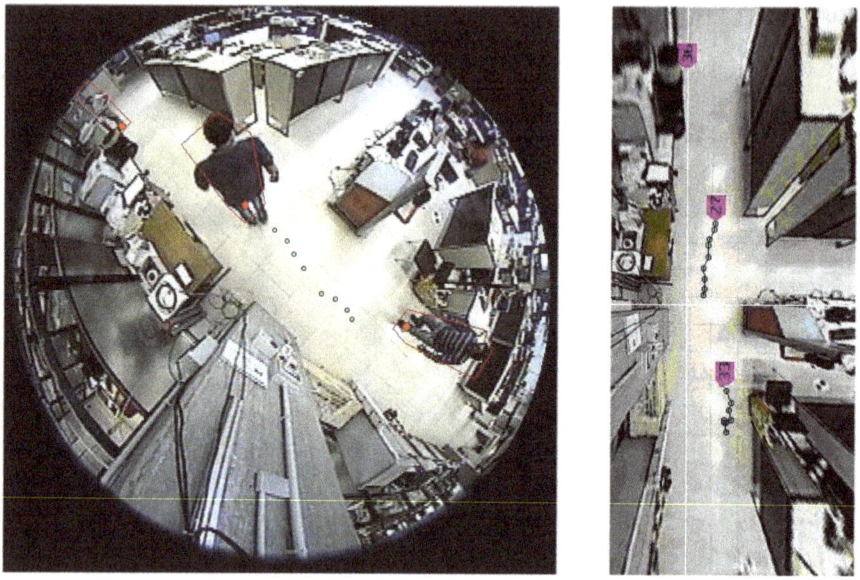

Figure 7. 3D model fitting in a panoramic fish-eye camera view.

Due to the fish-eye perspective distortion of the target, the only feature that can be used in this tracking process is the density of the foreground texture map within the projected 3D model outline, since color and edge gradient information are changing too much according to the variable appearance of the human body.

5.4. *Foot-ball player tracking*

The analysis of trajectories of video recorded sport events is another interesting application that has received a great deal of attention in the last few years, in different fields (foot-ball, tennis, hockey, etc.) [21] [22].

Fig. 8 shows an example of tracking obtained with the previously described technique. The input video sequence comes from the recording of a foot-ball match in the Italian first league (Serie A) in the year 2006-2007. The camera resolution was 1440x1080 pixels, from a pair of commercial digital camcorders. The video sequences are synchronized to achieve the same time-baseline during the match. Fig. 8 shows the tracked position of the players (with id code number) after a few hundred frames of the play, and the ground map depicts the trajectory of the tracked players and the referee in the last 15 frames of the sequence. The ball position has been estimated with a simple projection on the ground surface, without a control of its 3D position in space.

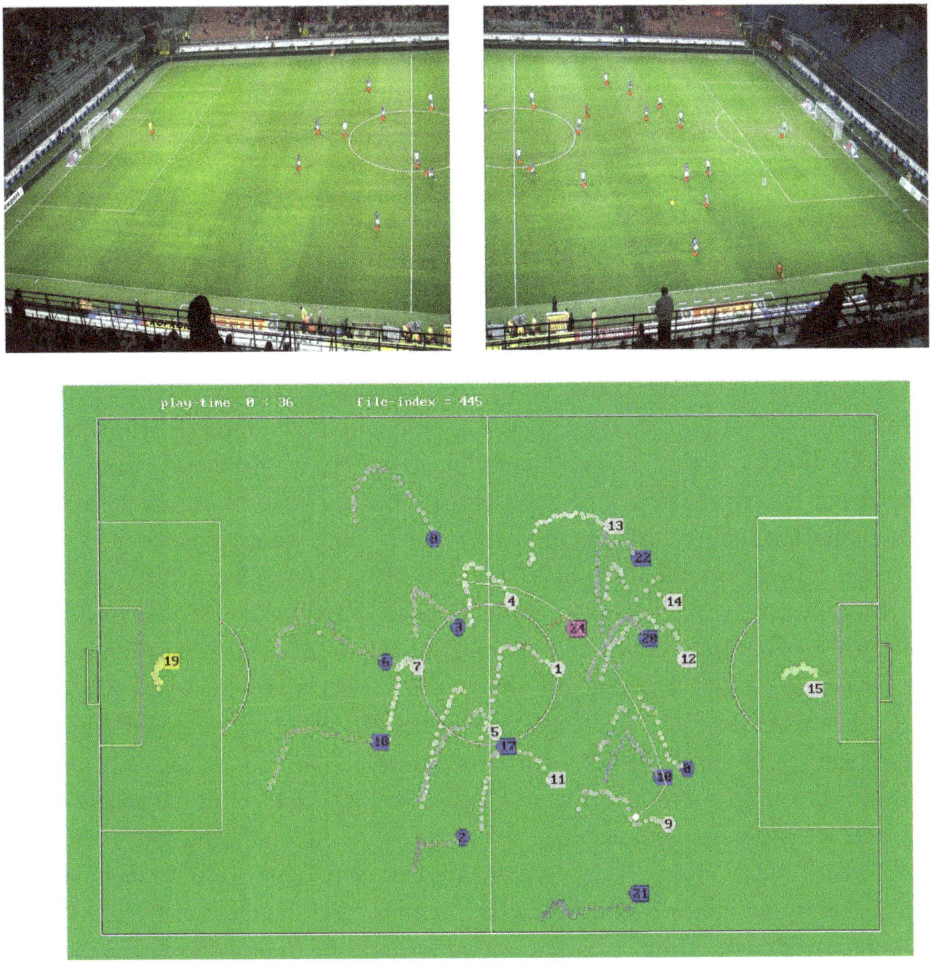

Figure 8. An example of foot-ball tracking

This is a quite representative example to show the main advantages of the proposed 3D model based tracking approach. The smooth transition of the players from the two cameras is achieved by a small amount of overlap between them. Of course a greater overlap between the cameras would allow multi-view check, to better manage the frequent situations of occlusion among the players. A stereo-arrangement would be even more helpful to better localize the players in the 3D space.

Currently, 3D localization of the players is managed only by the 3D model prediction & tracking process and by the low-level data fusion step of trajectory analysis. In this case, the different colors between the teams are nice features, at the price of a high level of confusion among the players of the same team. On the other hand the shape of the players' body is continuously changing, during the different phases of the play, and the correction of the vertical orientation (head position) turns out to be an essential preprocessing step, before feature extraction and matching.

6. Conclusions

The automatic detection of abnormal behavior represents one of the main objectives of any new intelligent video surveillance system to drive the human attention only when relevant actions take place.

This chapter refers an example of people detection and tracking to show the advantages that come from the use of a 3D model based approach. The analysis is carried directly in the 3D world coordinates by projecting the expected target appearance into the different camera images that are used to verify the correct hypothesis. The main advantage of this approach is the scale independency of the projected target, provided all the cameras are suitably calibrated in the world reference system. Adaptive pose correction and target size normalization are additional mechanisms to achieve an effective data representation to be used by the most appropriate learning and classification techniques. Around this 3D model based framework, the main algorithms for human tracking have been briefly summarized and discussed, from camera calibration, image and motion analysis, and their impact into the quickly evolving field of Intelligent Visual Surveillance systems. The accuracy of target localization, in the 3D reference coordinate system, strongly depends by the selected target resolution in wide-field cameras and it may be definitely improved by multi-camera configurations.

Further research efforts will be devoted to improve the robustness of the full processing chain and optimize the processing speed to approach video-rate (25 fps). The final goal is the design of an easy to use, almost adaptive and self-controlled system to be managed by security personnel, for real world applications, in both indoor and out-door environments.

References

[1] G. Garibotto, "Auto-Detector: Mobile Automatic Number Plate Recognition", Handbook of Pattern Recognition and Computer Vision, ed. C.H. Chen & P.S.P. Wang, chapter 5.6, 2005, pp. 601-618

[2] Special Issue on Modeling People: Vision based Understanding of a Person's shape, appearance, movement and behavior, Computer Vision and Image Understanding, vol 104, N.2-3, Nov/Dec. 2006.

[3] B. Leibe, E. Seemann, B. Schiele, "Pedestrian Detection in Crowded Scenes", IEEE Conf on Computer Vision and Pattern Recognition, 2005, CVPR 2005, vol. 1, pp 878-885, Jun. 2005.

[4] R.Y. Tsai, "An Efficient and Accurate Camera Calibration Technique for 3D Machine Vision", Proc. of the IEEE Conf. on Computer Vision and Pattern Recognition, Miami, 1986, pp. 364-374.
[5] F. Lv, T. Zhao, R. Nevatia, "Self-Calibration of a camera from video of a walking human", Proc. of the Int. Conf. on Pattern Recognition, vol 1, pp. 562-567, 2002.
[6] G. Garibotto, "A New Optical Distortion Model for Multi-Camera Calibration", Proc. of the AVSS 2009, Genova, Sep. 2009
[7] F. Devernay, O. Faugeras, "Straight lines have to be straight : automatic calibration and removal of distortion from scenes of structured environments", Machine Vision and Applications, (2001) 13: pp 14-24.
[8] B. Caprile, V. Torre, "Using Vanishing Points for Camera Calibration", Journal of Computing Vision 4(2) pp. 127-140. 1990.
[9] O.D. Faugeras, Three Dimensional Computer Vision: a Geometric Viewpoint, MIT Press, 1993
[10] G. Garibotto, "3-D model-based people detection & tracking", Proc. of the AVSS 2007, London, 5-7 Sep. 2007, pp. 579-584
[11] M. W. Lee, R. Nevatia, "Human Pose Tracking in Monocular Sequence using Multilevel Structured Models", IEEE Trans. On PAMI, vol. 31, n.1, pp. 27-38, Jan. 2009
[12] G. Rogez, J.J. Guerrero, C. Orrite, "View-invariant Human Feature Extraction for Video Surveillance Applications", Proc. Of the AVSS 2007, London, 5-7 Sep. 2007, pp.
[13] H. Wang, D. Suter, "Background Subtraction Based on a Robust Consensus Method, Proc. of the 18th Int. Conf. on Pattern Recognition, ICPR'06, 2006.
[14] R. Cucchiara, C. Grana, M. Piccardi, and A. Prati, "Detecting moving objects, ghosts and shadows in video streams", IEEE Trans. on Patt. Anal. and Machine Intell., vol. 25, no. 10, Oct. 2003, pp. 1337-1342.
[15] C. Wren, A. Azarbayejani, T. Darrell, and A. Pentland, "Pfinder:Real-time Tracking of the Human Body," IEEE Trans. on Patt. Anal. and Machine Intell., vol. 19, no. 7, pp. 780-785, 1997.
[16] N. Dalal, B. Triggs, "Histograms of Oriented Gradients for Human Detection", Proc. CVPR'05, vol.1, pp.886-893, 2005.
[17] T. Zhao, R. Nevatia, "Tracking Multiple Humans in Complex Situations", IEEE Trans. on Pattern Analysis and Machine Intelligence, PAMI, vol 26, n.9, Sep. 2004, pp. 1208-1221.
[18] PETS2006, (Int. Conf. on Performance Evaluation of Tracking and Surveillance, June 18, 2006, New York, www.pets2006.net.
[19] PETS2007, (Int. Conf. on Performance Evaluation of Tracking and Surveillance, Oct. 14, 2007, Rio de Janeiro, www.pets2007.net.
[20] G. Garibotto, C. Cibei, "3D Scene Analysis by Real-Time Stereovision", Proc of the IEEE Int. Conf. On Image Processing, ICIP-05, Genova, Vol.2, Sep. 2005.
[21] K. Okuma, J.J. Little, D. Lowe, "Automatic Acquisition of Motion Trajectories: tracking Hockey Players", Proceedings of the SPIE, vol. 5304, pp. 202-213, 2003.
[22] Xu Ming, J. Orwell, G. Jones, "Tracking football players with multiple cameras", IEEE Int Conf. on Image Processing ICIP'4, pp. 2909-2912, Vol.5, Oct. 2004.

Index

2D, 595
3D, 595
3D face reconstruction, 596
3D faces, 596
3D model, 755
3D morphable model, 596
3D morphological model, 595
3D supported 2D model, 597
3D surface, 607
3DMM, 617
8-point algorithm, 225

accurate, 30
active contour model (ACM), 497, 507, 518, 519, 673
active contours, 537, 540, 542
adaptive classifier selection, 39
adaptive resonance theory, 309
affective patterns, 415
affine invariance, 238
affine invariant features, 235
affine transformation, 236
algebraic distance minimization, 225
alternated least squares (ALS), 8
Ambrosio and Tortorelli, 297
American sign language recognition, 477
    with manual information, 477
    with non-manual information, 477
analysis-by-synthesis, 597
angiography, 498
annotation of biological images, 625
approximate entropy, 423
Arabic handwriting recognition, 380
association of classifiers, 24
atomic number, 647
atomic number image, 651
automated segmentation, 497
automated threat detection, 655
automatic annotation, 627
autoregressive model, 500
average accuracy, 323

background initialization, 182
background maintenance, 182
background modeling, 181, 182
backprojecting, 652
Bayesian approach, 24
Bayesian networks, 129, 130, 131, 146
beam hardening, 650
belief theory, 33
bending-invariant canonical form, 595, 596, 609
best emotion-relevant features, 429
binary tree, 314
binocular stereo matching, 165
bioinformatics, 590
biosensors, 415
bivariate Gaussian, 318
bivariate Nakagami, 318, 319
bivariate Weibull, 318, 320
bone image, 651

bounding boxes, 358
breathing rate variability, 424
B-spline, 509
bundle adjustment, 207

camera calibration, 750, 753
camera matrix, 220
candidate locations, 359
canonical correlation analysis, 14
carbon nanotube, 663
cascade error projection (CEP) algorithm, 112
category proliferation, 73
catheter, 499
cDNA microarrays, 557
cell segmentation, 537, 538, 539, 542, 545, 551, 552
CENPARMI Arabic Isolated Numerals Database, 386
certified EDS, 645
certified, 646
CFAR, 656
CFNS, 208
checked baggage screening, 645, 663
checked baggage screening process, 645
cheirality, 221
circular Hough transform (CHT), 503
classification, 73
classification of handwritten characters, 74
clique, 313
clustering, 358
combination of classifiers, 28
common pin-hole camera model, 672
component analysis, 3, 5
compound object, 656, 661
Compton projection, 651
Compton scatter, 648
computation of the edit distance, 52
computed tomography, 645
computer vision, 748
conditional density propagation (ConDensation), 501, 504
conditional random fields, 129, 133, 146
conditional topology, 42
confusion matrix, 614
connected component labeling, 659
constrained decomposition method, 654
context, 294
contextual classification, 307
contour, 618
cooperation of classifiers, 37
corner points, 224
coronary heart disease (CHD), 497
correct classification ratio, 427
correlation, 365
correlation similarity measure, 365
correspondence, 597
critical surface, 205
cubic spline polynomial, 622
cupping effect, 653
curve evolution, 297
    discrete, 295
    efficient alternative, 297
    Kimia, Tanenbaum and Zucker, 297
curve fitting, 506

data dependent classifier selection, 39
data fusion, 24, 307
decision space, 25
deformable object, 610
deformable template model (DTM), 497, 507
Dempster-Shafer evidence theory, 310
Dempster-Shafer evidential reasoning, 24
density image, 651

density measurements, 647
dependence tree, 307, 308, 309, 317
derivative of Gaussian (DoG), 508, 509
destreaking, 654
diagonal components, 614, 615
diffusion, 297, 299
digital watermarking, 687, 688, 689, 690, 692, 693, 696, 699, 701, 702, 707, 708, 713, 719
displacement, 604
dissimilarity embedding, 47
dissimilarity function, 362
dissimilarity representation, 49
distance map, 609, 610
distance transform, 286, 297
diverse, 30
documents classification, 736
dominant dimensions of facial expressions, 486
dominant plane, 257, 258
    hierarchy of, 268
DSCT, 652
dual energy, 645
dual energy CT, 647, 663
dual energy decomposition, 650, 653
dual energy index, 650
dynamic Bayesian networks, 129, 130, 131, 137
dynamic classes, 25
dynamic classifier selection, 39
dynamic programming, 163
dynamic time warping algorithm, 365

edit path, 51
EER, 615
effective atomic number, 647, 648, 654
EFNS, 202, 208, 209
eigenfaces, 599
eigenvalues, 599
eigenvectors, 599
electrocardiogram (ECG), 415, 419
electromyogram (EMG), 415, 419
emergent structures, 284
emotion recognition, 415
emotion-specific classification, 417
emotion-specific multilevel dichotomous classification (EMDC), 415
endocytosis, 537, 549, 550, 551
energy, 313
    boundary-based, 284
    minimizer, 284
    non-local, *see* non-local
    region-based, 284
energy resolving detectors, 649
ensemble, 25
ensemble learning, 726
epipolar equation, 205
epipoles, 213
equivalent number of looks, 319
erosions, 656
error correcting output codes, 580
essential matrix, 211
evolutionary algorithms, 557
evolutionary classes, 26
evolutionary computation techniques, 81
evolutionary computing (EC), 688, 703, 704, 707, 708, 709, 719
exact graph matching, 50
explosive trace detection, 646
explosives detection systems, 645

F_1, 740
face, 595
face template protection, 455, 460, 461
facial expression recognition, 129, 137, 152
facial expressions, 610

facial reconstruction, 595
factor graph, 130, 149
failure modes, 222
false alarm rates, 646, 656
false alarms, 646
fast marching method, 595, 596, 609
FDR (false discovery rate), 549, 551
feature extraction, 102, 238
 adaptive color, 111
 bio-inspired, 103
 color, 101, 102
 dynamic, 108
 shape, 101, 102
 static, 105
feature point definition table, 598
feature points, 224, 595
feature selection, 360, 425
feature space, 25
feature-based image-to-image matching, 364
features, 360
 boundary-based, 283
 global, 284, 295
 local, 284, 295
 region-based, 283
first rank measurement (FRM), 384
first two ranks measurement (FTRM), 385
fish-eye camera, 761
fixating, 214
fixed focal length method, 213, 214
flexible neural tree classifier, 584
FNS, 208, 210, 227
focal length, 204
food size measurement, 672, 674
foot-ball tracking, 762
foreground detection, 182
forward-projecting, 652
FPDT, 622
free focal length method, 213, 229
fundamental matrix, 205
fusion methodology, 25
fuzzy ARTMAP neural networks, 73
fuzzy logic study, 678

Gamma, 319
generalized eigenvalue problem, 5, 228
generalized Gaussian-Rayleigh, 319
generalized inverse, 230, 231
generalized Rayleigh, 319
generalized singular valued decomposition, 12
generic face model, 611
generic mean face, 598
genetic algorithms, 81
geodesic path, 610
geodesic propagation, 610
geodesic uniform sampling, 611
GEPs, 8
Gibbs form, 313
GMF, 618
gradient descent search, 510
gradient vector field (GVF) snakes, 522, 525, 526
grammar scale factor, 408
graph, 50
graph based pattern representation, 50
graph data sets, 63
graph edit distance, 51
graph embedding, 49, 55
graph embedding kernel, 57
graph matching, 50
graphics processor unit, 663
gray-scale images, 360
gridding, 557
guidewire, 498

hand search, 646
handwriting recognition, 397, 410
 baseline, 401
 classification, 405

corpus line, 401
feature extraction, 402
normalization, 400
offline recognition, 397
online recognition, 397
postprocessing, 406
preprocessing, 399
recognition system, 398
handwritten documents, 355
handwritten numerals, 377
Harris operator, 224
head, 598
head motion, 487
heart rate variability (HRV), 422
HEIV, 227
helical scanning, 646
HeroSVM, 381
hidden Markov models (HMM), 367, 405, 478
high voltage power supply, 653
histogram based invariants, 249
HMM network, 367
home made explosives, 648
human detection, 749
human visual system, 283
human-computer interaction (HCI), 415
hybrid graphical model, 145, 148, 149
hybrid methods, 40
hybrid model, 368
hybrid PGM, 129, 130
hybrid probabilistic graphical model, 148
hyper-parameter, 75

ICIA, 617
ICP, 613
image analysis, 557
image artifacts, 652
image format, 360
image labeling frameworks, 159
image matching, 363
image quality, 655, 656
image reconstructor, 645, 651
image representations, 360
image resolution, 655
image segmentation, 129, 145, 150, 153
image-warping, 609
in situ fruitfly gene expression, 634
in situ gene expression, 638
inconsistency, 652
independent component analysis, 258
for dominant plane detection, 264
of optical flow field, 263
inexact graph matching, 50
information retrieval, 354
integration of color and shape feature based approach, 116
intelligent video analysis, 747
intelligent watermarking, 687, 688, 698, 699, 707, 709, 710, 719
interaction, 283
non-local, *see* non-local
pairwise, 285
international patent classification (IPC), 736
intravascular ultrasound (IVUS), 497
intrinsic shape signatures, 609
invariant features, 239
iterated conditional mode, 315
iterative closest point, 609

K, 309, 319
K nearest neighbors, 309
Kalman filtering, 501, 503
kernel CCA, 14
kernel density estimation, 182, 188
kernel linear discriminant analysis, 13
kernel methods, 329, 331, 348

kernel principal component analysis, 11
kernel-based methods, 310
keypoints, 224
keywords, 354
Klein-Nishina cross section, 648
KLT, 225
k-means, 16
k-nearest neighbor, 729
Kronecker product, 226
Kruppa equations, 229
Kruskal's algorithm, 318
Kullback-Leibler divergence, 318
kVp switching, 649, 652, 663

Lagrangian, 602
landmarks, 596
large-scale, 725
LDA measurement (LDAM), 385
least squares, 225
least-squares weighted kernel reduced rank regression, 6
leave-one-out cross-validation, 427
level curve, 297
level-set theory, 673
LibSVM, 382
limitations, 613
line of sight, 204
line oriented approaches, 359
line processes, 325
linear discriminant analysis, 12, 415
liquid explosives, 648
LiST, 617
LLE, 613
LMedS, 225
local symmetry, *see* medial axis
lognormal, 309, 319

macro-F_1, 741
manual inspection, 647
Markov random fields, 129, 133, 309
mass density, 648
master-salve framework, 731
match tracking, 73
matching levels, 368
mathematical morphology, 436
matrix of intrinsic parameters, 220
maximum *a posteriori* probability, 33
maximum likelihood, 205
maximum-*a-posteriori*, 311
MDS, 613
mean shift algorithm, 662
mean shift method, 656
medial axis, 283, 284, 294
  Aslan and Tari, 296
  Blum, 286
  disconnected skeleton, 300
  grass-fire, 286
  Tari, Shah and Pien, 296
mesh surface, 610
method of log-cumulants, 319
micro-F_1, 741
mid-point method, 218
min-max modular network (M^3-network), 726
min-redundancy max-relevance (mRMR), 630
mixed pixels, 324
mixture of Gaussians, 182, 184
modulation transfer function, 655
module combination, 730
MoLC, 321
moment based invariants, 249
Monte Carlo, 652
morphological, 599
morphological operations, 656
morphology, 297
motion parameters, 211, 217
moving-average filter, 323
MSER, 225
multiclass classification, 415
multi-energy CT, 663

multilayer neural networks, 729
multilayer perceptron (MLP), 441, 700, 701, 702, 703, 707, 709, 720
multi-objective annotation, 628
multiple classifier systems, 24
multiple linear regression, 4
multiscale affine invariants, 243
multiscale autoconvolution, 244
multiscale framework, 241
multiscale sample entropy, 420, 423
multisensor data, 308
multisensor optical-SAR, 307
multisensor optical-SAR image classification, 308
multispectral, 307
multivariate Gaussian, 309
Mumford and Shah, 297
musical emotions, 415

Nakagami, 319
Nakagami-Gamma, 309
natural language processing, 407
neighborhood system, 313
neural networks, 309
Newton's method, 320, 595, 602
Newton-Raphson, 654
n-gram language models, 407
NIST SD19 data, 84
NIST SD19 handwritten character recognition, 73
non-feature points, 604
non-local, 283, 284, 295
nonparametric, 309
normalization, 238
normalized cuts, 17
NTCIR, 736

object recognition, 101, 102
object-based methods, 310
objective dietary physical activity assessment, 668
off-diagonal components, 615
one-versus-one, 728
one-versus-rest, 728
on-screen threat resolution, 646, 655
operational costs, 646
optical character recognition, 354
optical flow, 258, 260
optical flow estimation, 166
optical lens distortion, 751
optimization, 601, 689, 696, 698, 700, 703, 704, 705, 706, 707, 709, 713, 717, 718, 719, 720
optimization technique, 560
optimizing the parameters in rejection, 377
outliers, 225
overall accuracy, 323
over-segmentation, 162
oversketching snakes, 524, 525
overtraining, 73

page segmentation, 360
painting process, 605
parallel implementation, 731
parallel pipeline, 735
parallelism, 726
parametric, 309
partial volume effect, 656, 657
particle filtering, 503
particle swarm optimization (PSO), 73, 583, 688, 689, 703, 704, 705, 707, 708, 709, 712, 718, 719, 720
parts of visual form, 283
Parzen algorithm, 309
pattern recognition, 74, 645
PCA learning, 104
PDE, 296
  curve evolution, *see* curve evolution
  Laplace operator, 296
  Poisson equation, 296

people tracking, 749, 754, 757
permutation, 317
perspective projection, 204
photoelectric effect, 648, 651
photoelectric projection, 651
physical activity evaluation, 681
physiological signals, 416, 417
pin-hole camera model, 204
pixel-by-pixel matching, 364
Poincaré geometry, 422
points of interest, 224
post-reconstruction decomposition, 650
potentials, 313
Potts model, 312, 313
power spectral densities, 424
precision, 369
pre-processing, 358
pre-reconstruction decomposition, 650
P-R-graph, 369
principal component analysis, 9, 661
principal point, 204
probabilistic graphical models, 129
probability density estimation, 307
probability of detection, 646, 655
probability of false alarm, 655
projection data, 645
projection matrix, 220
projective Gauss-Newton iterations, 227
protein fold recognition, 579
    with bio-physics based method, 579
    with sequence-based method, 579
    with structure based method, 579
protein labeling map, 638
prototype graphs, 54
prototype selection, 59
pseudoinverse, 231
pseudoinverse LDA, 427
pyramid-based multiresolution analysis, 260

QRS complex, 421

radial basis function, 596, 606
random-forest, 310
range of interest, 657
RANSAC, 225
Rayleigh, 309
Rayleigh quotient, 8, 227
real-time implementation, 663
real-world data, 75
Recall, 369
reconstruction, 595
reduced-rank regression, 6
reference-density, 324
region adjacency graph, 161
region tree, 163
region-based classification, 308, 311
region-based methods, 310
registration, 506
regula-falsi method, 321
rejection measurements, 384
rejection rate, 378
relational distributions, 477
relaxation labeling, 309
reliability, 378
reliability factors, 309
remote sensing, 329, 330, 344, 345, 348, 349, 352
renormalization, 201
reprojection error, 205
respiration (RSP), 415, 419
robot navigation, 257
ROI annotation, 636
rule of Bayes, 32

SAIMC, 617

saliency, 293
sample entropy, 423
sandwich detector, 649
scatter, 652, 653
screening, 537, 538, 542, 552
scripts, 355
segmentation, 297, 307, 308, 310, 656
selection of classifiers, 38
semisupervised classification, 345
set marginal graph, 59
set median graph, 59
shape, 283,597
    boundary texture, 283
    evolution, 297
    global, 284
    gross structure, 283
    minima rule, 285
    occlusion, 288
    parsing, 294
    parts, 283
        semantically meaningful, 288
        the least deformable, 283
    PDE based representation, 283, 284
    peripheral structure, 283
    prickly, 294
    primitives, 283
    topological change, 299
    topology, 297
    with holes, 288
shape alignment, 595
shape alignment algorithm, 602
shape codes, 365
shape context, 365
shape correction, 596
shape measure, 596
shape parameters, 595
shape similarity, 609
shortest-spanning-tree problem, 318
SIFT, 225
similarity function, 362
similarity measure, 613
similarity measurement algorithms, 362
simulated annealing, 315
single 2D image, 596
single Gaussian, 182, 184
singular value decomposition (SVD), 230, 602
singular values, 230
skeleton, *see* medial axis
skin conductance response (SCR), 425
skin conductivity (SC), 415, 419
SLH algorithm, 364
slice sensitivity profile, 655
snakes, 518, 520
SNO, 617
Sobel edge detection, 505
solar flare detection, 438
solar image features, 435
space alternating search, 510
spatial reasoning method, 257
speckle, 308
spectral calibration, 655
spectral clustering, 16
spectral drift, 652, 653, 654
spectrum estimation, 651
split gain, 315
spot detection, 537, 538, 545, 548, 551
spot-segmentation, 557
stabilized inverse diffusion equation (SIDE), 443
stacked-vector, 309
static classifier selection, 39
statistical methods, 309
statistical pattern recognition, 47
step-and-shoot, 646
stochastic optimization, 81
stratification, 309

streaks, 653
stroke-based, 366
structural pattern recognition, 48
structure from motion, 201
subband spectral entropies, 424
subspace iteration, 8
subspace learning, 182, 190
supervised classification, 330, 332, 345, 351
support vector classifiers, 381
support vector domain description, 310
support vector machines (SVMs), 309, 310, 329, 350, 351, 352, 438, 441, 446, 582, 700, 702, 709, 720, 729
  with linear kernel, 442
  with non-linear kernel, 442
SUSAN, 225
SVM^{light,} 740
synthesis, 595
synthesized image, 597
synthetic aperture radar, 307
systems of classifiers, 25

task decomposition, 728
Taubin method, 205, 226
template matching, 364, 539, 546
template security, 455, 456, 459, 460, 473
term filtering, 739
term indexing, 739
texture, 596
texture recovery, 597
threat analysis, 645
throughput rates, 646
time delay neural networks, 405
tissue image, 651
tissue ultrasound, 498
tokenization, 739
training-based matching, 367
training-free, 363
transmedia machine, 366
tree-structured Markov random field, 307, 311
tree-structured MRF, 313, 314
triangle-facet, 598
triangulated domain, 609
triangulation, 207, 218
tunneling probability, 501

user interaction, 517, 530, 534
UV space, 595

vanishing point, 753
variability, 371
variance stabilizing transform (VST), 548, 549
vascular profiling, 498
vertex, 612
vertices and facets, 598
video analytics, 747
video surveillance, 748
viewpoint, 204
voting methods, 31

watershed, 287, 539
wavelet transform, 548, 549
wavelet-embryo, 629
Weibull, 309, 319
whiteboard, 397
word images, 367
word insertion penalty, 408
word spotting, 353
word spotting systems, 357
workload balance, 735

www.ingramcontent.com/pod-product-compliance
Lightning Source LLC
LaVergne TN
LVHW081351011026
849161LV00013BA/2787
* 9 7 8 9 8 1 4 2 7 3 3 8 1 *